谨以此书祝贺

王昆扬教授 80 寿辰！

北京师范大学数学家文库

李仲来◎主编

王昆扬文集

逼近与正交和

王昆扬◎著

BIJIN YU ZHENGJIAOHE

2021 · 北京

图书在版编目（CIP）数据

逼近与正交和：王昆扬文集 / 王昆扬著，李仲来主编．- 北京：北京师范大学出版社，2021.7

（北京师范大学数学家文库）

ISBN 978-7-303-26825-2

Ⅰ．①逼… Ⅱ．①王… ②李… Ⅲ．①逼近 - 文集 ②正交 - 文集 Ⅳ．① O174.41-53 ② O224-53

中国版本图书馆 CIP 数据核字 (2021) 第 020162 号

逼 近 与 正 交 和： 王 昆 扬 文 集

BIJIN YU ZHENGJIAOHE：WANG KUNYANG WENJI

出版发行：北京师范大学出版社 www.bnup.com
北京市西城区新街口外大街 12-3 号
邮政编码：100088
印　　刷：鸿博昊天科技有限公司
经　　销：全国新华书店
开　　本：710 mm ×1 000 mm　1/16
印　　张：43.5
插　　页：4
字　　数：668 千字
版　　次：2021 年 7 月第 1 版
印　　次：2021 年 7 月第 1 次印刷
定　　价：158.00 元

策划编辑：岳昌庆　　责任编辑：岳昌庆
美术编辑：李向昕　　装帧设计：李向昕
责任校对：段立超　　责任印制：马　洁

◀ 1985年，在北京师范大学数学楼与导师合影（左起，下同）。陆善镇、孙永生、王昆扬。

▶ 1987年，俄罗斯学者访问北京师范大学，在数学楼合影。王昆扬、陆善镇、捷里亚科夫斯基（孙永生的师弟）、赵桢、斯杰奇金（孙永生的导师）、孙永生、李占柄。

◀ 1997年，在北京师范大学数学楼，俄罗斯学者苏伯金讲学后合影。苏伯金、唐娉、王昆扬、□□□。

▶ 1998年，俄罗斯学者访问北京师范大学，在孙永生家合影。王昆扬、尤金、孙永生、安德列耶夫、房艮孙、曾美美。

◀ 2004年，在乌鲁木齐市，参加表示论与调和分析国际研讨会，和大学老师邓东皋合影。

▶ 2004年，在莫斯科，访问俄罗斯科学院斯捷克洛夫数学研究所。

2005 年，访问加拿大 Alberta 大学。张英伯、王昆扬、戴峰（东道主，王昆扬的博士生）。

2006 年，在北京师范大学珠海分校讲座后，与学生合影。

2007 年，访问澳大利亚悉尼大学。

▶ 2008年，在第4届高等学校教学名师奖表彰大会上。

◀ 2008年，在北京师范大学数学楼，为本科生讲课。

▶ 2008年，在北京师范大学数学楼，指导本科生。

自序

本人1978年考入北京师范大学数学系读研究生，师从孙永生教授．1982年留校任教之后继续读在职博士研究生．期间并得到陆善镇教授的指导．

入学之后，导师给我选定的研究方向是多元周期函数用Fourier级数逼近．本文集把这一方向的主要论文作为第一部分．这方面的基础知识和我的一些其他研究论文收入了专著《Bochner-Riesz平均》(与陆善镇教授合著，北京师范大学出版社出版)，曾作为硕士研究生的教材和参考书．

1991年，孙先生建议我做些多维球面上的可积函数用Fourier-Laplace级数逼近的研究．其后我的工作，转向了多维球面．本文集把这一方向的主要论文作为第二部分．我和师弟李落清教授收集整理了这方面的基础知识，以及我们自己的研究成果，写成了专著《Harmonic Analysis and Approximation on the Unit Sphere》(科学出版社出版)．

后来，我参加了悉尼大学Brown教授的关于正交和的正性的研究项目，合作撰写了若干这方面的论文．本文集把这一方向的论文作为第三部分．

自从1982年硕士研究生毕业留校之后，我一直把教学工作作为自己的首要任务，认真体验“得天下英才而教育之，三乐也”(孟子语)．对于教学内容有些思考和实践，编写了本科生教材《简明数学分析》《数学分析简明教程》《实变函数论讲义》，还修订了孙永生教授的《泛函分析讲义》，与陆善镇教授合写了研究生教材《实分析》．在教学实践的基础上，写了一些论文．其中几篇，收入本文集的第四部分．

导师孙永生教授已于2006年去世．我在不同场合发表过几篇文章介绍他，纪念他，祝贺他寿辰的同名文章《孙永生》．本文集的第五部分收录其中一篇．

孙永生教授，在学术上是我的启蒙者，领路人．做人处事更是我衷心崇敬的典范．孙先生精心指导我完成了我的第一篇论文《二元连续周期函数用其Marcinkiewicz型和强性逼近的估计式》，我曾在文末由衷地多写了几句感谢的话，先生批示“不必如此，写一句就够了”．在我取得硕士学位之后，孙先生安排陆善镇教授指导我继续做多元Fourier分析．我的博士论文《Approximation for continuous periodic functions of several variables and its conjugate functions by Riesz means on a set of total measure》，就是在陆先生指导下完成的．对陆先生表示衷心感谢．

退休之后十分怀念在北京师范大学的函数论研究集体中的生活，感谢众师兄弟以及学生们对我的帮助．

王昆扬

2018年10月

目　录

一、经典 Fourier 分析

二、球面上的 Fourier-Laplace 分析

三、正交和的正性

四、大学数学系分析类课程内容的改进

五、纪念导师

附录

Contents

Ⅰ. Classical Fourier Analysis

Ⅱ. Fourier-Laplace Analysis on the Sphere/

Ⅴ. In Memory of My Supervisor

Appendix

一、经典 Fourier 分析

I. Classical Fourier Analysis

北京师范大学学报(自然科学版)，1981，(1)：7－22.

二元连续周期函数用其 Marcinkiewicz 型和强性逼近的估计式①

Some Estimates for the Strong Approximation of Continuous Periodic Functions of Two Variables by Its Sums of Marcinkiewicz Type

§1. 引言

用 $L_p(R)(1\leqslant p\leqslant+\infty)$ 表示在 $R=[-\pi,\ \pi;\ -\pi,\ \pi]$ 上 p 次幂可和(当 $1\leqslant p<+\infty$)或在 R 上连续(当 $p=+\infty$)，对每个变量都以 2π 为周期的二元函数空间，记 $L_\infty(R)=C(R)$. 用 $S_{k,k}(f;\ x,\ y)(k\in\mathbf{N})$ 表示可和函数 $f(x,\ y)$ 的 Fourier 级数的对每个变元都是 k 阶的矩形部分和.

J. Marcinkiewicz[1] 最先考察了如下形式的和

$$\sigma_n(f;x,y)=\frac{1}{n+1}\sum_{k=1}^{n}S_{k,k}(f;x,y),\quad n\in\mathbf{N},$$

对于 $f\in C(R)$ 的情形，他证明了

$$\lim_{n\to+\infty}\|f(x,\ y)-\sigma_n(f;\ x,\ y)\|=0.$$

Л. В. Жижиашвили[2] 把下面的和

$$\sigma_n^\alpha(f;x,y)=\frac{1}{A_n^\alpha}\sum_{k=1}^{n}A_{n-k}^{\alpha-1}S_{k,k}(f;x,y)$$

叫作二重 Fourier 级数 $\sigma(f)$ 的 $(C,\ \alpha)$ 平均，其中 A_n^α 为 α 阶 Cesàro 系数(参见[3]，第三章§1). 他证明了对于 $f\in L(R)$，$\sigma(f)$ 的正阶 Cesàro 平均几乎处处收敛到 $f(x,\ y)$.

很自然地，还可以考虑序列 $\{S_{k,k}(f;\ x,\ y),\ k\in\mathbf{N}\}$ 的更一般的线性平均. 定义下三角矩阵

---

① 收稿日期：1980-02-09.

$$\boldsymbol{M} = \left\{\mu_k^{(n)} \mid_{k=0,1,2,\cdots,n} ; \sum_{k=0}^{n} \mu_k^{(n)} = 1 ; n \in \mathbf{N}\right\} \tag{1.1}$$

做和

$$L_n(f;x,y;\boldsymbol{M}) = \sum_{k=0}^{n} \mu_k^{(n)} S_{k,k}(f;x,y),$$

把它叫作 $\sigma(f)$ 的相应于矩阵 $\boldsymbol{M}$(或说求和法 $\boldsymbol{M}$)的 Marcinkiewicz 型和，把它对 $f(x, y)$ 的逼近度(按 L_p 度量)记为

$$R_n(f;\ \boldsymbol{M})_{L_p} = \|L_n(f;\ x,\ y;\ M) - f(x,\ y)\|_{L_p}$$

$\boldsymbol{M}$ 求和法概括了某些重要的线性求和法. 例如当 $\boldsymbol{M}$ 如下定义时，

$$\boldsymbol{M} = \left\{\frac{A_{n-k}^{\alpha-1}}{A_n^{\alpha}},\ k=0,1,2,\cdots,n;\ n\in\mathbf{N}\right\} \tag{1.2}$$

$L_n(f;\ x,\ y;\ \boldsymbol{M})$就是$(C,\ \alpha)$平均 $\sigma_n^{\alpha}(f;\ x,\ y)$. 又如，当

$$\boldsymbol{M} = \left\{\mu_k^{(n)} = \frac{(k+1)^r - k^r}{(n+1)^r},\ k=0,1,2,\cdots,n;\ n\in\mathbf{N}\right\} \quad (r>0) \tag{1.3}$$

时，$L_n(f;\ x,\ y;\ \boldsymbol{M})$即 Zygmund 典型平均.

М. Ф. Тиман 和 В. Г. Пономаренко[4] 对于 Zygmund 典型平均(r 取正整数时)，得到了估计式

$$R_n(f;\boldsymbol{M})_{L_p} \leqslant \frac{C_r}{(n+1)^r} \sum_{k=0}^{n} (k+1)^{r-1} E_{k,k}(f)_{L_p}, \tag{1.4}$$

其中 $E_{k,k}(f)_{L_p}$ 表示 $f(x,\ y)$ 按 L_p 度量的$(k,\ k)$阶最佳三角多项式逼近，即

$$E_{k,k}(f)_{L_p} = \min_{T_{\mu,\nu},\mu,\nu\leqslant k} \|f(x,\ y) - T_{\mu,\nu}(x,\ y)\|_{L_p},$$

C_r 是只与 r 有关的常数，$T_{\mu,\nu}$表$(\mu,\ \nu)$阶三角多项式.

本文所考虑的是连续函数的强性 Marcinkiewicz 型逼近. 即对于 $f(x,\ y)\in C(R)$，由(1.1)定义的 $\boldsymbol{M}$ 及任意的指数，$q>0$，考虑由下式定义的逼近度

$$H_n^q(f;\boldsymbol{M}) = \left\| \sum_{k=0}^{n} \mu_k^{(n)} \left| S_{k,k}(f;x,y) - f(x,y) \right|^q \right\|,$$

此处及下面所涉及的范数都是按 $C(R)$ 度量的范数. 主要的结果是下面两个定理.

定理 1 设 $q>0$，$\{\mu_n,\ n\in\mathbf{N}\}$是一个非负的单调数列. 若有常数 $K\geqslant1$使得

$$\mu_{2n} \leqslant K\mu_n,\ n\in\mathbf{N},$$

则存在只与 q 及 K 有关的常数 $C_{q,K}>0$，使对任意的 $f(x, y)\in C(R)$ 和 $n\in\mathbf{N}$ 有

$$\left\|\sum_{k=0}^{n}\mu_k\left|S_{k,k}(f;x,y)-f(x,y)\right|^q\right\|\leqslant C_{q,K}\sum_{k=0}^{n}\mu_k E_{k,k}^q(f) \quad (1.5)$$

对于 $C(R)$ 的情形，定理 1 把(1.4)改进为强性的不等式，从而加强了 Тиман 和 Пономаренко 的结果.

这是因为对于由(1.3)定义的 $\boldsymbol{M}$ 及任意的 $q>0$，从定理 1 可以推出

$$H_n^q(f;\boldsymbol{M})\leqslant\frac{C_r}{(n+1)^r}\sum_{k=0}^{n}(k+1)^{r-1}E_{k,k}^q(f) \quad (1.6)$$

(其中的 r 不必只取自然数). 当 $q=1$，r 取自然数时，这就是(1.4)在强性逼近时的情形($p=+\infty$时).

由定理 1 还得出当 $\alpha\geqslant1$ 时强性的(C, α)逼近的上界估计. 即对于由(1.2)定义的 $\boldsymbol{M}$，当 $\alpha\geqslant1$ 时，

$$\begin{aligned}H_n(f;\boldsymbol{M})&=\left\|\frac{1}{A_n^\alpha}\sum_{k=0}^{n}A_{n-k}^{\alpha-1}\left|f(x,y)-S_{k,k}(f;x,y)\right|\right\|\\&\leqslant\frac{C_\alpha}{A_n^\alpha}\sum_{k=0}^{n}A_{n-k}^{\alpha-1}E_{k,k}(f),\end{aligned} \quad (1.7)$$

对于 $\alpha\in(0, 1)$的情形，(1.7)不能从定理 1 推出. 此时有

定理 2 设 $q>0$，$\alpha\in(0, 1)$，矩阵 $\boldsymbol{M}$ 由(1.2)定义，则存在只与 q 和 α 有关的常数 $C_{q,\alpha}>0$，使对一切 $f\in C(R)$ 和 $n\in\mathbf{N}$ 有

$$\begin{aligned}H_n^q(f;M)&=\left\|\sum_{k=0}^{n}\frac{A_{n-k}^{\alpha-1}}{A_n^\alpha}\left|S_{k,k}(f;x,y)-f(x,y)\right|^q\right\|\\&\leqslant\frac{C_{q,\alpha}}{A_n^\alpha}\sum_{k=0}^{n}A_{n-k}^{\alpha-1}E_{k,k}^q(f).\end{aligned} \quad (1.8)$$

对于一元连续函数的情形，类似(1.6)和(1.7)的估计式曾为施咸亮[5]证明，不过那里分别限定 $q\leqslant1$ 及 $\alpha>\frac{1}{2}$.

把定理 1 用于满足一定条件的一类线性求和法，在一定的函数类上可以得到逼近的阶的准确估计.

Г. А. Фомин[6]曾研究了满足以下条件的线性求和法(下三角矩阵)：

$$\begin{cases}\boldsymbol{\Lambda}=\{\lambda_k^{(n)},k=0,1,2,\cdots,n;\lambda_{n+1}^{(n)}=0;\quad n\in\mathbf{N}\},\\(n+1)^{\alpha-1}\sum_{k=0}^{n}\left|\Delta\lambda_k^{(n)}\right|^\alpha=O(1),\quad n>1,\end{cases} \quad (1.9)$$

其中 $\Delta\lambda_k^{(n)}=\lambda_k^{(n)}-\lambda_{k+1}^{(n)}(k=0,1,2,\cdots,n)$，相当于我们这里的 $\mu_k^{(n)}$. 用本文的符号，条件(1.9)成为

$$(n+1)^{\alpha-1}\sum_{k=0}^{n}|\mu_k^{(n)}|^{\alpha}\leqslant K,\quad \alpha>1,K>0\text{ 为常数},\tag{1.10}$$

其中$\{\mu_k^{(n)}\}$由(1.1)定义. 把满足(1.10)的全体下三角矩阵 $\boldsymbol{M}$ 记作 M^*. 设 $\omega(t)$是连续模，函数类

$$H^{\omega}=\{f\mid f\in C(R),\ \omega(f;\ t)\leqslant\omega(t),\quad 0\leqslant t\leqslant\pi\},$$

其中 $\omega(f;\ t)$表示 $f(x,\ y)$的完全连续模在 t 点的值. 又设$\{F_n,\ n\in\mathbf{N}\}$是一个单调趋于零的正数列，函数类

$$C(F)=\{f\mid f\in C(R),\ E_{k,k}(f)\leqslant F_k,\ k\in\mathbf{N}\},$$

定义

$$A_n(\omega,\ M^*)=\sup_{f\in H^{\omega},\boldsymbol{M}\in M^*}R_n(f;\ \boldsymbol{M}),$$
$$A_n(F,\ M^*)=\sup_{f\in C(F),\boldsymbol{M}\in M^*}R_n(f;\ \boldsymbol{M}),$$

其中 $R_n(f;\ \boldsymbol{M})=\|L_n(f;\ x,\ y;\ \boldsymbol{M})-f(x,\ y)\|$，$L_n$ 表示相应于矩阵 $\boldsymbol{M}$ 的 Marcinkiewicz 型和.

B. A. Баскаков[7]对于一元函数的情形证明了

$$A_n(\omega,M^*)\asymp\left\{\frac{1}{n+1}\sum_{k=0}^{n}\omega^{\beta}\left(\frac{1}{k+1}\right)\right\}^{\frac{1}{\beta}};A_n(F,M^*)\asymp\left\{\frac{1}{n+1}\sum_{k=0}^{n}F_k^{\beta}\right\}^{\frac{1}{\beta}},$$

其中 $\beta=\frac{\alpha}{\alpha-1}$.

利用定理1，Баскаков 的结果得以推广于二元 Marcinkiewicz 型逼近的情形. 即有

定理 3 按前述记号，存在常数 $C_{\alpha,k}\geqslant C'_{\alpha,k}>0$，只与 α 和 K 有关，使得

$$C'_{\alpha,k}\left\{\frac{1}{n+1}\sum_{k=0}^{n}\omega^{\beta}\left(\frac{1}{k+1}\right)\right\}^{\frac{1}{\beta}}\leqslant A_n(\omega,M^*)\leqslant C_{\alpha,k}\left\{\frac{1}{n+1}\sum_{k=0}^{n}\omega^{\beta}\left(\frac{1}{k+1}\right)\right\}^{\frac{1}{\beta}},\tag{1.11}$$

$$C'_{\alpha,k}\left\{\frac{1}{n+1}\sum_{k=0}^{n}F_k^{\beta}\right\}^{\frac{1}{\beta}}\leqslant A_n(F,M^*)\leqslant C_{\alpha,k}\left\{\frac{1}{n+1}\sum_{k+1}^{n}F_k^{\beta}\right\}^{\frac{1}{\beta}},\tag{1.12}$$

其中 $\beta=\frac{\alpha}{\alpha-1}$，$n\in\mathbf{N}$.

§2. 引理

全部证明的关键在于以下的引理.

引理 设 $q\geqslant 2$，$m\in\mathbf{N}^*$，$f\in C(R)$，那么存在只与 q 有关的常数 $C_q>0$，使

$$J_m\equiv\sum_{k=m}^{2m-1}|S_{k,k}(f;x,y)|^q\leqslant C_q m\|f\|^q. \tag{2.1}$$

证 写出 $S_{k,k}(f;\ x,\ y)$的积分表达式

$$\begin{aligned}S_{k,k}(f;x,y)&=\frac{1}{\pi^2}\int_{-\pi}^{\pi}\int_{-\pi}^{\pi}f(x+u,y+v)D_k(u)D_k(v)\mathrm{d}u\mathrm{d}v\\&=\frac{1}{\pi^2}\int_0^{\pi}\int_0^{\pi}4\Phi(u,v)D_k(u)D_k(v)\mathrm{d}u\mathrm{d}v,\end{aligned}$$

其中 $\Phi(u,\ v)=\frac{1}{4}\{f(x+u,\ y+v)+f(x-u,\ y+v)+f(x+u,\ y-v)+f(x-u,\ y-v)\}$，$D_k(t)$表示 Dirichlet 核，$D_k(t)=\dfrac{\sin\left(k+\frac{1}{2}\right)t}{2\sin\frac{t}{2}}$.

因为

$$4D_k(u)D_k(v)=\frac{\sin ku\cdot\sin kv}{\tan\frac{u}{2}\cdot\tan\frac{v}{2}}+\frac{\sin ku}{\tan\frac{u}{2}}\cos kv+\cos ku\frac{\sin kv}{\tan\frac{v}{2}}+$$

$\cos ku\cos kv$，所以可把积分分成四个. 记

$$I_1=\frac{1}{\pi^2}\int_0^{\pi}\int_0^{\pi}\frac{\Phi(u,v)}{\tan\frac{u}{2}\tan\frac{v}{2}}\sin ku\cdot\sin kv\mathrm{d}u\mathrm{d}v,$$

$$I_2=\frac{1}{\pi^2}\int_0^{\pi}\int_0^{\pi}\frac{\Phi(u,v)}{\tan\frac{u}{2}}\sin ku\cdot\cos kv\mathrm{d}u\mathrm{d}v,$$

$$I_3=\frac{1}{\pi^2}\int_0^{\pi}\int_0^{\pi}\frac{\Phi(u,v)}{\tan\frac{v}{2}}\cos ku\cdot\sin kv\mathrm{d}u\mathrm{d}v,$$

$$I_4=\frac{1}{\pi^2}\int_0^{\pi}\int_0^{\pi}\Phi(u,v)\cos ku\cdot\cos kv\mathrm{d}u\mathrm{d}v.$$

在以下的推导中，为了简化符号，约定用 C 表示绝对常数，用带脚标的 C(例如 C_q)表示只与其脚标有关的常数，对于同一性质的常数总用同一

符号表示，而不计较其大小，就如同使用符号 $O(1)$ 一样．那么

$$|I_4|\leqslant\frac{1}{\pi^2}\int_0^\pi\int_0^\pi|\Phi(u,v)|\,\mathrm{d}u\mathrm{d}v\leqslant\|f\|,$$

$$|I_3|\leqslant\frac{1}{\pi^2}\int_0^\pi\left|\int_0^\pi\frac{\Phi(u,v)}{\tan\frac{v}{2}}\sin kv\mathrm{d}v\right||\cos ku|\,\mathrm{d}u.$$

$$\leqslant\frac{1}{\pi^2}\int_0^\pi\left(\left|\int^{m^{-1}}\frac{\Phi(u,v)}{\tan\frac{v}{2}}\sin kv\mathrm{d}v\right|+\left|\int_{m^{-1}}^\pi\frac{\Phi(u,v)}{\tan\frac{v}{2}}\sin kv\mathrm{d}v\right|\right)\mathrm{d}u.$$

当 $m\leqslant k<2m$ 时，

$$\left|\int_0^{m^{-1}}\frac{\Phi(u,v)}{\tan\frac{v}{2}}\sin kv\mathrm{d}v\right|\leqslant C\cdot k\|f\|\cdot m^{-1}\leqslant C\|f\|.$$

设 $q'=\dfrac{q}{q-1}$，则 $q'\in(1,\ 2]$．定义 v 的函数

$$\varphi(v)=\begin{cases}\dfrac{\Phi(u,\ v)}{\tan\frac{v}{2}}, & m^{-1}\leqslant v\leqslant\pi,\\ 0, & 0\leqslant v<m^{-1},\\ -\varphi(-v), & -\pi<v<0.\end{cases}$$

显然，$\varphi(v)\in L^{q'}[-\pi,\ \pi]$．根据 Hausdorff-Young 不等式(参见[3]，Ⅻ章§2)

$$\sum_{k=m}^{2m-1}\left|\int_{m^{-1}}^\pi\frac{\Phi(u,v)}{\tan\frac{v}{2}}\sin kv\mathrm{d}v\right|^q=\sum_{k=m}^{2m-1}\left|\int_{-\pi}^\pi\varphi(v)\sin kv\mathrm{d}v\right|^q\frac{1}{2^q}$$

$$\leqslant C_q\left(\int_{-\pi}^\pi|\varphi(v)\|^{q'}\mathrm{d}v\right)^{\frac{q}{q'}}\leqslant C_q\|f\|^q\left\{\int_{m^{-1}}^\pi\left(\frac{1}{v}\right)^{q'}\mathrm{d}v\right\}^{\frac{q}{q'}}$$

$$\leqslant C_q m\|f\|^q.$$

使用 Hölder 不等式(参见[3] Ⅰ章§9)，代入上面的结果，推得

$$\sum_{k=m}^{2m-1}|I_3|^q\leqslant C_q\left\{m\|f\|^q+\sum_{k=m}^{2m-1}\left[\int_0^\pi\left|\int_{m^{-1}}^\pi\frac{\Phi(u,v)}{\tan\frac{v}{2}}\sin kv\mathrm{d}v\right|\mathrm{d}u\right]^q\right\}$$

$$\leqslant C_q m\|f\|^q+C_q\sum_{k=m}^{2m-1}\left(\int_0^\pi 1^{q'}\mathrm{d}u\right)^{\frac{q}{q'}}\cdot\int_0^\pi\left|\int_{m^{-1}}^\pi\frac{\Phi(u,v)}{\tan\frac{v}{2}}\sin kv\mathrm{d}v\right|^q\mathrm{d}u$$

$$\leqslant C_q m\|f\|^q + C_q\int_0^{\pi}\left(\sum_{k=m}^{2m-1}\left|\int_{m^{-1}}^{\pi}\frac{\Phi(u,v)}{\tan\frac{v}{2}}\sin kv\mathrm{d}v\right|^q\right)\mathrm{d}u$$

$$\leqslant C_q m\|f\|^q.$$

同样地

$$\sum_{k=m}^{2m-1}|I_2|^q \leqslant C_q\cdot m\|f\|^q.$$

将 I_1 分成四个积分.

$$I_1 = \frac{1}{\pi^2}\left(\int_0^{m^{-1}}\int_0^{m^{-1}} + \int_{m^{-1}}^{\pi}\int_0^{m^{-1}} + \int_0^{m^{-1}}\int_{m^{-1}}^{\pi} + \int_{m^{-1}}^{\pi}\int_{m^{-1}}^{\pi}\right)\cdot$$

$$\frac{\Phi(u,v)}{\tan\frac{u}{2}\tan\frac{v}{2}}\sin ku\sin kv\mathrm{d}u\mathrm{d}v$$

$$= I_{11} + I_{12} + I_{13} + I_{14}.$$

当 $m\leqslant k<2m$ 时

$$|I_{11}|\leqslant C\|f\|,$$

同计算 $\sum_{k=m}^{2m-1}|I_3|^q$ 类似可得

$$\sum_{k=m}^{2m-1}|I_{12}|^q \leqslant \sum_{k=m}^{2m-1}\left\{\frac{1}{\pi^2}\int_0^{m^{-1}}\left|\frac{\sin ku}{\tan\frac{u}{2}}\right|\left|\int_{m^{-1}}^{\pi}\frac{\Phi(u,v)}{\tan\frac{v}{2}}\sin kv\mathrm{d}v\right|\mathrm{d}u\right\}^q$$

$$\leqslant C_q\sum_{k=m}^{2m-1}\left(\int_0^{m^{-1}}\left|\frac{\sin ku}{\tan\frac{u}{2}}\right|^{q'}\mathrm{d}u\right)^{\frac{q}{q'}}\cdot\int_0^{m^{-1}}\left|\int_{m^{-1}}^{\pi}\frac{\Phi(u,v)}{\tan\frac{v}{2}}\sin kv\mathrm{d}v\right|^q\mathrm{d}u$$

$$\leqslant C_q\sum_{k=m}^{2m-1}\left(\int_0^{m^{-1}}m^{q'}\mathrm{d}u\right)^{\frac{q}{q'}}\int_0^{m^{-1}}\left|\int_{m^{-1}}^{\pi}\frac{\Phi(u,v)}{\tan\frac{v}{2}}\sin kv\mathrm{d}v\right|^q\mathrm{d}u$$

$$\leqslant C_q\cdot m\cdot\int_0^{m^{-1}}\left(\sum_{k=m}^{2m-1}\left|\int_{m^{-1}}^{\pi}\frac{\Phi(u,v)}{\tan\frac{v}{2}}\sin kv\mathrm{d}v\right|^q\right)\mathrm{d}u$$

$$\leqslant C_q\cdot m\|f\|^q,$$

$$\sum_{k=m}^{2m-1}|I_{13}|^q \leqslant C_q m\|f\|^q.$$

下面估计 $\sum_{k=m}^{2m-1}|I_{14}|^q$.

因为 $\sin ku \cdot \sin kv=\dfrac{1}{2}[\cos k(u-v)-\cos k(u+v)]$，所以

$$I_{14}=\frac{1}{2\pi^2}\int_{m^{-1}}^{\pi}\int_{m^{-1}}^{\pi}\frac{\Phi(u,v)}{\tan\frac{u}{2}\cdot\tan\frac{v}{2}}\cdot[\cos k(u-v)-\cos k(u+v)].$$

记 $F(u,v)=\dfrac{\Phi(u,v)}{\tan\frac{u}{2}\cdot\tan\frac{v}{2}}$，$(m^{-1}\leqslant u, v\leqslant\pi)$，则 $|F(u,v)|\leqslant C\dfrac{\|f\|}{u\cdot v}$.

$$\begin{aligned}I_{14}&=\frac{1}{2\pi^2}\int_{m^{-1}}^{\pi}\int_{m^{-1}}^{\pi}F(u,v)\cos k(u-v)\mathrm{d}u\mathrm{d}v-\\&\quad\frac{1}{2\pi^2}\int_{m^{-1}}^{\pi}\int_{m^{-1}}^{\pi}F(u,v)\cos k(u,v)\mathrm{d}u\mathrm{d}v\\&=\frac{1}{2\pi^2}\tau_1-\frac{1}{2\pi^2}\tau_2.\end{aligned}$$

作坐标变换，令 $u-v=\xi$，$u+v=\eta$，其中积分区域 D 由直线

$$\eta=\xi+2m^{-1},\quad \eta=-\xi+2\pi,$$
$$\eta=\xi+2\pi,\qquad \eta=-\xi+2m^{-1}$$

围成(如图 1).

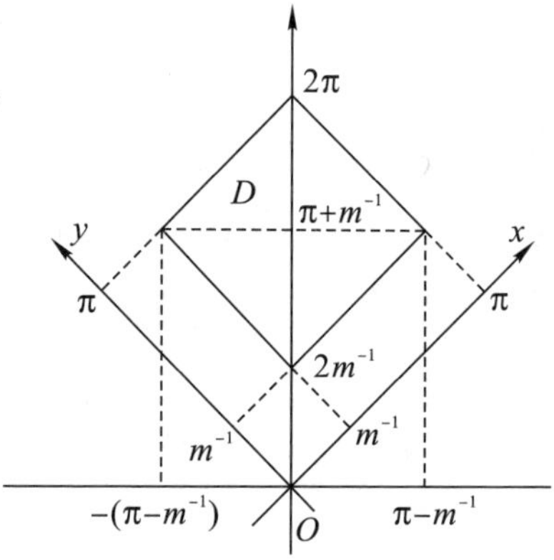

图 1

$$\begin{aligned}\tau_1&=\int_{m^{-1}}^{\pi}\int_{m^{-1}}^{\pi}F(u,v)\cos k(u-v)\mathrm{d}u\mathrm{d}v\\&=\frac{1}{2}\iint_D F\left(\frac{\eta+\xi}{2},\frac{\eta-\xi}{2}\right)\cos k\xi\mathrm{d}\xi\mathrm{d}\eta\\&=\frac{1}{2}\int_{-(\pi-m^{-1})}^{0}\left[\int_{-\xi+2m^{-1}}^{2\pi+\xi}F\left(\frac{\eta+\xi}{2},\frac{\eta-\xi}{2}\right)\mathrm{d}\eta\right]\cos k\xi\mathrm{d}\xi+\\&\quad\frac{1}{2}\int_{0}^{\pi-m^{-1}}\left[\int_{\xi+2m^{-1}}^{2\pi-\xi}F\left(\frac{\eta+\xi}{2},\frac{\eta-\xi}{2}\right)\mathrm{d}\eta\right]\cos k\xi\mathrm{d}\xi\\&=\frac{1}{2}\int_{0}^{\pi-m^{-1}}\left[\int_{\xi+2m^{-1}}^{2\pi-\xi}\left(F\left(\frac{\eta+\xi}{2},\frac{\eta-\xi}{2}\right)+F\left(\frac{\eta-\xi}{2},\frac{\eta+\xi}{2}\right)\right)\mathrm{d}\eta\right]\cos k\xi\mathrm{d}\xi.\end{aligned}$$

记

$$g(\xi)=\int_{\xi+2m^{-1}}^{2\pi-\xi}\left(F\left(\frac{\eta+\xi}{2},\frac{\eta-\xi}{2}\right)+F\left(\frac{\eta-\xi}{2},\frac{\eta+\xi}{2}\right)\right)\mathrm{d}\eta,$$
$$0\leqslant\xi\leqslant\pi-m^{-1}.$$

由于 $|F(u,v)|\leqslant C\dfrac{\|f\|}{uv}$，所以

$$\begin{aligned}|g(\xi)| &\leqslant C\|f\|\int_{\xi+2m^{-1}}^{2\pi-\xi}\frac{\mathrm{d}\eta}{(\eta+\xi)(\eta-\xi)}\\ &= C\|f\|\frac{1}{2\xi}\int_{\xi+2m^{-1}}^{2\pi-\xi}\left(\frac{1}{\eta-\xi}-\frac{1}{\eta+\xi}\right)\mathrm{d}\eta\\ &= C\|f\|\frac{1}{2\xi}\left[\ln\left(1-\frac{\xi}{\pi}\right)-\ln\frac{2m^{-1}}{2m^{-1}+2\xi}\right]\\ &= C\|f\|\frac{\ln(1+m\xi)}{\xi}.\end{aligned}$$

用 Hausdorff-Young 不等式，

$$\begin{aligned}\sum_{k=m}^{2m-1}|\tau_1|^q &= \frac{1}{2^q}\sum_{k=m}^{2m-1}\left|\int_0^{\pi-m^{-1}} g(\xi)\cos k\xi\mathrm{d}\xi\right|^q\\ &\leqslant C_q\left(\int_0^{\pi-m^{-1}}|g(\xi)|^{q'}\mathrm{d}\xi\right)^{\frac{q}{q'}}\\ &\leqslant C_q\|f\|^q\left(\int_0^{\pi-m^{-1}}\left|\frac{\ln(1+m\xi)}{\xi}\right|^{q'}\mathrm{d}\xi\right)^{\frac{q}{q'}}\\ &= C_q\|f\|^q\left\{\int_0^{m\pi-1}\left|\frac{\ln(1+t)}{\xi}\right|^{q'}m^{q'-1}\mathrm{d}t\right\}^{\frac{q}{q'}}\\ &\leqslant C_q m\cdot\|f\|^q\left\{\int_0^{+\infty}\left|\frac{\ln(1+t)}{t}\right|^{q'}\mathrm{d}t\right\}^{\frac{q}{q'}}.\end{aligned}$$

因为 $q'=\dfrac{q}{q-1}>1,\displaystyle\int_0^{+\infty}\left|\frac{\ln(1+t)}{t}\right|^{q'}\mathrm{d}t<+\infty$，所以

$$\sum_{k=m}^{2m-1}|\tau_1|^q\leqslant C_q m\|f\|^q.$$

也用累次积分来估计 τ_2：

$$\begin{aligned}\tau_2 &= \int_{m^{-1}}^{\pi}\int_{m^{-1}}^{\pi}F(u,v)\cos k(u+v)\mathrm{d}u\mathrm{d}v\\ &= \frac{1}{2}\iint_D F\left(\frac{\eta+\xi}{2},\frac{\eta-\xi}{2}\right)\cos k\eta\mathrm{d}\xi\mathrm{d}\eta\\ &= \frac{1}{2}\int_{2m^{-1}}^{2\pi}\left[\int_{-\eta+2m^{-1}}^{\eta-2m^{-1}}F\left(\frac{\eta+\xi}{2},\frac{\eta-\xi}{2}\right)I_D(\xi,\eta)\mathrm{d}\xi\right]\cos k\eta\mathrm{d}\eta,\end{aligned}$$

其中 $I_D(\xi,\ \eta)$ 表示 D 的特征函数(即在 D 上取值 1，在 D 处取值 0)．记

$$h(\eta)=\int_{-\eta+2m^{-1}}^{\eta-2m^{-1}}F\left(\frac{\eta+\xi}{2},\frac{\eta-\xi}{2}\right)I_D(\xi,\eta)\mathrm{d}\xi,\quad 2m^{-1}\leqslant\eta\leqslant 2\pi.$$

因为 $\left|F\left(\dfrac{\eta+\xi}{2},\dfrac{\eta-\xi}{2}\right)I_D(\xi,\eta)\right|\leqslant C\|f\|\dfrac{1}{(\eta+\xi)(\eta-\xi)}$，所以

$$|h(\eta)|\leqslant C\|f\|\frac{1}{\eta}\int_{-\eta+2m^{-1}}^{\eta-2m^{-1}}\left(\frac{1}{\eta-\xi}+\frac{1}{\eta+\xi}\right)\mathrm{d}\xi$$

$$\leqslant C\|f\|\frac{\ln(m\eta-1)}{\eta},\quad 2m^{-1}\leqslant\eta\leqslant 2\pi.$$

再用 Hausdorff-Young 不等式，得

$$\sum_{k=m}^{2m-1}|\tau_2|^q=\frac{1}{2^q}\sum_{k=m}^{2m-1}\left|\int_{2m^{-1}}^{2\pi}h(\eta)\cos k\eta\mathrm{d}\eta\right|^q$$

$$\leqslant C_q\|f\|^q\left\{\int_{2m^{-1}}^{2\pi}\left|\frac{\ln(m\eta-1)}{\eta}\right|^{q'}\mathrm{d}\eta\right\}^{\frac{q}{q'}}$$

$$=C_q\|f\|^q\left(\int_2^{2m\pi}\left|\frac{\ln(t-1)}{t}\right|^{q'}m^{q'-1}\mathrm{d}t\right)^{\frac{q}{q'}}$$

$$\leqslant C_q\cdot m\|f\|^q\left\{\int_2^{+\infty}\left|\frac{\ln(t-1)}{t}\right|^{q'}\mathrm{d}t\right\}^{\frac{q}{q'}}.$$

对于 $q'=\frac{q}{q-1}>1$，$\left\{\int_2^{+\infty}\left|\frac{\ln(t-1)}{t}\right|^{q'}\mathrm{d}t\right\}^{\frac{q}{q'}}$ 也是只与 q 有关的常数，所以

$$\sum_{k=m}^{2m-1}|\tau_2|^q\leqslant C_qm\|f\|^q,$$

总起来，用 Hölder 不等式，

$$J_m=\sum_{k=m}^{2m-1}|S_{k,k}(f;x,y)|^q\leqslant\sum_{k=m}^{2m-1}4^{q-1}(|I_1|^q+|I_2|^q+|I_3|^q+|I_4|^q)$$

$$\leqslant C_q\sum_{k=m}^{2m-1}(|I_{11}|^q+|I_{12}|^q+|I_{13}|^q+|\tau_1|^q+|\tau_2|^q+|I_2|^q+$$

$$|I_3|^q+|I_4|^q),$$

代入前面的结果，就得到

$$J_m\leqslant C_q\cdot m\|f\|^q.\quad\square$$

§3. 定理 1 和定理 2 的证明

定理 1 的证

对于正整数 m，记

$$\sum\nolimits_m=\sum_{k=m}^{2m-1}\mu_k|S_{k,k}|(f;xy)-f(x,y)|^q.$$

设 $T_{m,m}(x,y)$是 $f(x,y)$ (m,m)阶最佳逼近三角多项式. 取数 $q'>1$，使 $qq'\geqslant2$. 由 Hölder 不等式，

$$\sum_m \leqslant \left(\sum_{k=m}^{2m-1}\mu_k^{\frac{q'}{q'-1}}\right)^{1-\frac{1}{q'}}\left(\sum_{k=m}^{2m-1}|S_{k,k}(f)-f|^{qq'}\right)^{\frac{1}{q'}}.$$

根据条件，$\{\mu_k\}$单调且 $0\leqslant\mu_{2k}\leqslant K\mu_k(k\in\mathbf{N})$，$k\geqslant1$. 可见，当 $k=m$，$m+1$，…，$2m$ 时，有

$$\mu_k\leqslant K\mu_m,$$

从而

$$\sum_m \leqslant Km^{1-\frac{1}{q'}}\cdot\mu_m\cdot\left\{\sum_{k=m}^{2m-1}C_{qq'}\left[|S_{k,k}|(f)-T_{m,m}|^{qq'}+|T_{m,m}-f|^{qq'}\right]\right\}^{\frac{1}{q'}}.$$

因为当 $k\geqslant m$ 时，$S_{k,k}(f)-T_{m,m}=S_{k,k}(f-T_{m,m})$，使用引理及 $\|f-T_{m,m}\|=E_{m,m}(f)$，得

$$\sum_m \leqslant C_{qq'}K\cdot m^{1-\frac{1}{q'}}\mu_m\cdot\{m\cdot E_{m,m}^{qq'}(f)\}^{\frac{1}{q'}}.$$

注意 q'由 q 决定，把 $C_{qq'}\cdot K$ 记为 $C_{q,K}$，得

$$\sum_m \leqslant C_{q,K}\cdot m\cdot\mu_m E_{m,m}^{q}(f). \tag{3.1}$$

很明显，

$$|S_{0,0}(f)-f|^q\leqslant[2E_{0,0}(f)]^q,$$

而对于正整数 n，有正整数 l 满足 $2^{l-1}\leqslant n<2^l$，于是

$$\sum_{k=0}^{n}\mu_k|S_{k,k}(f)-f|^q\leqslant\mu_0|S_{0,0}(f)-f|^q+\sum_{j=0}^{l-1}\sum_{k=2^j}^{2^{j+1}-1}\mu_k|S_{k,k}(f)-f|^q.$$

根据(3.1)，

$$\sum_{2^j}\leqslant C_{q,K}2^j\mu_{2^j}E_{2^j,2^j}^q(f),$$

代入前式，得

$$\sum_{k=0}^{n}\mu_k|S_{k,k}(f)-f|^q\leqslant C_{q,K}\left\{\mu_0E_{0,0}^q(f)+\sum_{j=0}^{l-1}2^j\mu_{2^j}E_{2^j,2^j}^q(f)\right\}$$

$$=C_{q,K}\left\{\mu_0E_{0,0}^q+\mu_1E_{1,1}^q+\sum_{j=0}^{l-1}2^j\mu_{2^j}E_{2^j,2^j}^q\right\},$$

由$\{\mu_k\}$单调及 $0\leqslant\mu_{2k}\leqslant K\mu_k$，$(k\in\mathbf{N})$ $K\geqslant1$，得知

$$2^{j-1}\mu_{2^j}\leqslant K\sum_{k=2^{j-1}}^{2^j-1}\mu_k,\quad j\in\mathbf{N}^*,$$

再注意到 $E_{k,k}(f)$是单调降的，就得到

$$\sum_{k=0}^{n}\mu_k|S_{k,k}(f)-f|^q\leqslant C_{q,K}\left\{\mu_0E_{0,0}^q+\mu_1E_{1,1}^q+2K\sum_{j=1}^{l-1}\sum_{k=2^{j-1}}^{2^j-1}\mu_kE_{k,k}^q\right\}$$

$$\leqslant C_{q,K}\sum_{k=0}^{n}\mu_kE_{k,k}^q(f).\quad\square$$

注　对于单调非负数列$\{\mu_n\}$，条件 $\mu_{2n}\leqslant K\mu_n$ 与条件 $\frac{1}{n+1}\sum_{k=0}^{2n}\mu_k\leqslant K\mu_n$ 等价.

推论 1　若由(1.1)定义的 $\boldsymbol{M}$ 的每行都是非负单调降的，则对于 $f\in C(R)$，$q>0$，有

$$H_n^q(f;\boldsymbol{M})\leqslant C_q\sum_{k=0}^{n}\mu_k^{(n)}E_{k,k}^q(f).$$

证　对 M 的每行取 $K=1$ 用定理 1 便得.

特别地，由于(1.2)定义的 $\boldsymbol{M}$，当 $\alpha\geqslant1$ 时，满足此推论的条件，从而(1.7)成立.

推论 2　设 $\boldsymbol{M}$ 由(1.3)定义，则对任意的实数 $q>0$ 和 $r>0$，(1.6)成立.

证　因为 $\mu_0^{(n)}=\frac{1}{(n+1)^r}$，而当 $k\geqslant1$ 时，有 $\theta_k\in(0,1)$使得

$$\mu_k^{(n)}=\frac{(k+1)^r-k^r}{(n+1)^r}=\frac{r(k+\theta_k)^{r-1}}{(n+1)^r}\leqslant r\cdot2^{|1-r|}\frac{(k+1)^{r-1}}{(n+1)^r},\qquad r>0,$$

所以 $\mu_k^{(n)}\leqslant C_r\frac{(k+1)^{r-1}}{(n+1)^r}$，$(k=0,1,2,\cdots,n)$. 于是

$$H_n^q(f;M)=\left\|\sum_{k=0}^{n}\mu_k^{(n)}|S_{k,k}(f)-f|^q\right\|$$

$$\leqslant\frac{C_r}{(n+1)^r}\left\|\sum_{k=0}^{n}(k+1)^{r-1}|S_{k,k}(f)-f|^q\right\|.$$

数列$\{(k+1)^{r-1},k\in\mathbf{N}\}$，当 $r\leqslant1$ 时单调降，当 $r>1$ 时单调增且满足

$$(2k+1)^{r-1}=2^{r-1}\left(k+\frac{1}{2}\right)^{r-1}<2^{r-1}(k+1)^{r-1},$$

所以它总满足定理 1 的条件. 因此

$$\left\|\sum_{k=0}^{n}(k+1)^{r-1}|S_{k,k}(f)-f|^q\right\|\leqslant C_r\sum_{k=0}^{n}(k+1)^{r-1}E_{k,k}^q(f).\qquad(3.2)$$

由此即得(1.6).

注　(3.2)对于 $q>0$ 和任意实数 r 成立. 对于一元函数的情形，施

咸亮[5]证明了与(3.2)同样形状的不等式，(不过限制 $q\geqslant 1$，$r\geqslant 0$).

下面证明定理2.

对于 $n=0,1,2,\cdots,8$，存在常数 $C_q>0$，使

$$H_n^q(f;M)=\left\|\sum_{k=0}^{n}\frac{A_{n-k}^{\alpha-1}}{A_n^\alpha}\left|f(x,y)-S_{k,k}(f;x,y)\right|^q\right\|$$

$$\leqslant C_q\sum_{k=0}^{n}\frac{A_{n-k}^{\alpha-1}}{A_n^\alpha}E_{k,k}^q(f).$$

设 $n>8$，令 $\nu=\left[\frac{n}{2}\right]$，那么

$$\frac{1}{A_n^\alpha}\sum_{k=0}^{n}A_{n-k}^{\alpha-1}\left|S_{k,k}(f)-f\right|^q=\frac{1}{A_n^\alpha}\Big(\sum_{k=0}^{\nu}+\sum_{k=\nu+1}^{n}\Big)=\sigma_1+\sigma_2.$$

当 $k=0,1,2,\cdots,\nu$ 时，

$$\frac{A_{n-k}^{\alpha-1}}{A_n^\alpha}=\frac{\alpha(n-k+1)\cdots n}{(\alpha+n-k)\cdots(\alpha+n)}\leqslant\frac{\alpha}{\alpha+n-k}\leqslant\frac{\alpha}{\alpha+\frac{n}{2}}<\frac{2}{n+1},$$

所以

$$\sigma_1\leqslant\frac{2}{n+1}\sum_{k=0}^{\nu}\left|S_{k,k}(f)-f\right|^q\leqslant 2H_n^{q,1}.$$

其中 $H_n^{q,1}=\left\|\frac{1}{n+1}\sum_{k=0}^{n}\left|S_{k,k}(f)-f\right|^q\right\|$. 用Abel变换算 σ_2.

$$\sigma_2=\frac{1}{A_n^\alpha}\sum_{k=0}^{n-\nu-1}A_k^{\alpha-1}\left|S_{n-k,n-k}(f)-f\right|^q$$

$$=\frac{1}{A_n^\alpha}\Big\{\sum_{k=0}^{n-\nu-2}(-A_{k+1}^{\alpha-2})\sum_{j=0}^{k}\left|S_{n-j,n-j}(f)-f\right|^q+$$

$$A_{n-\nu-1}^{\alpha-1}\sum_{k=0}^{n-\nu-1}\left|S_{n-k,n-k}(f)-f\right|^q\Big\}.$$

因为 $\alpha\in(0,\ 1)$时有

$$0<\frac{-A_{k+1}^{\alpha-2}}{A_n^\alpha}\leqslant C_\alpha\cdot\frac{(k+1)^{\alpha-2}}{n^\alpha},\quad \frac{A_{n-\nu-1}^{\alpha-1}}{A_n^\alpha}\leqslant C_\alpha\frac{1}{n+1},$$

所以

$$\sigma_2\leqslant C_\alpha\Big\{\frac{1}{n^\alpha}\sum_{k=0}^{n-\nu-2}(k+1)^{\alpha-2}\sum_{j=0}^{k}\left|S_{n-j,n-j}(f)-f\right|^q+$$

$$\frac{1}{n+1}\sum_{k=0}^{n}\left|S_{k,k}(f)-f\right|^q\Big\}$$

$$\leqslant C_\alpha\cdot\sigma+C_\alpha H_n^{q,1},$$

其中

$$\sigma=\frac{1}{n^{\alpha}}\sum_{k=0}^{n-\nu-2}(k+1)^{\alpha-2}\sum_{j=0}^{k}|S_{n-j,n-j}(f)-f|^{q}.$$

设 $T_{\nu,\nu}(x, y)$ 为 $f(x, y)$ 的 ν，ν 阶最佳逼近三角多项式，并记 $g(x, y)=f(x, y)-T_{\nu,\nu}(x, y)$. 得

$$\begin{aligned}\sigma&\leqslant C_{q}\frac{1}{n^{\alpha}}\sum_{k=0}^{n-\nu-2}(k+1)^{\alpha-2}\Big\{\sum_{j=0}^{k}|S_{n-j,n-j}(g)|^{q}+(k+1)E_{\nu,\nu}^{q}(f)\Big\}\\&\leqslant C_{q}\Big[\frac{1}{n^{\alpha}}\sum_{k=0}^{n-\nu-2}(k+1)^{\alpha-1}\Big]E_{\nu,\nu}^{q}(f)+\\&\quad C_{q}\frac{1}{n^{\alpha}}\sum_{k=0}^{n-\nu-2}(k+1)^{\alpha-2}\sum_{j=0}^{k}|S_{n-j,n-j}(g)|^{q}\\&\leqslant C_{q,\alpha}E_{\nu,\nu}^{q}(f)+C_{q}\frac{1}{n^{\alpha}}\sum_{k=0}^{n-\nu-2}(k+1)^{\alpha-2}\sum_{j=0}^{k}|S_{n-j,n-j}(g)|^{q}.\end{aligned}$$

取 $q'=\frac{1}{\alpha}+\frac{2}{q}$，则 $q'>1$ 且 $qq'>2$. 使用 Hölder 不等式及引理，得知当 $k=0,1,2,\cdots,n-\nu-2$ 时，有

$$\begin{aligned}\sum_{j=0}^{k}|S_{n-j,n-j}(g)|^{q}&\leqslant(k+1)^{1-\frac{1}{q'}}\Big(\sum_{j=0}^{k}|S_{n-j,n-j}(g)|^{qq'}\Big)^{\frac{1}{q'}}\\&\leqslant(k+1)^{1-\frac{1}{q'}}\Big(\sum_{j=0}^{n-\nu-2}|S_{n-j,n-j}(g)|^{qq'}\Big)^{\frac{1}{q'}}\\&=(k+1)^{1-\frac{1}{q'}}\Big(\sum_{j=\nu+2}^{n}|S_{j,j}(g)|^{qq'}\Big)^{\frac{1}{q'}}\\&\leqslant(k+1)^{1-\frac{1}{q'}}\Big(\sum_{j=\nu+2}^{2(\nu+2)-1}|S_{j,j}(g)|^{qq'}\Big)^{\frac{1}{q'}}\\&\leqslant(k+1)^{1-\frac{1}{q'}}\cdot C_{q,\alpha}\big[(\nu+2)\|g\|^{qq'}\big]^{\frac{1}{q'}}\\&\leqslant C_{q,\alpha}(k+1)^{1-\frac{1}{q'}}\cdot n^{\frac{1}{q'}}\cdot E_{\nu,\nu}^{q}(f).\end{aligned}$$

于是

$$\begin{aligned}&\frac{1}{n^{\alpha}}\sum_{k=0}^{n-\nu-2}(k+1)^{\alpha-2}\sum_{j=0}^{k}|S_{n-j,n-j}(g)|^{q}\\\leqslant&\frac{1}{n^{\alpha}}\sum_{k=0}^{n-\nu-2}(k+1)^{\alpha-2}\cdot C_{q,\alpha}(k+1)^{1-\frac{1}{q'}}\cdot n^{\frac{1}{q'}}\cdot E_{\nu,\nu}^{q}(f)\\\leqslant&C_{q,\alpha}\frac{1}{n^{\alpha-\frac{1}{q'}}}\sum_{k=0}^{n}(k+1)^{\alpha-1-\frac{1}{q'}}\cdot E_{\nu,\nu}^{q}.\end{aligned}$$

注意 $\alpha-1-\dfrac{1}{q'}>-1$，求得

$$\sigma\leqslant C_{q,\alpha}E^{q}_{\nu,\nu}(f)\leqslant C_{q,\alpha}\frac{1}{\nu+1}\sum_{k=0}^{\nu}E^{q}_{k,k}(f)\leqslant C_{q,\alpha}\frac{1}{n+1}\sum_{k=0}^{n}E^{q}_{k,k}(f).$$

由定理 1 的推论 1 已有　$H_n^{q,1}\leqslant\dfrac{C_q}{n+1}\displaystyle\sum_{k=0}^{n}E^{q}_{k,k}(f).$

总起来得到

$$\frac{1}{A_n^{\alpha}}\sum_{k=0}^{n}A_{n-k}^{\alpha-1}\,|S_{k,k}(f)-f|^{q}=\sigma_1+\sigma_2$$

$$\leqslant C_{q,\alpha}\Big[H_n^{q,1}+\frac{1}{n+1}\sum_{k=0}^{n}E^{q}_{k,k}(f)\Big]\leqslant C_{q,\alpha}\frac{1}{n+1}\sum_{k=0}^{n}E^{q}_{k,k}(f).$$

又因为当 $\alpha\in(0,1)$时，$\dfrac{A_{n-k}^{\alpha-1}}{A_n^{\alpha}}\geqslant\dfrac{\alpha}{\alpha+n}>\dfrac{\alpha}{n+1}$，所以

$$H_n^{q}(f;M)\leqslant C_{q,\alpha}\frac{1}{n+1}\sum_{k=0}^{n}E^{q}_{k,k}(f)\leqslant C_{q,\alpha}\cdot\frac{1}{\alpha}\sum_{k=0}^{n}\frac{A_{n-k}^{\alpha-1}}{A_n^{\alpha}}E^{q}_{k,k}(f).\quad\square$$

§4. 量 $A_n(\omega, M^*)$ 和 $A_n(F, M^*)$ 的阶

现在证明定理 3.

因为　$R_n(f;\boldsymbol{M})=\|L_n(f;x,y;M)-f(x,y)\|$

$$=\Big\|\sum_{k=0}^{n}\mu_k^{(n)}(S_{k,k}(f;x,y)-f(x,y))\Big\|$$

$$\leqslant\Big\|\sum_{k=0}^{n}|\mu_k^{(n)}|\,|(S_{k,k}(f;x,y)-f(x,y))|\Big\|,$$

关于指数 $\alpha>1$ 及 $\beta=\dfrac{\alpha}{\alpha-1}$用 Hölder 不等式于上式，得

$$R_n(f;\boldsymbol{M})\leqslant\Big\{\sum_{k=0}^{n}|\mu_k^{(n)}|^{\alpha}\Big\}^{\frac{1}{\alpha}}\Big\|\sum_{k=0}^{n}|S_{k,k}(f)-f|^{\beta}\Big\|^{\frac{1}{\beta}}$$

$$=\Big\{(n+1)^{\alpha-1}\sum_{k=0}^{n}|\mu_k^{(n)}|^{\alpha}\Big\}^{\frac{1}{\alpha}}\Big\|\frac{1}{n+1}\sum_{k=0}^{n}|S_{k,k}(f)-f|^{\beta}\Big\|^{\frac{1}{\beta}}.$$

仍用 $H_n^{\beta,1}$ 表示带指数β 的强性$(C,1)$逼近，代入条件(1.10)，得到

$$R_n(f;\boldsymbol{M})\leqslant K^{\frac{1}{\alpha}}(H_n^{\beta,1})^{\frac{1}{\beta}},\quad M\in M^*,$$

用定理 1 的推论 1. 得到　$R_n(f;\boldsymbol{M})\leqslant C_{\alpha,K}\Big\{\dfrac{1}{n+1}\displaystyle\sum_{k=0}^{n}E^{\beta}_{k,k}(f)\Big\}^{\frac{1}{\beta}}.$

本文定义的完全连续模是指

$$\omega(f;\ t)=\sup_{\substack{\sqrt{u^2+v^2}\leqslant t\\(x,y)}}|f(x+u,\ y+v)-f(x,\ y)|,\quad 0\leqslant t\leqslant\pi,$$

它与[8]中定义的 $\omega(f;\ t;\ 0)$，$\omega(f;\ 0,\ t)$有如下关系：

$\max\{\omega(f;\ t,\ 0),\ \omega(f;\ 0,\ t)\}\leqslant\omega(f;\ t)\leqslant\omega(f;\ t,\ 0)+\omega(f;\ 0,\ t).$

故由 Jackson 定理(参见[8]5. 3. 1)，得

$$E_{k,k}(f)\leqslant C\omega\left(f;\ \frac{1}{k+1}\right),\quad K\in\mathbf{N}.$$

因此，对于 $f\in H^\omega$ 有

$$R_n(f;\boldsymbol{M})\leqslant C_{\alpha,K}\left\{\frac{1}{n+1}\sum_{k=0}^{n}\omega^\beta\left(\frac{1}{k+1}\right)\right\}^{\frac{1}{\beta}},\quad M\in M^*,$$

由此推出(1. 11)的右半式：

$$A_n(\omega,\boldsymbol{M}^*)=\sup_{f\in H^\omega,M\in M^*}R_n(f;\boldsymbol{M})\leqslant C_{\alpha,K}\left(\frac{1}{n+1}\sum_{k=0}^{n}\omega^\beta\left(\frac{1}{k+1}\right)\right\}^{\frac{1}{\beta}}.$$

同时得到(1. 12)式的右半式：

$$A_n(F,\boldsymbol{M}^*)=\sup_{f\in C(F),M\in M^*}R_n(f;\boldsymbol{M})\leqslant C_{\alpha,K}\left(\frac{1}{n+1}\sum_{k=0}^{n}F_k^\beta\right\}^{\frac{1}{\beta}}.$$

为了得到下界估计，只要利用 Баскаков[7] 对一元函数得到的结果，而把一元作为二元的特例即可．也就是说，若 $f(x)\in C_{2\pi}$，则 $F(x,y)\equiv f(x)$属于 $C(R)$，且 $S_{k,k}(F;\ x,\ y)=S_k(f;\ x)$，$E_{k,k}(F)=E_k(f)$，$\omega(F;\ t)=\omega(f;\ t)$．可见，对一元线性求和成立的下界估计式，对二元的 Marcinkiewicz 型线性求和也成立．由此得(1. 11)(1. 12)的左半式．从而定理 3 得证．

本文是在孙永生老师指导下完成的，谨致衷心的感谢．

参考文献

[1] Marcinkiewicz J. Sur une méthode remarquable de sommation des séries doubles de Fourier. Annal. della R scuola N sup. di Pisa，1939，8：149—160.

[2] Жижиашвили Л В. Обобщение одной теоремы Марценкевича Изв. АН СССР，сер. матем.，1968，32：1 112—1 122.

[3] Zygmund A. Trigonometric series. Cambridge，1959.

[4] Тиман М Ф，Пономаренко В Г. О приближении периодических функций двух переменных суммами типа Марценкевича. Изв. вузов. Матем.，

1975，(9)：59－67.

[5] 施咸亮. 关于傅里叶级数强性求和的几个不等式. 数学学报，1966，16：233－252.

[6] Фомин Г А. О линейных методах суммирования рядов Фуръе. Матем. сб. 1964，65(107)：144－152.

[7] Баскаков В А. О порядке приблиижения нецрерывных функций некоторыми линейными среднияmи их рядов Фурье. Матем. зам. 1971，9：617－627.

[8] Тиман А Ф.，Теория приближения функций действительного переменного. Москва，1960.

Abstract Let $C(R)$ be the linear normed space of functions of two variables which are continuous over $R=[-\pi, \pi; -\pi, \pi]$ and 2π-periodic in each variable. The norm of $f\in C(R)$ is defined by $\|f\|=\max\limits_{(x,y)\in R}|f(x, y)|$. We denote the partial sums of Fourier series of $f\in C(R)$ by $S_{k,j}(f; x, y)(k, j\in\mathbf{N})$. Let $E_{k,j}(f)=\inf\limits_{T_{k,j}}\|f(x, y)-T_{k,j}(x, y)\|$, where $T_{k,j}(x, y)$ denotes the trigonometric polynomials of degree$\leqslant(k, j)$.

The main results in this paper are the following theorems.

Theorem 1 Let $\{\mu_k\}_0^{+\infty}$ be a non-negative monotone sequence, $q>0$ and $f\in C(R)$. If $\{\mu_k\}_0^{+\infty}$ satisfies the condition $\mu_{2k}\leqslant K\mu_k$, $k\in\mathbf{N}$,

Where $K\geqslant 1$ is a constant, then

$$\left\|\sum_{k=0}^{n}\mu_k|S_{k,k}(f;x,y)-f(x,y)|^q\right\|\leqslant C_{q,K}\sum_{k=0}^{n}\mu_k E_{k,k}^q(f),$$

Where $C_{q,K}$ is a positive number, depending only on q and K.

Theorem 2 Let $f\in C(R)$, $\alpha>0$, $q>0$ and A_m^β denote the Cesàro numbers. Then

$$\left\|\frac{1}{A_n^\alpha}\sum_{k=0}^{n}A_{n-k}^{\alpha-1}|S_{k,k}(f;x,y)-f(x,y)|^q\right\|\leqslant C_{\alpha,q}\frac{1}{A_n^\alpha}\sum_{k=0}^{n}A_{n-k}^{\alpha-1}E_{k,k}^q(f),$$

Where $C_{\alpha,q}$ is a positive number, depending only on α and q.

From Theorem 1 we derive some inequalities and estimates for the strong approximation of continuous periodic functions of two variables by its sums of Marcinkiewicz type.

数学年刊，1982，3(6)：789−802.

多重 de la Vallée Poussin 方形余项的估计[①]

The Estimation for the Multiple de la Vallée Poussin Square Remainders

§1. 引言

设 E_N 为 N 维欧氏空间，其中的点记为 $\bar{x}=(x_1, x_2, \cdots, x_N)$. 用 Q^N 表示 E_N 中的基本方体：$Q^N=\{\bar{x}\in E_N \mid -\pi\leqslant x_k<\pi, k=1,2,\cdots,N\}$. 用 $L(Q^N)$表示在 Q^N 可和，对每个变元都以 2π 为周期的 N 元函数在平均范数下所成的线性赋范空间；用 $C(Q^N)$表示 $L(Q^N)$中全体连续函数在一致范数下所成的线性赋性空间. 一般地，用 $X(Q^N)$代表 $L(Q^N)$和 $C(Q^N)$.

全体非负整数的集合记为 $\mathbf{N}$. N 重脚标记作 $\bar{\boldsymbol{v}}=(v_1, v_2, \cdots, v_N)$, $(v_i\in\mathbf{N}, i=1,2,\cdots,N)$. 若 $v_1=v_2=\cdots=v_N=v$，则把 $\bar{\boldsymbol{v}}$ 简写作 $\boldsymbol{v}$.

设 $f(\bar{x})\in X(Q^N)$，$S_{\bar{n}}(f; \bar{x})$为其 Fourier 部分和，$E_{\bar{n}}(f)_X$ 为 f 在 X 度量下的 $\bar{n}$ 阶最佳三角多项式逼近. 对于任意的 n，$l\in\mathbf{N}$，如果脚标 $\bar{n}=(n, n, \cdots, n)$，$\overline{n+l}=(n+l, n+l, \cdots, n+l)$，那么把如下 de la Vallée Poussin 和

$$\frac{1}{(l+1)^N}\sum_{\bar{v}=\bar{n}}^{\overline{n+l}}S_{\bar{v}}(f;\bar{x})$$

$$=\frac{1}{\pi^N}\int_{Q^N}f(\bar{x}+\bar{t})\left\{\frac{1}{(l+1)^N}\prod_{i=1}^{N}\left(\sum_{v_i=n}^{n+l}D_{v_i}(t_i)\right)\right\}\mathrm{d}\bar{t}$$

---

① 收稿日期：1980-09-09.

叫作方形的，阶数是$(n, n+l)$，并记之为$V_n^{n+l}(f; \bar{x})$.（式中$D_k(t)=\dfrac{\sin\left(k+\dfrac{1}{2}\right)t}{2\sin\dfrac{t}{2}}$是 Dirichlet 核）. 差$R_{n,l}(f; \bar{x})=f(\bar{x})-V_n^{n+l}(f; \bar{x})$叫作$f$的 de la Vallée Poussin 方形$(n, n+l)$阶余项. 显然，当$l=0$时，$V_n^n(f; \bar{x})$就是 Fourier 方形部分和$S_n(f; \bar{x})$，相应的方形余项$R_{n,0}(f, \bar{x})$简记作$R_n(f; \bar{x})$.

本文的目的是在一元函数已有的结果的基础上，求出当$N>1$时$\|R_{n,l}(f)\|_X$用方形最佳逼近数列$\{E_v(f)\}_{v=0}^{+\infty}$来表示的阶的估计式. 因为$l=0$的情形是本质的，所以我们着重对 Fourier 方形余项的估计做详细的讨论，然后，只需做些细节的改动，就可推广到一般的 de la Vallée Poussin 的情形.

当$N=1$时，对于 Fourier 余项的估计，Осколков，К. И.[1,2]已得到完善的结果，孙永生[3]也进行了研究. Стечкин С. Б[4]，Dahmen W[5]和 Байбородов С П[6]完成了向 de la Vallée Poussin 情形的推广. 他们的结果总括起来可写成

定理 A 设$\{\varepsilon\}_{n=0}^{+\infty}$是单调下降到零的数列，函数类

$$X(\varepsilon)=\{f\in X(Q^1)\,|\,E_v(f)_X\leqslant\varepsilon_v, v\in\mathbf{N}\},$$

则存在常数$C_1>C_1'>0$，使对一切n，$l\in\mathbf{N}$有

$$C_1\sum_{v=0}^{n+l}\frac{\varepsilon_{n+v}}{v+l+1}\geqslant\sup_{f\in X(\varepsilon)}\|R_{n,l}(f)\|_X\geqslant C_1'\sum_{v=0}^{n+1}\frac{\varepsilon_{n+v}}{v+l+1}.\qquad(1.1)$$

Стечкин[7]还对于$n=0$，即 Féjer 余项的情形，当$X=C$时得到了使(1.1)式右半式成立而与l无关的极值函数.

对于$N=2$，$X=C$，$l=0$的情形，孙永生[8]证明了

定理 B 有绝对常数$C_2>0$，使对于$f\in C(Q^2)$有

$$\|R_n(f;x,y)\|_C\leqslant C_2\ln n\sum_{v=n}^{2n}\frac{E_v(f)_C}{v-n+1},\quad n=2,3,\cdots.\qquad(1.2)$$

定理 C 对于相应于任意的单调趋于零的非负数列$\{\varepsilon_n\}_{n=0}^{+\infty}$的函数类

$$C(\varepsilon)=\{f\in C(Q^2)\,|\,E_n(f)_C\leqslant\varepsilon_n,\quad n\in\mathbf{N}\},$$

存在绝对常数$C_2'>0$，使

$$\sup_{f\in C(\varepsilon)}\|R_n(f)\|_C\geqslant C_2'\sum_{v=n}^{2n}\frac{\varepsilon_v\ln(v-n+2)}{v-n+1},\quad n\in\mathbf{N}.\qquad(1.3)$$

[8]中预料(1.2)能改进成

$$\|R_n(f;x,y)\|_C \leqslant C_2 \sum_{v=n}^{2n} \frac{E_n(f)_C \ln(v-n+2)}{v-n+1}, \quad n \in \mathbf{N}. \tag{1.4}$$

本文的结果证实了这个推测. 本文的结果是

定理 存在只与维数 N 有关的常数 $C_N > C'_N > 0$，使对于一切 n，$l \in \mathbf{N}$和相应于任意的单调趋于零的非负数列$\{\varepsilon_n\}_{n=0}^{+\infty}$的函数类 $X(\varepsilon) = \{f \in X(Q_N) \mid E_v(f)_X \leqslant \varepsilon_v,\ v \in \mathbf{N}\}$，成立着

$$C'_N \sum_{v=0}^{n+l} \frac{\varepsilon_{n+v}}{v+l+1} \ln^{N-1}\left(3+\frac{v}{l+1}\right) \leqslant \sup_{f \in X(\varepsilon)} \|R_{n,l}(f)\|_X$$

$$\leqslant C_N \sum_{v=0}^{n+l} \frac{\varepsilon_{n+v}}{v+l+1} \ln^{N-1}\left(3+\frac{v}{l+1}\right). \tag{1.5}$$

§2. Fourier 方形余项的上方估计

(1.5)的右半式当 $l=0$ 时可等价地表述为

定理 1 存在只与维数 N 有关的常数 $C_N > 0$，使对一切 $f \in X(Q^N)$，有

$$\|R_n(f)\|_X \leqslant C_N \sum_{v=n}^{2n} \frac{E_v(f)_X \ln^{N-1}(v-n+3)}{v-n+1}, \quad n \in \mathbf{N}. \tag{2.1}$$

证 对维数 N 用数学归纳法.

当 $N=1$ 时，(2.1)是定理 A 的结果，设 $N=k$ 时(2.1)成立. 下面设 $N=k+1$. 设 $\bar{x}'=(x_1, x_2, \cdots, x_k)$. 那么

$$S_n(f;\bar{x}') = S_n(S_n(f;x_{k+1});\bar{x}') = S_n(g;\bar{x}').$$

记$g(\bar{x}')=S_n(f;\ x_{k+1})$，是 $f(\bar{x})$关于变元 x_{k+1}的一元 n 阶 Fourier 部分和，而 $S_n(g;\ \bar{x}')$是 $g(\bar{x})$关于变元 $\bar{x}'$ 的 Fourier 方形 n 阶部分和. 用 $E_v^{(\bar{x}')}(g)$表示$_X g(\bar{x})$关于变元 $\bar{x}'$ 的方形 v 阶偏最佳逼近(见[10]42～44页)，则由归纳假设

$$\|S_n(g;\bar{x}')-g(\bar{x}')\|_X \leqslant C_k \sum_{v=n}^{2n} \frac{E_v^{(\bar{x}')}(g)_X}{v-n+1} \ln^{k-1}(v-n+3). \tag{2.2}$$

设 $T_v^*(\bar{x})$是 $f(\bar{x})$的方形 v 阶最佳逼近三角多项式. 置

$$T_v(\bar{x}',x_{k+1}) = \frac{1}{\pi}\int_{-\pi}^{\pi} T_v^*(\bar{x}',x_{k+1}+t)(D_n(t)-D_v(t))\mathrm{d}t + T_v^*(\bar{x}).$$

由偏最佳逼近的定义可知 $E_v^{(\bar{x}')}(g)_X \leqslant \|g-T_v\|_X$. 所以

$$E_v^{(\bar{x}')}(g)_X \leqslant \left\| \frac{1}{\pi}\int_{-\pi}^{\pi} f(\bar{x}', x_{k+1}+t)(D_n(t)-D_v(t))\mathrm{d}t + S_v(f;x_{k+1}) - T_v \right\|_X$$

$$\leqslant E_v(f)_X \frac{1}{\pi}\int_{-\pi}^{\pi} |D_n(t)-D_v(t)|\,\mathrm{d}t +$$

$$\|S_v(f;x_{k+1})-f\|_X + E_v(f)_X. \tag{2.3}$$

对(2.3)式右端第二项用一元已有结果，得

$$\|S_v(f;x_{k+1})-f\|_X \leqslant C_1 \sum_{j=v}^{2v} \frac{E_j^{(x_{k+1})}(f)_X}{j-v+1} \leqslant C_1 \sum_{j=v}^{2v} \frac{E_j(f)_X}{j-v+1}. \tag{2.4}$$

(式中 $E_j^{(x_{k+1})}(f)_X$ 表示 f 关于变元 x_{k+1} 的偏最佳逼近). 因为

$$\frac{1}{\pi}\int_{-\pi}^{\pi} |D_n(t)-D_v(t)|\,\mathrm{d}t \leqslant 4(1+\ln^+|n-v|),$$

所以

$$E_v^{(\bar{x}')}(g)_X \leqslant E_v(f)_X \cdot 5(1+\ln^+|n-v|) + C_1 \sum_{j=v}^{2v} \frac{E_j(f)_X}{j-v+1}. \tag{2.5}$$

将此代入(2.2)，得

$$\|S_n(g;\bar{x}')-g(\bar{x})\|_X \leqslant$$

$$C_{k+1}\left\{ \sum_{v=n}^{2n} \frac{\ln^k(v-n+3)}{v-n+1} E_v(f)_X + \sum_{v=n}^{2n} \frac{\ln^{k-1}(v-n+3)}{v-n+1} \sum_{j=v}^{2v} \frac{E_j(f)_X}{j-v+1} \right\}.$$

因为

$$\sum_{v=n}^{2n} \frac{\ln^{k-1}(v-n+3)}{v-n+1} \sum_{j=v}^{2v} \frac{E_j(f)_X}{j-v+1}$$

$$\leqslant \sum_{v=n}^{4n} \sum_{j=v}^{4n} \frac{\ln^{k-1}(v-n+3)}{v-n+1} \frac{E_j(f)_X}{j-v+1}$$

$$= \sum_{j=n}^{4n} \sum_{v=n}^{j} \frac{\ln^{k-1}(v-n+3)}{(v-n+1)(j-v+1)} E_j(f)_X$$

$$= \sum_{j=n}^{4n} \frac{E_j(f)_X}{j-n+2} \sum_{v=n}^{j} \ln^{k-1}(v-n+3)\left(\frac{1}{v-n+1}+\frac{1}{j-v+1}\right)$$

$$\leqslant C\sum_{v=n}^{2n} \frac{E_v(f)_X \ln^k(v-n+3)}{v-n+1}, \tag{2.6}$$

其中 $C>0$ 为绝对常数，所以

$$\|S_n(g;\bar{x}')-g(\bar{x})\|_x \leqslant C_{k+1} \sum_{v=n}^{2n} \frac{E_v(f)_X \ln^k(v-n+3)}{v-n+1}, \tag{2.7}$$

再于(2.4)中代入 $v=n$，得到

$$\|g-f\|_X \leqslant C_1 \sum_{v=n}^{2n} \frac{E_v(f)_X}{v-n+1}. \tag{2.8}$$

(2.7)(2.8)合起来给出

$$\|S_n(f)-f\|_X \leqslant C_{k+1} \sum_{v=n}^{2n} \frac{E_v(f)_X \ln^k(v-n+3)}{v-n+1}. \quad \square \tag{2.9}$$

§3. 关于数列的一个命题

为了做出方形余项的下方估计，需要引入一个关于数列的命题.

取定正整数 n，把一切形如$\{u_0, u_1, \cdots, u_n, 0, \cdots\}$，且满足 $\Delta u_k=u_k-u_{k+1}\geqslant 0(k\in\mathbf{N})$的数列的集合记作 U_n，用 $\bar{u}$ 表示数列$\{u_k\}_{k=0}^{+\infty}$，并记 $\Delta\bar{u}=\{\Delta u_k\}_{k=0}^{+\infty}$.

U_n 中两个数列 $\bar{a}=\{\bar{a}_k\}_{k=0}^{+\infty}$和 $\bar{b}=\{b_k\}_{k=0}^{+\infty}$，说 $\bar{a}$ 不超过 $\bar{b}$，是指 $a_k\leqslant b_k(k\in\mathbf{N})$，记作 $\bar{a}\leqslant\bar{b}$. 对于 $\bar{a}\in U_n$，说 $\bar{a}^*=\{a_k^*\}_{k=0}^{+\infty}$是不超过 $\bar{a}$ 的最大凸序列，是指 $\bar{a}^*$ 有如下性质：

(1) $\bar{a}^*\in U_n$ 且 $\bar{a}^*\leqslant\bar{a}$；

(2) $\Delta\bar{a}^*\in U_n$，即 $\Delta^2 a_k^*\geqslant 0$，$k\in\mathbf{N}$；

并且 $\bar{a}^*$ 是具有这两条性质的序列中的最大者.

Осколков[2]指出了构作 $\bar{a}^*$ 的具体办法，从而也证实了 $\bar{a}^*$ 的存在性和唯一性，$\bar{a}^*$ 具有如下结构：存在数组

$$k_0=0<k_1<\cdots<k_v<k_{v+1}=n+1, \quad 0\leqslant v\leqslant n, \tag{3.1}$$

使 $a_{k_i}^*=a_{k_i}(i=0,1,2,\cdots,v+1)$，而且当 $k_i\leqslant k\leqslant k_{i+1}$ 时，$\{a_k^*\}$是线性的，即

$$a_k^*=\frac{k_{i+1}-k}{k_{i+1}-k_i}a_{k_i}+\frac{k-k_i}{k_{i+1}-k_i}a_{k_{i+1}}, \quad k_i\leqslant k\leqslant k_{i+1}, \tag{3.2}$$

由此推出 $\bar{a}^*$ 具有重要性质：

(3) $\bar{a}^*\cdot\bar{u}\geqslant C\bar{a}\cdot\bar{u}$，$\forall\bar{u}\in U_n$，

其中 $C>0$ 是与 $\bar{a}$，$\bar{u}$ 以及 n 无关的常数. 运算"·"表示内积，即

$$\bar{a}\cdot\bar{u}=\sum_{k=0}^{+\infty}a_k u_k.$$

我们的目的是拓广上述结果，证明

命题 $\forall \bar{a}\in U_n$，$\forall$ 正整数 m，$\exists \bar{a}^*\in U_n$，使

(1) $\bar{a}^*\leqslant\bar{a}$；

(2) $\Delta\bar{a}^*$，$\Delta^2\bar{a}^*$，…，$\Delta^m\bar{a}^*$ 都属于 U_n，即

$$\Delta^2 a_k\geqslant 0,\ \Delta^3 a_n\geqslant 0,\ \cdots,\ \Delta^{m+1}a_k\geqslant 0,\qquad k\in\mathbf{N};$$

(3) $\forall \bar{u}\in U_n$，有 $\bar{a}^*\cdot\bar{u}\geqslant C_m\bar{a}\cdot\bar{u}$ 成立，其中 C_m 是只与 m 有关的正数.

先证明一个辅助命题.

辅助命题 若$\{a_k\}_{k=0}^{+\infty}$是单调增的正数列，满足条件

$$a_{2k}\leqslant\lambda a_k,\qquad k\in\mathbf{N},\qquad \lambda>1,\tag{3.3}$$

则存在绝对常数 $\alpha>0$，使对一切整数 $0\leqslant p\leqslant q$ 有

$$\sum_{k=p}^{q-1}\frac{q-k}{q-p}a_k\geqslant(\alpha\lambda)^{-1}\sum_{k=p}^{q-1}a_k.\tag{3.4}$$

证 当 $q-p\leqslant 24$ 时，取 $\alpha=24$，不等式成立.

设 $q-p>24$，令 $r=\left[\dfrac{q-p}{4}\right]\geqslant 6$. 当 $q-r\leqslant k\leqslant q-1$ 时，记 $v_k=\left[\dfrac{k+1}{2}\right]$. 显然 $2v_k\geqslant k$. 由条件(3.3)及$\{a_k\}$单调增，得

$$a_k\leqslant a_{2v_k}\leqslant\lambda a_{v_k},\qquad q-r\leqslant k\leqslant q-1,\tag{3.5}$$

$$\sum_{k=q-r}^{q-1}a_k\leqslant\lambda\sum_{k=q-r}^{q-1}a_{v_k}.\tag{3.6}$$

又因为 $v_k\leqslant v_{q-1}=\left[\dfrac{q}{2}\right]\leqslant\dfrac{q+p}{2}=q-\dfrac{q-p}{2}\leqslant q-2r$，所以

$$\begin{aligned}\sum_{k=q-r}^{q-1}a_{v_k}&\leqslant ra_{q-2r}\leqslant\sum_{k=q-2r}^{q-r-1}a_k\leqslant\sum_{k=p}^{q-r-1}a_k=\frac{q-p}{r+1}\sum_{k=p}^{q-r-1}\frac{r+1}{q-p}a_k\\&\leqslant\frac{q-p}{r+1}\sum_{k=p}^{q-r-1}\frac{q-k}{q-p}a_k.\end{aligned}\tag{3.7}$$

将 $r+1>\dfrac{q-p}{4}$代入(3.7)，与(3.6)合并，得

$$\sum_{k=q-r}^{q-1}a_k\leqslant 4\lambda\sum_{k=p}^{q-1}\frac{q-k}{q-p}a_k.\tag{3.8}$$

另一方面，$r>\dfrac{q-p}{4}-1$，$5r>q-p-4+r>q-p$，因此

$$5\sum_{k=q-r}^{q-1}a_k\geqslant\sum_{k=p}^{q-1}a_k.\tag{3.9}$$

(3.8)和(3.9)合起来就得到

$$\sum_{k=p}^{q-1}\frac{q-k}{q-p}a_k \geqslant (24\lambda)^{-1}\sum_{k=p}^{q-1}a_k. \quad \square$$

为证命题，定义数列$\{S_k^{(m)}\}_{k=1}^{+\infty}(m\in\mathbf{N})$

$$S_k^{(0)}\equiv 1,\quad S_k^{(m)}=C_{k+m-1}^{m}=\frac{k\cdot(k+1)\cdots(k+m-1)}{m!},\quad m>0,\ k\in\mathbf{N}^*. \tag{3.10}$$

易见，当$m>0$时，$S_k^{(m)}=\sum_{j=1}^{k}S_j^{(m-1)}$，并且

$$\frac{S_{2k}^{(m)}}{S_k^{(m)}}=\frac{(2k)(2k+1)\cdots(2k+m-1)}{k(k+1)\cdots(k+m-1)}\leqslant 2^m,\quad k\in\mathbf{N}^*. \tag{3.11}$$

由此，根据辅助命题推出，对于一切整数k，p，q，只要满足$0\leqslant k\leqslant p<q$，就有

$$\sum_{j=p}^{q-1}\frac{q-j}{q-p}S_{j-k+1}^{(m)}\geqslant(24\times 2^m)^{-1}\sum_{j=p}^{q-1}S_{j-k+1}^{(m)},\quad m\in\mathbf{N}. \tag{3.12}$$

命题的证明　对m作归纳法.

当$m=1$时，取$\bar{a}^*$为不超过$\bar{a}$的最大下凸数列即可. 此时使(3)成立的常数C_1可取$\frac{1}{2}$(见[2]).

若对正整数m命题已证，就是说对于任意取定的$\bar{a}\in U_n$，$\exists\bar{b}\in U_n$，满足

(1) $\bar{b}\leqslant\bar{a}$;

(2) $\Delta^r b_k\geqslant 0$，$r=1,2,\cdots,m+1,k\in\mathbf{N}$;

(3) $\bar{b}\cdot\bar{u}\geqslant C_m\bar{a}\cdot\bar{u}$，$\quad\forall\,\bar{u}\in U_n$，

那么$\Delta^m\bar{b}\in U_n$. 作不超过$\Delta^m\bar{b}$的最大凸数列，记为$\bar{\delta}=\{\delta_k\}_{k=0}^{+\infty}$，$\bar{\delta}$具有(3.1)和(3.2)表出的结构. 即有一组号码$k_0=0<k_1<\cdots<k_v<k_{v+1}=n+1$，$(0\leqslant v\leqslant n)$，使

$$\delta_k=\frac{k_{i+1}-k}{k_{i+1}-k_i}\Delta^m b_{k_i}+\frac{k-k_i}{k_{i+1}-k_i}\Delta^m b_{k_{i+1}},\quad k_i\leqslant k\leqslant k_{i+1} \tag{3.13}$$

构作数列$\bar{a}^*=\{a_k^*\}_{k=0}^{+\infty}$如下：当$k>n$时，$a_k^*=0$；当$k\leqslant n$时，

$$a_k^*=\sum_{j=k}^{n}S_{j-k+1}^{(m-1)}\delta_j. \tag{3.14}$$

由(3.10)及$\bar{\delta}\leqslant\Delta^m\bar{b}$，当$k\leqslant n$时，

$$a_k^* = \sum_{j=k}^{n} C_{j-k+m-1}^{m-1}\delta_j \leqslant \sum_{j=k}^{n} C_{j-k+m-1}^{m-1}\Delta^m b_j = \sum_{j=k}^{n} C_{j-k+m-2}^{m-2}\Delta^{m-1} b_j$$

$$= \sum_{j=k}^{n} C_{j-k}^{0}\Delta b_j = b_k.$$

由此得知 $\bar{a}^* \leqslant \bar{b} \leqslant \bar{a}$，$\bar{a}^*$ 具有性质(1).

从(3.14)逐次作差，当 $k \leqslant n$ 时求得

$$\Delta a_k^* = \sum_{j=k}^{n} C_{j-k+m-1}^{m-1}\delta_j - \sum_{j=k+1}^{n} C_{j-k+m-2}^{m-1}\delta_j$$

$$= \delta_k + \sum_{j=k+1}^{n}(C_{j-k+m-1}^{m-1} - C_{j-k+m-2}^{m-1})\delta_j = \sum_{j=k}^{n} C_{j-k+m-2}^{m-2}\delta_j \geqslant 0,$$

……

$$\Delta^{m-1} a_k^* = \sum_{j=k}^{n}\delta_j \geqslant 0,$$

$$\Delta^m a_k^* = \delta_k \geqslant 0, \qquad \Delta^{m+1} a_k^* = \Delta\delta_k \geqslant 0,$$

$$\Delta^{m+2} a_k^* = \Delta^2\delta_k \geqslant 0.$$

这就是说，$\bar{a}^*$ 具有性质(2)(关于 $m+1$).

现在验证 $\bar{a}^*$ 具有性质(3). 设 $k_i \leqslant k < k_{i+1}$ $(i=0,1,2,\cdots,v)$，由(3.14)，有

$$a_k^* = \sum_{j=k}^{k_{i+1}-1} S_{j-k+1}^{(m-1)}\delta_j + \sum_{\mu=i+1}^{v}\sum_{j=k_\mu}^{k_{\mu+1}-1} S_{j-k+1}^{(m-1)}\delta_j. \qquad (3.15)$$

由(3.13)，根据(3.12)及 $\Delta^m \bar{b} \in U_n$，知

$$\sum_{j=k_\mu}^{k_{\mu+1}-1} S_{j-k+1}^{(m-1)}\delta_j$$

$$\geqslant \sum_{j=k_\mu}^{k_{\mu+1}-1} \frac{k_{\mu+1}-j}{k_{\mu+1}-k_\mu} S_{j-k+1}^{(m-1)}\Delta^m b_{k_\mu} \geqslant (24\times 2^m)^{-1}\sum_{j=k_\mu}^{k_{\mu+1}-1} S_{j-k+1}^{(m-1)}\Delta^m b_{k_\mu}$$

$$\geqslant (24\times 2^m)^{-1}\sum_{j=k_\mu}^{k_{\mu+1}-1} S_{j-k+1}^{(m-1)}\Delta^m b_j,\ (\mu = i+1,\cdots,v) \qquad (3.16)$$

以及

$$\sum_{j=k}^{k_{i+1}-1} S_{j-k+1}^{(m-1)}\delta_j$$

$$\geqslant \sum_{j=k}^{k_{i+1}-1} \frac{k_{i+1}-j}{k_{i+1}-k_i} S_{j-k+1}^{(m-1)}\Delta^m b_{k_i} = \frac{k_{i+1}-k}{k_{i+1}-k_i}\sum_{j=k}^{k_{i+1}-1} \frac{k_{i+1}-j}{k_{i+1}-k} S_{j-k+1}^{(m-1)}\Delta^m b_{k_i}$$

$$\geqslant \frac{k_{i+1}-k}{k_{i+1}-k_i}(24\times 2^m)^{-1}\sum_{j=k}^{k_{i+1}-1} S_{j-k+1}^{(m-1)}\Delta^m b_j. \tag{3.17}$$

将(3.16)(3.17)代入(3.15)，得

$$\begin{aligned} a_k^* &\geqslant \frac{k_{i+1}-k}{k_{i+1}-k_i}(24\times 2^m)^{-1}\sum_{j=k}^{n} S_{j-k+1}^{(m-1)}\Delta^m b_j \\ &= (24\times 2^m)^{-1}\frac{k_{i+1}-k}{k_{i+1}-k_i}b_k,\quad k_i\leqslant k<k_{i+1}. \end{aligned}\tag{3.18}$$

$\forall \bar{u}\in U_n$，作内积，用(3.18)式，得

$$\bar{a}^*\cdot\bar{u}=\sum_{\mu=0}^{v}\sum_{k=k_\mu}^{k_{\mu+1}-1}a_k^*\cdot u_k\geqslant(24\times 2^m)^{-1}\sum_{\mu=0}^{v}\sum_{k=k_\mu}^{k_{\mu+1}-k}\frac{k_{\mu+1}-k}{k_{\mu+1}-k_\mu}b_k u_k. \tag{3.19}$$

置 $v_\mu=\left[\frac{k_\mu+k_{\mu+1}}{2}\right]$，则

$$\sum_{k=k_\mu}^{k_{\mu+1}-1}\frac{k_{\mu+1}-k}{k_{\mu+1}-k_\mu}b_k\cdot u_k\geqslant\sum_{k=k_\mu}^{v_\mu}\frac{k_{\mu+1}-k}{k_{\mu+1}-k_\mu}b_k u_k\geqslant\frac{1}{2}\sum_{k=k_\mu}^{v_\mu}b_k u_k\geqslant\frac{1}{4}\sum_{k=k_\mu}^{k_{\mu+1}-1}b_k u_k. \tag{3.20}$$

由(3.20)和(3.19)得

$$\bar{a}^*\cdot\bar{u}\geqslant(24\times 2^{m+2})^{-1}\sum_{k=0}^{n}b_k u_k. \tag{3.21}$$

由归纳假设，$\bar{b}\cdot\bar{u}\geqslant C_m\bar{a}\cdot\bar{u}$. 于是

$$\bar{a}^*\cdot\bar{u}\geqslant C_{m+1}\bar{a}\cdot\bar{u}. \tag{3.22}$$

归纳法完成，命题证毕.

注 由(3.21)可知(3.22)中的 C_{m+1} 可取 $2^{-(m+8)^2}$.

§4. Fourier 方形余项在类 $X(\varepsilon)$ 上的极值的下方估计

现把(1.5)的左半式当 $l=0$ 时重写作

定理 2 存在只与 N 有关的常数 $C_N'>0$，使得对于一切 $n\in\mathbf{N}$ 和相应于任意单调趋于零的非负数列 $\{\varepsilon_n\}_{n=0}^{+\infty}$ 的函数类 $X(\varepsilon)=\{f\in X(Q^N)\mid E_v(f)_X\leqslant\varepsilon_v,\ v\in\mathbf{N}\}$，成立着

$$\sup_{f\in X(\varepsilon)}\|R_n(f)\|_X\geqslant C_N'\sum_{v=n}^{2n}\frac{\varepsilon_n\ln^{N-1}(v-n+3)}{v-n+1}. \tag{4.1}$$

证 分别对于 $X=L$ 和 $X=C$ 的情形进行证明.

(1) $L(Q^N)$的情形.

为了避免冗繁，下面仅对维数 $N=2$ 的情形写出证明，对于一般情形证法完全适用.

设$\{\varepsilon_k\}_{k=0}^{+\infty}$是单调趋于零的非负数列，$L(\varepsilon)=\{f\mid f(x,y)\in L(Q^2)$，方形逼近 $E_k(f)_L\leqslant\varepsilon_k$，$k\in\mathbf{N}\}$. n 是任取的非负整数. 我们证明存在 $g_n(x,y)\in L(\varepsilon)$，使

$$\|R_n(g_n;x,y)\|_L\geqslant C_2'\sum_{v=n}^{2n}\frac{\varepsilon_v\ln(v-n+3)}{v-n+1}. \tag{4.2}$$

首先，当 $n=0$ 时，取 $g_0=\frac{\varepsilon_0}{8\pi}\cos x$，则 $g_0\in L(\varepsilon)$，且有 $R_0(g_0)=g_0$，$\|R_0(g_0)\|_L=\varepsilon_0$，取 $C_2'\leqslant\frac{1}{\ln 3}$，就有(4.2).

设 $n>0$. 由§3 的命题，对于数列 $\bar{\varepsilon}(n)=\{\varepsilon_n,\cdots,\varepsilon_{2n},0,\cdots\}$，存在数列 $\bar{\eta}=\{\eta_k\}_{k=0}^{+\infty}\in U_n$，使

(1) $\eta_k\leqslant\varepsilon_{k+n}$，$k\in\mathbf{N}$;

(2) $\Delta^r\eta_k\geqslant0$，$r=1,2,3,4,k\in\mathbf{N}$;

(3) $\bar{\eta}\cdot\bar{u}\geqslant C\bar{\varepsilon}(n)\cdot\bar{u}$，$\forall\,\bar{u}\in U_n$，$C>0$ 为常数.

作函数 $f(x,y)=\sum\limits_{i=0}^{n}\sum\limits_{j=0}^{n}\lambda_{i,j}\cdot\eta_{i,j}\cos ix\cdot\cos jy$，其中 $\lambda_{0,0}=\frac{1}{4}$，$\lambda_{0,j}=\lambda_{i,0}=\frac{1}{2}$，$\lambda_{i,j}=1(i,j\in\mathbf{N}^*)$；$\eta_{i,j}=\eta_{i+j}$，则

$$f(x,y)=\sum_{i=0}^{n}\sum_{j=0}^{n}\Delta^{2,2}\eta_{i,j}\sigma_i(x)\cdot\sigma_j(y),$$

其中 $\sigma_i(t)=(i+1)F_i(t)=\dfrac{\sin^2\frac{i+1}{2}t}{2\sin^2\frac{t}{2}}$，($F_i(t)$表示 Féjer 核). 估计 f 的方形 v 阶 L 最佳逼近.

设 $v<n$，令

$$f_1(x,y)=\sum_{i=v+1}^{n}\sum_{j=0}^{v}\Delta^{2,2}\eta_{i,j}\sigma_i(x)\cdot\sigma_j(y),$$

$$f_2(x,y)=\sum_{i=0}^{v}\sum_{j=v+1}^{n}\Delta^{2,2}\eta_{i,j}\sigma_i(x)\cdot\sigma_j(y),$$

$$f_3(x,y)=\sum_{i=v+1}^{n}\sum_{j=v+1}^{n}\Delta^{2,2}\eta_{i,j}\sigma_i(x)\cdot\sigma_j(y).$$

由于 $f-(f_1+f_2+f_3)$ 是 v 阶三角多项式，所以

$$E_v(f)_L\leqslant E_v(f_1)_L+E_v(f_2)_L+E_v(f_3)_L. \tag{4.3}$$

设 $T_v^{(i)}(x)$ 是 $\sigma_i(x)$ 的一元 L 最佳逼近三角多项式(v 阶)，那么(见[9])，

$$\int_0^{2\pi}|\sigma_i(x)-T_v^i(x)|\,\mathrm{d}x\leqslant 2\pi(i-v),\quad i>v. \tag{4.4}$$

于是

$$\begin{aligned}E_v(f_1)_L&\leqslant\sum_{i=v+1}^{n}\sum_{j=0}^{v}\Delta^{2,2}\eta_{i,j}\int_0^{2\pi}\int_0^{2\pi}|\sigma_i(x)-T_v^{(i)}(x)|\sigma_j(y)\mathrm{d}x\mathrm{d}y\\&\leqslant 2\pi^2\sum_{i=v+1}^{n}\sum_{j=0}^{v}\Delta^{2,2}\eta_{i,j}(i-v)(j+1)\leqslant 2\pi^2\sum_{i=v+1}^{n}\Delta^{2,0}\eta_{i,0}(i-v)\\&=2\pi^2\eta_{v+1,0}=2\pi^2\eta_{v+1}\leqslant 2\pi^2\varepsilon_{v+1+n}.\end{aligned} \tag{4.5}$$

同理

$$E_v(f_2)_L\leqslant 2\pi^2\varepsilon_{v+1+n}. \tag{4.6}$$

当 i，$j\geqslant v$ 时，方形最佳逼近

$$\begin{aligned}&E_v(\sigma_i(x)\cdot\sigma_j(y))_L\\&\leqslant\|\sigma_i(x)[\sigma_j(y)-T_v^{(j)}(y)]\|_L+\|[\sigma_i(x)-T_v^{(i)}(x)]T_v^{(j)}(y)\|_L\\&\leqslant\pi(i+1)2\pi(j-v)+2\pi(i-v)[\pi(j+1)+2\pi(j-v)]\\&\leqslant 4\pi^2[(i+1)(j-v)+(i-v)(j+1)].\end{aligned}$$

注意到 i，j 和 x，y 的对称性，求得

$$\begin{aligned}E_v(f_3)_L&\leqslant\sum_{i=v+1}^{n}\sum_{j=v+1}^{n}\Delta^{2,2}\eta_{i,j}E_v(\sigma_i(x)\cdot\sigma_j(y))_L\\&\leqslant 8\pi^2\sum_{i=v+1}^{n}\sum_{j=v+1}^{n}\Delta^{2,2}\eta_{i,j}(i+1)(j-v)\\&=8\pi^2\sum_{i=v+1}^{n}\Delta^{2,0}\eta_{i,v+1}(i+1)\\&\leqslant 8\pi^2\Big[\sum_{i=v+1}^{n}\Delta^{1,0}\eta_{i,v+1}+(v+1)\Delta^{1,0}\eta_{v+1,v+1}\Big]\\&\leqslant 8\pi^2(\eta_{v+1,v+1}+\eta_{0,v+1})\leqslant 16\pi^2\eta_{v+1}\leqslant 16\pi^2\varepsilon_{v+1+n}.\end{aligned} \tag{4.7}$$

(4.5)～(4.7)及(4.3)给出

$$E_v(f)_L\leqslant 20\pi^2\varepsilon_{v+1+n},\qquad v<n. \tag{4.8}$$

置 $g_n(x,\ y)=\dfrac{1}{20\pi^2}\cos(n+1)x\cdot\cos(n+1)y\cdot f(x,\ y)$，则当 $v\leqslant n$ 时方形逼近

$$E_v(g_n)_L \leqslant \|g_n\|_L \leqslant \frac{1}{20\pi^2}\|f\|_L = \frac{1}{20}\eta_{0,0} \leqslant \varepsilon_n \leqslant \varepsilon_v.$$

当 $v>2n$ 时，由于 g_n 是 $2n+1$ 阶三角多项式，显然 $E_v(g_n)_L=0$．当 $n<v\leqslant 2n$ 时，设 $T^*_{v-n-1}(x, y)$ 是 f 的方形 $v-(n+1)$ 阶最佳逼近三角多项式，令

$$\tau_v(x, y) = \frac{1}{20\pi^2}\cos(n+1)x\cos(n+1)y \cdot T^*_{v-(n+1)}(x, y),$$

则 τ_v 是 v 阶三角多项式．于是

$$E_v(g_n)_L \leqslant \|g_n - \tau_n\|_L \leqslant \frac{1}{20\pi^2}E_{v-n-1}(f)_L.$$

根据(4.8)，得 $E_v(g_n)_L \leqslant \varepsilon_v$，$n<v\leqslant 2n$.

证得 $g_n \in L(\varepsilon)$.

估计 $\|R_n(g_n)\|_L$．先写出 g_n 的表达式.

$$\begin{aligned}
&g_n(x,y)\\
=&\frac{1}{80\pi^2}\eta_{0,0}\cos(n+1)x \cdot \cos(n+1)y + \\
&\frac{1}{80\pi^2}\cos(n+1)x \cdot \sum_{j=1}^{n}\eta_{0,j}[\cos(n+1-j)y + \cos(n+1+j)y] + \\
&\frac{1}{80\pi^2}\cos(n+1)y\sum_{i=1}^{n}\eta_{i,0}[\cos(n+1-i)x + \cos(n+1+i)x] + \\
&\frac{1}{80\pi^2}\sum_{i=1}^{n}\sum_{j=1}^{n}\eta_{i,j}\{[\cos(n+1-i)x + \cos(n+1+i)x] + \\
&[\cos(n+1-j)y + \cos(n+1+j)y]\}.
\end{aligned}$$

顺次记右边四项为 $h_1(x, y)$，$h_2(x, y)$，$h_3(x, y)$，$h_4(x, y)$．易见

$$R_n(g_n) = h_1 + h_2 + h_3 + R_n(h_4).$$

设 $\psi(x,y) = \sin(n+1)x \cdot \sin(n+1)y\sum_{i=1}^{n}\sum_{j=1}^{n}\frac{\sin ix \cdot \sin jy}{i \cdot j}$，展开之

$$\begin{aligned}
\psi(x,y) = \frac{1}{4}\sum_{i=1}^{n}\sum_{j=1}^{n}\frac{1}{i \cdot j}\{&[\cos(n+1-i)x - \cos(n+1+i)x] + \\
&[\cos(n+1-j)y - \cos(n+1+j)y]\}.
\end{aligned}$$

根据三角函数系的正交性，得到

$$\int_0^{2\pi}\int_0^{2\pi}R_n(g_n)\psi \mathrm{d}x\mathrm{d}y = \int_0^{2\pi}\int_0^{2\pi}R_n(h_4)\psi \mathrm{d}x\mathrm{d}y$$

$$=-\frac{1}{320\pi^2}\cdot\pi^2\sum_{i=1}^{n}\sum_{j=1}^{n}\frac{\eta_{i,j}}{i\cdot j}. \qquad (4.9)$$

取 $\varphi(x, y)=\cos(n+1)x\cdot\cos(n+1)y-\psi(x, y)$. 由于(见[11]，第二章)

$$\sup_m\left\|\sum_{k=1}^{m}\frac{\sin kx}{k}\right\|_C=\varkappa<+\infty, \qquad (4.10)$$

可见有常数 $C>0$，使 $\|\varphi(x, y)\|_C<C$. 从而

$$\|R_n(g_n)\|_L\geqslant\frac{1}{C}\int_0^{2\pi}\int_0^{2\pi}R_n(g_n)\varphi\,\mathrm{d}x\mathrm{d}y$$

$$=\frac{1}{C}\int_0^{2\pi}\int_0^{2\pi}[h_1\varphi-R_n(g_n)\psi]\mathrm{d}x\mathrm{d}y.$$

代入 $\int_0^{2\pi}\int_0^{2\pi}h_1\varphi\,\mathrm{d}x\mathrm{d}y=\frac{1}{80}\eta_{0,0}$ 及(4.9)，得

$$\|R_n(g_n)\|_L$$

$$\geqslant\frac{1}{320C}\left[4\eta_{0,0}+\sum_{i=1}^{n}\sum_{j=1}^{n}\frac{\eta_{i,j}}{i\cdot j}\right]=\frac{1}{320C}\left[4\eta_0+\sum_{v=2}^{n+1}\sum_{k=1}^{v-1}\frac{\eta_v}{k(v-k)}\right]$$

$$=\frac{1}{320C}\left[4\eta_0+\sum_{v=2}^{n+1}\frac{\eta_v}{v}\sum_{k=1}^{v-1}\left(\frac{1}{k}+\frac{1}{v-k}\right)\right]$$

$$\geqslant\frac{1}{160C}\left(2\eta_0+\sum_{v=2}^{n}\frac{\eta_v\ln v}{v}\right).$$

最后和式中缺少 $v=1$ 的项，但显然可用第一项($v=0$)来弥补，从而取适当的常数 $C'>0$，得有

$$\|R_n(g_n)\|_L\geqslant C'\sum_{v=n}^{2n}\frac{\eta_v\ln(v-n+3)}{v-n+1}.$$

由于$\left\{\frac{\ln 3}{1}, \frac{\ln 4}{2}, \cdots, \frac{\ln(n+3)}{n+1}, 0, \cdots\right\}\in U_n$，所以根据 $\bar{\eta}$ 的性质 3 得到(4.2).

(2) $C(Q^N)$的情形

也以 $N=2$ 为例写出证明，一般的情况，证法完全适用.

用 $L(\varepsilon)$的情形一样，只需考虑 n 为正整数的情形. 取数列

$$\bar{\varepsilon}(n)=\{\varepsilon_n, \varepsilon_{n+1}, \cdots, \varepsilon_{2n}, 0, \cdots\}.$$

用 § 3 的命题，存在 $\bar{\eta}=\{\eta_0, \eta_1, \cdots, \eta_n, 0, \cdots\}\in U_n$，满足

(1) $\eta_k\leqslant\varepsilon_{k+n}$，$k\in\mathbf{N}$;

(2) $\Delta\eta_k\geqslant0$，$\Delta^2\eta_k\geqslant0$，$k\in\mathbf{N}$;

(3) $\bar{\eta}\cdot\bar{u}\geqslant C_2\bar{\varepsilon}(n)\cdot\bar{u}$，$\forall\,\bar{u}\in U_n$，($C_2>0$ 为常数)

置 $\eta_{i,j}=\eta_{i+j-2}$，取 $f(x,y)=\sum\limits_{i=1}^{n+1}\sum\limits_{j=1}^{n+1}\eta_{i,j}\dfrac{\sin ix}{i}\cdot\dfrac{\sin jy}{j}$. 显然，

$$f(x,y)=\sum_{i=1}^{n+1}\sum_{j=1}^{n+1}\Delta^{1,1}\eta_{i,j}S_i(x)\cdot S_j(y),$$

其中 $S_v(t)=\sum\limits_{k=1}^{v}\dfrac{\sin kt}{k}$，$v\in\mathbf{N}^*$. 根据(4.10)

$$\|f\|_C\leqslant\varkappa^2\eta_{1,1}=\varkappa^2\eta_0\leqslant\varkappa^2\varepsilon_n \tag{4.11}$$

当 $0<v\leqslant n$ 时，设

$$f_1=\sum_{i=1}^{v}\sum_{j=v+1}^{n+1}\Delta^{1,1}\eta_{i,j}S_i(x)S_j(y),$$

$$f_2=\sum_{i=v+1}^{n+1}\sum_{j=1}^{v}\Delta^{1,1}\eta_{i,j}S_i(x)S_j(y),$$

$$f_3=\sum_{i=v+1}^{n+1}\sum_{j=v+1}^{n+1}\Delta^{1,1}\eta_{i,j}S_i(x)S_j(y),$$

由于 $f-(f_1+f_2+f_3)$是 v 阶三角多项式，所以方形最佳逼近

$$\begin{aligned}E_v(f)_C&\leqslant\|f_1\|_C+\|f_2\|_C+\|f_3\|_C\leqslant\varkappa^2(\eta_{1,v+1}+\eta_{v+1,1}+\eta_{v+1,v+1})\\&\leqslant3\varkappa^2\eta_v\leqslant3\varkappa^2\varepsilon_{v+n},\end{aligned}\tag{4.12}$$

取$g_n(x,y)=\dfrac{1}{3\varkappa^2}\sin nx\cdot\sin ny\cdot f(x,y)$，是 $2n+1$ 阶三角多项式. 当 $v>2n$ 时，$E_v(g_n)_C=0$；而当 $v<n$ 时，由(4.11)$E_v(g_n)_C\leqslant\dfrac{1}{3\varkappa^2}\|f\|_C\leqslant\varepsilon_n\leqslant\varepsilon_v$；当 $n<v\leqslant2n$ 时，取 $T^*_{v-n}(x,y)$为 f 的方形 $v-n$ 阶最佳逼近三角多项式，由(4.12)得

$$\begin{aligned}E_v(g_n)_C&\leqslant\left\|g_n(x,y)-\frac{1}{3\varkappa^2}\sin nx\cdot\sin ny\cdot T^*_{v-n}(x,y)\right\|_C\\&\leqslant\frac{1}{3\varkappa^2}E_{v-n}(f)_C\leqslant\varepsilon_v.\end{aligned}$$

总之，$g_n\in C(\varepsilon)$.

估计 $R_n(g_n)$在原点的值，得

$$\begin{aligned}\|R_n(g_n)\|_C&\geqslant|R_n(g_n;0,0)|=\left|\frac{1}{12\varkappa^2}\sum_{i=1}^{n+1}\sum_{j=1}^{n+1}\frac{\eta_{i,j}}{i\cdot j}(-1)\right|\\&=\frac{1}{12\varkappa^2}\sum_{i=1}^{n+1}\sum_{j=1}^{n+1}\frac{\eta_{i+j-2}}{i\cdot j}\geqslant\frac{1}{6\varkappa^2}\sum_{v=2}^{n+2}\frac{\eta_{v-2}}{v}\ln v\end{aligned}$$

$$\geqslant \frac{1}{24\varkappa^2}\sum_{v=0}^{n}\frac{\eta_v\ln(v+3)}{v+1}.$$

根据 $\overline{\eta}$ 的性质(3)得到

$$\|R_n(g_n)\|_C \geqslant C_2'\sum_{v=n}^{2n}\frac{\varepsilon_v\ln(v-n+3)}{v-n+1}. \quad \square \tag{4.13}$$

注 在定理 2 的证明中，求得的极值元 $g_n(\overline{x})$ 是与 n 有关的. 仿照[8]. 可以把那里的定理 3 拓广，求得 $f_0(\overline{x})\in X(Q^N)$，满足

(1) $f_0\in X(\varepsilon)$，$\widetilde{f}_0\in X(\varepsilon)$；

(2) $\overline{\lim\limits_{n\to+\infty}}\dfrac{\|R_n(f_0)\|_X}{K_n(\varepsilon)}\geqslant\gamma_N$，$\overline{\lim\limits_{n\to+\infty}}\dfrac{\|R_n(\widetilde{f}_0)\|_X}{K_n(\varepsilon)}\geqslant\gamma_N$，

其中 $K_n(\varepsilon)=\sum\limits_{v=n}^{2n}\dfrac{\varepsilon_v\ln^{N-1}(v-n+3)}{v-n+1}$，$\gamma_N>0$ 是只与 N 有关的常数.

§ 5. de la Vallée Poussin 方形余项的估计

完全类似于定理 1，可对维数 N 用归纳法证得

$$\|R_{n,l}(f)\|_X \leqslant C_N\sum_{v=0}^{n+l}\frac{E_v(f)_X\ln^{N-1}\left(\frac{v}{l+1}+3\right)}{v+l+1}. \tag{5.1}$$

证明的步骤同定理 1，只需把其中的 $T_v(\overline{x})$ 换为

$$T_v(\overline{x})=\frac{1}{\pi}\int_{-\pi}^{\pi}T_v^*(\overline{x}',x_{k+1}+t)W_{n,v}^l(t)\mathrm{d}t+T_v^*(\overline{x}),$$

式中代替 $D_n(t)-D_v(t)$ 的是 $W_{n,v}^l(t)=\dfrac{1}{l+1}\sum\limits_{\mu=v}^{v+l}[D_{n+\mu-v}(t)-D_\mu(t)]$，

把 S_n 换成 V_n^{n+l}，并注意到 $\dfrac{1}{\pi}\int_{-\pi}^{\pi}|W_{n,v}^l(t)|\mathrm{d}t\leqslant 4\left(1+\ln^{-1}\dfrac{|n-v|}{l+1}\right)$.

(5.1)与(1.5)的右半式是等价的.

(1.5)的左半式的证明，分两种情形.

(1) $X=C$ 的情形.

引入 Стечкин[7] 给出的一元极值函数

$$f_0(x_1)=\sum_{v=1}^{+\infty}\Delta\varepsilon_{v-1}\cos vx_1.$$

对于 n，$l\in\mathbf{N}^*$，有(见[7])

$$\|R_{n,l}(f_0)\|_X=f_0(0)-V_n^{n+l}(f;0)=\frac{1}{l+1}\sum_{v=n}^{n+l}\varepsilon_v. \tag{5.2}$$

由此可知当 $n=0$ 时(1.5)的左半式成立.

设 $n>0$. 据§3的命题，对于数列 $\{\varepsilon_n, \varepsilon_{n+1}, \cdots, \varepsilon_{2n+l}, 0, \cdots\}$ 存在

$$\{\eta_k\}_{k=0}^{+\infty}=\{\eta_0, \eta_1, \cdots, \eta_{n+l}, 0, \cdots\},$$

具有下列性质：

(1) $\eta_k\leqslant\varepsilon_{n+k}$, $k\in\mathbf{N}$;

(2) $\Delta^r\eta_k\geqslant0$, $\gamma=1,2,\cdots,N$, $k\in\mathbf{N}$;

(3) 对于任意的单调降非负数列 $\{u_k\}_{k=0}^{+\infty}$

$$\sum_{v=0}^{n+l}\eta_v\cdot u_v\geqslant\alpha_N\sum_{v=0}^{n+l}\varepsilon_{v+n}\cdot u_v,$$

其中 $\alpha_N>0$ 只有 N 有关.

令 $\eta_{\bar{v}}=\eta_{v_1,v_2,\cdots,v_N}=\eta_{v_1+v_2+\cdots+v_N-N}$，作函数

$$f_n(\bar{x})=\beta_N\Big(\prod_{i=1}^{N}\sin nx_i\Big)\sum_{\bar{v}=\bar{1}}^{\overline{2n}}\eta_{\bar{v}}\prod_{i=1}^{N}\frac{\sin v_ix_i}{v_i},$$

其中 $2\bar{n}=(2n, 2n, \cdots, 2n)$，$\bar{\boldsymbol{1}}=(1, 1, \cdots, 1)$，求和是 N 重的，$\beta_N>0$ 是只与 N 有关的适当的常数，使得 $f_n\in C(\varepsilon)$——这是做得到的，参见定理2的证明. f_n 展开为

$$f_n(\bar{x})=\frac{1}{2^N}\beta_N\sum_{\bar{v}=\bar{1}}^{\overline{2n}}\eta_{\bar{v}}\prod_{i=1}^{N}\frac{\cos(n-v_i)x_i-\cos(n+v_i)x_i}{v_i}.$$

可见

$$\begin{aligned}&V_n^{n+l}(f_n;\bar{x})\\&=\frac{1}{2^N}\beta_N\sum_{\bar{v}=\bar{1}}^{\overline{2n}}\eta_{\bar{v}}\prod_{i=1}^{n}\frac{1}{v_i}\Big[\cos(n-v_i)x_i+\Big(1-\frac{\tilde{v}_i}{l+1}\Big)\cos(n+v_i)x_i\Big]\\&=\frac{\beta_n}{2^N}\sum_{\bar{v}=\bar{1}}^{\overline{2n}}\eta_{\bar{v}}\prod_{i=1}^{N}\Big\{\frac{\cos(n-v_i)x_i-\cos(n+v_i)x_i}{v_i}+\\&\quad\frac{\tilde{v}_i}{v_i(l+1)}\cos(n+v_i)x_i\Big\}.\end{aligned}\tag{5.3}$$

式中

$$\tilde{v}_i=\begin{cases}v_i, & v_i\leqslant l,\\ l+1, & v_i>l,\end{cases}\tag{5.4}$$

于是

$$\|R_{n,l}(f_n)\|_C\geqslant V_n^{n+l}(f_n;\mathbf{0})-f_n(\mathbf{0})=\frac{\beta_N}{2^n}\sum_{\bar{v}=\bar{1}}^{\overline{2n}}\eta_{\bar{v}}\prod_{i=1}^{N}\frac{\tilde{v}_i}{v_i(l+1)}$$

$$\geqslant \frac{\beta_N}{2^N} \sum_{\bar{v}=\bar{1}}^{\overline{2n}} \eta_{\bar{v}} \prod_{i=1}^{N} \frac{1}{v_i + l + 1}. \tag{5.5}$$

因为

$$\sum_{\bar{v}=\bar{1}}^{\overline{2n}} \eta_{\bar{v}} \prod_{i=1}^{N} \frac{1}{v_i + l + 1}$$

$$\geqslant \sum_{v_1=N}^{2n+N-1} \sum_{v_2=N-1}^{v_1-1} \cdots \sum_{v_N=1}^{v_{N-1}-1} \eta_{v_1-N} \frac{1}{v_1 - v_2 + l + 1} \cdots \frac{1}{v_{N-1} - v_N + l + 1} \cdot$$

$$\frac{1}{v_N + l + 1} \geqslant C_N \sum_{v=0}^{2n-1} \eta_v \frac{\ln^{N-1}\left(1 + \frac{v+1}{l+1}\right)}{v + l + 1}, \tag{5.6}$$

其中 $C_N>0$ 只与 N 有关，所以，存在常数 $\beta'_N>0$ 使

$$V_n^{n+l}(f_n;\mathbf{0}) - f_n(\mathbf{0}) \geqslant \beta'_N \sum_{v=0}^{2n-1} \eta_v \frac{\ln^{N-1}\left(1 + \frac{v+1}{l+1}\right)}{v + l + 1}. \tag{5.7}$$

取 $f^*(\bar{x})=\frac{1}{2}(f_0(x_1)-f_n(\bar{x}))$，则 $f^*\in C(\varepsilon)$，且由(5.2)和(5.7)得

$$\|R_{n,l}(f^*)\|_C \geqslant f^*(0) - V_n^{n+l}(f^*;\mathbf{0})$$

$$\geqslant \frac{1}{2}\left\{\frac{1}{l+1}\sum_{v=0}^{l} \varepsilon_{v+n} + \beta'_N \sum_{v=0}^{2n-1} \frac{\eta_v \ln^{N-1}\left(1 + \frac{v+1}{l+1}\right)}{v + l + 1}\right\}. \tag{5.8}$$

由此，兼顾 $l<n$ 和 $l\geqslant n$ 两种情形并利用 $\{\eta_k\}_{k=0}^{+\infty}$ 的性质(3)，适当取 $C'_N>0$，得到

$$\|R_{n,l}(f^*)\|_C \geqslant C'_N \sum_{v=0}^{n+l} \frac{\varepsilon_{v+n}}{v + l + 1}\left(1 + \ln \frac{v + l + 1}{l + 1}\right)^{N-1}. \tag{5.9}$$

显然，因子 $\left(1+\ln\frac{v+l+1}{l+1}\right)^{N-1}$ 可以换为 $\ln^{N-1}\left(3+\frac{v}{l+1}\right)$.

从证明可见，所取的极值元 f^* 只与 n 有关而与 l 无关. 这是由于 f_0 与 l 无关，它控制了 $n\leqslant l$ 的情形.

(2) $X=L$ 的情形.

若 $l\geqslant n$，利用 Байбородов С П[6] 关于一元函数的结果，知 N 元时有

$$\sup_{f\in L(\varepsilon)} \|f - V_n^{n+l}(f)\|_L \geqslant C'_1 \sum_{v=0}^{n+l} \frac{\varepsilon_{v+n}}{v + l + 1}$$

$$\geqslant \frac{C_1'}{(1+\ln 3)^{N-1}} \sum_{v=0}^{n+l} \frac{\varepsilon_{v+n}}{v+l+1}\left(1+\ln \frac{v+l+1}{l+1}\right)^{N-1}, \quad l \geqslant n. \tag{5.10}$$

设 $l<n$. 同(1)一样选一个 $\{\eta_k\}_{k=0}^{+\infty}$，其满足的条件(2)加强到 $r=1,2,\cdots,2N$. 然后令 $\eta_{\bar{v}}=\eta_{v_1+v_2+\cdots+v_N}$. 作函数(与 n 有关)

$$g_N(\bar{x})=\frac{\beta_N}{\pi^N}\left(\prod_{i=1}^{N}\cos(n+1)x_i\right)\sum_{\bar{v}=\bar{0}}^{\overline{2n}}\lambda_{\bar{v}}\eta_{\bar{v}}\prod_{i=1}^{N}\cos v_i x_i,$$

式中 $\lambda_{\bar{v}}=2^{-\theta}$，$\theta$ 是 $\bar{v}$ 的零分量的个数.

易见(参见§4)，$\beta_N>0$ 取的适当时，$g_N\in L(\varepsilon)$. 显然

$$V_n^{n+l}(g_N;\bar{x})=\frac{\beta_N}{(2\pi)^N}\sum_{\bar{v}=\bar{0}}^{\overline{2n}}\lambda_{\bar{v}}\cdot\eta_{\bar{v}}\prod_{i=1}^{N}\left\{\left(1-\frac{\tilde{v}_i}{l+1}\right)\cos(n+1-v_i)x_i+\left(1-\frac{\tilde{\tilde{v}}_i+1}{l+1}\right)\cos(n+1+v_i)x_i\right\},$$

其中

$$\tilde{v}_i=\begin{cases}0, & v_i=0,\\ 1, & v_i>0,\end{cases}\qquad \tilde{\tilde{v}}_i=\begin{cases}v_i, & v_i<l,\\ l, & v_i\geqslant l.\end{cases}$$

再取

$$\varphi_N(\bar{x})=\frac{1}{\varkappa^N}\left(\prod_{i=1}^{N}\sin(n+1)x_i\right)\sum_{\bar{v}=\bar{1}}^{\overline{2n}}\left(\prod_{i=1}^{N}\frac{\sin v_i x_i}{v_i}\right),$$

那么，$\|\varphi_N\|_C\leqslant 1$. 于是

$$\begin{aligned}\|R_{n,l}(g_N)\|_L &\geqslant \int_{Q^N}\{g_N(\bar{x})-V_n^{n+l}(g_N;\bar{x})\}\varphi_N(\bar{x})\mathrm{d}\bar{x}\\ &=-\int_{Q^N}V_n^{n+l}(g_N;\bar{x})\cdot\varphi_N(\bar{x})\mathrm{d}\bar{x}\\ &=\frac{\beta_N}{(4\pi\varkappa)^N}\sum_{\bar{v}=\bar{1}}^{\overline{2n}}\eta_{\bar{v}}\prod_{i=1}^{N}\frac{\tilde{\tilde{v}}_i+1}{l+1}\frac{\pi}{v_i}\\ &\geqslant \frac{\beta_N}{(4\varkappa)^N}\sum_{\bar{v}=\bar{1}}^{\overline{2n}}\eta_{\bar{v}}\prod_{i=1}^{N}\frac{1}{v_i+l+1}.\end{aligned}$$

利用(5.6)得知有常数 $\gamma_N>0$ 使

$$\int_{Q^N}R_{n,l}(g_N)\varphi_N\mathrm{d}\bar{x}\geqslant\gamma_N\sum_{v=0}^{2n-1}\frac{\eta_{v+N}\ln^{N-1}\left(1+\frac{v+1}{l+1}\right)}{v+l+1}. \tag{5.11}$$

取 $g^*(\bar{x})=\frac{1}{2}(g_1(x_1)+g_N(\bar{x}))$，$\varphi(\bar{x})=\frac{1}{3}\{\cos(n+1)x_1+\varphi_1(x_1)+$

$\varphi_N(\bar{x})\}$，那么$g^* \in L(\varepsilon)$，$\|\varphi\|_C \leqslant 1$. 利用(5.11)得

$$\|R_{n,l}(g^*)\|_L$$

$$\geqslant \int_{Q^N} R_{n,l}(g^*)\varphi \mathrm{d}\bar{x} \geqslant \frac{\beta_1}{12(l+1)}(2\pi)^{N-1}\eta_0 +$$

$$\frac{\gamma_1}{6}(2\pi)^{N-1}\sum_{v=0}^{2n-1}\frac{\eta_{v+1}}{v+l+1} + \frac{\gamma_N}{6}\sum_{v=0}^{2n-1}\frac{\eta_{v+N}\ln^{N-1}\left(1+\frac{v+1}{l+1}\right)}{v+l+1}. \quad (5.12)$$

注意到$l<n$，使用$\{\eta_k\}_{k=0}^{+\infty}$的性质(c)，知只有与$N$有关的常数$C_N'>0$，使

$$\|R_{n,l}(g^*)\|_L \geqslant C_N' \sum_{v=0}^{n+l}\frac{\varepsilon_{v+n}}{v+l+1}\left(1+\ln\frac{v+l+1}{l+1}\right)^{N-1}. \quad (5.13)$$

由(5.1)(5.9)(5.10)和(5.13)证得§1的定理，即(1.5).

本文是在孙永生老师指导下完成的，谨表示衷心的感谢.

参考文献

[1] Осколков К И. К неравенству Лебега в равномерной метрике и на множестве полной меры. Мамем. замемкц，1975，18(4)：515—526.

[2] Осколков К И. К неравенству Лебега в среднем. Мамем. замемкц，1979，25(4)：551—555.

[3] 孙永生. 周期函数用其傅里叶部分和的平均逼近. 数学年刊，1980，1(2)：181—190.

[4] Stečkin S B. On the approximation of periodic functions by de la Vallée Poussin sums. Analysis mathematica，1978，4：61—74.

[5] Дамен В. О наилучшем приближении и суммах Валле Пуссена. Мамем замемкц，1978，23(5)：671—683.

[6] Ьайбородов С П. Приближение функций суммами Валле Пуссена，Мамем. замемкц，1980，27(1)：33—48.

[7] Стечкин С Б. О приближении периодических фунский суммами Фейера. Тр. мамем цкн-маименц В. А. Смеклова АН СССР，1961，62：48—60.

[8] 孙永生. 二元周期连续函数用它的方形傅里叶部分和的一致逼近. 北京师范大学学报(自然科学版)，1979，(3)：16—35.

[9] 孙永生. 关于一个周期函数类用三角多项式近迫问题的一点注记. 北京师范大学学报(自然科学版)，1959，(2)：36—45.

[10] Тиман А Ф. Теория приближения функдий действительного переменного. Москва，1960.

[11] Zygmund A. Trigonometric series，Volume I. Cambridge，1959.

Abstract Let $Q^N=\{\bar{x}=(x_1, x_2, \cdots, x_N)\mid -\pi\leqslant x_i<\pi, i=1, 2,\cdots,N\}$ and $X(Q^N)$ denote $L(Q^N)$ and $C(Q^N)$. The square de la Vallée Poussin sums of $f\in X(Q^N)$ are defined by

$$V_n^{n+l}(f;\bar{x})=\frac{1}{\pi^N}\int_{Q^N}f(\bar{x}+\bar{t})\prod_{i=1}^{N}\left(\frac{1}{l+1}\sum_{v=n}^{n+l}D_v(t_i)\right)\mathrm{d}\bar{t},(n,l\in\mathbf{N}),$$

where $D_v(t)=\dfrac{\sin\left(v+\frac{1}{2}\right)t}{2\sin\frac{t}{2}}$. The differences $R_{n,l}(f;\bar{x})=f(\bar{x})-V_n^{n+l}(f;\bar{x})$ are called square remainders. We denote by $E_k(f)_X$ the best approximation of the function $f\in X(Q^N)$ by N-multiple trigonometric polynomials of order K.

Theorem Let $\{\varepsilon_k\}_{k=0}^{+\infty}$ be a sequence such that $\varepsilon_n\downarrow 0\ (n\to+\infty)$, the class $X(\varepsilon)=\{f\in X(Q^N)\mid E_K(f)_X\leqslant\varepsilon_k, k\in\mathbf{N}\}$. Then

$$C_N'\sum_{v=0}^{n+l}\frac{\varepsilon_{v+n}\ln^{N-1}\left(3+\frac{v}{l+1}\right)}{v+l+1}$$

$$\leqslant\sup_{f\in X(\varepsilon)}\|R_{n,l}(f)\|_X\leqslant C_N\sum_{v=0}^{n+l}\frac{\varepsilon_{v+n}\ln^{N-1}\left(3+\frac{v}{l+1}\right)}{v+l+1},$$

where $C_N>C_N'>0$ are constants depending only on N.

北京师范大学学报(自然科学版)，1983，(2)：1—12.

关于多重 Fourier 级数的线性求和①

On the Summability of Multiple Fourier Series

§1. 引言

设 $Q=\{(x, y)\mid -\pi\leqslant x, y\leqslant\pi\}$. $L(Q)$表示在 Q 上可和且对于每个变元都以 2π 为周期的函数的全体. 设 $\lambda(x, y)$是支集(函数取值不为零的点的集合的闭包)有界的二元连续函数，由 $\lambda(x, y)$产生一个线性求和法 $\tau_{m,n}^{\lambda}(m, n\in\mathbf{N})$. 它是 $L(Q)$到(m, n)阶三角多项式集的线性算子：

$$\tau_{m,n}^{\lambda}f(x,y)=\frac{1}{4\pi^2}\int_Q f(x+s,y+t)\sum_{\mu,\nu}\lambda\left(\frac{\mu}{m+1},\frac{\nu}{n+1}\right)e^{-i(\mu s+\nu t)}\,dsdt.$$

我们考虑 $\tau_{m,n}^{\lambda}f(x, y)$在 f 的 Lebesgue 点集上的收敛性. 关于这个问题，Э. С. Белинский[1]得到了一些通过$\lambda(x, y)$的 Fourier 变换$\tilde{\lambda}(x, y)$的积分性质表示的可求和的充要条件. 先写出(参见[1])

定义 1 设 $f\in L(Q)$. 点(x, y)叫作 f 的强 Lebesgue 点，若

$$\lim_{h_1,h_2\to0^+}\frac{1}{h_1\cdot h_2}\int_{-h_1}^{h_1}\int_{-h_2}^{h_2}|f(x+s,y+t)-f(x,y)|\,dsdt=0,$$

$$\sup_{h_1,h_2>0}\frac{1}{4h_1\cdot h_2}\int_{-h_1}^{h_1}\int_{-h_2}^{h_2}|f(x+s,y+t)|\,dsdt\equiv Mf(x,y)<+\infty.$$

f 在 Q 中的全体强 Lebesgue 点的集合记作 $SE(f)$.

定义 2 设 $f\in L(Q)$. 点(x, y)叫作 f 的弱 Lebesgue 点，若

① 收稿日期：1982-06-11.

$$\lim_{h\to 0^+}\frac{1}{h^2}\int_{-h}^{h}\int_{-h}^{h}|f(x+s,y+t)-f(x,y)|\,\mathrm{d}s\mathrm{d}t=0.$$

f 在 Q 中的全体弱 Lebesgue 点的集合记作 $WE(f)$.

Белинский[1]得到

定理 A 等式

$$\lim_{m,n\to+\infty}\tau_{m,n}^{\lambda}f(x,y)=f(x,y) \tag{1.1}$$

对于一切 $f\in(Q)$ 在 $SE(f)$ 处处成立的充要条件是

(1) $\lambda(0,0)=1$, (2) $\int_0^{+\infty}\int_0^{+\infty}\sup\limits_{x<|u|,y<|v|}|\tilde{\lambda}(u,v)\mathrm{d}x\mathrm{d}y<+\infty$.

定理 B 等式

$$\lim_{n\to+\infty}\tau_{n,n}^{\lambda}f(x,y)=f(x,y) \tag{1.2}$$

对于一切 $f\in L(Q)$ 在 $WE(f)$ 处处成立的充要条件是

(1) $\lambda(0,0)=1$, (2) $\int_0^{+\infty}\rho\sup\limits_{\rho<\sqrt{x^2+y^2}}|\tilde{\lambda}(x,y)|\,\mathrm{d}\rho<+\infty$.

从定理 A 的证明可以看出，定理 A 的条件对于(1.2)式在 $SE(f)$ 成立(一切 $f\in L(Q)$)也是必要的.

我们的主要兴趣是研究“沿对角线”的求和. 设 $\varphi(r)$ 是支集含于[0，1]的连续函数. 对于 $f\in L(Q)$，定义

$$\mathfrak{S}_n^{\varphi}f(x,y)=\sum_{k=0}^{n}\left[\varphi\left(\frac{k}{n+1}\right)-\varphi\left(\frac{k+1}{n+1}\right)\right]S_{k,k}(f;x,y),\quad n\in\mathbf{N},$$

其中 $S_{k,k}(f;x,y)=\frac{1}{\pi^2}\int_Q f(x+s,y+t)D_k(s)D_k(t)\mathrm{d}s\mathrm{d}t$ 是 f 的方形 Fourier 部分和. $\mathfrak{S}_n^{\varphi}f(x,y)$ 叫作 f 的(由 φ 产生的)Marcinkiewicz 型线性平均. 此种求和方式为 M 氏于 1939 年首先提出[2]，故以 M 型名之，亦可称为“沿对角线”的求和.

倘支集含于[−1，1]×[−1，1]的连续函数 $\lambda(x,y)$ 满足

$$\lambda(x,y)=\lambda(\max(|x|,|y|),\max(|x|,|y|))\triangleq\varphi(\max(|x|,|y|)) \tag{1.3}$$

就得到

$$\tau_{n,n}^{\lambda}f=\mathfrak{S}_n^{\varphi}f, \tag{1.4}$$

这样一来，似乎只要把定理 A，B 的条件，通过(1.3)用加于 φ 的条件表出，就得到了关于 M 型求和的结果. 可是，我们在 § 3 证明，这样做得不到正面的结果，即

定理 1　对于任何支集在[0，1]上的连续函数 $\varphi(r)$，总存在 $f\in L(Q)$使得

$$\lim_{n\to+\infty}\mathfrak{S}_n^{\varphi}f(x,y)=f(x,y) \tag{1.5}$$

在 $SE(f)$的真子集上不成立.

可见，M 型求和法对于依定义 1 和定义 2 确定的 Lebesgue 点集，从整体上看来是不适用的. 于是在 § 2 中，我们以稍强的限制给出另一种强 Lebesgue 点和加强的弱 Lebesgue 点的定义. 这两种点集分别对于 $L\ln^{+}L(Q)$和 $L(Q)$的函数是满测度的. 然后求出在这样定义的强 Lebesgue 点集上(1.1)(1.2)成立的充要条件(定理 2)，以及在加强的弱 Lebesgue 点集上(1.2)成立的充分条件(定理 3). § 2 的结果不难推广到多于二元的情形. 最后在 § 3 中，通过(1.3)(1.4)把 § 2 的结果应用于 M 型求和，求得沿对角线求和的充分条件(定理 4，5)及一些推论. 这些推论确定了一些熟知的 M 型线性平均收敛的点集.

§ 2. 星形 Lebesgue 点上的求和

把实平面 $\mathbf{R}^2$ 分成 8 个角形区域，定义

$A_1=\{(x,\ y)\mid 0\leqslant y\leqslant x<+\infty\}$，　$A_2=\{(x,\ y)\mid (y,\ x)\in A_1\}$，

$A_3=\{(x,\ y)\mid (y,\ -x)\in A_1\}$，　$A_4=\{(x,\ y)\mid (-x,\ y)\in A_1\}$，

$A_5=\{(x,\ y)\mid (-x,\ -y)\in A_1\}$，　$A_6=\{(x,\ y)\mid (-y,\ -x)\in A_1\}$，

$A_7=\{(x,\ y)\mid (-y,\ x)\in A_1\}$，　$A_8=\{(x,\ y)\mid (x,\ -y)\in A_1\}$.

若点$(x,\ y)\in A_\nu(\nu=1,2,\cdots,8)$，以$(x,\ y)$为顶点作两条分别与 A_ν 的边平行(同方向)的射线，所成的锐角形区域(在 A_ν 内)记作 $A(x,\ y)$.

任给 $\mathbf{R}^2$ 上一个函数(可取复值)，定义一个函数 $Af(x,\ y)$，它在诸角 A_ν 内的值用下式定义

$$Af(x,y)=\sup_{(u,v)\in A(x,y)}|f(u,v)|.$$

姑且把 Af 叫作 f 的角形优函数.

显然，当点$(x,\ y)$落在角 A_ν 的边上(但不是原点)时，对应的角形区域 $A(x,\ y)$必是两个，因此 $Af(x,\ y)$可有两个值. 至于 $Af(0,\ 0)$，它可取 8 个值.

任取正数 a，b，记平行四边形

$$D(a,\ b)=\{(x,\ y)\mid 0\leqslant x-y\leqslant a,\ 0\leqslant y\leqslant b\}.$$

任意一点$(x, y)\in \mathbf{R}^2$，对应一个8点(或4点或1点)的集合$P(x, y)$：

$$P(x, y)=\{(x, y), (-x, y), (x, -y), (-x, -y), (y, x), (-y, x), (y, -x), (-y, -x)\}.$$

定义原点的星形邻域

$$E(a, b)=\{(x, y)\mid P(x, y)\cap D(a, b)\neq\varnothing\}.$$

显然有如下的测度的等式

$$|D(a, b)|=ab, \quad |E(a, b)|=8ab.$$

定义3 设$f\in L(Q)$. 点(x, y)叫作f的星形强Lebesgue点，若

$$\begin{cases}\lim\limits_{a,b\to 0^+}\dfrac{1}{ab}\displaystyle\int_{E(a,b)}|f(x+s,y+t)-f(x,y)|\,\mathrm{d}s\mathrm{d}t=0, \\ \sup\limits_{a,b>0}\dfrac{1}{8ab}\displaystyle\int_{E(a,b)}|f(x+s,y+t)|\,\mathrm{d}s\mathrm{d}t\equiv M_* f(x,y)<+\infty.\end{cases}$$

f在Q中的全体星形强Lebesgue点的集合记作$S_* E(f)$.

定义4 设$f\in L(Q)$，$p>0$. 点(x, y)叫作f的p阶加强的星形弱Lebesgue点(简称p阶弱L-点)，若

$$\lim_{\delta\to 0^+}\left\{\sup_{\mu,\nu\in\mathbf{N}}\frac{1}{2^{(1+p)(\mu+\nu)}\delta^2}\int_{E(2^\mu\delta,2^\nu\delta)}|f(x+s,y+t)-f(x,y)|\,\mathrm{d}s\mathrm{d}t\right\}=0.$$

f在Q中的全体p阶弱L-点的集合记作$W_* E^p(f)$.

易见，当$0<p<q<+\infty$时，

$$SE(f)\subset W_* E^p(f)\subset W_* E^q(f)\subset WE(f).$$

定理2 等式(1.1)对于一切$f\in L(Q)$在$S_* E(f)$上成立的充要条件是

$$(1)\ \lambda(0,0)=1, \qquad (2)\int_{\mathbf{R}^2}A\tilde{\lambda}(x,y)\mathrm{d}x\mathrm{d}y<+\infty.$$

而且为使等式(1.2)对于一切$f\in L(Q)$在$S_* E(f)$上成立，上述条件也是必要的.

此定理与定理A相当，证法一样，其证从略.

下面给出一个在$W_* E^p(f)$上求和的充分条件. 先定义一个函数ρ. 设$(x, y)\in A_\nu(1\leqslant\nu\leqslant 8)$. 它到$A_\nu$的两边的距离分别为$d_1$和$d_2$. 定义

$$\rho(x, y)=(d_1+1)(d_2+1)$$

(显然ρ在角的边上的值仍是单值的).

定理3 设$p>0$. 若以下三条满足

(1) $\lambda(0, 0)=1$,

(2) $\int_{\mathbf{R}^2}\rho^p(x,y)\cdot A\tilde{\lambda}(x,y)\mathrm{d}x\mathrm{d}y<+\infty$,

(3) $\rho^p(x, y)A\tilde{\lambda}(x, y)$在诸角$A_\nu$的边上的一维积分取有限值(函数在每条边上有两个分支)，则等式(1.2)对于一切$f\in L(Q)$在$W_* E^p(f)$上成立.

证　由条件$\lambda(0, 0)=1$知，只需证明当$(0, 0)\in W_* E^p(f)$，且$f(0, 0)=0$时，

$$\lim_{n\to+\infty}\tau_{n,n}^{\lambda}f(0, 0)=0.$$

像[1]中证定理A那样，使用Poisson积分公式，得到

$$\begin{aligned}|\tau_{n-1,n-1}^{\lambda}f(0,0)|&\leqslant C\int_{\mathbf{R}^2}\left|f\left(\frac{s}{n},\frac{t}{n}\right)\right||\tilde{\lambda}(s,t)|\mathrm{d}s\mathrm{d}t\\&=C\sum_{\nu=1}^{8}\int_{A_\nu}\left|f\left(\frac{s}{n},\frac{t}{n}\right)\right||\tilde{\lambda}(s,t)|\mathrm{d}s\mathrm{d}t,\quad n\in\mathbf{N}^*.\end{aligned}\tag{2.1}$$

其中$C>0$为绝对常数(以下我们总用字母C表示正的常数，它只可能与明确标明的参量有关，不计其大小). (2.1)右端诸A_ν上的积分的估计办法是一样的，下面只估算A_1上的积分，依照前面规定的记号，令

$$D_{\mu,\nu}=D(2^\mu, 2^\nu),\ F_\mu=D_{\mu,0}\setminus D_{\mu-1,0},\ G_\nu=D_{0,\nu}\setminus D_{0,\nu-1},$$
$$H_{\mu,\nu}=D_{\mu,\nu}\setminus(D_{\mu-1,\nu}\cup D_{\mu,\nu-1}),\quad \mu,\ \nu\in\mathbf{N}.$$

并记

$$\alpha_n^p=\sup_{\mu,\nu\in\mathbf{N}}\frac{n^2}{2^{(1+p)(\mu+\nu)}}\int_{E\left(\frac{2^\mu}{n},\frac{2^\nu}{n}\right)}|f(s,t)|\mathrm{d}s\mathrm{d}t.\tag{2.2}$$

那么，

$$\begin{aligned}I_1&\triangleq\int_{A_1}\left|f\left(\frac{s}{n},\frac{t}{n}\right)\right||\tilde{\lambda}(s,t)|\mathrm{d}s\mathrm{d}t=\\&=\left\{\int_{D(1,1)}+\sum_{\mu=1}^{+\infty}\left(\int_{F_\mu}+\int_{G_\mu}\right)+\sum_{\mu,\nu=1}^{+\infty}\int_{H_{\mu,\nu}}\right\}\left|f\left(\frac{s}{n},\frac{t}{n}\right)\tilde{\lambda}(s,t)\right|\mathrm{d}s\mathrm{d}t.\end{aligned}\tag{2.3}$$

其中

$$\begin{aligned}&\int_{D(1,1)}\left|f\left(\frac{s}{n},\frac{t}{n}\right)\tilde{\lambda}(s,t)\right|\mathrm{d}s\mathrm{d}t\\&\leqslant\sup_{(s,t)\in D(1,1)}|\tilde{\lambda}(t,t)|\cdot n^2\int_{D\left(\frac{1}{n},\frac{1}{n}\right)}|f(s,t)|\mathrm{d}s\mathrm{d}t\leqslant C(\lambda)\cdot\alpha_n^p.\end{aligned}\tag{2.4}$$

另外，

$$\int_{F_k}\left|f\left(\frac{s}{n},\frac{t}{n}\right)\tilde{\lambda}(s,t)\right|\mathrm{d}s\mathrm{d}t\leqslant\sup_{(s,t)\in F_k}|\tilde{\lambda}(s,t)|\int_{E(2^k,1)}\left|f\left(\frac{s}{n},\frac{t}{n}\right)\right|\mathrm{d}s\mathrm{d}t$$

$$=\sup_{(s,t)\in F_k}|\tilde{\lambda}(s,t)|\cdot 2^{(1+p)k}\frac{n^2}{2^{(1+p)k}}\int_{E(\frac{2^k}{n},\frac{1}{n})}|f(s,t)|\mathrm{d}s\mathrm{d}t$$

$$\leqslant C\cdot\alpha_n^p\cdot\sup_{(s,t)\in F_k}|\tilde{\lambda}(s,t)|\cdot 2^{pk}|F_{k-1}|.$$

当$(x,\ 0)\in\bar{F}_{k-1}$时，

$$\sup_{(s,t)\in F_k}|\tilde{\lambda}(s,\ t)|2^{pk}\leqslant C(p)\cdot\rho^p(x,\ 0)\cdot A\tilde{\lambda}(x,\ +0).$$

其中$A\tilde{\lambda}(x,\ +0)$表示$A\tilde{\lambda}$的一个值，它是由在A_1内取上确界得到的.

于是

$$\int_{F_k}\left|f\left(\frac{s}{n},\frac{t}{n}\right)\tilde{\lambda}(s,t)\right|\mathrm{d}s\mathrm{d}t\leqslant C(p)\alpha_n^p\cdot\int_{2^{k-2}}^{2^{k-1}}\rho^p(x,0)A\tilde{\lambda}(x,+0)\mathrm{d}x,$$

$$\sum_{k=1}^{+\infty}\int_{F_k}\left|f\left(\frac{s}{n},\frac{t}{n}\right)\tilde{\lambda}(s,t)\right|\mathrm{d}s\mathrm{d}t\leqslant C(p)\cdot\alpha_n^p\int_{\frac{1}{2}}^{+\infty}\rho^p(x,0)A\tilde{\lambda}(x,+0)\mathrm{d}x.$$

由条件(3)得

$$\sum_{k=1}^{+\infty}\int_{F_k}\left|f\left(\frac{s}{n},\frac{t}{n}\right)\tilde{\lambda}(s,t)\right|\mathrm{d}s\mathrm{d}t\leqslant C(p,\lambda)\alpha_n^p.\tag{2.5}$$

同理可证

$$\sum_{k=1}^{+\infty}\int_{G_k}\left|f\left(\frac{s}{n},\frac{t}{n}\right)\tilde{\lambda}(s,t)\right|\mathrm{d}s\mathrm{d}t\leqslant C(p,\lambda)\alpha_n^p.\tag{2.6}$$

最后，

$$\int_{H_{\mu,\nu}}\left|f\left(\frac{s}{n},\frac{t}{n}\right)\tilde{\lambda}(s,t)\right|\mathrm{d}s\mathrm{d}t\leqslant\sup_{(s,t)\in H_{\mu,\nu}}|\tilde{\lambda}(s,t)|\cdot\int_{E(2^\mu,2^\nu)}\left|f\left(\frac{s}{n},\frac{t}{n}\right)\right|\mathrm{d}s\mathrm{d}t$$

$$=\sup_{(s,t)\in H_{\mu,\nu}}\left|\tilde{\lambda}(s,t)\cdot 2^{(1+p)(\mu+\nu)}\frac{n^2}{2^{(1+p)(\mu+\nu)}}\int_{E(2^\mu n^{-1},2^\nu n^{-1})}\right|f(s,t)\mathrm{d}s\mathrm{d}t$$

$$\leqslant C\alpha_n^p\cdot\sup_{(s,t)\in H_{\mu,\nu}}|\tilde{\lambda}(s,t)|\cdot 2^{p(\mu+\nu)}|H_{\mu-1,\nu-1}|$$

$$\leqslant C(p)\cdot\alpha_n^p\cdot\int_{H_{\mu-1,\nu-1}}\rho^p(x,y)A\tilde{\lambda}(x,y)\mathrm{d}x\mathrm{d}y,(\mu,\nu\in\mathbf{N}^*).$$

所以，根据条件(2)，

$$\sum_{\mu=1}^{+\infty}\sum_{\nu=1}^{+\infty}\int_{H_{\mu,\nu}}\left|f\left(\frac{s}{n},\frac{t}{n}\right)\tilde{\lambda}(s,t)\right|\mathrm{d}s\mathrm{d}t$$

$$\leqslant C(p)\alpha_n^p\cdot\sum_{\mu=0}^{+\infty}\sum_{\nu=0}^{+\infty}\int_{H_{\mu,\nu}}\rho^p(x,y)A\tilde{\lambda}(x,y)\mathrm{d}x\mathrm{d}y\leqslant C(p,\lambda)\cdot\alpha_n^p.\tag{2.7}$$

由(2.3)～(2.7)得

$$I_1 \leqslant C(p, \lambda)\alpha_n^p.$$

从而

$$|\tau_{n,n}^{\lambda} f(0, 0)| \leqslant C(p, \lambda) \cdot \alpha_{n+1}^p, \quad n \in \mathbf{N}. \tag{2.8}$$

由于$(0, 0) \in W_* E^p(f)$，$\alpha_n^p \to 0$ $(n \to +\infty)$，故得定理.

由 О. Д. Габисония[3]得知，当 $f \in L\ln^+ L(Q)$时，$|SE(f)| = |Q|$. 由此便可推出$|S_* E(f)| = |Q|$. 而当 $f \in L(Q)$时，对于任意的 $p > 0$，也可由[3]的结果推出$|W_* E^p(f)| = |Q|$.

我们指出，本节的结果不难推广到多于二元的情形.

§3. Marcinkiewicz 型求和

设(1.3)成立，则(1.4)成立：

$$\begin{aligned}\tau_{n,n}^{\lambda} f(x,y) &= \mathfrak{S}_n^{\varphi} f(x,y) \\ &= \frac{1}{\pi^2}\int_Q f(x+s, y+t)\left(\sum_{k=0}^{n} \varphi\left(\frac{k}{n+1}\right) B_k(s,t)\right) \mathrm{d}s\mathrm{d}t.\end{aligned}$$

其中 $B_0 = \frac{1}{4}$，

$$B_k(s, t) = \cos ks \cdot \frac{\sin kt}{2\tan\frac{t}{2}} + \cos kt \, \frac{\sin ks}{2\tan\frac{s}{2}}, \quad k \in \mathbf{N}^*.$$

计算 $\tilde{\lambda}(x, y)$，得

$$\begin{aligned}\tilde{\lambda}(x,y) &= \int_{-1}^{1}\int_{-1}^{1} \lambda(u,v) \mathrm{e}^{-\mathrm{i}(xu+yv)} \mathrm{d}u\mathrm{d}v \\ &= 4\int_0^1 \varphi(r)\left(\cos rx \, \frac{\sin ry}{y} + \cos ry \, \frac{\sin rx}{x}\right) \mathrm{d}r.\end{aligned} \tag{3.1}$$

若记 $\hat{\varphi}(u) = \int_0^1 \varphi(r) \sin ru \, \mathrm{d}r$，则得到

$$\tilde{\lambda}(x, y) = \frac{2}{xy}\{(x+y)\hat{\varphi}(x+y) + (y-x)\hat{\varphi}(x-y)\}. \tag{3.2}$$

定理 1 的证 如果定理 A 的条件成立，那么 $\varphi(0) = 1$，且对于任意的 $a \geqslant 0$，

$$\int_{\substack{a \leqslant x-y < +\infty \\ 0 \leqslant y,}} \sup_{\substack{x < |u| \\ y < |v|}} |\tilde{\lambda}(u,v)| \mathrm{d}x\mathrm{d}y < +\infty. \tag{3.3}$$

但据(3.2)，当 $a \leqslant x-y < +\infty$且 $y \geqslant 0$ 时，

$$\sup_{\substack{x<|u|\\ y<|v|}}|\tilde{\lambda}(u,\ v)|\geqslant|\tilde{\lambda}(x,\ x-a)|$$

$$=\frac{2}{x(x-a)}|(2x-a)\hat{\varphi}(2x-a)-a\hat{\varphi}(a)|. \tag{3.4}$$

于(3.4)中，先取 $a=0$，代入(3.2)，得

$$\int_0^{+\infty}\mathrm{d}x\cdot\int_0^x\left|\frac{4x\hat{\varphi}(2x)}{x^2}\right|\mathrm{d}y=2\int_0^{+\infty}|\hat{\varphi}(x)|\mathrm{d}x<+\infty. \tag{3.5}$$

再取 $a<0$，从(3.3)～(3.5)得

$$\int_{\substack{a<x-y<+\infty\\ 0<y}}\left|\frac{2a\hat{\varphi}(a)}{x(x-a)}\right|\mathrm{d}x\mathrm{d}y$$

$$\leqslant\int_0^{+\infty}\int_0^{+\infty}\sup_{\substack{x<|u|\\ y<|v|}}|\tilde{\lambda}(u,v)|\mathrm{d}x\mathrm{d}y+\int_{\substack{a<x-y<+\infty\\ 0<y}}\frac{2(2x-a)|\hat{\varphi}(2x-a)|}{x(x-a)}\mathrm{d}x\mathrm{d}y$$

$$=\int_0^{+\infty}\int_0^{+\infty}\sup_{\substack{x<|u|\\ y<|v|}}|\tilde{\lambda}(u,v)|\mathrm{d}x\mathrm{d}y+2\int_{2a}^{+\infty}\frac{u-a}{u}|\hat{\varphi}(u-a)|\mathrm{d}u<+\infty.$$

由此可见，必有

$$\hat{\varphi}(a)=0,\qquad a>0. \tag{3.6}$$

于是 $\varphi(r)$ 几乎处处等于零．这与 φ 连续且 $\varphi(0)=1$ 矛盾．这说明定理A的条件不能成立． □

下面假定 $\varphi(r)$ 在 $[0,\ 1]$ 连续而且有界变差，$\varphi(0)=1$，$\varphi(1)=0$．那么

$$\hat{\varphi}(u)=\frac{1}{u}+\frac{1}{u}\int_0^1\cos ru\,\mathrm{d}\varphi(r). \tag{3.7}$$

记 $\Phi(u)=\int_0^1\cos ru\,\mathrm{d}\varphi(r)$，由(3.4)(3.7)得

$$\tilde{\lambda}(x,\ y)=\frac{2}{x\cdot y}(\Phi(x+y)-\Phi(x-y)). \tag{3.8}$$

在这些假定下，我们证明

定理4 若

$$\int_1^{+\infty}\frac{1}{x}\{\sup_{\xi>x}|\Phi'(\xi)|+\ln x\cdot\sup_{\xi>x}|\Phi(\xi)|\}\mathrm{d}x<+\infty, \tag{3.9}$$

则对于一切 $f\in L(Q)$，等式(5)在 $S_*E(f)$ 上成立．

证 根据定理2，只需验证当(3.9)成立时，对于由(3.8)表示的 $\tilde{\lambda}$，有

$$\int_{A_\nu}A\tilde{\lambda}(x,y)\mathrm{d}x\mathrm{d}y<+\infty,\quad \nu=1,2,\cdots,8. \tag{3.10}$$

由于对各角 A_ν 上的积分的估算办法都一样，下面只对 $\nu=1$ 来证明(3.10).

$$\int_{A_1} A\tilde{\lambda}(x,y)\mathrm{d}x\mathrm{d}y$$

$$=\Big(\int_{0<y<x<2}+\int_{\substack{2<x<+\infty\\0<y<1}}+\int_{\substack{2<x<+\infty\\1<y<x-1}}+\int_{\substack{2<x<+\infty\\x-1<y<x}}\Big)A\tilde{\lambda}(x,y)\mathrm{d}x\mathrm{d}y$$

右边四个积分顺次记为 J_1，J_2，J_3，J_4. 显然

$$J_1\leqslant 2\sup_{(x,y)}|\tilde{\lambda}(x,\ y)|<+\infty.$$

从(3.8)知，当 $2<x<+\infty$，$0<y<1$ 时

$$A\tilde{\lambda}(x,y)=\sup_{\substack{x-y<u-v\\y<v}}\left|\frac{2}{u\cdot v}(\Phi(u+v)-\Phi(u-v))\right|$$

$$=4\sup_{\substack{x-y<u-v\\y<v}}\left|\frac{1}{u}\Phi'(u+\theta v)\right|\leqslant\frac{4}{x}.\sup_{x-1<\xi}|\Phi'(\xi)|. \tag{3.11}$$

式中 θ 与$(u,\ v)$有关，但$|\theta|<1$. 由(3.9)和(3.11)知

$$J_2\leqslant 4\int_2^{+\infty}\frac{1}{x}\lim_{x-1<\xi}|\Phi'(\xi)|\,\mathrm{d}x<+\infty.$$

由(3.8)知，当 $2<x<+\infty$，$0<y<\frac{x}{2}$时，

$$A\tilde{\lambda}(x,\ y)\leqslant\frac{4}{xy}\sup_{\xi>\frac{x}{2}}|\Phi(\xi)|. \tag{3.12}$$

$$\int_2^{+\infty}\mathrm{d}x\int_1^{\frac{x}{2}}\mathrm{d}yA\tilde{\lambda}(x,y)\leqslant 4\int_2^{+\infty}\frac{1}{x}\sup_{\xi>\frac{x}{2}}|\Phi(\xi)|\,\mathrm{d}x\int_1^{\frac{x}{2}}\frac{\mathrm{d}y}{y}$$

$$=4\int_1^{+\infty}\frac{\ln x}{x}\sup_{x<\xi}|\Phi(\xi)|\,\mathrm{d}x.$$

而当 $2<x<+\infty$，$\frac{x}{2}<y<x-1$ 时，

$$A\tilde{\lambda}(x,\ y)\leqslant\frac{8}{x^2}\sup_{x-y<\xi}|\Phi(\xi)|. \tag{3.13}$$

$$\int_2^{+\infty}\mathrm{d}x\int_{\frac{x}{2}}^{x-1}\mathrm{d}yA\tilde{\lambda}(x,y)\leqslant 8\int_2^{+\infty}\frac{\mathrm{d}x}{x^2}\int_{\frac{x}{2}}^{x-1}\sup_{x-y<\xi}|\Phi(\xi)|\,\mathrm{d}y$$

$$=8\int_2^{+\infty}\frac{\mathrm{d}x}{x^2}\int_1^{\frac{x}{2}}\sup_{\xi>t}|\Phi(\xi)|\,\mathrm{d}t\leqslant C\int_1^{+\infty}\frac{1}{x}\sup_{\xi>x}|\Phi(\xi)|\,\mathrm{d}x.$$

所以，

$$J_3\leqslant C\int_1^{+\infty}(1+\ln x)\frac{1}{x}\sup_{\xi>x}|\Phi(\xi)|\,\mathrm{d}x<+\infty.$$

最后，当 $2<x<+\infty$，$x-1<y<x$ 时，

$$
\begin{aligned}
A\tilde{\lambda}(x,y) &\leqslant \frac{2}{x(x-1)} \sup_{\substack{x-y<u-v\\ y<v}} |\Phi(u+v)-\Phi(u-v)| \\
&\leqslant \frac{4}{x(x-1)} \sup_{\xi>0} |\Phi(\xi)| \leqslant \frac{4}{x(x-1)} \bigvee_0^1(\varphi).
\end{aligned} \tag{3.14}
$$

恒有 $J_4<+\infty$.

总之，当(3.9)成立时(3.10)成立. □

推论 1 当 $\varphi(r)$ 取下列函数时，定理 4 的结论成立：

(1) $\varphi(r)=1-r^{\alpha}$，$(\alpha>0)$，$r\in[0,1]$；

(2) $\varphi(r)=(1-r)^{\alpha}$，$(\alpha>0)$，$r\in[0,1]$；

(3) $\varphi(r)=\cos\frac{\pi}{2}r$，$r\in[0,1]$.

证 (1) $\Phi(\xi)=\int_0^1 \cos\xi r\,\mathrm{d}\varphi(r)=-\alpha\int_0^1 r^{\alpha-1}\cos\xi r\,\mathrm{d}r.$

$$|\Phi(\xi)|\leqslant C(\alpha)\frac{1}{|\xi|^{\bar{\alpha}}},\ \bar{\alpha}=\min(\alpha,1),\quad |\xi|\to+\infty.$$

$$|\Phi'(\xi)|\leqslant(\alpha)\frac{1}{|\xi|}.$$

(2)
$$
\begin{aligned}
\Phi(\xi) &= -\alpha\int_0^1 (1-r)^{\alpha-1}\cos\xi r\,\mathrm{d}r \\
&= -\alpha\int_0^1 u^{\alpha-1}\cos\xi(1-u)\,\mathrm{d}u \\
&= -\alpha\cos\xi\cdot\int_0^1 u^{\alpha-1}\cos\xi u\,\mathrm{d}u-\alpha\sin\xi\int_0^1 u^{\alpha-1}\sin\xi u\,\mathrm{d}u.
\end{aligned}
$$

$$|\Phi(\xi)|\leqslant C(\alpha)\cdot\frac{1}{|\xi|^{\bar{\alpha}}},\quad \bar{\alpha}=\min(\alpha,1).$$

$$|\Phi'(\xi)|\leqslant C(\alpha)\frac{1}{|\xi|^{\bar{\alpha}}},\quad \bar{\alpha}=\min(\alpha,1).$$

(3)
$$
\begin{aligned}
\Phi(\xi) &= -\frac{\pi}{2}\int_0^1 \sin\frac{\pi}{2}r\cos\xi r\,\mathrm{d}r \\
&= -\frac{\pi}{2}\int_0^1 \left\{\sin\left(\frac{\pi}{2}+\xi\right)r+\sin\left(\frac{\pi}{2}-\xi\right)r\right\}\mathrm{d}r.
\end{aligned}
$$

$$|\Phi(\xi)|\leqslant C\frac{1}{|\xi|},$$

$$|\Phi'(\xi)|\leqslant C\frac{1}{|\xi|}.$$

易见，在三种情况下，(3.9)都成立. □

依然保留前面对 $\varphi(r)$ 的假设，则有

定理 5 设 $0<p<\frac{1}{2}$. 若

$$\int_1^{+\infty}\frac{1}{x^{1-p}}\cdot\{x^p\cdot\sup_{\xi>x}|\Phi(\xi)|+\sup_{\xi>x}|\Phi'(\xi)|\}\mathrm{d}x<+\infty, \tag{3.15}$$

则对于一切 $f\in L(Q)$ 等式(1.5)在 $W_*E^p(f)$ 上成立.

证 只需验证当(3.15)成立时，定理 3 的条件(2)和(3)成立. 先验证条件(3).

在角 A_1 的边、射线 $\overrightarrow{Ox}$ 上，当 $x\geqslant 1$ 时，由(3.8)

$$\rho^p(x,\ 0)A\tilde{\lambda}(x,\ 0)=\left(1+\frac{x}{\sqrt{2}}\right)^p\cdot\sup_{\substack{x<u-v\\0<v}}\left|\frac{2}{uv}(\Phi(u+v)-\Phi(u-v))\right|$$

$$\leqslant C(p)\cdot\frac{1}{x^{1-p}}\sup_{\xi>x}|\Phi'(\xi)|.$$

而在 A_1 的另一条边 $y=x\geqslant 0$ 上，当 $y\geqslant 1$ 时

$$\rho^p(y,\ y)\cdot A\tilde{\lambda}(y,\ y)=(1+y)^p\cdot\sup_{\substack{0<u-v\\y<v}}\left|\frac{2}{u\cdot v}(\Phi(u+v)-\Phi(u-v))\right|$$

$$\leqslant C(p)\cdot\frac{1}{y^{2-p}}\bigvee_0^1(\varphi).$$

可见定理 3 的条件(3)在 A_1 的边上满足，同理在其他各角的边上也满足.

现在验证

$$\int_{A_\nu}\rho^p(x,y)A\tilde{\lambda}(x,y)\mathrm{d}x\mathrm{d}y<+\infty,\quad \nu=1,2,\cdots,8. \tag{3.16}$$

显然只需对 $\nu=1$ 证明. 与证(3.10)类似，有

$$\int_{A_1}\rho^pA\tilde{\lambda}\mathrm{d}x\mathrm{d}y=\left(\int_{0<y<x<2}+\int_{\substack{2<x<+\infty\\0<y<1}}+\int_{\substack{2<x<+\infty\\1<y<x-1}}+\int_{\substack{2<x<+\infty\\x-1<y<x}}\right)\rho^pA\tilde{\lambda}\mathrm{d}x\mathrm{d}y$$

$$=J_1'+J_2'+J_3'+J_4'.$$

显然，

$$J_1'\leqslant C(p)\cdot\sup_{(x,y)}|\tilde{\lambda}(x,\ y)|<+\infty.$$

由 $\rho(x,\ y)$ 之定义，用(3.11)，知当 $2<x<+\infty$，$0<y<1$ 时，

$$\rho^p(x,\ y)A\tilde{\lambda}(x,\ y)\leqslant\left[1+\frac{x-y}{\sqrt{2}}\right]^p\cdot(1+y)^p\cdot\frac{2}{x}\cdot\sup_{\xi>x-1}|\Phi'(\xi)|$$

$$\leqslant C(p)\cdot\frac{1}{x^{1-p}}\sup_{\xi>x-1}|\Phi'(\xi)|.$$

$$J'_2\leqslant C(p)\int_1^{+\infty}\frac{1}{x^{1-p}}\sup_{\xi>x}|\Phi'(\xi)|\mathrm{d}x<+\infty.$$

用(3.12)，知当 $2<x<+\infty$，$1<y<\dfrac{x}{2}$时，

$$\rho^p(x,\ y)A\tilde{\lambda}(x,\ y)\leqslant C(p)\cdot\frac{1}{x^{1-p}}\frac{1}{y^{1-p}}\sup_{\xi>\frac{x}{2}}|\Phi(\xi)|.$$

$$\int_2^{+\infty}\mathrm{d}x\int_1^{\frac{x}{2}}\mathrm{d}y\rho^pA\tilde{\lambda}\leqslant C(p)\int_0^{+\infty}\frac{1}{x^{1-p}}\sup_{\xi>\frac{x}{2}}|\Phi(\xi)|\cdot\int_1^{\frac{x}{2}}\frac{\mathrm{d}y}{y^{1-p}}$$

$$=C(p)\cdot\int_1^{+\infty}\frac{1}{x^{1-2p}}\sup_{\xi>x}|\Phi(\xi)|\mathrm{d}x<+\infty.$$

用(3.13)，得知当 $2<x<+\infty$，$\dfrac{x}{2}<y<x-1$ 时，

$$\rho^p(x,\ y)A\tilde{\lambda}(x,\ y)\leqslant\left(1+\frac{x-y}{\sqrt{2}}\right)^p(1+y)^p\frac{8}{x^2}\sup_{\xi>x-y}|\Phi(\xi)|$$

$$\leqslant C(p)\cdot\frac{(x-y)^p}{x^{2-p}}\sup_{\xi>x-y}|\Phi(\xi)|.$$

从而

$$\int_{\substack{2<x<+\infty\\ \frac{x}{2}<y<x-1}}\rho^pA\tilde{\lambda}\mathrm{d}x\mathrm{d}y\leqslant C(p)\cdot\int_2^{+\infty}\mathrm{d}x\int_{\frac{x}{2}}^{x-1}\mathrm{d}y\frac{(x-y)^p}{x^{2-p}}\sup_{\xi>x-y}|\Phi(\xi)|$$

$$=C(p)\int_2^{+\infty}\frac{\mathrm{d}x}{x^{2-p}}\int_1^{\frac{x}{2}}t^p\sup_{\xi>t}|\Phi(\xi)|\mathrm{d}t$$

$$=C(p)\cdot\lim_{M\to+\infty}\left\{\frac{-1}{1-p}\frac{1}{x^{1-p}}\int_1^{\frac{x}{2}}t^p\cdot\sup_{\xi>t}|\Phi(\xi)|\mathrm{d}t\Big|_{x=2}^{x=M}+\right.$$

$$\left.\frac{1}{1-p}\cdot\frac{1}{2}\int_2^M\frac{1}{x^{1-p}}\left(\frac{x}{2}\right)^p\sup_{\xi>\frac{x}{2}}|\Phi(\xi)|\mathrm{d}x\right\}$$

$$\leqslant C(p)\int_1^{+\infty}\frac{1}{x^{1-2p}}\sup_{\xi>x}|\Phi(\xi)|\mathrm{d}x<+\infty.$$

于是

$$J'_3\leqslant C(p)\int_1^{+\infty}\frac{1}{x^{1-2p}}\sup_{\xi>x}|\Phi(\xi)|\mathrm{d}x<+\infty.$$

用(3.14)，得到当 $2<x<+\infty$，$x-1<y<x$ 时，

$$\rho^p(x,\ y)\cdot A\tilde{\lambda}(x,\ y)\leqslant C(p)\frac{x^p}{x(x-1)}\bigvee_0^1(\varphi).$$

可见

$$J_4' \leqslant C(p) \bigvee_0^1 (\varphi) \cdot \int_1^{+\infty} \frac{\mathrm{d}x}{x^{2-p}} < +\infty.$$

总之，(3.16)式成立，即定理 3 的条件(2)满足. □

利用定理 5 可得

推论 2 设 $0<p<\frac{1}{2}$，$\alpha>2p$，则对于

(1) $\varphi(r)=1-r^{\alpha}$，　$r\in[0, 1]$，

(2) $\varphi(r)=(1-r)^{\alpha}$，　$r\in[0, 1]$，

(3) $\varphi(r)=\cos\frac{\pi}{2}$，　$r\in[0, 1]$

定理 5 的结论成立.

参考文献

[1] Белинский Э С. В кн: Теория функций. функционалъньгй анализ и их приложения，1975，23：3－12.

[2] Marcinkiewicz. J. Sur une méthode remarquable de sommation des séries doubles de Fourier，Collected papers，Warszawa. 1964，527－538.

[3] Габнсония. О Д. Изб. буз. мамемамцка. 1972，5：29－37.

Abstract In this paper the summability of multiple Fourier series on a set of points，at which the integral of function has certain differentiable properties，is discussed.

Some conditions of summability for double Fourier series by the method of Marcinkiewicz are obtained.

数学年刊，1984，5A(2)：133－152.

二元周期函数用其 Marcinkiewicz 型 Cesàro 平均逼近与绝对求和[①]

Approximation and Absolute Summation for Periodic Functions of Two Variables by Cesàro Means of Marcinkiewicz Type

摘要 设 $f\in L(Q)$，$Q=\{(x,y)\mid -\pi\leqslant x<\pi,\ -\pi\leqslant y<\pi\}$. f 的 Marcinkiewicz 型 Cesàro 平均定义为

$$\sigma_n^\alpha(f;x,y)=\frac{1}{A_n^\alpha}\sum_{k=0}^{n}A_{n-k}^{\alpha-1}S_{k,k}(f;x,y),\quad n\in\mathbf{N},$$

其中 $S_{k,k}(f)$是 f 的 Fourier 部分和. 本文将一元周期函数$(C,\alpha)(\alpha>0)$逼近及$|C,\alpha|(\alpha>0)$求和的一些结果推广到 M 型的情形，得到了 M 型$|C,\alpha|(\alpha>0)$可求和的一些充分条件.

§1. 引言

为写法上方便起见，我们用 $L^p(Q)(1\leqslant p\leqslant+\infty)$表示在 $Q=\{(x,y)\mid -\pi\leqslant x<\pi,\ -\pi\leqslant y<\pi\}$上 p 次幂可和(当 $p<+\infty$)或连续(当 $p=+\infty$)且对每个变元都以 2π 为周期的函数的 Banach 空间. $f\in L^p(Q)$的范数是

$$\|f\|_p=\left\{\int_Q|f(x,y)|^p\mathrm{d}x\mathrm{d}y\right\}^{\frac{1}{p}},\quad 1\leqslant p<+\infty,$$

$$\|f\|_C=\|f\|_\infty=\max_{(x,y)\in Q}|f(x,y)|,\quad p=+\infty.$$ [②]

f 的 Fourier 部分和是

---

① 收稿日期：1981-10-13；收修改稿日期：1983-07-07.

② 我们用$\|f\|_\infty$表示$\|f\|_C$，为的是以后写法上的方便.

$$S_{i,j}(f;x,y)=\frac{1}{\pi^2}\int_Q f(x-s,y-t)D_i(s)D_j(t)\mathrm{d}s\mathrm{d}t,\quad i,j\in\mathbf{N},$$

其中 $D_\nu(t)=\dfrac{\sin\left(\nu+\frac{1}{2}\right)t}{2\sin\frac{t}{2}}$ 为 Dirichlet 核.

依照 J. Marcinkiewicz[1] 1939 年提出的求和法，f 的 M 型 Cesàro 平均是

$$\begin{aligned}\sigma_n^\alpha(f;x,y)&=\frac{1}{A_n^\alpha}\sum_{k=0}^{n}A_{n-k}^{\alpha-1}S_{k,k}(f;x,y)\\&=\frac{1}{\pi^2}\int_Q f(x-s,y-t)K_n^\alpha(s,t)\mathrm{d}s\mathrm{d}t,\quad n\in\mathbf{N},\end{aligned}$$

其中 A_ν^α 表示通常 Cesàro 平均中的系数，核

$$K_n^\alpha(s,t)=\frac{1}{A_n^\alpha}\sum_{k=0}^{n}A_{n-k}^{\alpha-1}D_k(s)D_k(t).$$

M 型线性求和是对 Fourier 部分和序列 $\{S_{k,k}(f)\}_{k=0}^{+\infty}$ 进行线性平均，其求和方式与一元函数一样. 因而自然想到将一元 Fourier 级数线性求和的讨论，保持原有形式推广到二元 M 型求和的问题.

本文讨论的是正阶 M 型 Cesàro 平均逼近的阶及正阶 M 型 Cesàro 绝对求和的问题.

在一元情形，郭竹瑞[2]得到在 $C_{2\pi}$ 中

$$\sigma_n^\alpha(f;x)-f(x)=\frac{\alpha}{\pi}\frac{1}{n}\int_{\frac{1}{n}}^{\pi}\frac{\varphi_x(u)}{2\sin^2\frac{u}{2}}\mathrm{d}u+O\left(\omega_2\left(f;\frac{1}{n}\right)\right),$$

$$\alpha>0;\ n\to+\infty,\tag{1.1}$$

其中 $\varphi_x(u)=f(x+u)+f(x-u)-2f(x)$，$\omega_2(f;\ u)$ 是 f 的二阶连续模. 由此立即推出

$$\|\sigma_n^\alpha(f)-f\|_C\leqslant\frac{C(\alpha)}{n+1}\int_{\frac{1}{n+1}}^{\pi}\frac{\omega_2(f;u)}{u^2}\mathrm{d}u\leqslant\frac{C(\alpha)}{n+1}\sum_{k=0}^{n}E_k(f)_C,\tag{1.2}$$

其中 $E_k(f)_C$ 为 f 的 k 阶最佳三角多项式逼近(按 C 范数)，$C(\alpha)>0$ 只与 α 有关.

我们在 § 2 推广这些结果. 设 $f\in L^p(Q)(1\leqslant p\leqslant+\infty)$. 记 f 的二阶偏连续模(L^p 尺度)为 $\omega_2(f;\ u,\ 0)_p$ 和 $\omega_2(f;\ 0,\ u)_p$，偏最佳逼近记为 $E_{k,\infty}(f)_p$ 和 $E_{\infty,k}(f)_p$(参见 A. Φ. Тиман[3])，当 $p=+\infty$时，由于约定 $L_\infty(Q)$表示 $C(Q)$，所以下标∞亦写作 c. 定义

$$\widetilde{\omega}(f;\ u)_p=\omega_2(f;\ u,\ 0)_p+\omega_2(f;\ 0,\ u)_p,$$
$$\widetilde{E}_k(f)_p=E_{k,\infty}(f)_p+E_{\infty,k}(f)_p.$$

§2 的结果是

定理 1 设 $1\leqslant p\leqslant+\infty$，$\alpha>0$，则存在只与 α 有关的常数 $C(\alpha)>0$，使对一切 $f\in L^p(Q)$有

$$\|\sigma_n^\alpha(f)-f\|_p\leqslant\frac{C(\alpha)}{n+1}\int_{\frac{1}{n+1}}^{\pi}\frac{\widetilde{\omega}(f;u)_p}{u^2}\mathrm{d}u,\quad n\in\mathbf{N}. \tag{1.3}$$

由此导出

$$\|\sigma_n^\alpha(f)-f\|_p\leqslant\frac{C(\alpha)}{n+1}\sum_{k=0}^{n}\widetilde{E}_k(f)_p,\quad n\in\mathbf{N}. \tag{1.4}$$

说 $f\in L(Q)$（或者说 f 的 Fourier 级数 $\sigma(f)$）在点$(x,\ y)$处可以 M 型$|C,\ \alpha|$求和是指

$$\sum_{n=0}^{+\infty}|\sigma_n^\alpha(f;x,y)-\sigma_{n+1}^\alpha(f;x,y)|<+\infty.$$

为了研究正阶$|C,\ \alpha|$求和问题，需要考虑 Cesàro 阶数 $\beta\in(-1,\ 0]$时的逼近情况. 在§3 中，对于 $f\in L^p(Q)$，$1<p\leqslant2$ 及 $p=+\infty$的情形，求得了量

$$\left\|\frac{1}{n+1}\sum_{k=n}^{2n}|\sigma_k^\beta(f;x,y)-f(x,y)|\right\|_p$$

用 f 的二阶偏连续模（及偏最佳逼近）控制的不等式，即定理 2、定理 3 及其推论.

§4 给出了 $L^p(Q)(1<p\leqslant2)$中函数几乎处处 M 型$|C,\ \alpha|(\alpha>0)$可求和的一个充分条件，同时也给出了 $C(Q)$中函数一致$|C,\ \alpha|(\alpha>0)$求和的充分条件. 即

定理 4 设 $\alpha>0$，$1<p\leqslant2$ 或 $p=+\infty$，$f\in L^p(Q)$. 则

$$\left\|\sum_{k=0}^{n}|\sigma_k^\alpha(f)-\sigma_{k+1}^\alpha(f)|\right\|_p\leqslant C(\alpha,p)\sum_{k=0}^{n}-\frac{\mu_k(\alpha,p)}{k+1}\widetilde{E}_k(f)_p, \tag{1.5}$$

$$\left\|\sum_{k=n}^{+\infty}|\sigma_k^\alpha(f)-\sigma_{k+1}^\alpha(f)|\right\|_p\leqslant C(\alpha,p)\left\{\sum_{k=n}^{+\infty}\frac{\mu_k(\alpha,p)}{k+1}\widetilde{E}_k(f)_p+\right.$$
$$\left.\frac{\mu_n(\alpha,p)}{(n+1)^2}\sum_{k=0}^{n}(k+1)\widetilde{E}_k(f)_p+\frac{1}{(n+1)^{\bar{\alpha}}}\sum_{k=0}^{n}(k+1)^{\bar{\alpha}-1}\widetilde{E}_k(f)_p\right\}, \tag{1.6}$$

其中 $\bar{\alpha}=\min(1,\ \alpha)$，

$$
\mu_k(\alpha, p)=\begin{cases}1, & \alpha>\frac{1}{p}, \quad 1<p\leqslant 2,\\ \ln^{1+\frac{1}{p}}(k+2), & \alpha=\frac{1}{p}, \quad 1<p\leqslant 2,\\ (k+1)^{\frac{1}{p}-\alpha}\ln(k+1), & 0<\alpha<\frac{1}{p}, \quad 1<p\leqslant 2,\\ \mu_k(\alpha, 2), & p=+\infty.\end{cases}
$$

这个结果，当 $p=2$ 时可与 ПономаренкоЮА[4] 关于多重矩形 Cesàro 求和的相应于二元情形的结果(见§4)进行比较，彼处所给的条件较强. 而当 $p=+\infty$ 时，定理 4 可以看作是施咸亮[5] 关于一元函数的结果(见§4)的推广.

最后(§5)，我们顺便推广王福春[6] 关于一元函数的定理于 M 型的情形，得到

定理 5 设 $f(x, y)\in L^2(Q)$，$\alpha\geqslant\frac{1}{2}$. 定义

$$
B_k(f;x,y)=\frac{1}{\pi^2}\int_Q f(x-s,y-t)\left(\frac{\sin ks}{2\sin\frac{s}{2}}\cos kt+\cos ks\frac{\sin kt}{2\sin\frac{t}{2}}\right)\mathrm{d}s\mathrm{d}t,
$$

$$
\gamma_k(f)=\|B_k(f)\|_2^2, \quad k\in\mathbf{N}^*.
$$

设正数列 $\{\varphi(n)\}_{n=1}^{+\infty}$ 满足 $\sum\limits_{n=1}^{+\infty}\frac{1}{\varphi(n)n^2}<+\infty$. 置

$$
r_k=k^2\sum_{n=k}^{+\infty}\frac{\varphi(n)}{n(n-k+1)}, \quad k\in\mathbf{N}^*.
$$

若

$$
\sum_{k=1}^{+\infty}\gamma_k(f)r_k<+\infty,
$$

则 $\sigma(f)$ 几乎处处 $|C, \alpha|$ 可求和.

§2. M 型正阶 Cesàro 逼近的阶

设 $f\in L^p(Q)$，$1\leqslant p\leqslant+\infty$，$\alpha>0$. 有

$$
\sigma_n^\alpha(f;x,y)-f(x,y)=\frac{1}{\pi^2}\int_0^\pi\int_0^\pi\varphi_{x,y}(s,t)K_n^\alpha(s,t)\mathrm{d}s\mathrm{d}t,
$$

式中

$$
\varphi_{x,y}(s, t)=f(x+s, y+t)+f(x-s, y+t)+f(x+s, y-t)+
$$

$$f(x-s,\ y-t)-4f(x,\ y).$$

显然

$$\|\varphi_{x,y}(s,t)\|_p=\left\{\int_Q|\varphi_{x,y}(s,t)|^p\mathrm{d}x\mathrm{d}y\right\}^{\frac{1}{p}}$$
$$\leqslant C(\omega_2(f;|s|,0)_p+\omega_2(f;0,|t|)_p)$$
$$\leqslant C(\tilde{\omega}(f;|s|)_p+\tilde{\omega}(f;|t|)_p).$$

为证定理 1，只需考虑 $n\geqslant 8$ 的情形．把积分区域划分成

$$E_1=\left\{(s,\ t)\ |\ 0<t<s<\frac{2\pi}{n}\right\},$$

$$E_2=\left\{(s,\ t)\ |\ 0<s-t<\frac{\pi}{n},\ \frac{2\pi}{n}<s<\pi\right\},$$

$$E_3=\left\{(s,\ t)\ |\ 0<t<\frac{\pi}{n},\ \frac{2\pi}{n}<s<\pi\right\},$$

$$E_4=\left\{(s,\ t)\ |\ \frac{2\pi}{n}<t+\frac{\pi}{n}<s<\pi\right\},$$

$$E_5=\{(s,\ t)\ |\ 0<s<t<\pi\}.$$

记

$$I_j(x,y)=\int_{E_j}\varphi_{x,y}(s,t)K_n^\alpha(s,t)\mathrm{d}s\mathrm{d}t,\quad j=1,2,3,4,5.$$

对于一切 s，t：$0<s<\pi$，$0<t<\pi$，

$$|K_n^\alpha(s,\ t)|\leqslant\left(n+\frac{1}{2}\right)^2,\quad |K_n^\alpha(s,\ t)|\leqslant\frac{C}{st}.$$

所以，由广义 Minkowski 不等式(见[7])可知

$$\|I_1\|_p=\left\{\int_Q|I_1(x,y)|^p\mathrm{d}x\mathrm{d}y\right\}^{\frac{1}{p}}\leqslant C\tilde{\omega}\left(f;\frac{1}{n}\right)_p,\tag{2.1}$$

$$\|I_2\|_p\leqslant C\int_{E_2}\frac{\|\varphi_{x,y}(s,t)\|_p}{s\cdot t}\mathrm{d}s\mathrm{d}t\leqslant\frac{C}{n}\int_{\frac{1}{n}}^{\pi}\frac{\tilde{\omega}(f;s)_p}{s^2}\mathrm{d}s.\tag{2.2}$$

把 M 型 Cesàro 核 $K_n^\alpha(s,\ t)$写成

$$K_n^\alpha(s,t)=\frac{1}{8A_n^\alpha\sin\frac{s}{2}\cdot\sin\frac{t}{2}}\mathrm{Re}\left\{\sum_{k=0}^{n}A_{n-k}^{\alpha-1}\left[\mathrm{e}^{\mathrm{i}\left(k+\frac{1}{2}\right)(s-t)}-\mathrm{e}^{\mathrm{i}\left(k+\frac{1}{2}\right)(s+t)}\right]\right\}.$$

记 $\xi=s-t$，$\eta=s+t$．利用下面的等式(见[8]60 页)

$$\sum_{k=0}^{n}A_k^{\alpha-1}\mathrm{e}^{-\mathrm{i}k\theta}=-\mathrm{e}^{-\mathrm{i}(n+1)\theta}\sum_{\nu=1}^{l}A_n^{\alpha-\nu}\frac{1}{(1-\mathrm{e}^{-\mathrm{i}\theta})^\nu}+\frac{1}{(1-\mathrm{e}^{-\mathrm{i}\theta})^\alpha}-$$

$$\frac{\sum\limits_{k=n+1}^{+\infty} A_k^{\alpha-l-1} \mathrm{e}^{-\mathrm{i}k\theta}}{(1-\mathrm{e}^{-\mathrm{i}\theta})^l} \tag{2.3}$$

其中 l 只要保证级数收敛即可．我们取 $l=[\alpha]+2$，得

$$K_n^\alpha(s,\ t)=J_n^{(1)}(s,\ t)+J_n^{(2)}(s,\ t)+J_n^{(3)}(s,\ t), \tag{2.4}$$

$$\begin{aligned} &J_n^{(1)}(s,t)\\ &=\frac{1}{8A_n^\alpha \sin\frac{s}{2}\sin\frac{t}{2}}\mathrm{Re}\left\{\sum_{\nu=1}^{l} A_n^{\alpha-\nu}\left[\frac{-\mathrm{e}^{-\frac{1}{2}\mathrm{i}\xi}}{(1-\mathrm{e}^{-\mathrm{i}\xi})^\nu}+\frac{\mathrm{e}^{-\frac{1}{2}\mathrm{i}\eta}}{(1-\mathrm{e}^{-\mathrm{i}\eta})^\nu}\right]\right\}\\ &=\frac{1}{8A_n^\alpha \sin\frac{s}{2}\cdot\sin\frac{t}{2}}\cdot\sum_{\nu=2}^{l} A_n^{\alpha-\nu}\left[\frac{\cos\left(\frac{\nu-1}{2}\eta-\frac{\nu\pi}{2}\right)}{\left(2\sin\frac{\eta}{2}\right)^\nu}-\frac{\cos\left(\frac{\nu-1}{2}\xi-\frac{\nu\pi}{2}\right)}{\left(2\sin\frac{\xi}{2}\right)^\nu}\right], \end{aligned} \tag{2.5}$$

$$\begin{aligned} &J_n^{(2)}(s,\ t)\\ &=\frac{1}{8A_n^\alpha \sin\frac{s}{2}\cdot\sin\frac{t}{2}}\left\{\frac{\cos\left[\left(n+\frac{1+\alpha}{2}\right)\xi-\frac{\alpha\pi}{2}\right]}{\left(2\sin\frac{\xi}{2}\right)^\alpha}-\frac{\cos\left[\left(n+\frac{1+\alpha}{2}\right)\eta-\frac{\alpha\pi}{2}\right]}{\left(2\sin\frac{\eta}{2}\right)^\alpha}\right\}, \end{aligned} \tag{2.6}$$

$$\begin{aligned} &J_n^{(3)}(s,\ t)\\ &=\frac{-1}{8A_n^\alpha \sin\frac{s}{2}\cdot\sin\frac{t}{2}}\cdot\left\{\frac{\sum\limits_{k=n+1}^{+\infty} A_k^{\alpha-l-1}\cos\left[\left(n+\frac{1+l}{2}-k\right)\eta-\frac{l\pi}{2}\right]}{\left(2\sin\frac{\eta}{2}\right)^l}-\right.\\ &\left.\frac{\sum\limits_{k=n+1}^{+\infty} A_k^{\alpha-l-1}\cos\left[\left(n+\frac{1+l}{2}-k\right)\xi-\frac{l\pi}{2}\right]}{\left(2\sin\frac{\xi}{2}\right)^l}\right\}. \end{aligned} \tag{2.7}$$

以上各式当 $0<t<\pi$，$0<s<\pi$ 时成立．为方便，记

$$\gamma_n=\frac{1}{n+1}\int_{\frac{1}{n+1}}^{\pi}\frac{\widetilde{\omega}(f;u)_p}{u^2}\mathrm{d}u,$$

$$g_n(s,\ t)=\frac{\cos\left(n+\frac{\alpha+1}{2}\right)t}{8A_n^\alpha \sin\frac{s}{2}\cdot\sin\frac{t}{2}}\left\{\frac{1}{\left(2\sin\frac{\xi}{2}\right)^\alpha}-\frac{1}{\left(2\sin\frac{\eta}{2}\right)^\alpha}\right\},$$

$$h_n(s,\ t)=\frac{\sin\left(n+\frac{\alpha+1}{2}\right)t}{8A_n^\alpha\sin\frac{s}{2}\cdot\sin\frac{t}{2}}\left\{\frac{1}{\left(2\sin\frac{\xi}{2}\right)^\alpha}+\frac{1}{\left(2\sin\frac{\eta}{2}\right)^\alpha}\right\},$$

那么

$$J_n^{(2)}(s,\ t)=g_n(s,\ t)\cdot\cos\left[\left(n+\frac{\alpha+1}{2}\right)s-\frac{\alpha\pi}{2}\right]+$$
$$h_n(s,\ t)\sin\left[\left(n+\frac{\alpha+1}{2}\right)s-\frac{\alpha\pi}{2}\right].$$

我们有

引理 1 当 $\alpha>0$ 时，

$$\left\|\int_{E_3}\varphi_{x,y}(s,t)\,g_n(s,t)\cos\left[\left(n+\frac{\alpha+1}{2}\right)s-\frac{\alpha\pi}{2}\right]\mathrm{d}s\mathrm{d}t\right\|_p\leqslant C(\alpha)\gamma_n,\tag{2.8}$$

$$\left\|\int_{E_3}\varphi_{x,y}(s,t)h_n(s,t)\sin\left[\left(n+\frac{\alpha+1}{2}\right)s-\frac{\alpha\pi}{2}\right]\mathrm{d}s\mathrm{d}t\right\|_p\leqslant C(\alpha)\gamma_n,\tag{2.9}$$

式中$\|\cdot\|_p$是关于变元$(x,\ y)$的$L^p(Q)$范数，$n\geqslant8$.

引理 2 当 $\alpha>0$ 时，

$$\left\|\int_{E_4}\varphi_{x,y}(s,t)\,g_n(s,t)\cos\left[\left(n+\frac{\alpha+1}{2}\right)s-\frac{\alpha\pi}{2}\right]\mathrm{d}s\mathrm{d}t\right\|_p\leqslant C(\alpha)\gamma_n,\tag{2.10}$$

$$\left\|\int_{E_4}\varphi_{x,y}(s,t)h_n(s,t)\sin\left[\left(n+\frac{\alpha+1}{2}\right)s-\frac{\alpha\pi}{2}\right]\mathrm{d}s\mathrm{d}t\right\|_p\leqslant C(\alpha)\gamma_n,\tag{2.11}$$

引理 1 的证明比较简单，略去. 下面证明引理 2. 由于(2.10)和(2.11)的证明是一样的，我们只证(2.11).

记$\tilde{n}=n+\frac{1+\alpha}{2}$，$\delta=\frac{\pi}{\tilde{n}}$，$\psi(s)=\psi_{x,y}(s,\ t)=\varphi_{x,y}(s,\ t)h_n(s,\ t)$. 把区域$E_4$平移$(\delta,\ 0)$记为$E_4'$($E_4'=E_4+(\delta,\ 0)$)，记$E_4'\cap E_4=E_4^{(1)}$；$E_4^{(1)}+(\delta,\ 0)=E_4^{(1)\prime}$，$E_4^{(1)\prime}\cap E_4^{(1)}=E_4^{(2)}$；$E_4^{(2)}+(\delta,\ 0)=E_4^{(2)\prime}$，$E_4^{(2)\prime}\cap E_4^{(2)}=E_4^{(3)}$.

又设

$$J(x,y)=\int_{E_4}\psi(s)\sin\left(\tilde{n}s-\frac{\alpha\pi}{2}\right)\mathrm{d}s\mathrm{d}t,$$

那么经变量替换得

$$J(x,y)=\int_{E_4+(\delta,0)}\psi(s-\delta)\sin\left(\tilde{n}s-\frac{\alpha\pi}{2}-\pi\right)\mathrm{d}s\mathrm{d}t$$

$$=-\int_{E_4'}\psi(s-\delta)\sin\left(\tilde{n}s-\frac{\alpha\pi}{2}\right)\mathrm{d}s\mathrm{d}t$$

$$=-\int_{E_4^{(1)}}\psi(s-\delta)\sin\left(\tilde{n}s-\frac{\alpha\pi}{2}\right)\mathrm{d}s\mathrm{d}t-$$

$$\int_{E_4'\setminus E_4}\psi(s-\delta)\sin\left(\tilde{n}s-\frac{\alpha\pi}{2}\right)\mathrm{d}s\mathrm{d}t.$$

另一方面，由 $J(x,y)$的定义式知

$$J(x,y)=\left(\int_{E_4\setminus E_4^{(1)}}+\int_{E_4^{(1)}}\right)\psi(s)\sin\left(\tilde{n}s-\frac{\alpha\pi}{2}\right)\mathrm{d}s\mathrm{d}t.$$

将此式与前式相加除以 2 得

$$J(x,y)=\frac{1}{2}\int_{E_4\setminus E_4^{(1)}}\psi(s)\sin\left(\tilde{n}s-\frac{\alpha\pi}{2}\right)\mathrm{d}s\mathrm{d}t+$$

$$\frac{1}{2}\int_{E_4^{(1)}}[\psi(s)-\psi(s-\delta)]\sin\left(\tilde{n}s-\frac{\alpha\pi}{2}\right)\mathrm{d}s\mathrm{d}t-$$

$$\frac{1}{2}\int_{E_4'\setminus E_4}\psi(s-\delta)\sin\left(\tilde{n}s-\frac{\alpha\pi}{2}\right)\mathrm{d}s\mathrm{d}t.$$

当$(s,t)\in E_4\setminus E_4'$时，

$$|h_n(s,t)|\leqslant\frac{C(\alpha)}{s^2},$$

$$\|\varphi_{x,y}(s,t)\|_p\leqslant C(\tilde{\omega}(f;|s|)_p+\tilde{\omega}(f;|t|)_p)\leqslant C\tilde{\omega}(f;s)_p,$$

所以

$$\left\|\int_{E_4\setminus E_4'}\psi_{x,y}(s)\sin\left(\tilde{n}s-\frac{\alpha\pi}{2}\right)\mathrm{d}s\mathrm{d}t\right\|_p\leqslant C(\alpha)\frac{1}{n}\int_{\frac{\pi}{n}}^{\pi}\frac{\tilde{\omega}(f;s)_p}{s^2}\mathrm{d}s\leqslant C(\alpha)\gamma_n. \tag{2.12}$$

当$(s,t)\in E_4'\setminus E_4$ 时，

$$|h_n(s,t)|\leqslant\frac{C(\alpha)}{n^{\alpha}t(s-t)^{\alpha}},$$

$$\|\varphi_{x,y}(s-\delta,t)\|_p\leqslant C\tilde{\omega}(f;\pi)_p,$$

所以

$$\left\|\int_{E_4'\setminus E_4}\psi_{x,y}(s-\delta)\sin\left(\tilde{n}s-\frac{\alpha\pi}{2}\right)\mathrm{d}s\mathrm{d}t\right\|_p$$

$$\leqslant C(\alpha)\tilde{\omega}(f;\pi)_p\cdot\int_{\pi}^{\pi+\delta}\mathrm{d}s\int_{\frac{\pi}{n}}^{s-\delta}\frac{\mathrm{d}t}{n^{\alpha}t(s-t)^{\alpha}}$$

$$\leqslant C(\alpha)\frac{\ln n}{n^{1+\bar{\alpha}}}\tilde{\omega}(f;\ \pi)_p\leqslant C(\alpha)\gamma_n, \tag{2.13}$$

式中 $\bar{\alpha}=\min(\alpha,\ 1)$，$\alpha>0$.

于是

$$\|J(x,y)\|_p\leqslant C(\alpha)\gamma_n+\left\|\int_{E_4^{(1)}}\Delta\psi_{(x,y)}(s)\sin\left(\tilde{n}s-\frac{\alpha\pi}{2}\right)\mathrm{d}s\mathrm{d}t\right\|_p,$$

式中 $\Delta\psi_{(x,y)}(s)=\psi_{x,y}(s,\ t)-\psi_{x,y}(s-\delta,\ t)$.

重复上面的步骤求得

$$\begin{aligned}&\int_{E_4^{(1)}}\Delta\psi_{x,y}(s)\sin\left(\tilde{n}s-\frac{\alpha\pi}{2}\right)\mathrm{d}s\mathrm{d}t\\&=\frac{1}{2}\int_{E_4^{(1)}\setminus E_4^{(1)'}}\Delta\psi_{x,y}(s)\sin\left(\tilde{n}s-\frac{\alpha\pi}{2}\right)\mathrm{d}s\mathrm{d}t+\\&\quad\frac{1}{2}\int_{E_4^{(2)}}\Delta^2\psi_{x,y}(s)\sin\left(\tilde{n}s-\frac{\alpha\pi}{2}\right)\mathrm{d}s\mathrm{d}t-\\&\quad\frac{1}{2}\int_{E_4^{(1)'}\setminus E_4^{(1)}}\Delta\psi_{x,y}(s-\delta)\sin\left(\tilde{n}s-\frac{\alpha\pi}{2}\right)\mathrm{d}s\mathrm{d}t,\end{aligned}$$

式中

$$\Delta^2\psi_{x,y}(s)=\Delta\psi_{x,y}(s)-\Delta\psi_{x,y}(s-\delta)=\psi(s)-2\psi(s-\delta)+\psi(s-2\delta).$$

同推导(2.12)一样地算出

$$\begin{aligned}&\left\|\int_{E_4^{(1)}\setminus E_4^{(1)'}}\Delta\psi_{x,y}(s)\sin\left(\tilde{n}s-\frac{\alpha\pi}{2}\right)\mathrm{d}s\mathrm{d}t\right\|_p\\&\leqslant\int_{E_4^{(1)}\setminus E_4^{(1)'}}(\|\psi_{x,y}(s,t)\|_p+\|\psi_{x,y}(s-\delta,t)\|_p)\mathrm{d}s\mathrm{d}t\\&\leqslant C(\alpha)\gamma_n;\end{aligned}$$

同推导(2.13)一样地算出

$$\begin{aligned}&\left\|\int_{E_4^{(1)'}\setminus E_4^{(1)}}\Delta\psi_{x,y}(s-\delta)\sin\left(\tilde{n}s-\frac{\alpha\pi}{2}\right)\mathrm{d}s\mathrm{d}t\right\|\\&\leqslant\int_{E_4^{(1)'}\setminus E_4^{(1)}}(\|\psi_{x,y}(s-\delta)\|_p+\|\psi_{x,y}(s-2\delta)\|_p)\mathrm{d}s\mathrm{d}t\\&\leqslant C(\alpha)\frac{\ln n}{n^{1+\bar{\alpha}}}\tilde{\omega}(f;\ \pi)_p\leqslant C(\alpha)\gamma_n.\end{aligned}$$

于是

$$\|J(x,y)\|_p\leqslant C(\alpha)\gamma_n+\left\|\int_{E_4^{(2)}}\Delta^2\psi_{x,y}(s)\sin\left(\tilde{n}s-\frac{\alpha\pi}{2}\right)\mathrm{d}s\mathrm{d}t\right\|.$$

再重复一次上述步骤，得

$$\int_{E_4^{(2)}} \Delta^2 \psi_{x,y}(s)\sin\left(\tilde{n}s-\frac{\alpha\pi}{2}\right)\mathrm{d}s\mathrm{d}t$$
$$=\frac{1}{2}\int_{E_4^{(2)}\setminus E_4^{(2)'}} \Delta^2 \psi_{x,y}(s)\sin\left(\tilde{n}s-\frac{\alpha\pi}{2}\right)\mathrm{d}s\mathrm{d}t+$$
$$\frac{1}{2}\int_{E_4^{(3)}} \Delta^3 \psi_{x,y}(s)\sin\left(\tilde{n}s-\frac{\alpha\pi}{2}\right)\mathrm{d}s\mathrm{d}t-$$
$$\frac{1}{2}\int_{E_4^{(2)'}\setminus E_4^{(2)}} \Delta^2 \psi_{x,y}(s-\delta)\sin\left(\tilde{n}s-\frac{\alpha\pi}{2}\right)\mathrm{d}s\mathrm{d}t,$$

式中

$$\Delta^3\psi_{x,y}(s)=\Delta^2\psi_{x,y}(s)-\Delta^2\psi_{x,y}(s-\delta)$$
$$=\psi_{x,y}(s,\ t)-3\psi_{x,y}(s-\delta,\ t)+3\psi_{x,y}(s-2\delta,\ t)-\psi_{x,y}(s-3\delta,\ t).$$

仍同(2.12)一样地算出

$$\left\|\int_{E_4^{(2)}\setminus E_4^{(2)'}} \Delta^2 \psi_{x,y}(s)\sin\left(\tilde{n}s-\frac{\alpha\pi}{2}\right)\mathrm{d}s\mathrm{d}t\right\|_p$$
$$\leqslant\int_{E_4^{(2)}\setminus E_4^{(2)'}}(\|\psi_{x,y}(s,t)\|_p+2\|\psi_{x,y}(s-\delta,t)\|_p+\|\psi_{x,y}(s-2\delta,t)\|_p)\mathrm{d}s\mathrm{d}t$$
$$\leqslant C(\alpha)\gamma_n;$$

同(2.13)一样地算出

$$\left\|\int_{E_4^{(2)'}\setminus E_4^{(2)}} \Delta^2 \psi_{x,y}(s-\delta)\sin\left(\tilde{n}s-\frac{\alpha\pi}{2}\right)\mathrm{d}s\mathrm{d}t\right\|_p$$
$$\leqslant\int_{E_4^{(2)'}\setminus E_4^{(2)}}(\|\psi_{x,y}(s-\delta,t)\|_p+2\|\psi_{x,y}(s-2\delta,t)\|_p+$$
$$\|\psi_{x,y}(s-3\delta,t)\|_p)\mathrm{d}s\mathrm{d}t$$
$$\leqslant C(\alpha)\frac{\ln n}{n^{1+\bar{\alpha}}}\tilde{\omega}(f;\pi)_p\leqslant C(\alpha)\gamma_n;$$

终于得到

$$\|J(x,y)\|_p\leqslant C(\alpha)\gamma_n+\int_{E_4^{(3)}}\|\Delta^3\psi_{x,y}(s)\|_p\mathrm{d}s\mathrm{d}t.$$

因为

$$\|\Delta^3\psi_{x,y}(s)\|_p\leqslant C(\alpha)\left\{\frac{\tilde{\omega}\left(f;\ \frac{1}{n}\right)_p}{n^\alpha st(s-t)^\alpha}+\frac{\tilde{\omega}(f;\ s)_p\delta^2}{n^\alpha st(s-t)^{\alpha+2}}\right\},$$

$$\int_{E_4^{(3)}}\frac{\mathrm{d}s\mathrm{d}t}{n^\alpha s\cdot t(s-t)^\alpha}\leqslant\int_{\frac{2\pi}{n}}^{\pi}\mathrm{d}s\int_{\frac{\pi}{n}}^{s-\frac{\pi}{n}}\frac{\mathrm{d}t}{n^\alpha st(s-t)^\alpha}$$
$$=\int_{\frac{2\pi}{n}}^{\pi}\mathrm{d}s\left(\int_{\frac{\pi}{n}}^{\frac{s}{2}}+\int_{\frac{s}{2}}^{s-\frac{\pi}{n}}\right)\frac{\mathrm{d}t}{n^\alpha s\cdot t(s-t)^\alpha}$$

$$\leqslant \frac{C(\alpha)}{n^{\alpha}}\int_{\frac{2\pi}{n}}^{\pi}\left(\frac{\ln ns}{s^{1+\alpha}}+\frac{1}{s^{2}}\int_{\frac{\pi}{n}}^{\frac{s}{2}}\frac{\mathrm{d}u}{u^{\alpha}}\right)\mathrm{d}s\leqslant C(\alpha),$$

以及

$$\int_{E_4^{(3)}}\frac{\widetilde{\omega}(f;s)_p}{n^{\alpha+2}s\cdot t(s-t)^{2+\alpha}}\mathrm{d}s\mathrm{d}t$$

$$\leqslant \int_{\frac{2\pi}{n}}^{\pi}\frac{\widetilde{\omega}(f;s)_p}{s}\left(\int_{\frac{\pi}{n}}^{s-\frac{\pi}{n}}\frac{\mathrm{d}t}{n^{\alpha+1}t(s-t)^{\alpha+2}}\right)\mathrm{d}s$$

$$\leqslant C(\alpha)\int_{\frac{2\pi}{n}}^{\pi}\frac{\widetilde{\omega}(f;s)_p}{s}\frac{1}{n^{\alpha+2}}\left(\frac{\ln ns}{s^{\alpha+2}}+\frac{n^{\alpha+2}}{s}\right)\mathrm{d}s$$

$$\leqslant C(\alpha)\gamma_n,$$

所以$\|J(x,\ y)\|_p\leqslant C(\alpha)\gamma_n$. □

由引理 1 和引理 2 得到

$$\left\|\int_{E_2\cup E_4}\psi_{x,y}(s,t)J_n^{(2)}(s,t)\mathrm{d}s\mathrm{d}t\right\|_p\leqslant C(\alpha)\gamma_n. \tag{2.14}$$

另一方面，在 E_3 上

$$\left|\frac{\cos\left(\frac{\nu-1}{2}\eta-\frac{\nu\pi}{2}\right)}{\left(2\sin\frac{\eta}{2}\right)^{\nu}}-\frac{\cos\left(\frac{\nu-1}{2}\xi-\frac{\nu\pi}{2}\right)}{\left(2\sin\frac{\xi}{2}\right)^{\nu}}\right|$$

$$\leqslant\left|\frac{1}{\left(2\sin\frac{\eta}{2}\right)^{\nu}}-\frac{1}{\left(2\sin\frac{\xi}{2}\right)^{\nu}}\right|+\frac{\nu-1}{2}t\left|\frac{1}{\left(2\sin\frac{\eta}{2}\right)^{\nu}}+\frac{1}{\left(2\sin\frac{\xi}{2}\right)^{\nu}}\right|$$

$$\leqslant C(\alpha)\frac{t}{s^{1+\nu}},$$

所以在 E_3 上有

$$|J_n^{(1)}(s,t)|\leqslant\frac{C(\alpha)}{n^2s^4},$$

$$\left\|\int_{E_3}\varphi_{x,y}(s,t)J_n^{(1)}(s,t)\mathrm{d}s\mathrm{d}t\right\|_p\leqslant C(\alpha)\int_{E_3}\frac{\widetilde{\omega}(f;s)_p}{n^2s^4}\mathrm{d}s\mathrm{d}t$$

$$\leqslant C(\alpha)\widetilde{\omega}\left(f;\frac{1}{n}\right)\int_{E_3}\frac{\mathrm{d}s\mathrm{d}t}{s^2}$$

$$\leqslant C(\alpha)\gamma_n. \tag{2.15}$$

同时，在 E_3 上

$$|J_n^{(3)}(s,t)|\leqslant\sum_{k=n+1}^{+\infty}|A_k^{\alpha-l-1}|\cdot\left\{\left|\frac{1}{\left(2\sin\frac{\eta}{2}\right)^{l}}-\frac{1}{\left(2\sin\frac{\xi}{2}\right)^{l}}\right|+\right.$$

$$\left|\sin\left(n+\frac{1+l}{2}-k\right)t\right|\left|\frac{1}{\left(2\sin\frac{\eta}{2}\right)^{l}}+\frac{1}{\left(2\sin\frac{\xi}{2}\right)^{l}}\right|\Bigg\}$$

$$\leqslant C(\alpha)n^{\alpha-l+1}\frac{t}{s^{l}}.$$

所以

$$\left\|\iint_{E_3}\varphi_{x,y}(s,t)J_n^{(3)}(s,t)\mathrm{d}s\mathrm{d}t\right\|_p\leqslant C(\alpha)\int_{E_3}\frac{\widetilde{\omega}(f;s)_p n^{\alpha-l+1}t}{n^{\alpha}s\cdot t\cdot s^{l}}\mathrm{d}s\mathrm{d}t\leqslant C(\alpha)\gamma_n. \tag{2.16}$$

而在 E_4 上，有

$$|J_n^{(1)}(s,\ t)|\leqslant\frac{C(\alpha)}{n^2st(s-t)^2},$$

$$|J_n^{(3)}(s,\ t)|\leqslant\frac{C(\alpha)}{n^lst(s-t)^l}\leqslant\frac{C(\alpha)}{n^2st(s-t)^2},$$

所以

$$\left\|\iint_{E_4}\varphi_{x,y}(s,t)(J_n^{(1)}+J_n^{(3)})\mathrm{d}s\mathrm{d}t\right\|_p$$

$$\leqslant\frac{C(\alpha)}{n^2}\int_{\frac{2\pi}{n}}^{\pi}\mathrm{d}s\int_{\frac{\pi}{n}}^{s-\frac{\pi}{n}}\frac{\widetilde{\omega}(f;s)_p}{st(s-t)^2}\mathrm{d}t\leqslant C(\alpha)\gamma_n. \tag{2.17}$$

把(2.14)～(2.17)合起来，得到

$$\|I_3+I_4\|_p\leqslant C(\alpha)\gamma_n. \tag{2.18}$$

再与(2.1)(2.2)合起来就得到

$$\left\|\iint_{0<t<s<\pi}\varphi_{x,y}(s,t)K_n^{\alpha}(s,t)\mathrm{d}s\mathrm{d}t\right\|_p\leqslant C(\alpha)\gamma_n. \tag{2.19}$$

对于函数$g(x,\ y)=f(y,\ x)$用(2.19)，由于 $K_n^{\alpha}(s,\ t)=K_n^{\alpha}(t,\ s)$，就得到

$$\|I_5\|_p\leqslant C(\alpha)\gamma_n. \tag{2.20}$$

这样，我们完成了定理 1 的证明.

推论 1 设 $\alpha>0$，$1\leqslant p\leqslant+\infty$，则对于一切 $f\in L^p(Q)$，(1.4)成立，即

$$\|\sigma_n^{\alpha}(f)-f\|_p\leqslant\frac{C(\alpha)}{n+1}\sum_{k=0}^{n}\widetilde{E}_k(f)_p,\quad n\in\mathbf{N}.$$

为证此式，只需把(1.3)右端的积分写成求和并使用不等式(见[3] 363 页)

$$\widetilde{\omega}\left(f;\frac{1}{k+1}\right)_p\leqslant C(p)\frac{1}{(k+1)^2}\sum_{\nu=0}^{k}(\nu+1)\widetilde{E}_{\nu}(f)_p. \tag{2.21}$$

§3. Cesàro 阶数 $\beta\in(-1,0]$时的辅助不等式

定理 2 设$-1<\beta\leqslant 0$，$1<p\leqslant 2$，$f\in L^p(Q)$，则

$$\left\|\frac{1}{n+1}\sum_{k=n}^{2n}|\sigma_k^\beta(f)-f|\right\|_p\leqslant\frac{C(\beta,p)}{(n+1)^{1+\beta}}\int_{\frac{1}{n+1}}^{\pi}\frac{\tilde{\omega}(f;u)_p}{u^{2+\beta}}\mathrm{d}u+$$

$$C(\beta,p)\mu_n(\beta+1,p)\tilde{\omega}\left(f;\frac{1}{n+1}\right)_p,\tag{3.1}$$

式中$\mu_n(\beta+1,p)$的定义见§1定理4.

证 只需考虑$n\geqslant 8$. 仍如§2那样划分区域$\{(s,t)\mid 0<s<\pi,0<t<\pi\}$. 记

$$I_k^{(j)}(x,y)=\int_{E_j}\varphi_{x,y}(s,t)K_k^\beta(s,t)\mathrm{d}s\mathrm{d}t,\quad j=1,2,3,4,5.$$

有

$$\sigma_k^\beta(f;x,y)-f(x,y)=\frac{1}{\pi^2}\sum_{j=1}^{5}I_k^{(j)}(x,y).\tag{3.2}$$

因为仍有$|K_k^\beta(s,t)|\leqslant C(\beta)(k+1)^2$，所以

$$\|I_k^{(1)}\|_p\leqslant C(\beta)(k+1)^2\int_{E_1}\tilde{\omega}\left(f;\frac{1}{n}\right)_p\mathrm{d}s\mathrm{d}t$$

$$\leqslant C(\beta)\tilde{\omega}\left(f;\frac{1}{n}\right)_p,\quad k=n,n+1,\cdots,2n.\tag{3.3}$$

先在(2.3)中代入$l=1$，由(2.4)～(2.7)可见

$$|K_k^\beta(s,t)|\leqslant\frac{C(\beta)}{k^\beta st}\left(\frac{1}{\xi^\beta}+\frac{1}{\eta^\beta}+k^\beta\right),\qquad -1<\beta\leqslant 0.$$

于是当$(s,t)\in E_2$时，$|K_k^\beta(s,t)|\leqslant\frac{C(\beta)}{k^\beta st\eta^\beta}$. 从而

$$\|I_k^{(2)}(x,y)\|_p\leqslant\frac{C(\beta)}{n^\beta}\int_{E_2}\frac{\tilde{\omega}(f;s)_p}{st(s+t)^\beta}\mathrm{d}s\mathrm{d}t$$

$$\leqslant\frac{C(\beta)}{n^{1+\beta}}\int_{\frac{\pi}{n}}^{\pi}\frac{\tilde{\omega}(f;s)_p}{s^{2+\beta}}\mathrm{d}s,\quad k=n,n+1,\cdots,2n.\tag{3.4}$$

现于(2.3)中取$l=2$，如(2.4)～(2.7)写出

$K_k^\beta(s,t)=J_k^{(1)}(s,t)+J_k^{(2)}(s,t)+J_k^{(3)}(s,t)$，

$$J_k^{(1)}(s,t)=\frac{A_k^{\beta-2}}{8A_k^\beta}\frac{1}{\sin\frac{s}{2}\cdot\sin\frac{t}{2}}\left(\frac{\cos\xi}{4\sin^2\frac{\xi}{2}}-\frac{\cos\eta}{4\sin^2\frac{\eta}{2}}\right),$$

$$J_k^{(2)}(s,\ t)=\frac{1}{8A_k^\beta \sin\frac{s}{2}\sin\frac{t}{2}}\mathrm{Re}\left\{\frac{\mathrm{e}^{\mathrm{i}\left(k+\frac{1}{2}\right)\xi}}{(1-\mathrm{e}^{-\mathrm{i}\xi})^\beta}-\frac{\mathrm{e}^{\mathrm{i}\left(k+\frac{1}{2}\right)\eta}}{(1-\mathrm{e}^{-\mathrm{i}\eta})^\beta}\right\},$$

$$J_k^{(3)}(s,t)=\frac{1}{32A_k^\beta \sin\frac{s}{2}\cdot\sin\frac{t}{2}}\left\{\frac{\sum\limits_{j=k+1}^{+\infty}A_j^{\beta-3}\cos\left(k+\frac{3}{2}-j\right)\xi}{\sin^2\frac{\xi}{2}}-\frac{\sum\limits_{j=k+1}^{+\infty}A_j^{\beta-3}\cos\left(k+\frac{3}{2}-j\right)\eta}{\sin^2\frac{\eta}{2}}\right\}.$$

记 $\tau_k^\nu(x,\ y)=\int_{E_\nu}\varphi_{x,y}(s,\ t)(J_k^{(1)}(s,\ t)+J_k^{(3)}(s,\ t))\mathrm{d}s\mathrm{d}t(\nu=3,4)$.
容易算出，当$(s,\ t)\in E_3$，$n\leqslant k\leqslant 2n$ 时，

$$|J_k^{(1)}(s,\ t)|\leqslant\frac{C(\beta)}{s^2},\quad |J_k^{(3)}(s,\ t)|\leqslant\frac{C(\beta)}{s^2}.$$

从而

$$\|\tau_k^{(3)}\|_p\leqslant C(\beta)\int_{E_3}\frac{\tilde{\omega}(f;s)_p}{s^2}\mathrm{d}s\mathrm{d}t\leqslant C(\beta)\gamma_n. \tag{3.5}$$

而当$(s,\ t)\in E_4$，$n\leqslant k\leqslant 2n$ 时，

$$|J_k^{(1)}(s,\ t)|+|J_k^{(3)}(s,\ t)|\leqslant\frac{C(\beta)}{k^2st(s-t)^2},$$

从而

$$\|\tau_k^{(4)}\|_p\leqslant C(\beta)\gamma_n. \tag{3.6}$$

由(3.3)～(3.6)得，当 $k=n,\ n+1,\ \cdots,\ 2n$ 时，

$$\|I_k^{(1)}\|_p+\|I_k^{(2)}\|_p+\|\tau_k^{(3)}\|_p+\|\tau_k^{(4)}\|_p\leqslant\frac{C(\beta)}{n^{1+\beta}}\int_{\frac{1}{n}}^{\pi}\frac{\tilde{\omega}(f;u)_p}{u^{2+\beta}}\mathrm{d}u,$$

所以

$$\begin{aligned}&\frac{1}{n+1}\sum_{k=n}^{2n}(\|I_k^{(1)}\|_p+\|I_k^{(2)}\|_p+\|\tau_k^{(3)}\|_p+\|\tau_k^{(4)}\|_p)\\ \leqslant&\frac{C(\beta)}{n^{1+\beta}}\int_{\frac{1}{n}}^{\pi}\frac{\tilde{\omega}(f;u)_p}{u^{2+\beta}}\mathrm{d}u.\end{aligned} \tag{3.7}$$

令

$$W_k^{(\nu)}(x,y)=\int_{E_\nu}\varphi_{x,y}(s,t)J_k^{(2)}(s,t)\mathrm{d}s\mathrm{d}t,\quad \nu=3,4.$$

为估计

$$\left\|\frac{1}{n+1}\sum_{k=n}^{2n}\left|W_k^{(\nu)}(x,y)\right|\right\|_p,\quad \nu=3,4,$$

先证明

引理 3 设 $F(s)\in L_{2\pi}^p$，$(1<p\leqslant 2)$，F 的 Fourier 级数为

$$\sigma(F;s)=\sum_{k=-\infty}^{+\infty}C_k\mathrm{e}^{\mathrm{i}ks},\quad q=\frac{p}{p-1},$$

则

$$\left\{\sum_{k=n}^{2n}(|C_K|^q+|C_{-k}|^q)\right\}^{\frac{1}{q}}\leqslant C(p)\left\{\int_{-\pi}^{\pi}|\Delta_{\frac{\pi}{2u}}^4F(s)|^p\mathrm{d}s\right\}^{\frac{1}{p}},\quad n\in\mathbf{N}^*,\tag{3.8}$$

式中算符"Δ_h^r"定义为

$$\Delta_h^rF(s)=\sum_{\nu=0}^{r}(-1)^{r-\nu}C_r^\nu F(s+\nu h),\quad r\in\mathbf{N}^*.\tag{3.9}$$

证 显然有

$$\sigma(F(\cdot+h);s)=\sum_{k=-\infty}^{+\infty}C_k\mathrm{e}^{\mathrm{i}ks}\cdot\mathrm{e}^{\mathrm{i}kh},$$

$$\begin{aligned}\sigma(\Delta_h^2F(\cdot-h);s)&=\sigma(F(\cdot+h)+F(\cdot-h(-2F(\cdot));s))\\&=\sum_{k=-\infty}^{+\infty}C_k\mathrm{e}^{\mathrm{i}ks}2(\cos kh-1),\end{aligned}$$

$$\sigma(\Delta_h^4F(\cdot-2h);s)=16\sum_{k=-\infty}^{+\infty}C_k\sin^4\frac{kh}{2}\cdot\mathrm{e}^{\mathrm{i}ks}.$$

由 Hausdorff-Young 不等式(见[8])

$$\begin{aligned}&\sum_{k=n}^{2n}\left(\left|C_k\sin^2\frac{kh}{2}\right|^q+\left|C_{-k}\sin^4\frac{kh}{2}\right|^q\right)\\&\leqslant C(p)\left\{\int_{-\pi}^{\pi}|\Delta_h^4F(s-2h)|^p\mathrm{d}s\right\}^{\frac{q}{p}}=C(p)\left\{\int_{-\pi}^{\pi}|\Delta_h^4F(s)|^p\mathrm{d}s\right\}^{\frac{q}{p}}.\end{aligned}$$

代入 $h=\frac{\pi}{2n}$，由于 $\sin\frac{kh}{2}\geqslant\frac{1}{\sqrt{2}}$，$(n\leqslant k\leqslant 2n)$，就得到(3.8). □

因为 $J_k^{(2)}(s,t)$是 s 的偶函数，所以

$$W_k^{(3)}(x,y)=\frac{1}{2}\int_0^{\frac{\pi}{n}}\mathrm{d}t\int_{\frac{2n}{n}\leqslant|s|\leqslant\pi}\varphi_{x,y}(s,t)J_k^{(2)}(s,t)\mathrm{d}s,\tag{3.10}$$

其中

$$
\begin{aligned}
&J_k^{(2)}(s,\ t)\\
=&\frac{1}{8A_k^\beta\sin\frac{s}{2}\cdot\sin\frac{t}{2}}\mathrm{Re}\left\{\frac{\mathrm{e}^{\mathrm{i}\left(k+\frac{1}{2}\right)\xi}}{(1-\mathrm{e}^{-\mathrm{i}\xi})^\beta}-\frac{\mathrm{e}^{\mathrm{i}\left(k+\frac{1}{2}\right)\eta}}{(1-\mathrm{e}^{-\mathrm{i}\eta})^\beta}\right\}\\
=&\frac{1}{16A_k^\beta\sin\frac{s}{2}\cdot\sin\frac{t}{2}}\left\{\frac{\mathrm{e}^{\mathrm{i}\left(k+\frac{1}{2}\right)\xi}}{(1-\mathrm{e}^{-\mathrm{i}\xi})^\beta}+\frac{\mathrm{e}^{-\mathrm{i}\left(k+\frac{1}{2}\right)\xi}}{(1-\mathrm{e}^{\mathrm{i}\xi})^\beta}-\frac{\mathrm{e}^{\mathrm{i}\left(k+\frac{1}{2}\right)\eta}}{(1-\mathrm{e}^{-\mathrm{i}\eta})^\beta}-\frac{\mathrm{e}^{-\mathrm{i}\left(k+\frac{1}{2}\right)\eta}}{(1-\mathrm{e}^{\mathrm{i}\eta})^\beta}\right\}.
\end{aligned}
\tag{3.11}
$$

设

$$
\chi(s)=\begin{cases}0, & |s|<\frac{2\pi}{n},\\ 1, & \frac{2\pi}{n}\leqslant|s|\leqslant\pi,\\ \chi(s+2\pi), & 一切\ s,\end{cases}
$$

$$
u_k(x,y;t)=\int_{-\pi}^{\pi}\varphi_{x,y}(s,t)\chi(s)J_k^{(2)}(s,t)\mathrm{d}s.
$$

再令

$$
g(s,\ t)=\frac{\mathrm{e}^{\mathrm{i}\frac{s}{2}}}{\sin\frac{s}{2}\cdot\sin\frac{t}{2}}\left\{\frac{1}{(1-\mathrm{e}^{-\mathrm{i}\xi})^\beta}-\frac{1}{(1-\mathrm{e}^{-\mathrm{i}\eta})^\beta}\right\},
$$

$$
h(s,\ t)=\frac{\mathrm{e}^{\mathrm{i}\frac{s}{2}}}{\sin\frac{s}{2}}\frac{1}{(1-\mathrm{e}^{-\mathrm{i}\eta})^\beta},
$$

$$
a_k(x,y;t)=\int_{-\pi}^{\pi}\varphi_{x,y}(s,t)\chi(s)\cdot g(s,t)\mathrm{e}^{\mathrm{i}ks}\mathrm{d}s,
$$

$$
b_k(x,y;t)=\int_{-\pi}^{\pi}\varphi_{x,y}(s,t)\chi(s)h(s,t)\mathrm{e}^{\mathrm{i}ks}\mathrm{d}s.
$$

易见

$$
u_k(x,\ y;\ t)=\frac{a_k(x,\ y;\ t)}{8A_k^\beta}\mathrm{e}^{-\mathrm{i}\left(k+\frac{1}{2}\right)t}-\frac{\mathrm{i}b_k(x,\ y;\ t)\sin\left(k+\frac{1}{2}\right)t}{4A_k^\beta\sin\frac{t}{2}}.
$$

我们略去证明而写出容易从引理 3 推出的

引理 4 按上述记号，当 $n\geqslant 8$，$0<t<\frac{\pi}{n}$ 时，

$$\left\|\left\{\sum_{k=n}^{2n}|a_k(\cdot,\ \cdot;t)|^q\right\}^{\frac{1}{q}}\right\|_p\leqslant C(\beta,p)n^{2+\beta-\frac{1}{p}}\tilde{\omega}\left(f;\frac{1}{n}\right)_p,\qquad(3.12)$$

$$\left\|\left\{\sum_{k=n}^{2n}|b_k(\cdot,\ \cdot;t)|^q\right\}^{\frac{1}{q}}\right\|_p\leqslant C(\beta,p)\cdot\lambda_n(\beta,p)\tilde{\omega}\left(f;\frac{1}{n}\right)_p,\qquad(3.13)$$

其中

$$\lambda_n(\beta,\ p)=\begin{cases}n^{1+\beta-\frac{1}{p}}, & \beta>\frac{1}{p}-1,\\ \ln^{\frac{1}{p}}n, & \beta=\frac{1}{p}-1,\\ 1, & \beta<\frac{1}{p}-1.\end{cases}\qquad(3.14)$$

由于

$$W_k^{(3)}(x,y)=\frac{1}{2}\int_0^{\frac{\pi}{n}}u_k(x,y;t)\mathrm{d}t,$$

所以(暂隐去变元$(s,\ y)$)有

$$|W_k^{(3)}|\leqslant C(\beta)\int_0^{\frac{\pi}{n}}\left(\frac{1}{k^\beta}|a_k(t)|+k^{1-\beta}|b_k(t)|\right)\mathrm{d}t,$$

$$\frac{1}{n+1}\sum_{k=n}^{2n}|W_k^{(3)}|\leqslant\frac{C(\beta)}{(n+1)^{1+\beta}}\int_0^{\frac{\pi}{n}}\sum_{k=n}^{2n}(|a_k(t)|+|nb_k(t)|)\mathrm{d}t.$$

于右端用 Hölder 不等式，得

$$\frac{1}{n+1}\sum_{k=n}^{2n}|W_k^{(3)}|$$

$$\leqslant\frac{C(\beta)}{(n+1)^{1+\beta}}\int_0^{\frac{\pi}{2}}\left\{\left(\sum_{k=n}^{2n}|a_k(t)|^q\right)^{\frac{1}{q}}\cdot n^{\frac{1}{p}}+\left(\sum_{k=n}^{2n}|b_k(t)|^q\right)^{\frac{1}{q}}\cdot n^{1+\frac{1}{p}}\right\}\mathrm{d}t.$$

再用引理 4，得

$$\left\|\frac{1}{n+1}\sum_{k=n}^{2n}|W_k^{(3)}(x,y)|\right\|_p$$

$$\leqslant\frac{C(\beta,p)}{(n+1)^{2+\beta}}\{n^{2+\beta-\frac{1}{p}}\cdot n^{\frac{1}{p}}+\lambda_n(\beta,p)n^{1+\frac{1}{p}}\}\tilde{\omega}\left(f;\frac{1}{n}\right)_p$$

$$\leqslant\frac{C(\beta,p)}{(n+1)^{1+\beta-\frac{1}{p}}}\lambda_n(\beta,p)\tilde{\omega}\left(f;\frac{1}{n}\right)_p,\quad n\geqslant 8.\qquad(3.15)$$

现在来处理 $W_k^{(4)}$，定义

$$\chi(s,\ t)=\begin{cases}0, & 0\leqslant|s|<|t|+\dfrac{\pi}{n},\\ 1, & |t|+\dfrac{\pi}{n}\leqslant|s|\leqslant\pi,\\ \chi(s+2\pi,\ t), & \text{一切 } s,\end{cases}$$

而 t 的取值范围是 $\frac{\pi}{n}\leqslant|t|\leqslant\pi-\frac{\pi}{n}$. 设

$$e_k(x,y;t)\equiv e_k(t)=\int_{-\pi}^{\pi}\frac{\varphi_{x,y}(s,t)\chi(s,t)\mathrm{e}^{\mathrm{i}\frac{s}{2}}}{\sin\frac{s}{2}\cdot\sin\frac{t}{2}(1-\mathrm{e}^{-\mathrm{i}\xi})^{\beta}}\mathrm{e}^{\mathrm{i}ks}\mathrm{d}s.$$

从(3.11)可知

$$W_k^4(x,y)=\frac{1}{8A_k^{\beta}}\int_{\frac{\pi}{n}}^{\pi-\frac{\pi}{n}}(e_k(t)\mathrm{e}^{-\mathrm{i}\left(k+\frac{1}{2}\right)t}+e_k(-t)\mathrm{e}^{\mathrm{i}\left(k+\frac{1}{2}\right)t})\mathrm{d}t.$$

从而

$$\begin{aligned}&\frac{1}{n+1}\sum_{k=n}^{2n}|W_k^{(4)}|\\ \leqslant&\frac{C(\beta)}{n^{1+\beta}}\int_{\frac{\pi}{n}}^{\pi-\frac{\pi}{n}}\sum_{k=n}^{2n}(|e_k(t)|+|e_k(-t)|)\mathrm{d}t\\ \leqslant&\frac{C(\beta)}{n_{1+\beta-\frac{1}{p}}}\int_{\frac{\pi}{n}}^{\pi-\frac{\pi}{n}}\left\{\left(\sum_{k=n}^{2n}|e_k(t)|^q\right)^{\frac{1}{q}}+\left(\sum_{k=n}^{2n}|e_k(-t)|^q\right)^{\frac{1}{q}}\right\}\mathrm{d}t.\end{aligned}\tag{3.16}$$

引理 5 按上面的记号，当 $n\geqslant8$，$\frac{\pi}{n}\leqslant|t|\leqslant\pi-\frac{\pi}{n}$ 时，有

$$\left\|\left(\sum_{k=n}^{2n}|e_k(t)|^q\right)^{\frac{1}{q}}\right\|_p\leqslant\frac{C(\beta,p)}{n^{\frac{1}{p}}}\frac{\tilde{\omega}(f;|t|)_p}{|t|^{2+\beta}}+C(\beta,p)\lambda_{\beta,p}(t)\tilde{\omega}\left(f;\frac{1}{n}\right)_p.\tag{3.17}$$

式中

$$\lambda_{\beta,p}(t)=\begin{cases}\dfrac{1}{|t|^{2+\beta-\frac{1}{p}}}, & \beta>\dfrac{1}{p}-1,\\ \dfrac{1}{|t|}\cdot\ln^{\frac{1}{p}}\dfrac{\pi}{|t|}, & \beta=\dfrac{1}{p}-1,\\ \dfrac{1}{|t|}, & \beta<\dfrac{1}{p}-1.\end{cases}$$

证 记 $\psi(s)=\varphi_{x,y}(s,\ t)\chi(s,\ t)\dfrac{\mathrm{e}^{\mathrm{i}\frac{s}{2}}}{\sin\frac{s}{2}\cdot\sin\frac{t}{2}(1-\mathrm{e}^{-\mathrm{i}\xi})^{\beta}}$.

据引理 3

$$\left(\sum_{k=n}^{2n}|e_k(t)|^q\right)^{\frac{1}{q}}\leqslant C(p)\left\{\int_{-\pi}^{\pi}|\Delta_{\frac{\pi}{2n}}^4\psi(s)|^p\mathrm{d}s\right\}^{\frac{1}{p}}$$

从而

$$\begin{aligned}&\left\|\left(\sum_{k=n}^{2n}|e_k(t)|^q\right)^{\frac{1}{q}}\right\|_p\\&\equiv\left\{\int_Q\left(\sum_{k=n}^{2n}|e_k(t)|^q\right)^{\frac{p}{q}}\mathrm{d}x\mathrm{d}y\right\}^{\frac{1}{p}}\\&\leqslant C(p)\cdot\left\{\int_Q\left(\int_{-\pi}^{\pi}|\Delta_{\frac{\pi}{2n}}^4\psi(s)|^p\mathrm{d}s\right)\mathrm{d}x\mathrm{d}y\right\}^{\frac{1}{p}}\\&\leqslant C(p)\cdot\left\{\int_{-\pi}^{\pi}\left(\int_Q|\Delta_{\frac{\pi}{2n}}^4\psi(s)|^p\mathrm{d}x\mathrm{d}y\right)\mathrm{d}s\right\}^{\frac{1}{p}}\\&=C(p)\left\{\left(\int_{|s|\leqslant|t|+\frac{4\pi}{n}}+\int_{|t|+\frac{4\pi}{n}\leqslant|s|\leqslant\pi}\right)\left(\int_Q|\Delta_{\frac{\pi}{2n}}^4\psi(s)|\mathrm{d}x\mathrm{d}y\right)\mathrm{d}s\right\}^{\frac{1}{p}}.\end{aligned}\tag{3.18}$$

先算出

$$\begin{aligned}&\int_{-(|t|+\frac{4\pi}{n})}^{|t|+\frac{4\pi}{n}}\left(\int_Q|\Delta_{\frac{\pi}{2n}}^4\psi(s)|^p\mathrm{d}x\mathrm{d}y\right)\mathrm{d}s\\&\leqslant C(\beta,p)\int_{|t|+\frac{\pi}{n}\leqslant|s|\leqslant|t|+\frac{6\pi}{n}}\left\{\frac{\tilde{\omega}(f;|s|)_p}{|st|(|s|+|t|)^\beta}\right\}^p\mathrm{d}s\\&\leqslant C(\beta,p)\frac{1}{n}\left(\frac{\tilde{\omega}(f;|t|)_p}{|t|^{2+\beta}}\right)^p.\end{aligned}\tag{3.19}$$

$$\begin{aligned}&\int_{|t|+\frac{4\pi}{n}\leqslant|s|\leqslant\pi}\left(\int_Q|\Delta_{\frac{\pi}{2n}}^4\psi(s)|^p\mathrm{d}x\mathrm{d}y\right)\mathrm{d}s\\&\leqslant C(\beta,p)\int_{|t|+\frac{2\pi}{n}}^{\pi}\left\{\frac{\tilde{\omega}\left(f;\frac{1}{n}\right)_p}{|st|(s+|t|)^\beta}+\frac{\tilde{\omega}(f;s)_p}{|st|(s-|t|)^{3+\beta}n^3}\right\}^p\mathrm{d}s.\end{aligned}\tag{3.20}$$

(3.20)右端第一项用估计式

$$\int_{|t|+\frac{2\pi}{n}}^{\pi}\left(\frac{1}{|st|(s+|t|)^\beta}\right)^p\mathrm{d}s\leqslant\frac{C(\beta,p)}{|t|^p}\int_{|t|}^{\pi}\frac{\mathrm{d}s}{s^{(1+\beta)p}}.\tag{3.21}$$

第二项的积分分成两段来算，有

$$\int_{|t|+\frac{\pi}{n}}^{2|t|}\left\{\frac{\tilde{\omega}(f;s)_p}{s|t|(s-|t|)^{3+\beta}n^3}\right\}^p\mathrm{d}s$$

$$\leqslant C(p)\,\frac{(\widetilde{\omega}(f;|t|)_p)^p}{n^{3p}\cdot|t|^{2p}}\int_{\frac{\pi}{n}}^{|t|}\frac{\mathrm{d}u}{u^{(\beta+3)p}}$$

$$\leqslant C(\beta,p)\cdot\left(\frac{\widetilde{\omega}(f;|t|)_p}{|t|^2}n^{\beta-\frac{1}{p}}\right)^p. \tag{3.22}$$

$$\int_{2|t|}^{\pi}\left\{\frac{\widetilde{\omega}(f;s)_p}{s|t|(s-|t|)^{\beta+3}n^3}\right\}^p\mathrm{d}s$$

$$\leqslant\int_{2|t|}^{\pi}\left(\frac{\widetilde{\omega}(f;s)_p}{s^{\beta+4}|t|n^3}\right)^p\mathrm{d}s$$

$$\leqslant C(\beta,p)\left[\frac{\widetilde{\omega}\left(f;\frac{1}{n}\right)_p}{n|t|}\right]^p\int_{|t|}^{\pi}\frac{\mathrm{d}s}{s^{(\beta+2)p}}$$

$$\leqslant C(\beta,p)\left[\frac{\widetilde{\omega}\left(f;\frac{1}{n}\right)_p}{n|t|^{\beta+3-\frac{1}{p}}}\right]^p. \tag{3.23}$$

把(3.21)～(3.23)代入(3.20)再与(3.19)合并，就得到

$$\left\{\int_{-\pi}^{\pi}\left(\int_Q|\Delta_{\frac{\pi}{2n}}^4\psi(s)|^p\mathrm{d}x\mathrm{d}y\right)\mathrm{d}s\right\}^{\frac{1}{p}}$$

$$\leqslant C(\beta,p)\left\{\frac{1}{n^{\frac{1}{p}}}\frac{\widetilde{\omega}(f;|t|)_p}{|t|^{2+\beta}}+\frac{\widetilde{\omega}\left(f;\frac{1}{n}\right)_p}{|t|}\left(\int_{|t|}^{\pi}\frac{\mathrm{d}s}{s^{(1+\beta)p}}\right)^{\frac{1}{p}}+\frac{\widetilde{\omega}\left(f;\frac{1}{n}\right)_p}{n|t|^{\beta+3-\frac{1}{p}}}\right\}. \tag{3.24}$$

注意到

$$\left(\int_{|t|}^{\pi}\frac{\mathrm{d}s}{s^{(1+\beta)p}}\right)^{\frac{1}{p}}\leqslant\begin{cases}\dfrac{C(\beta,p)}{|t|^{1+\beta-\frac{1}{p}}}, & \beta>\dfrac{1}{p}-1,\\ \left(\ln\dfrac{\pi}{|t|}\right)^{\frac{1}{p}}, & \beta=\dfrac{1}{p}-1,\\ C(\beta,p), & \beta<\dfrac{1}{p}-1,\end{cases}$$

将(3.24)代入(3.18)就得(3.17). □

从(3.16)和(3.17)得

$$\left\|\frac{1}{n+1}\sum_{k=n}^{2n}|W_k^{(4)}(x,y)|\right\|_p$$

$$\leqslant\frac{C(\beta)}{n^{1+\beta-\frac{1}{p}}}\int_{\frac{\pi}{n}}^{\pi-\frac{\pi}{n}}\left\{\frac{\widetilde{\omega}(f;|t|)_n}{n^{\frac{1}{p}}|t|^{2+p}}+\lambda_{\beta,p}(t)\widetilde{\omega}\left(f;\frac{1}{n}\right)_p\right\}\mathrm{d}t. \tag{3.25}$$

显然

$$\frac{1}{n^{1+\beta-\frac{1}{p}}}\int_{\frac{\pi}{n}}^{\pi-\frac{\pi}{n}}\lambda_{\beta,p}(t)\mathrm{d}t\leqslant C(\beta,p)\cdot\begin{cases}1, & \beta>\frac{1}{p}-1,\\ \ln^{1+\frac{1}{p}}n, & \beta=\frac{1}{p}-1,\\ n^{-\beta-\frac{1}{p}-1}\ln n, & \beta<\frac{1}{p}-1,\end{cases}$$

以此代入(3.25)，证得

$$\left\|\frac{1}{n+1}\sum_{k=n}^{2n}|W_k^{(4)}(x,y)|\right\|_p$$

$$\leqslant C(\beta,p)\left\{\frac{1}{n^{1+\beta}}\int_{\frac{1}{n}}^{\pi}\frac{\tilde{\omega}(f;t)_p}{t^{2+\beta}}\mathrm{d}t+\mu_n(\beta+1,p)\tilde{\omega}\left(f;\frac{1}{n}\right)_p\right\}. \qquad (3.26)$$

把(3.7)(3.15)及(3.26)合起来，再处理$\left\|\frac{1}{n+1}\sum_{k=n}^{2n}|I_k^{(5)}(x,y)|\right\|_p$，就得到(3.1). □

从定理2的证明中明显可以看出，成立

定理3 设$-1<\beta\leqslant 0$，$f(x, y)\in L^{\infty}(Q)$(即$C(Q)$—约定)，则

$$\left\|\frac{1}{n+1}\sum_{k=n}^{2n}|\sigma_k^{\beta}(f)-f|\right\|_{\infty}$$

$$\leqslant C(\beta)\left\{\frac{1}{(n+1)^{1+\beta}}\int_{\frac{1}{n+1}}^{\pi}\frac{\tilde{\omega}(f;u)_{\infty}}{u^{2+\beta}}\mathrm{d}u+\mu_n(\beta,2)\tilde{\omega}\left(f;\frac{1}{n}\right)_{\infty}\right\}. \quad (3.27)$$

从定理2和定理3，把积分写成求和并使用(2.21)，就得到

推论2 设$-1<\beta\leqslant 0$，$1<p\leqslant 2$，$f\in L^p(Q)$，则

$$\left\|\frac{1}{n+1}\sum_{k=n}^{2n}|\sigma_k^{\beta}(f)-f|\right\|_p$$

$$\leqslant C(\beta,p)\left\{\frac{1}{(n+1)^{1+\beta}}\sum_{k=0}^{n}(k+1)^{\beta}\widetilde{E}_k(f)_p+\right.$$

$$\left.\frac{\mu_n(\beta+1,p)}{(n+1)^2}\sum_{k=0}^{n}(k+1)\widetilde{E}_k(f)_p\right\}. \qquad (3.28)$$

推论3 设$-1<\beta\leqslant 0$，$f\in C(Q)$，则

$$\left\|\frac{1}{n+1}\sum_{k=n}^{2n}|\sigma_k^{\beta}(f)-f|\right\|_c$$

$$\leqslant C(\beta)\left\{\frac{1}{(n+1)^{1+\beta}}\sum_{k=0}^{n}(k+1)^{\beta}\widetilde{E}_k(f)_c+\frac{\mu_n(\beta+1,2)}{(n+1)^2}\sum_{k=0}^{n}(k+1)\widetilde{E}_k(f)_c\right\}.$$

(3.29)

§4. 正阶 M 型 $|C, \alpha|$ 求和的充分条件

定理 4 的证　当 $\alpha>0$ 时，

$$\sigma_k^\alpha(f)-\sigma_{k+1}^\alpha(f)=\frac{\alpha}{k+1}\{\sigma_{k+1}^\alpha(f)-\sigma_{k+1}^{\alpha-1}(f)\}$$

$$=\frac{\alpha}{k+1}\{(\sigma_{k+1}^\alpha(f)-f)-(\sigma_{k+1}^{\alpha-1}(f)-f)\}. \tag{4.1}$$

所以

$$\sum_{k=0}^{n}|\sigma_k^\alpha(f)-\sigma_{k+1}^\alpha(f)|\leqslant\alpha\sum_{k=0}^{n}\left\{\frac{|\sigma_{k+1}^\alpha(f)-f|}{k+1}+\frac{|\sigma_{k+1}^{\alpha-1}(f)-f|}{k+1}\right\}. \tag{4.2}$$

由(1.4)得知

$$\left\|\sum_{k=0}^{n}\frac{|\sigma_{k+1}^\alpha(f)-f|}{k+1}\right\|_p\leqslant C(\alpha)\sum_{k=0}^{n}\frac{1}{(k+1)^2}\sum_{\nu=0}^{k+1}\widetilde{E}_\nu(f)_p$$

$$\leqslant C(\alpha)\sum_{k=0}^{n}\frac{1}{k+1}\widetilde{E}_k(f)_p, \tag{4.3}$$

$$\left\|\sum_{k=n}^{+\infty}\frac{|\sigma_{k+1}^\alpha(f)-f|}{k+1}\right\|_p$$

$$\leqslant C(\alpha)\sum_{k=n}^{+\infty}\frac{1}{(k+1)^2}\sum_{\nu=0}^{k+1}\widetilde{E}_\nu(f)_p$$

$$\leqslant C(\alpha)\left\{\sum_{\nu=n+1}^{+\infty}\left(\widetilde{E}_\nu(f)_p\cdot\sum_{k=\nu-1}^{+\infty}\frac{1}{(k+1)^2}\right)+\sum_{\nu=0}^{n}\widetilde{E}_\nu(f)_p\cdot\sum_{k=n}^{+\infty}\frac{1}{(k+1)^2}\right\}$$

$$\leqslant C(\alpha)\left\{\sum_{k=n}^{+\infty}\frac{1}{k+1}\widetilde{E}_k(f)_p+\frac{1}{n+1}\sum_{k=0}^{n}\widetilde{E}_k(f)_p\right\}. \tag{4.4}$$

另一方面，(3.28)(3.29)可合写成

$$\left\|\sum_{k=n}^{2n}\frac{1}{k+1}|\sigma_k^\beta(f)-f|\right\|_p$$

$$\leqslant C(\beta,p)\sum_{k=n}^{2n}\frac{1}{(k+1)^{2+\beta}}\sum_{\nu=0}^{k}(\nu+1)^\beta\widetilde{E}_\nu(f)_p+$$

$$C(\beta,p)\sum_{k=n}^{2n}\frac{\mu_k(1+\beta,p)}{(k+1)^3}\sum_{\nu=0}^{k}(\nu+1)\widetilde{E}_\nu(f)_p,\quad -1<\beta\leqslant 0.$$

若 $0<\alpha\leqslant 1$，以 $\beta=\alpha-1$ 代入上式，然后求得

$$\left\|\sum_{k=0}^{n}\frac{|\sigma_{k+1}^{\alpha-1}(f)-f|}{k+1}\right\|_p$$

$$\leqslant \sum_{j=0}^{\lfloor \log_2(n+1)\rfloor+1}\left\|\sum_{k=2^j-1}^{2^{j+1}-2}\frac{|\sigma_{k+1}^{\alpha-1}(f)-f|}{k+1}\right\|_p$$

$$\leqslant C(\alpha,p)\sum_{j=0}^{[\log_2(n+1)]+1}\sum_{k=2^j-1}^{2^{j+1}-2}\left\{\frac{1}{(k+1)^{1+\alpha}}\sum_{\nu=0}^{k}(\nu+1)^{\alpha-1}\widetilde{E}_\nu(f)_p+\right.$$

$$\left.\frac{1}{(k+1)^3}\mu_k(\alpha,p)\sum_{\nu=0}^{k}(\nu+1)\widetilde{E}_\nu(f)_p\right\}$$

$$\leqslant C(\alpha,p)\sum_{k=0}^{n}\left\{\frac{1}{(k+1)^{1+\alpha}}\sum_{\nu=0}^{k}(\nu+1)^{\alpha-1}\widetilde{E}_\nu(f)_p+\frac{\mu_k(\alpha,p)}{(k+1)^3}\sum_{\nu=0}^{k}(\nu+1)\widetilde{E}_\nu(f)_p\right\}$$

$$\leqslant C(\alpha,p)\sum_{k=0}^{n}\frac{\mu_k(\alpha,p)}{k+1}\widetilde{E}_k(f)_p, \tag{4.5}$$

以及

$$\left\|\sum_{k=n}^{+\infty}\frac{|\sigma_{k+1}^{\alpha-1}(f)-f|}{k+1}\right\|_p$$

$$\leqslant\left\|\sum_{k=n}^{2n}\frac{|\sigma_{k+1}^{\alpha-1}(f)-f|}{k+1}\right\|_p+\sum_{j=0}^{+\infty}\left\|\sum_{k=2^j(n+1)}^{2^{j+1}(n+1)}\frac{|\sigma_{k+1}^{\alpha-1}(f)-f|}{k+1}\right\|_p$$

$$\leqslant C(\alpha,p)\left(\sum_{k=n}^{2n}+\sum_{j=0}^{+\infty}\sum_{k=2^j(n+1)}^{2^{j+1}(n+1)}\right)\times$$

$$\left\{\frac{1}{(k+1)^{1+\alpha}}\sum_{\nu=0}^{k}(\nu+1)^{\alpha-1}\widetilde{E}_\nu(f)_p+\frac{\mu_k(\alpha,p)}{(k+1)^3}\sum_{\nu=0}^{k}(\nu+1)\widetilde{E}_\nu(f)_p\right\}$$

$$\leqslant C(\alpha,p)\left(\sum_{\nu=0}^{n}\sum_{k=n}^{+\infty}+\sum_{\nu=n+1}^{+\infty}\sum_{k=\nu-1}^{+\infty}\right)\left\{\frac{(\nu+1)^{\alpha-1}}{(k+1)^{\alpha+1}}+\frac{\mu_k(\alpha,p)(\nu+1)}{(k+1)^3}\right\}\widetilde{E}_\nu(f)_p$$

$$\leqslant C(\alpha,p)\left\{\sum_{\nu=0}^{n}\left[\frac{(\nu+1)^{\alpha-1}}{(n+1)^{\alpha}}+\frac{\mu_n(\alpha,p)}{(n+1)^2}(\nu+1)\right]\widetilde{E}_\nu(f)_p+\right.$$

$$\left.\sum_{\nu=n}^{+\infty}\frac{\mu_\nu(\alpha,p)}{\nu+1}\widetilde{E}_\nu(f)_p\right\}. \tag{4.6}$$

由(4.2)(4.3)(4.5)得(1.5)，由(4.1)(4.4)(4.6)得(1.6). □

由定理4得到正阶M型$|C,\alpha|$可和的两个判别条件，即

推论4 设$\alpha>0$，$1<p\leqslant 2$，$f\in L^p(Q)$. 若

$$\sum_{k=0}^{+\infty}\frac{\mu_k(\alpha,p)}{k+1}\widetilde{E}_k(f)_p<+\infty,$$

则$\sigma(f)$几乎处处可以M型$|C,\alpha|$求和.

推论5 设$\alpha>0$，$f\in C(Q)$. 若

$$\sum_{k=0}^{+\infty}\frac{\mu_k(\alpha,2)}{k+1}\widetilde{E}_K(f)_c<+\infty,$$

则 $\sigma(f)$一致地可以 M 型$|C,\alpha|$求和.

Ю. А. Пономаренко[4]的前三个定理可以合起来写成

定理 A 设 2π 周期函数 $f(x_1, x_2, \cdots, x_k)\in L^2$，满足

$$\sum_{n=1}^{+\infty}\lambda(n,\beta_\nu)\omega_k^{(\nu)}\left(f;\frac{1}{n}\right)_{L_2}<+\infty,\quad \nu=1,2,\cdots,k,$$

其中

$$\omega_k^{(\nu)}(f;h)_{L_2}=\sup_{|t|\leqslant h}\left\|\sum_{\mu=0}^{k}(-1)^{k-\mu}C_k^\mu f(x_1,x_2,\cdots,x_{\nu+\mu}t,\cdots,x_k)\right\|_{L_2},$$

$$\lambda(n,\beta_\nu)=\begin{cases}n^{\frac{k}{2}-\beta_\nu-1}, & -1<\beta_\nu<\frac{1}{2},\nu=1,2,\cdots,k,\\ n^{\frac{k-3}{2}}\sqrt{\ln n}, & \beta_\nu=\frac{1}{2},\qquad \nu=1,2,\cdots,k,\\ n^{k-2}, & \beta_\nu>\frac{1}{2},\qquad \nu=1,2,\cdots,k,\end{cases}$$

则 $\sigma(f)$几乎处处$|C;\beta_1,\cdots,\beta_k|$可和.

当 $k=2$，$\beta_1=\beta_2=\alpha$ 时，定理 A 的条件中乘数

$$\lambda(n,\alpha)=\begin{cases}1, & \alpha>\frac{1}{2},\\ \sqrt{n^{-1}\ln n}, & \alpha=\frac{1}{2},\\ n^{-\alpha}, & \alpha<\frac{1}{2},\quad \alpha>-1.\end{cases}$$

相应情形下，当 $p=2$ 时推论 4 中的相应的数是

$$\frac{\mu_n(\alpha,2)}{n+1}=\begin{cases}\frac{1}{n+1}, & \alpha>\frac{1}{2},\\ \frac{1}{n+1}\ln^{\frac{3}{2}}(n+2), & \alpha=\frac{1}{2},\\ \frac{1}{(n+1)^{\frac{1}{2}+\alpha}}\ln(n+2), & 0<\alpha<\frac{1}{2}.\end{cases}$$

可见推论 4 的条件比较好.

当 $p=+\infty$ 时，即连续函数的情形，施咸亮[5]关于一元函数$|C,\alpha|_q$求和的结果是

定理 B 设 $f(x)\in C_{2\pi}$，$\alpha>0$，$q\geqslant1$，则

$$\sum_{n=1}^{N}|\sigma_n^\alpha(f;x)-\sigma_{n-1}^\alpha(f;x)|^q n^{q-1}\leqslant C(\alpha,q)\sum_{n=0}^{N}\frac{p_n(q,\alpha)}{n+1}E_n^q(f),$$

$$\sum_{n=N}^{+\infty}|\sigma_n^\alpha(f;x)-\sigma_{n-1}^\alpha(f;x)|^q n^{u-1}$$
$$\leqslant C(\alpha,q)\left\{\sum_{n=N}^{+\infty}\frac{p_n(q,\alpha)}{n}E_n^q(f)+\frac{1}{N+1}\sum_{n=0}^{N}E_n^q(f)\right\},$$

其中

$$p_n(q,\ \alpha)=\begin{cases}1, & q\geqslant 2,\ \alpha>1-\dfrac{1}{q},\\ \{\ln(n+2)\}^{q-1}, & q\geqslant 2,\ \alpha=1-\dfrac{1}{q},\\ (n+1)^{q(1-\alpha)-1}, & q\geqslant 2,\ 0<\alpha<1-\dfrac{1}{q},\\ \{p_n(2,\ \alpha)\}^{\frac{q}{2}}, & 1\leqslant q<2.\end{cases}$$

当 $q=1$ 时，与 M 型结果比较，易见相应的 $|C,\ \alpha|$ 求和条件当 $\alpha>\frac{1}{2}$ 时形式是一样的，此时 $p_n(1,\ \alpha)=1=\mu_n(\alpha,\ +\infty)$；而当 $0<\alpha\leqslant\frac{1}{2}$ 时，$\mu_n(\alpha,\ +\infty)=p_n(1,\ \alpha)\cdot\ln(n+2)$，此时 M 型条件比一元条件劣.

§5. 用 Fourier 系数给出的 $|C,\ \alpha|$ 求和条件

定理 5 的证 由定义知

$$\sigma_n^\alpha(f)-\sigma_{n-1}^\alpha(f)=\frac{1}{nA_n^\alpha}\sum_{k=0}^{n}A_{n-k}^{\alpha-1}kB_k(f),\quad n\in\mathbf{N}^*,$$

$$\|\sigma_n^\alpha(f)-\sigma_{n-1}^\alpha(f)\|_2=\frac{1}{nA_n^\alpha}\left\{\sum_{k=1}^{n}(A_{n-k}^{\alpha-1}k)^2\gamma_k\right\}^{\frac{1}{2}},\quad n\in\mathbf{N}^*.\quad(5.1)$$

用 Hölder 不等式得

$$\sum_{n=1}^{N}\|\sigma_n^\alpha(f)-\sigma_{n-1}^\alpha(f)\|_2$$
$$=\sum_{k=1}^{N}\frac{1}{\sqrt{\varphi(n)}n}\cdot\frac{\sqrt{\varphi(n)}}{A_n^\alpha}\left\{\sum_{k=1}^{n}(A_{n-k}^{\alpha-1}k)^2\gamma_k\right\}^{\frac{1}{2}}$$
$$\leqslant\left\{\sum_{k=1}^{N}\frac{1}{\varphi(n)n^2}\right\}^{\frac{1}{2}}\left\{\sum_{k=1}^{N}\frac{\varphi(n)}{(A_n^\alpha)^2}\sum_{k=1}^{n}(A_{n-k}^{1-\alpha}k)^2\gamma_k\right\}^{\frac{1}{2}}.$$

由于 $\sum\limits_{n=1}^{+\infty}\frac{1}{\varphi(n)n^2}=C\in\mathbf{R}$，所以代入 $A_n^\alpha\sim n^\alpha$ 得到

$$\left\{\sum_{n=1}^{N}\|\sigma_n^\alpha(f)-\sigma_{n-1}^\alpha(f)\|_2\right\}^2$$
$$\leqslant C\sum_{n=1}^{N}\frac{\varphi(n)}{n^{2\alpha}}\sum_{k=1}^{n}(n-k+1)^{2(\alpha-1)}k^2\gamma_k$$
$$\leqslant C\sum_{k=1}^{N}k^2\gamma_k\sum_{n=k}^{N}\frac{\varphi(n)}{n(n-k+1)}\cdot\left(\frac{n-k+1}{n}\right)^{2\alpha-1}.$$

由于 $\alpha\geqslant\frac{1}{2}$ 时，$\left(\frac{n-k+1}{n}\right)^{2\alpha-1}\leqslant 1$，$(n\geqslant k\geqslant 1)$，所以

$$\sum_{n=1}^{N}\|\sigma_n^\alpha(f)-\sigma_{n-1}^\alpha(f)\|_2\leqslant C\left\{\sum_{k=1}^{N}\gamma_k\cdot k^2\sum_{n=k}^{+\infty}\frac{\varphi(n)}{n(n-k+1)}\right\}^{\frac{1}{2}}$$
$$\leqslant C\left(\sum_{k=1}^{+\infty}\gamma_k\cdot r_k\right)^{\frac{1}{2}}<+\infty,$$

从而 $\sum_{n=1}^{+\infty}|\sigma_n^\alpha(f)-\sigma_{n-1}^\alpha(f)|\in L^2(Q)$，当然 a. e. 有限.　□

如果 $\sigma(f)\sim\sum_{m,n=-\infty}^{+\infty}C_{m,n}\mathrm{e}^{\mathrm{i}(mx+ny)}$，那么

$$\gamma_k(f)=\sum_{\max(|m|,|n|)=k}|C_{m,n}|^2.$$

所以定理 5 是在 $L^2(Q)$ 中由 Fourier 系数给出的 a. e. $|C,\alpha|$ 求和 $\left(\alpha\geqslant\frac{1}{2}\right)$ 的一个充分条件. 由它得

推论 6　设 $f\in L^2(Q)$，其 Fourier 系数满足条件

$$\sum_{n=3}^{+\infty}\gamma_n(f)\ln n(\ln\ln n)^p<+\infty, p>1, \tag{5.2}$$

则当 $\alpha\geqslant\frac{1}{2}$ 时，$\sigma(f)$ 几乎处处 M 型 $|C,\alpha|$ 可和.

证　取

$$\varphi(n)=\begin{cases}2, & n=1,2,\\ \frac{1}{n}\ln n\cdot(\ln\ln n)^p, & n\geqslant 3,\end{cases}$$

因为

$$\sum_{n=1}^{+\infty}\frac{1}{\varphi(n)n^2}\leqslant C+\sum_{n=3}^{+\infty}\frac{1}{n\ln n\cdot(\ln\ln n)^p}<+\infty, \tag{5.3}$$

$$k^2\sum_{n=k}^{+\infty}\frac{\varphi(n)}{n(n-k+1)}\triangleq r_k\leqslant C(p)\ln k\cdot(\ln\ln k)^p,\quad k\geqslant 3, \tag{5.4}$$

所以，根据(5.2)，定理5的条件全部满足，用定理5就得所需的结论.

推论6是王福春[6]关于一元情形的结果的直接推广. 他的结果是

定理C 设 $f\in L_{2\pi}$，f 的 Fourier 系数是 a_n，$b_n(n\in\mathbf{N}^*)$，则当

$$\sum_{n=2}^{+\infty}(a_n^2+b_n^2)(\ln n)^p<+\infty, p>1$$

时，$\sigma(f)$ 对于 $\alpha>\frac{1}{2}$ 几乎处处 $|C,\alpha|$ 可和. p 不可改成1，条件 $\alpha>\frac{1}{2}$ 不可改成 $0<\alpha<\frac{1}{2}$.

作为定理 C 的直接推广，推论6中的 $p>1$ 也是不能改成 $p=1$ 的，而且 $\alpha>\frac{1}{2}$ 也不可改成 $0<\alpha<\frac{1}{2}$. 当然推论6比定理C的一个改进之处是包括了 $\alpha=\frac{1}{2}$ 的情形.

作者对孙永生老师的指导表示衷心的感谢.

参考文献

[1] Marcinkiewicz J. Collected papers. Warszawa，1964，527—538.

[2] 郭竹瑞. 连续函数用它的 Fourier 级数的蔡查罗平均数来逼近. 数学学报，1962，12(3)：320—329.

[3] Тиман А Ф. Теория дриблиЖения функций действительного переменного. Москва，1960.

[4] Пономаренко Ю А. Доклады. АН СССР，1963，152(6)：1 305—1 307.

[5] 施咸亮. 关联着连续函数富里叶级数的蔡查罗绝对求和的种种估计. 杭州大学学报，1964，2(1)：83—97.

[6] Wang Futraing. The absolute Cesàro Summability of trigonometrical series. Duke Math. J.，1942，9：567—572.

[7] Hardy G H，Littlewood J E，Polya G. Inequalities. Cambridge Univ. Press，1934.

[8] Zygmund A. Trigonometric series. Warszawa，1935.

数学学报，1984，27(6)：811－816.

二重 Fourier 级数及其共轭级数的 M 型(H, q)求和①

M(H, q) Summability of Double Fourier Series and of Their Conjugate Series

§1. 引言

设$Q=\{(x, y)\mid -\pi\leqslant x<\pi, -\pi\leqslant y<\pi\}$，$L(Q)$表示在$Q$上可和，对每个变元都以$2\pi$为周期的二元函数的全体，$L\ \ln^+ L(Q)$表示$L(Q)$中满足$|f|\ln^+|f|\in L(Q)$的函数$f$所成的子类. $f(x, y)$的 Fourier 部分和记为$S_{m,n}(f; x, y)$.

J. Marcinkiewicz[1] (1939)首倡对$\{S_{k,k}\}_{k=0}^{+\infty}$作线性平均的求和法，并证明了$(C, 1)$平均$\sigma_n(f; x, y)=\frac{1}{n+1}\sum\limits_{k=0}^{n}S_{k,k}(f; x, y)$当$f$连续时一致收敛，而当$f\in L\ \ln^+ L(Q)$时几乎处处收敛. Л. В. Жижиашвили[2] (1968)证明，只要$f\in L(Q)$，就有

$$\lim_{n\to+\infty}\sigma_n(f; x, y)\overset{\text{a. e.}}{=\!=}f(x, y). \tag{1.1}$$

我们把这种类型的求和及强性求和都叫作是M型的. 本文考虑的是当$f\in L\ \ln^+ L(Q)$时，f的 Fourier 级数及其共轭级数的M型的带有正指数q的强求和((H, q)求和)问题.

周知，当$f\in L\ \ln^+ L(Q)$时，f的三个共轭函数都是几乎处处有限的(引自[3])，记之为

① 收稿日期：1982-03-10；收精简稿日期：1984-02-09.

$$\overline{f}_1(x,y)=\lim_{\varepsilon\to0^+}-\frac{1}{\pi}\int_{\varepsilon<|s|<\pi}\frac{f(x+s,y)}{2\tan\frac{s}{2}}\mathrm{d}s,$$

$$\overline{f}_2(x,y)=\lim_{\eta\to0^+}-\frac{1}{\pi}\int_{\eta<|t|<\pi}\frac{f(x,y+t)}{2\tan\frac{t}{2}}\mathrm{d}t,$$

$$\overline{f}_3(x,y)=\lim_{\varepsilon\to0^+,\eta\to0^+}\frac{1}{\pi^2}\int_{\varepsilon<|s|<\pi,\eta<|t|<\pi}\frac{f(x+s,y+t)}{2\tan\frac{s}{2}\cdot2\tan\frac{t}{2}}\mathrm{d}s\mathrm{d}t.$$

f 的 Fourier 级数的三个共轭级数的部分和相应地分别记为 $S_{m,n}^{(1)}(f)$，$S_{m,n}^{(2)}(f)$及 $S_{m,n}^{(3)}(f)$. 为方便，记 f 为$\overline{f}_0$，$S_{m,n}(f)$为 $S_{m,n}^{(0)}(f)$.

本文的结果是

定理 若 $f(x,y)\in L\ \ln^+L(Q)$，$q>0$，则

$$\lim_{n\to+\infty}\frac{1}{n+1}\sum_{k=0}^{n}\left|S_{k,k}^{(\nu)}(f;x,y)-\overline{f}_\nu(x,y)\right|^q\overset{\text{a.e.}}{=\!=}0,\quad \nu=0,1,2,3. \tag{1.2}$$

多重矩形(H, q)求和的相应结果可见于[4].

§2. 记号及引理

设 $f\in L(Q)$，$\mathbf{N}$ 为正整数集，$n\in\mathbf{N}$，定义

$$K_n^{(1)}(s,t)=\frac{\sin ns}{2\tan\frac{s}{2}}\cdot\cos nt,\quad K_n^{(2)}(s,t)=K_n^{(1)}(t,s),$$

$$K_n^{(3)}(s,t)=\sin ns\frac{\sin nt}{2\tan\frac{t}{2}},$$

$$K_n^{(5)}(s,t)=K_n^{(3)}(t,s),\quad K_n^{(4)}(s,t)=\frac{1-\cos ns}{2\tan\frac{s}{2}}\cos nt,$$

$$K_n^{(6)}(s,t)=K_n^{(4)}(t,s),\quad K_n^{(7)}(s,t)=\frac{1-\cos ns}{2\tan\frac{s}{2}}\sin nt,$$

$$K_n^{(8)}(s,t)=K_n^{(7)}(t,s);$$

$$A_n^{(\nu)}(f;x,y)=\frac{1}{\pi^2}\int_Q f(x-s,y-t)K_n^{(\nu)}(s,t)\mathrm{d}s\mathrm{d}t,\quad \nu=1,2,\cdots,8.$$

那么，方形部分和与 M 型$(C, 1)$平均的差是

$$S_{n,n}^{(\nu)}(f)-\sigma_n^{(\nu)}(f)=\frac{1}{n+1}\sum_{k=1}^{n}k(A_k^{(2\nu+1)}(f)+A_k^{(2\nu+2)}(f)),$$
$$\nu=0,1,2,3, \tag{2.1}$$

其中

$$\sigma_n^{(\nu)}(f)=\frac{1}{n+1}\sum_{k=0}^{n}S_{k,k}^{(\nu)}(f).$$

由[3]得知，当 $f\in L\ \ln^+L(Q)$时，

$$\lim_{n\to+\infty}\sigma_n^{(\nu)}(f;\ x,\ y)\overset{\text{a. e.}}{=\!=}\overline{f}_\nu(x,\ y),\quad \nu=0,1,2,3. \tag{2.2}$$

所以，欲证定理只需证明

$$\lim_{n\to+\infty}\frac{1}{n+1}\sum_{k=1}^{n}|S_{k,k}^{(\nu)}(f;x,y)-\sigma_k^{(\nu)}(f;x,y)|^q\overset{\text{a. e.}}{=\!=}0,$$
$$\nu=0,1,2,3. \tag{2.3}$$

引入复数 $z=re^{i\theta}$，$r\in[0,\ 1)$，$\theta\in(-\infty,\ +\infty)$. 记 $\delta=1-r$. 对于 $f\in L(Q)$，定义单位圆内的解析函数

$$F_\nu(z)=\sum_{n=1}^{+\infty}A_n^{(\nu)}(f;x,y)z^n,\quad \nu=1,2,\cdots,8.$$

设 $q\geqslant2$，$p=\dfrac{q}{q-1}$. 定义算子 T_v，S_ν 和 R_μ：

$$T_\nu f(x,y)=\sup_{0<\delta\leqslant\frac{1}{2}}\left\{\delta^{2p-1}\int_{-\pi}^{\pi}\left|\frac{zF'_\nu(z)}{1-z}\right|^p\mathrm{d}\theta\right\}^{\frac{1}{p}},\quad \nu=1,2,\cdots,8,$$

$$S_\nu f(x,y)=\sup_{n\in\mathbf{N}}\left\{\frac{1}{(n+1)^{q+1}}\sum_{k=1}^{n}\left|\sum_{j=1}^{k}jA_j^{(\nu)}(f;x,y)\right|^q\right\}^{\frac{1}{q}},$$
$$\nu=1,2,\cdots,8,$$

$$R_\mu f(x,y)=\sup_{n\in\mathbf{N}}\left\{\frac{1}{n+1}\sum_{k=1}^{n}|S_{k,k}^{(\mu)}(f;x,y)-\sigma_k^{(\mu)}(f;x,y)|^q\right\}^{\frac{1}{q}},$$
$$\mu=0,1,2,3.$$

引理 1　设$\{a_n\}_{n=0}^{+\infty}$是非负数列，$q>0$，则

$$\sup_{n\in\mathbf{N}}\left\{\frac{1}{n+1}\sum_{k=0}^{n}a_k\right\}\leqslant C_q\sup_{n\in\mathbf{N}}\left\{\frac{1}{(n+1)^{q+1}}\sum_{k=0}^{n}(k+1)^qa_k\right\}. \tag{2.4}$$

(此简单命题载于[4]，但其证明细节有误.)

引理 2　设 $f\in L(Q)$，$q\geqslant2$，$p=\dfrac{q}{q-1}$，则

$$R_\mu f\leqslant C_q\sum_{\nu=1}^{8}S_\nu f\leqslant C_q\sum_{\nu=1}^{8}T_\nu f,\quad \mu=0,1,2,3.$$

证 由(3)知

$$\left\{\frac{1}{n+1}\sum_{k=1}^{n}\mid S_{k,k}^{(\mu)}-\sigma_k^{(\mu)}\mid^q\right\}^{\frac{1}{q}}\leqslant\left(\frac{1}{n+1}\sum_{k=1}^{n}\left|\frac{1}{k+1}\sum_{j=1}^{k}jA_j^{(2\mu+1)}\right|^q\right)^{\frac{1}{q}}+$$

$$\left(\frac{1}{n+1}\sum_{k=1}^{n}\left|\frac{1}{k+1}\sum_{j=1}^{k}jA_j^{(2\mu+2)}\right|^q\right)^{\frac{1}{q}}\leqslant\sum_{\nu=1}^{8}\left(\frac{1}{n+1}\sum_{k=1}^{n}\left|\frac{1}{k+1}\sum_{j=1}^{k}jA_i^{(\nu)}\right|^q\right)^{\frac{1}{q}}$$

$$\leqslant\sum_{\nu=1}^{8}\sup_{n\in\mathbf{N}}\left\{\frac{1}{n+1}\sum_{k=1}^{n}\left|\frac{1}{k+1}\sum_{j=1}^{k}jA_j^{(\nu)}\right|^q\right\}^{\frac{1}{q}},\quad \mu=0,1,2,3.$$

用引理 1，上式右端不超过 $C_q\sum_{\nu=1}^{8}S_\nu f$. 从而

$$R_\mu f\leqslant C_q\sum_{\nu=1}^{8}S_\nu f,\quad \mu=0,1,2,3. \tag{2.5}$$

另一方面，

$$\frac{zF'_\nu(z)}{1-z}=\sum_{n=1}^{+\infty}\left(\sum_{k=1}^{n}kA_k^{(\nu)}(f;x,y)\right)z^n.$$

据 Hausdorff-Young 不等式(见[5]，Ⅻ § 2)，

$$\left\{\sum_{n=1}^{+\infty}\left|\sum_{k=1}^{n}kA_k^{(\nu)}(f;x,y)\right|^q r^{nq}\right\}^{\frac{1}{q}}\leqslant C_q\left\{\int_{-\pi}^{\pi}\left|\frac{zF'_\nu(z)}{1-z}\right|^p\mathrm{d}\theta\right\}^{\frac{1}{p}}.$$

代入 $r=\frac{n}{n+1}$，据 $r^m=\left(1-\frac{1}{n+1}\right)^m\geqslant \mathrm{e}^{-1}$，($m\leqslant n$)及 T_ν 之定义得

$$\left\{\frac{1}{(n+1)^{q+1}}\sum_{m=1}^{n}\left|\sum_{k=1}^{m}kA_k^{(\nu)}(f)\right|^q\right\}^{\frac{1}{q}}\leqslant C_qT_\nu f,$$

$$S_\nu(f)(x,\ y)\leqslant C_qT_\nu f(x,\ y),\qquad \nu=1,2,\cdots,8. \tag{2.6}$$

(7)(8)合起来，证得引理 2.

引理 3 设 $f(x)$是周期 2π 且在$(0,\ 2\pi)$可和的函数，$p\in(1,\ 2]$. f 的 Poisson 积分记作

$$f(r,x)=\frac{1}{\pi}\int_{-\pi}^{\pi}f(x-t)P(r,t)\mathrm{d}t,$$

核 $P(r,\ t)=\frac{1-r^2}{2(1-2r\cos t+r^2)}$，$(0\leqslant r<1)$. 定义算子 T：

$$Tf(x)=\sup_{0<\delta\leqslant 1}\left\{\delta^{p-1}\int_{-\pi}^{\pi}\left|\frac{f(r,x+\theta)}{1-z}\right|^p\mathrm{d}\theta\right\}^{\frac{1}{p}},\quad \delta=1-r.$$

则对任何 $\lambda\in(0,\ 1)$都有

$$\left\{\int_0^{2\pi}(Tf(x))^\lambda\mathrm{d}x\right\}^{\frac{1}{\lambda}}\leqslant C_{p,\lambda}\int_0^{2\pi}|f(x)|\mathrm{d}x.$$

引理 3 的证明实质上载于[5](Vol. Ⅱ，p. 184)，可参阅[4].

引理 4 设 $f(x)\in(L\ \ln^+ L)_{2\pi}$. 定义算子"$\Lambda$":

$$\hat{f}(x)=\sup_{m\in\mathbf{N}}\left|\int_{m^{-1}}^{\pi}\frac{f(x+t)-f(x-t)}{2\tan\frac{t}{2}}\mathrm{d}t\right|.$$

则 $\hat{f}\in L_{2\pi}$且

$$\int_0^{2\pi}\hat{f}(x)\mathrm{d}x\leqslant C\left(\int_0^{2\pi}|f(x)|\ln^+|f(x)|\mathrm{d}x+1\right). \tag{2.7}$$

证明根据[6]的结果及[5](Vol. Ⅱ, p. 119).

§ 3. 定理的证明

为估计算子 T_ν 先算 $F'_\nu(z)$. 定义

$$K(\tau,\ z)=\frac{\mathrm{e}^{\mathrm{i}2\tau}}{(1-z\mathrm{e}^{\mathrm{i}2\tau})^2},$$

并记 $\xi=\frac{s-t}{2}$, $\eta=\frac{s+t}{2}$. 经初等计算得

$$F'_1(z)=\frac{1}{\pi^2}\int_Q\frac{f(x-s,y-t)}{2\tan\frac{s}{2}}\frac{1}{4\mathrm{i}}[K(\xi,z)-K(-\xi,z)+$$

$$K(\eta,\ z)-K(-\eta,\ z)]\mathrm{d}s\mathrm{d}t,$$

$F'_\nu(z)$当 $\nu=2,3,\cdots,8$ 时亦有类似形状的表达式. 现设

$$Q_m=\left\{(s,\ t)\left|\begin{array}{l}m^{-1}\leqslant|s|\leqslant\pi,\\ 0<|t|<\pi\end{array}\right.\right\},\ m\in\mathbf{N},$$

并记

$$\tau_1=\frac{t}{2}=-\tau_2,\ \tau_3=\xi=-\tau_4,\ \tau_5=\eta=-\tau_6,$$

定义

$$I_\mu(x,y,z)=\varlimsup_{m\to+\infty}\left|\int_{Q_m}\frac{f(x-s,y-t)}{2\tan\frac{s}{2}}K(\tau_\mu,z)\mathrm{d}s\mathrm{d}t\right|,\quad \mu=1,2,\cdots,6.$$

由积分的绝对连续性推出, 对于 $\nu=1,4,5,7$,

$$|F'_\nu(z)|\leqslant\sum_{\mu=1}^{6}I_\mu(x,y,z). \tag{3.1}$$

再定义

$$\varphi_1(x,y)=\hat{f}_1(x,y)=\sup_{m\in\mathbf{N}}\left|\int_{m^{-1}<|s|<\pi}\frac{f(x-s,y)}{2\tan\frac{s}{2}}\mathrm{d}s\right|,$$

$$\varphi_2(x,y)=\sup_{m\in\mathbf{N}}\left|\int_{m^{-1}<|\tau|<\pi}\frac{f(x-\tau,y-\tau)}{2\tan\frac{\tau}{2}}\mathrm{d}\tau\right|,$$

$$\varphi_3(x,y)=\sup_{m\in\mathbf{N}}\left|\int_{m^{-1}<|\tau|<\pi}\frac{f(x-\tau,y+\tau)}{2\tan\frac{\tau}{2}}\mathrm{d}\tau\right|.$$

由引理 4 直接得到

$$\int_0^{2\pi}\varphi_1(x,y)\mathrm{d}x\leqslant C\left(\int_0^{2\pi}|f(x,y)|\ln^+|f(x,y)|\mathrm{d}x+1\right).\tag{3.2}$$

设 $u=\frac{1}{2}(x-y)$，$v=\frac{1}{2}(x+y)$，并记 $g(u,v)=f(x,y)$. 则由 $f\in L\ln^+L(Q)$知 $g\in L\ln^+L(Q)$. 且用引理 4 的记号可写

$$\varphi_2(x,y)=\hat{g}_2(u,v)=\sup_{m\in\mathbf{N}}\left|\int_{m^{-1}<|\tau|<\pi}\frac{g(u,v-\tau)}{2\tan\frac{\tau}{2}}\mathrm{d}\tau\right|,$$

$$\varphi_3(x,y)=\hat{g}_1(u,v)=\sup_{m\in\mathbf{N}}\left|\int_{m^{-1}<|\tau|<\pi}\frac{g(u-\tau,v)}{2\tan\frac{\tau}{2}}\mathrm{d}\tau\right|.$$

故由引理 4 得

$$\int_0^{2\pi}\varphi_2(x,y)\mathrm{d}v=\int_0^{2\pi}\hat{g}_2(u,v)\mathrm{d}v\leqslant C\left(\int_0^{2\pi}|g(u,v)|\ln^+|g(u,v)|\mathrm{d}v+1\right),\tag{3.3}$$

$$\int_0^{2\pi}\varphi_3(x,y)\mathrm{d}u=\int_0^{2\pi}\hat{g}_1(u,v)\mathrm{d}v\leqslant C\left(\int_0^{2\pi}|g(u,v)|\ln^+|g(u,v)|\mathrm{d}u+1\right).\tag{3.4}$$

把 $\varphi_\nu(x,y)$关于第二个变元的 Poisson 积分记为 $\varphi_\nu^{(x)}(r,y)$，$(\nu=1,2,3)$. 得

$$\begin{aligned}I_1(x,y,z)&\leqslant\int_{-\pi}^{\pi}\varphi_1(x,y-t)\frac{\mathrm{d}t}{|1-z\mathrm{e}^{\mathrm{i}t}|^2}\\&\leqslant C\frac{1}{\delta}\int_{-\pi}^{\pi}\varphi_1(x,y-t)P(r,\theta+t)\mathrm{d}t\\&=C\frac{1}{\delta}\varphi_1^{(x)}(r,y+\theta),\quad\delta=1-r,\end{aligned}\tag{3.5}$$

$$I_2(x,y,z)\leqslant C\frac{1}{\delta}\varphi_1^{(x)}(r,y-\theta).\tag{3.6}$$

由函数的周期性，并进行坐标变换$\begin{cases}s=\xi+\eta,\\t=-\xi+\eta,\end{cases}$算出

$$\int_{Q_m} \frac{f(x-s,y-t)}{2\tan\frac{s}{2}} K(\xi,z)\,\mathrm{d}s\mathrm{d}t$$

$$=\int_{m^{-1}<|s|<\pi,\, s<t<s+2\pi} \frac{f(x-s,y-t)}{2\tan\frac{s}{2}} K(\xi,z)\,\mathrm{d}s\mathrm{d}t$$

$$=2\int_{-\pi}^{0}\left\{\int_{m^{-1}<|\xi+\eta|<\pi} \frac{f(x-\xi-\eta,y-\eta+\xi)}{2\tan\frac{\xi+\eta}{2}}\mathrm{d}\eta\right\}K(\xi,z)\,\mathrm{d}\xi$$

$$=2\int_{-\pi}^{0}\left\{\int_{m^{-1}<|\tau|<\pi} \frac{f(x-\tau,y-\tau+2\xi)}{2\tan\frac{\tau}{2}}\mathrm{d}\tau\right\}K(\xi,z)\,\mathrm{d}\xi.$$

从而

$$\begin{aligned} I_3(x,y,z) &\leqslant 2\int_{-\pi}^{0}\varphi_2(x,y+2\xi)\,|K(\xi,z)|\,\mathrm{d}\xi \\ &\leqslant \frac{C}{\delta}\int_{-\pi}^{\pi}\varphi_2(x,y+\tau)P(r,\theta+\tau)\,\mathrm{d}\tau \\ &= \frac{C}{\delta}\varphi_2^{(x)}(r,y-\theta). \end{aligned} \tag{3.7}$$

同理

$$I_4(x,\ y,\ z)\leqslant\frac{C}{\delta}\varphi_2^{(x)}(r,\ y+\theta). \tag{3.8}$$

同样地算出

$$I_5(x,\ y,\ z)\leqslant\frac{C}{\delta}\varphi_3^{(x)}(r,\ y+\theta),\ I_6(x,\ y,\ z)\leqslant\frac{C}{\delta}\varphi_3^{(x)}(r,\ y-\theta). \tag{3.9}$$

使用引理 3 的算子 T，由(3.1)(3.5)～(3.9)推出

$$T_\nu f(x,y)\leqslant C\sum_{\mu=1}^{3}T\varphi_\mu^{(x)}(y),\quad \nu=1,4,5,7. \tag{3.10}$$

再用引理 3，得知对一切 $\lambda\in(0,\ 1)$，$\mu=1,2,3$，有

$$\int_Q\{T\varphi_\mu^{(x)}(y)\}^\lambda\mathrm{d}x\mathrm{d}y\leqslant C_{p,\lambda}\left\{\int_Q\varphi_\mu(x,y)\,\mathrm{d}x\mathrm{d}y\right\}^\lambda.$$

将(3.2)～(3.4)代入上式，然后代入(3.10)得 $\nu=1,4,5,7$ 时，

$$\left\{\int_Q(T_\nu f)^\lambda\mathrm{d}x\mathrm{d}y\right\}^{\frac{1}{\lambda}}\leqslant C_{p,\lambda}\left(\int_Q|f|\ln^+|f|\,\mathrm{d}x\mathrm{d}y+1\right). \tag{3.11}$$

同样可证(3.11)对于 $\nu=2,3,6,8$ 亦真．总起来得

引理 5 设 $f(x, y)\in L\ \ln^{+}L(Q)$，$q\geqslant 2$，$p=\frac{q}{q-1}$，则对于一切 $\lambda\in(0, 1)$，不等式(3.11)当 $\nu=1,2,\cdots,8$ 时成立.

定理的证 欲证(2.3)当 $q>0$ 成立，只需证它当 $q>2$ 时成立. 再根据引理 2 及引理 5，使用典型的逼近办法证得(2.3). □

根据这个定理，对求和法稍作推广，可得

推论 设 $\Lambda=\{\lambda_k^{(n)}{}_{n\in\mathbf{N}}^{k=0,1,\cdots,n+1}\lambda_{n+1}^{(n)}=0\}$满足

$$(1)\ \lim_{n\to+\infty}\lambda_k^{(n)}=0,\ k\in\mathbf{N},\ (2)\ \sum_{k=0}^{n}(k+1)\left|\lambda_k^{(n)}-\lambda_{k+1}^{(n)}\right|\leqslant C,\ n\in\mathbf{N}.$$

则对于 $f\in L\ \ln^{+}L(Q)$，当 $q>0$ 时有

$$\lim_{n\to+\infty}\sum_{k=0}^{n}\lambda_k^{(n)}\left|S_{k,k}^{(v)}(f;x,y)-\hat{f}_\nu(x,y)\right|^q\overset{\text{a. e.}}{=\!=}0,\quad \nu=0,1,2,3. \tag{3.12}$$

我们指出，正阶 Cesàro 数阵及正阶 Zygmund 典型平均数阵都符合推论的条件.

致谢 本文是在孙永生老师指导下完成的，谨表示衷心的感谢.

参考文献

[1] Marcinkiewicz J. Sur une méthode remarquable des séries doubles de Fourier. Collected papers. Warszawa，1964，527－538.

[2] Жижиащвили Л В. Обобщение одной теоремы Марцинкевича. Изв. АН СССР，сер. матем. 1968，32：1 112－1 122.

[3] Жижиащвили Л В. О некоторых вопросах из теории простых и кратных тригонометрических и ортогональных рядов，У. М. Н. 1973，28 (2)：65－119.

[4] Гоголадзе Л Д. О($н$，$к$)-суммируемости кратных тригонометрических рядов Фурье. ИЗВ. АН СССР，сер. матем. 1977，41(4)：937－958.

[5] Zygmund A. Trigonometric series，Vol. Ⅰ and Ⅱ. Cambridge，1979.

[6] Stein E M. On limits of sequences of operators. Annals of Math.，1961，74(1)：140－170.

北京师范大学学报(自然科学版)，1985，(1)：11—16.

关于 Walsh-Fourier 级数的几乎处处收敛问题的一点注记①

A Note on the Almost Everywhere Convergence of Walsh-Fourier Series

摘要　本文证明空间 $C_q(1<q\leqslant+\infty)$中函数的 Walsh-Fourier 级数几乎处处收敛. 函数空间 C_q 是 M. Taibleson 和 G. Weiss 在[5]中引入的.

L. Garleson[1] 证明了 $L^2_{2\pi}$ 中的 Fourier 级数 a. e. 收敛. 随后，P. Billard[2] 将此结果推广到 $L^2(0,1)$的 Walsh-Fourier 级数. 接着 R. A. Hunt[3]证明了 $L^p_{2\pi}(p>1)$中的 Fourier 级数 a. e. 收敛. P. Sjölin[4] 又将此结果推广到 Walsh-Fourier 级数，证明了

定理 A　若$(0,1)$上的可测函数 f 属于 $L\ln^+L\ln\ln^+L$，则 $f(x)$的 Walsh-Fourier 级数 a. e. 收敛到 $f(x)$.

M. H. Taibleson 和 G. Weiss[5] 引入了函数空间 $C_q(1<q\leqslant+\infty)$，用块分解的办法很简洁地证明了 C_q 中的 Fourier 级数 a. e. 收敛. 此处我们用[5]的块分解办法简洁地证明 C_q 中 Walsh-Fourier 级数 a. e. 收敛. 由此可以再次看出块分解方法的功效.

至于空间 C_q 的大小，

1) [5]中指出 $J\subsetneqq C_\infty\subsetneqq C_q(1<q<+\infty)$，其中 J 表示熵有限的函数类；

2) Sato[6]指出，$J\setminus L\ln^+L\ln^+\ln^+L\neq\varnothing$. 因此，

$$C_q\setminus L\ln^+L\ln^+\ln^+L\neq\varnothing,\ (1<q\leqslant+\infty).$$

① 收稿日期：1984-04-17.

由此可见本文的结果并不包含在定理A中.

1° C_q 空间 $(1<q\leqslant+\infty)$

定义1 (q-block)设函数 $b(x)$ 满足

(1) Supp. $b\subset I$,I 是含在$[0,1)$中的区间,

(2) $\|b\|_g\leqslant|I|^{\frac{1}{g}-1}$,

其中$\|\ \|_q$ 为$(0,1)$上的 L^q 范数,$|I|$表 I 的 Lebesgue 测度,则称 $b(x)$为q-block(即 q 块).

定义2 用m表示数列$\{m_k\}_{k=1}^{+\infty}$. 当$\sum\limits_{l=1}^{+\infty}|m_l|<+\infty$时,定义$N(m)=\sum\limits_{k=1}^{+\infty}|m_k|\left(1+\ln\frac{1}{|m_k|}\sum\limits_{l=1}^{+\infty}|m_l|\right)$,求和式中若某 $m_k=0$,则认为 $|m_k|\left(1+\ln\frac{1}{|m_k|}\sum\limits_{l=1}^{+\infty}|m_l|\right)=0$;当$\sum\limits_{l=1}^{+\infty}|m_l|=+\infty$时亦认为$N(m)=+\infty$. 设$[0,1)$上的可积函数 $f(x)$ 能按 $L(0,1)$ 度量表成 $f(x)=\sum\limits_{k=1}^{+\infty}m_kb_k(x)$,其中 b_k 是q-block且 $m=\{m_k\}_{k=1}^{+\infty}$ 满足 $N(m)<+\infty$. 这样的函数 f 的全体叫作空间C_q.

2° Walsh-Fourier 级数

设 $f\in L(0,1)$,$\{w_k\}_{k=0}^{+\infty}$是 Walsh 函数系,f 的 W-F 级数是

$$\sigma_w(f;x)\sim\sum_{k=0}^{+\infty}c_k(f)w_k(x),$$

其中 W-F 系数 $c_k(f)=\int_0^1 f(x)w_k(x)\mathrm{d}x$. W-F 级数的部分和是

$$\begin{aligned}S_n(f;x)&=\sum_{k=0}^{n-1}c_k(f)w_k(x)=\int_0^1 f(t)\sum_{k=0}^{n-1}w_k(t)w_k(x)\mathrm{d}t\\&=\int_0^1 f(t)\cdot D_n(x\oplus t)\mathrm{d}t,\qquad n\in\mathbf{N}^*,\end{aligned}$$

其中核 $D_n(t)=\sum\limits_{k=0}^{n-1}w_k(t)$,“$\oplus$”为伪加运算.

关于核 D_n,有估计式

$$|D_n(t)|\leqslant n,\qquad t\in[0,1),$$

$$|D_n(t)|\leqslant\frac{2}{t},\qquad 0<t<1.$$

关于这些基本知识可参阅[7].

3° **引理** 设 I 是 $[0,1)$ 中的区间，$\delta\in(0,1)$. 若 $|I|\leqslant\delta$，则

$$\left|\{x\in(0,1)\mid \inf_{t\in I}\{x\oplus t\}<\delta\}\right|\leqslant c\delta, \tag{1}$$

其中 $c>0$ 是绝对常数.

证 设整数 μ 使 $2^{-\mu-1}\leqslant\delta\leqslant2^{-\mu}$，无妨认为 δ 较小，使 $\mu\geqslant3$(否则取 $c=2^4$ 便使不等式(1)成立). 由于 $|I|\leqslant\delta<2^{-\mu}$，所以存在 $\nu\in\{0,1,\cdots,2^{\mu}-2\}$ 使得 $I\subset[\nu2^{-\mu},(\nu+2)2^{-\mu})$. 写 $I_1=[\nu2^{-\mu},(\nu+1)2^{-\mu})$，$I_2=[(\nu+1)2^{-\mu},(\nu+2)2^{-\mu})$ 及 $I_0=[\nu2^{-\mu},(\nu+2^{-\mu}))$. 令

$$e_i=\{x\in[0,1)\mid \inf_{t\in I_i}\{x\oplus t\}<\delta\},\quad i=0,1,2.$$

设 $\nu2^{-\mu}$ 的二进表示是 $0.a_1\cdots a_\mu$(即 $\nu2^{-\mu}=\sum_{j=1}^{\mu}a_j2^{-j}$，其中 a_j 或取 1 或取 0). 则 I_1 中的点 t 得表为 $t=0.a_1\cdots a_\mu t_{\mu+1}\cdots$(集 $\{t_{\mu+1},t_{\mu+2},\cdots\}\neq\{1\}$). 设 $x\in[0,1)$，$x=0.x_1x_2\cdots$，则 $x\oplus t=0.\ a_1\oplus x_1\ \ a_2\oplus x_2\cdots a_\mu\oplus x_\mu\ \ t_{\mu+1}\oplus x_{\mu+1}\cdots$

如果 $x\notin I_1$，那么 $x\oplus t\geqslant2^{-\mu}>\delta$. 可见 $e_1\subset I_1$. 同理 $e_2\subset I_2$. 因此 $|e_0|=|e_1\cup e_2|\leqslant|I_0|=2^{-\mu+2}\leqslant8\delta$. □

4° **定理** 设 $f\in C_q$，$1<q\leqslant+\infty$，则

$$\lim_{n\to+\infty}S_n(f;x)\overset{\text{a. e.}}{=\!=}f(x),$$

其中 S_n 表 Walsh-Fourier 第 n 部分和.

证 定义 $L(0,1)$ 上的算子 M：

$$(Mf)(x)=\sup_{n\in\mathbf{N}}|S_n(f;x)|,\quad f\in L(0,1).$$

显然我们只需考虑 $1<q<+\infty$ 的情形，那么由 Sjölin[4] 知，M 是强 (q,q) 型的.

现设 $b(x)$ 是 q-block，supp. b 含于区间 $I\subset[0,1)$，$\|b\|_q\leqslant|I|^{\frac{1}{q}-1}$. 任取 $\lambda>0$，作集 $E_\lambda=\{x\in(0,1)\mid(Mb)(x)>\lambda\}$. 于是存在只与 q 有关的常数 c_q 使 $|E_\lambda|\leqslant c_q\left(\frac{\|b\|_q}{\lambda}\right)^q\leqslant c_q(|I|\lambda)^{1-q}\cdot\frac{1}{\lambda}$. 如果 $\lambda\geqslant|I|^{-1}$ 的话，由此得到

$$|E_\lambda|\leqslant c_q\frac{1}{\lambda}. \tag{2}$$

现设 $\lambda<|I|^{-1}$，并记 $e=\{x\in[0,1)\mid\inf_{t\in I}\{x\oplus t\}<|I|\}$.

那么 $E_\lambda=(E_\lambda\cap e)\cup(E_\lambda\cap e^c)$，其中 $e^c=[0,1)\setminus e=\{x\in[0,1)\mid$

$\inf\limits_{t\in I}\{x\oplus t\}\geqslant|I|\}$. 显然，$|E_\lambda|\leqslant|e|+|E_\lambda\cap e^c|$. 由引理知

$$|e|<c|I|<\frac{c}{\lambda}. \tag{3}$$

而当 $x\in e^c$ 时，

$$|S_n(b;\ x)|=\left|\int_0^1 b(t)D_n(x\oplus t)\mathrm{d}t\right|\leqslant$$

$$\leqslant\sup_{t\in I}|D_n(x\oplus t)|\,\|b\|_1\leqslant 2/\inf_{t\in I}\{x\oplus t\}.$$

从而

$$(Mb)(x)\leqslant 2/\inf_{t\in I}\{x\oplus t\}.$$

因此

$$(E_\lambda\cap e^c)\subset\left\{x\in[0,\ 1)\,\middle|\,\frac{2}{\inf\limits_{t\in I}\{x\oplus t\}}>\lambda\right\}=\left\{x\in[0,\ 1)\,\middle|\,\inf_{t\in I}\{x\oplus t\}<\frac{2}{\lambda}\right\}.$$

再用引理得

$$|E_\lambda\cap e^c|\leqslant\left|\left\{x\in[0,\ 1)\,\middle|\,\inf_{t\in I}\{x\oplus t\}<\frac{2}{\lambda}\right\}\right|\leqslant\frac{c}{\lambda}, \tag{4}$$

由(2)～(4)得知 $|E_\lambda|<\frac{c_q}{\lambda}(\forall\lambda>0)$.

现设 $f\in C_q\,(1<q<+\infty)$. 据定义 2，f 有 block 分解式 $f(x)=\sum\limits_{k=1}^{+\infty}m_kb_k(x)$ 按($L(0,\ 1)$度量)，系数满足条件

$$\sum_{k=1}^{+\infty}|m_k|\left(1+\ln\frac{1}{|m_k|}\sum_{l=1}^{+\infty}|m_l|\right)<+\infty. \tag{5}$$

显然，我们可以认为诸 m_k 都是正的，置 $A=\sum\limits_{k=1}^{+\infty}m_k$ 及 $m'_k=\frac{m_k}{A}$.

任给 $\varepsilon>0$，存在正整数 N，使

$$\sum_{k=N}^{+\infty}m'_k\left(1+\ln\frac{1}{m'_k}\right)<\varepsilon. \tag{6}$$

令 $g(x)=\sum\limits_{k=1}^{N}m_kb_k(x), h(x)=f(x)-g(x)$. 显然，$g\in L^q(0,\ 1)$. 所以，根据定理 A 知

$$\lim_{n\to+\infty}|S_n(g;\ x)-g(x)|=0\quad\text{a. e.}, \tag{7}$$

于是

$$\overline{\lim_{n\to+\infty}}|S_n(f;\ x)-f(x)|\leqslant\overline{\lim_{n\to+\infty}}|S_n(g;\ x)-g(x)|+$$

$$\sup_{n\in\mathbf{N}}|S_n(h;\ x)|+|h(x)|\leqslant(Mh)(x)+|h(x)|\quad\text{a. e.}. \tag{8}$$

任取正数λ，定义集合

$$E_1(\lambda)=\{x\in(0,\ 1)\mid \overline{\lim_{n\to+\infty}}\,|S_n(f;\ x)-f(x)|>\lambda\},$$
$$E_2(\lambda)=\{x\in(0,\ 1)\mid (Mh)(x)>\lambda\},$$
$$E_3(\lambda)=\{x\in(0,\ 1)\mid |h(x)|>\lambda\}.$$

显然

$$|E_3(\lambda)|\leqslant\frac{1}{\lambda}\int_0^1|h(x)|\,\mathrm{d}x\leqslant\frac{1}{\lambda}\sum_{k=N+1}^{+\infty}m_k\int_0^1|b_k(x)|\,\mathrm{d}x$$
$$\leqslant\frac{1}{\lambda}\sum_{k=N+1}^{+\infty}m_k\|b_k\|_q\cdot|I_k|^{1-\frac{1}{q}}\leqslant\frac{1}{\lambda}\sum_{k=N+1}^{+\infty}m_k,$$

式中 I_k 表示如定义 1 所要求的包含 b_k 的支集的区间．进而使用(6)，得

$$|E_3(\lambda)|\leqslant\frac{A}{\lambda}\varepsilon. \tag{9}$$

为估计 $E_2(\lambda)$的测度，我们引述迭加原理(见[8]的引理 2.3)．

迭加原理　设 $g_k(x)$是一个测度空间上的非负可测函数($k\in\mathbf{N}^*$)满足$|\{x\mid g_k(x)>s\}|<\dfrac{1}{s}$，$(s>0)$．设$\{c_k\}$是正数列使 $\sum c_k=1$，又置

$$K=\sum_k c_k\ln\frac{1}{c_k},$$

则

$$\left|\left\{x\,\middle|\,\sum_k c_kg_k(x)>s\right|\right\}<\frac{2(K+2)}{s}.$$

由于按 $L(0,\ 1)$尺度 $h(x)=\lim\limits_{p\to+\infty}\sum\limits_{k=N+1}^{p}m_kb_k(x)$，所以几乎处处成立着 $S_n(h;x)=\sum\limits_{k=N+1}^{+\infty}m_kS_n(b_k;x)$．因此

$$(Mh)(x)\leqslant\sum_{k=N+1}^{+\infty}m_k(Mb_k)(x)\quad\text{a. e.}. \tag{10}$$

前面已经证明，对于 q-block b_k，

$$|\{x\in(0,\ 1)\mid (Mb_k)(x)>\lambda\}|<\frac{c_q}{\lambda}.$$

令 $g_k=\dfrac{1}{c_q}(Mb_k)$，用迭加原理，就得到

$$\left|\left\{x\in(0,1)\,\middle|\,\sum_{k=N+1}^{+\infty}m'_kg_k(x)>\lambda\right\}\right|<\frac{2}{\lambda}\sum_{k=N+1}^{+\infty}m'_k\ln\frac{1}{m'_k}.$$

据(6)，上式右端小于$\dfrac{2}{\lambda}\varepsilon$．所以

$$\left|\left\{x \in (0,1) \,\middle|\, \sum_{k=N+1}^{+\infty} m_k(Mb_k)(x) > \lambda\right\}\right| < \frac{2c_q A}{\lambda}\varepsilon.$$

再据(10)，得

$$|E_2(\lambda)| < \frac{2c_q A}{\lambda}\varepsilon. \tag{11}$$

由(9)(11)，得到

$$|E_2(\lambda) \cup E_3(\lambda)| < \frac{A}{\lambda}(1+2c_q)\varepsilon. \tag{12}$$

由(8)知$|E_1(\lambda)| \leqslant \left|E_2\left(\frac{\lambda}{2}\right) \cup E_3\left(\frac{\lambda}{2}\right)\right|$，故由(12)得

$$|E_1(\lambda)| < \frac{2A}{\lambda}(1+2c_q)\varepsilon. \tag{13}$$

集$E_1(\lambda)$是与ε无关的，故由ε之任意性，从(13)推出$|E_1(\lambda)|=0$. 再由λ的任意性，推出

$$\overline{\lim_{n\to+\infty}} |S_n(f; x) - f(x)| = 0 \quad \text{a. e.}.$$

这就完成了定理的证明.

作者谨对陆善镇老师的指导表示衷心感谢.

参考文献

[1] Carleson L. Acta Math., 1966, 116: 135—157.
[2] Billard P. Studia Math., 1967, 28: 363—388.
[3] Hunt R A. Proc. conf. Edwardsvillε, 1967, SIU Press, Garbondale Illinois, 1968, 235—255.
[4] Sjölin P. Arkiv för Matem., 1968, 7: 551—570.
[5] Taibleson M H, Weiss G. Procecdings of the Conference on Harmonic Analysis. Univ. of Chicago, Vol. Ⅰ, 1982, 95—113.
[6] Shuichi Sato. Tôhoku Math. J., 1981, 33(4): 593—597.
[7] 郑维行，等. 沃尔什函数理论与应用. 上海：上海科学技术出版社，1983.
[8] Stein F M, Weiss N J. Trans. of the Amer. Math. Soc., 1969, 140: 35—54.

Abstract It is proved that the walsh-Fourier series of any function in $C_q(1<q\leqslant+\infty)$ is convergent almost everywhere, where C_q is the function space introduced by M. Taibleson and G. Weiss.

数学学报，1986，29(2)：156－175.

多重共轭 Fourier 级数的强求和①

Strong Summability of Conjugate Multiple Fourier Series

§1. 引言

设正整数 $k\geqslant 2$，E_k 为 k 维欧氏空间，$Q^k=\{x\in E_k \mid x=(x_1, x_2, \cdots, x_k), -\pi\leqslant x_j<\pi, j=1,2,\cdots,k.\}$. $L(Q)$表示在 Q 上可积对每个变元都以 2π 为周期的函数的全体. $L(E_k)$表示在 E_k 可积的函数全体. 设 $P(x)$是 k 元 n 阶齐次多项式，$n\geqslant 1$，满足 Laplace 方程 $\Delta P(x)=0$. 核 $K(x)=P(x)|x|^{-k-n}(x\neq 0)$. 下面先简要地叙述一下基本概念和定义，然后再说明本文的目的和结果. 所叙述的概念可参阅[1].

(1) 共轭级数与共轭函数. 设 $f\in L(Q)$. f 的 Fourier 级数是

$$\sigma(f)\sim\sum_{m\in\mathbf{Z}^k}a_m e^{imx}, \tag{1.1}$$

其中 Fourier 系数

$$a_m=(2\pi)^{-k}\int_Q f(x)e^{-imx}dx. \tag{1.2}$$

f 关于核 K 的共轭 Fourier 级数是

$$\tilde{\sigma}(f)\sim\sum_{m\in\mathbf{Z}^k}a_m\hat{K}(m)e^{imx}, \tag{1.3}$$

这里 $\hat{K}(y)$是主值意义下的 Fourier 变换，即

---

① 收稿日期：1983-06-11.

$$\hat{K}(y) = (2\pi)^{-k} \lim_{\varepsilon \to 0^+} \lim_{\rho \to +\infty} \int_{\varepsilon \leqslant |x| \leqslant \rho} K(x) e^{-ixy} dx. \tag{1.4}$$

它是处处存在的，并且

$$\hat{K}(y) = \frac{(-i)^n}{c(k, n)} P\left(\frac{y}{|y|}\right)(y \neq 0), \ \hat{K}(0) = 0, \tag{1.5}$$

其中

$$c(k, n) = 2^k \pi^{\frac{k}{2}} \frac{\Gamma\left(\frac{k+n}{2}\right)}{\Gamma\left(\frac{n}{2}\right)}. \tag{1.6}$$

共轭级数(1.3)的 γ 阶 Riesz 球形平均是

$$\widetilde{S}_R^\gamma(f;x) = \sum_{|m|<R} a_m \hat{K}(m) \left(1 - \frac{|m|^2}{R^2}\right)^\gamma e^{imx}. \tag{1.7}$$

代入 a_m 的表达式(1.2)就得

$$\widetilde{S}_R^\gamma(f;x) = (2\pi)^{-k} \int_Q f(x-y) \widetilde{D}_R^\gamma(y) dy, \tag{1.8}$$

其中共轭 Dirichlet 核

$$\begin{aligned} \widetilde{D}_R^\gamma(y) &= \sum_{|m|<R} \hat{K}(m) \left(1 - \frac{|m|^2}{R^2}\right)^\gamma e^{imy} \\ &= \frac{(-i)^n}{c(k,n)} \sum_{0<|m|<R} P\left(\frac{m}{|m|}\right) \left(1 - \frac{|m|^2}{R^2}\right)^\gamma e^{imy}. \end{aligned} \tag{1.9}$$

核 $K(x)$ 生成的周期核定义为

$$K^*(x) = \sum_{m \in \mathbf{Z}^k} (K(x - 2\pi m) - I_m), \tag{1.10}$$

其中

$$I_m = (2\pi)^{-k} \int_{Q+2\pi m} K(y) dy. \tag{1.11}$$

$K^*(x)$在集$\{2\pi m \mid m \in \mathbf{Z}^k\}$之外连续. f 关于核 K 的共轭函数定义为

$$f^*(x) = \lim_{\varepsilon \to 0^+} (2\pi)^{-k} \int_{Q \setminus B(0,\varepsilon)} K^*(y) f(x-y) dy. \tag{1.12}$$

我们写

$$f_\varepsilon^*(x) = (2\pi)^{-k} \int_{Q \setminus B(0,\varepsilon)} K^*(y) f(x-y) dy. \tag{1.13}$$

(2) 共轭积分与 Hilbert 变换. 设 $g \in L(E_k)$. 形式上定义 g 关于核 K 的共轭 Fourier 积分为

$$\int_{E_k}\hat{K}(y)\hat{g}(y)\mathrm{e}^{\mathrm{i}xy}\mathrm{d}y,$$

其中 $\hat{g}$ 是 g 的 Fourier 变换，即

$$\hat{g}(y)=(2\pi)^{-k}\int_{E_k}g(x)\mathrm{e}^{-\mathrm{i}yx}\mathrm{d}x.$$

g 的共轭积分的 γ 阶 Riesz 球形平均是

$$\tilde{\sigma}_R^{\gamma}(g;x)=\int_{|y|<R}\hat{K}(y)\hat{g}(y)\left(1-\frac{|y|^2}{R^2}\right)^{\gamma}\mathrm{e}^{\mathrm{i}xy}\mathrm{d}y. \tag{1.14}$$

写成卷积形式即

$$\tilde{\sigma}_R^{\gamma}(g;x)=(2\pi)^{-k}\int_{E_k}g(x-y)\widetilde{H}_R^{\gamma}(y)\mathrm{d}y, \tag{1.15}$$

其中核

$$\begin{aligned}\widetilde{H}_R^{\gamma}(y)&=\int_{|u|<R}\hat{K}(u)\left(1-\frac{|u|^2}{R^2}\right)^{\gamma}\mathrm{e}^{\mathrm{i}uy}\mathrm{d}u\\&=\frac{(-\mathrm{i})^n}{c(k,n)}\int_{|u|<R}P\left(\frac{u}{|u|}\right)\left(1-\frac{|u|^2}{R^2}\right)^{\gamma}\mathrm{e}^{\mathrm{i}uy}\mathrm{d}u.\end{aligned} \tag{1.16}$$

g 关于核 K 的 Hilbert 变换是

$$\hat{g}(x)=\lim_{\varepsilon\to 0^+}(2\pi)^{-k}\int_{|y|>\varepsilon}K(y)g(x-y)\mathrm{d}y. \tag{1.17}$$

我们写

$$\hat{g}_{\varepsilon}(x)=(2\pi)^{-k}\int_{|y|>\varepsilon}K(y)g(x-y)\mathrm{d}y. \tag{1.18}$$

(3) 本文的目的. 关于 $\widetilde{S}_R^{\gamma}(f;\ x)$的收敛性，当 $\gamma>\frac{k-1}{2}$时，Shapiro 得到了一些结果，可参阅[2]. 而当 $\gamma=\frac{k-1}{2}$(临界指数)时，Lippman[3]证明了 $\widetilde{S}_R^{\frac{k-1}{2}}(f;\ x)$的收敛不具有局部化性质，即存在于点 x 的邻域内消失的函数 $f\in L(Q)$，使

$$\overline{\lim_{R\to+\infty}}\left|\widetilde{S}_R^{\frac{k-1}{2}}(f;\ x)\right|=+\infty.$$

但对于$(c,\ 1)$平均$\frac{1}{R}\int_0^R\widetilde{S}_u^{\frac{k-1}{2}}(f;x)\mathrm{d}u$，我们已证明局部化定理依然成立①，此量 a. e. 收敛到 $f^*(x)$. 现在我们要进一步考虑强性求和的问

① 见作者：多重共轭 Fourier 级数的$(c,\ 1)$平均，科学通报，1984，24：1 473—1 477.

题. 我们先求得局部化的结果，即

定理 1 设 $f\in L(Q)$，f 在 $x+B(0,\delta)$上消失($\delta>0$)，$q>0$. 那么

$$\lim_{R\to+\infty}\frac{1}{R}\int_0^R|\widetilde{S}_u^{\frac{k-1}{2}}(f;x)-f^*_{\frac{1}{u}}(x)|^q\mathrm{d}u=0. \tag{1.19}$$

然后我们求出一个关于 $\tilde{\sigma}_R^{\frac{k-1}{2}}(g;x)$定点强求和的定理，即

定理 2 设 $g\in L(E_k)$，在点 x 处

$$\int_0^t\left(\int_{|y|=1}g(x-\tau y)P(y)\mathrm{d}\sigma(y)\right)\mathrm{d}\tau=o(t),\quad t\to0. \tag{1.20}$$

又设 g 在 $x+B(0,\delta)$上 p 次可积，$\delta>0$，$p>1$，且

$$\int_0^t\left|\int_{|y|=1}g(x-\tau y)P(y)\mathrm{d}\sigma(y)\right|^p\mathrm{d}\tau=O(t),\quad t\to0. \tag{1.21}$$

那么，对于任意的 $q>0$，

$$\lim_{R\to+\infty}\frac{1}{R}\int_0^R|\tilde{\sigma}_u^{\frac{k-1}{2}}(g;x)-\hat{g}_{\frac{1}{u}}(x)|^q\mathrm{d}u=0. \tag{1.22}$$

周知，Chang[4]的主要结果是

定理 A 设 $f\in L\ln^+L(Q)$，$g=f\chi_Q$(χ_Q 为 Q 的特征函数)，则在 Q 内部的闭集上

$$\lim_{R\to+\infty}(\widetilde{S}_R^{\frac{k-1}{2}}(f;x)-\tilde{\sigma}_R^{\frac{k-1}{2}}(g;x))=C(x) \tag{1.23}$$

一致成立，其中

$$C(x)=(2\pi)^k\int_Q f(y)(K^*(x-y)-K(x-y))\mathrm{d}y. \tag{1.24}$$

根据定理 1 和定理 2，使用定理 A 就得到本文的主要结果. 即

定理 3 设 $f\in L(Q)$，f 在 $x+B(0,\delta)$上 p 次可积，$\delta>0$，$p>1$. 若在点 x 处 f 满足条件(1.20)和(1.21)，则对于任意的 $q>0$，

$$\lim_{R\to+\infty}\frac{1}{R}\int_0^R|\widetilde{S}_u^{\frac{k-1}{2}}(f;x)-f^*_{\frac{1}{u}}(x)|^q\mathrm{d}u=0. \tag{1.25}$$

§ 2. $\gamma>\frac{k-1}{2}$时，$\widetilde{S}_R^\gamma(f;x)$的估计式

我们证明定理 1 的途径是根据

$$\lim_{\beta\to0^+}\frac{1}{R}\int_0^R|\widetilde{S}_u^{\frac{k-1}{2}+\beta}(f;x)-f^*_{\frac{1}{u}}(x)|^q\mathrm{d}u$$
$$=\frac{1}{R}\int_0^R|\widetilde{S}_u^{\frac{k-1}{2}}(f;x)-f^*_{\frac{1}{u}}(x)|^q\mathrm{d}u,\quad q>0, \tag{2.1}$$

把问题转化为证明

$$\overline{\lim_{R\to+\infty}}\left(\sup_{\beta\in(0,1)}\frac{1}{R}\int_0^R|\widetilde{S}_u^{\frac{k-1}{2}+\beta}(f;x)-f_{\frac{1}{u}}^*(x)|^q\mathrm{d}u\right)=0. \tag{2.2}$$

所以下面设 $\gamma=\frac{k-1}{2}+\beta$，$\beta>0$，先求出 $\widetilde{S}_R^\gamma(f;x)$的渐近估计.

首先，像在[4]中那样写出核 $\widetilde{D}_R^\gamma(y)$的渐近展开式. 对于函数

$$P\left(\frac{x}{|x|}\right)\left(1-\frac{|x|^2}{R^2}\right)^\gamma\chi_{B(0,R)}(x),\qquad R>\pi^{-1},$$

使用 Poisson 球形求和公式$\left(这里用到 \gamma>\frac{k-1}{2}\right)$，得

$$\sum_{0<|m|<R}P\left(\frac{m}{|m|}\right)\left(1-\frac{|m|^2}{R^2}\right)^\gamma\mathrm{e}^{\mathrm{i}my}=\sum_{j=0}^{+\infty}\sum_{m\in S_j}\Phi(y+2\pi m), \tag{2.3}$$

其中

$$\Phi(y)=\int_{Ek}P\left(\frac{x}{|x|}\right)\left(1-\frac{|x|^2}{R^2}\right)^\gamma\chi_{B(0,R)}(x)\mathrm{e}^{\mathrm{i}xy}\mathrm{d}x, \tag{2.4}$$

$$S_j=\{u\in E_k\cap\mathbf{Z}^k\,|\,j\leqslant|u|<j+1\},\qquad j\in\mathbf{N}. \tag{2.5}$$

然后用[4]的引理

$$\int_{|x|<R}P\left(\frac{x}{|x|}\right)\left(1-\frac{|x|^2}{R^2}\right)^{\frac{k-1}{2}+s}\mathrm{e}^{\mathrm{i}xy}\mathrm{d}x=\mathrm{i}^nR^{n+k}E(k,s,n,R|y|), \tag{2.6}$$

其中复数 s 的实部大于$-\frac{k+1}{2}$，而 $E(k,\ s,\ n,\ z)$是 Chang 引入的变量 z 的整函数

$$E(k,s,n,z)=\frac{\pi^{\frac{k}{2}}}{2^n}\sum_{j=0}^{+\infty}\frac{(-1)^j\Gamma\left(j+\frac{k+n}{2}\right)\Gamma\left(\frac{k+1}{2}+s\right)}{j\,!\,\Gamma\left(j+\frac{k}{2}+n\right)\Gamma\left(j+k+\frac{n+1}{2}+s\right)}\left(\frac{z}{2}\right)^{2j}. \tag{2.7}$$

于是(1.9)(2.3)(2.4)和(2.6)得

$$\widetilde{D}_R^{\frac{k-1}{2}+\beta}(y)=\frac{1}{c(k,n)}R^{k+n}\sum_{j=0}^{+\infty}\sum_{m\in S_j}P(y+2\pi m)E(k,\beta,n,R|y+2\pi m|),$$
$$(\beta>0,R>\pi^{-1}). \tag{2.8}$$

Chang[4]得到当 $u\geqslant1$ 时，

$$E(k,\ s,\ n,\ u)=\frac{c(k,\ n)}{u^{k+n}}+\frac{R^{(1)}(k,\ n,\ s,\ u)}{u^{k+n+1}}+\frac{R^{(2)}(k,\ n,\ s,\ u)}{u^{k+n+1+s}}+$$

$$A(k,\ s)\frac{\cos\left(u-\frac{\pi(k+n+s)}{2}\right)}{u^{k+n+s}},\tag{2.9}$$

其中 $c(k,\ n)$ 如(1.6)，$R^{(1)}$，$R^{(2)}$ 当 $\operatorname{Re} s\in\left[-\frac{k+1}{2}+h,\ G\right](0<h<G<+\infty)$ 时有界：

$$|R^{(1)}|+|R^{(2)}|\leqslant A_{k,n}(h,\ G)\mathrm{e}^{\frac{3\pi}{2}|\operatorname{Im} s|},\tag{2.10}$$

$A(k,\ s)$ 是常数：

$$A(k,\ s)=\pi^{\frac{k-1}{2}}2^{k+s}\Gamma\left(\frac{k+1}{2}+s\right).\tag{2.11}$$

现设 $f\in L(Q)$，$0<\delta<1$，f 在 $x+B(0,\ \delta)$ 上消失．那么对于 $R\geqslant\delta^{-1}>1$，当 $y\in Q$ 且 $|y|>\delta\geqslant R^{-1}$ 时，由于

$$R|y|>1,\ R|y+2\pi m|>1,$$

可将(2.9)代入(2.8)，于是得到

$$\begin{aligned}\widetilde{D}_R^{\gamma}(y)=&\frac{1}{c(k,n)}R^{k+n}\sum_{j=0}^{+\infty}\sum_{m\in S_j}P(y+2\pi m)\cdot\left\{\frac{c(k,n)}{R(|y+2\pi m|)^{k+n}}+\right.\\&\frac{R^{(1)}(k,n,\beta,R|y+2\pi m|)}{(R|y+2\pi m|)^{k+n+1}}+\frac{R^{(2)}(k,n,\beta,R|y+2\pi m|)}{(R|y+2\pi m|)^{k+n+1+\beta}}+\\&\left.\frac{A(k,\beta)\cos\left(R|y+2\pi m|-\frac{\pi(k+n+\beta)}{2}\right)}{(R|y+2\pi m|)^{k+n+\beta}}\right\}\\=&K^*(y)+\frac{1}{R}\left\{\frac{1}{c(k,n)}\sum_{j=0}^{+\infty}\sum_{m\in S_j}\left(\frac{R^{(1)}(k,n,\beta,R|y+2\pi m|)}{|y+2\pi m|^{k+n+1}}+\right.\right.\\&\left.\left.\frac{R^{(2)}(k,n,\beta,R|y+2\pi m|)}{R^{\beta}|y+2\pi m|^{k+n+1+\beta}}\right)\right\}+\frac{A(k,\beta)}{c(k,n)R^{\beta}}\times\\&\sum_{j=0}^{+\infty}\sum_{m\in S_j}\frac{P(y+2\pi m)\cos\left(R|y+2\pi m|-\frac{\pi}{2}(k+n+\beta)\right)}{|y+2\pi m|^{k+n+\beta}}.\end{aligned}\tag{2.12}$$

这里我们使用了等式 $K^*(y)=\sum_{j=0}^{+\infty}\sum_{m\in S_j}\frac{P(y+2\pi m)}{|y+2\pi m|^{k+n}}$．把右端第二项记作 $\frac{1}{R}M_R^{\beta}(y)$，并让 $\beta\in(0,\ 1)$，则据(2.10)，

$$|M_R^{\beta}(y)|\leqslant M,\quad M\text{ 只与 }k,\ n\text{ 有关}.\tag{2.13}$$

把(2.12)右端第三项记为 $\Phi_R^{\beta}(y)$．得

$$\widetilde{D}_R^{\gamma}(y)=K^*(y)+\frac{1}{R}M_R^{\beta}(y)+\Phi_R^{\beta}(y). \tag{2.14}$$

从(1.12)～(2.14)得，当 $R\geqslant\delta^{-1}$ 时，

$$\widetilde{S}_R^{\gamma}(f;\ x)-f_{\frac{1}{R}}^*(x)=\widetilde{S}_R^{\frac{k-1}{2}+\beta}(f;\ x)-f^*(x)$$

$$=(2\pi)^{-k}\int_Q f(x-y)\Phi_R^{\beta}(y)\mathrm{d}y+\frac{1}{R(2\pi)^k}\int_Q f(x-y)M_R^{\beta}(y)\mathrm{d}y. \tag{2.15}$$

显然第二项被控制于 $M\|f\|_{L(Q)}\cdot\frac{1}{R}=o(1)(R\to+\infty)$. 所以当 $\beta\in(0,1)$ 时一致地有

$$\widetilde{S}_R^{\frac{k-1}{2}+\beta}(f;x)-f_{\frac{1}{R}}^*(x)=(2\pi)^{-k}\int_0 f(x-y)\Phi_R^{\beta}(y)\mathrm{d}y+o(1), \tag{2.16}$$

其中

$$\Phi_R^{\beta}(y)=\frac{A(k,\beta)}{c(k,n)}\frac{1}{R^{\beta}}\sum_{j=0}^{+\infty}\sum_{m\in S_j}\frac{P(y+2\pi m)}{|y+2\pi m|^{k+n+\beta}}\cos\left(R|y+2\pi m|-\frac{\pi}{2}(k+n+\beta)\right). \tag{2.17}$$

写

$$I^{\beta}(R)=\int_Q f(x-y)\Phi_R^{\beta}(y)\mathrm{d}y \tag{2.18}$$

则关于 $\beta\in(0,1)$ 一致地有

$$\widetilde{S}_R^{\frac{k-1}{2}+\beta}(f;x)-f_{\frac{1}{R}}^*(x)=I^{\beta}(R)(2\pi)^{-k}+o(1),\quad R\to+\infty. \tag{2.19}$$

§3. 定理1的证明

由(2.19)知，只需证(当然无妨认为 $\delta<1$)

$$\overline{\lim_{R\to+\infty}}\sup_{\beta\in(0,1)}\frac{1}{R}\int_{\delta^{-1}}^{R}|I^{\beta}(u)|^q\mathrm{d}u=0. \tag{3.1}$$

引入以下记号：$m\in\mathbf{Z}^k$，

$$S_m(\rho)=\{u\in Q+2\pi m\,|\,|u|=\rho\},\quad \rho>0, \tag{3.2}$$

$$\rho_m^-=\inf_{y\in Q}\{|y+2\pi m|\},\ \rho_m^+=\sup_{y\in Q}\{|y+2\pi m|\}, \tag{3.3}$$

$$\varphi_m(\rho)=\int_{S_m(\rho)}f(x-y)P\left(\frac{y}{|y|}\right)\mathrm{d}\sigma(y),\varphi_m(\rho)=0,\quad \rho\leqslant0, \tag{3.4}$$

$$F_m^\beta(u)=\int_Q f(x-y)\frac{P(y+2\pi m)\cos\left(u|y+2\pi m|-\frac{\pi}{2}(k+n+\beta)\right)}{u^\beta|y+2\pi m|^{k+n+\beta}}\mathrm{d}y$$

$$=\int_{\rho_m^-}^{\rho_m^+}\varphi_m(\rho)\frac{\cos\left(u_\rho-\frac{\pi}{2}(k+n+\beta)\right)}{\rho^{k+\beta}u^\beta}\mathrm{d}\rho,\quad u>\delta^{-1}. \tag{3.5}$$

由(2.17)(2.18)知

$$I^\beta(u)=\frac{A(k,\beta)}{c(k,n)}\sum_{j=0}^{+\infty}\sum_{m\in S_j}F_m^\beta(u),\quad u>\delta^{-1}. \tag{3.6}$$

引理 1 对于 $m\in\mathbf{Z}^k$,

$$\max\{\pi|m|,\ 2\pi|m|-\sqrt{k}\pi\}\leqslant\rho_m^-<\rho_m^+\leqslant 2\pi|m|+\sqrt{k}\pi. \tag{3.7}$$

引理 2 设 $M=\max|P(y)|$,则

$$\int_{\rho_m^-}^{\rho_m^+}|\varphi_m(\rho)|\mathrm{d}\rho=\int_{-\infty}^{+\infty}|\varphi_m(\rho)|\mathrm{d}\rho\leqslant M\int_Q|f(y)|\mathrm{d}y. \tag{3.8}$$

引理 3 $\forall\varepsilon>0$,$\exists\eta>0$,只要 $0<\rho_2-\rho_1<\eta$,就有

$$\int_{\rho_1}^{\rho_2}|\varphi_m(\rho)|\mathrm{d}\rho<\varepsilon,\quad\forall m\in\mathbf{Z}^k. \tag{3.9}$$

引理 4 对于 $m\in\mathbf{Z}^k$,

$$\lim_{u\to+\infty}\sup_{\beta\in(0,1)}|F_m^\beta(u)|=0. \tag{3.10}$$

引理 1 和引理 2 都是明显的. 此处说明,当 $\rho\notin[\rho_m^-,\ \rho_m^+]$ 时,$S_m(\rho)=\varnothing$,从而 $\varphi_m(\rho)=0$.

引理 3 之成立是根据 $S_m(\rho)$ 的 $k-1$ 维测度关于 m 和 ρ 的一致有界性及 f 在周期上积分的绝对连续性.

我们只对 $m=0$ 使用(3.10),所以我们写出 $m=0$ 时它的证明,$m\neq 0$ 时的证明完全类似.

由于 f 在 $B(x,\delta)$ 上消失,所以当 $0\leqslant\rho<\delta$ 时,$\varphi_0(\rho)=0$,故

$$F_0^\beta(u)=\int_\delta^{\rho_0^+}\varphi_0(\rho)\frac{\cos u\rho}{\rho^{k+\beta}}\mathrm{d}\rho\cdot\frac{\cos\left(\frac{\pi}{2}(k+n+\beta)\right)}{u^\beta}+$$

$$\int_\delta^{\rho_0^+}\varphi_0(\rho)\frac{\sin u\rho}{\rho^{k+\beta}}\mathrm{d}\rho\frac{\sin\frac{\pi}{2}(k+n+\beta)}{u^\beta},\quad u>\delta^{-1}>1.$$

分部积分算出

$$\int_\delta^{\rho_0^+}\varphi_0(\rho)\cos u\rho\frac{\mathrm{d}\rho}{\rho^{k+\beta}}=\left(\frac{1}{\rho_0^+}\right)^{k+\beta}\cdot\int_\delta^{\rho_0^+}\varphi_0(\rho)\cos u\rho\mathrm{d}\rho+$$

$$(k+\beta)\int_{\delta}^{\rho_0^+}\frac{1}{\rho^{k+\beta+1}}\left(\int_{\delta}^{\rho}\varphi_0(t)\cos ut\,\mathrm{d}t\right)\mathrm{d}\rho.$$

$$\int_{\delta}^{\rho_0^+}\varphi_0(\rho)\sin u\rho\frac{\mathrm{d}\rho}{\rho^{k+\beta}}=\left(\frac{1}{\rho_0^+}\right)^{k+\beta}\cdot\int_{\delta}^{\rho_0^+}\varphi_0(\rho)\sin u\rho\mathrm{d}\rho+$$

$$(k+\beta)\int_{\delta}^{\rho_0^+}\frac{1}{\rho^{k+\beta+1}}\left(\int_{\delta}^{\rho}\varphi_0(t)\sin ut\,\mathrm{d}t\right)\mathrm{d}\rho.$$

得到

$$\sup_{\beta\in(0,1)}|F_0^\beta(u)|\leqslant M\cdot\left(\left|\int_{\delta}^{\rho_0^+}\varphi_0(\rho)\cos u\rho\mathrm{d}\rho\right|+\left|\int_{\delta}^{\rho_0^+}\varphi_0(\rho)\sin u\rho\mathrm{d}\rho\right|\right)+$$

$$M\cdot\left(\int_{\delta}^{\rho_0^+}\left|\int_{\delta}^{\rho}\varphi_0(t)\cos ut\,\mathrm{d}t\right|\mathrm{d}\rho+\int_{\delta}^{\rho_0^+}\left|\int_{\delta}^{\rho}\varphi_0(t)\sin ut\,\mathrm{d}t\right|\mathrm{d}\rho\right),$$

$$(u>\delta^{-1}).$$

故由 Riemann-Lebesgue 引理及积分号下取极限得

$$\lim_{u\to+\infty}\sup_{\beta\in(0,1)}|F_0^\beta(u)|=0.$$

写

$$J^\beta(u)=\sum_{j=1}^{+\infty}\sum_{m\in S_j}F_m^\beta(u),\quad u>\delta^{-1}.\tag{3.11}$$

由于 $A(k,\beta)$，当 $\beta\in(0,1)$ 时有界，故由(3.6)(3.10)知，欲证(3.1)，只需证

$$\overline{\lim_{R\to+\infty}}\sup_{\beta\in(0,1)}\frac{1}{R}\int_{\delta^{-1}}^{R}|J^\beta(u)|^q\mathrm{d}u=0,\quad q>0.\tag{3.12}$$

且由 Hölder 不等式可知，只需对于 $q=2^\lambda(\lambda\in\mathbf{N})$ 这样的数来进行证明. 再因

$$F_m^\beta(u)=\int_{\rho_m^-}^{\rho_m^+}\frac{\varphi_m(\rho)}{\rho^{k+\beta}}\frac{\cos u\rho}{u^\beta}\mathrm{d}\rho\cdot\cos\frac{\pi}{2}(k+n+\beta)+$$

$$\int_{\rho_m^-}^{\rho_m^+}\frac{\varphi_m(\rho)}{\rho^{k+\beta}}\frac{\sin u\rho}{u^\beta}\mathrm{d}\rho\cdot\sin\frac{\pi}{2}(k+n+\beta),\tag{3.13}$$

故只需证明

引理 5 设 $q=2^\lambda$，$\lambda\in\mathbf{N}$，则

$$\overline{\lim_{R\to+\infty}}\sup_{\beta\in(0,1)}\frac{1}{R}\int_{\delta^{-1}}^{R}\left(\sum_{j=1}^{+\infty}\sum_{m\in S_j}\int_{\rho_m^-}^{\rho_m^+}\frac{\varphi_m(\rho)}{\rho^{k+\beta}}\frac{\cos u\rho}{u^\beta}\mathrm{d}\rho\right)^q\mathrm{d}u=0,\tag{3.14}$$

$$\overline{\lim_{R\to+\infty}}\sup_{\beta\in(0,1)}\frac{1}{R}\int_{\delta^{-1}}^{R}\left(\sum_{j=1}^{+\infty}\sum_{m\in S_j}\int_{\rho_m^-}^{\rho_m^+}\frac{\varphi_m(\rho)}{\rho^{k+\beta}}\frac{\sin u\rho}{u^\beta}\mathrm{d}\rho\right)^q\mathrm{d}u=0.\tag{3.15}$$

两式的证明是一样的，下面证明(3.14). 写

$$h^{\beta}(R)=\int_{\delta^{-1}}^{R}\left(\sum_{j=1}^{+\infty}\sum_{m\in S_j}\int_{\rho_m^-}^{\rho_m^+}\frac{\varphi_m(\rho)}{\rho^{k+\beta}}\frac{\cos u\rho}{u^{\beta}}\mathrm{d}\rho\right)^q\mathrm{d}u$$

$$=\int_{\delta^{-1}}^{R}\left\{\sum_{j_1=1}^{+\infty}\cdots\sum_{j_q=1}^{+\infty}\sum_{m_1\in S_{j_1}}\cdots\sum_{m_q\in S_{j_q}}\int_{\rho_{m_1}^-}^{\rho_{m_1}^+}\frac{\varphi_{m_1}(\rho_1)}{\rho_1^{k+\beta}}\cdots\right.$$

$$\left.\int_{\rho_{m_q}^-}^{\rho_{m_q}^+}\frac{\varphi_{m_q}(\rho_q)}{\rho_q^{k+\beta}}\cdot\frac{1}{u^{q\beta}}\cos u\rho_1\cdots\cos u\rho_q\mathrm{d}\rho_1\mathrm{d}\rho_2\cdots\mathrm{d}\rho_q\right\}\mathrm{d}u$$

$$=\sum_{j_1=1}^{+\infty}\cdots\sum_{j_q=1}^{+\infty}\sum_{m_1\in S_{j_1}}\cdots\sum_{m_q\in S_{j_q}}\int_{\rho_{m_1}^-}^{\rho_{m_1}^+}\int_{\rho_{m_2}^-}^{\rho_{m_2}^+}\cdots\int_{\rho_{m_q}^-}^{\rho_{m_q}^+}\left\{\frac{\varphi_{m_1}(\rho_1)\varphi_{m_2}(\rho_2)\cdots\varphi_{m_q}(\rho_q)}{(\rho_1\rho_2\cdots\rho_q)^{k+\beta}}\cdot\right.$$

$$\left.\int_{\delta^{-1}}^{R}\frac{\cos u\rho_1\cos u\rho_2\cdots\cos u\rho_q}{u^{q\beta}}\mathrm{d}u\right\}\mathrm{d}\rho_1\mathrm{d}\rho_2\cdots\mathrm{d}\rho_q. \qquad (3.16)$$

把坐标为1或−1的 q 维向量排成一列 $\{e_l\}_{l=1}^{2q}$，其中

$$\boldsymbol{e}_l=(e_{l1},\ e_{l2},\ \cdots,\ e_{lq}),\ e_{lj}=1\text{ 或 }-1,\qquad j=1,2,\cdots,q.$$

又写 $\boldsymbol{e}_l\cdot\boldsymbol{\rho}=e_{l1}\rho_1+e_{l2}\rho_2+\cdots+e_{lq}\rho_q$，则

$$\cos u\rho_1\cos u\rho_2\cdots\cos u\rho_q=\frac{1}{2^q}\sum_{l=1}^{2^q}\cos u(\boldsymbol{e}_l\cdot\boldsymbol{\rho}).$$

用积分中值定理得

$$\int_{\delta^{-1}}^{R}\frac{\cos u(\boldsymbol{e}_l\cdot\boldsymbol{\rho})}{u^{q\beta}}\mathrm{d}u=\delta^{q\beta}\cdot\int_{\delta^{-1}}^{\xi}\cos u(\boldsymbol{e}_l\cdot\boldsymbol{\rho})\mathrm{d}u$$

$$=\frac{\delta^{q\beta}}{\boldsymbol{e}_l\cdot\boldsymbol{\rho}}(\sin\xi(\boldsymbol{e}_l\cdot\boldsymbol{\rho})-\sin\delta^{-1}(\boldsymbol{e}_l\cdot\boldsymbol{\rho})),\quad \xi\in[\delta^{-1},R].$$

令

$$\sup_{\beta\in(0,1)}\left|\int_{\delta^{-1}}^{R}\frac{\cos u(\boldsymbol{e}_l\cdot\boldsymbol{\rho})}{u^{q\beta}}\mathrm{d}u\right|=\frac{1}{|\boldsymbol{e}_l\cdot\boldsymbol{\rho}|}M_R(\boldsymbol{e}_l\cdot\boldsymbol{\rho}), \qquad (3.17)$$

则见

$$0\leqslant M_R(\boldsymbol{e}_l\cdot\boldsymbol{\rho})\leqslant M(\forall R,\ l), \qquad (3.18)$$

$$0\leqslant M_R(\boldsymbol{e}_l\cdot\boldsymbol{\rho})\leqslant M\cdot R|\boldsymbol{e}_l\cdot\boldsymbol{\rho}|,\qquad \forall R,\ l. \qquad (3.19)$$

定义

$$\tau_{m_1m_2\cdots m_q}^{l}(R)=$$

$$\int_{\rho_{m_1}^-}^{\rho_{m_1}^+}\int_{\rho_{m_2}^-}^{\rho_{m_2}^+}\cdots\int_{\rho_{m_q}^-}^{\rho_{m_q}^+}|\varphi_{m_1}(\rho_1)\varphi_{m_2}(\rho_2)\cdots\varphi_{m_q}(\rho_q)|\frac{M_R(\boldsymbol{e}_l\cdot\boldsymbol{\rho})}{|\boldsymbol{e}_l\cdot\boldsymbol{\rho}|}\mathrm{d}\rho_1\mathrm{d}\rho_2\cdots\mathrm{d}\rho_q.$$

$$(3.20)$$

那么，由(3.7)得知

$$\sup_{\beta\in(0,1)}\left|h^{\beta}(R)\right|$$
$$\leqslant \sum_{j_1,j_2,\cdots,j_q=1}^{+\infty}\ \sum_{m_1\in S_{j_1},m_2\in S_{j_2},\cdots,m_q\in S_{j_q}} \frac{1}{(|m_1||m_2|\cdots|m_q|)^k}\sum_{l=1}^{2^q}\tau^l_{m_1m_2\cdots m_q}(R). \tag{3.21}$$

从(3.16)和(3.21)可知，只需证明对于任意一个 $l(l=1,2,\cdots,2^q)$，如下定义的量

$$\sigma(R)=\sum_{j_1,j_2,\cdots,j_q=1}^{+\infty}\ \sum_{m_1\in S_{j_1},m_2\in S_{j_2},\cdots,m_q\in S_{j_q}} \frac{\tau^l_{m_1m_2\cdots m_q}(R)}{(|m_1||m_2|\cdots|m_q|)^k}, \tag{3.22}$$

当 $R\to+\infty$ 时是 $o(R)$ 即可.

无妨认为

$$\boldsymbol{e}_l=(\overbrace{1,\ \cdots,\ 1}^{\alpha个},\ \overbrace{-1,\ \cdots,\ -1}^{q-\alpha个}).$$

先证明

(1) 对于 $(m_1,\ m_2,\ \cdots,\ m_q)\in S_{j_1}\times S_{j_2}\times\cdots\times S_{j_q}$ 一致有

$$\tau^l_{m_1,m_2,\cdots,m_q}=o(R),\quad R\to+\infty.$$

(2) 当 $|j_1+j_2+\cdots+j_\alpha-(j_{\alpha+1}+j_{\alpha+2}+\cdots+j_q)|>4\pi q\sqrt{k}$ 时，对于 $(m_1,\ m_2,\ \cdots,\ m_q)\in S_{j_1}\times S_{j_2}\times\cdots\times S_{j_q}$ 有

$$\tau^l_{m_1,m_2,\cdots,m_q}(R)\leqslant C|j_1+j_2+\cdots+j_\alpha-(j_{\alpha+1}+j_{\alpha+2}+\cdots+j_q)|^{-1}.$$

做变量变换

$$\begin{pmatrix}\rho_1\\ \rho_2\\ \vdots\\ \rho_q\end{pmatrix}=\overset{\alpha列}{\begin{pmatrix}1 & -1\cdots-1 & 1\cdots1\\ 1 & & \\ & \ddots & 0\\ 0 & & \ddots\\ & & 1\end{pmatrix}}\begin{pmatrix}t_1\\ t_2\\ \vdots\\ t_q\end{pmatrix},$$

即

$$\begin{pmatrix}t_1\\ t_2\\ \vdots\\ t_q\end{pmatrix}=\overset{\alpha列}{\begin{pmatrix}1\cdots1 & -1 & \cdots-1\\ & 1 & \\ & \ddots & 0\\ 0 & & \ddots\\ & & 1\end{pmatrix}}\begin{pmatrix}\rho_1\\ \rho_2\\ \vdots\\ \rho_q\end{pmatrix},$$

先对 t_1 求积，再对 t_2，t_3，…，t_q 求积，得

$$\tau^l_{m_1m_2\cdots m_q}(R)=\int_{\rho^-_{m_2}}^{\rho^+_{m_2}}\cdots\int_{\rho^-_{m_q}}^{\rho^+_{m_q}}|\varphi_{m_2}(t_2)\cdots\varphi_{m_q}(t_q)|\cdot I_{m_1}(t_2,t_3,\cdots,t_q;R)\mathrm{d}t_2\mathrm{d}t_3\cdots\mathrm{d}t_q, \tag{3.23}$$

其中

$$I_{m_1}(t_2,t_3,\cdots,t_q;R)=\int_{-\infty}^{+\infty}|\varphi_{m_1}(t_1-(t_2+t_3+\cdots+t_\alpha)+(t_{\alpha+1}+\cdots+t_q))|\cdot\frac{M_R(t_1)}{|t_1|}\mathrm{d}t_1. \tag{3.24}$$

令 $s=-(t_2+t_3+\cdots+t_\alpha)+(t_{\alpha+1}+\cdots+t_q)$，则

$$I_{m_1}=\int_{-\infty}^{+\infty}|\varphi_{m_1}(t_1+s)|\frac{M_R(t_1)}{|t_1|}\mathrm{d}t_1.$$

根据引理 3，$\forall\varepsilon>0$，$\exists\eta>0$，只要 $0<\rho_2-\rho_1\leqslant2\eta$，就有

$$\int_{\rho_1}^{\rho_2}|\varphi_m(\rho)|\mathrm{d}\rho<\varepsilon,\quad\forall m\in\mathbf{Z}^k.$$

再用(3.18)(3.19)，得知当 $R>\frac{1}{\eta}$（且 $R>\frac{1}{\delta}$）时，

$$\begin{aligned}I_{m_1}(t_2,\cdots,t_q;R)&\leqslant\int_{|t_1|<\eta}|\varphi_{m_1}(t_1+s)|M\cdot R\mathrm{d}t_1+\int_{\eta<|t_1|<+\infty}|\varphi_{m_1}(t_1+s)|\cdot\frac{M}{\eta}\mathrm{d}t_1\\&\leqslant\int_{-\eta+s}^{\eta+s}|\varphi_{m_1}(t)|\mathrm{d}t\cdot MR+\frac{M}{\eta}\int_{-\infty}^{+\infty}|\varphi_{m_1}(t)|\mathrm{d}t\\&\leqslant MR\cdot\varepsilon+\frac{1}{\eta}M\|f\|_{L(Q)},\quad R>\frac{1}{\eta}+\frac{1}{\delta}.\end{aligned} \tag{3.25}$$

这里用了(3.8)．以(3.25)代入(3.23)，再用(3.8)就得

$$\tau^l_{m_1m_2\cdots m_q}(R)\leqslant M\cdot\|f\|^{q-1}_{L(Q)}\left(R\varepsilon+\frac{1}{\eta}\|f\|_{L(Q)}\right),$$

从而(1)的结论成立．

根据(3.7)，对于 $r=1,2,\cdots,q$，$\rho_r\in(\rho^-_{m_r},\ \rho^+_{m_r})$，$m_r\in S_{j_r}$ 有

$$\begin{aligned}2\pi j_r-\sqrt{k}\pi&\leqslant2\pi|m_r|-\sqrt{k}\pi\leqslant\rho^-_{m_r}<\rho_r<\rho^+_{m_r}\\&\leqslant2\pi|m_r|+\sqrt{k}\pi<2\pi(j_r+1)+\sqrt{k}\pi,\end{aligned}$$

$$\begin{aligned}2\pi(j_1+j_2+\cdots+j_\alpha)-\sqrt{k}\alpha\pi&<\rho_1+\rho_2+\cdots+\rho_\alpha\\&<2\pi(j_1+j_2+\cdots+j_\alpha)+(2+\sqrt{k})\pi\alpha,\end{aligned}$$

$$2\pi(j_{\alpha+1}+\cdots+j_q)-\sqrt{k}(q-\alpha)\pi<\rho_{\alpha+1}+\cdots+\rho_q$$
$$<2\pi(j_{\alpha+1}+\cdots+j_q)+(2+\sqrt{k})\pi(q-\alpha).$$

因此，在(2)的条件下有

$$\begin{aligned}&|\rho_1+\rho_2+\cdots+\rho_\alpha-(\rho_{\alpha+1}+\cdots+\rho_q)|\\ \geqslant&2\pi|j_1+j_2+\cdots+j_\alpha-(j_{\alpha+1}+\cdots+j_q)|-(2+\sqrt{k})q\pi\\ >&|j_1+j_2+\cdots+j_\alpha-(j_{\alpha+1}+\cdots+j_q)|.\end{aligned}\tag{3.26}$$

将(3.18)(3.26)代入(3.20)，就得到(2)的结论.

记 $h=[4\pi q\sqrt{k}]$. 把(3.22)的求和分成两部分，一部分是

$$\sum_{j_1,j_2,\cdots,j_q=1,\ |j_1+j_2+\cdots+j_\alpha-(j_{\alpha+1}+\cdots+j_q)|\leqslant h}^{+\infty},$$

记作 $\sigma_1(R)$，另一部分记为 $\sigma_2(R)$. 注意到 S_{j_r} 的定义，S_{j_r} 所含整点 m_r 的数目不超过 $C\cdot j_r^{k-1}$. 代入(1)的估计，得

$$\begin{aligned}\sigma_1(R)&=o(R)\cdot\sum_{|j_1+j_2+\cdots+j_\alpha-(j_{\alpha+1}+\cdots+j_q)|\leqslant h,j_1,j_2,\cdots,j_q\geqslant1}\frac{1}{j_1j_2\cdots j_q}\\&=o(R)\cdot\sum_{\mu=\alpha}^{+\infty}\sum_{\mu_1=\alpha-1}^{\mu-1}\cdots\sum_{\mu_{\alpha-1}=1}^{\mu_{\alpha-2}-1}\cdot\sum_{v\geqslant q-\alpha,\ |v-\mu|\leqslant h}\cdot\sum_{v_1=q-\alpha-1}^{v-1}\cdots\cdot\\&\quad\sum_{v_{q-\alpha-1}=1}^{v_{q-\alpha-2}-1}\left(\frac{1}{\mu-\mu_1}\cdot\frac{1}{\mu_1-\mu_2}\cdots\frac{1}{\mu_{\alpha-2}-\mu_{\alpha-1}}\cdot\frac{1}{\mu_{\alpha-1}}\right)\cdot\\&\quad\left(\frac{1}{v-v_1}\cdot\frac{1}{v_1-v_2}\cdots\frac{1}{v_{q-\alpha-2}-v_{q-\alpha-1}}\cdot\frac{1}{v_{q-\alpha-1}}\right).\end{aligned}\tag{3.27}$$

容易算出，当 $\alpha>1$ 时($\mu\geqslant\alpha$)，

$$\begin{aligned}&\sum_{\mu_j=\alpha-1}^{\mu-1}\cdots\sum_{\mu_{\alpha-1}=1}^{\mu_{\alpha-2}-1}\frac{1}{\mu-\mu_1}\cdot\frac{1}{\mu_1-\mu_2}\cdots\frac{1}{\mu_{\alpha-2}-\mu_{\alpha-1}}\cdot\frac{1}{\mu_{\alpha-1}}\\&\leqslant C\frac{\ln^{\alpha-1}(1+\mu)}{\mu}.\end{aligned}\tag{3.28}$$

于(3.27)中用(3.28)得

$$\begin{aligned}\sigma_1(R)&=o(R)\cdot\sum_{\mu=\alpha}^{+\infty}\left(\frac{\ln^{\alpha-1}(1+\mu)}{\mu}\cdot\sum_{v\geqslant q-\alpha,\ |v-\mu|\leqslant h}\frac{\ln^{q-\alpha-1}(1+v)}{v}\right)\\&=o(R)\cdot\sum_{\mu=1}^{+\infty}\frac{\ln^{q-2}(1+\mu)}{\mu^2}=o(R).\end{aligned}\tag{3.29}$$

另一方面，在 $\sigma_2(R)$ 的和式中代入(2)的估计得

$$\sigma_2(R)\leqslant C\sum_{j_1,j_2,\cdots,j_q\geqslant1,\ |j_1+j_2+\cdots+j_\alpha-(j_{\alpha+1}+\cdots+j_q)|>h}\frac{1}{j_1,j_2,\cdots,j_q}\cdot$$

$$\frac{1}{|j_1+j_2+\cdots+j_\alpha-(j_{\alpha+1}+\cdots+j_q)|}$$

$$=C\sum_{\mu=\alpha}^{+\infty}\sum_{u_1=\alpha-1}^{\mu-1}\cdots\sum_{\mu_{\alpha-1}=1}^{\mu_{\alpha-2}-1}\sum_{v\geqslant q-\alpha,\ |v-\mu|>h}\sum_{v_1=q-\alpha-1}^{v-1}\cdots\sum_{v_{q-\alpha-1}=1}^{v_{q-\alpha-2}-1}\cdot$$

$$\left(\frac{1}{\mu-\mu_1}\cdots\frac{1}{\mu_{\alpha-2}-\mu_{\alpha-1}}\frac{1}{\mu_{\alpha-1}}\right)\cdot\left(\frac{1}{v-v_1}\cdots\frac{1}{v_{q-\alpha-2}-v_{q-\alpha-1}}\frac{1}{v_{q-\alpha-1}}\right)\cdot\frac{1}{|\mu-v|}.$$

用(3.28)于上式得

$$\sigma_2(R)\leqslant C\sum_{\mu=\alpha}^{+\infty}\frac{\ln^{\alpha-1}(1+\mu)}{\mu}\cdot\sum_{v\geqslant q-\alpha,\ |v-\mu|>h}\frac{1}{|\mu-v|}\cdot\frac{\ln^{q-\alpha-1}(1+v)}{v}$$

$$\leqslant C\sum_{\mu=\alpha}^{+\infty}\frac{\ln^{\alpha-1}(1+\mu)}{\mu}\Big(\sum_{v>\mu+h}+\sum_{q-\alpha\leqslant v<\mu-h}\Big)\frac{1}{|\mu-v|}\cdot\frac{\ln^{q-\alpha-1}(1+v)}{v}.\tag{3.30}$$

由于(当 $\mu\geqslant 1$ 时)

$$\sum_{v>\mu+h}\frac{1}{|\mu-v|}\frac{\ln^{q-\alpha-1}(1+v)}{v}$$

$$\leqslant\Big(\sum_{v=\mu+1}^{2\mu}+\sum_{v=2\mu+1}^{+\infty}\Big)\frac{1}{v-\mu}\cdot\frac{\ln^{q-\alpha-1}(1+v)}{v}$$

$$\leqslant C\left\{\frac{\ln^{q-\alpha-1}(1+\mu)}{\mu}\sum_{v=\mu+1}^{2\mu}\frac{1}{v-\mu}+\sum_{v=\mu}^{+\infty}\frac{1}{v^2}\ln^{q-\alpha-1}(1+v)\right\}$$

$$\leqslant C\frac{1}{\mu}\ln^{q-\alpha}(1+\mu),$$

$$\sum_{q-\alpha\leqslant v<\mu-h}\frac{1}{|\mu-v|}\frac{\ln^{q-\alpha-1}(1+v)}{v}\leqslant\ln^{q-\alpha-1}(1+\mu)\cdot\sum_{v=1}^{\mu-1}\frac{1}{v(\mu-v)}$$

$$\leqslant C\frac{1}{\mu}\ln^{q-\alpha}(1+\mu),$$

所以

$$\sigma_2(R)\leqslant C\sum_{\mu=\alpha}^{+\infty}\frac{\ln^{q-1}(1+\mu)}{\mu^2}<+\infty.\tag{3.31}$$

把(3.29)(3.31)合起来，得 $\sigma(R)=o(R)(R\to+\infty)$. 于是证得(3.14). 引理 5 证毕. 从而完成了定理 1 的证明.

§ 4. 共轭 Fourier 积分的强求和

本节证明定理 2.

设 $g\in L(E_k)$. 无妨认为所考虑的点 $x=0$. 设 $p>1$，$\delta>0$，$f\in$

$L^p(B(0, \delta))$，且满足条件(1.20)(1.21)．定义

$$\psi(t) = \int_{|y|=1} g(-y)P(y)\mathrm{d}\sigma(y), \Psi(t) = \int_0^1 \psi(\tau)\mathrm{d}\tau. \tag{4.1}$$

条件(1.20)和(1.21)分别是

$$\Psi(t)=o(t), \quad t\to 0, \tag{4.2}$$

$$\int_0^t |\psi(\tau)|^p \mathrm{d}\tau = O(t), \quad 0 < t < \delta. \tag{4.3}$$

先对条件(4.3)做一点说明．当 $f\in L^p(B(0, \delta))$时，对于任何$p'\in[1, p)$都有 $f\in L^{p'}(B(0, \delta))$．当(4.3)对指数 $p>1$ 成立时，由 Hölder 不等式可知对于指数 $p'\in[1, p)$有

$$\int_0^t |\psi(\tau)|^{p'} \mathrm{d}\tau \leqslant \left(\int_0^t |\psi(\tau)|^{p'} \cdot \frac{p}{p'}\mathrm{d}\tau\right)^{\frac{p'}{p}} \cdot t^{1-\frac{p'}{p}}$$

$$=O(t^{\frac{p'}{p}}) \cdot t^{1-\frac{p'}{p}}=O(t).$$

由此可知，无妨认为条件中的 $p<2$，并且只要对于 p 的共轭数 $q=\frac{p}{p-1}$证明了(1.22)成立，事实上就证明了(1.22)对于一切 $q>0$ 成立．所以下面只对 $q=\frac{p}{p-1}(1<p<2)$来进行证明．

由(1.15)(1.18)得

$$\tilde{\sigma}_R^{\frac{k-1}{2}}(g;0) - \tilde{g}_{\frac{1}{R}}(0) = (2\pi)^{-k}\int_{|y|>R^{-1}} g(-y)(\widetilde{H}_R^{\frac{k-1}{2}}(y) - K(y))\mathrm{d}y +$$

$$(2\pi)^{-k}\int_{|y|<R^{-1}} g(-y)\widetilde{H}_R^{\frac{k-1}{2}}(y)\mathrm{d}y$$

$$=(2\pi)^{-k}(I_1(R)+I_2(R)). \tag{4.4}$$

由 $\widetilde{H}_R^{\frac{k-1}{2}}(y)$的表达式(1.16)，用 Chang[4]的引理 5 得

$$\widetilde{H}_R^{\frac{k-1}{2}}(y)=\frac{(-\mathrm{i})^n}{C(k, n)}\mathrm{i}^n R^{n+k}P(y)E(k, 0, n, R|y|)$$

$$=\frac{1}{C(k, n)}R^{k+n}P(y)E(k, 0, n, R|y|). \tag{4.5}$$

由此得

$$I_2(R) = \frac{C}{C(k,n)}R^{k+n}\int_0^{R^{-1}} \psi(t)t^{n+k-1}E(k,0,n,Rt)\mathrm{d}t$$

$$= \frac{C}{C(k,n)}R^{k+n}\left\{\Psi(t)t^{k+n-1}E(k,0,n,Rt)\Big|_0^{R^{-1}} - \right.$$

$$\int_0^{R^{-1}} \Psi(t) \frac{\mathrm{d}}{\mathrm{d}t}(t^{k+n-1}E(k,0,n,Rt))\mathrm{d}t\Big\}$$

$$= o(1) - \frac{CR^{k+n}}{C(k,n)}\int_0^{R^{-1}} \Psi(t)(n+k-1)t^{k+n-2}E(k,0,n,Rt)\mathrm{d}t -$$

$$\frac{CR^{k+n}}{C(k,n)}\int_0^{R^{-1}} \Psi(t)t^{k+n-1}\left(\frac{\mathrm{d}}{\mathrm{d}t}E(k,0,n,Rt)\right)\mathrm{d}t. \tag{4.6}$$

使用公式(见[4]p. 102，Sublemma 14.2)

$$\frac{\mathrm{d}}{\mathrm{d}z}E(k,\ s,\ p,\ z) = -\frac{z}{2\pi}E(k+2,\ s-1,\ p,\ z), \tag{4.7}$$

由于$|Rt|<1$时$E(k,\ 0,\ n,\ Rt)$和$E(k+2,\ -1,\ n,\ Rt)$都有界，再利用条件(4.2)就得到

$$I_2(R) = o(1), \quad R\to+\infty. \tag{4.8}$$

将(4.5)代入$I_1(R)$的积分式中，分部积分得

$$I_1(R) = \int_{R^{-1}}^{+\infty}\left(\frac{R^{k+n}t^n}{C(k,n)}E(k,0,n,Rt) - \frac{1}{t^k}\right)\int_{|y|=t} g(-y)P\left(\frac{y}{|y|}\right)\mathrm{d}\sigma(y)\mathrm{d}t$$

$$= C\int_{R^{-1}}^{+\infty}\left(\frac{R^{k+n}t^{k+n-1}}{C(k,n)}E(k,0,n,Rt) - \frac{1}{t}\right)\psi(t)\mathrm{d}t$$

$$= C\Psi(t)\left(\frac{R^{k+n}t^{k+n-1}}{C(k,n)}E(k,0,n,Rt) - \frac{1}{t}\right)\Bigg|_{R^{-1}}^{+\infty} -$$

$$C\int_{R^{-1}}^{+\infty}\Psi(t)\frac{\mathrm{d}}{\mathrm{d}t}\left(\frac{R^{k+n}t^{k+n-1}}{C(k,n)}E(k,0,n,Rt) - \frac{1}{t}\right)\mathrm{d}t. \tag{4.9}$$

从展开式(2.9)、$|\psi(t)|\leqslant M<+\infty$及条件(4.2)可知(4.9)右端第一项是$o(1)(R\to+\infty)$. 为估计第二项，用公式(4.7)并代入(2.9)，得

$$\frac{\mathrm{d}}{\mathrm{d}t}\left(\frac{R^{k+n}t^{k+n-1}}{C(k,\ n)}E(k,\ 0,\ n,\ Rt) - \frac{1}{t}\right)$$

$$= \frac{R^{k+n}}{C(k,\ n)}\left(-\frac{R^2t^{k+n}}{2\pi}E(k+2,\ -1,\ n,\ Rt) +\right.$$

$$\left.(k+n-1)t^{k+n-2}E(k,\ 0,\ n,\ Rt)\right) + \frac{1}{t^2}$$

$$= \frac{R^{k+n}}{C(k,\ n)}\Bigg\{-\frac{R^2t^{k+n}}{2\pi}\Bigg[\frac{C(k+2,\ n)}{(Rt)^{n+k+2}} + \frac{R^{(1)}(k+2,\ n,\ -1,\ Rt)}{(Rt)^{k+n+3}} +$$

$$\frac{R^{(2)}(k+2,\ n,\ -1,\ Rt)}{(Rt)^{k+n+2}} + \frac{A(k+2,\ -1)\cos\left(Rt - \frac{\pi}{2}(k+n+1)\right)}{(Rt)^{k+n+1}}\Bigg] +$$

$$(k+n-1)t^{k+n-2}\Bigg[\frac{C(k,\ n)}{(Rt)^{k+n}} + \frac{R^{(1)}(k,\ n,\ 0,\ Rt) + R^{(2)}(k,\ n,\ 0,\ Rt)}{(Rt)^{k+n+1}} +$$

$$\left.\left.\frac{A(k,\ 0)\cos\left(Rt-\frac{\pi}{2}(k+n)\right)}{(Rt)^{k+n}}\right]\right\}+\frac{1}{t^2}$$

$$=\left[-\frac{C(k+2,\ n)}{2\pi C(k,\ n)}+(k+n-1)+1\right]\frac{1}{t^2}+\frac{-R^{(1)}(k+2,\ n,\ -1,\ Rt)}{2\pi C(k,\ n)Rt^3}+$$

$$\frac{-R^{(2)}(k+2,\ n,\ -1,\ Rt)}{2\pi C(k,\ n)t^2}+$$

$$\frac{-A(k+2,\ -1)\sin\left(Rt-\frac{\pi}{2}(k+n)\right)}{2\pi C(k,\ n)t}R+$$

$$\frac{(k+n-1)[R^{(1)}(k,\ n,\ 0,\ Rt)+R^{(2)}(k,\ n,\ 0,\ Rt)]}{C(k,\ n)Rt^3}+$$

$$\frac{(k+n-1)A(k,\ 0)\cos\left(Rt-\frac{\pi}{2}(k+n)\right)}{C(k,\ n)t^2}$$

$$=\frac{G_1(Rt)}{Rt^3}-\frac{R^{(2)}(k+2,\ n,\ -1,\ Rt)}{2\pi C(k,\ n)t^2}-$$

$$\frac{A(k+2,\ -1)}{2\pi C(k,\ n)}\frac{R\sin\left(Rt-\frac{\pi}{2}(k+n)\right)}{t}+$$

$$\frac{(k+n-1)A(k,\ 0)}{C(k,\ n)}\frac{1}{t^2}\cos\left(Rt-\frac{\pi}{2}(k+n)\right). \tag{4.10}$$

其中

$$G_1(Rt)=\frac{-R^{(1)}(k+2,\ n,\ -1,\ Rt)}{2\pi C(k,\ n)}+$$

$$\frac{(k+n-1)}{C(k,\ n)}[R^{(1)}(k,\ n,\ 0,\ Rt)+R^{(2)}(k,\ n,\ 0,\ Rt)]$$

当 $t>R^{-1}$ 时是有界的.

注意到 $R^{(2)}$ 的表达式(见[4])有

$$R^{(2)}(k+2,n,-1,Rt)$$

$$=\sum_{j=1}^{2q}b_jG(k+2,j)\frac{\cos\left(Rt-\frac{\pi(k+2+n+j)}{2}\right)}{(Rt)^{j-1}}+$$

$$\sum_{j=1}^{2q}b_jR^*(k+2,n,j,Rt)\frac{1}{(Rt)^j}$$

$$=-b_1G(k+2,\ 1)\sin\left(Rt-\frac{\pi}{2}(k+n)\right)+$$

$$\frac{1}{Rt}\left\{\sum_{j=2}^{2q} b_j G(k+2,j)\frac{\cos\left(Rt-\frac{\pi}{2}(k+2+n+j)\right)}{(Rt)^{j-2}}+\right.$$

$$\left.\sum_{j=1}^{2q} b_j R^*(k+2,n,j,Rt)\frac{1}{(Rt)^{j-1}}\right\}, \tag{4.11}$$

式中 b_j，$G(k+2, j)$是常数，q 是正整数(与我们的求和指数是两码事)，而

$$R^*(k+2, n, j, Rt)=2^{\frac{k+1}{2}+j}\Gamma\left(1+\frac{k+1}{2}+j\right)R\left(k+2+n+j-\frac{1}{2}, Rt\right),$$

$$R(v, z)=J_v(z)-\left(\frac{2}{\pi}\right)^{\frac{1}{2}}z^{-\frac{1}{2}}\cos\left(z-\frac{\pi v}{2}-\frac{\pi}{4}\right).$$

把(4.11)花括号中的函数写作 $G_2(Rt)$，它当 $t>R^{-1}$ 时有界，而后将(4.11)代入(4.10)，得到

$$\frac{\mathrm{d}}{\mathrm{d}t}\left(\frac{R^{k+n}t^{k+n-1}}{C(k, n)}E(k, 0, n, Rt)-\frac{1}{t}\right)$$

$$=\frac{G(Rt)}{Rt^3}+A_1\frac{\sin Rt}{t^2}+A_2\frac{\cos Rt}{t^2}+A_3\frac{R\sin Rt}{t}+A_4\frac{R\cos Rt}{t},$$

$$t>R^{-1}, \tag{4.12}$$

其中 A_1，A_2，A_3，A_4 是与 k，n 有关的常数，函数

$$G(\xi)=G_1(\xi)-\frac{1}{2\pi C(k, n)}G_2(\xi)$$

当 $|\xi|\geqslant 1$ 时有界，而且是连续的.

将(4.12)代入(4.9)得

$$I_1(R)=-C\int_{R^{-1}}^{+\infty}\Psi(t)\left\{\frac{G(Rt)}{Rt^3}+(A_1\sin Rt+A_2\cos Rt)\frac{1}{t^2}+\right.$$

$$\left.(A_3\sin Rt+A_4\cos Rt)\frac{R}{t}\right\}\mathrm{d}t+o(1), \quad R\to+\infty. \tag{4.13}$$

从条件(4.2)及 $G(Rt)$的有界性得到

$$\int_{R^{-1}}^{+\infty}\Psi(t)\frac{G(Rt)}{Rt^3}\mathrm{d}t=o(1), \quad R\to+\infty. \tag{4.14}$$

经分部积分并用(4.2)得

$$\int_{R^{-1}}^{+\infty}\Psi(t)\frac{\sin Rt}{t^2}\mathrm{d}t=-\frac{\cos Rt}{Rt^2}\Psi(t)\Big|_{R^{-1}}^{+\infty}+\frac{1}{R}\int_{R^{-1}}^{+\infty}\left(\frac{\varphi(t)}{t^2}-\frac{2\Psi(t)}{t^3}\right)\cos Rt\,\mathrm{d}t$$

$$=o(1)+\frac{1}{R}\int_{R^{-1}}^{+\infty}\frac{\Psi(t)}{t^2}\cos Rt\,\mathrm{d}t. \tag{4.15}$$

同样地，

$$\int_{R^{-1}}^{+\infty}\Psi(t)\,\frac{\cos Rt}{t^2}\mathrm{d}t=o(1)-\frac{1}{R}\int_{R^{-1}}^{+\infty}\frac{\varphi(t)}{t^2}\sin Rt\,\mathrm{d}t. \tag{4.16}$$

又算出

$$\begin{aligned}
&R\int_{R^{-1}}^{+\infty}\Psi(t)\,\frac{A_3\sin Rt+A_4\cos Rt}{t}\mathrm{d}t\\
&=\frac{\Psi(t)}{t}(-A_3\cos Rt+A_4\sin Rt)\Big|_{R^{-1}}^{+\infty}-\\
&\quad\int_{R^{-1}}^{+\infty}\left(\frac{\varphi(t)}{t}-\frac{\Psi(t)}{t^2}\right)(-A_3\cos Rt+A_4\sin Rt)\mathrm{d}t\\
&=o(1)+\int_{R^{-1}}^{+\infty}\frac{\varphi(t)}{t}(A_3\cos Rt+A_4\sin Rt)\mathrm{d}t+\\
&\quad\int_{R^{-1}}^{+\infty}\frac{\Psi(t)}{t^2}(-A_3\cos Rt+A_4\sin Rt)\mathrm{d}t.
\end{aligned} \tag{4.17}$$

由(4.13)～(4.17)得

$$\begin{aligned}
I_1(R)=&B_1\int_{R^{-1}}^{+\infty}\frac{\varphi(t)}{t}\cos Rt\,\mathrm{d}t+B_2\int_{R^{-1}}^{+\infty}\frac{\varphi(t)}{t}\sin Rt\,\mathrm{d}t+\\
&B_3\int_{R^{-1}}^{+\infty}\frac{\varphi(t)}{Rt^2}\cos Rt\,\mathrm{d}t+B_4\int_{R^{-1}}^{+\infty}\frac{\varphi(t)}{Rt^2}\sin Rt\,\mathrm{d}t+o(1).
\end{aligned} \tag{4.18}$$

把此式右端的四个积分顺次记为 $J_1(R)$，$J_2(R)$，$J_3(R)$，$J_4(R)$. 从(4.4)(4.8)(4.18)知要证定理 2((1.22)式)只需证

$$\lim_{R\to+\infty}\frac{1}{R}\int_0^R|J_v(u)|^q\mathrm{d}u=0,\quad v=1,2,3,4. \tag{4.19}$$

我们只写出 $v=1$ 时的证明，$v=2,3,4$ 证明完全类似. 令

$$\alpha=\overline{\lim_{R\to+\infty}}\frac{1}{R}\int_0^R|J_1(u)|^q\mathrm{d}u.$$

任取 $\varepsilon\in(0,1)$有

$$\begin{aligned}
\int_{\frac{1}{u}}^{\frac{1}{\varepsilon u}}\frac{\varphi(t)}{t}\cos ut\,\mathrm{d}t&=\Psi(t)\,\frac{\cos ut}{t}\Big|_{\frac{1}{u}}^{\frac{1}{\varepsilon u}}-\int_{\frac{1}{u}}^{\frac{1}{\varepsilon u}}\Psi(t)\left(\frac{-\cos ut}{t^2}-\frac{u\sin ut}{t}\right)\mathrm{d}t\\
&=o(1)+o(1)\ln\frac{1}{\varepsilon}+o(1)\,\frac{1}{\varepsilon}=o(1),\quad u\to+\infty.
\end{aligned} \tag{4.20}$$

所以

$$\alpha\leqslant 2^q\,\overline{\lim_{R\to+\infty}}\frac{1}{R}\int_0^R\left|\int_{\frac{1}{u}}^{\frac{1}{\varepsilon u}}\frac{\varphi(t)}{t}\cos ut\,\mathrm{d}t\right|^q\mathrm{d}u+$$

$$2^q \overline{\lim_{R\to+\infty}} \frac{1}{R}\int_0^R \left|\int_{\frac{1}{\varepsilon u}}^{+\infty} \frac{\varphi(t)}{t}\cos ut\,\mathrm{d}t\right|^q \mathrm{d}u$$

$$= 2^q \overline{\lim_{R\to+\infty}} \frac{1}{R}\int_0^R \left|\int_{\frac{1}{\varepsilon u}}^{+\infty} \frac{\varphi(t)}{t}\cos ut\,\mathrm{d}t\right|^q \mathrm{d}u \tag{4.21}$$

记 $\Phi(u,t)=\int_0^t \varphi(\tau)\cos u\tau\,\mathrm{d}\tau$. 显然

$$\Phi(u,t) = \Psi(t)\cos ut + u\int_0^t \Psi(\tau)\sin u\tau\,\mathrm{d}\tau$$

$$=o(t)+uo(t^2),\quad t\to 0, \tag{4.22}$$

$$|\Phi(u,t)|\leqslant \int_0^t |\varphi(\tau)|\,\mathrm{d}\tau \leqslant M<+\infty,\quad \forall u,t. \tag{4.23}$$

将(4.21)右端的积分分部计算，用(4.22)得

$$\alpha \leqslant C_q \overline{\lim_{R\to+\infty}} \frac{1}{R}\int_0^R \left|\int_{\frac{1}{\varepsilon u}}^{+\infty} \frac{\Phi(u,t)}{t^2}\mathrm{d}t\right|^q \mathrm{d}u$$

$$\leqslant C_q \overline{\lim_{R\to+\infty}} \frac{1}{R}\int_0^R \left(\int_{\frac{1}{\varepsilon R}}^{+\infty} \frac{|\Phi(u,t)|}{t^2}\mathrm{d}t\right)^q \mathrm{d}u$$

$$\leqslant C_q \overline{\lim_{R\to+\infty}} \frac{1}{R}\int_0^R \left(\int_{\frac{1}{\varepsilon R}}^{\delta} \frac{|\Phi(u,t)|}{t^2}\mathrm{d}t\right)^q \mathrm{d}u +$$

$$C_q \overline{\lim_{R\to+\infty}} \frac{1}{R}\int_0^R \left(\int_{\delta}^{+\infty} \frac{|\Phi(u,t)|}{t^2}\mathrm{d}t\right)^q \mathrm{d}u \tag{4.24}$$

由于$\frac{|\Phi(u,\ t)|}{t^2}\leqslant\frac{M}{t^2}$，并据 Riemann-Lebesgue 引理

$$\lim_{u\to+\infty}|\Phi(u,t)| = \lim_{u\to+\infty}\left|\int_0^t \varphi(\tau)\cos u\tau\,\mathrm{d}\tau\right| = 0,$$

积分号下取极限得

$$\lim_{u\to+\infty}\int_{\delta}^{+\infty} \frac{|\Phi(u,t)|}{t^2}\mathrm{d}t = \int_{\delta}^{+\infty} \lim_{u\to+\infty}|\Phi(u,t)|\frac{1}{t^2}\mathrm{d}t = 0,$$

可见

$$\overline{\lim_{R\to+\infty}} \frac{1}{R}\int_0^R \left(\int_{\delta}^{+\infty} \frac{|\Phi(u,t)|}{t^2}\mathrm{d}t\right)^q \mathrm{d}u = 0. \tag{4.25}$$

代入(4.24)，再用广义 Minkowski 不等式，得

$$\alpha \leqslant C_q \overline{\lim_{R\to+\infty}} \frac{1}{R}\left\{\int_{\frac{1}{\varepsilon R}}^{\delta} \frac{1}{t^2}\left(\int_0^R |\Phi(u,t)|^q \mathrm{d}u\right)^{\frac{1}{q}} \mathrm{d}t\right\}^q$$

$$= C_q \overline{\lim_{R\to+\infty}} \frac{1}{R}\left\{\int_{\frac{1}{\varepsilon R}}^{\delta} \frac{1}{t^2}\left(\int_0^R \left|\int_0^t \varphi(\tau)\cos u\tau\,\mathrm{d}\tau\right|^q \mathrm{d}u\right)^{\frac{1}{q}} \mathrm{d}t\right\}^q. \tag{4.26}$$

用关于 Fourier 变换的 Hausdorff-Young 不等式(见[5]第五章)，得

$$\left(\int_0^R\left|\int_0^t \varphi(\tau)\cos u\tau\mathrm{d}\tau\right|^q\mathrm{d}u\right)^{\frac{1}{q}}$$

$$\leqslant C_q\left(\int_0^t|\varphi(\tau)|^p\mathrm{d}\tau\right)^{\frac{1}{p}},\quad (0<t<\delta) \tag{4.27}$$

于此使用条件(4.3)并代入(4.26)，求得

$$\alpha\leqslant C_q\varlimsup_{R\to+\infty}\frac{1}{R}\left(\int_{\frac{1}{\varepsilon R}}^{\delta}\frac{1}{t^2}\cdot t^{\frac{1}{p}}\mathrm{d}t\right)^q$$

$$\leqslant C_q\varlimsup_{R\to+\infty}\frac{1}{R}(\varepsilon R)^{(1-\frac{1}{p})q}=C_q\cdot\varepsilon. \tag{4.28}$$

由 ε 之任意性知 $\alpha=0$. 这就证明了(4.19). □

§5. 定理 3 的证明

设 $f\in L(Q)$满足定理 3 的条件. 定义

$$\xi(y)=\begin{cases}f(x+y)\chi_{B(0,\delta)}(y), & y\in Q,\\ \text{有周期 }2\pi, & y\in R.\end{cases}$$

$$\eta(y)=f(x+y)-\xi(y),$$

$$g(y)=\xi(y)\chi_Q(y),\qquad y\in E_k.$$

那么 g 在原点处满足定理 2 的条件，所以

$$\lim_{R\to+\infty}\frac{1}{R}\int_0^R|\tilde{\sigma}_u^{\frac{k-1}{2}}(g;0)-\tilde{g}_{\frac{1}{u}}(0)|^q\mathrm{d}u=0. \tag{5.1}$$

而 η 在原点处满足定理 1 的条件，所以

$$\lim_{R\to+\infty}\frac{1}{R}\int_0^R|\widetilde{S}_u^{\frac{k-1}{2}}(\eta;0)-\eta^*_{\frac{1}{u}}(0)|^q\mathrm{d}u=0. \tag{5.2}$$

根据定理 A(见(1.23))

$$\lim_{u\to+\infty}|\widetilde{S}_u^{\frac{k-1}{2}}(\xi;\ 0)-\tilde{\sigma}_u^{\frac{k-1}{2}}(g;\ 0)-C(0)|=0, \tag{5.3}$$

其中函数(见(1.24))

$$C(x)=(2\pi)^{-k}\int_Q\xi(y)(K^*(x-y)-K(x-y))\mathrm{d}y$$

是连续的. 代入 $x=0$ 得

$$C(0)=(2\pi)^{-k}\int_Q\xi(y)(K^*(-y)-K(-y))\mathrm{d}y$$

$$=\lim_{\varepsilon\to0^+}(2\pi)^{-k}\int_{Q\setminus B(0,\varepsilon)}\xi(y)(K^*(-y)-K(-y))\mathrm{d}y.$$

注意到(1.13)(1.18)知

$$C(0)=\lim_{\varepsilon\to 0^+}(\xi_\varepsilon^*(0)-\tilde{\eta}_\varepsilon(0)). \tag{5.4}$$

由于

$$\begin{aligned}\widetilde{S}_u^{\frac{k-1}{2}}(f;\ x)-f_{\frac{1}{u}}^*(x)&=\widetilde{S}_u^{\frac{k-1}{2}}(\xi+\eta;\ 0)-(\xi+\eta)_{\frac{1}{u}}^*(0)\\&=\widetilde{S}_u^{\frac{k-1}{2}}(\xi;\ 0)-\xi_{\frac{1}{u}}^*(0)+\widetilde{S}_u^{\frac{k-1}{2}}(\eta;\ 0)-\eta_{\frac{1}{u}}^*(0)\\&=\widetilde{S}_u^{\frac{k-1}{2}}(\xi;\ 0)-\xi_{\frac{1}{u}}^*(0)-(\tilde{\sigma}_u^{\frac{k-1}{2}}(g;\ 0)-\tilde{g}_{\frac{1}{u}}(0))+\\&\quad\tilde{\sigma}_u^{\frac{k-1}{2}}(g;\ 0)-\tilde{g}_{\frac{1}{u}}(0)+\widetilde{S}_u^{\frac{k-1}{2}}(\eta;\ 0)-\eta_{\frac{1}{u}}^*(0),\end{aligned} \tag{5.5}$$

再根据(5.1)～(5.4)，就得(1.25). □

致谢 本文是在陆善镇老师的指导下完成的，谨表示衷心的感谢.

参考文献

[1] Calderón A P, Zygmund A. Singular integral and periodic functions. Studia Mathematica, 1954, 14(2): 249—271.

[2] Shapiro V L. Fourier series in several variables. Bulletin of the Amer Math. Soc, 1964, 70(1): 48—93.

[3] Lippman G E. Spherical summability of conjugate multiple Fourier series and integrals at the critical index. SIAM J. Math. Anal., 1973, 4(4): 681—695.

[4] Chang C P. On certain exponential sums arising in conjugate multiple Fourier series. Ph. D. dissertation, University of Chicago, Chicago, 1964.

[5] Stein E M, Weiss G. Introduction to Fourier analysis on Euclidean spaces. Priceton University press, 1971.

数学季刊，1987，2(2)：59－68.

多重 Fourier 级数及其线性平均[①]

Multiple Fourier Series and Its Linear Means

§1. 引言

设 n 为正整数. $\mathbf{R}^n$ 为 n 维欧氏空间. Q^n为 $\mathbf{R}^n$ 中的方体：$Q^n=\{(x_1,\ x_2,\ \cdots,\ x_n)=x \mid -\pi\leqslant x_j<\pi,\ j=1,2,\cdots,n\}$. $\mathbf{R}^n$ 中的点 $x=(x_1,\ x_2,\ \cdots,\ x_n)$与 $y=(y_1,\ y_2,\ \cdots,\ y_n)$的欧氏内积记作 $xy=x_1y_1+x_2y_2+\cdots+x_ny_n$，欧氏范数是$|x|=\sqrt{x_1^2+x_2^2+\cdots+x_n^2}$.

$L(Q)$表示在 Q 上 Lebesgue 可积，对每个变元都以 2π 为周期的 n 元函数的空间. 设 $f\in L(Q)$，它的 Fourier 系数是

$$C_m(f)=\hat{f}(m)=(2\pi)^{-n}\int_{Q^n}f(x)\mathrm{e}^{-\mathrm{i}mx}\,\mathrm{d}x,\quad m\in\mathbf{Z}^n.$$

($\mathbf{Z}^n$ 为 n 维整点的集合). f 的 Fourier 级数记作 $\sigma(f)(x)$. 即

$$\sigma(f)(x)\sim\sum_{m\in\mathbf{Z}^n}\hat{f}(m)\mathrm{e}^{\mathrm{i}mx}.$$

我们讨论的问题是多元周期函数的 Fourier 级数的部分和或某种方式的线性平均对函数的逼近性质或收敛性质.

此处涉及三种求和方式：矩形、球形和对角线型(或叫作 Marcinkiewicz 型).

关于矩形方式，那些可以用累次降维数的方式直接把问题归结到一维情形的问题不具有本质的意义，不是我们讨论的重点. 矩形方式的极

① 收稿日期：1986-08-10.

限还有限制型与非限制型之分. 限制型的极限指的是第一卦限的整点 $m=(m_1, m_2, \cdots, m_n)$ 的最小分量 $\min(m_1, m_2, \cdots, m_n)$ 在条件 $\max(m_1, m_2, \cdots, m_n)\leqslant\lambda \min(m_1, m_2, \cdots, m_n)$ 的限制下趋于正无穷，此处 λ 是预先任给的不小于 1 的数. 我们记此极限过程为"$\lim\limits_{\substack{m\to+\infty\\(\lambda)}}$".

非限制型矩形极限指的是 $\min(m_1, m_2, \cdots, m_n)$ 趋于正无穷而无其他限制. 记之为"$\lim\limits_{m\to+\infty}$"(或"$\lim\limits_{\substack{m\to+\infty\\(\infty)}}$": 相当于限制型定义中 $\lambda=\infty$ 的情况). 极限"$\lim\limits_{\substack{m\to+\infty\\(1)}}$"又叫作方形极限，在限制型极限中，它往往是有代表性的.

至于球形方式，指的是 $|m|=(m_1^2+m_2^2+\cdots+m_n^2)^{\frac{1}{2}}$ 趋于正无穷的极限方式. 记作"$\lim\limits_{|m|\to+\infty}$".

对角线型求和是 Marcinkiewicz 于 1939 年研究二重 Fourier 级数的线性求和问题时首先提出的. 指的是对 Fourier 级数的方形部分和(在二元情形)即 $S_{k,k}(f)(k\in\mathbf{N})$ 的序列做线性平均的一种求和方式，此时的极限过程是一维的，处理的对象是 Fourier 部分和的脚标沿第一卦限的对角线取值的一个子序列.

本文的目的是扼要叙述笔者感兴趣的一些方面的已有的结果及一些看来有思考的可能的未解决的问题，所以并不求材料的完备.

§2. 多重 Fourier 级数的收敛问题

2.1 平均收敛与几乎处处收敛

周知，当 $n=1$ 时，$\exists f\in L_{2\pi}$ 使 $\overline{\lim\limits_{k\to+\infty}}\|S_k(f)-f\|=+\infty$; 而若 $f\in L\ln^+L$，则 $\lim\limits_{k\to+\infty}\|S_k(f)-f\|_1=0$. 这个结果的多元推广是

定理 1 (见 Л. В. Жижиащвили[1]) 若 $f(x_1, x_2, \cdots, x_n)\in L(\ln^+L)^n(Q)$，则 $\sigma(f)$ 及其一切共轭级数皆以矩形方式(无限制) L 收敛.

另外，还有以 L 连续模刻画的收敛条件，即如下的定理(亦见[1]).

定理 2 设 $f\in L(Q)$. 若 $\omega_j(f, \delta)_L=O(\ln\delta^{n+\varepsilon})\varepsilon>0$, $j=1,2,\cdots,n$，则 $\sigma(f)$ 及其共轭级数都以矩形方式(无限制) L 收敛及 a. e. 收敛.

这里 ω_j 表示对第 j 个变元的偏连续模，所说共轭级数，是经典意义的，在 Жижиащвили 的书[2]中有基本的介绍.

从定理 2，自然提出如下问题.

[问题 1] 若 $\omega_j(f,\delta)_L=O\left(\left(\ln\frac{1}{\delta}\right)^{-n}\right)$，$j=1,2,\cdots,n$ 能否断定 $\sigma(f)$以矩形方式 a. e. 收敛？关于共轭级数情形如何？

当 $n=1$ 时，这就是经典的 Zygmund 问题，见[3]XIII(3.10).

我们知道，$n=1$ 时，部分和 $S_k(f,x)$依测度收敛到 f. 文献[4]中提出下列问题.

[问题 2] $n>1$ 时矩形部分和 $S_m(f,x)$是否依测度收敛到 $f(f\in L(Q))$?

在方形的情形，Psjölin[5]推广 Carleson，Hunt 等关于一元的结果，得到

定理 3 若 $f(x_1,x_2,\cdots,x_n)\in L(\ln^+L)^n\ln^+\ln^+L(Q)$，则 $S_{m,m,\cdots,m}(f,x)$a. e. 收敛到 $f(x)$. $(m\to+\infty)$.

对于 L^p 尺度$(1<p<+\infty)$，矩形部分和是与同阶数最佳逼近多项式达到同样的逼近阶的，这与一元的情形一样. 所以矩形情形已不需考虑. 而球形的情形却不一样. 当 $n>1$ 时，对于球形方式的求和，我们代替球形部分和 $S_R(f,x)=\sum\limits_{|m|<R}\hat{f}(m)\mathrm{e}^{\mathrm{i}mx}(R>0)$ 而考虑球形 a 阶 Riesz 平均 $S_R^a(f,x)=\sum\limits_{|m|<R}\left(1-\frac{|m|^2}{R^2}\right)^a\hat{f}(m)\mathrm{e}^{\mathrm{i}mx}(R>0)$. 在 $L^p(1<p<+\infty)$ 尺度下，指数 $a_p=\frac{n-1}{2}\left|\frac{2}{p}-1\right|$ 具有临界的意义. E. M. Stein[6]的结果是

定理 4 若 $f\in L^p(Q)$，$\delta>a_p=\frac{n-1}{2}\left|\frac{2}{p}-1\right|$，则 $S_R^\delta(f,x)\xrightarrow{L^p}f(x)$.

后来，Д. С. Бели[7]求得当 $\delta>a_p$ 时 S_R^δ 对 f 的 L^p 逼近的结果，这在 §4 还要提到.

$p=1$ 时，Stein[6]还得到

定理 5 若 $f\in L\ln^+L(Q)$，则 $S_R^{\frac{n-1}{2}}(f)\xrightarrow{L}f$.

与逼近问题相联系(见 §4[问题 11])，可考虑

[问题 3] 对于一般的维数 n，一般的 $p\in(1,+\infty)(p\neq2)$，当 $\delta\leqslant a^p$ 时，对于 $S_R^\delta(f)$的 L^P 收敛性能说明什么？

对于 L^1 和 L^∞，Riesz 球形平均 S_R^δ 以 $\delta_0=\frac{n-1}{2}$为临界指数. 关于 a. e. 收敛有[6].

定理 6 若 $f\in L(\ln^+L)^2$，则 $S_R^{\frac{n-1}{2}}(f,\ x)\xrightarrow{\text{a. e.}}f(x)$.

当考虑球形求和时，宜于考虑按 Calderón-Zygmund 意义引入的共轭函数及共轭 Fourier 级数(见[8]). 对于相应于球调和核的共轭级数，上述 $L^p(1<p<+\infty)$结果(定理 4)仍然成立. 而定理 5 和定理 6 当将条件中的 $\ln^+L$ 升高一次幂时，结论依然成立，这只要使用 Calderon-Zygmund 关于奇异积分的一般理论便可推出.

关于 a. e. 收敛，还有 Chang Chaoping[9] 的结果，给出了 Marcinkiewicz 型的收敛条件.

定理 7 若 $f\in L\ln^+L(Q)$且满足条件.

$$\frac{1}{h^n}\int_{|\xi|<h}|f(x+\xi)-f(x)|\,\mathrm{d}\xi=O\left(\frac{1}{\ln\frac{1}{h}}\right),\forall x\in Q,$$

则
$$S_R^{\frac{n-1}{2}}(f,x)\xrightarrow{\text{a. e.}}f(x).$$

当然，定理 7 的原来形式是在任意一个正测度集上不必在整个 Q 上.

陆善镇[10]引入了球形积分

$$F(x,\gamma)=\int_{|\xi|>r}(f(x+\xi)-f(x))\,\mathrm{d}\xi,$$

得到 Salem 型收敛条件.

定理 8 若 $f\in L\ln^+L(Q)$且存在 $\gamma_0>0$ 使对于每个 x，关于 $\gamma\in(h,\gamma_0)(0<h<\gamma_0)$一致成立着

$$\{F(x,\ r+2h)+F(x,\ r)-2F(x,\ r+h)\}=O\left(\frac{h}{\ln\frac{1}{h}}\right)(h\to0)\text{，则}$$

$$S_R^2(f,\ x)\xrightarrow{\text{a. e.}}f(x).$$

Lu Shan-Zhen. M. H. Taibleson 和 G. Weiss[11]证明了

定理 9 若 $f\in L(Q)$，f 的熵有限，则

$$S_R^{\frac{n-1}{2}}(f,\ x)\xrightarrow{\text{a. e.}}f(x),\qquad R\to+\infty.$$

这里，函数的熵是 R. Fellerman 1978 年引入的一个概念. f 的熵记作 $J(f)$. 且 $J(f)<+\infty\Rightarrow f\in L\ln L(Q)$. 定理 9 给出了在 $L\ln L(Q)$中一个使 $S_R^{\frac{n-1}{2}}$ a. e. 收敛的子类.

事实上，从[11]的证明中还可以看到稍强的结论. 即

定理 9′　$f\in C_q\cap L\ln^+L(Q)(1<q\leqslant+\infty)\Rightarrow S_R^{\frac{n-1}{2}}(f,\ x)\xrightarrow{\text{a. e.}}f(x)$.

这里，C_q 是由 M. H. Taibleson 和 G. Weiss 于 1981 年引入的块(block)生成的空间. 若 $J(f)<+\infty$，则 $f\in C_q(\forall 1<q\leqslant+\infty)$.

块理论的一大优点是，通过迭加原理，把问题归结到一个块(相当于分子)上，从而使问题的讨论变得十分简捷.

稍后，Lu[12]继续对依 Calderón-Zygmund 意义，相应于球调和核的共轭级数做出了同样的结果. 即

定理 10　若 $f\in C_q\cap L\ln^+L(Q)$，则 $\widetilde{S}_R^{\frac{n-1}{2}}(f,\ x)\xrightarrow{\text{a. e.}}f^*(x)$.

这里 f^* 是共轭函数(参见[8]).

当指数 a 低于$\dfrac{n-1}{2}$时，E. M. Stein[6]的结果是

定理 11　若 $f\in L^p(Q)$，$1<p\leqslant2$，$\dfrac{n-1}{2}\geqslant a>\dfrac{n-1}{2}\left(\dfrac{2}{p}-1\right)$，则

$$S_R^a(f,\ x)\xrightarrow{\text{a. e.}}f(x).$$

这里要求 $f\in L^p$，对于 L^p 的“临界指数”$a_p=\dfrac{n-1}{2}\left(\dfrac{2}{p}-1\right)$，定理 11 的结论还是处在高于“临界指数”的情形，由此自然产生.

[问题 4] 当 $p\in(1,\ 2]$，$a_p=\dfrac{n-1}{2}\left(\dfrac{2}{p}-1\right)$，$f\in L^p(Q)$时，$S_R^{a_p}(f,\ x)$是否总是 a. e. 收敛到 $f(x)$?

如果 $p=2$，那么 $a_2=0$，S_R^0 就是原原本本的球形部分和，这就回到 $L^2(Q)$上球形部分和是否 a. e. 收敛这样一个经典的，然而至今没有解决的问题.

有些人求得一些 L^2 上 S_Ra. e. 收敛的充分条件，如苏联的 Голубов[13]得到

定理 12　若 $f\in L^2(Q)$且

$$\omega(f,\ \delta)L^2=O\left\{\left(\ln\frac{1}{\delta}\left(\ln\ln\frac{1}{\delta}\right)^{\frac{1}{2}+\varepsilon}\right)^{-1}\right\},$$

则 $S_R(f,\ x)=S_R^0(f,\ x)\xrightarrow{\text{a. e.}}f(x)$.

我们也可以继续沿着这个方向，考虑

[问题 5] 寻求使 $S_R^{a^p}(f)$在 L^p 类上 a. e. 收敛的尽可能低的充分条

件，如果 $p=1$，也可以考虑构造容纳 $L(\ln^+ L)^2$ 在内的 $L(Q)$ 的子类使 $S_R^{\frac{n-1}{2}}$ 在这个类上 a. e. 收敛.

2.2　局部化问题

一元的 Riemann 局部化准则是众所周知的，多元情形完全两样.

$n>1$ 时，对于方形极限，[4]中叙述了 Ильин 的结果，对于函数类加了非常强的限制才得到局部化性质. 文[4]认为，对于方形和寻求终极的局部化条件是有意义的.

对于球形情形. 以 Riesz 平均取代 Fourier 部分和，高于临界指数 $\frac{n-1}{2}$ 时，局部化性质成立而对于临界阶 $\frac{n-1}{2}$ $(n>1)$ 的情形. S. Bochner[14] 早已证明

定理 13　存在 $f\in L(Q)(n>1)$，f 在原点附近消失，但

$$\varlimsup_{R\to+\infty} S_R^{\frac{n-1}{2}}(f, 0)=+\infty.$$

E. M. Stein[6] 证明

定理 14　在 $L\ln^+ L(Q)$ 上对于 $S_R^{\frac{n-1}{2}}$，收敛的局部化原理成立.

G. E. Lippman[15] 对于相应于球调和核的共轭级数得到了与定理 13 类似的结果. 我们可以问

[问题 6] 定理 14 的结论是否是终极的? 也就是说，能不能找到 $L(Q)$ 的子类，它真包含 $L\ln^+ L(Q)$，而使 $S_R^{\frac{n-1}{2}}$ 的收敛只是局部性质?

又，当 $a\in\left(\left(\frac{n-3}{2}\right)_+, \frac{n-1}{2}\right)(n>1)$ 时，在怎样的函数类上 $S_R^0(f, x)$ 的收敛性只是 f 的局部性质?

通过对局部化问题的研究，可能会对 Riesz 平均的性质达到更深刻的了解.

我们指出，当 Stein[16] 于 1961 年揭示了 Fourier 级数和 Fourier 积分(关于 Riesz 平均)的联系(在 $L\ln^+ L(Q)$)之后，利用关于 Fourier 积分的结果，定理 14 的结论就极易得出了.

2.3　一致收敛问题

关于矩形极限，我们列出用偏连续模刻画的收敛充分条件，可在[2]中找到.

定理 15　若 $f\in C(Q)$，满足

$$\omega_j(f,\ \delta)c=O\left(\left(\ln\frac{1}{\delta}\right)^{-n-\varepsilon}\right),\ \varepsilon>0,\ j=1,2,\cdots,n,$$

则 $\sigma(f)$及其一切共轭级数(按经典意义)都以矩形方式一致收敛.

对于临界指数的球形 Riesz 平均，Голубов[13]得到了 Dini 型收敛条件，即

定理 16 若 $f\in C(Q)$满足

$$\sup_{x\in Q^n}\ \sup_{0<h<\delta}\ \sup_{t>0}|f_x(t+h)-f_x(t)|=o\left(\left(\ln\frac{1}{\delta}\right)^{-1}\right),$$

则$\|S_R^{\frac{n-1}{2}}(f)-f\|c\to 0(R\to+\infty)$. 式中 $f_x(t)=\int\limits_{|\xi|=1}f(x+t\xi)\mathrm{d}\sigma(\xi)$.

条件"o"不能减弱为"O".

陆善镇[10]加强了上面的结果，得到

定理 17 若 $f\in C(Q)$满足：存在 $r_0>0$，使

$$\sup_{h<r<r_0}\ \sup_{x\in Q^n}\frac{1}{r^{n-1}}\{F(x,\ r+2h)+F(x,\ r)-2F(x,\ r+h)\}=o\left(\frac{h}{\ln\frac{1}{h}}\right)$$

$(h\to 0^+)$，则$\|S_R^{\frac{n-1}{2}}(f)-f\|\to 0\ (R\to+\infty)$.

§3. 多重 Fourier 级数的线性求和及强求和

3.1 矩形及方形求和

Cesáro 求和是常用的方法. 周知，(见[3]14 章)早在 1935 年就已建立 Jensen，Marcinkiewicz 及 Zygmund 的

定理 18 设 $\boldsymbol{a}=(a_1,\ a_2,\ \cdots,\ a_n)>(0,\ 0,\ \cdots,\ 0)=\boldsymbol{0}$. 若 $f\in L(\ln^+L)^{n-1}(Q^n)$，则 $\sigma(f)(x)$以矩形方式几乎处处可以$(C,\ \sigma)$求和.

根据 S. Saks[17]的结果，此定理的结论在下述意义之下是终极的. 任给 $\omega(t)\uparrow_0^{+\infty}(t\uparrow_0^{+\infty})$，$\omega(t)=o(\ln^+t)^{n-1}(t\to+\infty)$，存在 $f\in L\omega(L)$，但 $\sigma(f)$以矩形方式 a. e. 不可$(C,\ 1)$求和.

问题的关键是，多元的"强"Lebesgue 点不必是几乎处处的(可以几乎处处不是). 所谓"强"L 点指的是使

$$\lim_{(h_1,h_2,\cdots,h_n)\to(0^+,0^+,\cdots,0^+)}\frac{1}{h_1h_2\cdots h_n}\int_{-h_1}^{h_1}\cdots\int_{-h_n}^{h_n}|f(x+\xi)-f(x)|\mathrm{d}\xi=0$$

成立的点 x，这里极限是非限制型的.

1977 年，Л. Д. Гоголадзе[18]将定理 18 的结论加强为强性的，得到

定理 19 若 $f\in L(\ln^+L)^{n-1}(Q^n)$，则 $\sigma(f)$ 及其一切共轭级数皆以矩形方式 a. e. $(H, q)(q>0)$ 可知

对于限制型求和，定理 18 的条件可以减弱到 $f\in L(Q)$（见[3]，14 章§3）. 可是对于强求和，当考虑限制型极限时，定理 19 的条件能否减弱？我们提出（就二元情形）：

[问题 7] 在怎样的函数类上能保证

$$\lim_{m\to+\infty}\frac{1}{m^2}\sum_{j=0}^{m}\sum_{i=0}^{m}|S_{i,j}^{i}(f,x,y)-f(x,y)|\xlongequal{\text{a. e.}}0.$$

3.2 球形情形（关于 $S_R^{\frac{n-1}{2}}(f)$ 的线性及强性平均）

当 $n>1$ 时，$S_R^{\frac{n-1}{2}}$ 的地位与一元 Fourier 部分和的地位相当，所以我们把 $\frac{1}{R}\int_0^R S_r^{\frac{n-1}{2}}(f,x)\mathrm{d}r$ 叫作 Fourier 级数的$(C, 1)$平均.

笔者[19]考虑了共轭级数的$(C, 1)$平均，证明了对于按 Calderán-Zygmund 意义相应于球调和核的共轭级数成立着：

定理 20 设 $f\in L(Q)$，则 $\frac{1}{R}\int_0^R \widetilde{S}_r^{\frac{n-1}{2}}(f,x)\mathrm{d}r\xrightarrow{\text{a. e.}}f^*(x)$，

这里 f^* 是相应于球调和核的共轭函数.

关于强求和，E. M. Stein 在[6]中曾提到对于 $f\in L(Q)$.

$$\lim_{R\to+\infty}\frac{1}{R}\int_0^R|S_r^{\frac{n-1}{2}}(f,x)-f(x)|^2\mathrm{d}r\xlongequal{\text{a. e.}}0. \tag{3.1}$$

但在 1984 年 9 月北京国际分析学讨论会上问及 Stein，他说(3.1)尚待证明. 于是我们提出

[问题 8] 在 $L(Q)$ 上，(3.1)是否成立？或者更一般地把指数 2 换为 $q>0$，情况如何？又，对于共轭情形，(H, q) 可和性又如何？

关于定点的(H, q)可和充分条件，陆善镇[20]证明了

定理 21 若 $f\in L(Q)$，f 在 $|x-x_0|<\eta$ 内属于 L^p，$p>1$，$q>1$，且满足条件

$$\int_0^t\tau^{k-1}\{f_{x_0}(\tau)-f(x_0)\}\mathrm{d}\tau=o(t^k),\quad t\to0, \tag{3.2}$$

$$\int_0^t|f_{x_0}(\tau)-f(x_0)|^p\mathrm{d}\tau=o(t),\quad t\to0, \tag{3.3}$$

则
$$\lim_{R\to+\infty}\frac{1}{R}\int_0^R|S_u^{\frac{n-1}{2}}(f,x_0)-f(x_0)|^q\mathrm{d}u=0.$$

对于共轭的情形，笔者证明了类似的结论(见[21]第21页).

此处提出

[问题9] 定理21的条件(3.3)可怎样减弱？能否能得到终极(不可再改进)的结果？

3.3 Marcinkiewicz型求和

关于M型求和，在[22]中有较详细的叙述，此处只叙述一个(C,α)求和和一个(H,q)求和的结果. 关于(C,α)求和，Л. В. Жижиащвили[23]得到

定理22 若$f(x,y)\in L(Q)(n=2)$，$\alpha>0$，则

$$\lim_{R\to+\infty}\frac{1}{A_k^{\alpha}}\sum_{j=0}^{k}A_{k-j}^{\alpha-1}S_{j,j}(f,x,y)\overset{\text{a. e.}}{=\!=}f(xy).$$

笔者[24]得到的(H,q)求和的结果是

定理23 若$f(x,y)\in L\ln^{+}L(Q)(n=2)$，则

$$\lim_{k\to+\infty}\frac{1}{k+1}\sum_{j=0}^{k}|S_{j,j}(f,x,y)-f(x,y)|^{q}\overset{\text{a. e.}}{=\!=}0,\quad q>0.$$

且此式对于f的三种共轭级数也成立.

[问题10] 定理23的条件能不能减弱为$f(x,y)\in L(Q)$？

§4. 用多重Fourier级数的线性平均逼近函数

4.1 矩形方式

许多一元的结果可以用降维数的办法推广到多元. 此处只提一下一元Осколков不等式的多元推广. 结论是

$$\|f-S_{k,\cdots,k}(f)\|_{x}\leqslant C_n\sum_{\gamma=0}^{k}\frac{\ln^{n-1}(\gamma+2)}{\gamma+1}E_{k+\gamma}(f)_X,\qquad(4.1)$$

式中X表示尺度L或C. $E_{k+\gamma}(f)_X$表示函数f在空间X中的$k+\gamma$阶方形最佳三角逼近. 对于方形de le Vallée Poussin余项也有类似的结果. 可见于С. Л. Баиборoдов[25]及笔者的文章[26]. 在证明(4.1)时，完全可以用归纳的(即降维数的)办法，这与一元的证法相比就不能说有本质的不同，[25]没有看到这一点.

重要的是做出下方估计，即

定理24 设$\{\varepsilon_n\}_{n=0}^{+\infty}$是单调下降到零的数列，函数类$X(\varepsilon)=\{f\in X(Q_n)\,|\,E_\gamma(f)_X|\leqslant\varepsilon_\gamma\quad\gamma\in\mathbf{N}^*\}$. 存在只与维数$n$有关的常数$C>0$，使

$$\sup_{f\in X(\varepsilon)}\|f-S_{k,\cdots,k}(f)\|_X\geqslant C\sum_{\gamma=0}^{k}\frac{\ln^{n-1}(\gamma+2)}{\gamma+1}\varepsilon_{k+\gamma}.$$

[25]和[26]各自独立地用不同的方法证明了定理 24，关对 de la Vallée Poussin 余项作出了类似的估计.

4.2 球形方式

在 L^p 尺度，有 Белинский[7]的结果

定理 25 设 $1\leqslant p\leqslant 2$，$\alpha_p=\frac{n-1}{2}\left(\frac{2}{p}-1\right)$，$f\in L^p(Q^n)$. 若 $\alpha>\alpha_p$，

则
$$\|f-S_R^\alpha(f)\|_p\leqslant C_{\alpha,p}\omega_2\left(f,\frac{1}{R}\right)_p.$$

$\omega_2(f,t)_p(t>0)$表示 f 的二阶 L^p 连续模.

[问题 11]当 $\alpha\leqslant\alpha_p$ 时，对于一般的维数 $n>1$，$\|f-S_R^\alpha(f)\|_p$ 的阶如何用 L^p 连续模刻画?(相应的收敛性问题见[问题 3]).

关于球形逼近的较详细的叙述，可参阅笔者毕业论文的序言部分.

最近，蒋迅证明了(见[21]，p. 35～36)

定理 26 任给连续模 $\omega(r)$，存在正的常数 C_1，C_2 使

$$\frac{C_1}{R}\int_0^R\omega\left(\frac{1}{r}\right)\mathrm{d}r\leqslant\sup_{f\in H_C^\omega}\left\|f-\frac{1}{R}\int_0^R S_r^{\frac{n-1}{2}}(f)\,\mathrm{d}r\right\|_C\leqslant\frac{C_2}{R}\int_0^R\omega\left(\frac{1}{r}\right)\mathrm{d}r.$$

并对于共轭(相应于球调和核)情形，得到

定理 27 设 $n\geqslant 2$，$f\in C(Q)$，则

$$\left\|\frac{1}{R}\int_0^R(\widetilde{S}_r^{\frac{n-1}{2}}(f)-f_{1/r}^*)\,\mathrm{d}r\right\|_C=O\left(\frac{1}{R}\int_1^R\omega\left(f,\frac{1}{r}\right)\mathrm{d}r\right)+O\left(\frac{\|f\|_C}{R^{n-1}}\right).$$

蒋迅还考虑了有关球形 Riesz 平均算子的饱和类的问题.

王昆扬在毕业论文中考虑了 $S_R^{\frac{n-1}{2}}$ 在全测度集上的逼近问题(见[21]，p. 23～24)(包括共轭情形)所得结果包含下述定理

定理 28 设 $f\in \mathrm{Lip}\,\delta$，$0<\delta<1$，则

$$|f(x)-S_R^{\frac{n-1}{2}}(f,x)|+|\widetilde{S}_R^{\frac{n-1}{2}}(f,x)-f^*(x)|\leqslant C(x)\frac{\ln\ln R}{R^\delta},$$

$(R\geqslant 3)$，其中 $C(x)$a. e. 有限且

$$|\{x\in Q^n\,|\,C(x)>\lambda\}|\leqslant A\mathrm{e}^{-\frac{\lambda}{A}},\quad\forall\lambda<0.$$

A 只与维数 n 及共轭球调和核有关.

但所得的结果是否终极的，尚未结论. 也就是说，例如针对

Lip $\delta(0<\delta<1)$类可提出

[问题 12]当维数 $n>1$ 时，能不能证明

“存在函数 $f_0\in\text{Lip}\,\delta\ (0<\delta<1)$，使得

$$\overline{\lim_{R\to+\infty}}\frac{\max(|f_0(x)-S_R^{\frac{n-1}{2}}(f_0,\ x)|,\ |f_0^*(x)-\widetilde{S}_R^{\frac{n-1}{2}}(f_0,\ x)|)}{\ln\ln R/R^\delta}>0$$

几乎处处成立”?

关于对角线方式逼近，此处不拟详述，可参阅[22](p. 40～70).

§ 5. 关于 Lebesgue 常数

在 Fourier 级数的线性平均的收敛问题及逼近(特别是 L 尺度和 C 尺度)问题的研究中，线性平均算子的 Lebesgue 常数是一个很重要的量. 一些苏联学者使用 Fourier 变换的理论，对一些线性平均算子的 Lebesgue 常数做了一般性的研究，取得了一些普遍性的结果，应用到一些特殊的算子，得到了比较精确的对 Lebesgue 常数的量的估计. 此处列出一些文章以备查考.

① Белинскнй Б С，Лифлянд И Р. Констанмы Лебела и интегри руецосми иреобразования Фурье радиальнй. Докл. АНУССР，1980，A，6：5—10.

② Bëlins'kil E S. Behavior ob Lebesgue constants in certain methods on swumation of multiple Fourier series，Mr 58. ＃29816.

③ Triguf R M. Summability of multiple Fourier series. Growth of Lebesgue constants，Analysis Mathematica，1980，6：255—267.

参考文献

[1] Жижиащвили Л В. У. М. Н.，1973，28(2)：65—119.

[2] Жижиащвили Л В. Соиряжённые функции и Тригономет рические. Тбилиси，1969.

[3] Zygmund A. Trigonometric series. Combridge，1979.

[4] Алилов Ш А. Ильин ВА Никищин Е М. У. М. Н. 1976，31(6)：28—83.

[5] Sjölin P. Ark för Matematik，1971，9(9)：65—90.

[6] Stein E M. Acta Math. 1958，100(1—2)：93—147.

[7] Белінський Е С. Доповіді АНУРСР. 1975，7：579—581.

[8] Calderón A P，Zygmund A. Studia Math. 1954，14(2)：249—271.
[9] Chang C P. Studia Math.，1956，26：25—66.
[10] 陆善镇. 数学学报，1980，23(4)：609—623.
[11] Lu Shan-zhen. Taibleson M H，Weiss G. Proceedings of the hazmonic analysis confezence，Univ of Minnesota，1981.
[12] Lu Shan-zhen. A note on the almost everywhere Convergence of Boc Bochner-Riesz means of multiple conjugate Fourier series. Lecture Notes in Math.，908，1982.
[13] Голубов Ь И. Матем，сб.，1975，96(2)：189—211.
[14] Bochner S. Trans. Amer. Math. Soc.，1936，40：175—207.
[15] Lippman G E. SIAM J. Math. Anal.，1973，4(4)：681—695.
[16] Stein E M. Ann. Math.，1961，73(1)：87—109.
[17] Saks S. Fundamenta Mathematicae，1934，22：257—261.
[18] Гоголпдзе Л Д. Изв. АНСССР，сер，Матем.，1977，41(4)：937—958.
[19] Wang Kun-Yang. Kexue Tongbao，1984，29(12)：19—24.
[20] 陆善镇. 中国科学，1984，14A(6)：492—503.
[21] 北师大数学系. 逼近论与调和分析论文集(Ⅳ)，1984，6.
[22] 北师大数学系. 调和分析与逼近论文集(Ⅲ)，1982，10.
[23] Жижиашвили Л В. Изв. АНСССР，сер. матем，1968，32：1 112—1 122.
[24] 王昆扬. 数学学报，1984，27(6)：811—816.
[25] Байбородов С П. Матем Заметки，1981，29(5)：711—730.
[26] 干昆扬. 数学年刊，1982，3(6)：789—802.
[27] 程民德，邓东皋. 多元调和分析的近代发展. 函数论专辑，1978.

Abstract This paper gives a short summary of those results in the theory of multiple Fourier series in which the author is interested. These results deal with convergence，linear summability and strong summability，approximation by linear means and the esitmates of Lebesgue's constants. Someopen problems are also posed.

北京师范大学学报(自然科学版)，1987，(4)：1—8.

一类奇异积分算子的逼近性质[①]

Approximation Behaviour of Certain Singular Integral Operators

摘要 对于 n 元连续周期函数及其共轭函数，由

$$\Gamma_R(f)(x)=\int_{|y|>1} f\left(x-\frac{y}{R}\right)|y|^{-n-1}\mathrm{d}y,\ (R>0)$$

定义的算子 Γ_R 在全测度集上的逼近性态被讨论且所得的结果被用来得到对于用

$$S_R(f)(x)=\sum_{|m|<R}\left(1-\frac{|m|}{R}\right)^{\frac{n-1}{2}}a_m(f)\mathrm{e}^{imx}$$

定义的广义 Riesz 平均的逼近的估计.

关键词 多重 Fourier 级数；多重共轭 Fourier 级数；Riesz 平均；逼近.

§1. 定义及结果

设 $n\in\mathbf{N}$，$n\geqslant2$，函数

$$\Gamma(t)=\frac{\gamma}{t^{n+1}},\ 0<t<+\infty,\ \gamma=\frac{\Gamma\left(\frac{n}{2}\right)}{2\pi^{\frac{n}{2}}} \tag{1.1}$$

在 $L(Q^n)$上定义算子 $\Gamma_R(R>0)$：

$$\Gamma_R(f)(x)=\int_{|y|>1}f\left(x-\frac{y}{R}\right)\Gamma(|y|)\mathrm{d}y. \tag{1.2}$$

---

① 国家自然科学基金资助项目.
收稿日期：1986-07-08.

下面，我们只讨论$C(Q^n)$中不满足 Lip 1 条件的函数．这样的函数的连续模ω满足

$$\overline{\lim_{t\to 0^+}}\ \omega(t)t^{-1}=+\infty.$$

设$f\in C(Q^n)-\mathrm{Lip}\,1$，$\omega$为$f$的连续模．定义

$$\delta_0=1,\ \delta_{k+1}=\min\left\{\delta:\ \max\left(\frac{\omega(\delta)}{\omega(\delta_k)},\ \frac{\omega(\delta_k)}{\omega(\delta)}\frac{\delta}{\delta_k}\right)=\frac{1}{6}\right\},$$

$$\Omega(\delta)=6^{-k}\text{当}\delta\in(\delta_{k+1},\ \delta_k],\ k\in\mathbf{N}^*.$$

并定义

$$\rho(f)(x)=\sup_{R>1}\frac{|\Gamma_R(f)(x)-f(x)|}{\omega\left(\frac{1}{R}\right)\ln\ln\frac{9}{\Omega\left(\frac{1}{R}\right)}},\tag{1.3}$$

$$L(f)(x)=\overline{\lim_{R\to+\infty}}\frac{|\Gamma_R(f)(x)-f(x)|}{\omega\left(\frac{1}{R}\right)\ln\ln\frac{9}{\Omega\left(\frac{1}{R}\right)}}.\tag{1.4}$$

取定一个p次齐次调和多项式$P(x)$，这里$p\geqslant 1$，并于核$K(x)=\dfrac{P(x)}{|x|^{n+p}}$引入 Calderón-Zygmund 意义下的共轭函数

$$f^*(x)=\lim_{\varepsilon\to 0,\rho\to+\infty}\frac{1}{(2\pi)^n}\int_{\varepsilon<|y|<\rho}f(x-y)K(y)\mathrm{d}y\tag{1.5}$$

并定义

$$\tilde{\rho}(f)(x)=\sup_{R>1}\frac{|\Gamma_R(f^*)(x)-f^*(x)|}{\omega\left(\frac{1}{R}\right)\left[\ln\ln\frac{9}{\Omega\left(\frac{1}{R}\right)}\right]^2}.\tag{1.6}$$

$$\widetilde{L}(f)(x)=\overline{\lim_{R\to+\infty}}\frac{|\Gamma_R(f^*)(x)-f^*(x)|}{\omega\left(\frac{1}{R}\right)\left[\ln\ln\frac{9}{\Omega\left(\frac{1}{R}\right)}\right]^2}.\tag{1.7}$$

结果是

定理 1 (1) $|\{x\in Q^n|\rho(f)(x)>\lambda\}|\leqslant A\mathrm{e}^{-\frac{\lambda}{A}}$，$\lambda>0$；式中$A$与$f$及$\lambda$无关；

(2) $L(f)(x)\leqslant A$ a. e. A与f无关.

定理 2 (1) $|\{x\in Q^n|\tilde{\rho}(f)(x)>\lambda\}|\leqslant A\mathrm{e}^{-\frac{\sqrt{\lambda}}{A}}$，$\lambda>0$，式中$A$与$f$

及λ无关；

(2) $\widetilde{L}(f)(x)\leqslant A$　a. e. A与f无关.

使用定理1、定理2推出一个广义Riesz平均在全测度集上逼近的结果，即§4定理3.

§2. 定理1的证明

定义集合

$$E_R(\lambda)=\left\{x\in Q^n \,\middle|\, |\Gamma_R(f)(x)-f(x)|>\lambda\omega\left(\frac{1}{R}\right)\ln\ln\frac{9}{\Omega\left(\frac{1}{R}\right)}\right\},$$

$$\Delta_k(\lambda)=\bigcup_{R\in[\delta_k^{-1},\delta_{k+1}^{-1})}E_R(\lambda),\qquad k\in\mathbf{N}^*,\ \lambda>0.$$

我们来估计$\Delta_k(\lambda)$的测度. 分两种情况.

第一种情况，$\omega(\delta_{k+1})=\frac{1}{6}\omega(\delta_k)$. 取$g$为$f$的$\delta_k^{-1}$阶最佳一致逼近三角多项式. 当$R\in[\delta_k^{-1},\ \delta_{k+1}^{-1})$时，有

$$\Gamma_R(f)(x)-f(x)=\Gamma_R(f-g)(x)+\Gamma_R(g)(x)-g(x)+(g-f)(x). \tag{2.1}$$

因此　$|\Gamma_R(f)(x)-f(x)|\leqslant 2\|f-g\|_C+\Gamma_R(g)(x)-g(x)|$

$$\leqslant c\omega\left(\frac{1}{R}\right)+|\Gamma_R(g)(x)-g(x)|.$$

记　$g_x(t)=\int_{|\xi|=1}[g(x-t\xi)-g(x)]\mathrm{d}\sigma(\xi),(t\geqslant 0)$. 那么

$$\begin{aligned}\Gamma_R(g)(x)-g(x)&=\gamma\int_1^{+\infty}g_x\left(\frac{t}{R}\right)t^{-2}\mathrm{d}t\\&=\gamma\left\{g_x\left(\frac{t}{R}\right)\frac{-1}{t}\,\Big|_1^{+\infty}+\int_1^{+\infty}\left(g_x\left(\frac{t}{R}\right)\right)'\frac{1}{t}\mathrm{d}t\right\}\\&=\gamma\left\{g_x\left(\frac{1}{R}\right)+\frac{1}{R}\int_1^{+\infty}g_x'\left(\frac{t}{R}\right)t^{-1}\mathrm{d}t\right\},\end{aligned}\tag{2.2}$$

其中

$$\begin{aligned}g_x'(s)&=\frac{\mathrm{d}}{\mathrm{d}s}\int_{|\xi|=1}[g(x-s\xi)-g(x)]\mathrm{d}\sigma(\xi)\\&=\sum_{j=1}^{n}\int_{|\xi|=1}g_j(x-s\xi)(-\xi_j)\mathrm{d}\sigma(\xi),\end{aligned}\tag{2.3}$$

g_j表示g对第j变元的偏导数，ξ_j为ξ的第j分量. 于是我们看到

$$|\Gamma_R(g)(x)-g(x)|\leqslant c\omega\left(\frac{1}{R}\right)+\frac{1}{R}\sum_{j=1}^{n}\left|\int_{|y|>\frac{1}{R}}g_j(x-y)K_j(y)\mathrm{d}y\right|,$$

其中核 $K_j(y)=\frac{\gamma y_j}{|y|^{n+1}}$. 令

$$R_j^*(g_j)(x)=\sup_{R>0}\left|\int_{|y|>\frac{1}{R}}g_j(x-y)K_j(y)\mathrm{d}y\right|$$

(事实上 R_j^* 是第 j 极大 Riesz 变换，在周期情形它是相应于 K_j 的极大共轭算子的常数倍). 注意到

$$\|g_j\|_C\leqslant c\delta_k^{-1}\omega(\delta_k)\leqslant cR\omega\left(\frac{1}{R}\right)^{[2]}$$

推出

$$|\Gamma_R(g)(x)-g(x)|\leqslant c\omega\left(\frac{1}{R}\right)\left\{1+\sum_{j=1}^{n}R_j^*\left(\frac{g_j}{\|g_j\|_C}\right)(x)\right\},R\in[\delta_k^{-1},\delta_{k+1}^{-1}].$$

由关于共轭极大算子的强型估计[3]，得

$$|\{x\in Q^n\mid R_j^*(g_j/\|g_j\|_C)(x)>\lambda\ln(k+2)\}|\leqslant A(k+2)^{-\frac{\lambda}{A}},$$

$$j=1,2,\cdots,n.$$

从而推出
$$|\Delta_k(\lambda)|\leqslant A(k+2)^{-\frac{\lambda}{A}}.\qquad(2.4)$$

在第二种情况下，$\frac{\delta_{k+1}}{\omega(\delta_{k+1})}=\frac{1}{6}\frac{\delta_k}{\omega(\delta_k)}$. 这时取 g 为 f 的 δ_{k+1}^{-1} 阶最佳一致逼近三角多项式，经类似的推导，同样得到(2.4).

于是，当 $\lambda>2A$ 时有

$$|\{x\in Q^n:\rho(f)(x)>\lambda\}|\leqslant A\sum_{k=0}^{+\infty}(k+2)^{-\frac{\lambda}{A}}\leqslant A\mathrm{e}^{-\frac{\lambda}{A}}.$$

这就是结论(1).

同时，因为对于任何 $m\in\mathbf{N}$ 有

$$\{x\in Q^n:L(f)(x)>2A\}\subset\bigcup_{k=m}^{+\infty}\Delta_k(2A),$$

所以 $L(f)(x)\leqslant 2A$ a. e. 这就证得结论(2).

§3. 定理 2 的证明

用 T 表示相应于核 K 的共轭算子，即 $T(f)=f^*$(参见(1.5)). 将(2.1)两边用 T 作用，得

$$\Gamma_R(f^*)(x)-f^*(x)=(\Gamma_R\circ T)(f-g)(x)+$$

$$\Gamma_R(g^*)(x)-g^*(x)+T(g-f)(x), \text{ a. e.} \tag{3.1}$$

显然

$$|(\Gamma_R \circ T)(f-g)(x)| \leqslant \|f-g\|_C \cdot \sup_{u>0} \left|(\Gamma_u \circ T)\left(\frac{f-g}{\|f-g\|_C}\right)(x)\right|$$

$$\leqslant c\omega\left(\frac{1}{R}\right)(HL \circ T)\left(\frac{f-g}{\|f-g\|_C}\right)(x), \quad R \in [\delta_k^{-1}, \delta_{k+1}^{-1}], \tag{3.2}$$

式中 HL 表示 Hardy-Littlewood 极大算子. 并且

$$|T(g-f)(x)| \leqslant \|g-f\|_C \left|T\left(\frac{g-f}{\|g-f\|_C}\right)(x)\right|$$

$$\leqslant c\omega\left(\frac{1}{R}\right)\left|T\left(\frac{g-f}{\|g-f\|_C}\right)(x)\right|, \quad R \in [\delta_k^{-1}, \delta_{k+1}^{-1}]. \tag{3.3}$$

现考察 $\Gamma_R(g^*)-g^*$. 在(2.2)(2.3)两边作用算子 T(注意到 g 是三角多项式)得

$$\Gamma_R(g^*)(x)-g^*(x)$$

$$=\gamma\left\{g_x^*\left(\frac{1}{R}\right)+\frac{1}{R}\sum_{j=1}^n\int_0^{+\infty}\frac{1}{t}\int_{|\xi|=1}g_j^*(x-s\xi)(-\xi_j)\mathrm{d}\sigma(\xi)\mathrm{d}t\right\}, \tag{3.4}$$

其中

$$g_x^*(t)=\int_{|\xi|=1}[g^*(x-t\xi)-g^*(x)]\mathrm{d}\sigma(\xi).$$

为了估计 $g_x^*(t)$, 我们单独证明下述引理.

引理 设 g 是 n 元三角多项式, g^* 是它关于核 K 的共轭函数, $g_j=\dfrac{\partial g}{\partial x_j}$, $j=1,2,\cdots,n$. 那么

$$|g_x^*(t)| \leqslant 2t\sum_{j=1}^n \|g_j\|_C(R_j^* \circ T)\left(\frac{g_j}{\|g_j\|_C}\right)(x), \quad t \geqslant 0.$$

证 记

$$h_j(\tau)=\frac{1}{\tau}\int_{|\xi|=1}g_j^*(x-\tau\xi)(-\xi_j)\mathrm{d}\sigma(\xi), \quad \tau>0.$$

那么

$$\frac{\mathrm{d}}{\mathrm{d}\tau}(g_x^*(\tau))=\tau\sum_{j=1}^n h_j(\tau), \quad \tau>0.$$

从而

$$g_x^*(t)=\int_0^t\tau\sum_{j=1}^n h_j(\tau)\mathrm{d}\tau$$

$$= \sum_{j=1}^{n} \left\{ \tau \int_0^\tau h_j(s)\mathrm{d}s \ \Big|_0^t - \int_0^t \int_0^\tau h_j(s)\mathrm{d}s\mathrm{d}\tau \right\}.$$

由于

$$\left| \int_0^t h_j(\tau)\mathrm{d}\tau \right| \leqslant 2\sup_{\varepsilon>0} \left| \int_\varepsilon^{+\infty} h_j(\tau)\mathrm{d}\tau \right|$$

$$= 2R_j^*(g_j^*)(x) = 2\|g_j\|_C (R_j^* \circ T)\left(\frac{g_j}{\|g_j\|_C}\right)(x),$$

所以引理成立.

现在我们接着证定理 2.

用引理，得

$$\left| g_x^*\left(\frac{1}{R}\right) \right| \leqslant c\omega\left(\frac{1}{R}\right) \sum_{j=1}^{n} (R_j^* \circ T)\left(\frac{g_j}{\|g_j\|_C}\right)(x), R \in [\delta_k^{-1}, \delta_{k+1}^{-1}].$$

同时，

$$\left| \frac{1}{R}\int_0^{+\infty} \frac{1}{t}\int_{|\xi|=1} g_j^*(x-t\xi)(-\xi_j)\mathrm{d}\sigma(\xi)\mathrm{d}t \right|$$

$$\leqslant \frac{1}{R}R_j^*(g_j^*)(x) = \frac{1}{R}\|g_j\|_C (R_j^* \circ T)\left(\frac{g_j}{\|g_j\|_C}\right)(x)$$

$$\leqslant c\omega\left(\frac{1}{R}\right)(R_j^* \circ T)\left(\frac{g_j}{\|g_j\|_C}\right)(x),\ R \in [\delta_k^{-1},\ \delta_{k+1}^{-1}).$$

故由(3.4)推出，当 $R \in [\delta_k^{-1},\ \delta_{k+1}^{-1})$ 时

$$|\Gamma_R(g^*)(x) - g^*(x)| \leqslant c\omega\left(\frac{1}{R}\right) \sum_{j=1}^{n} (R_j^* \circ T)\left(\frac{g_j}{\|g_j\|_C}\right)(x). \tag{3.5}$$

由(3.1)～(3.5)推出，当 $R \in [\delta_k^{-1},\ \delta_{k+1}^{-1})$ 时，

$$|\Gamma_R(f^*)(x) - f^*(x)| \leqslant c\omega\left(\frac{1}{R}\right)(HL \circ T)\left(\frac{f-g}{\|f-g\|_C}\right)(x) +$$

$$\left| T\left(\frac{f-g}{\|f-g\|_C}\right)(x) \right| + \sum_{j=1}^{n} (R_j^* \circ T)\left(\frac{g_j}{\|g_j\|_C}\right)(x). \tag{3.6}$$

我们说，一个$(p,\ p)$型算子 S $(1<p<+\infty)$，如果当 $p\geqslant 2$ 时满足

$$\|S(h)\|_p \leqslant cp^2 \|h\|_p,\ \ \forall h \in L^p(Q^n),$$

(其中 c 与 h 及 p 无关)，那么对于 $h \in L^\infty(Q^n)$ 有

$$|\{x \in Q^n : |S(h)(x)| > \lambda\}| \leqslant A\mathrm{e}^{-\frac{1}{A}\sqrt{\frac{\lambda}{\|h\|_\infty}}},\ (\lambda>0). \tag{3.7}$$

显然，(3.6)右端出现的算子 $HL \circ T$，$|T|$ 及 $R_j^* \circ T$ 都是这样的

算子，因而都满足(3.7).

现定义

$$\widetilde{E}_R(\lambda)=\left\{x\in Q^n \mid |\Gamma_R(f^*)(x)-f^*(x)|>\lambda\omega\left(\frac{1}{R}\right)\left[\ln\ln\frac{9}{\Omega\left(\frac{1}{R}\right)}\right]^2\right\},$$

$$\widetilde{\Delta}_k(\lambda)=\bigcup_{R\in[\delta_k^{-1},\delta_{k+1}^{-1}]}\widetilde{E}_R(\lambda),\ k\in\mathbf{N}^*.$$

那么，由(3.6)(3.7)，并注意到当 $R\in[\delta_k^{-1},\ \delta_{k+1}^{-1}]$时，$\ln\dfrac{9}{\Omega\left(\frac{1}{R}\right)}\geqslant k+2$，我们推出 $|\widetilde{\Delta}_k(\lambda)|\leqslant A\mathrm{e}^{-\frac{1}{A}\sqrt{\lambda}\ln(k+2)}=A(k+2)^{-\frac{\sqrt{\lambda}}{A}}$，$k\in\mathbf{N}^*$.

由此推出定理 2 的结论.

§4. 一个广义 Riesz 平均逼近定理

在[3]中我们已详细讨论过多重 Fourier 级数

$$\sigma(f)\sim\sum a_m(f)\mathrm{e}^{\mathrm{i}mx}$$

的广义 Riesz 平均(临界阶)

$${}^lS_R^{\frac{n-1}{2}}(f)(x)=\sum_{|m|<R}\left[1-\left(\frac{m}{R}\right)^l\right]^{\frac{n-1}{2}}a_m(f)\mathrm{e}^{\mathrm{i}mx},\quad R>0,$$

对连续函数及其共轭函数在全测度集上的逼近问题，但所得的结论只适用于 $l\geqslant2$ 的情形．一般说来，l 越大算子${}^lS_R^{\frac{n-1}{2}}$的逼近性能越好，现在我们容易利用前两节的结果得到 $l=1$ 时的结论.

把算子${}^lS_R^{\frac{n-1}{R}}$简记为 S_R．仍设 $f\in C(Q^n)-\mathrm{Lip}\ 1$．定义

$$d(f)(x)=\sup_{R>1}\frac{|S_R(f)(x)-f(x)|}{\omega\left(\frac{1}{R}\right)\ln\ln\dfrac{9}{\Omega\left(\frac{1}{R}\right)}},$$

$$l(f)(x)=\varlimsup_{R\to+\infty}\frac{|S_R(f)(x)-f(x)|}{\omega\left(\frac{1}{R}\right)\ln\ln\dfrac{9}{\Omega\left(\frac{1}{R}\right)}},$$

$$\widetilde{d}(f)(x)=\sup_{R>1}\frac{|S_R(f^*)(x)-f^*(x)|}{\omega\left(\frac{1}{R}\right)\left[\ln\ln\dfrac{9}{\Omega\left(\frac{1}{R}\right)}\right]^2},$$

$$\tilde{l}(f)(x)=\varlimsup_{R\to+\infty}\frac{|S_R(f^*)(x)-f^*(x)|}{\omega\left(\frac{1}{R}\right)\left[\ln\ln\frac{9}{\Omega\left(\frac{1}{R}\right)}\right]^2}.$$

定理 3 存在与 f 及 λ 无关的正数 A，使

(1) $|\{x\in Q^n \mid d(f)(x)>\lambda\}|\leqslant Ae^{-\frac{\lambda}{A}}$，$\lambda>0$；

(2) $l(f)(x)\leqslant A$ a. e.；

(3) $|\{x\in Q^n \mid \tilde{d}(f)(x)>\lambda\}|\leqslant Ae^{-\frac{\sqrt{\lambda}}{A}}$；

(4) $\tilde{l}(f)(x)\leqslant A$ a. e..

为证定理 3，需援引[3]的结果. 由[3]的定理 5，

$${}^lS_R^\alpha(f;x)=\sum_{j=0}^{2q+1}b_jS_R^{\alpha+j}(f;x)+{}^l\Gamma_R^\alpha(f)(x),\mathrm{Re}\,\alpha>-1,$$
其中 $S_j^{\alpha+j}$ 表示 $\alpha+j$ 阶 Riesz 平均，$l\in\mathbf{N}$，${}^l\Gamma_R^\alpha$ 由[3]中定义 3 给出：

$${}^l\Gamma_R^\alpha(f)(x)=\int_{R^n}f(x-y){}^l\gamma_n^\alpha(R|y|)R^n\mathrm{d}y,$$

式中${}^l\gamma_n^\alpha$ 当 l 为奇数时有如下表达式(见[3]，引理 3，(14))：

$${}^l\gamma_n^\alpha(t)=\frac{1}{(2\pi)^{\frac{n}{2}}}\left(\frac{c_0A_0}{t^{n+l}}+\frac{(-c_0+c_1)A_1}{t^{n+l+1}}\right)+O\left(\frac{1}{t^{n+l+2}}\right).$$

特别，当 $l=1$ 时由[3]的定理 2 的证明中见到

$${}^l\gamma_n^\alpha(t)=\frac{\alpha\Gamma(n+1)}{2^n\pi^{\frac{n}{2}}\Gamma\left(\frac{n}{2}+1\right)}t^{-(n+1)}+O(t^{-(n+3)}).$$

根据这些结果，利用关于 $S_R^{\frac{n-1}{2}}$(及 $S_R^{\frac{n-1}{2}+j}$，$j>0$)的已有结论[2]，使用本文前两节证明的定理 1 和定理 2 以及引理，就可分别推出定理 3 的结论(1)～(4). 其推导的细节此处不拟赘述.

顺序指出，从[3]的论证及本文的讨论可以看出，当 $l>1$ 时，不必限制 l 为整数，用算子${}^lS_R^{\frac{n-1}{2}}$ 对连续函数及其共轭函数在全测度集上逼近可以得到与 $l=2$ 时(见[2]或[3])同样的结果.

参考文献

[1] Stein E M. Singular Integrals and Differentiability Properties of Functions. Princeton Univ. Press，1970.

[2] Wang Kunyang. Approximation Theory and Its Applications，1985，

1(4): 19.

[3] Wang Kunyang. Acta Mathematica Sinica, 1986, 2(2): 178.

Abstract For continuous periodic functions of n variables ($n \geqslant 2$) and their conjugate functions the approximation behaviour on set of full measure of the operators Γ_R defined by

$$\Gamma_R(f)(x) = \int_{|y|>1} f\left(x - \frac{y}{R}\right) |y|^{-n-1} \mathrm{d}y \quad (R > 0)$$

is discussed and the result is applied to get the estimate for approximation by generalized Riesz means S_R defined as

$$S_R(f)(x) = \sum_{|m|<R} \left(1 - \frac{m}{R}\right)^{\frac{n-1}{2}} \alpha_m(f) \mathrm{e}^{\mathrm{i}mx}.$$

Keywords multiple Fourier Series; multiple conjugate Fourier Series; Riesz means, approximation.

科学通报，1987，32(15)：1 124—1 127.

再论多重共轭 Fourier 级数的强求和①

Further Discussion on Strong Summability of Multiple Conjugate Fourier Series

本文是文献[1]的继续，其目的是改进文献[1]关于多重共轭 Fourier 级数强求和的结果.

沿用文献[1]的记号. 设 $Q^k=\{x=(x_1, x_2, \cdots, x_k) \mid -\pi\leqslant x_j<\pi, j=1,2,\cdots,k\}$. $L(Q)$表示在 Q 上可积的函数的集合，设 $P(x)$是一个$n\geqslant1$ 次的 k 元齐次调和多项式. 对于 $f\in L(Q)$我们用 f_ε^* ($\varepsilon>0$)表示它关于核 $K(x)=\dfrac{P(x)}{|x|^{n+k}}$ 的截断共轭函数而 $\widetilde{S}u^{\frac{k-1}{2}}(f)$表示 f 的共轭 Fourier 级数的临界阶 Bochner-Riesz 平均. 现将文献[1]的主要结果叙述如下：

定理 A 设 $f\in L(Q)$. 设 f 在点 x 的一个邻域 $B(x, \delta)$上 L^p 可积，$\delta>0$，$p>1$. 若 f 满足下述条件：

$$\int_0^t \psi(\tau)\mathrm{d}\tau = o(t), \quad t\to 0, \tag{1}$$

$$\int_0^t |\psi(\tau)|^p\mathrm{d}\tau = O(t), \quad t\in(0,\delta), \tag{2}$$

其中

$$\psi(\tau)=\int_{|y|=1} f(x-\tau y)P(y)\mathrm{d}\sigma(y),$$

则对于任何 $q>0$，

---

① 国家自然科学基金资助项目.

收稿日期：1986-04-16.

$$\lim_{R\to+\infty}\frac{1}{R}\int_0^R \left|\widetilde{S}_u^{\frac{k-1}{2}}(f;x)-f_{\frac{1}{u}}^*(x)\right|^q \mathrm{d}u=0. \tag{3}$$

受王福春[2]的启发，我们现在减弱(2)式中的 L^p 条件，得到了一个 $L\ln^+L$ 条件以取代之.

定理 1　设 $f\in L(Q)$. 设 $f\ln^+|f|$ 在 $B(x,\delta)(\delta>0)$上可积. 若 f 满足条件(1)及下述条件：

$$\int_0^t |\psi(\tau)|(1+\ln^+|\psi(\tau)|)\mathrm{d}\tau\leqslant At \text{ 对于 } t\in(0,\delta), \tag{4}$$

其中 A 是与 t 无关的常数(我们可以认为 $A\geqslant 1$)，则等式(3)成立.

我们指出，我们的论证同样适用于非共轭情形. 于是我们得到下述定理，它改进了陆善镇[3]的主要结果.

定理 2　设 $f\in L(Q)$. 设 $f\ln^+|f|$ 在 $B(x,\delta)(\delta>0)$上可积. 置

$$\varphi(\tau)=\int_{|y|=1}[f(x-\tau y)-f(x)]\mathrm{d}\sigma(y).$$

若 f 满足下列条件：

$$\int_0^t \varphi(\tau)\mathrm{d}\tau=o(t),\quad t\to 0, \tag{5}$$

$$\int_0^t |\varphi(\tau)|(1+\ln^+|\varphi(\tau)|)\mathrm{d}\tau\leqslant At,\quad t\in(0,\delta), \tag{6}$$

则对于任何 $q>0$，

$$\lim_{R\to+\infty}\frac{1}{R}\int_0^R \left|S_u^{\frac{k-1}{2}}(f;x)-f(x)\right|^q \mathrm{d}u=0, \tag{7}$$

其中 $S_u^{\frac{k-1}{2}}(f)$表示 f 的 Fourier 级数的临界阶 Bochner-Riesz 平均.

我们从文献[1]的(4.19)(4.21)得知，为证定理 1 只需证明量

$$\beta(\varepsilon)\xlongequal{\text{def.}}\varlimsup_{R\to+\infty}\frac{1}{R}\int_0^R\left|\int_{\frac{1}{\varepsilon u}}^{+\infty}\frac{\psi(t)}{t}\cos ut\right|^q \mathrm{d}u,$$

当 $\varepsilon\to 0^+$ 时是个无穷小量.

注意到

$$\left|\int_{\frac{1}{\varepsilon R}}^{\frac{1}{\varepsilon u}}\frac{\psi(t)}{t}\cos ut\ \mathrm{d}t\right|^q=o(1)\left(\ln\frac{R}{u}\right)^q,\quad R\geqslant u\to+\infty,$$

我们见到

$$\begin{aligned}&\frac{1}{R}\int_0^R\left|\int_{\frac{1}{\varepsilon R}}^{\frac{1}{\varepsilon u}}\frac{\psi(t)}{t}\cos ut\mathrm{d}t\right|^q\mathrm{d}u\\&=\int_0^{+\infty}o(1)t^{-2}(\ln t)^q\mathrm{d}t=o(1),\quad R\to+\infty.\end{aligned}$$

于是我们得到

$$\beta(\varepsilon)\leqslant c_q \overline{\lim_{R\to+\infty}}\frac{1}{R}\int_0^R\left|\int_{\frac{1}{\varepsilon R}}^{\frac{\delta}{\varepsilon}}\frac{\psi(t)}{t}\cos ut\,\mathrm{d}t\right|^q\mathrm{d}u.$$

为了方便起见，我们只对于 $q=2$ 的情形来叙述证明．我们的方法同样适用于一般情形，但细节将十分冗繁．

现设 $q=2$．显然，我们可以认为 $\delta=1$，$\varepsilon<\mathrm{e}^{-100}$ 且

$$R>\frac{1}{\varepsilon}\mathrm{e}^{100}.$$

我们有

$$\begin{aligned}
&\int_0^R\left|\int_{\frac{1}{\varepsilon R}}^{\mathrm{e}^{-1}}\frac{\psi(t)}{t}\cos ut\ \mathrm{d}t\right|^2\mathrm{d}u\\
=&\int_{\frac{1}{\varepsilon R}}^{\mathrm{e}^{-1}}\int_{\frac{1}{\varepsilon R}}^{\mathrm{e}^{-1}}\frac{\psi(s)\psi(t)}{st}\frac{1}{2}\left[\frac{\sin R(s+t)}{s+t}+\frac{\sin R(s-t)}{s-t}\right]\mathrm{d}s\mathrm{d}t\\
\leqslant&\left(\int_{\frac{1}{\varepsilon R}}^{\mathrm{e}^{-1}}\frac{|\psi(t)|}{t^{\frac{3}{2}}}\mathrm{d}t\right)^2+\int_{\frac{1}{\varepsilon R}}^{\mathrm{e}^{-1}}\frac{|\psi(s)|}{s^2}\int_{\frac{1}{\varepsilon R}}^{s}|\psi(t)|\left|\frac{\sin R(s-t)}{s-t}\right|\mathrm{d}t\mathrm{d}s.
\end{aligned}$$

根据条件(4)，我们得到

$$\left(\int_{\frac{1}{\varepsilon R}}^{\mathrm{e}^{-1}}\frac{|\psi(t)|}{t^{\frac{3}{2}}}\mathrm{d}t\right)^2\leqslant C\varepsilon R,$$

其中 C 代表一个与 ε 和 R 无关的常数．

我们记

$$H_R(u)=\left|\frac{\sin Ru}{u}\right|,$$

且我们只用到 H_R 的下述性质：

$$0\leqslant H_R(u)\leqslant R,\quad |u|\leqslant\frac{1}{R},$$

$$0\leqslant H_R(u)\leqslant\frac{1}{u},\quad u>\frac{1}{R}.$$

现在我们看到

$$\beta(\varepsilon)\leqslant C\left\{\varepsilon+\overline{\lim_{R\to+\infty}}\frac{1}{R}\int_{\frac{1}{\varepsilon R}}^{\mathrm{e}^{-1}}\frac{|\psi(s)|}{s^2}\int_{\frac{1}{\varepsilon R}}^{s}|\psi(t)|H_R(s-t)\mathrm{d}t\mathrm{d}s\right\}.\qquad(8)$$

令 $\quad k_0=\left[\ln\frac{1}{\varepsilon}\right](k_0\geqslant100)$ 及 $k_1=[\ln R]-1(k_1\geqslant199)$，

设 $v_k=R^{-1}\mathrm{e}^k$，$k=k_0$，k_0+1，…，k_1+1．

引理 1 设

$$P(s)=\frac{1}{s^2}\int_{v_{k_0}}^{s}|\psi(t)|H_R(s-t)\mathrm{d}t, s\in(v_{k_0},1),$$

则
$$P(s)\leqslant 4AR[s\ln(Rs)]^{-1}.$$

证 置

$$I(s)=\int_{v_{k_0}}^{s}H_R(s-t)\mathrm{d}t,$$

则
$$I(s)=\int_{0}^{s-v_{k_0}}H_R(u)\mathrm{d}u\leqslant 1+\ln(Rs)<2\ln(Rs).$$

用 $J(t)=t(1+\ln^+t)$，$t\in[0,+\infty)$来定义 J．根据 Jensen 公式[4]，我们得到

$$J\left(\frac{p(s)s^2}{I(s)}\right)\leqslant\int_{v_{k_0}}^{s}J(|\psi(t)|)\frac{H_R(s-t)\mathrm{d}t}{I(s)},$$

$$s^2P(s)+\left(1+\ln^+\frac{P(s)s^2}{I(s)}\right)\leqslant\int_{v_{k_0}}^{s}J(|\psi(t)|)H_R(s-t)\mathrm{d}t.$$

据条件(4)我们见到右端不大于 ARs．那么，如果

$$P(s)>\frac{2R}{s\ln(Rs)}$$

的话，就有

$$\frac{P(s)s^2}{I(s)}>\frac{Rs}{\ln^2(Rs)}>1,$$

$$1+\ln^+\left[\frac{P(s)s^2}{I(s)}\right]>\ln Rs-2\ln\ln Rs>\frac{1}{4}\ln Rs.$$

从而我们得到了所需要的不等式.

引理 2 写

$$a_k=\int_{v_k}^{v_{k+1}}P(s)\mathrm{d}s,\quad k_0\leqslant k\leqslant k_1,$$

则
$$a_k\leqslant 4ARk\mathrm{e}^{-k}.$$

证 我们有

$$\begin{aligned}a_k&=\int_{v_{k_0}}^{v_k}|\psi(t)|\int_{v_k}^{v_{k+1}}\frac{1}{s^2}H_R(s-t)\mathrm{d}s\mathrm{d}t+\\&\quad\int_{v_k}^{v_{k+1}}|\psi(t)|\int_{t}^{v_{k+1}}\frac{1}{s^2}H_R(s-t)\mathrm{d}s\mathrm{d}t\\&\leqslant\frac{1}{v_k^2}\int_{v_{k_0}}^{v_{k+1v}}|\psi(t)|\int_{0}^{v_{k+1}}H_R(s)\mathrm{d}s\mathrm{d}t\\&\leqslant\frac{1}{v_k^2}Av_{k+1}\left(1+\int_{\frac{1}{k}}^{\frac{1}{k}\mathrm{e}^{k+1}}\frac{1}{s}\mathrm{d}s\right)<4ARk\mathrm{e}^{-k}.\end{aligned}$$

引理 3 令

$$b_k=\int_{v_k}^{v_{k+1}}|\psi(s)|P(s)\mathrm{d}s,\quad k_0\leqslant k\leqslant k_1,$$

则
$$b_k\leqslant 72A^2Rk^{-2}.$$

证 用 Jensen 公式得

$$J\left(\frac{b_k}{a_k}\right)\leqslant\frac{1}{a_k}\int_{v_k}^{v_{k+1}}J(|\psi(s)|)P(s)\mathrm{d}s.$$

根据引理 1 及条件(4)我们得到

$$\int_{v_k}^{v_{k+1}}J(|\psi(s)|)P(s)\mathrm{d}s\leqslant\int_{v_k}^{v_{k+1}}J(|\psi(s)|)\frac{4AR}{s\ln(Rs)}\mathrm{d}s$$

$$\leqslant 4AR\frac{1}{v_k\ln(Rv_k)}\int_{v_k}^{v_{k+1}}J(|\psi(s)|)\mathrm{d}s\leqslant 4\mathrm{e}A^2Rk^{-1}.$$

因此

$$b_k\left[1+\ln^+\left(\frac{b_k}{a_k}\right)\right]<12A^2Rk^{-1}.$$

若 $b_k>4ARk^{-2}$，则据引理 2，我们有

$$\ln\left(\frac{b_k}{a_k}\right)>\ln(k^{-3}\mathrm{e}^k)=k-3\ln k>\frac{1}{6}\ln k.$$

于是我们得到
$$b_k<72A^2Rk^{-2},$$
从而完成了证明.

引理 4

$$\beta(\varepsilon)\leqslant C\left(\varepsilon+\frac{1}{\ln\frac{1}{\varepsilon}}\right),\quad \varepsilon<\mathrm{e}^{-100}.$$

证 置

$$I_R(\varepsilon)=\int_{\frac{1}{\varepsilon R}}^{\mathrm{e}^{-1}}\frac{|\psi(s)|}{s^2}\int_{\frac{1}{\varepsilon R}}^{s}|\psi(t)|H_R(s-t)\mathrm{d}t\mathrm{d}s,$$

那么
$$I_R(\varepsilon)\leqslant\sum_{k=k_0}^{k_1}b_k.$$

用引理 3 并注意到

$$k_0=\left[\ln\frac{1}{\varepsilon}\right]\geqslant 100,$$

我们得到

$$I_R(\varepsilon)\leqslant\sum_{k=k_0}^{k_1}72A^2Rk^{-2}\leqslant 72A^2R\int_{k_0-1}^{+\infty}\frac{\mathrm{d}t}{t^2}$$

$$= 72A^2R\frac{1}{k_0-1} < 100A^2R\frac{1}{\ln\frac{1}{\varepsilon}}.$$

将此不等式代入(8)我们就得到引理 4 的结论.

现在我们已经证得定理 1.

参考文献

[1] 王昆扬. 多重共轭 Fourier 级数的强求和. 数学学报，1986，29(2)：156—175.

[2] Wang Fu-Traing. Strong summability of Fourier series. Duke Math. J.，1945，12：77—87.

[3] 陆善镇. Bochner-Riesz 球形平均的强性求和. 中国科学，A 辑，1984，(6)：492—503.

[4] Hardy G H，Littlewood J E，Pólya G. Inequalities. Cambridge，1934.

北京师范大学学报(自然科学版)，1988，(2)：18—27.

关于多重 Fourier 级数的收敛性[①]

Convergence of Multiple Fourier Series

摘要 经典的 Rogosinsky 恒等式推广到多元情形并用来求得一致收敛和 a. e. 收敛的判别条件. 所得结果推广了一元级数的 Salem-Стечкин 定理.

关键词 多重 Fourier 级数；收敛.

§0. 引言

在讨论多重 Fourier 级数时，对于可积类或连续类的函数，将临界阶的 Bochner-Riesz 平均看作一元 Fourier 部分和的类比，这一观点是恰当的. 我们的讨论完全基于这一观点.

在引入部分和的高次算术平均的概念后，研究了这种线性平均在 Lebesgue 点上的收敛性及一致收敛性. 并把一元的 Rogosinsky 恒等式推广到多元情形. 使用这一公式及有关结果求得 C 类函数部分和一致收敛及 L 类函数部分和 a. e. 收敛的充分条件. 全部论证适用于一元情形. 所得的收敛条件在一元时蕴含了 Salem-Стечкин 判别法.

这里，a. e. 收敛是在 L 类上讨论的，而以往这方面的讨论都限制在 $L\ln^+L$ 类上. 另外，从我们关于一致收敛的讨论可得到一致逼近阶的估计.

① 收稿日期：1987-04-10.

§1. 部分和的高次算术平均

设 $f\in L(Q^n)$，$Q^n=\{(x_1,\ x_2,\ \cdots,\ x_n)\mid -\pi\leqslant x_j<\pi,\ j=1,2,\cdots,n\}$. 将 f 的临界阶 Bochner-Riesz 平均简写作 $S_R(f)$，即

$$S_R(f)(x)=\sum_{|m|<R}\left(1-\frac{|m|^2}{R^2}\right)^{\frac{n-1}{2}}c_m(f)\mathrm{e}^{\mathrm{i}mx},\quad R>0. \tag{1.1}$$

$(S_0(f)(x)=c_0(f))$. 对于 $0<r\leqslant R$，定义

$$S_{R,r}(f)(x)=\sum_{|m|<r}\left(1-\frac{|m|^2}{R^2}\right)^{\frac{n-1}{2}}c_m(f)\mathrm{e}^{\mathrm{i}mx},$$

$(S_{R,0}(f)(x)=c_0(f))$.

在往下的讨论中，$R>0$ 暂且是固定的，x 也是固定的. 我们简记

$$a_m=\left(1-\frac{|m|^2}{R^2}\right)^{\frac{n-1}{2}}c_m(f)\mathrm{e}^{\mathrm{i}mx},$$

$$S_{R,r}(f)=\sum_{|m|<r}a_m,\quad 0<r\leqslant R.$$

对于正整数 k，定义 $S_R(f)$的 k 次算术平均为

$$\begin{aligned}
\sigma_{R,r}^k(f)&=\frac{k!}{r^k}\int_0^r\int_0^{r_1}\cdots\int_0^{r_{k-1}}S_{R,r_k}(f)\mathrm{d}r_k\cdots\mathrm{d}r_1\\
&=\frac{k!}{r^k}\sum_{m\in\mathbf{Z}^n}\int_{D\triangleq\left\{\begin{subarray}{l}(r_1,r_2,\cdots,r_k):\\ 0<r_k<\cdots<r_1<r\end{subarray}\right\}}a_m\chi_D(|m|)\mathrm{d}r_k\cdots\mathrm{d}r_1\\
&=\frac{k!}{r^k}\sum_{|m|<r}a_m\cdot\int_{|m|}^r\int_{|m|}^{r_1}\cdots\int_{|m|}^{r_{k-1}}\mathrm{d}r_k\cdots\mathrm{d}r_1\\
&=\frac{1}{r^k}\sum_{|m|<r}a_m(r-|m|)^k\\
&=\sum_{|m|<r}a_m\left(1-\frac{|m|}{r}\right)^r,\quad 0<r\leqslant R.
\end{aligned}\tag{1.2}$$

并令 $\sigma_{R,0}^k(f)=\lim\limits_{r\to0^+}\sigma_{R,r}^k(f)=a_0=c_0(f)$. 特别地写 $\sigma_{R,R}^k(f)=\sigma_R^k(f)$，即

$$\sigma_R^k(f)(x)=\sum_{|m|<R}\left(1-\frac{|m|}{R}\right)^k\left(1-\frac{|m|^2}{R^2}\right)^{\frac{n-1}{2}}c_m(f)\mathrm{e}^{\mathrm{i}mx} \tag{1.3}$$

定义核

$$K_R^k(y)=\sum_{|m|<R}\left(1-\frac{|m|}{R}\right)^k\left(1-\frac{|m|^2}{R^2}\right)^{\frac{n-1}{2}}\mathrm{e}^{\mathrm{i}my},$$

那么

$$\sigma_R^k(f)(x) = \frac{1}{(2\pi)^n}\int_{Q^n} f(x-y)\cdot K_R^k(y)\mathrm{d}y. \tag{1.4}$$

只考虑 $k=2,3,\cdots,x+1$ 的情形.

定义一元函数

$$\varphi(t)=\begin{cases}(1-t)^k(1-t^2)^{\frac{n-1}{2}}, & 0\leqslant t<1,\\ 0, & t\geqslant 1.\end{cases}$$

置 $\Phi(x)=\varphi(|x|)$，$x\in\mathbf{R}^n$. 那么

$$K_R^k(y) = \sum_{m\in\mathbf{Z}^n}\Phi\left(\frac{m}{R}\right)\mathrm{e}^{\mathrm{i}my}. \tag{1.5}$$

易见

$$\hat{\Phi}(x) = \frac{1}{(2\pi)^n}\int_{\mathbf{R}^n}\Phi(y)\mathrm{e}^{-\mathrm{i}xy}\mathrm{d}y = \frac{1}{(2\pi)^{\frac{n}{2}}}\int_0^1\varphi(t)t^{n-1}\frac{J_{\frac{n}{2}-1}(t|x|)}{(t|x|)^{\frac{n}{2}-1}}\mathrm{d}t. \tag{1.6}$$

式中 J_v 表示第一类 Bessel 函数.

记 $u_0(t)=\frac{1}{(2\pi)^{\frac{n}{2}}}\varphi(t)$，并令 $l=\left[\frac{n+3}{2}\right]$. 那么 $t=1$ 是 u_0 的 $\frac{n+3}{2}$ 级零点，且 $u_0\in C^l[0, 1]$，

$$u_0(1)=u_0'(1)=\cdots=u_0^{(l-1)}(1)=0.$$

令(对于 $t\in(0, 1)$)，

$$u_1(t)=u_0'(t)t^{-1},\quad u_{\mu+1}(t)=u_\mu'(t)t^{-1},\quad \mu\in\mathbf{N}.$$

那么，容易归纳出表达式

$$u_\mu(t) = \sum_{j=1}^{\mu} b_{\mu,j}u_0^{(\mu)}t^{-2\mu+j}, \mu\in\mathbf{N}.$$

于是

$$\lim_{t\to 0^+}t^{2\mu-1}u_\mu(t)=0,\quad \mu\in\mathbf{N}.$$

并且，当 $\mu=0,1,\cdots,l-1$ 时，

$$u_\mu(1)=\lim_{t\to 1^-}u_\mu(t)=0.$$

定义

$$v_\nu(t)=t^{\frac{n}{2}-1+\nu}J_{\frac{n}{2}-1+\nu}(t|x|),\qquad \nu\in\mathbf{Z}.$$

式中 $|x|$ 是正数(即 $x\neq 0$). 那么，有递推式

$$v_\nu'(t)=|x|tv_{\nu-1}(t).$$

并且有估计式

$$|v_\nu(t)|\leqslant Mt^{n-2+2\nu},\qquad t\geqslant 0,$$

$$|v_\nu(t)|\leqslant Mt^{\frac{n-3}{2}+\nu}\frac{1}{\sqrt{|x|}},\qquad t\geqslant 0,\ x\in\mathbf{R}^n,\ x\neq 0.$$

其中 M 是只与 n，ν 有关的常数[1].

根据上述事实，分部积分得到

$$\hat{\Phi}(x)=(-1)^l|x|^{-\frac{n}{2}-l+1}\int_0^1 tu_l(t)v_l(t)\mathrm{d}t.$$

于是得到

$$|\hat{\Phi}(x)|\leqslant\frac{M}{1+|x|^{n+1}},\ k\in\{2,\ 3,\ \cdots,\ n+1\}.\tag{1.7}$$

式中常数 M 只与 n 有关.

令 $(2\pi)^n\hat{\Phi}(x)=\Psi(x)$，则 $\Psi\in L(\mathbf{R}^n)$，$\hat{\Psi}=\Phi$. 再定义 $\Psi_R(x)=R^n\Psi(Rx)(R>0)$. 那么

$$\hat{\Psi}_R(y)=\hat{\Psi}\left(\frac{y}{R}\right)=\Phi\left(\frac{y}{R}\right),\ y\in\mathbf{R}^n.$$

用 Poisson 求和公式[2]得到

$$K_R^k(y)=\sum_{m\in\mathbf{Z}^n}\Psi_R(x+2\pi m),\quad k\geqslant 2.\tag{1.8}$$

由此推出

引理 1　设 $f\in L(Q^n)$，$R>0$，则当 $k\in\{2,\ 3,\ \cdots,\ n+1\}$ 时，$S_R(f)$ 的次平均 $\sigma_R^k(f)$ 有如下表达式：

$$\sigma_R^k(f)(x)=\frac{1}{(2\pi)^n}\int_0^{+\infty}\int_{|\xi|=1}f\left(x-\frac{t}{R}\xi\right)\mathrm{d}\sigma(\xi)t^{n-1}\psi(t)\mathrm{d}t.\tag{1.9}$$

式中 ψ 由等式 $\psi(|x|)=\Psi(x)(x\in\mathbf{R}^n)$ 定义.

由此引理立即推出两个定理.

定理 1　设 $f\in L(Q^n)$. 若 x 是 f 点 Lebesgue 点，则当 $k\in\{2,3,\cdots,n+1\}$ 时，

$$\lim_{R\to+\infty}\sigma_R^k(f)(x)=f(x).$$

定理 2　设 $f\in C(Q^n)$，

$$\omega_2(f;\delta)=\sup\{|f(x+h)+f(x-h)-2f(x)|:x\in\mathbf{R}^n,|h|\leqslant\delta\}.$$

那么，存在只与 n 有关的常数 M 使

$$\|\sigma_R^k(f)-f\|_C\leqslant\frac{M}{R}\int_0^R\omega_2\left(f;\frac{1}{t}\right)\mathrm{d}t,\quad k\in\{2,3,\cdots,n+1\}.$$

现在我们来考察

$$\sigma_{R,r}^{n+1}(f)(x)=\sum_{|m|<r}\left(1-\frac{|m|}{r}\right)^{n+1}\cdot\left(1-\frac{|m|^2}{R^2}\right)^{\frac{n-1}{2}}c_m(f)\mathrm{e}^{\mathrm{i}mx}.$$

定义

$$\varphi^a(t)=\begin{cases}(1-t)^{n+1}\cdot(1-at^2)^{\frac{n-1}{2}}, & 0\leqslant t<1,\\ 0, & t\geqslant 1.\end{cases}$$

参数 $a\in(0,1]$. 置 $\Phi^a(x)=\varphi^a(|x|)$，$x\in\mathbf{R}^n$. 有

$$\hat{\Phi}^a(x)=\frac{1}{(2\pi)^{\frac{n}{2}}}\int_0^1\varphi^a(t)t^{n-1}\frac{J_{\frac{n}{2}-1}(t|x|)}{(t|x|)^{\frac{n}{2}-1}}\mathrm{d}t.$$

类似从(1.6)到(1.7)的推导，注意到 $a\in[0,1]$，我们推出

$$|\hat{\Phi}^a(x)|\leqslant\frac{M}{1+|x|^{n+1}}.$$

常数 M 只与 n 有关而与 a 无关. 用 Poisson 求和公式，得

$$\sum_{m\in\mathbf{Z}^n}r^n\hat{\Phi}^a(r(x+2\pi m))=\frac{1}{(2\pi)^n}\sum_{m\in\mathbf{Z}^n}\Phi^a\left(\frac{m}{r}\right)\mathrm{e}^{\mathrm{i}mx}$$

$$=\frac{1}{(2\pi)^n}\sum_{|m|<r}\left(1-\frac{|m|}{r}\right)^{n+1}\left(1-a\frac{|m|^2}{r^2}\right)^{\frac{n-1}{2}}\mathrm{e}^{\mathrm{i}mx}$$

$$\overset{\mathrm{df}}{=}K_r(x;a),\quad 0<r<+\infty.$$

于是

$$\sigma_{R,r}^{n+1}(f)(x)=\int_Q f(x-y)\cdot K_r\left(y;\frac{r^2}{R^2}\right)\mathrm{d}y$$

$$=\int_{\mathbf{R}^n}f(x-y)\cdot r^n\hat{\Phi}^{\left(\frac{r}{R}\right)^2}(ry)\mathrm{d}y,$$

依等式 $\varphi^a(|x|)=\hat{\Phi}^a(x)$ 定义 $\psi^a(t)\quad(t\geqslant 0)$. 得

$$\sigma_{R,r}^{n+1}(f)(x)=\int_0^{+\infty}\int_{|\xi|=1}f\left(x-\frac{t}{r}\xi\right)\mathrm{d}\sigma(\xi)t^{n-1}\varphi\left(\frac{r}{R}\right)^2(t)\mathrm{d}t.$$

其中 $\varphi^{\left(\frac{r}{R}\right)^2}(t)$ 满足估计式

$$\left|\varphi^{\left(\frac{r}{R}\right)^2}(t)\right|\leqslant\frac{M}{1+t^{n+1}},\quad 0<r\leqslant R.$$

由此立即得到

定理 3 设 $f\in L(Q^n)$. 若 x 是 f 的 Lebesgue 点，则在 $0<r\leqslant R$ 条件下，关于 R 一致成立

$$\lim_{r\to+\infty}\sigma_{R,r}^{n+1}(f)(x)=f(x).$$

定理 4 设 $f\in C(Q^n)$，则关于 $R\geqslant r$ 一致有

$$\|\sigma_{R,r}^{n+1}(f)-f\|_C\leqslant M\cdot\frac{1}{r}\int_0^r\omega_2\left(f;\frac{1}{t}\right)\mathrm{d}t.$$

其中 M 只与 n 有关.

§ 2. Rogosinsky 恒等式

用 ω_n 表示 $\mathbf{R}^n$ 中单位球面的面积，即 $\omega_n=\dfrac{2\pi^{\frac{n}{2}}}{\Gamma\left(\frac{n}{2}\right)}$；用 V_n 表单位球体的体积，即 $V_n=\dfrac{1}{n}\omega_n$.

设 $f\in L(Q^n)$，$\alpha>0$. 定义(对于 $R>0$)

$$\begin{aligned}\mu_R(f)(x;\alpha)&=\frac{1}{\omega_n}\int_{|\xi|=1}S_R(f)(x+\alpha\xi)\mathrm{d}\sigma(\xi)\\&=\sum_{|m|<R}\left(1-\frac{|m|^2}{R^2}\right)^{\frac{n-1}{2}}c_m(f)\mathrm{e}^{imx}\cdot2^{\frac{n}{2}-1}\Gamma\left(\frac{n}{2}\right)\cdot\frac{J_{\frac{n}{2}-1}(\alpha|m|)}{(\alpha|m|)^{\frac{n}{2}-1}}.\end{aligned}$$

对于 $h>0$，

$$\begin{aligned}&\frac{n}{h^n}\int_0^k\alpha^{n-1}\mu_R(f)(x;\alpha)\mathrm{d}\alpha\\&=\frac{1}{V_nh^n}\int_{|\xi|<h}S_R(f)(x+\xi)\mathrm{d}\xi\\&=\sum_{|m|<R}\left(1-\frac{|m|^2}{R^2}\right)^{\frac{n-1}{2}}c_m(f)\mathrm{e}^{imx}\cdot2^{\frac{n}{2}}\Gamma\left(\frac{n}{2}+1\right)\cdot\frac{J_{\frac{n}{2}}(h|m|)}{(h|m|)^{\frac{n}{2}}}.\end{aligned}$$

取 λ 为 $J_{\frac{n}{2}+1}$ 的最小正根. 定义

$$T_R(f)(x)=\frac{1}{V_n\lambda^n}R^n\int_{|\xi|<\frac{\lambda}{R}}S_R(f)(x+\xi)\mathrm{d}\xi,\tag{2.1}$$

并令

$$H(t)=2^{\frac{n}{2}}\Gamma\left(\frac{n}{2}+1\right)\frac{J_{\frac{n}{2}}(t)}{t^{\frac{n}{2}}}.\tag{2.2}$$

那么

$$T_R(f)(x)=\sum_{|m|<R}\left(1-\frac{|m|^2}{R^2}\right)^{\frac{n-1}{2}}c_m(f)\mathrm{e}^{imx}H\left(\frac{\lambda}{R}|m|\right).\tag{2.3}$$

对于固定的 $R>0$，沿用 § 1 的记号，可将 $T_R(f)$ 表为如下的关于变元 r 的 Riemann-Stieltjes 积分

$$T_R(f) - S_{R,0} = \int_0^R H\left(\frac{\lambda}{R}r\right)\mathrm{d}S_{R,r}(f).$$

分部积分，注意到 $H(0)=1$，得到

$$T_R(f) = H(\lambda)S_R(f) - \int_0^R S_{R,r}(f)\mathrm{d}H\left(\frac{\lambda}{R}r\right).$$

$$T_R(f) - f = H(\lambda)(S_R(f) - f) + \int_0^R (f - S_{R,r}(f))\mathrm{d}H\left(\frac{\lambda}{R}r\right)$$

$$= H(\lambda)(S_R(f) - f) + \frac{\lambda}{R}\int_0^R H'\left(\frac{\lambda}{R}r\right)\mathrm{d}[r(f - \sigma_{R,r}^1(f))]$$

$$= H(\lambda)(S_R(f) - f) + \lambda H'(\lambda)(f - \sigma_{R,R}^1(f)) -$$

$$\left(\frac{\lambda}{R}\right)^2\int_0^R r[f - \sigma_{R,r}(f)]H''\left(\frac{\lambda}{R}r\right)\mathrm{d}r$$

$$= H(\lambda)(S_R(f) - f) + \lambda H'(\lambda)(f - \sigma_R^1(f)) -$$

$$\left(\frac{\lambda}{R}\right)^2\int_0^R H''\left(\frac{\lambda}{R}r\right)\mathrm{d}\left[\frac{1}{2}r^2(f - \sigma_{R,r}^2(f))\right]$$

$$= H(\lambda)(S_R(f) - f) + \lambda H'(\lambda)(f - \sigma_R^1(f)) -$$

$$\frac{1}{2}\lambda^2 H''(\lambda)(f - \sigma_R^2(f)) + \frac{1}{2}\left(\frac{\lambda}{R}\right)^3\int_0^R H'''\left(\frac{\lambda}{R}r\right)r^2(f - \sigma_{R,r}^2(f))\mathrm{d}r. \tag{2.4}$$

继续做分部积分，一直到求得

$$T_R(f) - f$$

$$= H(\lambda)(S_R(f) - f) + \sum_{j=1}^{n+1}\frac{(-1)^{j-1}}{j!}\lambda^j H^{(j)}(\lambda)(f - \sigma_R^j(f)) +$$

$$(-1)^{n+1}\left(\frac{\lambda}{R}\right)^{n+2}\frac{1}{(n+1)!}\int_0^R H^{(n+2)}\left(\frac{\lambda}{R}r\right)r^{n+1}(f - \sigma_{R,r}^{n+1}(f))\mathrm{d}r.$$

因为 λ 是 $J_{\frac{n}{2}+1}$ 的最小正根，所以

$$H'(\lambda) = 0,\ H(\lambda) \neq 0. \tag{2.5}$$

定义

$$\gamma_R(f)(x) = \frac{1}{H(\lambda)}\sum_{j=2}^{n+1}\frac{(-1)^j}{j!}\lambda^j H^{(j)}(\lambda)[f(x) - \sigma_R^j(f)(x)] +$$

$$\frac{(-1)^n}{(n+1)!}\frac{\lambda^{n+2}}{H(\lambda)}\frac{1}{R^{n+2}}\int_0^R H^{(n+2)}\left(\frac{\lambda}{R}r\right)r^{n+1}[f(x) - \sigma_{R,r}^{n+1}(f)(x)]\mathrm{d}r. \tag{2.6}$$

那么

$$S_R(f)(x)-f(x)=\frac{1}{H(\lambda)}(T_R(f)(x)-f(x))+\gamma_R(f)(x). \tag{2.7}$$

我们称此式为 Rogosinsky 恒等式，把它作为一元形式(见[3]第 288 页)的推广.

定义 设 $f\in L(Q^n)$，$R>0$.

$$F_R(x)=F_R(f)(x)=\frac{1}{V_n\lambda^n}R^n\int_{|\xi|<\frac{\lambda}{R}}f(x+\xi)\mathrm{d}\xi. \tag{2.8}$$

显然 $F_R\in C(Q^n)$. 容易算出 F_R 的 Fourier 系数

$$c_m(F_R)=c_m(f)\cdot\frac{R^n}{V_n\lambda^n}\int_{|\xi|<\frac{\lambda}{R}}\mathrm{e}^{im\xi}\mathrm{d}\xi,$$

与(2.4)比较，可知

$$T_R(f)(x)=S_R(F_R)(x).$$

那么

$$T_R(f)(x)-f(x)=S_R(F_R)(x)-F_R(x)+F_R(x)-f(x).$$

由此可见，Rogosinsky 恒等式(2.7)可改写为

$$S_R(f)(x)-f(x)=$$

$$\frac{1}{H(\lambda)}[S_R(F_R)(x)-F_R(x)]+\frac{1}{H(\lambda)}[F_R(x)-f(x)]+\gamma_R(f)(x). \tag{2.9}$$

§3. 一致收敛及 a.e. 收敛

先考虑一致收敛问题.

设 $f\in C(Q^n)$. 令

$$\varepsilon_R=\frac{1}{R}\int_0^R\omega_2\left(f;\frac{1}{t}\right)\mathrm{d}t.$$

由定理 2 知，对于 $k\in\{2, 3, \cdots, n+1\}$，

$$\|\sigma_R^k(f)-f\|_C\leqslant M\varepsilon_k,\quad M\text{ 只与 }n\text{ 有关}. \tag{3.1}$$

由定理 4 知，对于 $R\geqslant r>0$，

$$\|\sigma_{R,r}^{n+1}(f)-f\|_C\leqslant M\varepsilon_r,\quad M\text{ 只与 }n\text{ 有关}. \tag{3.2}$$

据(3.1)(3.2)及(2.6)，得

$$\|\gamma_R(f)\|_c\leqslant M\left(\varepsilon_R+\frac{1}{R^{n+2}}\int_0^R r^{n+1}\varepsilon_r\mathrm{d}r\right)$$

$$\leqslant M\varepsilon_R \tag{3.3}$$

式中 M 只与 n 有关($J_{\frac{n}{2}+1}$的最小正根 λ 亦只由 n 决定).

由 F_R 的定义(见 2.8),显然有

$$\|F_R-f\|_C\leqslant M\varepsilon_R.$$

于是,从(2.9)推出下述定理.

定理 5 设 $f\in C(Q^n)$,则

$$S_R(f)(x)-f(x)=\frac{1}{H(\lambda)}[S_R(F_R)(x)-F_R(x)]+O(\varepsilon_R).$$

由[4]的定理 1 知,

$$\|S_R(F_R)-F_R\|_C\leqslant M\ln R\omega_2\left(F_R;\ \frac{1}{R}\right),$$

式中 M 是只与 n 有关的常数. 于是得到下述

推论 1 设 $f\in C(Q^n)$. 若

$$\omega_2\left(F_R;\ \frac{1}{R}\right)=o\left(\frac{1}{\ln R}\right),\quad R\to+\infty, \tag{3.4}$$

则 $S_R(f)$一致收敛到 f.

现在我们来讨论几乎处处收敛问题.

设 $f\in L(Q^n)$. 若 x 是 f 的 Lebesgue 点,则由定理 1 及定理 3 知

$$f(x)-\sigma_R^j(f)(x)=o(1),\ j=2,3,\cdots,n+1,\quad R\to+\infty,$$

$$f(x)-\sigma_{R,r}^{n+1}(f)(x)=o(1),\quad R\geqslant r\to+\infty.$$

其中第二式的 $o(1)$关于 $R\geqslant r$ 是一致的,即

$$\sup_{R\geqslant r}\{|f(x)-\sigma_{R,r}^{n+1}(f)(x)|\}=o(1),\quad r\to+\infty.$$

据这些结果及(2.6),得 $\gamma_R(f)(x)=o(1)$. 于是以(2.9)推出

定理 6 设 $f\in L(Q^n)$. 若 x 是 f 的 Lebesgue 点,则

$$S_R(f)(x)-f(x)=\frac{1}{H(\lambda)}[S_R(F_R)(x)-F_R(x)]+o(1).$$

推论 2 设 $f\in L(Q^n)$. 若条件(3.4)成立,则 $S_R(f)$几乎处处收敛到 f.

我们把条件(3.4)看作一元的 Salem-Стечкин 条件的推广. 下面分析一下条件(3.4).

据 F_R 的定义

$$F_R(x+h)+F_R(x-h)-2F(x)$$

$$=\frac{R^n}{V_n\lambda^n}\int_{|\xi|<\frac{\lambda}{R}}[f(x+h+\xi)+f(x-h+\xi)-2f(x+\xi)]\mathrm{d}\xi.$$

对于 $h\in\mathbf{R}^n$，$h\neq0$，令 $h^0=\frac{h}{|h|}$. 定义 $n-1$ 维圆盘

$$\Sigma_x=\left\{\xi\,\middle|\,|\xi-x|\leqslant\frac{\lambda}{R},\ (\xi-x)\cdot h=0\right\}.$$

对于 $\xi\in\Sigma_x$，令

$$t_\xi=\left[\left(\frac{\lambda}{R}\right)^2-|\xi-x|^2\right]^{\frac{1}{2}}.$$

那么

$$F_R(x+h)+F_R(x-h)-2F(x)$$
$$=\frac{R^n}{V_n\lambda^n}\int_{\Sigma_x}\int_{-t_\xi}^{t_\xi}[f(\xi+h+th^0)+f(\xi-h+th^0)-2f(\xi+th^0)]\mathrm{d}t\mathrm{d}\sigma(\xi).$$

易见

$$\int_{-t_\xi}^{t_\xi}[f(\xi+h+th^0)-f(\xi+th^0)]\mathrm{d}t$$
$$=\int_{-t_\xi+\frac{1}{2}|h|}^{t_\xi+\frac{1}{2}|h|}\left[f\left(\xi+\frac{1}{2}h+sh^0\right)-f\left(\xi+\frac{1}{2}h-sh^0\right)\right]\mathrm{d}s,$$
$$\int_{-t_\xi}^{t_\xi}[f(\xi+th^0)-f(\xi-h+th^0)]\mathrm{d}t$$
$$=\int_{-t_\xi+\frac{1}{2}|h|}^{t_\xi+\frac{1}{2}|h|}\left[f\left(\xi-\frac{1}{2}h+sh^0\right)-f\left(\xi-\frac{1}{2}h-sh^0\right)\right]\mathrm{d}s.$$

定义

$$F(u,h^0,t)=\frac{1}{t}\int_0^t[f(u+sh^0)-f(u-sh^0)]\mathrm{d}s,$$

其中 $u\in\mathbf{R}^n$，$h^0\in\mathbf{R}^n$ 且 $|h^0|=1$，$t>0$. 如果 f 是连续函数，或者是 $n=1$ 的情形，那么条件

$$\sup\left\{\,|F(u+\delta h^0,h^0,t)-F(u,h^0,t)|\,\middle|\,u\in\mathbf{R}^n,\ |h^0|=1,-\frac{1}{R}\leqslant\delta\leqslant\frac{1}{R},0<t\leqslant\frac{\lambda}{R}\right\}=o\left(\frac{1}{\ln R}\right)\qquad(3.5)$$

足以保证条件(3.4)成立.

周知，一元经典的 Salem-Стечкин 条件是(见[3]293 页)

$$\frac{1}{h}\int_0^h [f(x+t)-f(x-t)]\mathrm{d}t = o\left(\frac{1}{|\ln h|}\right),(\text{关于 } x \text{ 一致}). \tag{3.6}$$

从(3.6)(3.5)看到，条件(3.4)的确是条件(3.6)的推广．用于一维时，(3.4)略弱于(3.6)，因此推论 1 和推论 2 略优于经典的结果．

顺便指出，陆善镇[5]用完全不同的方法研究了一致收敛和几乎处处收敛的问题．以完全不同的形式成功地推广了 Salen 判别法．彼处的讨论基于 Stein 关于级数与积分同收敛的著名论断．由于使用这一结果，必须限定 $f\in L\ln^+L(Q^n)$．而且在一致尺度时无法考察逼近的阶．现在我们使用 Rogosinsky 恒等式来处理问题就避免了这两方面的缺陷．

参考文献

[1] Watson G N. A Treatise on the Theory of Bessel Functions. Cambridge Press，1952.

[2] Stein E M，Weiss G. Introduction to Fourier Analysis on Euclidean Space. Princeton University Press，1971.

[3] Бари Н К. Тригонометрические ряды. Москва，ФИЗМАТГИЗ，1961.

[4] Wang Kunyang. Approximation Theory and Its Applications，1985，1(4)：19.

[5] 陆善镇．数学学报，1980，23(4)：609.

Abstract The classical Rogosinsky identical relation is extended to the case of several variables and the extension is used to get the conditions for uniform convergence and for a. e. convergence. The results extend the Salem-Stechkin theorems of single series.

Keywords multiple Fourier series；convergence.

Chinese Science Bulletin，1985，30(2)：160－165.

多重共轭 Fourier 级数的(C，1)求和

(C，1) Summability of Multiple Conjugate Fourier Series①

§1. Introduction

Let E_k be k-dimensional Euclidean space $k\geqslant 2$, and $Q^k=\{(x_1, x_2, \cdots, x_k)\in E_k \mid -\pi\leqslant x_j<\pi,\ j=1,2,\cdots,k\}$. $L(Q^k)$ denotes the space of functions which are integrable on Q^k and have a period 2π with respect to each variable element.

Let $P(x)$ be a homogeneous harmonic polynomial of degree n, $n\geqslant 1$. The function $K(x)=P\left(\frac{x}{|x|}\right)|x|^{-k}(x\neq 0)$ is called the spherical harmonic kernel. The periodic function

$$K^*(x)=\sum_{m\in\mathbf{Z}^k}\left\{K(x+2\pi m)-|Q^k|^{-1}\int_{Q^k+2\pi m}K(y)\mathrm{d}y\right\}$$

is called the periodic kernel generated by $K(x)$ (see Ref. [1]).

If $f\in L(Q^k)$, its Fourier series is $\quad \sigma(f)\sim\sum_m a_m \mathrm{e}^{\mathrm{i}mx}, \qquad (1.1)$

then the series
$$\tilde{\sigma}(f)\sim\sum_m a_m\hat{K}(m)\mathrm{e}^{\mathrm{i}mx}, \qquad (1.2)$$

is called conjugate Fourier series, with respect to the kernel K, of f,

where
$$\hat{K}(y)=\frac{1}{(2\pi)^k}\lim_{\rho\to+\infty}\lim_{\varepsilon\to 0}\int_{0<\varepsilon<|x|<\rho}K(x)\mathrm{e}^{-\mathrm{i}xy}\mathrm{d}x. \qquad (1.3)$$

① Received：1983-06-13.

$\hat{K}(y)$ exists everywhere and $\hat{K}(y)=\frac{(-\mathrm{i})^n}{c(k,\ n)}P\left(\frac{y}{|y|}\right)$, $y\neq 0$, $\hat{K}(0)=0$,

$$c(k,n)=\frac{(2\pi)^k\Gamma\left(\frac{k+n}{2}\right)}{\pi^{\frac{k}{2}}\Gamma\left(\frac{n}{2}\right)}. \tag{1.4}$$

The conjugate function, with respect to the kernel K, of f is defined by

$$f^*(x)=\lim_{\varepsilon\to 0^+}\frac{1}{(2\pi)^k}\int_{Q^k\setminus B(0,\varepsilon)}f(x-y)K^*(y)\mathrm{d}y. \tag{1.5}$$

We write

$$f_\varepsilon^*(x)=\frac{1}{(2\pi)^k}\int_{Q^k\setminus B(0,\varepsilon)}f(x-y)K^*(y)\mathrm{d}y. \tag{1.6}$$

The spherical Riesz means of order γ of $\tilde{\sigma}(f)$ is defined by

$$\widetilde{S}_R^\gamma(f;x)=\sum_{|m|<R}a_m\hat{K}(m)\left(1-\frac{|m|^2}{R^2}\right)^\gamma \mathrm{e}^{\mathrm{i}mx}. \tag{1.7}$$

At the index $\gamma>\frac{K-1}{2}$, Shapiro obtained a convergence theorem for Riesz means of conjugate Fourier series with respect to more general kernels K (see Ref. [2]), namely,

Theorem A Let $f\in L(Q^k)$, $K(x)$ be a Calderón-Zygmund kernel in class C^{k+4}, index α be defined as follows:

$$\frac{K-1}{2}<\alpha\leqslant K-1+\frac{1}{2},\qquad \text{if } K \text{ is even},$$

$$\frac{K-1}{2}<\alpha\leqslant \frac{K-1}{2}+1,\qquad \text{if } K \text{ is odd}.$$

Suppose that

$$|B(x_0,h)|^{-1}\int_{B(x_0,h)}|f(x)-f(x_0)|\mathrm{d}x\to 0,\ \text{as } h\to 0. \tag{1.8}$$

Then

$$\lim_{R\to+\infty}\left\{\widetilde{S}_R^\alpha(f;x_0)-\lim_{\lambda\to+\infty}\frac{1}{(2\pi)^k}\int_{B(0,\lambda)\setminus B(0,R^{-1})}f(x_0-y)K(y)\mathrm{d}y\right\}=0. \tag{1.9}$$

When $\gamma=\frac{K-1}{2}$, Lippman[3] proved the opposite result, namely,

Theorem B Let $0<\delta<1$ be given. There exists a function $f\in L(Q^k)$ which vanishes in $B(0,\ \delta)$ but

$$\overline{\lim_{R\to+\infty}}\left|\widetilde{S}_R^{\frac{k-1}{2}}(f;\ 0)\right|=+\infty. \tag{1.10}$$

In this note, we investigate the convergence of $(C,\ 1)$ means

$$\frac{1}{R}\int_0^R \widetilde{S}_u^{\frac{k-1}{2}}(f;x)\mathrm{d}u.$$

We first obtain a localization theorem for $\widetilde{S}_R^{\gamma}(f;\ x)$ with $\gamma>\frac{k-1}{2}$, that is

Theorem 1 Let $f\in L(Q^k)$,

$$\psi_x(t)=\int_{|y|=1} f(x-ty)P(y)\mathrm{d}\sigma(y). \tag{1.11}$$

If
$$\int_0^t |\psi_x(\tau)|\mathrm{d}\tau=o(t),\quad t\to 0, \tag{1.12}$$

then for $\gamma>\frac{k-1}{2}$,
$$\lim_{R\to+\infty}(\widetilde{S}_R^{\gamma}(f;\ x)-f_{\frac{1}{R}}^{*}(x))=0. \tag{1.13}$$

If we only consider the spherical harmonic kernel defined as above, then Theorem 1 improves Theorem A, because the condition (1.12) is weaker than (1.8).

Using the prove of Theorem 1 we lead to the critical index and obtain our main result.

Theorem 2 Let $f\in L(Q^k)$. If (1.12) is satisfied, then

$$\lim_{R\to+\infty}\frac{1}{R}\int_0^R(\widetilde{S}_u^{\frac{k-1}{2}}(f;x)-f_{\frac{1}{u}}^{*}(x))\mathrm{d}u=0. \tag{1.14}$$

Because (1.12) holds a. e. and $f^*(x)$ exists a. e., so we derive from Theorem 2 that

$$\lim_{R\to+\infty}\frac{1}{R}\int_0^R\widetilde{S}_u^{\frac{k-1}{2}}(f;x)\mathrm{d}u=f^*(x)\quad \text{a. e.} \tag{1.15}$$

§ 2. Proof of theorem 1

By (1.1)(1.4) and (1.7),

$$\widetilde{S}_R^{\gamma}(f;x)=\frac{1}{(2\pi)^k}\int_{Q^k}f(x-y)\widetilde{D}_R^{\gamma}(y)\mathrm{d}y, \tag{2.1}$$

where
$$\widetilde{D}_R^{\gamma}(y)=\frac{(-\mathrm{i})^n}{c(k,n)}\sum_{0<|m|<R}P\left(\frac{m}{|m|}\right)\left(1-\frac{|m|^2}{R^2}\right)\mathrm{e}^{\mathrm{i}my}. \tag{2.2}$$

we write $\gamma=\frac{k-1}{2}+\beta$, $\beta>0$. By Lemmas of Ref. [4], we have

$$\widetilde{D}_R^{\frac{k-1}{2}+\beta}(y)=-\frac{1}{c(k,n)}R^{k+n}\sum_{j=0}^{+\infty}\sum_{m\in S_j}P(y+2\pi m)E(k,\beta,n,R|y+2\pi m|),$$

$$\beta>0,\quad R>\frac{1}{\pi},\tag{2.3}$$

Where $\quad S_j=\{y\in E_k\cap \mathbf{Z}^k \mid j\leqslant|y|<j+1\}$, $j\in\mathbf{N}$,

and $E(k, s, n, z)$ is an entire function of variable z which has been introduced by chang[4],

$$E(k,s,n,z)=\frac{\pi^{\frac{k}{2}}}{2^n}\sum_{j=0}^{+\infty}\frac{(-1)^j\Gamma\left(j+\frac{k+n}{2}\right)\Gamma\left(\frac{k+1}{2}+s\right)}{j!\Gamma\left(j+\frac{k}{2}+n\right)\Gamma\left(j+k+\frac{n+1}{2}+s\right)}\left(\frac{z}{2}\right)^{2j},$$

$$\operatorname{Re} s>-\frac{k+1}{2}.\tag{2.4}$$

In our discussion k, n are constants, so we simply write $E(\beta, z)$ instead of $E(k, \beta, n, z)$ and sometimes omit k, n in other notations.

By the lemma of chang's[4], we have for $u\geqslant 1$,

$$E(\beta, u)=$$

$$\frac{c(k, n)}{u^{k+n}}+\frac{R^{(1)}(\beta, u)}{u^{k+n+1}}+\frac{R^{(2)}(\beta, u)}{u^{k+n+1+\beta}}+A(k, \beta)\frac{\cos\left(u-\frac{\pi(k+n+\beta)}{2}\right)}{u^{k+n+\beta}},\tag{2.5}$$

where $c(k, n)$ is as in (1.4), $R^{(1)}$, $R^{(2)}$ are bounded for $\beta\in\left[-\frac{k+1}{2}+h, G\right](0<h<G)$ and

$$A(k, \beta)=\pi^{\frac{k-1}{2}}2^{k+\beta}\Gamma\left(\frac{k+1}{2}+\beta\right).\tag{2.6}$$

When $|y|>\frac{1}{R}$, $y\in Q^k$ and $R>\frac{1}{\pi}$, from (2.3) and (2.5) we deduce that

$$\widetilde{D}_R^{\frac{k-1}{2}+\beta}(y)=\left\{\sum_{j=0}^{+\infty}\sum_{m\in S_j}\frac{P(y+2\pi m)}{|y+2\pi m|^{k+n}}\right\}+$$

$$\left\{\frac{1}{c(k,n)}\left(\frac{R^{(1)}(\beta,R|y|)}{R|y|^{k+n+1}}+\frac{R^{(2)}(\beta,R|y|)}{R^{1+\beta}|y|^{k+n+1+\beta}}\right)P(y)\right\}+$$

$$\left\{\frac{1}{c(k,n)}\sum_{j=1}^{+\infty}\sum_{m\in S_j}\frac{P(y+2\pi m)}{|y+2\pi m|^{k+n+1}}\left(\frac{R^{(1)}(\beta,R|y+2\pi m|)}{R}+\right.\right.$$

$$\left.\left.\frac{R^{(2)}(\beta,R|y+2\pi m|)}{R^{1+\beta}|y+2\pi m|^{\beta}}\right)\right\}+$$

$$\left\{\frac{A(k,\beta)}{c(k,n)}\frac{1}{R^{\beta}}\frac{P(y)}{|y|^{k+n+\beta}}\cos\left(R|y|-\frac{\pi(k+n+\beta)}{2}\right)\right\}+$$

$$\left\{\frac{A(k,\beta)}{c(k,n)}\frac{1}{R^{\beta}}\sum_{j=1}^{+\infty}\sum_{m\in S_j}\frac{P(y+2\pi m)}{|y+2\pi m|^{k+n+\beta}}\cdot\right.$$

$$\left.\cos\left(R|y+2\pi m|-\frac{\pi(k+n+\beta)}{2}\right)\right\}. \tag{2.7}$$

We notice that the first term on the right-hand side is just $K^*(y)$, and we denote the rest four terms by $\Phi_R^{(1)}(\beta, y)$, $\Phi_R^{(2)}(\beta, y)$, $\Phi_R^{(3)}(\beta, y)$ and $\Phi_R^{(4)}(\beta, y)$ respectively. Then we have

$$\widetilde{S}_R^{\frac{k-1}{2}+\beta}(f;x)-f^*_{\frac{1}{R}}(x)=\frac{1}{(2\pi)^k}\int_{B(0,\frac{1}{R})}f(x-y)\widetilde{D}_R^{\frac{k-1}{2}+\beta}(y)\mathrm{d}y+$$

$$\sum_{v=1}^{4}\frac{1}{(2\pi)^k}\int_{Q^k\setminus B(0,\frac{1}{R})}f(x-y)\Phi_R^{(v)}(\beta,y)\mathrm{d}y=\triangleq I_0^{\beta}(R)+\sum_{v=1}^{4}I_V^{\beta}(R). \tag{2.8}$$

On account of the boundedness of $R^{(1)}$, $R^{(2)}$ and the condition (1.2), we get

$$|I_1^{\beta}(R)|\leqslant M_{\beta}\int_{\frac{1}{R}}^{\pi}\frac{|\psi_x(t)|}{Rt^2}\mathrm{d}t=M_{\beta}\cdot o(1),\quad R\to+\infty, \tag{2.9}$$

$$|I_2^{\beta}(R)|\leqslant\int_{Q^k\setminus B(0,\frac{1}{R})}|f(x-y)||\Phi_R^{(2)}(\beta,y)|\mathrm{d}y$$

$$\leqslant\frac{M_{\beta}}{R}\|f\|_{L(Q^k)}=M_{\beta}\cdot o(1), \tag{2.10}$$

where the constant M_{β} is bounded for $\beta\in(0, G](\forall G>0)$.

Considering $|y|<\frac{1}{R}$ in (2.3), we have

$$\widetilde{D}_R^{\frac{k-1}{2}+\beta}(y)=\frac{1}{c(k,n)}R^{k+n}P(y)E(\beta,R|y|)+$$

$$\sum_{j=1}^{+\infty}\sum_{m\in S_j}\frac{P(y+2\pi m)}{|y+2\pi m|^{k+n}}+\Phi_R^{(2)}(\beta,y)+\Phi_R^{(4)}(\beta,y). \tag{2.11}$$

We notice that the second term on right-hand side of (2.11) is equal to $K^*(y)-K(y)$, which is bounded on Q^k; the third term $\Phi_R^{(2)}$ is bounded on Q^k uniformly for $\beta\in(0, G](\forall G>0)$; and $\sup\limits_{\beta>0}\max\limits_{0\leqslant u\leqslant 1}|E(\beta, u)|=$

$M<+\infty$. Hence

$$I_0^\beta(R)=\frac{1}{(2\pi)^k}\int_{B(0,R^{-1})}f(x-y)\Phi_R^{(4)}(\beta,y)\mathrm{d}y+M_\beta\cdot o(1),$$
$$R\to+\infty. \tag{2.12}$$

Substituting (2.9)(2.10) and (2.12) into (2.8), we obtain

$$\widetilde{S}_R^{\frac{k-1}{2}+\beta}(f;x)-f_{\frac{1}{R}}^*(x)-\frac{1}{(2\pi)^k}\left(I_3^\beta(R)+\int_{Q^k}f(x-y)\Phi_R^{(4)}(\beta,y)\mathrm{d}y\right)$$
$$=M_\beta\cdot o(1),\quad R\to+\infty. \tag{2.13}$$

Finally,

$$|I_3^\beta(R)|\leqslant\frac{A(k,\beta)}{c(k,n)}\frac{1}{R^\beta}\int_{\frac{1}{R}}^{\pi}\frac{|\psi_x(t)|\mathrm{d}t}{t^{1+\beta}}=C_\beta\cdot o(1), \tag{2.14}$$

$$\left|\int_{Q^k}f(x-y)\Phi_R^{(4)}(\beta,y)\mathrm{d}y\right|\leqslant C_\beta\|f\|_{L(Q^k)}\frac{1}{R^\beta}, \tag{2.15}$$

(where the constant C_β may be unbounded for $\beta\in(0,G]$).

Thus Theorem 1 is deduced from (2.13)~(2.15).

§ 3. Proof of theorem 2

From (2.13), we see

$$\left|\frac{1}{R}\int_{\frac{1}{\pi}}^{R}\left(\widetilde{S}_u^{\frac{k-1}{2}+\beta}(f;x)-f_{\frac{1}{u}}^*(x)\right)\mathrm{d}u\right|$$
$$\leqslant\frac{1}{R}\left|\int_{\frac{1}{\pi}}^{R}I_3^\beta(u)\mathrm{d}u\right|+\frac{1}{R}\left|\int_{\frac{1}{\pi}}^{R}\left\{\int_{Q^k}f(x-y)\Phi_u^{(4)}(\beta,y)\mathrm{d}y\right\}\mathrm{d}u\right|+M_\beta\cdot o(1)$$
$$\leqslant\frac{1}{R}\left|\int_{\frac{1}{\pi}}^{R}\left(\int_{Q^k\setminus B(0,\pi)}f(x-y)\Phi_u^{(3)}(\beta,y)\mathrm{d}y\right)\mathrm{d}u\right|+$$
$$\frac{1}{R}\left|\int_{\frac{1}{\pi}}^{R}\left(\int_{B(0,\pi)\setminus B(0,1/u)}f(x-y)\Phi_u^{(3)}(\beta,y)\mathrm{d}y\right)\mathrm{d}u\right|+$$
$$\frac{1}{R}\left|\int_{\frac{1}{\pi}}^{R}\left(\int_{Q^k}f(x-y)\Phi_u^{(4)}(\beta,y)\mathrm{d}y\right)\mathrm{d}u\right|+M_\beta\cdot o(1)$$
$$\triangleq J_1(\beta,R)+J_2(\beta,R)+J_3(\beta,R)+M_\beta\cdot o(1). \tag{3.1}$$

We define that $\quad M_R(\beta,t,\tau)=t\int_\tau^R\frac{1}{(ut)^\beta}\cos\left(ut-\frac{\pi(k+n+\beta)}{2}\right)\mathrm{d}u,$

$$\beta>0,\ \tau\in(0,+\infty),\ t\tau\geqslant1. \tag{3.2}$$

By mean value theorem, we have

$M_R(\beta,t,\tau)$

$$= \frac{t}{(t\tau)^{\beta}}\int_{\tau}^{\xi}\cos\left(ut - \frac{\pi(k+n+\beta)}{2}\right)\mathrm{d}u = \frac{1}{(\tau t)^{\beta}}\int_{t\tau}^{t\xi}\cos\left(u - \frac{\pi(k+n+\beta)}{2}\right)\mathrm{d}u.$$

We see that

$$|M_R(\beta,\ t,\ \tau)| \leqslant 2 \quad (\beta>0,\ \tau\in(0,\ +\infty),\ t\tau\geqslant 1). \tag{3.3}$$

Then by the definition of $\Phi_u^{(3)}$ (see (2.7)) we get

$$\left|\int_{\frac{1}{\pi}}^{R}\Phi_u^{(3)}(\beta,y)\mathrm{d}u\right| \leqslant C\frac{|P(y)|}{|y|^{k+n+1}} \leqslant M, (|y| \geqslant \frac{1}{\pi}, \beta\in(0,1)),$$

$$\sup_{\beta\in(0,1)} J_1(\beta,R) \leqslant M\cdot\|f\|_{L(Q^k)}\cdot\frac{1}{R}; \tag{3.4}$$

$$\begin{aligned} J_2(\beta,R) &= \frac{C}{R}\left|\int_{\frac{1}{R}}^{\pi}t^{k-1}\left[\int_{|y|=1}f(x-ty)\left(\int_{\frac{1}{t}}^{R}\Phi_u^{(3)}(\beta,ty)\mathrm{d}u\right)\mathrm{d}\sigma(y)\right]\mathrm{d}t\right| \\ &= \frac{CA(k,\beta)}{c(k,n)}\left|\int_{\frac{1}{R}}^{\pi}\frac{\psi_x(t)}{t^2}M_R(\beta,t,t^{-1})\mathrm{d}t\right|, \end{aligned}$$

$$\sup_{\beta\in(0,1)} J_2(\beta,R) \leqslant M\cdot\frac{1}{R}\int_{\frac{1}{R}}^{\pi}\frac{|\psi_x(t)|}{t^2}\mathrm{d}t = o(1),\quad R\to+\infty. \tag{3.5}$$

Finally, by (3.3) and the definition of $\Phi_u^{(4)}$ (see (2.7)) we get

$$\begin{aligned} &\int_{\frac{1}{\pi}}^{R}\Phi_u^{(4)}(\beta,y)\mathrm{d}u \\ &= \frac{A(k,\beta)}{c(k,n)}\sum_{j=1}^{+\infty}\sum_{m\in S_j}\frac{P(y+2\pi m)}{|y+2\pi m|^{k+n+1}}M_R\left(\beta, |y+2\pi m|, \frac{1}{\pi}\right), \end{aligned}$$

$$\sup_{\beta\in(0,1)}\left|\int_{\frac{1}{\pi}}^{R}\Phi_u^{(4)}(\beta,y)\mathrm{d}u\right| \leqslant C\sum_{j=1}^{+\infty}\sum_{m\in S_j}\frac{1}{|m|^{k+1}} < +\infty,$$

$$\begin{aligned} &\sup_{\beta\in(0,1)} J_3(\beta,R) \\ &\leqslant \frac{1}{R}\int_{Q^k}|f(x-y)|\sup_{\beta\in(0,1)}\left|\int_{\frac{1}{\pi}}^{R}\Phi_u^{(4)}(\beta,y)\mathrm{d}u\right|\mathrm{d}y \leqslant \frac{m\|f\|_{L(Q^k)}}{R}. \end{aligned} \tag{3.6}$$

On account of (3.4)～(3.6), taking limit as $\beta\to 0^+$ in (3.1) we deduce Theorem 2.

References

[1] Calderón A P, Zygmund A. Studia Mathematica, 1954, 14: 249—271.
[2] Shapiro V L. Bulletin of Amer. Math. Soc., 1964, 70(1): 48—93.
[3] Lippman G E. SIAM J. Math. Anal., 1973, 4(4): 681—695.
[4] Chang C P. Ph. D. Dissertation, Univ. of Chicago, Chicago, 1964.

Approximation Theory and Its Applications，1985，1(4)：19—56.

多元连续周期函数及其共轭函数用 Riesz 平均在全测度集上逼近

Approximation for Continuous Periodic Function of Several Variables and Its Conjugate Function by Riesz Means on Set of Total Measure[①]

§ 1. Introduction

Let $Q=Q^k=\{(x_1, x_2, \cdots, x_k) \mid -\pi\leqslant x_j<\pi, j=1,2,\cdots,k\}$.

Suppose $f\in L(Q)$. The Fourier series of f is denoted by

$$\sigma(f)(x) \sim \sum_{m\in \mathbf{Z}^k} a_m(f)\mathrm{e}^{imx},$$

where $\mathbf{Z}^k$ denotes set of all k-dimensional integers. The Riesz means of order α of f are defined by

$$S_R^\alpha(f;x) = \sum_{|m|<R}\left(1-\left|\frac{m}{R}\right|^2\right)^\alpha a_m(f)\mathrm{e}^{imx}, \quad R>0.$$

The concept of "conjugate" is introduced in Calderon-Zygmund sense (see [1]). Let $p(x)$ be a homogeneous harmonic polynomial of degree $p\geqslant 1$. The function $K(x)=p(x)|x|^{-k-p}$ $(x\neq 0)$ is called a spherical harmonic kernel. The conjugate Fourier series of f with respect to the kernel K is defined by

$$c(f)(x) \sim \sum_{m\in \mathbf{Z}^k} a_m(f)\hat{K}(m)\mathrm{e}^{imx},$$

where $\hat{K}$ denotes the principal-valued Fourier transform of K, i. e.

① Received：1984-09-22.

$$\hat{K}(x) = \lim_{\varepsilon \to 0^+, \rho \to +\infty} \frac{1}{(2\pi)^k} \int_{\varepsilon < |y| < \rho} K(y) e^{-jyx} dy.$$

It is well known that

$$\hat{K}(x) = (-1)^p \frac{\Gamma\left(\frac{p}{2}\right)}{2^k \pi^{\frac{k}{2}} \Gamma\left(\frac{k+p}{2}\right)} P\left(\frac{x}{|x|}\right), \quad x \neq 0,$$

$$\hat{K}(0) = 0.$$

Define the periodization of K by

$$K^*(x) = \sum_{m \in \mathbf{Z}^k} (K(x + 2\pi m) - I_m), \quad x \neq 2\pi m, \quad m \in \mathbf{Z}^k,$$

where

$$I_m = \frac{1}{|Q|} \int_Q K(x + 2\pi m) dx, \quad m \in \mathbf{Z}^k,$$

and when $m = (0, 0, \cdots, 0) = 0$ the integral defining I_0 is in principal value sense, that is

$$I_0 = \frac{1}{|Q|} \lim_{\varepsilon \to 0^+} \int_{y \in Q, |y| > \varepsilon} K(y) dy.$$

The truncated conjugate function of f with respect to the kernel K is defined by

$$f_\varepsilon^*(x) = \frac{1}{(2\pi)^k} \int_{\gamma \in Q, |y| \geqslant \varepsilon} f(x - y) K^*(y) dy, \quad 0 < \varepsilon < \sqrt{k}\pi;$$

$$f_\varepsilon^*(x) = 0, \quad \varepsilon \geqslant \sqrt{k}\pi.$$

And the conjugate function is

$$f^*(x) = \lim_{\varepsilon \to 0^+} f_\varepsilon^*(x),$$

In face (see [1])

$$f^*(x) = \lim_{\varepsilon \to 0^+, \rho \to +\infty} \frac{1}{(2\pi)^k} \int_{\varepsilon < |y| < \rho} f(x - y) K(y) dy.$$

It is also known that

$$\| f^* \|_q \leqslant A_k(p) \frac{q^2}{q-1} \| f \|_q \tag{1.1}$$

provided $1 < q < +\infty$ and $f \in L^q(Q)$, where $A_k(p)$ is a constant depending only on k and p (see [1]).

The Riesz means of order α of the conjugate Fourier series is

$$\widetilde{S}_R^\alpha(f;x) = \sum_{|m|<R}\left(1-\left|\frac{m}{R}\right|^2\right)^\alpha a_m(f)\hat{K}(m)e^{imx}, \quad R>0.$$

If $f^* \in L(Q)$ then

$$\tilde{\sigma}(f)=\sigma(f^*), \quad \widetilde{S}_R^\alpha(f)=S_R^\alpha(f^*).$$

Now suppose $f\in C(Q)$. Define the first modulus of continuity of f by

$$\omega(f;\ \delta)=\omega_1(f;\ \delta)=\sup_{x\in Q}\sup_{|h|<\delta}|f(x+h)-f(x)|,$$

and the second modulus of continuity of f by

$$\omega_2(f;\ \delta)=\sup_{x\in Q}\sup_{|h|<\delta}|f(x+h)+f(x-h)-2f(x)|.$$

If $\omega(f;\ \delta)=0(\delta)$ then we say $f\in W^1L^\infty$ of $f\in \mathrm{Lip}\ 1$. If f has all first partial derivatives in W^1L^∞ then we say $f\in W^2L^\infty$. For a function $f\in C(Q)$ if $f\notin W^1L$ then it must hold that

$$\lim_{\delta\to 0^+}\frac{\omega(f;\ \delta)}{\delta}=+\infty. \tag{1.2}$$

This can be easily proved by a property of the modulus of continuity. And the version is also true.

The main purpose of the present paper is to discuss the problem of the approximation for the functions in $C(Q)\setminus W^1L^{+\infty}$ and their conjugate functions by their Riesz means of critical order $\left(\alpha=\frac{k-1}{2}\right)$ on set of total measure. The result obtained here (Theorem 5) is parallel with the corresponding result of Oskolkov's paper [2] in the case of $k=1$.

In § 2 the problem of the uniform approximation is discussed. The conclusion is the following

Theorem 1 Let $f\in C(Q)$. If $\alpha>\frac{k-3}{2}$, then

$$\|f-S_R^\alpha(f)\|_C\leqslant A_{k,\alpha}(L_R^\alpha+1)\omega_2\left(f;\ \frac{1}{R}\right), \tag{1.3}$$

where L_R^α denotes the Lebesgue constant:

$$L_R^\alpha=\begin{cases} O(1), & \alpha>\frac{k-1}{2}; \\ O(\ln R), & \alpha=\frac{k-1}{2}; \\ O(R^{\frac{k-1}{2}-\alpha}), & \frac{k-3}{2}<\alpha<\frac{k-1}{2}. \end{cases}$$

In order to discuss the problem of the approximation on set of total measure it is needed to find the estimates for the maximal partial sum operators and some connected maximal operators. This will be done in § 3. The lemmas in that section will be the important tools to solve the main problem.

In § 4 the problem of the approximation on set of total measure is discussed. First we prove the following two theorems.

Theorem 2　If $f\in W^2L^\infty$ then

$$|S_{R^2}^{\frac{k-1}{2}}(f;\ x)-f(x)|\overset{\text{a. e.}}{=\!=}O\left(\frac{1}{R^2}\right),$$

$$\|S_{R^2}^{\frac{k-1}{2}}(f)-f\|_C=O\left(\frac{\ln R}{R}\right).$$

and

$$\|\widetilde{S}_{R^2}^{\frac{k-1}{2}}(f)-f^*\|_C=O\left(\frac{\ln R}{R^2}\right).$$

Theorem 3　If $f\in W^1L^\infty$ then

$$|S_{R^2}^{\frac{k-1}{2}}(f;\ x)-f(x)|\overset{\text{a. e.}}{=\!=}O\left(\frac{1}{R}\right),$$

$$|\dot{S}_{R^2}^{\frac{k-1}{2}}(f;\ x)-f(x)|\overset{\text{a. e.}}{=\!=}O\left(\frac{1}{R}\right).$$

Then we prove our main theorems.

Theorem 4　Let $f\in \operatorname{Lip}\alpha$, $0<\alpha<1$, i. e. $\omega(f;\ t)=O(t^\alpha)$. Then the inequality

$$\max\left(|S_{R^2}^{\frac{k-1}{2}}(f;\ x)-f(x)|,\ |S_{R^2}^{\frac{k-1}{2}}(f;\ x)-f^*(x)|\right)$$
$$\leqslant C(x)\frac{\ln\ln R}{R^2},\quad R\geqslant 9$$

holds almost everywhere, where $C(x)$ is finite almost everywhere.

This theorem is a particular case of the following

Theorem 5　Let $f\in C(Q)\setminus W^1L^\infty$. That means the condition (1. 2) is satisfied. Write simply $\omega(\delta)$ to denote $\omega(f;\ \delta)$ and set $\bar{\omega}(\delta)=\frac{\delta}{\omega(\delta)}\omega(1)$, $0<\delta\leqslant 1$. Define

$$\delta_0=1,\quad \delta_{n+1}=\min\left\{\delta\,\middle|\,\max\left(\frac{\omega(\delta)}{\omega(\delta_n)},\ \frac{\bar{\omega}(\delta)}{\bar{\omega}(\delta_n)}\right)=\frac{1}{6}\right\}$$

and a function $\Omega(\delta)$: $\Omega(\delta)=6^{-n}$ as $\delta\in(\delta_{n+1}, \delta_n]$, $n\in\mathbf{N}$.

Write

$$\rho_R(f;\ x)=\max(|S_R^{\frac{k-1}{2}}(f;\ x)-f(x)|,\ |\widetilde{S}_R^{\frac{k-1}{2}}(f;\ x)-f^*(x)|).$$

Then

(1) $$\rho_R(f;\ x)\leqslant C(x)\omega\left(\frac{1}{R}\right)\ln\ln\frac{9}{\Omega\left(\frac{1}{R}\right)},\quad R\geqslant 1$$

Where $C(x)$ is finite almost everywhere and

$$|\{x\in Q|C(x)>\lambda\}|<A_k(p)\mathrm{e}^{-\frac{\lambda}{A_k(p)}},\quad \lambda>0;$$

(2) $$\varlimsup_{R\to+\infty}\rho_R(f;\ x)\left\{\omega\left(\frac{1}{R}\right)\ln\ln\frac{9}{\Omega\left(\frac{1}{R}\right)}\right\}^{-1}\leqslant A_k(p),\ \text{a. e.}$$

Finally, a theorem of approximation which is in the non-conjugate case and is in the terms of the second modulus of continuity is stated by the way (Theorem 6).

§ 2. Some results about uniform approximation

From E. M. Stein[3], we know that the degree of the Lebesgue constant $L_R^\alpha=\int_Q|D_R^\alpha(y)|\mathrm{d}y$ is as in Theorem 1. Here

$$D_R^\alpha(y)=\sum_{|m|<R}\left(1-\left|\frac{m}{R}\right|^2\right)^\alpha\mathrm{e}^{imy}.$$

Suppose $U(x)$ is a trigonometric polynomial of degree R ($R>0$). From S. B. Stečkin[5] we know that

$$\left\|\frac{\partial^2U}{\partial x_j^2}\right\|_C\leqslant\left(\frac{R}{2}\right)^2\omega_2^{(j)}\left(U;\ \frac{\pi}{R}\right)\leqslant CR^2\omega_2\left(U;\ \frac{1}{R}\right)\tag{2.1}$$

where $\omega_2^{(j)}$ denotes the second partial modulus of continuity in the j-th Variable and C denotes an absolute constant. If $U(x)$ is the trigonometric polynomial of best approximation of $f\in C(Q)$ then from (1.3) and (2.1) we get

$$\|\Delta U\|_C\leqslant A_kR^2\omega_2\left(U;\ \frac{1}{R}\right)\leqslant A_kR^2\omega_2\left(f;\ \frac{1}{R}\right).\tag{2.2}$$

Proof of theorem 1 We first assume $\alpha>\frac{k-1}{2}$. Then according to the Bochner's formula (see [6]) we have

$$S_R^{\alpha}(f;x)-f(x)=C_{k,\alpha}\int_0^{+\infty}f_x(t)R^kt^{k-1}\frac{J_{\frac{k}{2}+\alpha}(Rt)}{(Rt)^{\frac{k}{2}+\alpha}}\mathrm{d}t, \qquad (2.3)$$

where $c_{k,\alpha}=\dfrac{2^{\alpha+1}\Gamma(\alpha+1)}{(2\pi)^{\frac{k}{2}}}$,

$$f_x(t)=\int_{S^+}[f(x+t\xi)+f(x-t\xi)-2f(x)]\mathrm{d}\sigma(\xi),$$

$$S^+=\{\xi \mid |\xi|=1,\ \xi_1>0\}, \qquad (2.4)$$

$$|f_x(t)|\leqslant A_k\omega_2(f;\ t),$$

and J_ν denotes the Bessel function of order ν.

From (2.3) and (2.4) it is derived that

$$\begin{aligned}&S_R^{\alpha}(f;x)-f(x)\\&=c_{k,\alpha}\left(\int_0^1+\int_1^{+\infty}\right)f_x\left(\frac{t}{R}\right)t^{k-1}\frac{J_{\frac{k}{2}+\alpha}(t)}{t^{\frac{k}{2}+\alpha}}\mathrm{d}t\\&=O\left(\omega_2\left(f;\frac{1}{R}\right)\right)+c_{k,\alpha}\int_1^{+\infty}f_x\left(\frac{t}{R}\right)t^{k-1}\frac{J_{\frac{k}{2}+\alpha}(t)}{t^{\frac{k}{2}+\alpha}}\mathrm{d}t. \qquad (2.5)\end{aligned}$$

We make use of the asymptotic expansion formula for the Bessel functions (see [7] 7.21, (1), p. 199)

$$J_\nu(z)=\sqrt{\frac{2}{\pi z}}\left\{\cos\left(z-\frac{\nu}{2}\pi-\frac{\pi}{4}\right)-\frac{\left(\nu+\frac{1}{2}\right)\left(\nu-\frac{1}{2}\right)}{2z}\sin\left(z-\frac{\nu}{2}\pi-\frac{\pi}{4}\right)+O\left(\frac{1}{z^2}\right)\right\}, z\to+\infty \qquad (2.6)$$

Then we get

$$\begin{aligned}&\int_1^{+\infty}f_x\left(\frac{t}{R}\right)\frac{J_{\frac{k}{2}+\alpha}(t)}{t^{\frac{k}{2}+\alpha}}\mathrm{d}t\\&=\int_1^{+\infty}f_x\left(\frac{t}{R}\right)\sqrt{\frac{2}{\pi}}\frac{1}{t^{\alpha-\frac{k}{2}+1}}\left\{\cos\left[t-\frac{\pi}{2}\left(\frac{k}{2}+\alpha\right)-\frac{\pi}{4}\right]-\right.\\&\left.\left(\frac{k+1}{2}+\alpha\right)\left(\frac{k-1}{2}+\alpha\right)\frac{1}{t}\sin\left[t-\frac{\pi}{2}\left(\frac{k}{2}+\alpha\right)-\frac{\pi}{4}\right]+O\left(\frac{1}{t^2}\right)\right\}\mathrm{d}t\\&\overset{\text{def.}}{=\!=}I_1+I_2+I_3.\end{aligned}$$

Set $h(t)=f_x\left(\frac{t}{R}\right)t^{-(1+\delta)}$, $\delta=\alpha-\frac{k-1}{2}>0$ and $\beta=-\frac{\pi}{2}\left(\frac{k}{2}+\alpha\right)-\frac{\pi}{4}$. Then

$$\begin{aligned} I_1 &= \sqrt{\frac{2}{\pi}}\int_1^{+\infty} h(t)\cos(t+\beta)\mathrm{d}t \\ &= \frac{1}{8}\sqrt{\frac{2}{\pi}}\int_1^{+\infty}(\Delta_\pi^3 h(t))\cos(t+\beta)\mathrm{d}t + \\ &\int_1^{1+\pi}\sqrt{\frac{2}{\pi}}\left[\frac{1}{8}\Delta_\pi^2 h(t)+\frac{1}{4}\Delta_\pi h(t)+\frac{1}{2}h(t)\right]\cos(t+\beta)\mathrm{d}t, \end{aligned}$$

where "Δ_π^r" is defined by

$$\Delta_\pi^r h(t) = \sum_{\nu=0}^{r}(-1)^\nu C_r^\nu h(t+\nu\pi),\quad r\in\mathbf{N}^*.$$

Applying (2.4) wc get

$$\int_1^{1+\pi}\left[\frac{1}{8}\Delta_\pi^2 h(t)+\frac{1}{4}\Delta_\pi h(t)+\frac{1}{2}h(t)\right]\cos(t+\beta)\mathrm{d}t = O\left(\omega_2\left(f;\frac{1}{R}\right)\right).$$

We make use of the following formula:

$$\begin{aligned} \Delta_\pi^3(u(t)v(t)) = &(\Delta_\pi^3 u(t))v(t)+3(\Delta_\pi^2 u(t+\pi))\Delta_\pi v(t)+ \\ &3(\Delta_\pi u(t+2\pi))\Delta_\pi^2 v(t)+u(t+3\pi)\Delta_\pi^3 v(t). \end{aligned}$$

Substituting $u(t)=f_x\left(\frac{t}{R}\right)$ and $v(t)=\frac{1}{t^{1+\delta}}$ into the above equation, then noticing that

$$|\Delta_\pi^3 u(t)| = O(|\Delta_\pi^2 u(t)|) = O\left(\omega_2\left(f;\ \frac{1}{R}\right)\right),$$

$$|\Delta_\pi u(t)| = O(|u(t)|) = O\left(\omega_2\left(f;\ \frac{t}{R}\right)\right) = O\left((1+t^2)\omega_2\left(f;\ \frac{1}{R}\right)\right),$$

$$|\Delta_\pi v(t)| = O\left(\frac{1}{t^{2+\delta}}\right),\qquad |\Delta_\pi^2 v(t)| = O\left(\frac{1}{t^{3+\delta}}\right),$$

$$|\Delta_\pi^3 v(t)| = O\left(\frac{1}{t^{4+\delta}}\right),$$

we get

$$\int_1^{+\infty}(\Delta_\pi^3 h(t))\cos(t+\beta)\mathrm{d}t = O\left[\int_1^{+\infty}\frac{\omega_2\left(f;\frac{1}{R}\right)}{t^{1+\delta}}\mathrm{d}t\right] = O\left(\omega_2\left(f;\frac{1}{R}\right)\right).$$

Hence $I_1=O\left(\omega_2\left(f;\ \frac{1}{R}\right)\right)$. Similarly $I_2=O\left(\omega_2\left(f;\ \frac{1}{R}\right)\right)$.

Finally,

$$I_3=\int_1^{+\infty} f_x\left(\frac{t}{R}\right)O\left(\frac{1}{t^{3+\delta}}\right)\mathrm{d}t=O\left(\omega_2\left(f;\frac{1}{R}\right)\right).$$

Substituting all these results into (2.5) we obtain

$$\|S_R^\alpha(f)-f\|_C\leqslant A_{k,\alpha}\omega_2\left(f;\ \frac{1}{R}\right),\quad \alpha>\frac{k-1}{2}. \tag{2.7}$$

Now we consider the case of $\frac{k-3}{2}<\alpha\leqslant\frac{k-1}{2}$. Let $U(x)$ be the trigonometric polynomial of best approximation of degree R of f. From the equation

$$S_R^\alpha(f)-f=S_R^\alpha(f-U)-(f-U)+S_R^\alpha(U)-U,$$

we see that

$$\|S_R^\alpha(f)-f\|_C\leqslant(L_R^\alpha+1)E_R(f)+\|S_R^\alpha(U)-U\|_C,$$

where $E_R(f)$ denotes the best uniform approximation of function f by trigonometrical polynomial of order m, $|m|<R$. Since

$$\begin{aligned}S_R^\alpha(U)-U&=S_R^\alpha(U)-S_R^{\alpha+1}(U)+S_R^{\alpha+1}(U)-U\\&=-\frac{1}{R^2}S_R^\alpha(\Delta U)+S_R^{\alpha+1}(U)-U\end{aligned}$$

and $\alpha+1>\frac{k-1}{2}$, we can use (2.7) and obtain that

$$\|S_R^\alpha(U)-U\|_C\leqslant\frac{1}{R^2}L_R^\alpha\|\Delta U\|_C+A_{k,\alpha}\omega_2\left(U;\ \frac{1}{R}\right).$$

Using (2.2) we get

$$\|S_R^\alpha(U)-U\|_C\leqslant A_{k,\alpha}(L_R^\alpha+1)\omega_2\left(f;\ \frac{1}{R}\right).$$

Hence using (1.3) we complete the proof.

Remark 1 From the proof we can see that in the particular case of $\alpha=\left[\frac{k-1}{2}\right]$ we can get

$$\|f-S_R^{\frac{k-1}{2}}(f)\|_C\leqslant(L_R^{\frac{k-1}{2}}+1)E_R(f)+A_k\omega_2\left(f;\ \frac{1}{R}\right).$$

Therefore if k is odd then

$$\|f-S_R^{\frac{k-1}{2}}(f)\|_C\leqslant A_k\left(\ln(R+2)E_R(f)+\omega_2\left(f;\ \frac{1}{R}\right)\right).$$

Remark 2 The conclusion of Theorem 1 holds also for the case of $L^p(Q)(1\leqslant p\leqslant+\infty)$.

§ 3. Estimates for partial sum maximal operators

Suppose $f\in L(Q)$. Define $S^\alpha_*(f;\ x)=\sup\limits_{R>0}|S^\alpha_R(f;\ x)|$ and $S^\alpha_*(f;\ x)=\sup\limits_{R>0}|\widetilde{S}^\alpha_R(f;\ x)|$ with respect to a given spherical harmonic ernel $K(x)=\dfrac{p(x)}{|x|^{k+p}}$.

We use the symbol HL to denote the Hardy-Littlewood maximal operator:

$$HL(f)(x)=\sup_{r>0}\frac{1}{\gamma^k}\int_{|y|<x}|f(x-y)|\,\mathrm{d}y.$$

Moreover we define the conjugate maximal operator N as follows:

$$N(f)(x)=\sup_{0<\varepsilon\leqslant 1}|f^*_\varepsilon(x)|. \tag{3.1}$$

It is well known that for $1<q<+\infty$

$$\|HL(f)\|_q\leqslant A_k\frac{q}{q-1}\|f\|_q, \tag{3.2}$$

$$\|N(f)\|_q\leqslant A_k(p)\frac{q^2}{q-1}\|f\|_q. \tag{3.3}$$

The formula (3.2) is derived from [8, pp. 14~18]; and (3.3) is taken from [1] (In the reference [1] it was indicated that the coefficient $A_k(p)\dfrac{q^2}{q-1}$ in the right-hand side of (3.3) can be found in Cotlar's thesis).

For the operator $S_*^{\frac{k-1}{2}}$ of critical order, the estimates of strong type are also well known (see [9]):

$$\|S_*^{\frac{k-1}{2}}(f)\|_q\leqslant A_k\frac{q^3}{(q-1)^2}\|f\|_q,\quad 1<q<+\infty. \tag{3.4}$$

In the conjugate case there are the same estimates for the operator $\widetilde{S}_*^{\frac{k-1}{2}}$ as for $S_*^{\frac{k-1}{2}}$. For this we give a proof as follows.

Let β be a complex number. Write

$$\widetilde{S}_R^{\frac{k-1}{2}+\beta}(f;x) = \frac{1}{|Q|}\int_Q f(x-y)\widetilde{D}_R^{\frac{k-1}{2}+\beta}(y)\mathrm{d}y,$$

$$\widetilde{D}_R^{\frac{k-1}{2}+\beta}(x) = \sum_{|m|<R}\hat{K}(m)\left(1-\left|\frac{m}{R}\right|^2\right)^{\frac{k-1}{2}+\beta}\mathrm{e}^{\mathrm{i}mx}. \tag{3.5}$$

When Re $\beta>0$, we have the following expression (see [10]):

$$\widetilde{D}_R^{\frac{k-1}{2}+\beta}(y) = \frac{R^{k+p}}{c(k,p)}\sum_{j=0}^{+\infty}\sum_{m\in S_j}p(y+2\pi m)\cdot$$

$$E(k,\ \beta,\ p,\ R|y+2\pi m|),\quad R>\pi^{-1}, \tag{3.6}$$

where

$$c_{k,p}=2^k\pi^{\frac{k}{2}}\frac{\Gamma\left(\frac{k+p}{2}\right)}{\Gamma\left(\frac{p}{2}\right)}\quad (p \text{ is the degree of the polynomial } p);$$

$$S_j=\{m\in\mathbf{Z}^k\,|\,j\leqslant|m|<j+1\},\quad j\in\mathbf{N};$$

$$E(k,\beta,p,u) = \frac{(2\pi)^{\frac{k}{2}}}{u^{\frac{k}{2}+p-1}}\int_0^1(1-t^2)^{\frac{k-1}{2}+\beta}t^{\frac{k}{2}}\cdot J_{\frac{k}{2}+p-1}(ut)\mathrm{d}t.$$

It is due to [10] that when $u\geqslant1$,

$$E(k,\ \beta,\ p,\ u)=\frac{c(k,\ p)}{u^{k+p}}+\frac{R^{(1)}(\beta,\ u)}{u^{k+p+1}}+\frac{R^{(2)}(\beta,\ u)}{u^{k+p+1+\beta}}+$$

$$A(k,\ \beta)\frac{\cos\left(u-\frac{\pi}{2}(k+p+\beta)\right)}{u^{k+p+\beta}}, \tag{3.7}$$

where

$$A(k,\ \beta)=\pi^{\frac{k-1}{2}}2^{k+\beta}\Gamma\left(\frac{k+1}{2}+\beta\right),$$

$$|R^{\nu}(\beta,\ u)|=|R^{(\nu)}(k,\ \beta,\ p,\ u)|\leqslant A_{k,p}\mathrm{e}^{\frac{3}{2}\cdot\pi|\mathrm{Im}\beta|},\quad \nu=1,2.$$

Because we only consider the case of Re $\beta=\sigma>0$, we can rewrite (3.7) to be as the following:

$$E(k,\ \beta,\ p,\ u)=\frac{c(k,\ p)}{u^{k+p}}+\frac{R^{(3)}(\beta,\ u)}{u^{k+p+1}}+A(k,\ \beta)\frac{\cos\left(u-\frac{\pi}{2}(k+p+\beta)\right)}{u^{k+p+\beta}},$$

$$u\geqslant1, \tag{3.8}$$

where $R^{(3)}$ satisfies the same estimate as for $R^{(1)}$ or $R^{(2)}$.

Proposition If Re $\beta>0$, then

$$\widetilde{S}_R^{\frac{k-1}{2}+\beta}(f;x)=\frac{1}{|Q|c(k,p)}\int_0^{+\infty}\psi_x\left(\frac{t}{R}\right)t^{k+p-1}E(k,\beta,p,t)\mathrm{d}t, \qquad (3.9)$$

where $\int_0^{+\infty}$ denotes $\lim\limits_{\rho\to+\infty}\int_0^{\rho}$, $\psi_x(s)=\int_{|\xi|=1}f(x-t\xi)p(\xi)\mathrm{d}\sigma(\xi)$.

Proof According to the uniform convergence of the limit " $\sum\limits_{j=0}^{+\infty}\sum\limits_{m\in S_j}$ " in (3.6) and the periodicity of f we conclude immediately.

Lemma 1 Suppose $\beta=\sigma+\mathrm{i}\tau$, $\sigma>0$, $\tau\in\mathbf{R}$. Define

$$M^{\beta}(f)(x)=\sup_{R\geqslant1}|f_{\frac{1}{R}}^{*}(x)-\widetilde{S}_R^{\frac{k-1}{2}+\beta}(f;\ x)|.$$

Then

$$M^{\beta}(f)(x)\leqslant A_k(p)\mathrm{e}^{\frac{3}{2}\cdot\pi|\tau|}\frac{2^{\sigma}(1+\sigma)\Gamma\left(\frac{k+1}{2}+\sigma\right)}{\sigma}HL(f)(x). \qquad (3.10)$$

Proof By the definition we have that for $R\geqslant1$,

$$\begin{aligned}
f_{\frac{1}{R}}^{*}(x)&=\frac{1}{|Q|}\int_{y\in Q,\,|y|>\frac{1}{R}}f(x-y)K^{*}(y)\mathrm{d}y\\
&=\frac{1}{|Q|}\int_{y\in Q,\,|y|>\frac{1}{R}}f(x-y)(K(y)-I_0)\mathrm{d}y+\\
&\quad\frac{1}{|Q|}\int_{y\in Q,\,|y|>\frac{1}{R}}f(x-y)\sum_{m\neq0}[K(y+2\pi m)-I_m]\mathrm{d}y\\
&=\frac{1}{|Q|}\int_{y\in Q,\,|y|>\frac{1}{R}}f(x-y)K(y)\mathrm{d}y-\\
&\quad I_0\frac{1}{|Q|}\int_{y\in Q,\,|y|>\frac{1}{R}}f(x-y)\mathrm{d}y+\\
&\quad\frac{1}{|Q|}\int_{Q}f(x-y)\sum_{m\neq0}[K(y+2\pi m)-I_m]\mathrm{d}y-\\
&\quad\frac{1}{|Q|}\int_{|y|<\frac{1}{R}}f(x-y)\sum_{m\neq0}[K(y+2\pi m)-I_m]\mathrm{d}y.
\end{aligned}$$

Since

$$\sum_{m\neq0}|K(y+2\pi m)-I_m|\leqslant A_k(p),\quad y\in Q,$$

we have

$$\frac{1}{|Q|}\int_{Q}f(x-y)\sum_{m\neq0}[K(y+2\pi m)-I_m]\mathrm{d}y$$

$$= \sum_{j=1}^{+\infty} \sum_{m \in S_j} \left\{ \frac{1}{|Q|} \int_Q f(x-y) K(y+2\pi m) \mathrm{d}y - I_m \frac{1}{|Q|} \int_Q f(x-y) \mathrm{d}y \right\}.$$

The right-hand side can be divided into two sums. We have

$$\begin{aligned} & \sum_{j=1}^{+\infty} \sum_{m \in S_j} \frac{1}{|Q|} \int_Q f(x-y) K(y+2\pi m) \mathrm{d}y \\ = & \sum_{j=1}^{+\infty} \sum_{m \in S_j} \frac{1}{|Q|} \int_{Q+2\pi m} f(x-y) K(y) \mathrm{d}y \\ = & \lim_{\rho \to +\infty} \frac{1}{|Q|} \int_{B(0,\rho) \setminus Q} f(x-y) K(y) \mathrm{d}y, \end{aligned}$$

and

$$\begin{aligned} & \sum_{j=1}^{+\infty} \sum_{m \in S_j} (-I_m) \frac{1}{|Q|} \int_Q f(x-y) \mathrm{d}y \\ = & \frac{1}{|Q|} \int_Q f(x-y) \mathrm{d}y \Big(- \sum_{j=1}^{+\infty} \sum_{m \in S_j} I_m \Big) \\ = & \frac{1}{|Q|} \int_Q f(x-y) \mathrm{d}y \cdot I_0. \end{aligned}$$

Here and below we denote by $B(x, r)$ the ball with center at x and radius r. Hence

$$\begin{aligned} f^*_{\frac{1}{R}}(x) = & \lim_{\rho \to +\infty} \frac{1}{|Q|} \int_{B(0,\rho) \setminus B\left(0,\frac{1}{R}\right)} f(x-y) K(y) \mathrm{d}y + \\ & I_0 \frac{1}{|Q|} \int_Q f(x-y) \mathrm{d}y - \frac{1}{|Q|} \int_{B\left(0,\frac{1}{R}\right)} f(x-y) \cdot \\ & \sum_{m \neq 0} [K(y+2\pi m) - I_m] \mathrm{d}y. \end{aligned}$$

Write

$$\gamma_R(f;x) = \frac{1}{|Q|} \int_{B\left(0,\frac{1}{R}\right)} f(x-y) [I_0 - \sum_{m \neq 0} K(y+2\pi m) - I_m] \mathrm{d}y. \tag{3.11}$$

We get

$$\begin{aligned} f^*_{\frac{1}{R}}(x) & = \frac{1}{|Q|} \int_{R^{-1}}^{+\infty} \psi_x(t) \frac{1}{t} \mathrm{d}t + \gamma_R(f;x) \\ & = \frac{1}{|Q|} \int_1^{+\infty} \psi_x\left(\frac{t}{R}\right) \frac{1}{t} \mathrm{d}t + \gamma_R(f;x). \end{aligned} \tag{3.12}$$

It is easy to see that

$$|\gamma_R(f;x)| \leqslant A_k(P) \int_{B\left(x,\frac{1}{R}\right)} |f(y)| \mathrm{d}y.$$

By virtue to the absolute continuity of integral we see that $\gamma_R(f;\ x)\to 0$ (as $R\to+\infty$) uniformly and by the definition of HL we get

$$|\gamma_R(f,\ x)|\leqslant A_k(P)\frac{1}{R^k}HL(f)(x). \tag{3.13}$$

From (3.19) and (3.12) it follows that

$$\begin{aligned}
&\widetilde{S}_R^{\frac{k-1}{2}+\beta}(f;\ x)-f^*_{\frac{1}{R}}(x)\\
&=\frac{1}{c(k,\ p)}\frac{1}{|Q|}\int_0^1\psi_x\left(\frac{t}{R}\right)t^{k+p-1}E(k,\beta,p,t)\mathrm{d}t+\\
&\quad\frac{1}{c(k,\ p)}\frac{1}{|Q|}\int_1^{+\infty}\psi_x\left(\frac{t}{R}\right)\left[E(k,\beta,p,t)-\frac{c(k,p)}{t^{k+p}}\right]t^{t+p-1}\mathrm{d}t-\gamma_R(f;x).
\end{aligned} \tag{3.14}$$

Since $|E(k,\ \beta,\ p,\ t)|\leqslant A_{k,p}(\mathrm{Re}\,\beta\geqslant 0)$ so

$$\left|\int_0^1\psi_x\left(\frac{t}{R}\right)t^{k+p-1}E(k,\beta,p,t)\mathrm{d}t\right|\leqslant A_{k,p}\int_0^1\left|\psi_x\left(\frac{t}{R}\right)\right|t^{k+p-1}\mathrm{d}t.$$

Substituting the inequality into the right-hand side of the above ineouality we get

$$\left|\int_0^1\psi_x\left(\frac{t}{R}\right)t^{k+p-1}E(k,\beta,p,t)\mathrm{d}t\right|\leqslant A_k(p)HL(f)(x). \tag{3.15}$$

Using (18) we get

$$\begin{aligned}
&\left|\int_1^{+\infty}\psi_x\left(\frac{t}{R}\right)t^{k+p-1}\left(E(k,\beta,p,t)-\frac{c(k,p)}{t^{k+p}}\right)\mathrm{d}t\right|\\
&\leqslant A_{k,p}\mathrm{e}^{\frac{3}{2}\cdot\pi|\tau|}\int_1^{+\infty}\left|\psi_x\left(\frac{t}{R}\right)\right|\frac{\mathrm{d}t}{t^2}+|A(k,\beta)|\int_1^{+\infty}\left|\psi_x\left(\frac{t}{R}\right)\right|\frac{\mathrm{d}t}{t^{1+\sigma}}.
\end{aligned} \tag{3.16}$$

It is easy to see that

$$\begin{aligned}
\int_1^{+\infty}\left|\psi_x\left(\frac{t}{R}\right)\right|\frac{\mathrm{d}t}{t^{1+\sigma}}&=\sum_{j=0}^{+\infty}\int_{2^j}^{2^{j+1}}\left|\psi_x\left(\frac{t}{R}\right)\right|\frac{\mathrm{d}t}{t^{1+\sigma}}\\
&\leqslant A_k(P)\sum_{j=0}^{+\infty}\frac{1}{2^{\sigma j}}HL(f)(x)\\
&\leqslant A_k(P)\frac{1+\sigma}{\sigma}HL(f)(x).
\end{aligned} \tag{3.17}$$

Now the conclusion of Lemma 1 is derived from (3.14)~(3.17).

Lemma 2 Suppose $\beta=\sigma+\tau\mathrm{i}$, $\sigma\in\mathbf{R}$, $\tau\in\mathbf{R}$. If $\frac{k+1}{2}>\sigma>-\frac{k+1}{2}$ then

$$\|\widetilde{S}_*^{\frac{k-1}{2}+\beta}(f)\|_2 \leqslant A_k(P)\,\frac{e^{\pi|\tau|}}{\sigma+\frac{k-1}{2}}\|f\|_2.$$

Proof Let $f\in L^2(Q)$. According to [1] we know that $f^*\in L^2(Q)$ and $\|f^*\|_2\leqslant A_k(P)\|f\|_2$. Now $\widetilde{S}_*^{\frac{k-1}{2}+\beta}(f;\ x)=S_*^{\frac{k-1}{2}+\beta}(f^*;\ x)$. From [9] (Theorem 7 in [9]) we get

$$\|S_*^{\frac{k-1}{2}+\beta}(f^*)\|_2 \leqslant A_k e^{\pi|\tau|}\,\frac{1}{\sigma+\frac{k-1}{2}}\|f^*\|_2.$$

Hence the lemma is established.

Lemma 3 $\|\widetilde{S}_*^{\frac{k-1}{2}}(f)\|_q\leqslant A_k(P)\dfrac{q^3}{(q-1)^2}\|f\|_q,\quad 1<q<+\infty.$

Proof When $k=1$ the conclusion is known. Suppose now $k>1$.

We first consider the case of $q\geqslant 3$. For any simple function $R(x)$ $(x\in Q)$ having only a finite number of distinct positive values, we can define an analytic family of linear operators $\{T_z\}$ by

$$T_z(f)(x)=\widetilde{S}_{R(x)}^{\delta(z)}(f;\ x),\quad f\in L(Q),$$

where $\delta(z)=\dfrac{k-1}{2}(1-z)+\left(\dfrac{q-\frac{3}{2}}{q-2}\right)\dfrac{k-1}{2}z,\ z=\sigma+\tau i,\ 0\leqslant\sigma\leqslant 1$. It is easy to see that $\{T_z\}$ is an "admissible" family (see [9] or [12] about this terminology). Since $\sigma(\tau i)=\dfrac{k-1}{4}+\dfrac{k-1}{4}\dfrac{q-1}{q-2}\tau i$, so by lemma 2 we have

$$\|T_{\tau i}(f)\|_2\leqslant A_k(P)e^{\frac{(k-1)\pi|\tau|}{2}}\frac{1}{k-1}\|f\|_2.$$

On the other hand, $\delta(1+\tau i)=\dfrac{q-\frac{3}{2}}{q-2}\,\dfrac{k-1}{2}+\dfrac{k-1}{4}\,\dfrac{q-1}{q-2}\tau i$. Hence by

Lemma 1 and (3.2)(3.3) we have

$$\|T_{1+\tau i}(f)\|_{2q}\leqslant A_k(P)e^{A_k|\tau|}q\|f\|_{2q}.$$

Using the interpolation theorem (see [12] Chapter V, Theorem 4.1) we get for $t=\dfrac{q-2}{q-1}$,

$$\|T_t(f)\|_q=\|\widetilde{S}_{R(x)}^{\frac{k-1}{2}}(f;\ x)\|_q\leqslant A_k(P)q\|f\|_q,\quad q\leqslant 3.$$

Then taking supremum over all simple functions $R=R(x)$ we get

$$\|\widetilde{S}_*^{\frac{k-1}{2}}(f)\|_q\leqslant A_k(p)q\|f\|_q,\quad q\geqslant 3.$$

We have just proved the conclusion of the lemma in the case of $q\geqslant 3$. For the case of $1<q<3$ the argument is similar.

The proof of Lemma 3 is complete.

Definition Suppose $P(x)$ is a homogeneous harmonic polynomial $P(x)$ of degree p ($p\in\mathbf{N}$). Let $f\in L(Q)$. Write

$$P(f)(x;t)=\int_{|\xi|=1}f(x-t\xi)P(\xi)\mathrm{d}\sigma(\xi).\tag{3.18}$$

The operator TP is defined by

$$TP(f)(x)=\sup_{R\geqslant 1}\left\{\overline{\lim_{\beta\to 0^+}}\left|\int_1^{+\infty}P(f)\left(x;\frac{t}{R}\right)\frac{\cos\left(t-\frac{\pi}{2}(k+p+\beta)\right)}{t^{1+\beta}}\mathrm{d}t\right|\right\}.\tag{3.19}$$

Lemma 4 If $1<q<+\infty$, $f\in L^q(Q)$, then

$$\|TP(f)\|_q\leqslant A_k(P)\frac{q^3}{(q-1)^2}\|f\|_q.$$

Proof We first consider the case of $p>0$ (We recall that p is the degree of p which can be zero in our lemma). Take arbitrarily $\beta\in(0,1)$. According to (3.14) we have

$$\widetilde{S}_R^{\frac{k-1}{2}+\beta}(f;x)-\Big[f_{\frac{1}{R}}^*(x)+\frac{1}{c(k,p)|Q|}\int_0^1P(f)\left(x;\frac{t}{R}\right)t^{k+p-1}\cdot$$

$$E(k,\beta,p,t)\mathrm{d}t-\gamma_R(f;x)\Big]$$

$$=\frac{1}{c(k,p)|Q|}\int_1^{+\infty}P(f)\left(x;\frac{t}{R}\right)t^{k+p-1}\left(E(k,\beta,p,t)-\frac{c(k,p)}{t^{k+p}}\right)\mathrm{d}t.$$

From this, using (3.13) and (3.15) we get

$$|\hat{S}_R^{\frac{k-1}{2}+\beta}(f;x)|+N(f)(x)+A_k(p)HL(f)(x)\geqslant$$

$$\frac{1}{c(k,p)|Q|}\left|\int_1^{+\infty}P(f)\left(x;\frac{t}{R}\right)t^{k+p-1}\left(E(k,\beta,p,t)-\frac{c(k,p)}{t^{k+p}}\right)\mathrm{d}t\right|.$$

Using the asymptotic formula (3.8) we get

$$\left|\int_1^{+\infty}P(f)\left(x;\frac{t}{R}\right)t^{k+p-1}\left(E(k,\beta,p,t)-\frac{c(k,p)}{t^{k+p}}\right)\mathrm{d}t\right|$$

$$\geqslant A(k,\beta)\left|\int_1^{+\infty}P(f)\left(x;\frac{t}{R}\right)\frac{\cos\left(t-\frac{\pi}{2}(k+p+\beta)\right)}{t^{1+\beta}}\mathrm{d}t\right|-$$

$$A_{k,p}\int_1^{+\infty}\left|P(f)\left(x;\frac{t}{R}\right)\frac{1}{t^2}\right|\mathrm{d}t.$$

From (3. 17) we see that

$$\int_1^{+\infty}\left|P(f)\left(x;\frac{t}{R}\right)\right|\frac{1}{t^2}\mathrm{d}t\leqslant A_k(P)HL(f)(x).$$

So we get

$$A(k,\beta)\left|\int_1^{+\infty}P(f)\left(x;\frac{t}{R}\right)\frac{\cos\left(t-\frac{\pi}{2}(k+p+\beta)\right)}{t^{1+\beta}}\mathrm{d}t\right|$$

$$\leqslant|\widetilde{S}_R^{\frac{k-1}{2}+\beta}(f;x)|+N(f)(x)+A_k(P)HL(f)(x),\quad \beta\in(0,1),$$

where $A(k,\beta)=2^{k+\beta}\pi^{\frac{k-1}{2}}\Gamma\left(\frac{k+1}{2}+\beta\right)$. Letting $\beta\to 0^+$ and then taking "$\sup\limits_{R\geqslant 1}$" we obtain that

$$TP(f)(x)\leqslant A_k(P)HL(f)(x)+A_k(N(f(x)+\widetilde{S}_*^{\frac{k-1}{2}}(f;\ x))).$$

From this, by Lemma 3 and (3. 2) (3. 3) we obtain the conclusion of Lemma 4 (for $p>0$).

For $p=0$, we replace $\widetilde{S}_R^{\frac{k-1}{2}+\beta}$ by $S_R^{\frac{k-1}{2}+\beta}$, then a similar argument yields the desired result.

The proof of lemma 4 is finished.

Lemma 5 If an operator T defined on $L(Q)$ satisfies the condition

$$\|T(f)\|_q\leqslant A_0q\cdot\|f\|_q,\quad 2<q<+\infty,$$

then there exists a constant A depending only on A_0 such that for any $f\in L^\infty(Q)$ and any $\lambda>0$,

$$|\{x\in Q\mid T(f)(x)>\lambda\}|\leqslant A\mathrm{e}^{-\frac{\lambda}{A\|f\|_\infty}}.$$

Proof Let $f\in L(Q)$. Write $E_\lambda=\{x\in Q\mid T(f)(x)>\lambda\}$, $\lambda>0$. Then for any $q>2$,

$$|E_\lambda|\leqslant\int_{E_\lambda}\left|\frac{T(f)(x)}{\lambda}\right|^q\mathrm{d}x\leqslant\left(\frac{A_0q\|f\|_q}{\lambda}\right)^q.$$

If $\lambda>2A_0\|f\|_\infty\mathrm{e}$, putting $q=\frac{\lambda}{A_0\|f\|_\infty\mathrm{e}}$ in the above inequality we get

$$|E_\lambda| \leqslant e^{-\frac{\lambda}{A_0 \|f\|_\infty e}}.$$

Hence the constant $A = |Q| e^2 + A_0 e$ will satisfy the requirement. This finishes the proof.

§ 4. Theorems of the approximation on set of total measure

Lemma 6 Let $f \in W^1 L^\infty$. Then

$$|S_R^{\frac{k-1}{2}}(f;x) - f(x)| \leqslant \frac{2\Gamma\left(\frac{k+1}{2}\right)}{\pi^{\frac{k}{2}}} \frac{1}{R}\left\{\sum_{j=1}^{k} TP_j(f_j)(x) + \theta_R(f)(x)\right\},$$

$$R \geqslant 1,$$

where TP_j is an operator defined by (3.18) and (3.19) with $P = P_j(x) = x_j$, $f_j = \frac{\partial f}{\partial x_j} (j = 1, 2, \cdots, k)$, the remainder $\theta_R(f)(x)$ satisfies the following two conditions:

(1) $0 \leqslant \theta_R(f)(x) \leqslant A_k \max\{\|f_j\|_\infty | j = 1, 2, \cdots, k\}$,

(2) $\lim\limits_{R \to +\infty} \theta_R(f)(x) = 0$, a. e.

Proof Let $\beta \in (0, 1)$. From (2.3) we get

$$S_R^{\frac{k-1}{2}+\beta}(f;x) - f(x) = \frac{2^{\frac{k+1}{2}+\beta}\Gamma\left(\frac{k+1}{2}+\beta\right)}{(2\pi)^{\frac{k}{2}}} \cdot \int_0^{+\infty} f_x\left(\frac{t}{R}\right) \frac{J_{\frac{k-1}{2}+\beta}(t)}{t^{\frac{1}{2}+\beta}} dt.$$

Set $F_\beta(t) = \int_t^{+\infty} \frac{J_{\frac{k-1}{2}+\beta}(s)}{s} ds$. Using (2.6) we get for $t \geqslant 1$,

$$F_\beta(t) = \sqrt{\frac{2}{\pi}} \int_t^{+\infty} \left\{ \frac{\cos\left(s - \frac{\pi(k+\beta)}{2}\right)}{s^{1+\beta}} - A_{k,\beta} \frac{\sin\left(s - \frac{\pi(k+\beta)}{2}\right)}{s^{2+\beta}} + O\left(\frac{1}{s^3}\right) \right\} ds$$

$$= \sqrt{\frac{2}{\pi}} \frac{\cos\left(t - \frac{\pi(k+1+\beta)}{2}\right)}{t^{1+\beta}} + O\left(\frac{1}{t^2}\right), \quad t \to +\infty, \tag{4.1}$$

where $O\left(\frac{1}{t^2}\right)$ is uniform for $\beta \in (0, 1)$. When $0 \leqslant t < 1$ we have

$$F_\beta(t) = \int_t^1 \frac{J_{\frac{k-1}{2}+\beta}(s)}{s^{\frac{1}{2}+\beta}} ds + F_\beta(1) = O(1). \tag{4.2}$$

Integrating by parts we have

$$\int_0^{+\infty} f_x\left(\frac{t}{R}\right)\frac{J_{\frac{k-1}{2}+\beta}(t)}{t^{\frac{1}{2}+\beta}}\mathrm{d}t = -\int_0^{+\infty} F_\beta(t)\frac{\mathrm{d}}{\mathrm{d}t}f_x\left(\frac{t}{R}\right)\mathrm{d}t$$

$$= -\left(\int_0^1 + \int_1^{+\infty}\right)F_\beta(t)\frac{\mathrm{d}}{\mathrm{d}t}f_x\left(\frac{t}{R}\right)\mathrm{d}t$$

$$\stackrel{\text{def.}}{=\!=} I_1 + I_2.$$

From (4.2) we get

$$|I_1| \leqslant A_k\int_0^1\left|\frac{\mathrm{d}}{\mathrm{d}t}f_x\left(\frac{t}{R}\right)\right|\mathrm{d}t.$$

It is easy to see that

$$\frac{\mathrm{d}}{\mathrm{d}t}f_x\left(\frac{t}{R}\right) = \frac{1}{R}\int_{|\xi|=1}\sum_{j=1}^k f_j\left(x-\frac{t}{R}\xi\right)(-\xi_j)\mathrm{d}\sigma(\xi)$$

$$= -\frac{1}{R}\sum_{j=1}^k P_j(f_j)\left(x;\frac{t}{R}\right). \tag{4.3}$$

It is obvious that if $f_j(x)$ exists then

$$\left|P_j(f_j)\left(x;\frac{t}{R}\right)\right| = \left|\int_{|\xi|=1}\left[f_j\left(x-\frac{t}{R}\xi\right)-f_j(x)\right]\xi_j\mathrm{d}\sigma(\xi)\right|$$

$$\leqslant \int_{|\xi|=1}\left|f_j\left(x-\frac{t}{R}\xi\right)-f_j(x)\right|\mathrm{d}\sigma(\xi).$$

Suppose that x is a common Lebesgue point of all f_j $(j=1,2,\cdots,k)$. Write

$$\Phi_j(x;s) = \int_0^s\int_{|\xi|=t}|f_j(x-\xi)-f_j(x)|\mathrm{d}\sigma(\xi)\mathrm{d}t.$$

Then $\Phi_j(x;s)=o(s^k)$ (as $s\to 0^+$). Hence

$$\int_0^1\left|P_j(f_j)\left(x;\frac{t}{R}\right)\right|\mathrm{d}t \leqslant R\int_0^{\frac{1}{R}}\frac{1}{t^{k-1}}\int_{|\xi|=1}|f_j(x-\xi)-f_j(x)|\mathrm{d}\sigma(\xi)\mathrm{d}t$$

$$= R\left\{\frac{1}{t^{k-1}}\Phi_j(x;t)\Big|_0^{\frac{1}{R}} + (k-1)\int_0^{\frac{1}{R}}\frac{\Phi_j(x;t)}{t^k}\mathrm{d}t\right\}$$

$$= O(1), \quad R\to+\infty.$$

From this we get $I_1 = o\left(\frac{1}{R}\right)$.

Using (4.1) and (4.3) we get

$$I_2 = -\sqrt{\frac{2}{\pi}}\frac{1}{R}\sum_{j=1}^k\int_1^{+\infty}P_j(f_j)\left(x;\frac{t}{R}\right)\frac{\cos\left(t-\frac{\pi}{2}(k+1+\beta)\right)}{t^{1+\beta}}\mathrm{d}t + I_2',$$

$$
\begin{aligned}
|I_2'| &\leqslant A_k\int_1^{+\infty}\left|\frac{\mathrm{d}}{\mathrm{d}t}f_x\left(\frac{t}{R}\right)\right|\frac{1}{t^2}\mathrm{d}t \\
&\leqslant \frac{A_k}{R}\sum_{j=1}^{k}\int_{\frac{1}{R}}^{+\infty}\frac{1}{Rt^{k+1}}\int_{|\xi|=1}|f_j(x-\xi)-f_j(x)|\,\mathrm{d}\sigma(\xi)\cdot\mathrm{d}t \\
&\leqslant \frac{A_k}{R}\sum_{j=1}^{k}\int_{\frac{1}{R}}^{+\infty}\frac{\Phi_j(x;t)}{Rt^{k+2}}\mathrm{d}t=o\left(\frac{1}{R}\right).
\end{aligned}
$$

Now let $\beta\to0^+$ and define

$$
\theta_R(f;\ x)=\sqrt{\frac{\pi}{2}}R\varlimsup_{\beta\to0+}|I_1+I_2'|.
$$

Then we get

$$
|S_R^{\frac{k-1}{2}}(f;x)-f(x)|\leqslant\frac{2\Gamma\left(\frac{k+1}{2}\right)}{\pi^{\frac{k}{2}}}\frac{1}{R}\left\{\sum_{j=1}^{k}TP_j(f_j)(x)+\theta_R(f)(x)\right\},
$$

where $\theta_R(f)(x)$ is $o(1)$ at the common Lebesgue point of all f_j and it is obvious that $0\leqslant\theta_R(f)(x)\leqslant A_k\max\{\|f_j\|_\infty\,|\,j=1,2,\cdots,k\}$. □

In order to establish a similar lemma for the conjugate case we need to refine the asymptotic expansion formula for the function $E(k,\ \beta,\ p,\ u)$.

Lemma 7 Let $s=\sigma+\mathrm{i}\tau$, $-\frac{k+1}{2}+h\leqslant\sigma\leqslant G$ ($0<h<G<+\infty$, $\tau\in\mathbf{R}$). Then for $u\geqslant1$,

$$
\begin{aligned}
E(k,s,p,u)=&\frac{c(k,p)}{u^{k+p}}+\frac{G(k,s,p,u)}{u^{k+p+2}}+\frac{H(k,s,p,u)}{u^{k+p+2+s}}+ \\
&\frac{B(k,s)}{u^{k+p+s}}\left(\sqrt{u}\,J_{k+p+s-\frac{1}{2}}(u)+p\frac{k+1+s}{\sqrt{u}}J_{k+p+s+\frac{1}{2}}(u)\right),
\end{aligned}
\tag{4.4}
$$

where

$$
c(k,\ p)=\frac{2^k\pi^{\frac{k}{2}}\Gamma\left(\frac{k+p}{2}\right)}{\Gamma\left(\frac{p}{2}\right)};
$$

$$
B(k,\ s)=2^{\frac{k-1}{2}+s}\pi^{\frac{k}{2}}\Gamma\left(\frac{k+1}{2}+s\right);
$$

$$
|G(k,\ s,\ p,\ u)|\leqslant A_{k,p,h,G}\,\mathrm{e}^{3\pi\frac{|\tau|}{2}};
$$

$$
|H(k,\ s,\ p,\ u)|\leqslant A_{k,p,h,G}\,\mathrm{e}^{3\pi\frac{|\tau|}{2}}.
$$

Proof Define

$$f_\xi(u)=\int_0^{+\infty}\mathrm{e}^{-t}t^{\frac{k}{2}+\xi}J_{\frac{k}{2}+p-1}(ut)\mathrm{d}t,\quad(\xi\geqslant 0),$$

We know that Sublemma 7. 2 in [10] established the asymptotic expansion for f_ξ. Now we calculate anew to obtain more precise result. Define another function ϕ_ξ by

$$\phi_\xi(x)=\int_0^{+\infty}\mathrm{e}^{-xt}t^{\frac{k}{2}+\xi}J_{\frac{k}{2}+p-1}(t)\mathrm{d}t,\quad 0<x\leqslant 1.$$

It is clear that

$$f_\xi(u)=\frac{1}{u^{\frac{k}{2}+\xi+1}}\phi_\xi\left(\frac{1}{u}\right),\quad u>0.$$

Using a formula in [11] (p. 26, formula (6)) we get

$$f_\xi(u)=\left(\frac{1}{\sqrt{1+u^2}}\right)^{\frac{k}{2}+\xi+1}\Gamma(k+p+\xi)p_{\frac{k}{2}+\xi}^{-(\frac{k}{2}+p-1)}\left(\frac{1}{\sqrt{1+u^2}}\right),$$

where p_μ^ν denotes legendre function (see [11] p. 426). Therefore

$$\lim_{u\to+\infty}u^{\frac{k}{2}+\xi+1}f_\xi(u)=\Gamma(k+p+\xi)p_{\frac{k}{2}+\xi}^{-(\frac{k}{2}+p-1)}(0)=\varphi_\xi(0^+).\quad(4.5)$$

On the other hand we have obviously that for $x>0$.

$$\phi_\xi'(x)=-\int_0^{+\infty}\mathrm{e}^{-xt}t^{\frac{k}{2}+\xi+1}J_{\frac{k}{2}+p-1}(ut)\mathrm{d}t=-\phi_{\xi+1}(x).$$

Hence we get

$$\phi_\xi'(0^+)=-\phi_{\xi+1}(0^+).\quad(4.6)$$

Therefore by L'Hospital's rule it follows that

$$f_\xi(u)=\phi_\xi(0^+)u^{-(\frac{k}{2}+\xi+1)}+\phi_\xi'(0^+)u^{-(\frac{k}{2}+\xi+2)}+g_\xi(u)u^{-(\frac{k}{2}+\xi+3)},\quad(4.7)$$

where $|g_\xi(u)|\leqslant A_{k,p,\xi}$.

Let $\psi(t)$ be a function of the class $C_{[0,+\infty)}^{+\infty}$ satisfying the conditions

$$\psi(t)=1,\quad\text{if }0\leqslant t\leqslant\frac{1}{3},\quad\psi(t)=0,\quad\text{if }t\geqslant\frac{2}{3}.$$

Define a polynomial $Q(t)=\sum_{j=0}^{2q}a_jt^j$, $q=\left[\frac{k}{2}+p+2\right]$, such that

$$\frac{\mathrm{d}^n}{\mathrm{d}t^n}\{\mathrm{e}^{-t}Q(t)-(1-t^2)^{\frac{k-1}{2}+s}\psi(t)\}|_{t=0}=0,\quad n=0,1,2,\cdots,2q.$$

It is easy to see that $a_0=a_1=1$. Write

$$I_1(u)=\int_0^{+\infty}\psi(t)(1-t^2)^{\frac{k-1}{2}+s}t^{\frac{k}{2}}J_{\frac{k}{2}+p-1}(ut)\mathrm{d}t,$$

$$I_1^*(u)=\int_0^{+\infty}\mathrm{e}^{-t}Q(t)t^{\frac{k}{2}}J_{\frac{k}{2}+p-1}(ut)\mathrm{d}t,$$

$$\Delta_1(u)=I_1^*(u)-I_1(u).$$

Then we have as in [10] that

$$|\Delta_1(u)|\leqslant A_{k,p}(1+|s|^{4q})u^{-\left(\frac{k}{2}+p+3\right)}. \tag{4.8}$$

Now, $I_1^*(u)=f_0(u)+f_1(u)+\sum_{j=2}^{2q}a_jf_j(u)$. By (4.7),

$$f_0(u)=\phi_0(0^+)u^{-\left(\frac{k}{2}+1\right)}+\phi_0'(0^+)u^{-\left(\frac{k}{2}+2\right)}+g_0(u)u^{-\left(\frac{k}{2}+3\right)},$$

$$f_1(u)=\phi_1(0^+)u^{-\left(\frac{k}{2}+2\right)}+\phi_1'(0^+)u^{-\left(\frac{k}{2}+3\right)}+g_1(u)u^{-\left(\frac{k}{2}+4\right)},$$

and

$$\sum_{j=2}^{2q}a_jf_j(u)=G(u)u^{-\left(\frac{k}{2}+3\right)}.$$

It is not difficult to get the following estimate:

$$|G(u)|\leqslant A_{k,p,h,G}\mathrm{e}^{\frac{3\pi|\tau|}{2}}.$$

Hence we get

$$I_1^*(u)=\phi_0(0^+)u^{-\left(\frac{k}{2}+1\right)}+(\phi_0'(0^+)+\phi_1(0^+))u^{-\left(\frac{k}{2}+2\right)}+G_1(u)u^{-\left(\frac{k}{2}+3\right)},$$

$$|G_1(u)|\leqslant A_{k,p,h,G}\mathrm{e}^{\frac{3\pi|\tau|}{2}}. \tag{4.9}$$

According to (4.6) the equation $\phi_0'(0^+)+\phi_1(0^+)=0$ holds. Therefore

$$I_1^*(u)=\phi_0(0^+)u^{-\left(\frac{k}{2}+1\right)}+G_1(u)u^{-\left(\frac{k}{2}+3\right)}. \tag{4.10}$$

Now define another polynomial $\bar{Q}(t)=\sum_{j=0}^{2q}b_jt^j$ such that if we set $\varepsilon(t)=\bar{Q}(1-t^2)-t^{-p}(1-\psi(t))$, then $\varepsilon^{(n)}(1)=0$, $n=0,1,\cdots,2q$. It is clear that $b_0=1$ and $b_1=\frac{1}{2}p$. Then write

$$I_2(u)=\int_0^1(1-\psi(t))(1-t^2)^{\frac{k-1}{2}+s}t^{\frac{k}{2}}J_{\frac{k}{2}+p-1}(ut)\mathrm{d}t,$$

$$I_2^*(u)=\int_0^1\bar{Q}(1-t^2)t^{\frac{k}{2}+p}(1-t^2)^{\frac{k-1}{2}+s}J_{\frac{k}{2}+p-1}(ut)\mathrm{d}t,$$

$$\Delta_2(u)=I_2^*(u)-I_2(u).$$

We have as in [10] that

$$|\Delta_2(u)|\leqslant A_{k,p}(1+|s|^{4q})u^{-\left(\frac{k}{2}+p+3\right)}. \tag{4.11}$$

Now, $$I_2^*(u)=\sum_{j=0}^{2q}b_j\int_0^1(1-t^2)^{\frac{k-1}{2}+s+j}t^{\frac{k}{2}+p}J_{\frac{k}{2}+p-1}(ut)\mathrm{d}t.$$

Using a formula in [11] (p. 26, formula (33)) we have

$$\int_0^1(1-t^2)^{\frac{k-1}{2}+s+j}t^{\frac{k}{2}+p}J_{\frac{k}{2}+p-1}(ut)\mathrm{d}t$$

$$=2^{\frac{k-1}{2}+s+j}\Gamma\left(\frac{k-1}{2}+s+j\right)\frac{J_{\frac{k-1}{2}+p+s+j}(u)}{u^{\frac{k-1}{2}+s+j+1}},$$

Using again the asymptotic formula for Bessel function we get

$$I_2^*(u)=\frac{2^{\frac{k-1}{2}+s}}{u^{\frac{k+1}{2}+s}}\Gamma\left(\frac{k+1}{2}+s\right)(J_{\frac{k-1}{2}+p+s}(u)+$$

$$p\frac{\frac{k+1}{2}+s}{u}J_{\frac{k+1}{2}+p+s}(u))+\frac{H(u)}{u^{\frac{k}{2}+s+3}},$$

$$|H(u)|\leqslant A_{k,p,h,G}\mathrm{e}^{\frac{3\pi|\tau|}{2}}.$$

Hence we obtain

$$I_2(u)=\frac{2^{\frac{k-1}{2}+s}\Gamma\left(\frac{k+1}{2}+s\right)}{u^{\frac{k}{2}+1+s}}(\sqrt{u}\,J_{\frac{k-1}{2}+p+s}(u)+p\frac{\frac{k+1}{2}+s}{\sqrt{u}}J_{\frac{k+1}{2}+p+s}(u))+$$

$$\frac{H(u)}{u^{\frac{k}{2}+s+3}}-\Delta_2(u). \tag{4.12}$$

Combining (4.8) ~ (4.12) and noticing $E(k,s,p,t)=\frac{(2\pi)^{\frac{k}{2}}}{u^{\frac{k}{2}+p-1}}\cdot(I_1(u)+I_2(u))$ we get (4.4). □

From (4.4) we see that if $\beta\geqslant0$ then for $u\geqslant1$,

$E(k,\beta,p,u)=$

$$\frac{c(k,p)}{u^{k+p}}+\frac{I(k,\beta,p,u)}{u^{k+p+2}}+\frac{A(k,\beta)}{u^{k+p+\beta}}\cos\left(u-\frac{\pi}{2}(k+p+\beta)\right)+$$

$$\frac{A_1(k,\beta)}{u^{k+p+1+\beta}}\sin\left(u-\frac{\pi}{2}(k+p+\beta)\right), \tag{4.13}$$

where $I(k,\beta,p,u)\in C^{+\infty}_{[1,+\infty)}$ (as a function of variable u) and $|I(k,\beta,p,u)|\leqslant A_{k,p,\beta}$. Particularly, when $\beta\in[0,1]$, $|I(k,\beta,p,u)|\leqslant A_{k,p}$.

Lemma 8 There exists uniquely determinate system of homogeneous harmonic polynomials $\{P_{j,p+1-2l} \mid j=1,2,\cdots,k;\ l=0,1,2,\cdots,v_p\}$ corresponding to any homogeneous harmonic polynomial $P(x)$ of degree p, where $v_p=\frac{p+1}{2}$ and $P_{j,p+1-2l}(x)$ is of degree $p+1-2l$, such that

$$x_jP(x)=P_{j,p+1}(x)+P_{j,p-1}(x)|x|^2+\cdots+P_{j,p+1-2v_p}(x)|x|^{2v_p},$$

$j=1,\ 2,\ \cdots,\ k$. Particularly, it holds on the spherical surface $\{\xi \mid |\xi|=1\}$ that

$$\xi_jP(\xi)=\sum_{l=0}^{v_p}P_{j,p+1-2l}(\xi). \tag{4.14}$$

This lemma is taken from [12] (see Chapter IV, Theorem 2.1).

Now we assume a homogeneous harmonic polynomial $P(x)$ of degree p is fixed and p is a natural number. We consider the corresponding problem of conjugate approximation.

Lemma 9 If $f\in W^1L^\infty$ then

$$|\widetilde{S}_R^{\frac{k-1}{2}}(f;x)-f^*(x)|\leqslant\frac{A_{k,p}}{R}\Big(\sum_{j=1}^{k}\sum_{l=0}^{v_p}TP_{j,p+1-2l}(f_j)(x)+\tilde{\theta}_R(f)(x)\Big),$$

where $TP_{j,p+1-2l}$ denotes the operator defined by (3.18) and (3.19) with $P=P_{j,p+1-2l}$ which is the same as in Lemma 8, f_j denotes the partial derivative of f with respect to the j-th variable. The remainder $\tilde{\theta}_R(f)(x)$ satisfies the condition

$$0\leqslant\tilde{\theta}_R(f)(x)\leqslant A_k(P)\max\{\|f_j\|_\infty \mid j=1,2,\cdots,k\}.$$

Proof Let $\beta\in(0,\ 1)$ be arbitrary. From (3.9) we get

$$\widetilde{S}_R^{\frac{k-1}{2}+\beta}(f;x)=\frac{1}{c(k,p)|Q|}\int_0^{+\infty}P(f)\Big(x;\frac{t}{R}\Big)t^{k+p-1}E(k,\beta,p,t)\mathrm{d}t.$$

And from (3.11) (3.12) (or from [1]) we see that

$$f^*(x)=\frac{1}{|Q|}\int_0^{+\infty}P(f)(x;t)\frac{1}{t}\mathrm{d}t=\frac{1}{|Q|}\int_0^{+\infty}P(f)\Big(x;\frac{t}{R}\Big)\frac{1}{t}\mathrm{d}t.$$

Therefore

$$\widetilde{S}_R^{\frac{k-1}{2}+\beta}(f;x)-f^*(x)=\frac{1}{|Q|}\int_0^{+\infty}P(x)\Big(x;\frac{t}{R}\Big)t^{t+p-1}\cdot\Big(\frac{E(k,\ \beta,\ p,\ t)}{c(k,\ p)}-\frac{1}{t^{k+p}}\Big)\mathrm{d}t.$$

Set

$$G_\beta(t)=\int_t^{+\infty}S^{k+p-1}\left(\frac{E(k,\beta,p,s)}{c(k,p)}-\frac{1}{s^{k+p}}\right)\mathrm{d}s,\quad t>0.$$

Using (4. 13) we get for $t\geqslant 1$,

$$G_\beta(t)=\int_t^{+\infty}\frac{I(k,\beta,p,s)}{c(k,p)s^3}\mathrm{d}s+\frac{A(k,\beta)}{c(k,p)}\int_t^{+\infty}\frac{\cos\left(s-\frac{\pi}{2}(k+p+\beta)\right)}{s^{1+\beta}}\mathrm{d}s+$$

$$\frac{A_1(k,\beta)}{c(k,p)}\int_t^{+\infty}\frac{\sin\left(s-\frac{\pi}{2}(k+p+\beta)\right)}{s^{2+\beta}}\mathrm{d}s.$$

Integrating by parts we get

$$\int_t^{+\infty}\frac{\cos\left(s-\frac{\pi}{2}(k+p+\beta)\right)}{s^{1+\beta}}\mathrm{d}s$$

$$=-\frac{\sin\left(t-\frac{\pi}{2}(k+p+\beta)\right)}{t^{1+\beta}}+(1+\beta)\int_t^{+\infty}\frac{\sin\left(s-\frac{\pi}{2}(k+p+\beta)\right)}{s^{2+\beta}}\mathrm{d}s$$

$$=-\frac{\sin\left(t-\frac{\pi}{2}(k+p+\beta)\right)}{t^{1+\beta}}+(1+\beta)\frac{\cos\left(t-\frac{\pi}{2}(k+p+\beta)\right)}{s^{2+\beta}}-$$

$$(1+\beta)(2+\beta)\int_t^{+\infty}\frac{\cos\left(s-\frac{\pi}{2}(k+p+\beta)\right)}{s^{3+\beta}}\mathrm{d}s,$$

$$\int_t^{+\infty}\frac{\sin\left(s-\frac{\pi}{2}(k+p+\beta)\right)}{s^{2+\beta}}\mathrm{d}s$$

$$=\frac{\cos\left(t-\frac{\pi}{2}(k+p+\beta)\right)}{t^{2+\beta}}-(2+\beta)\int_t^{+\infty}\frac{\cos\left(s-\frac{\pi}{2}(k+p+\beta)\right)}{s^{3+\beta}}\mathrm{d}s.$$

Therefore

$$G_\beta(t)=-\frac{A(k,\beta)}{c(k,p)}\frac{\cos\left(t-\frac{\pi}{2}(k+p+\beta)\right)}{t^{1+\beta}}+$$

$$\left[(1+\beta)\frac{A(k,\beta)}{c(k,p)}+\frac{A_1(k,\beta)}{c(k,p)}\right]\cdot\frac{\cos\left(t-\frac{\pi}{2}(k+p+\beta)\right)}{t^{2+\beta}}+$$

$$\int_t^{+\infty}\frac{1}{s^{3+\beta}}\left\{\frac{I(k,\beta,p,s)}{c(k,p)s^{-\beta}}-\left[(1+\beta)(2+\beta)\frac{A(k,\beta)}{c(k,p)}+\right.\right.$$

$$(2+\beta)\frac{A_1(k,\beta)}{c(k,p)}\Big]\cdot\cos\Big(s-\frac{\pi}{2}(k+p+\beta)\Big)\Big\}\mathrm{d}s=-\frac{A(k,\beta)}{c(k,p)}. \tag{4.15}$$

where $\gamma(k, \beta, p, t)$ is continuous in (t, β) on $[1, +\infty)\times[0, 1]$ and bounded: $|\gamma(k, \beta, p, t)|\leqslant A_{k,p}$.

When $0<t<1$, we have

$$\begin{aligned}G_\beta(t)&=G_\beta(1)+\int_t^1 s^{k+p-1}\Big(\frac{E(k,\beta,p,s)}{c(k,p)}-\frac{1}{s^{k+p}}\Big)\mathrm{d}s\\&=O(1)+\ln t\end{aligned} \tag{4.16}$$

uniformly for $\beta\in[0, 1]$.

Write $M=\max\{\|f_j\|_\infty\,|\,j=1,2,\cdots,k\}$. We have

$$\begin{aligned}\Big|P(f)\Big(x;\frac{t}{R}\Big)\Big|&\leqslant\int_{|\xi|=1}\Big|f\Big(x-\frac{t}{R}\xi\Big)-f(x)\Big|\,|P(\xi)|\,\mathrm{d}\sigma(\xi)\\&\leqslant A_k(P)M\frac{t}{R}.\end{aligned} \tag{4.17}$$

Integrating by parts we get

$$\begin{aligned}&\widetilde{S}_R^{\frac{k-1}{2}+\beta}(f;\ x)-f^*(x)\\&=\frac{1}{|Q|}\int_0^{+\infty}\frac{\mathrm{d}}{\mathrm{d}t}\Big(P(f)\Big(x;\frac{t}{R}\Big)\Big)G_\beta(t)\mathrm{d}t\\&=\frac{1}{|Q|}\Big(\int_0^1+\int_0^{+\infty}\Big)\frac{\mathrm{d}}{\mathrm{d}t}\Big(P(f)\Big(x;\frac{t}{R}\Big)\Big)G_\beta(t)\mathrm{d}t\\&=I_1+I_2.\end{aligned}$$

It follows from (4.14) that

$$\begin{aligned}\frac{\mathrm{d}}{\mathrm{d}t}\Big(P(f)\Big(x;\frac{t}{R}\Big)\Big)&=-\frac{1}{R}\sum_{j=1}^k\int_{|\xi|=1}f_j\Big(x-\frac{t}{R}\xi\Big)\xi_jP(\xi)\mathrm{d}\sigma(\xi)\\&=-\frac{1}{R}\sum_{j=1}^k\sum_{l=0}^{v_p}P_{j,p+1-2l}(f_j)\Big(x;\frac{t}{R}\Big).\end{aligned} \tag{4.18}$$

It is clear that

$$\Big|\frac{\mathrm{d}}{\mathrm{d}t}P(f)\Big(x;\ \frac{t}{R}\Big)\Big|\leqslant\frac{1}{R}A_k(P)M. \tag{4.19}$$

Hence by (4.16) we get

$$|I_1|\leqslant\frac{A_k(P)}{R}M\int_0^1|\ln t|\,\mathrm{d}t=\frac{A_k(P)}{R}M. \tag{4.20}$$

On the other hand, from (4.15) and (4.18), we get

$$I_2=-\frac{1}{R}\frac{1}{|Q|}\frac{A(k,\beta)}{c(k,p)}\sum_{j=1}^{k}\sum_{l=0}^{v_p}\int_1^{+\infty}P_{j,p+1-2l}(f_j)\left(x;\frac{t}{R}\right)\cdot$$

$$\frac{\cos\left(t-\frac{\pi}{2}(k+p+1+\beta)\right)}{t^{1+\beta}}\mathrm{d}t+\frac{1}{|Q|}\int_1^{+\infty}\frac{\mathrm{d}}{\mathrm{d}t}\left(P(f)\left(x;\frac{t}{R}\right)\right)\cdot$$

$$\frac{1}{t^2}\gamma(k,\ \beta,\ p,\ t)\mathrm{d}t.$$

According to (3.18) (3.19) and (4.19) we get

$$\overline{\lim_{\beta\to0^+}}|I_2|\leqslant\frac{A_{k,p}}{R}\sum_{j=1}^{k}\sum_{l=0}^{v_p}TP_{j,p+1-2l}(f_j)(x)+\frac{A_k(P)}{R}M.\qquad(4.21)$$

The conclusion of Lemma 9 is now deduced from (4.20) and (4.21).

Lemma 10 (1) if $f\in W^1L^\infty$ then

$$\omega(f^*;\ \delta)=O\left(\delta\ln\frac{1}{\delta}\right),\quad \text{as } \delta\to0^+;$$

(2) if $f\in W^2L^\infty$ then

$$\omega_2(f^*;\ \delta)=O\left(\delta^2\ln\frac{1}{\delta}\right),\quad \text{as } \delta\to0^+.$$

Proof By the definition we have

$$f^*(x)=\frac{1}{|Q|}\int_Q f(x-y)K^*(y)\mathrm{d}y=\frac{1}{|Q|}\int_Q f(x-y)K(y)\mathrm{d}y+$$

$$\frac{1}{|Q|}\int_Q f(x-y)\Big[\sum_{m\neq0}(K(y+2\pi m)-I_m)-I_0\Big]\mathrm{d}y$$

$$=g_1(x)+g_2(x).$$

It is clear that $\left|\sum_{m\neq0}(K(y+2\pi m)-I_m)-I_0\right|\leqslant A_k(P)$ for all $y\in Q$.

Therefore

$$\omega(g_2;\ \delta)\leqslant A_k(P)\omega(f;\ \delta),$$

$$\omega_2(g_2;\ \delta)\leqslant A_k(P)\omega_2(f;\ \delta).$$

Now let $h\in Q$, $0<|h|\leqslant\frac{1}{2}$. Then

$$g_1(x)=\frac{1}{|Q|}\int_0^{|h|}P(f)(x;t)\frac{1}{t}\mathrm{d}t+\frac{1}{|Q|}\int_{|y|>|h|,y\in Q}f(x-y)K(y)\mathrm{d}y$$

$$=g_{1,1}(x)+g_{1,2}(x),$$

Integrating by parts we get

$$g_{1,1}(x)=\frac{1}{|Q|}\left\{P(f)(x;\ t)\ln t\Big|_0^{|h|}-\int_0^{|h|}\ln t\sum_{j=1}^{k}P_j(f_j)(x;t)\mathrm{d}t\right\}$$

$$=\frac{1}{|Q|}\left\{P(f)(x;\ |h|)\ln|h|-\sum_{j=1}^{k}\int_0^{|h|}\ln t\,P_j(f_j)(x;t)\mathrm{d}t\right\},$$

where $P_j(x)=-x_jP(x)$, $f_j=\dfrac{\partial f}{\partial x_j}$.

It is obvious that

$$|P(f)(x+h;\ t)-P(f)(x;\ t)|\leqslant A_k(P)_\omega(f;\ |h|),\quad \forall t>0,$$

$$|P(f)(x+h;\ t)+P(f)(x-h;\ t)-2P(f)(x;\ t)|$$

$$\leqslant A_k(P)_{\omega_2}(f;\ |h|),\quad \forall t>0.$$

Accordingly we get

$$\omega(g_{1,1};|h|)\leqslant A_k(P)_\omega(f;|h|)\ln\frac{1}{|h|}+A_k(P)M\cdot\int_0^{|h|}\ln\frac{1}{t}\mathrm{d}t$$

$$=O\left(|h|\ln\frac{1}{|h|}\right),\quad f\in W^1L^\infty.$$

where $M=\max\{\|f_j\|_\infty\ |j=1,2,\cdots,k\}$ as before. If $f\in W^2L^\infty$ then we have

$$\omega_2(g_{1,1};|h|)\leqslant A_k(P)_{\omega_2}(f;|h|)\ln\frac{1}{|h|}+A_k(P)\int_0^{|h|}|h|\ln\frac{1}{t}\mathrm{d}t$$

$$=O\left(|h|^2\ln\frac{1}{|h|}\right).$$

Concerning $g_{1,2}$, it is obvious that

$$\omega(g_{1,2};|h|)\leqslant\int_{|y|>|h|}\omega(f;|h|)|K(y)|\mathrm{d}y=O\left(|h|\ln\frac{1}{|h|}\right),$$

$$\omega_2(g_{1,2};|h|)\leqslant\int_{|y|>|h|}\omega_2(f;|h|)|K(y)|\mathrm{d}y$$

$$=O\left(\omega_2(f;\ |h|)\ln\frac{1}{|h|}\right).$$

From these results the conclusion of Lemma 10 is deduced.

Remark From the proof of Lemma 10 it is easy to see that for general kernel $K(x)=\dfrac{\Omega(x/|x|)}{|x|^k}\left(\int_{|\xi|=1}\Omega(\xi)\mathrm{d}\sigma(\xi)=0,\int_0^1\dfrac{\omega(\Omega;\delta)}{\delta}\mathrm{d}\delta<+\infty\right)$ the conclusion of Lemma 10 still holds.

Proof of Theorem 2 Let $f\in W^2L^\infty$. We have

$$S_R^{\frac{k-1}{2}}(f;x)-f(x)=S_R^{\frac{k-1}{2}}(f;x)-S_R^{\frac{k+1}{2}}(f;x)+S_R^{\frac{k+1}{2}}(f;x)-f(x)$$
$$=\frac{1}{R^2}S_R^{\frac{k-1}{2}}(-\Delta f;x)+S_R^{\frac{k+1}{2}}(f;x)-f(x), \quad (4.22)$$

where "Δ" denotes the Laplace operator as usual. Since

$$\|S_R^{\frac{k-1}{2}}(-\Delta f)\|_C\leqslant L_R^{\frac{k-1}{2}}\|\Delta f\|_\infty=O(\ln R),$$

and by Theorem 1,

$$\|S_R^{\frac{k+1}{2}}(f)-f\|_C=O\left(\omega_2\left(f;\frac{1}{R}\right)\right)=O\left(\frac{1}{R^2}\right),$$

so we have

$$\|S_R^{\frac{k-1}{2}}(f)-f\|_C=O\left(\frac{\ln R}{R^2}\right).$$

Furthermore

$$|S_R^{\frac{k-1}{2}}(-\Delta f;x)|\leqslant S_*^{\frac{k-1}{2}}(-\Delta f;x).$$

which is finite almost everywhere, Hence

$$|S_R^{\frac{k-1}{2}}(f;x)-f(x)|=O\left(\frac{1}{R^2}\right),\quad \text{a. e.}$$

In the conjugate case (4.22) holds also. Notice that the Lebesgue constant of $\widetilde{S}_R^{\frac{k-1}{2}}$ is also $o(\ln R)$ (see [10]) and

$$\|\widetilde{S}_R^{\frac{k+1}{2}}(f)-f^*\|_C=\|S_R^{\frac{k+1}{2}}(f^*)-f^*\|_C=O\left(\omega_2\left(f^*;\frac{1}{R}\right)\right).$$

An application of Lemma 10 yields

$$\|\widetilde{S}_R^{\frac{k-1}{2}}(f)-f^*\|_C=O\left(\frac{\ln R}{R^2}\right). \quad \square$$

Proof of Theorem 3 Let $f\in W^1L^\infty$ and $\varepsilon>0$ be arbitrary. Define the Steklov (Стеклов) function of f by

$$f_\varepsilon(x)=\frac{1}{(2\varepsilon)^k}\int_{(-\varepsilon,\varepsilon)^k}f(x+y)\mathrm{d}y.$$

Obviously we have

$$\frac{\partial f_\varepsilon}{\partial x_j}(x)=\frac{1}{(2\varepsilon)^k}\int_{(-\varepsilon,\varepsilon)^k}\frac{\partial}{\partial x_j}f(x+y)\mathrm{d}y$$
$$=\frac{1}{(2\varepsilon)^k}\int_{(-\varepsilon,\varepsilon)^{k-1}}[f(x+\bar{y}+\varepsilon e_j)-f(x+\bar{y}-\varepsilon e_j)]\mathrm{d}\bar{y},$$

where

$$\bar{y}=(y_1,\ y_2,\ \cdots,\ y_{j-1},\ 0,\ y_{j+1},\ \cdots,\ y_k);$$
$$e_j=(0,\ 0,\ \cdots,\ 0,\ 1,\ 0,\ \cdots,\ 0);$$
$$\mathrm{d}\bar{y}=\mathrm{d}y_1\mathrm{d}y_2\cdots\mathrm{d}y_{j-1}\mathrm{d}y_{j+1}\cdots\mathrm{d}y_k.$$

It is easy to see that $\dfrac{\partial f_\varepsilon}{\partial x_j}\in W^1L^\infty(j=1,2,\cdots,k)$ and

$$\left\|\frac{\partial^2 f_\varepsilon}{\partial x_j\,\partial x_l}\right\|_\infty\leqslant\frac{1}{\varepsilon}\max\left\{\left\|\frac{\partial f}{\partial x_j}\right\|_\infty\middle|j=1,2,\cdots,k\right\}.$$

According to Theorem 2 we have

$$R\|S_R^{\frac{k-1}{2}}(f_\varepsilon)-f_\varepsilon\|_C=O\left(\frac{\ln R}{R}\right).\tag{4.23}$$

We define an operator γ on $L(Q)$ by

$$\gamma(h)(x)=\varlimsup_{R\to+\infty}R\,|S_R^{\frac{k-1}{2}}(h;\ x)-h(x)|,\quad h\in L(Q).$$

Set $g=f-f_\varepsilon$. Since

$$R\,|S_R^{\frac{k-1}{2}}(f;\ x)-f(x)|\leqslant R\,|S_R^{\frac{k-1}{2}}(g;\ x)-g(x)|+$$
$$R\,|S_R^{\frac{k-1}{2}}(f_\varepsilon;\ x)-f_\varepsilon(x)|,$$

using (4.23) we get

$$\gamma(f)(x)\leqslant\gamma(g)(x).$$

Noticing $g\in W^1L^\infty$ and using Lemma 6 we get

$$\gamma(g)(x)\leqslant A_k\sum_{j=1}^{k}TP_j(g_j)(x),\quad \text{a. e.}$$

By Lemma 4 we have

$$\|TP_j(g_j)\|_2\leqslant A_k\|g_j\|_2.$$

Now from the equation

$$g_j(x)=\frac{\partial f(x)}{\partial x_j}-\frac{\partial f_\varepsilon(x)}{\partial x_j}=\frac{1}{(2\varepsilon)^k}\int_{(-\varepsilon,\varepsilon)^k}\left[\frac{\partial f(x)}{\partial x_j}-\frac{\partial f(x+y)}{\partial x_j}\right]\mathrm{d}y$$

we get

$$\|g_j\|_2\leqslant\frac{1}{(2\varepsilon)^k}\int_{(-\varepsilon,\varepsilon)^k}\sqrt{\int_Q|f_j(x)-f_j(x+y)|^2\mathrm{d}x}\ \mathrm{d}y\leqslant\omega(f_j;\sqrt{k}\varepsilon)_{L^2}.$$

Here we denote by $\omega(f_j;\ t)_{L^2}$ the $L^2(Q)$-modulus of continuity of f_j.

Hence

$$\|\gamma(f)\|_2\leqslant A_k\sum_{j=1}^{k}\omega(f_j;\sqrt{k}\varepsilon)_{L^2}.$$

Letting $\varepsilon \to 0^+$, we get $\gamma(f)(x)=0$, a. e. This yields

$$|S_R^{\frac{k-1}{2}}(f;\ x)-f(x)|=O\left(\frac{1}{R}\right),\quad \text{a. e.}$$

Concerning the conjugate case, an application of Lemma 9 yields

$$|\widetilde{S}_R^{\frac{k-1}{2}}(f;\ x)-f(x)|=O\left(\frac{1}{R}\right),\quad \text{a. e.}\quad \square$$

Proof of theorem 5 Set $Q_f=\{x\in Q|\lim_{R\to+\infty}\rho_R(f;\ x)=0\}$. It is clear that $|Q_f|=|Q|$. Let $\lambda>0$ be arbitrary. Define

$$G_R(\lambda)=\left\{x\in Q_f\,|\,\rho_R(f;\ x)>\lambda\omega\left(\frac{1}{R}\right)\ln\ln\frac{g}{\Omega\left(\frac{1}{R}\right)}\right\},\quad R\geqslant 1,$$

$$\Delta_n(\lambda)=\bigcup_{R\in[\delta_n^{-1},\delta_{n+1}^{-1})}G_R(\lambda),\quad n\in\mathbf{N}.$$

For any $n\in\mathbf{N}$ there are only two possibilities: (1) $\omega(\delta_{n+1})=\frac{1}{6}\omega(\delta_n)$ is true; (2) $\omega(\delta_{n+1})\neq\frac{1}{6}\omega(\delta_n)$ but $\frac{\delta_{n+1}}{\omega(\delta_{n+1})}=\frac{1}{6}\frac{\delta_n}{\omega(\delta_n)}$. We consider these cases separately.

Case (1), $\omega(\delta_{n+1})=\frac{1}{6}\omega(\delta_n)$. Suppose $g(x)$ is the trigonometric polynomial of best approximation of degree δ_n^{-1} of f. Set $f-g=h$. Let $R\in[\delta_n^{-1},\ \delta_{n+1}^{-1})$. Then $\|h\|_C=E_{\delta_n^{-1}}(f)\leqslant A_k\omega(\delta_n)\leqslant A_k\omega\left(\frac{1}{R}\right)$.

Obviously, $\rho_R(f;\ x)\leqslant\rho_R(g;\ x)+\rho_R(h;\ x)$,

$$\begin{aligned}\rho_R(h;\ x)&\leqslant\|h\|_C\rho_R\left(\frac{h}{\|h\|_C};\ x\right)\\&\leqslant A_k\omega\left(\frac{1}{R}\right)\left[S_*^{\frac{k-1}{2}}\left(\frac{h}{\|h\|_C};\ x\right)+\widetilde{S}_*^{\frac{k-1}{2}}\left(\frac{h}{\|h\|_C};\ x\right)\right],\ \forall x\in Q_f.\end{aligned}$$

And it is also obvious that

$$\begin{aligned}|S_R^{\frac{k-1}{2}}(g;\ x)-g(x)|&\leqslant|S_R^{\frac{k-1}{2}}(g;\ x)-S_R^{\frac{k+1}{2}}(g;\ x)|+\\&\quad|S_R^{\frac{k+1}{2}}(g;\ x)-g(x)|\\&\leqslant\frac{1}{R^2}|S_R^{\frac{k-1}{2}}(-\Delta g;\ x)|+A_k\omega\left(g;\ \frac{1}{R}\right)\\&\leqslant\frac{1}{R^2}\|\Delta g\|_C\cdot S_*^{\frac{k-1}{2}}\left(\frac{\Delta g}{\|\Delta g\|_C};\ x\right)+A_k\omega\left(\frac{1}{R}\right).\end{aligned}$$

According to (2. 2) we get $\|\Delta g\|_C \leqslant A_k \delta_n^{-2} \omega_2(\delta_n) \leqslant A_k R^2 \omega\left(\frac{1}{R}\right)$.

Hence

$$|S_R^{\frac{k-1}{2}}(g;\ x)-g(x)| \leqslant A_k \omega\left(\frac{1}{R}\right)\left(S_*^{\frac{k-1}{2}}\left(\frac{\Delta g}{\|\Delta g\|_C};\ x\right)+1\right). \quad (4.24)$$

In the conjugate case we similarly get

$$|\widetilde{S}_R^{\frac{k-1}{2}}(g;\ x)-g(x)|$$
$$\leqslant A_{k^\omega}\left(\frac{1}{R}\right)\widetilde{S}_R^{\frac{k-1}{2}}\left(\frac{\Delta g}{\|\Delta g\|_C};\ x\right)+|\widetilde{S}_R^{\frac{k+1}{2}}(g;\ x)-g^*(x)|.$$

Using (3. 9) and (4. 13) we get

$$\widetilde{S}_R^{\frac{k+1}{2}}(g;\ x)-g^*(x)$$
$$=\frac{1}{|Q|}\int_0^{+\infty} P(g)\left(x;\frac{t}{R}\right)t^{k+p-1}\left(\frac{E(k,1,p,t)}{c(k,p)}-\frac{1}{t^{k+p}}\right)\mathrm{d}t$$
$$=\frac{1}{|Q|}\int_0^1 P(g)\left(x;\frac{t}{R}\right)\left(\frac{t^{k+p-1}E(k,1,p,t)}{c(k,p)}-\frac{1}{t}\right)\mathrm{d}t+$$
$$\frac{1}{|Q|}\int_1^{+\infty} P(g)\left(x;\frac{t}{R}\right)\frac{A(k,1)}{t^2}\cos\left(t-\frac{\pi}{2}(k+p+1)\right)\mathrm{d}t+$$
$$\frac{1}{|Q|}\int_1^{+\infty} P(g)\left(x;\frac{t}{R}\right)\frac{1}{t^3}\Big[I(k,1,p,t)+$$
$$A_1(k,1)\sin\left(t-\frac{\pi}{2}(k+p+1)\right)\Big]\mathrm{d}t$$
$$=I_1+I_2+I_3.$$

Noticing $\quad P(g)\left(x;\frac{t}{R}\right)=\int_{|\xi|=1}\left[g\left(x-\frac{t}{R}\xi\right)-g(x)\right]P(\xi)\mathrm{d}\sigma(\xi)$

we have $\quad \left|P(g)\left(x;\ \frac{t}{R}\right)\right| \leqslant A_k(P)M\frac{t}{R},$

Where $M=\max\left\{\left\|\frac{\partial g}{\partial x_j}\right\|_C \,\middle|\, j=1,2,\cdots,k\right\}$. Hence $|I_1| \leqslant A_k(P)\frac{1}{R}M$.

We deal with I_2 now. We have

$$I_2=\frac{A(k,1)}{2|Q|}\left\{\int_1^{+\infty}\left[P(g)\left(x;\frac{t}{R}\right)-P(g)\left(x;\frac{t+\pi}{R}\right)\right]\cdot\right.$$
$$\frac{\cos\left(t-\frac{\pi}{2}(k+p+1)\right)}{t^2}\mathrm{d}t+\int_1^{+\infty}P(g)\left(x;\frac{t+\pi}{R}\right)\cdot$$

$$\left(\frac{1}{t^2}-\frac{1}{(t+\pi)^2}\right)\cos\left(t-\frac{\pi}{2}(k+p+1)\right)\mathrm{d}t+$$

$$\left.\int_1^{1+\pi}P(g)\left(x;\frac{t}{R}\right)\frac{\cos\left(t-\frac{\pi}{2}(k+p+1)\right)}{t^2}\mathrm{d}t\right\}.$$

Therefore we get

$$|I_2|\leqslant A_k(P)\omega\left(g;\frac{1}{R}\right)+A_k\int_1^{+\infty}\left|P(g)\left(x;\frac{t+\pi}{R}\right)\right|\frac{\mathrm{d}t}{t^3}+A_k\int_1^{1+\pi}\left|P(g)\left(x;\frac{t}{R}\right)\right|\frac{\mathrm{d}t}{t^2}.$$

The third term in right-hand side of the above inequality does not exceed $A(P)\omega\left(g;\ \frac{1}{R}\right)$ and

$$\int_1^{+\infty}\left|P(g)\left(x;\frac{t+\pi}{R}\right)\right|\frac{\mathrm{d}t}{t^3}\leqslant\int_1^{+\infty}\frac{A_k(P)\omega\left(g;\frac{t}{R}\right)}{t^3}\mathrm{d}t\leqslant A_k(P)\omega\left(g;\frac{1}{R}\right).$$

Hence we get $\quad |I_2|\leqslant A_k(P)\omega\left(g;\ \frac{1}{R}\right)\leqslant A_k(P)\omega\left(\frac{1}{R}\right).$

Simultaneously we get $\quad |I_3|\leqslant A_k(P)\omega\left(\frac{1}{R}\right).$

A combination of these results yields

$$\|\widetilde{S}_R^{\frac{k+1}{2}}(g)-g^*\|_C\leqslant A_k(P)\omega\left(\frac{1}{R}\right).$$

Therefore we get

$$|\widetilde{S}_R^{\frac{k-1}{2}}(g;x)-g^*(x)|\leqslant A_k(P)\omega\left(\frac{1}{R}\right)\left[\widetilde{S}_*^{\frac{k-1}{2}}\left(\frac{\Delta g}{\|\Delta g\|_C};x\right)+1\right]. \tag{4.25}$$

A combination of (4.24) and (4.25) yields

$$\rho_R(g;\ x)\leqslant A_k(P)\omega\left(\frac{1}{R}\right)\left[S_*^{\frac{k-1}{2}}\left(\frac{\Delta g}{\|\Delta g\|_C};\ x\right)\widetilde{S}_*^{\frac{k-1}{2}}\left(\frac{\Delta g}{\|\Delta g\|_C};\ x\right)+1\right].$$

Finally we get (for $x\in Q_f$, $R\in[\delta_n^{-1},\ \delta_{n+1}^{-1})$),

$$\rho_R(f;\ x)\leqslant A_k(P)\omega\left(\frac{1}{R}\right)\left\{S_*^{\frac{k-1}{2}}\left(\frac{h}{\|h\|_C};\ x\right)+\widetilde{S}_*^{\frac{k-1}{2}}\left(\frac{h}{\|h\|_C};\ x\right)+\right.$$

$$\left.S_*^{\frac{k-1}{2}}\left(\frac{\Delta g}{\|\Delta g\|_C};\ x\right)+\widetilde{S}_*^{\frac{k-1}{2}}\left(\frac{\Delta g}{\|\Delta g\|_C};\ x\right)+1\right\}. \tag{4.26}$$

For the convenience of writing we make use of the following definitions; $\quad T_1^*(x)=S_*^{\frac{k-1}{2}}\left(\frac{h}{\|h\|_C};\ x\right),\quad T_2^*(x)=\widetilde{S}_*^{\frac{k-1}{2}}\left(\frac{h}{\|h\|_C};\ x\right);$

$$T_3^*(x)=S_*^{\frac{k-1}{2}}\left(\frac{\Delta g}{\|\Delta g\|_C};\ x\right),\quad T_4^*(x)=\widetilde{S}_*^{\frac{k-1}{2}}\left(\frac{\Delta g}{\|\Delta g\|_C};\ x\right);$$

$$T_5^*(x)=1;$$

$$E_\nu=\left\{x\in Q_f\ \middle|\ T_\nu^*(x)>\frac{\lambda}{5A_k(P)}\cdot\ln(n+2)\right\},\quad \nu=1,2,3,4,5,$$

Where $A_k(P)$ is the same constant as in the right-hand side of (4.26).

Noticing $\Omega\left(\frac{1}{R}\right)=6^{-n}$, if $R\in[\delta_n^{-1},\ \delta_{n+1}^{-1})$, we get $G_R(\lambda)\subset\bigcup_{\nu=1}^{5}E_\nu$. According to (3.4), Lemma 3 and Lemma 5 we deduce that

$$|E_\nu|\leqslant A_k(P)(n+2)^{-\frac{\lambda}{A_k(P)}}.$$

Hence

$$|\Delta_n(\lambda)|\leqslant\sum_{\nu=1}^{5}|E_\nu|\leqslant A_k(P)(n+2)^{-\frac{\lambda}{A_k(P)}}.\tag{4.27}$$

Now we come to the case (2). In this case the equation $\frac{\delta_{n+1}}{\omega(\delta_{n+1})}=\frac{1}{6}\frac{\delta_n}{\omega(\delta_n)}$ holds. Hence according to the property of the modulus of continuity (see [5]) we have

$$\frac{\omega(\delta_{n+1})}{\delta_n}=6\frac{\omega(\delta_n)}{\delta_n}\leqslant 12R\omega\left(\frac{1}{R}\right),\ \text{for any } R\in[\delta_n^{-1},\ \delta_{n+1}^{-1}).$$

We suppose now $g(x)$ is the trigonometric polynomial of best approximation of degree δ_{n+1}^{-1} of f and set $h=f-g$. By an argument analogous to that of case (1) we get

$$\left|S_R^{\frac{k-1}{2}}(f;\ x)-f(x)\right|\leqslant A_k\omega\left(\frac{1}{R}\right)S_*^{\frac{k-1}{2}}\left(\frac{h}{\|h\|_C};\ x\right)+\left|S_R^{\frac{k-1}{2}}(g;\ x)-g(x)\right|,$$

for $x\in Q_f$. Using Lemma 6 we get

$$\left|S_R^{\frac{k-1}{2}}(g;x)-g(x)\right|\leqslant\frac{A_k}{R}\left\{\sum_{j=1}^{k}TP_j(g_j)(x)+\theta_R(g)(x)\right\},$$

where θ_R satisfies the inequality

$$|\theta_R(g)(x)|\leqslant M\leqslant A_k\delta_{n+1}^{-1}\omega(\delta_{n+1})\leqslant A_kR\omega\left(\frac{1}{R}\right).$$

On the other hand

$$TP_j(g_j)(x)\leqslant\|g_j\|_C TP_j\left(\frac{g_j}{\|g_j\|_C};\ x\right)\leqslant A_kR\omega\left(\frac{1}{R}\right)TP_j\left(\frac{g_j}{\|g_j\|_C};\ x\right).$$

Hence

$$|S_R^{\frac{k-1}{2}}(f;\ x)-f(x)|$$
$$\leqslant A_k\omega\left(\frac{1}{R}\right)\left\{S_*^{\frac{k-1}{2}}\left(\frac{h}{\|h\|_C};\ x\right)+\sum_{j=1}^{k}TP_j\left(\frac{g_j}{\|g_j\|_C};x\right)+1\right\}. \quad (4.28)$$

Similarly, we have

$$|\widetilde{S}_R^{\frac{k-1}{2}}(f;\ x)-f^*(x)|$$
$$\leqslant A_k\omega\left(\frac{1}{R}\right)\widetilde{S}_*^{\frac{k-1}{2}}\left(\frac{h}{\|h\|_C};\ x\right)+|\widetilde{S}_R^{\frac{k-1}{2}}(g;\ x)-g^*(x)|,$$

for $x\in Q_f$, By Lemma 9 we get

$$|\widetilde{S}_R^{\frac{k-1}{2}}(g;\ x)-g^*(x)|$$
$$\leqslant\frac{A_{k,p}}{R}\left(\sum_{j=1}^{k}\sum_{l=0}^{\nu_p}TP_{j,p+1-2l}(g_j)(x)+\tilde{\theta}_R(g)(x)\right)$$
$$\leqslant A_k(P)\frac{M}{R}\left\{\sum_{j=1}^{k}\sum_{l=0}^{\nu_p}TP_{j,p+1-2l}\left(\frac{g_j}{\|g_j\|_C}\right)(x)+1\right\},$$

where $M=\max\{\|g_j\|_C\,|\,j=1,2,\cdots,k\}\leqslant A_k\delta_{n+1}^{-1}\omega(\delta_{n+1})\leqslant A_kR\omega\left(\frac{1}{R}\right)$.

Hence we get

$$|\widetilde{S}_R^{\frac{k-1}{2}}(f;\ x)-f^*(x)|$$
$$\leqslant A_k(P)\omega\left(\frac{1}{R}\right)\left\{\widetilde{S}_*^{\frac{k-1}{2}}\left(\frac{h}{\|h\|_C};\ x\right)+\sum_{j=1}^{k}\sum_{l=0}^{\nu_p}TP_{j,p+1-2l}\left(\frac{g_j}{\|g_j\|_C}\right)(x)+1\right\}. \quad (4.29)$$

Using Lemma 4 and Lemma 5 from (4.28) and (4.29) it is deduced that the inequality (4.27) holds also in the case (2). Because the detail is completely similar to that of case (1) so we omit it.

Therefore by (4.27) we get

$$\left|\bigcup_{n=0}^{+\infty}\Delta_n(\lambda)\right|\leqslant A\sum_{n=0}^{+\infty}(n+2)^{-\frac{\lambda}{A}},\quad A=A_k(P).$$

If $\lambda\geqslant 3A$, then. $\displaystyle\int_2^{+\infty}x^{-\frac{\lambda}{A}}\mathrm{d}x=\left(\frac{\lambda}{A}-1\right)^{-1}2^{-\frac{\lambda}{A}+1}<\mathrm{e}^{-\frac{\lambda}{A}\ln 2}$.

An inspection of this inequality shows that $\displaystyle\sum_{n=0}^{+\infty}(n+2)^{-\frac{\lambda}{A}}\leqslant\mathrm{e}^{-\frac{\lambda}{A}}$.

Hence we get

$$\left|\bigcup_{n=0}^{+\infty}\Delta_n(\lambda)\right|\leqslant Ae^{-\frac{\lambda}{A}},\quad \text{for any } \lambda>0,\quad A=A_k(P). \tag{4.30}$$

We define $C(x)=\sup\limits_{R\geqslant 1}\dfrac{\rho_R(f;\ x)}{\omega\left(\frac{1}{R}\right)\ln\ln\left\{9\left[\Omega\left(\frac{1}{R}\right)\right]^{-1}\right\}}$.

Then for any $\lambda>0$ we have $\{x\in Q_f\mid C(x)>\lambda\}\subset\bigcup\limits_{n=0}^{+\infty}\Delta_n(\lambda)$.

Hence by (4.30) we get

$$\{x\in Q_f\mid C(x)>\lambda\}\mid=\mid\{x\in Q\mid C(x)>\lambda\}\mid<Ae^{-\frac{\lambda}{A}},\ A=A_k(P).$$

This proves the conclusion (1) of Theorem 5.

Now we rewrite the constant $A_k(P)$ appeared in (4.27) to be A_0. Let $\lambda_0=2A_0$. Then

$$\left\{x\in Q_f\,\middle|\,\varlimsup_{R\to+\infty}\frac{\rho_p(f;\ x)}{\omega\left(\frac{1}{R}\right)\ln\ln\left\{9\left[\Omega\left(\frac{1}{R}\right)\right]^{-1}\right\}}>\lambda_0\right\}\subset\varlimsup_{n\to+\infty}\Delta_n(\lambda_0).$$

Since $\left|\varlimsup\limits_{n\to+\infty}\Delta_n(\lambda_0)\right|=\lim\limits_{n\to+\infty}\left|\bigcup\limits_{j=n}^{+\infty}\Delta_j(\lambda_0)\right|\leqslant\lim\limits_{n\to+\infty}\sum\limits_{j=n}^{+\infty}\dfrac{A_0}{(j+2)^2}=0$,

So we get $\varlimsup\limits_{R\to+\infty}\dfrac{\rho_R(f;\ x)}{\omega\left(\frac{1}{R}\right)\ln\ln\left\{9\left[\Omega\left(\frac{1}{R}\right)\right]^{-1}\right\}}\leqslant 2A_0$, a. e.

This proves the conclusion (2). □

Finally we state an approximation theorem characterazed in terms of the second modulus of continuity.

Theorem 6 Let $f\in C(Q)$ satisfy the condition

$$\lim_{\delta\to 0^+}\frac{\omega_2(f;\ \delta)}{\delta^2}=+\infty.$$

Write $\omega_2(\delta)=\omega_2(f;\ \delta)$, $\bar{\omega}_2(\delta)=\dfrac{\delta^2}{\omega_2(\delta)}\omega(1)$ and define

$$\delta_0=1,\ \delta_{n+1}=\min\left\{\delta\,\middle|\,\max\left(\frac{\omega_2(\delta)}{\omega_2(\delta_n)},\ \frac{\bar{\omega}_2(\delta)}{\bar{\omega}_2(\delta_n)}\right)=\frac{1}{6}\right\},$$

$$\Omega(\delta)=6^{-n},\ \text{if } \delta\in(\delta_{n+1},\ \delta_n],\ n\in\mathbf{N}.$$

Then

(i) $\left|S_R^{\frac{k-1}{2}}(f;\ x)-f(x)\right|\leqslant C(x)\omega_2\left(\dfrac{1}{R}\right)\ln\ln\dfrac{9}{\Omega\left(\frac{1}{R}\right)}$, $R\geqslant 1$,

$$|\{x\in Q|C(x)>\lambda\}|<Ae^{-\frac{\lambda}{A}},\ \lambda>0,\ A=A_k;$$

(ii) $$\overline{\lim_{R\to+\infty}}\frac{|S_R^{\frac{k-1}{2}}(f;\ x)-f(x)|}{\omega_2\left(\frac{1}{R}\right)\ln\ln\frac{9}{\Omega\left(\frac{1}{R}\right)}}\leqslant A,\ \text{a. e.}$$

The author thanks Sun Yong-sheng and Lu Shan-zhen for their guidance.

References

[1] Calderon A P，Zygmund A. Studia Math.，1954，14(2)：249—271.

[2] Осколков К И. Изв. АН СССР，Сер. Матем.，1974，38(6)：1 393—1 407.

[3] Stein E M. Ann. of Math.，1961，73(1)：87—109.

[4] Тиман А Ф. Теория приблнжения функднй действительвого переменного，ФНЗМАТГИЗ，1960.

[5] Стечкия С Б. Нзв. АН СССР，Сер. Матем.，1951，15：219—242.

[6] Bochner S. Trans. Amer. Math. Soc.，1936，40：175—207.

[7] Watson G N. Theory of Bessel Functions，Cambridge Univ. Press，1952.

[8] Стейн И. Сингулярные Ннтегралы н Дифференциалвые Свойства Функций，Москва，1973.

[9] Stein E M. Acta Mathematica，1958，100：93—147.

[10] Chang C P. Dh. D. Dissertation. Univ. of Chicago，Chicago，1964.

[11] Bateman H. Tables of Integral Transforms，Vol. 2，New York，1954.

[12] Stein E M，Weiss G. Introduction to Fourier Analysis on Euclidean Spaces. Princeton Univ. Press，1971.

Acta Mathematica Sinica, New Series, 1986, 2(2): 178—202.

多重 Fourier 级数的广义球型 Riesz 平均

Generalized Spherical Riesz Means of Multiple Fourier Series①

§ 1. Introduction

Let $k\in\mathbf{N}$, $Q^n=Q=\{(x_1, x_2, \cdots, x_k): -\pi\leqslant x_j<\pi, j=1, 2,\cdots,k\}$, and $L(Q)$ be the space of all functions integrable on Q and periodic in each variable with period 2π. Suppose $f\in L(Q)$. The Fourier series of f is denoted by

$$\sigma(f)(x)\sim\sum_{m\in\mathbf{Z}^k}a_m(f)\mathrm{e}^{\mathrm{i}mx},$$

where $\mathbf{Z}^k$ denotes the set of all k-dimensional integers and $a_m(f)$ the Fourier coefficient of f. Define the generalized Riesz means by

$${}^lS_R^\alpha(f;x)=\sum_{|m|<R}\left(1-\left|\frac{m}{R}\right|^l\right)^\alpha a_m(f)\mathrm{e}^{\mathrm{i}mx},\quad R>0,$$

where $l\in\mathbf{N}$ and the index α is a complex number with $\operatorname{Re}\alpha>-1$. If $l=2$, the sum ${}^lS_R^\alpha={}^2S_R^\alpha$ is the usual Riesz mean and in this case we write S_R^α instead of ${}^2S_R^\alpha$.

Cheng and Chen[1] investigated the problem of the uniform approximation by the operators ${}^lS_R^\alpha$. They indicated that the saturation of

---

① Supported by the Science Fund of the Chinese Academy of Sciences.

Received: 1985-01-22; Revised: 1986-06-10.

$\{{}^lS_R^\alpha\}$ is R^{-l} and in the particular case of $\alpha=\left[\frac{k-1}{2}\right]+1$, they obtained the estimates of the degree of the uniform approximation for differentiable functions.

Our purpose is to investigate the operators ${}^lS_R^\alpha(l\neq 2)$ further on.

In § 2 we examine the kernel of ${}^lS_R^\alpha$, point out that the critical index for ${}^lS_R^\alpha(l\neq 2)$ is $\alpha_0=\frac{k-1}{2}$, and deduce the asymptotic estimates of the kernel for $\alpha>\alpha_0$. Then we generalize the Bochner formula to the case of $l\neq 2$(Theorem 1).

In § 3 the problem of uniform approximation by ${}^lS_R^\alpha$ $\left(\text{when } \alpha>\alpha_0=\frac{k-1}{2}\right)$ is discussed. When $l=1$ we get the result (Theorem 2) completely parallel with that of one-dimensional Cesàro approximation. When $l>1$, our result shows that the degree of approximation by ${}^lS_R^\alpha$ $(\alpha>\alpha_0)$ is the same as by S_R^α, that is $O(\omega_2(R^{-1}))$, where ω_2 denotes the second modulus of continuity which is defined by

$$\omega_2(f;\ t)=\sup\{|f(x+h)+f(x-h)-2f(x)|:\ |h|\leqslant t\},\qquad t\geqslant 0.$$

This is Theorem 3. We also use the modulus of continuity of first order which is defined by

$$\omega_1(f;\ t)=\omega(f;\ t)=\sup\{|f(x+h)-f(x)|:\ |h|\leqslant t\},\qquad t\geqslant 0.$$

For differentiable functions higher degree of approximation can be obatined if we use the operators ${}^lS_R^\alpha$ with large l. In this direction, Theorem 4 generalizes the result of [1].

In § 4 we make use of the result of § 2 and apply a theorem about analytic functions to obtain the relation between ${}^lS_R^\alpha$ and S_R^α (Theorem 5). This gives us a way to investigate the operators ${}^lS_R^\alpha$ for the index $\alpha\leqslant\alpha_0$. Particularly, we can directly derive the degree of the uniform approximation by ${}^lS_R^{\alpha_0}$ from the results obtained in case of $l=2$(Therorem 6).

Finally, in § 5, we discuss the problem of the approximation for continuous function and its conjugate function by the operator ${}^lS_R^{\alpha_0}$ $\left(\alpha_0=\frac{k-1}{2}\right)$ on set of total measure. The conclusion generalizes that of case $l=2$ obtained by the author in paper [6].

§ 2. On the kernel of operator ${}^lS_R^\alpha$

The kernel of ${}^lS_R^\alpha$ is

$$ {}^lD_R^\alpha(y) = \sum_{|m|<R}\left(1-\left|\frac{m}{R}\right|^l\right)^\alpha e^{imr}. \tag{2.1}$$

Set

$$\varphi(x)=\begin{cases}(1-|x|^l)^\alpha, & \text{if } |x|<1,\\ 0, & \text{if } |x|\geqslant 1.\end{cases} \tag{2.2}$$

We use the symbol $\mathscr{F}$ to denote the Fourier transform. We have

$$\begin{aligned}\mathscr{F}(\varphi)(y) &= \frac{1}{(2\pi)^k}\int_{\mathbf{R}^k}\varphi(x)e^{-ixy}dx\\ &= \frac{1}{(2\pi)^{\frac{k}{2}}}\int_0^1(1-r^l)^\alpha r^{\frac{k}{2}}J_{\frac{k}{2}-1}(|y|r)dr\cdot|y|^{\frac{k}{2}-1},\end{aligned} \tag{2.3}$$

where J_ν denotes the Bessel function of first kind. The derivation of this formula can be found in [2](Chapter Ⅳ). Define an entire function of variable s,

$$ {}^lH_k^\alpha(s) = (2\pi)^{-\frac{k}{2}}\int_0^1(1-t^l)^\alpha t^{\frac{k}{2}}\frac{\frac{k}{2}-1^{(st)}}{s^{\frac{k}{2}-1}}dt,\quad \operatorname{Re}\alpha>-1. \tag{2.4}$$

Definition 1 Let α be a complex number with $\operatorname{Re}\alpha>-1$, $q=\left[\frac{k+1}{2}\right]+4$. The polynomial $Q(t)=\sum_{j=0}^{2q+1}b_jt^j$ is defined by the following condition:

$$\frac{d^j}{dt^j}\left\{Q(1-t^2)-\left(\frac{1+t+\cdots+t^{l-1}}{1+t}\right)^\alpha\right\}\Bigg|_{t=1}=0,\ j=0,1,\cdots,2q+1. \tag{2.5}$$

we notice that the condition (2.5) determines uniquely the coefficients $b_j=b_j(l,\alpha)$, $j=0,1,2,\cdots,2q+1$, and a simple calculation shows that

$$b_0=\left(\frac{l}{2}\right)^\alpha,\ b_1=-\frac{\alpha}{2}\left(\frac{l}{2}\right)^\alpha\frac{l-2}{2}. \tag{2.6}$$

Definition 2 Assume $\mathrm{Re}\,\alpha>-1$, $q=\left[\frac{k+1}{2}\right]+4$ and $Q(t)$ is as in Definition 1. We define an entire function of variable s, ${}^l\gamma_k^\alpha(s)$, by

$$ {}^l\gamma_k^\alpha(s)=(2\pi)^{-\frac{k}{2}}\int_0^1[(1-t^l)^\alpha-Q(1-t^2)(1-t^2)^\alpha]t^{\frac{1}{2}k}\frac{J_{\frac{1}{2}k-1}k-1(st)\mathrm{d}t}{s^{\frac{1}{2}k-1}}. \tag{2.7}$$

We point out that when α variates in the domain $\mathrm{Re}\,\alpha>-1$ all coefficients $b_j(l,\alpha)$ of $Q(t)$ and ${}^l\gamma_k^\alpha(s)$ are analytic in α.

For convenience we set

$$V_\nu(z)=\frac{J_\nu(z)}{z^\nu}=\sum_{j=0}^{+\infty}\frac{(-1)^j}{j\,!\,\Gamma(j+\nu+1)}\left(\frac{1}{2}z\right)^{2j}2^{-\nu},\quad \mathrm{Re}\,\nu>-1.$$

Lemma 1 Let $\mathrm{Re}\,\alpha>-1$. If l is even then

$${}^lH_k^\alpha(s)=2^\alpha(2\pi)^{-\frac{1}{2}k}\sum_{j=0}^{2q+1}b_j2^j\Gamma(\alpha+j+1)V_{\frac{1}{2}k+\alpha+j}(s)+{}^l\gamma_k^\alpha(s),$$

and

$$|{}^l\gamma_k^\alpha(s)|\leqslant M(k,l,\alpha)s^{-(k+l+6)},\quad s\geqslant1, \tag{2.8}$$

where $M(k,l,\alpha)$ is a constant depending only on k, l, α.

Proof Let

$$\begin{aligned}\omega(t)&=[(1-t^l)^\alpha-Q(1-t^2)(1-t^2)^\alpha]\chi_{[0,1)}(t)\\&=\left[\left(\frac{1+t+\cdots+t^{l-1}}{1+t}\right)^\alpha-Q(1-t^2)\right](1-t^2)^\alpha\chi_{[0,1)}(t).\end{aligned}$$

By the definition of $Q(t)$ we observe that $\omega\in C^{2q}[0,+\infty)$. We denote by Δ the following operator

$$\Delta(t)=\omega''(t)+\frac{k-1}{t}\omega'(t),\quad \Delta^2\omega(t)=\Delta(\Delta\omega(t)),\ \cdots.$$

It is clear that Δ is just Laplacean operator on the radial function $\omega(|x|)$ of k variables and that $\Delta^q\omega\in C(0,+\infty)$ and $\Delta^q\omega(t)=0$ when $t\geqslant1$. We proceed to consider the asymptotic behavious of $\Delta^q\omega(t)$ as $t\to0^+$. For $t\in(0,1)$, $\omega(t)=(1-t^l)^\alpha-Q(1-t^2)(1-t^2)^\alpha$. It is easy to observe that for any $n\in\mathbf{N}$ and $\beta\in\mathbf{C}$ the function $\Delta^q(1-t^{2n})^\beta(t\in(0,1))$ is a linear combination of $(1-t^{2n})^\gamma t^{2m}(m\in\mathbf{N}^*,\gamma\in\mathbf{C})$. Now let l be a positive even number, it is evident that $\Delta^q\omega(t)$ is such a combination. Hence it has a finite limit as $t\to0^+$. Therefore

$$|\Delta^q \omega(t)| \leqslant M(k,\ l,\ \alpha),\ t \in (0,\ +\infty).$$

Thus by Lemma 6 in [3] we get

$$\left| \int_0^1 \omega(t) t^{k-1} V_{\frac{1}{2}k-1}(st)\,\mathrm{d}t \right| \leqslant M s^{-2q} (s \geqslant 1).$$

Since $2q > k+l+6$ so the inequality (2.8) holds.

On the other hand,

$$\begin{aligned} &\int_0^1 Q(1-t^2)(1-t^2)^\alpha t^{k-1} V_{\frac{1}{2}k-1}(st)\,\mathrm{d}t \\ &= \sum_{j=0}^{2q+1} b_j \int_0^1 (1-t^2)^{\alpha+j} t^{k-1} V_{\frac{1}{2}k-1}(st)\,\mathrm{d}t. \end{aligned} \tag{2.9}$$

According to [4](p. 26(33))

$$\begin{aligned} &\int_0^1 (1-t^2)^{\alpha+j} t^{k-1} V_{\frac{1}{2}k-1}(st)\,\mathrm{d}t \\ &= 2^{\alpha+j} \Gamma(\alpha+j+1) V_{\frac{1}{2}k+\alpha+j}(s). \end{aligned} \tag{2.10}$$

Hence

$$\begin{aligned} &\int_0^1 Q(1-t^2)(1-t^2)^\alpha t^{k-1} V_{\frac{1}{2}k-1}(st)\,\mathrm{d}t \\ &= \sum_{j=0}^{2q+1} b_j 2^{\alpha+j} \Gamma(\alpha+j+1) V_{\frac{1}{2}k+\alpha+j}(s). \end{aligned} \tag{2.11}$$

Since

$${}^l H_k^\alpha(s) = {}^l \gamma_k^\alpha(s) + (2\pi)^{-\frac{1}{2}k} \int_0^1 Q(1-t^2)(1-t^2)^\alpha t^{k-1} V_{\frac{1}{2}k-1}(st)\,\mathrm{d}t,$$

the desired result deduced from (2.9) and (2.11).

The proof of Lemma 1 is complete.

Lemma 2 Let $k,\ l \in \mathbf{N}$. Define

$$I_j(s) = \int_0^\infty \mathrm{e}^{-t} t^{\frac{1}{2}k+j} J_{\frac{1}{2}k+l-1}(st)\,\mathrm{d}t, \quad j \in \mathbf{N}.$$

Then

$$I_j(s) = \frac{A_j(k,\ l)}{s^{\frac{1}{2}k+j+1}} - \frac{A_{j+1}(k,\ 1)}{s^{\frac{1}{2}k+j+2}} + o\left(\frac{1}{s^{\frac{1}{2}k+j+3}}\right), \quad s \to +\infty,$$

where

$$A_j(k,\ l) = \frac{\sqrt{\pi}\,\Gamma(k+l+j)}{2^{\frac{1}{2}k+l-1} \Gamma\left(\frac{k+l+j+1}{2}\right) \Gamma\left(\frac{l-j}{2}\right)}, \quad j \in \mathbf{N},$$

and by Weierstrass' definition,

$$\frac{1}{\Gamma(z)} = z\mathrm{e}^{\gamma z}\prod_{n=1}^{+\infty}\left[\left(1+\frac{z}{n}\right)\mathrm{e}^{-\frac{z}{n}}\right], \qquad \gamma\text{—Euler constant.}$$

Proof From [4]((p. 29)(6)) we obtain that

$$I_j(s)=(1+s^2)^{-(\frac{1}{4}k+\frac{1}{2}j+\frac{1}{2})}\Gamma(k+l+j)P_{\frac{1}{2}k+j}^{-\frac{1}{2}k-l+1}((1+s^2)^{-\frac{1}{2}}),$$

where $P_\mu^\nu(.)$ is Legendre function (see [4] p. 4(26)). Define

$$\psi_j(\varepsilon)=(1+\varepsilon^2)^{-\frac{1}{4}k-\frac{1}{2}j-\frac{1}{2}}\Gamma(k+l+j)P_{\frac{1}{2}k+j}^{-\frac{1}{2}k-l+1}(\varepsilon(1+\varepsilon^2)^{-\frac{1}{2}}),$$

When $\varepsilon>0$ we observe

$$\psi_j(\varepsilon)=I_j\left(\frac{1}{\varepsilon}\right)\varepsilon^{-\frac{1}{2}k-j-1}.$$

Thus

$$\psi_j(\varepsilon)=\int_0^{+\infty}\mathrm{e}^{-\varepsilon t}t^{\frac{1}{2}k+j}J_{\frac{1}{2}k+l-1}(t)\,\mathrm{d}t, \quad \varepsilon>0.$$

By differentiation, we obtain

$$\psi_j'(\varepsilon)=-\int_0^{+\infty}\mathrm{e}^{-\varepsilon t}t^{\frac{1}{2}k+j+1}J_{\frac{1}{2}k+l-1}(t)\,\mathrm{d}t=-\psi_{j+1}(\varepsilon), \quad \varepsilon>0.$$

Since $\psi_j\in C^\infty[0,+\infty)$, we get from the above equation that

$$\psi_j'(0)=-\psi_{j+1}(0). \tag{2.12}$$

From the equation $\psi_j(\varepsilon)=\psi_j(0)+\psi_j'(0)\varepsilon+o(\varepsilon^2)$, (as $\varepsilon\to0$) and (2.12) we have

$$I_j(\varepsilon^{-1})=\{\psi_j(0)-\psi_{j+1}(0)+o(\varepsilon^2)\}\varepsilon^{\frac{1}{2}k+j+1}, \quad \text{as } \varepsilon\to0^+. \tag{2.13}$$

By definition,

$$\psi_j(0)=\Gamma(k+l+j)P_{\frac{1}{2}+j}^{-\frac{1}{2}k-l+1}(0),$$

and according to [5](p. 171)

$$P_\nu^\mu(0)=2^\mu\pi^{-\frac{1}{2}}\cos\frac{1}{2}\pi(\nu+\mu)\Gamma\left(\frac{1}{2}(1+\nu+\mu)\right)\left[\Gamma\left(1+\frac{1}{2}\nu-\frac{1}{2}\mu\right)\right]^{-1}.$$

Hence

$$\psi_j(0)=2^{-\frac{1}{2}k-l+1}\pi^{-\frac{1}{2}}\sin\frac{1}{2}\pi(1-j)\Gamma\left(1-\frac{1}{2}l+\frac{1}{2}j\right)\Gamma(k+l+j)\cdot$$
$$\left[\Gamma\left(\frac{1}{2}k+\frac{1}{2}j+\frac{1}{2}l+\frac{1}{2}\right)\right]^{-1}$$
$$=\frac{\pi^{\frac{1}{2}}\Gamma(k+l+j)}{\Gamma\left(\frac{1}{2}\mu+\frac{1}{2}l+\frac{1}{2}j+\frac{1}{2}\right)\Gamma\left(\frac{1}{2}l-\frac{1}{2}j\right)}\cdot 2^{-\frac{1}{2}k-l+1}$$

$$=A_j(k,\ l).$$

Substituting it into (2.13) and putting $\varepsilon=s^{-1}(s>0)$ we complete the proof of Lemma 2.

Lemma 3 Let Re $\alpha>-1$. if l is odd then

$$^{l}H_k^{\alpha}(s)=2^{\alpha-\frac{1}{2}k}\pi^{-\frac{1}{2}k}\sum_{j=0}^{2q+1}b_j 2^{\alpha+j}\Gamma(\alpha+j+1)V_{\frac{1}{2}k+\alpha+j}(s)+{}^{l}\gamma_k^{\alpha}(s)$$

and

$$^{l}\gamma_k^{\alpha}(s)=(2\pi)^{-\frac{1}{2}k}(c_0A_0s^{-k-1}+(-c_0+c_1)A_1s^{-k-l-1})+O(s^{-k-l-2}) \tag{2.14}$$

Where $A_j=A_j(k,\ l)(j=0,1)$ are the same as in Lemma 2 and $c_j=c_j(k,\ l,\ \alpha)(j=0,1)$ are certain constants depending only on $k,\ l,\ \alpha$.

Proof Because the equation (1.11) holds also for odd l, we need only prove (2.14).

Let $\varphi(t)$ be such that $\varphi\in C^{\infty}_{[0,+\infty)}$ and

$$\varphi(x)=\begin{cases}1, \text{ if } & 0\leqslant t\leqslant\frac{1}{3},\\ 0, \text{ if } & \frac{2}{3}\leqslant t<+\infty.\end{cases}$$

Set $\psi(t)=1-\varphi(t)$ and

$$\omega(t)=[(1-t^l)^{\alpha}-Q(1-t^2)(1-t^2)^{\alpha}]\chi_{[0,1)}(t).$$

It has been shown in the proof of Lemma 1 that $\omega\in C^{2q}_{(0,+\infty)}$. So it is clear that $\Delta^q(\omega(t)\psi(t))\in C_{[0,+\infty)}$ and $\Delta^q(\omega(t)\psi(t))=0$ if $t\geqslant 1$. Hence we obtain by the same way as in deducing (2.8) that

$$\left|\int_0^1\omega(t)\psi(t)t^{k-1}V_{\frac{1}{2}k-1}(st)\mathrm{d}t\right|\leqslant M(k,l,\alpha)s^{-k-l-6}. \tag{2.15}$$

Similarly,

$$\left|\int_0^1 Q(1-t^2)(1-t^2)^{\alpha}\varphi(t)t^{k-1}V_{\frac{1}{2}k-1}(st)\mathrm{d}t\right|\leqslant M(k,l,\alpha)s^{-k-l-6}. \tag{2.16}$$

Set $u_0(t)=(1-t^l)^{\alpha}\quad\varphi(t)$ and

$$U(s)=\int_0^1 u_0(t)t^{\frac{1}{2}k}J_{\frac{1}{2}k-1}(st)\mathrm{d}t.$$

Using $\frac{\mathrm{d}}{\mathrm{d}t}(t^{\nu}J_{\nu}(t))=t^{\nu}J_{\nu-1}(t)$ and integrating by parts, we get

$$U(s)=s^{-1}\int_0^1 t^{\frac{1}{2}k+1}J_{\frac{1}{2}k}(st)(-u_0'(t)t^{-1})\mathrm{d}t.$$

Put $u_1(t)=-u_0'(t)t^{-1}$. If $l>1$ we can integrate by parts again and get

$$U(s)=s^{-2}\int_0^1(-u_1'(t)t^{-1})t^{\frac{1}{2}k+2}J_{\frac{1}{2}k+1}(st)\mathrm{d}t.$$

Put $u_2(t)=-u_1'(t)t^{-1}$ and integrate by parts again if $l>2$. Continue this process until we get

$$U(s)=s^{-1}\int_0^1 u_l(t)t^{\frac{1}{2}k+l}J_{\frac{1}{2}k+l-1}(st)\mathrm{d}t. \tag{2.17}$$

It is easy to verify inductively that

$$u_l(t)=\sum_{j=0}^{l+1}a_j^{(l)}t^{-l-j}u_0^{(1-j)}(t).$$

For $t\in\left(0,\ \frac{1}{3}\right)$, we have the following expression:

$$t^{-j}u_0^{(l-j)}(t)=\sum_{\mu=0}^{l}\sum_{\nu=0}^{l^2}c_{\mu,\nu,j}^{(l)}(1-t^l)^{\alpha-\mu}t^{\nu},\qquad j=0,1,2,\cdots,l-1.$$

Therefore

$$u_l(t)t^l=\sum_{j=0}^{l-1}\sum_{\mu=0}^{l}\sum_{\nu=0}^{l^2}a_j^{(l)}c_{\mu,\nu,j}^{(l)}(1-t^l)^{\alpha-\mu}t^{\nu},\ t\in\left(0,\ \frac{1}{3}\right).$$

From this we observe that $u_l(t)t^l\in C^{\infty}_{\left[0,\frac{1}{3}\right]}$, furthermore $u_l(t)t^l\in C^{\infty}_{[0,+\infty)}$.

Let $P(t)=\sum_{j=0}^{2q}c_j t^j$ such that

$$\frac{\mathrm{d}^{\nu}}{\mathrm{d}t^{\nu}}\{u_l(t)t^l-\mathrm{e}^{-t}P(t)\}\big|_{t=0}=0,\ \nu=0,1,2,\cdots,2q. \tag{2.18}$$

$P(t)$ is uniquely determinated by the conditions (2.18).

Denote $\delta(t)=u_l(t)-\mathrm{e}^{-t}P(t)t^{-1}$. Obviously, $\Delta^q\delta(t)\in C^{\infty}_{(0,+\infty)}$ and

$$\Delta^q\delta(t)=-\Delta^q(\mathrm{e}^{-t}P(t)t^{-1})=0(t^{2q}\mathrm{e}^{-t})\quad \text{as}\quad t\to+\infty. \tag{2.19}$$

It is easy to prove that

$$\Delta^q\delta(t)=\sum_{j=0}^{2q}d_j^{(q)}\delta^{(j)}(t)t^{j-2q}.$$

By Leibniz formula

$$\delta^{(j)}(t)=\sum_{\nu=0}^{j}c_j^{\nu}\frac{\mathrm{d}^{\nu}}{\mathrm{d}t^{\nu}}[u_l(t)t^l-\mathrm{e}^{-t}P(t)]\frac{\mathrm{d}^{j-\nu}}{\mathrm{d}t^{j-\nu}}(t^{-l})$$

$$= \sum_{\nu=0}^{j} c_j^\nu o(t^{2q-\nu}) \cdot O(t^{\nu-l-j})$$
$$= O(t^{2q-1-j}) \quad \text{as} \quad t \to 0^+, \ j = 1, 2, \cdots, 2q.$$

Hence

$$\Delta^q \delta(t) = O(t^{-1}) \quad \text{as} \quad t \to 0^+. \tag{2.20}$$

From (2. 17) and the definition of $\delta(t)$ we have

$$U(s) = s^{-1}\int_0^{+\infty} \delta(t) t^{\frac{1}{2}k+1} J_{\frac{1}{2}k+l-1}(st)\mathrm{d}t + s^{-1}\int_0^{+\infty} \mathrm{e}^{-t} P(t) t^{\frac{1}{2}k} J_{\frac{1}{2}k+l-1}(st)\mathrm{d}t. \tag{2.21}$$

Now according to Lemma 6 in [3] it follows from (2. 19) and (2. 20) that

$$s^{-l}\int_0^{+\infty} \delta(t) t^{\frac{1}{2}k+l} J_{\frac{1}{2}k+l-1}(st)\mathrm{d}t = O(s^{-\frac{1}{2}k-l-1}).$$

And an application of Lemma 2 yields

$$s^{-1}\int_0^{+\infty} \mathrm{e}^{-t} P(t) t^{\frac{1}{2}k} J_{\frac{1}{2}k+l-1}(st)\mathrm{d}t = s^{-1}\sum_{j=0}^{2q} c_j I_j$$
$$= c_0 A_0 s^{-\frac{1}{2}k-l-1} + (-c_0 + c_1) A_1 s^{-\frac{1}{2}k-l-2} + O(s^{-\frac{1}{2}k-l-3}).$$

Substituting these two estimates into (2. 21), we obtain

$$U(s) = c_0 A_0 s^{-\frac{1}{2}k-l-1} + (-c_0 + c_1) A s^{-\frac{1}{2}k-l-2} + O(s^{-\frac{1}{2}k-l-3}). \tag{2.22}$$

By Definition 2 we have

$${}^l\gamma_k^\alpha(s) = (2\pi)^{-\frac{1}{2}k}\int_0^1 \omega(t)\psi(t) t^{k-1} V_{\frac{1}{2}k-1}(st)\mathrm{d}t +$$
$$(2\pi)^{-\frac{1}{2}k}\int_0^1 Q(1-t^2)(1-t^2)^\alpha \varphi(t) t^{k-1} V_{\frac{1}{2}k-1}(st)\mathrm{d}t +$$
$$(2\pi)^{-\frac{1}{2}k} s^{-\frac{1}{2}k+1} U(s). \tag{2.23}$$

An combination of (2. 15)(2. 16)(2. 22) and (2. 23) yields (2. 14). □

Remark If l is small the coefficient c_0 can be easily obtained. For example, if $l=1$ then

$$t u_1(t) = -u_0'(t) = \alpha(1-t)^{\alpha-1}, \quad 0 < t < \frac{1}{3}.$$

Hence $c_0 = \alpha$.

It is well known that for $l=2$,

$${}^2 H_k^\alpha(t) = 2^{\alpha-\frac{1}{2}k} \pi^{-\frac{1}{2}k} \Gamma(\alpha+1) V_{\frac{1}{2}k+\alpha}(t).$$

Therefore from Lemma 1 and Lemma 3 we obtain the relation between the kernel functions ${}^lH_k^\alpha$ and ${}^2H_k^\alpha$. That is

Proposition If Re $\alpha>-1$ then

$$ {}^lH_k^\alpha(t)=\sum_{j=0}^{2q+1} b_j\,{}^2H_k^{\alpha+j}(t)+{}^l\gamma_k^\alpha(t). \tag{2.24}$$

where ${}^l\gamma_k^\alpha$ is defined by (2.7) and satisfies the estimate (2.8) or (2.14) according as l is even or odd.

In virtue of this proposition we conclude that the necessary and sufficient condition for $t^{k-1}\cdot{}^lH_k^\alpha(t)\in L(0,+\infty)$ is Re $\alpha>\frac{1}{2}(k-1)$.

This condition is also necessary and sufficient for $\mathscr{F}(\varphi)(y)\in L(R^k)$, where φ is defined by (2.2). Hence $\alpha_0=\frac{1}{2}(k-1)$ is also the critical index for the operator ${}^lS_R^\alpha\,(l\neq 2)$. When $\alpha>\alpha_0$ we can use the Poission summation formula (See [2] Chap. Ⅶ) to derive the generalized Bochner formula, that is

Theorem 1 If Re $\alpha>\frac{1}{2}(k-1)$, $f\in L(Q)$ then

$$ {}^lS_R^\alpha(f;x)=\int_0^{+\infty}\varphi_x(tR^{-1})t^{k-1}\,{}^lH_k^\alpha(t)\mathrm{d}t,\ (R>0),$$

where
$$\varphi_x(t)=\int_{|\xi|=1} f(x-t\xi)\mathrm{d}\sigma(\xi).$$

From Proposition and Theorem 1 the following result is directly derived.

Corollary 1 Suppose Re $\alpha>\frac{1}{2}(k-1)$, $f\in L(Q)$. Then

$$ {}^lS_R^\alpha(f;x)=\sum_{j=0}^{2q+1} b_jS_R^{\alpha+j}(f;x)+\int_0^{+\infty}\varphi_x(tR^{-1})t^{k-1}\,{}^l\gamma_k^\alpha(t)\mathrm{d}t. \tag{2.25}$$

§ 3. On uniform approximation (when $\alpha>\frac{k-1}{2}$)

We notice that when $l=1$ ${}^l\gamma_k^\alpha(t)={}^1\gamma_k^\alpha(t)$ tends to 0 most slowly as $t\to+\infty$. So we discuss the case of $l=1$ particularly.

Theorem 2 Let $f\in C(Q)$. if $\alpha>\frac{1}{2}(k-1)$ then

$$ {}^1S_R^\alpha(f;x)-f(x)=\frac{\alpha\Gamma(k+1)}{2^k\pi^{\frac{1}{2}k}\Gamma\left(\frac{1}{2}k+1\right)}\int_{\frac{\pi}{R}}^{+\infty}f_x(t)R^{-1}t^{-2}\mathrm{d}t+O(\omega_2(f;R^{-1})), $$

where "O" is uniform for all x, and

$$ f_x(t)=\int_{|\xi|=1}(f(x-t\xi)-f(x))\mathrm{d}\sigma(\xi). $$

Proof In [6] we have proved that if $\alpha>\frac{1}{2}(k-1)$,

$$ \| S_R^\alpha(f)-f\|_c=O(\omega_2(f;\ R^{-1})). $$

Hence by (2.25) we observe that it is sufficient to prove the following formula

$$ \int_0^{+\infty}f_x(tR^{-1})t^{k-1}\,{}^1\gamma_k^\alpha(t)\mathrm{d}t $$

$$ =\frac{\alpha\Gamma(k+1)}{2^k\pi^{\frac{1}{2}k}\Gamma\left(\frac{1}{2}k+1\right)}\int_{\frac{\pi}{R}}^{+\infty}f_x(t)R^{-1}t^{-2}\mathrm{d}t+O(\omega_2(f;R^{-1})). \tag{3.1} $$

Obviously,

$$ f_x(t)=\int_{|\xi|=1,\xi_1>0}(f(x+t\xi)+f(x-t\xi)-2f(x))\mathrm{d}\sigma(\xi)=O(\omega_2(f;t)). $$

Hence

$$ \int_0^{\pi}f_x(tR^{-1})t^{k-1}\,{}^1\gamma_k^\alpha(t)\mathrm{d}t=O(\omega_2(f;R^{-1})). \tag{3.2} $$

By (2.14), for $t\geqslant1$,

$$ {}^1\gamma_k^\alpha(t)=(2\pi)^{-\frac{1}{2}k}[c_0A_0t^{-k-1}+(-c_0+c_1)A_1t^{-k-2}]+O(t^{-k-3}). $$

According to Lemma 2, in our case

$$ A_0=A_0(k,\ 1)=\frac{\Gamma(k+1)}{2^{\frac{1}{2}k}\Gamma\left(\frac{1}{2}k+1\right)},\qquad A_1=A_1(k,\ 1)=0. $$

From Remark of Lemma 3 we observe that $c_0=c_0(k,\ 1,\ \alpha)=\alpha$. Therefore

$$ {}^1\gamma_k^\alpha(t)=\frac{\alpha\Gamma(k+1)}{2^k\pi^{\frac{1}{2}k}\Gamma\left(\frac{1}{2}k+1\right)}t^{-k-1}+O(t^{-k-3}). $$

Hence

$$ \int_\pi^{+\infty}f_x(tR^{-1})t^{k-1}\,{}^1\gamma_k^\alpha(t)\mathrm{d}t $$

$$= \frac{\alpha\Gamma(k+1)}{2^k\pi^{\frac{1}{2}k}\Gamma\left(\frac{1}{2}k+1\right)}\int_{\pi}^{+\infty} f_x(tR^{-1})t^{-2}\mathrm{d}t + O\left(\int_{\pi}^{+\infty}\omega_2(f;tR^{-1})t^{-4}\mathrm{d}t\right)$$

$$= \frac{\alpha\Gamma(k+1)}{2^k\pi^{\frac{1}{2}k}\Gamma\left(\frac{1}{2}k+1\right)}\int_{\frac{\pi}{R}}^{+\infty} f_x(t)R^{-1}t^{-2}\mathrm{d}t + O(\omega_2(f;R^{-1})). \tag{3.3}$$

A combination of (3.2) and (3.3) yields (3.1). This completes the proof of Theorem 2.

The conclusion of Theorem 2 is perfectly parallel with the result obtained by Guo Zhurui[7] which is in respect to the approximation by Cesàro means of positive order. So we can say that when $k>1$ the operator ${}^1S_R^\alpha$ is indeed the generalization of the Cesàro operator.

When $l>1$ the following conclusion is obvious.

Theorem 3 Let $l>1$, $\alpha>\frac{1}{2}(k-1)$ and $f\in C(Q)$. Then

$$\| {}^lS_R^\alpha(f)-f\|_c = O(\omega_2(f;\ R^{-1})).$$

Now we generalize the result concerning differentiable functions in [1].

Theorem 4 Let $k>1$, $\alpha>\frac{1}{2}(k-1)$ and $f\in C^j(Q)$. Then

(i) $\| {}^lS_R^\alpha(f)-f\|_c \leqslant A_{k,l,\alpha}R^{-j}\omega^{(j)}(f;\ R^{-1})$, $\quad j=1,2,\cdots,l-2$; $\hfill (3.4)$

(ii) for $j=l-1$ if l is even, (3.4) still holds; if l is odd,

$$\| {}^lS_R^\alpha(f)-f\|_c \leqslant A_{k,l,\alpha}\ln(2+R)R^{-(l-1)}\omega^{(l-1)}(f;\ R^{-1}) \tag{3.5}$$

holds where

$$\omega^{(j)}(f;\ t)=\max\{\omega(D^{(s_1,s_2,\cdots,s_k)}f;\ t):\ s_1+s_2+\cdots+s_k=j\},\qquad t\geqslant 0,$$

$D^{(s_1,s_2,\cdots,s_k)}f=\frac{\partial^{s_1+s_2+\cdots+s_k}f}{\partial x_1^{s_1}\cdots\partial x_k^{s_k}}$, ($s_1$, s_2, $\cdots$, s_k are non-negative integers).

The result of [1] is the particular case of Theorem 4 for

$$x=\left[\frac{1}{2}(k-1)\right]+1.$$

Define $G_l(t)=t^{k-1}\,{}^lH_k^\alpha(t)$, $F_l^\alpha(t)=t^{k-1}\,{}^l\gamma_k^\alpha(t)$ and $J_l^\alpha(t)=\sum_{j=0}^{2q+1}b_jG_2^{\alpha+j}(t)$. It is known from (24) that

$$G_l^\alpha(t)=J_l^\alpha(t)+F_l^\alpha(t). \tag{3.6}$$

Lemma 4 Suppose $k>1$, $\alpha>\frac{1}{2}(k-1)$. and $l>1$. Define an operator I:

$$I(f)(t)=\int_t^{+\infty} f(s)\mathrm{d}s \quad f\in L(0,+\infty), \quad t\geqslant 0.$$

Then

(i) $\forall n\in\mathbf{N}$, $I^n(J_l^\alpha)(t)=O(t^{\frac{k-3}{2}-\alpha})$, $t\to+\infty$;

(ii) for $n=1,2,\cdots,l$, $I^n(F_l^\alpha)(t)$ is well-defined; if l is even $I^n(F_l)(t)=O(t^{-1-7+n})$, (as $t\to+\infty$); if l is odd,

$$I^n(F_l^\alpha)(t)=o(t^{-l-1+n}), \qquad t\to+\infty;$$

(iii) when $l\geqslant 3$,

$$I^3(G_l^\alpha)(0)=\cdots=I^{2\left[\frac{l+1}{2}\right]-1}(G_l^\alpha)(0)=0.$$

Proof *For any* $n\in\mathbf{N}$ it is known from the asymptotic expansion of the Bessel function that

$$\begin{aligned}G_2^\alpha(t)&=\frac{2^\alpha\Gamma(\alpha+1)}{(2\pi)^{\frac{k}{2}}}t^{k-1}V_{\frac{1}{2}k+\alpha}(t)\\&=t^{\frac{k-3}{2}-\alpha}\left\{\mathrm{Re}\left(\sum_{\nu=0}^{n+1}c_\nu \mathrm{e}^{\mathrm{i}t}t^{-\nu}\right)+o(t^{-n-2})\right\},\end{aligned} \tag{3.7}$$

Where c_ν denote certain complex valued constants. When $\beta>1$ we can integrate by parts and obtain that

$$\int_t^{+\infty}\mathrm{e}^{\mathrm{i}s}s^{-\beta}\mathrm{d}s=\mathrm{e}^{\mathrm{i}t}t^{-\beta}\sum_{\nu=0}^{n+1}t^{-\nu}\mathrm{d}_\nu+o(t^{-n-1-\beta}), \tag{3.8}$$

where d_ν are some complex numbers. From (3.7) and (3.8) it follows that $I^n(G_2^\alpha)(t)=O(t^{\frac{1}{2}k-\frac{3}{2}-\alpha})$ (as $t\to+\infty$) for any $n\in\mathbf{N}$. From this and the definition of J_l^α we observe that the conclusion (i) is true.

From (2.8) and (2.14) we know that the conclusion (ii) holds.

Now set $f(x)=\cos x_1$ (x_1 is the first coordinate of x) then we have ${}^lS_R^\alpha(f;x)=(1-R^{-l})^\alpha\cos x_1$ $(R>1)$. Using Theorem 1 and putting $x_1=0$,

$$\varphi_x=\mathrm{Re}\int_{|\xi|=1}\mathrm{e}^{\mathrm{i}(x_1-t\xi_1)}\mathrm{d}\sigma(\xi)=(2\pi)^{\frac{k}{2}}\frac{J_{\frac{k}{2}-1}(t)}{t^{\frac{k}{2}-1}}\cos x_1=(2\pi)^{\frac{k}{2}}V_{\frac{1}{2}k-1}(t)\xi(x),$$

we obtain that

$$(2\pi)^{-\frac{1}{2}k}(1-R^{-1})^{\alpha}=\int_{0}^{+\infty}V_{\frac{1}{2}k-1}(tR^{-1})G_{l}^{\alpha}(t)\mathrm{d}t,\quad R>1. \tag{3.9}$$

Since $V_{\frac{1}{2}k-1}(u)=\sum\limits_{j=0}^{+\infty}(-1)^{j}\left[j!\Gamma\left(j+\frac{1}{2}k\right)\right]^{-1}u^{2j}2^{1-\frac{1}{2}k-2j}$,

we have $V_{\frac{1}{2}k-1}^{(2j+1)}(0)=0$, $\quad j\in\mathbf{N}$,

$V_{\frac{1}{2}k-1}^{(2j)}(0)=(-1)^{j}(2j)!\ (j!\ (j+\frac{1}{2}k))^{-1}2^{1-2j-\frac{1}{2}k}\neq 0$, $\quad j\in\mathbf{N}$.

Noticing $V_{\nu}'(t)=-V_{\nu+1}(t)$, we can prove inductively that

$$V_{\nu}^{(2j)}=a_{1}^{(j)}V_{\nu+j}+a_{2}^{(j)}V_{\nu+j+1}t^{2}+\cdots+a_{j+1}^{(j)}V_{\nu+2j}t^{2j},$$

$$V_{\nu}^{(2j+1)}=t(b_{1}^{(j)}V_{\nu+j+1}+b_{2}^{(j)}V_{\nu+j+2}t^{2}+\cdots+b_{j+1}^{(j)}V_{\nu+2j+1}t^{2j}).$$

Hence $V_{\frac{1}{2}k-1}^{(j)}(t)=o(t^{\frac{1}{2}-\frac{1}{2}k})$ (as $t\to+\infty$). Intergrating the right-hand side of (3.9), we obtain that

$$\begin{aligned}&(2\pi)^{-\frac{1}{2}k}(1-R^{-l})^{\alpha}\\&=V_{\frac{1}{2}k-1}(0)I(G_{l}^{\alpha})(0)+R^{-1}\int_{0}^{+\infty}V_{\frac{1}{2}k-1}'(tR^{-1})I(G_{l}^{\alpha})(t)\mathrm{d}t,\end{aligned}$$

Letting $R\to+\infty$, noticing that $l>1$, $I(G_{l}^{\alpha})\in L(0,+\infty)$ and $V_{\frac{1}{2}k-1}'(tR^{-1})=O(1)$ we obtain that $V_{\frac{1}{2}k-1}(0)I(G_{l}^{\alpha})(0)=(2\pi)^{-\frac{1}{2}k}$. Hence

$$(2\pi)^{-\frac{1}{2}k}[(1-R^{-l})^{\alpha}-1]=R^{-1}\int_{0}^{+\infty}V_{\frac{1}{2}k-1}(tR^{-1})I(G_{l}^{\alpha})(t)\mathrm{d}t.$$

Integrating by parts, we obtain that

$$\begin{aligned}&(2\pi)^{-\frac{1}{2}k}[(1-R^{-l})^{\alpha}-1]\\&=R^{-1}V_{\frac{1}{2}k-1}'(0)I^{2}(G_{l}^{\alpha})(0)+R^{-2}\int_{0}^{+\infty}V_{\frac{1}{2}k-1}''(tR^{-1})\cdot I^{2}(G_{l})(t)\mathrm{d}t\\&=R^{-2}\int_{0}^{+\infty}V_{\frac{1}{2}k-1}''(tR^{-1})I^{2}(G_{l}^{\alpha})(t)\mathrm{d}t.\end{aligned}$$

If $l\geqslant 3$ then we can integrate by parts again and obtain that

$$\begin{aligned}&(2\pi)^{-\frac{1}{2}k}[(1-R^{-l})^{\alpha}-1]\\&=R^{-2}V_{\frac{1}{2}k-1}''(0)I^{3}(G_{l}^{\alpha})(0)+R^{-3}\int_{0}^{+\infty}V_{\frac{1}{2}k-1}'''(tR^{-1})I^{3}(G_{l}^{\alpha})(t)\mathrm{d}t.\end{aligned}$$

Notice that

$$\begin{aligned}&\int_{0}^{+\infty}V_{\frac{1}{2}k-1}'''(tR^{-1})I^{3}(G_{l}^{\alpha})(t)\mathrm{d}t\\&=\left(\int_{0}^{1}+\int_{1}^{R}+\int_{R}^{+\infty}\right)V_{\frac{1}{2}k-1}'''(tR^{-1})I^{3}(G_{l}^{\alpha})(t)\mathrm{d}t\\&=O(1)+\int_{1}^{R}o(t^{-1})\mathrm{d}t+\int_{R}^{+\infty}O[(Rt^{-1})^{\frac{1}{2}k-\frac{1}{2}}]t^{-1}\mathrm{d}t\end{aligned}$$

$$=o(\ln R),\ R>1,$$

(here the condition $\frac{1}{2}k>\frac{1}{2}$ is used). Multiply both sides of the above equation by R^2 and let $R\to+\infty$, we obtain that $0=V''_{\frac{1}{2}k-1}(0)I^3(G_l^\alpha)(0)$. Since $V''_{\frac{1}{2}k-1}(0)\neq0$, $I^3(G_l^3)(0)=0$ follows. If $l\geqslant4$ we can continue this argument and obtain the conclusion (iii). □

Proof of Theorem 4 Fix $R\geqslant1$, define the Steklov (Стеклов) function of f by

$$f_R(x)=(2R)^k\int_{(-R^{-1},R^{-1})^k}f(x+y)\mathrm{d}y.$$

Since $f\in C^j$, we have $f_R\in C^{j+1}$. Let n_1, n_2, $\cdots$, n_k be non-negative integers. If $n_1+n_2+\cdots+n_k\leqslant j$ then

$$\| D^{(n_1,n_2,\cdots,n_k)}(f_R-f)\|_c\leqslant\omega^{(\langle n\rangle)}(f;\ R^{-1}),\qquad(3.10)$$

where the symbol $\langle n\rangle$ denotes the sum $n_1+n_2+\cdots+n_k$ and here $\langle n\rangle\leqslant j$.

Set $g=f-f_R$. Clearly,

$${}^lS_R^\alpha(f;\ x)-f(x)={}^lS_R^\alpha(g;\ x)-g(x)+{}^lS_R^\alpha(f_R;\ x)-f_R(x).$$

Put $g_x(t)=\int_{|\xi|=1}[g(x-t\xi)-g(x)]\mathrm{d}\sigma(\xi)$. Then $g_x(0)=0$ and

$$(g_x)'(0)=-\int_{|\xi|=1}\sum_{\nu=1}^{k}\frac{\partial g(x-t\xi)}{\partial x_\nu}\bigg|_{t=0}\xi_\nu\mathrm{d}\sigma(\xi)=0.$$

Generally,

$$(g_x)^{(2\mu-1)}(0)=0,\qquad \mu=1,2,\cdots,\left[\frac{1}{2}j+\frac{1}{2}\right].\qquad(3.11)$$

By Theorem 1,

$${}^lS_R^\alpha(g;x)-g(x)=\int_0^{+\infty}g_x(tR^{-1})G_l^\alpha(t)\mathrm{d}t.\qquad(3.12)$$

Using Lemma 4 and (3.11), integrating the right-hand side of (3.12) by parts j times ($j\leqslant l-1$), we obtain that

$${}^lS_R^\alpha(g;x)-g(x)=R^{-j}\int_0^{+\infty}(g_x)^{(j)}(tR^{-1})I^j(G_l^\alpha)(t)\mathrm{d}t.\qquad(3.13)$$

From (3.10) it follows that $|(g_x)^{(j)}(u)|\leqslant A_k\omega^{(j)}(f;\ R^{-1})(u\geqslant0)$. And from Lemma 4 it is known that $I^j(G_l^\alpha)\in L(0,\ +\infty)$. Therefore, we obtain from (3.13) that

$$\| {}^{l}S_R^{\alpha}(g)-g \|_c \leqslant A_{k,l,\alpha} R^{-j} \omega^{(j)}(f;\ R^{-1}). \tag{3.14}$$

For the function f_R, the equation (3.13) holds also. This time we write $f_{R,x}(t) = \int\limits_{|\xi|=1} [f_R(x-t\xi) - f_R(x)] \mathrm{d}\sigma(\xi)$. Hence

$$ {}^{l}S_R^{\alpha}(f;x) - f_R(x) = R^{-j} \int_0^{+\infty} (f_{R,x})^{(j)}(tR^{-1}) I^j(G_l^{\alpha})(t) \mathrm{d}t. \tag{3.15}$$

When $j \leqslant l-2$, since $f_R \in C^{j+1}$ we can integrate the right-hand side of (3.15) by parts and obtain that

$$ {}^{l}S_R^{\alpha}(f_R;x) - f_R(x) = R^{-j-1} \int_0^{+\infty} (f_{R,x})^{j+1}(tR^{-1}) I^{j+1}(G_l^{\alpha})(t) \mathrm{d}t. \tag{3.16}$$

If $\langle n \rangle = n_1 + n_2 + \cdots + n_k = j$, then

$$\frac{\partial}{\partial x_\nu} D^{(n_1,n_2,\cdots,n_k)}(f_R)(x) = \frac{\partial}{\partial x_\nu} (2R)^k \int\limits_{(-R^{-1},R^{-1})^k} D^{(n_1,n_2,\cdots,n_k)}(f)(x+y)\mathrm{d}y$$

$$= (2R)^k \int\limits_{(-R^{-1},R^{-1})^k} [D^{(n_1,n_2,\cdots,n_k)}(f)(x+y_{(\nu)}^{+}) - D^{(n_1,n_2,\cdots,n_k)}(f)(x+y_{(\nu)}^{-})] \mathrm{d}y_{(\nu)}.$$

where

$$y_{(\nu)}^{+} = (y_1,\ \cdots,\ y_{\nu-1},\ R^{-1},\ y_{\nu+1},\ \cdots,\ y_k),$$
$$y_{(\nu)}^{-} = (y_1,\ \cdots,\ y_{\nu-1},\ -R^{-1},\ y_{\nu+1},\ \cdots,\ y_k),$$
$$dy_{(\nu)} = dy_1 \cdots dy_{\nu-1} dy_{\nu+1} \cdots dy_k.$$

Therefore

$$|(f_{R,x})^{(j+1)}(u)| \leqslant A_k R \omega^{(j)}(f;\ R^{-1}) \quad (j=1,2,\cdots,l-1). \tag{3.17}$$

Substituting it into (3.16) we conclude that

$$\| {}^{l}S_R^{\alpha}(f_R) - f_R \|_c \leqslant A_{k,1,\alpha} R^{-j} \omega^{(j)}(f;\ R^{-1}), \quad j=1,2,\cdots,l-2. \tag{3.18}$$

From (3.14) and (3.18) the estimate (3.4) follows. This is just the conclusion (i).

If l is even and $j=l-1$ the above argument keeps valid and (3.18) holds also.

Suppose now l is odd and $j=l-1$. Then since $I^{(l)}(G_l^{\alpha})(0)=0$, (by Lemma 4 (iii)) the equation (3.15) can be rewritten as

$$ {}^{l}S_R^{\alpha}(f;x) - f_R(x)$$

$$= R^{1-l}\int_0^{+\infty}[(f_{R,x})^{(l-1)}(tR^{-1})-(f_{R,x})^{(l-1)}(0)]\cdot I^{l-1}(G_l^\alpha)(t)\mathrm{d}t$$

$$= R^{1-l}\int_0^{+\infty}[(f_{R,x})^{(l-1)}(tR^{-1})-(f_{R,x})^{(l-1)}(0)]\cdot I^{l-1}(J_l^\alpha)(t)\mathrm{d}t+$$

$$R^{1-l}\int_0^{+\infty}[(f_{R,x})^{(l-1)}(tR^{-1})-(f_{R,x})^{(l-1)}(0)]\cdot I^{l-1}(F_l^\alpha)(t)\mathrm{d}t. \quad (3.19)$$

The first integral in right side can be calculated by parts and the result is $R^{-l}\int_0^{+\infty}(f_{R,x})^{(l)}(tR^{-1})I^l(J_l^\alpha)(t)\,\mathrm{d}t$. From (3.17) and the fact that $I^l(J_l^\alpha)\in L(0,+\infty)$ we observe that this quantity is equal to $O(R^{-l+1}\omega^{(l-1)}(f;R^{-1}))$. We split second integral in right side of (3.19) to three parts: $\int_0^1$, $\int_1^R$, $\int_R^{+\infty}$. We have

$$\int_0^1[(f_{Rx})^{(l-1)}(tR^{-1})-(f_{Rx})^{(l-1)}(0)]I^{l-1}(F_l^\alpha)(t)\mathrm{d}t$$

$$=O(\omega^{(l-1)}(f_R;R^{-1}));$$

$$\int_R^{+\infty}[(f_{Rx})^{(l-1)}(tR^{-1})-(f_{Rx})^{(l-1)}(0)]I^{l-1}(F_l^\alpha)(t)\mathrm{d}t$$

$$=O(R^{-1}\omega^{(l-1)}(f_R;1))=O(\omega^{(l-1)}(f_R;R^{-1})).$$

As for the integral $\int_1^R$, integrate by parts and obtain that

$$\int_1^R[(f_{Rx})^{(l-1)}(tR^{-1})-(f_{Rx})^{(l-1)}(0)]I^{l-1}(F_l^\alpha)(t)\mathrm{d}t$$

$$=[(f_{Rx})^{(l-1)}(0)-(f_{Rx})^{(l-1)}(tR^{-1})]I^l(F_l^\alpha)(t)\Big|_1^R+$$

$$R^{-1}\int_1^R(f_{Rx})^{(l)}(tR^{-1})I^l(F_l^\alpha)(t)\mathrm{d}t$$

$$=O(\ln R\omega^{(l-1)}(f,R^{-1})),\quad R\to+\infty.$$

From the definition of f_R we deduce that

$$\omega^{(l-1)}(f_R;R^{-1})=O(\omega^{(l-1)}(f;R^{-1})).$$

Substituting all these results into (3.19) we obtain (3.5). □

§ 4. The case of $\alpha\leqslant\alpha_0$

Set

$$h(\alpha,\ t)=[(1-t^l)^\alpha-Q(1-t^2)(1-t^2)^\alpha]\chi_{[0,1)}(t)$$
$$=\left[(1-t^l)^\alpha-\sum_{j=0}^{2q+1}b_j(l,\alpha)(1-t^2)^{\alpha+j}\right]\chi_{[0,1)}(t).$$

By Definition 2 (see (2.7)) we obtain that

$$^l\gamma_k^\alpha(s)=(2\pi)^{-\frac{1}{2}k}\int_0^1 h(\alpha,t)t^{k-1}V_{\frac{1}{2}k-1}(st)\mathrm{d}t,\quad \operatorname{Re}\alpha>-1. \tag{4.1}$$

According to the condition (2.5) we have

$$\frac{\partial^j}{\partial t^j}h(\alpha,\ t)\bigg|_{t=1}=0,\qquad j=0,1,2,\cdots,2q,$$

for any α such that $\operatorname{Re}\alpha>-1$. Hence

$$\frac{\partial^j}{\partial t^j}\left(\frac{\partial}{\partial\alpha}h(\alpha,\ t)\right)\bigg|_{t=1}=\frac{\partial}{\partial t}\left(\frac{\partial^j}{\partial t^j}h(\alpha,\ t)\bigg|_{t=1}\right)=0,\qquad j=0,1,2,\cdots,2q. \tag{4.2}$$

By this formula we can prove as in Lemma 3 to obtain

$$\int_0^1\frac{\partial}{\partial\alpha}h(\alpha,t)t^{k-1}V_{\frac{1}{2}k-1}(st)\mathrm{d}t=O(s^{-k-l}), \tag{4.3}$$

and it is evident that the "O" in the right-hand side is uniform for α variated in any compact set contained in the half plane $\operatorname{Re}\alpha>-1$.

Definition 3 Suppose $\operatorname{Re}\alpha>-1$, $^l\gamma_k^\alpha(t)$ is the same as in Definition 2. The operator $^l\Gamma_R^\alpha$ is defined by

$$^l\Gamma_R^\alpha(f)(x)=\int_{\mathbf{R}^k}f(x-y)^l\gamma_k^\alpha(R|y|)R^k\mathrm{d}y,\quad f\in L(Q). \tag{4.4}$$

It is obvious that

$$^l\Gamma_R^\alpha(f)(x)=\int_0^{+\infty}\varphi_x(tR^{-1})t^{k-1}\,{}^l\gamma_k^\alpha(t)\mathrm{d}t, \tag{4.5}$$

where $\varphi_x(t)=\int_{|\xi|=1}f(x-t\xi)\mathrm{d}\sigma(\xi)$.

When $\operatorname{Re}\alpha>-1$ from (4.3) we conclude that

$$\frac{\partial}{\partial\alpha}(^l\gamma_k^\alpha(t))=O(t^{-k-l}),\qquad t\to+\infty, \tag{4.6}$$

where "O" is uniform for α variated in any compact set contained in the half plane Re $\alpha>-1$. From (4.5) and (4.6) we obtain that

$$\frac{\partial}{\partial\alpha}{}^{l}\Gamma_R^{\alpha}(f)(x)=\int_0^{+\infty}\varphi_x(tR^{-1})t^{k-1}\left(\frac{\partial}{\partial\alpha}l\gamma_k^{\alpha}(t)\right)\mathrm{d}t.$$

Let $x\in Q$, $R>0$, $l\in\mathbf{N}$, $f\in L(Q)$ be all fixed. Then ${}^{l}\Gamma_R^{\alpha}(f)(x)$ as function of variable α is analytic in Re $\alpha>-1$.

Set

$$F(\alpha)=\sum_{j=0}^{2q+1}b_j(\alpha)S_R^{\alpha+j}(f;x)+{}^{l}\Gamma_R^{\alpha}(f)(x).$$

Because of $F(\alpha)$ and ${}^{l}S_R^{\alpha}(f;\ x)$ are all analytic in Re $\alpha>-1$, by Corollary 1 (see(2.25)), ${}^{l}S_R^{\alpha}(f;\ x)=F(\alpha)$ if Re $\alpha>\frac{1}{2}k-\frac{1}{2}$, thus we conclude that this equation holds in the half plane Re $\alpha>-1$ This is the following

Theorem 5 Let $f\in L(Q)$, $\alpha\in\mathbf{C}$. if Re $\alpha>-1$ then

$$ {}^{l}S_R^{\alpha}(f;x)=\sum_{j=0}^{2q+1}b_j(l,\alpha)S_R^{\alpha+j}(f;x)+{}^{l}\Gamma_R^{\alpha}(f)(x). \tag{4.7}$$

If we denote Hardy-Littlewood maximal operator by HL then from (4.5), (2.8) and (2.14) we obtain that

$$\sup_{R>0}|{}^{l}\Gamma_R^{\alpha}(f)(x)|\leqslant M(k,l,\alpha)HL(f)(x),\quad \text{Re }\alpha>-1. \tag{4.8}$$

Therefore we obtain the following corollary from Theorem 5.

Corollary 2 When Re $\alpha>-1$,

$${}^{l}S_*^{\alpha}(f)(x)\leqslant\sum_{j=0}^{2q+1}|b_j|S_*^{\alpha+j}(f)(x)+M(k,\ l,\ \alpha)HL(f)(x),$$

where ${}^{l}S_*^{\alpha}(f)(x)=\sup\limits_{R>0}|{}^{l}S_R^{\alpha}(f;\ x)|$.

When $\alpha=\alpha_0=\frac{1}{2}k-\frac{1}{2}$, from Theorem 5 we derive the following

Corollary 3 The lebesgue constant of the operator ${}^{l}S_R^{\frac{k-1}{2}}$ is

$${}^{l}L_R^{\frac{k-1}{2}}=o(\ln(2+R)).$$

These results show that many results for the operator S_R^{α} can directly generalized to the case of ${}^{l}S_R^{\alpha}$. Here we only state a theorem of uniform approximation for critical index.

Theorem 6 Suppose $f\in C(Q)$, if $l>1$, then

$$\| {}^lS_R^{\frac{k-1}{2}}(f)-f\|_c \leqslant A_{k,l}\ln(2+R)\omega_2(f;R^{-1});$$

if $l=1$ then

$$\| {}^lS_R^{\frac{k-1}{2}}(f)-f\|_c \leqslant A_k\ln(2+R)\omega_2(f;R^{-1})+A_kR^{-1}\int_{R^{-1}}^1 \frac{\omega_2(f;t)}{t^2}\mathrm{d}t$$

$$\leqslant A_k\ln(2+R)\omega_1(f;R^{-1}).$$

Notice that when $l=1$ the modulus of continuity used in right-hand side of the inequality is of first order.

For the proof we only need to quote a result in [6], that is

$$\| S_R^{\frac{k-1}{2}}(f)-f\|_c \leqslant A_k\ln(2+R)\omega_2(f;R^{-1}).$$

§ 5. For critical index the approximation of continuous function on set of total measure

Let $P(x)$ be a homogeneous harmonic polynomial of degree $p(p\in \mathbf{N})$. We denote by ${}^l\widetilde{S}_R^\alpha$ and f^* the conjugates, with respect to the kernel $K(x)=P(x)|x|^{-k-p}(x\neq 0)$, of the operator ${}^lS_R^\alpha$ and the function f respectively. Here the concept of "conjugate" is in Calderon-Zygmund sence (see [8] for the detail). In this section we want to generalize the result in [6] which is about the approximation on set of total measure by the operator $S_R^{\frac{k-1}{2}}$ and $\widetilde{S}_R^{\frac{k-1}{2}}$ to the case of ${}^lS_R^{\frac{k-1}{2}}$ and ${}^l\widetilde{S}_R^{\frac{k-1}{2}}$, $(l>1)$.

Consider $f\in C(Q)$ and suppose the modulus of continuity $\omega(f;\ t)$ satisfies the condition

$$\lim_{t\to 0^+}\omega(f;\ t)t^{-1}=+\infty, \tag{5.1}$$

which is equivalent to that $f\notin \mathrm{Lip}\ 1$.

Define $\delta_0=1$, $\delta_{n+1}=\min\left\{\delta:\ \max\left(\frac{\omega(f;\ \delta)}{\omega(f;\ \delta_n)},\ \frac{\omega(f;\ \delta_n)\delta}{\omega(f;\ \delta)\delta_n}\right)=\frac{1}{6}\right\}$

and $\Omega(\delta)=6^{-n}$ when $\delta\in(\delta_{n+1},\ \delta_n](n\in\mathbf{N})$.

Theorem 7 Let $f\in C(Q)$ satisfy the condition (5.1), $\{\delta_n\}$ and Ω be defined as the above, If $l>2$ then

(i) $\max(|{}^lS_R^{\frac{k-1}{2}}(f;\ x)-f(x)|,\ |{}^l\widetilde{S}_R^{\frac{k-1}{2}}(f;\ x)-f^*(x)|)$

$$\leqslant C(x)\omega(f;\ R^{-1})\ln\ln\left\{9\left[\Omega\left(\frac{1}{R}\right)\right]^{-1}\right\},\quad R\geqslant 1, \tag{5.2}$$

where $C(x)$ is finite almost everywhere and

$$|\{x\in Q:\ C(x)>\lambda\}|\leqslant Ae^{-\frac{\lambda}{A}} \tag{5.3}$$

for and $\lambda>0$.

(ii) $$\varlimsup_{R\to+\infty}\frac{\max\left(\left|{}^{l}S_R^{\frac{k-1}{2}}(f;\ x)-f(x)\right|,\ \left|{}^{l}\widetilde{S}_R^{\frac{k-1}{2}}(f;\ x)-f^*(x)\right|\right)}{\omega\left(f;\ \frac{1}{R}\right)\ln\ln\left[\frac{9}{\Omega\left(\frac{1}{R}\right)}\right]}$$

$$\leqslant A,\ \text{a. e.} \tag{5.4}$$

where the constant A depends only on k, l and P.

For convenience we define

$${}^{l}\rho_R^{\alpha}(f;\ x)=\max\left(\left|{}^{l}S_R^{\alpha}(f;\ x)-f(x)\right|,\ \left|{}^{l}\widetilde{S}_R^{\alpha}(f;\ x)-f^*(x)\right|\right)$$

and use ρ_R^{α} instead of ${}^{2}\rho_R^{\alpha}$. Let $\gamma_0=\int_{\mathbf{R}^k}{}^{l}\gamma_k^{\alpha}(|y|)\,\mathrm{d}y$. We can find that $\gamma_0=1-\sum_{j=0}^{2q+1}b_j$. Define

$$\theta_R^{l}(f)(x)=\max\left(\left|{}^{l}\Gamma_R^{\frac{k-1}{2}}(f)(x)-\gamma_0 f(x)\right|,\ \left|{}^{l}\Gamma_R^{\frac{k-1}{2}}(f^*)(x)-\gamma_0 f^*(x)\right|\right).$$

Lemma 5 Suppose $f\in C(Q)$ and the condition (5.1) holds. If $l>2$ then (5.2)~(5.4) hold for $\theta_R^{l}(f)(x)$ in place of ${}^{l}\rho_R^{\frac{k-1}{2}}(f)(x)$.

Proof When $k=1$, ${}^{l}\Gamma_R^{\frac{1}{2}k-\frac{1}{2}}=0$ identically. Hence we need only to consider the case of $k>1$. Because $l>2$ so we have

$${}^{l}\Gamma_R^{\frac{k-1}{2}}(f)(x)-\gamma_0 f(x)=\int_0^{+\infty}f_x(tR^{-1})t^{k-1}\,{}^{l}\gamma_K^{\frac{k-1}{2}}(t)\,\mathrm{d}t$$

$$=O(\omega_2(f;R^{-1}))+O\left(\int_1^{+\infty}\omega_2(f;tR^{-1})t^{-1-1}\,\mathrm{d}t\right)=O(\omega_2(f;R^{-1})), \tag{5.5}$$

where "O" depends only on k and l.

Now we use the letter T to denote the operator of "conjugate", that is $T(g)=g^*$ for any $g\in L(Q)$. Abbreviate Γ_R for ${}^{l}\Gamma_R^{\frac{1}{2}k-\frac{1}{2}}$ and define

$$E_R(\lambda)=\left\{x\in Q:\left|\Gamma_R(f^*)(x)-\gamma_0 f^*(x)\right|\right.$$

$$\left.>\lambda\omega\left(f;\frac{1}{R}\right)\ln\ln\left[\frac{9}{\Omega\left(\frac{1}{R}\right)}\right]\right\},\quad R\geqslant 1,\ \lambda>0,$$

$$\Delta_n(\lambda) \bigcup_{R\in[\delta_n^{-1},\delta_{n+1}^{-1})} E_R(\lambda), \quad n\in\mathbf{N}.$$

For fixed $n\in\mathbf{N}^*$, we consider the following two cases. Case (l), $\omega(f;\ \delta_{n+1})=6^{-1}\omega(f;\ \delta_n)$. Let $g(x)$ be the trigonometric polynomial of best approximation for f with degree δ_n^{-1}. Then

$$\Gamma_R(f^*)-\gamma_0 f^* =\Gamma_R(f^*-g^*)+\Gamma_R(g^*)-\gamma_0 g^*+\gamma_0(g^*-f^*).$$

Set $h=\dfrac{f-g}{\| f-g\|_c}$. We have

$$|(\Gamma_R(f^*)-\gamma_0 f^*)(x)|\leqslant \| f-g\|_c[|\Gamma_R(T(h))(x)|+ \gamma_0|T(h)(x)|]+|\Gamma_R(g^*)(x)-\gamma_0 g^*(x)|.$$

For $R\in[\delta_n^{-1},\ \delta_{n+1}^{-1})$,

$$\| f-g\|_c\leqslant A_k\omega(f;\ \delta_n)=6A_k\omega(f;\ \delta_{n+1})\leqslant A_k\omega(f;\ R^{-1}).$$

Hence by (4.8) we obtain that

$$|\Gamma_R(f^*)(x)-\gamma_0 f^*(x)| \leqslant A_{k,l}\omega(f;\ R^{-1})HL(T(h))(x)+|\Gamma_R(g^*)(x)-\gamma_0 g^*(x)|. \tag{5.6}$$

Integrating by parts we obtain that

$$\Gamma_R(g^*)(x)-\gamma_0 g^*(x)=R^{-1}\int_0^{+\infty}(g_x^*)'\left(\frac{t}{R}\right)I(F_l^{\frac{k-1}{2}})(t)\mathrm{d}t,$$

where I, $F_l^{\frac{1}{2}k-\frac{1}{2}}$ are the same as in (3.6) and Lemma 4. Since g is a trigonometric polynomial, $\dfrac{\partial T(g)}{\partial x_\nu}=T\left(\dfrac{\partial g}{\partial x_\nu}\right)$. Therefore

$$(g_x^*)'\left(\frac{t}{R}\right)=T\left(\sum_{\nu=1}^k\int_{|\xi|=1}g_\nu\left(x-\frac{t\xi}{R}\right)(-\xi_\nu)\mathrm{d}\sigma(\xi)\right),$$

where g_ν denotes $\dfrac{\partial g}{\partial x_\nu}$. Since $I(F_l^{\frac{1}{2}k-\frac{1}{2}})(t)$ is bounded for $t\in[0,\ 1]$ and $I(F_l^{\frac{1}{2}k-\frac{1}{2}})(t)=o(t^{-1})$ as $t\to+\infty$, we have

$$\left|\int_0^{+\infty}(g_x^*)'\left(\frac{t}{R}\right)I(F_l^{\frac{1}{2}k-\frac{1}{2}})(t)\mathrm{d}t\right| \leqslant A_{k,l}\sum_{\nu=1}^k HL[T(g_\nu)](x)\leqslant A_{A,l}\cdot\sum_{\nu=1}^k\| g_\nu\|_c HL\left[T\left(\frac{g_\nu}{\| g_\nu\|_c}\right)\right](x).$$

Since $\displaystyle\sum_{\nu=1}^k\| g_\nu\|_c\leqslant A_k\delta_n^{-1}\omega(f;\delta_n)\leqslant A_kR\omega\left(f;\frac{1}{R}\right)$,

therefore,

$$|\Gamma_R(g^*)(x)-\gamma_0 g^*(x)|\leqslant A_{k,l}\omega\left(f;\frac{1}{R}\right)\sum_{\nu=1}^{k}HL\left[T\left(\frac{g_\nu}{\|g_\nu\|_c}\right)\right](x).$$

Substituting this into (5.6) and writing $\overline{g}_\nu=\frac{g_\nu}{\|g_\nu\|_c}$ we obtain that

$$|\Gamma_R(f^*)(x)-\gamma_0 f^*(x)|$$
$$\leqslant A_{k,l}\omega(f;R^{-1})\{HL(T(h))(x)+\sum_{\nu=1}^{k}HL(T(\overline{g}_\nu))(x)\}$$
$$\text{for } R\in[\delta_n^{-1},\ \delta_{n+1}^{-1}). \tag{5.7}$$

Case (2) $\frac{\delta_{n+1}}{\omega(f;\ \delta_{n+1})}=\frac{\delta_n}{6\omega(f;\ \delta_n)}$. Let g be the trigonometric polynomial of best approximation of f with degree δ_{n+1}^{-1}. Then for $R\in[\delta_n^{-1}$, $\delta_{n+1}^{-1})$ we have

$$R\omega(f;\ R^{-1})\geqslant\frac{1}{2}\omega(f;\ \delta_n)\delta_n^{-1}=\frac{1}{12}\omega(f;\ \delta_{n+1})\delta_{n+1}^{-1},$$

and we derive (5.7) similarly as in case (1).

Now let $E(\lambda)=\{x\in Q:\ HL[T(h)](x)>\lambda\}$ and $F_\nu(\lambda)=\{x\in Q:\ HL[T(\overline{g}_\nu)](x)>\lambda\}$, $\nu=1,2,\cdots,k$, in both cases. Noticing $\Omega(R^{-1})=6^{-n}$ and $\ln\ln\left[\frac{9}{\Omega(R^{-1})}\right]\geqslant\ln(n+2)$ for $R\in[\delta_n^{-1},\ \delta_{n+1}^{-1})$, we derive from (5.7) that

$$E_R(\lambda)\subset E[A_{k,l}\lambda\ln(n+2)]\cup\left[\bigcup_{\nu=1}^{k}F_\nu(A_{k,l}\lambda\ln(n+2))\right],\quad\lambda>0. \tag{5.8}$$

For any function $\varphi(x)\in L^\infty(Q)$ and $q>3$ we have

$$\|HL[T(\varphi)]\|_q\leqslant A_k\|T(\varphi)\|_q\leqslant A_k(P)q\|\varphi\|q.$$

Consequently,

$$|\{x\in Q:\ HL[T(\varphi)](x)>\lambda\}|\leqslant A\exp\left(-\frac{\lambda}{A\|\varphi\|_\infty}\right),\qquad\lambda>0. \tag{5.9}$$

By (5.9) we obtain that

$$|E(A_{k,l}\ln(2+n))|\leqslant Ae^{-\left[\frac{\lambda}{A}\ln(2+n)\right]},$$
$$|E_\nu(A_{k,l}\ln(2+n))|\leqslant Ae^{-\left[\frac{\lambda}{A}\ln(2+n)\right]}.$$

Hence

$$|\Delta_n(\lambda)| \leqslant \left|\bigcup_{R\in[\delta_n^{-1},\delta_{n+1}^{-1})} E_R(\lambda)\right| \leqslant A_0(n+2)^{-\frac{\lambda}{A_0}},\quad \lambda>0, \tag{5.10}$$

where $A_0=A_{k,l}(P)$.

Define $C(x)=\sup\limits_{R>0}\frac{|\Gamma_R(f^*)(x)-\gamma_0 f^*(x)|}{\omega(f;\ R^{-1})\ln\ln[9(\Omega(\mathrm{R}^{-1}))^{-1}]}$, Then

$$\{x\in Q:\ C(x)>\lambda\}\subset\bigcup_{n=0}^{+\infty}\Delta_n(x).$$

If $\lambda>2A_0$ (A_0 is the constant in the right-hand side of (5.10)) then by (5.10).

$$|\{x\in Q: C(x)>\lambda\}| \leqslant A_0\sum_{n=0}^{+\infty}(n+2)^{-\frac{\lambda}{A_0}} \leqslant A\mathrm{e}^{-\frac{\lambda}{A}},$$

and

$$\left|\left\{x\in Q:\ \overline{\lim_{R\to+\infty}}\frac{|\Gamma_R(f^*)(x)-\gamma_0 f^*(x)|}{\omega(f;\ R^{-1})\ln\ln[9(\Omega(R^{-1}))^{-1}]}\geqslant 2A_0\right\}\right|$$

$$\leqslant \lim_{n\to+\infty}\sum_{j=n}^{+\infty}|\Delta_j(2A_0)|=0.$$

A combination of these results and (5.5) proves Lemma 5.

Lemma 6 Suppose function $f\in C(Q)$ satisfies the condition (5.1) and $\alpha\geqslant\frac{1}{2}k-\frac{1}{2}$. Then (5.2)~(5.4) hold for $\rho_R^\alpha(f)(x)$ in place of ${}^l\rho_R^{\frac{1}{2}k-\frac{1}{2}}(f)(x)$.

Proof In the case of $\alpha=\alpha_0=\frac{1}{2}k-\frac{1}{2}$ the conclusion is known (see [6].) Assume $\alpha=\alpha_0+\delta$ and $\delta>0$. We make use of Lemma 4 in [3] and obtain that

$$S_R^\alpha(f;x)=[B(\alpha_0+1,\delta)]^{-1}\int_0^{-1}(1-t)^{\delta-1}t^{\alpha_0}S_{R\sqrt{t}}^{\alpha_0}(f;x)\mathrm{d}t.$$

Therefore

$$\rho_R^\alpha(f)(x)\leqslant\frac{1}{B(\alpha_0+1,\delta)}\int_0^1(1-t)^{\delta-1}t^{\alpha_0}\rho_{R\sqrt{t}}^{\alpha_0}(f)(x)\mathrm{d}t.$$

Since we have already proved that

$$\rho_{R\sqrt{t}}^{\alpha_0}(f)(x)\leqslant C(x)\omega(f;\ R^{-1}t^{-\frac{1}{2}})\ln\ln[9(\Omega(R^{-1}t^{-\frac{1}{2}}))^{-1}]$$
$$\leqslant C(x)\omega(f;\ R^{-1})\ln\ln[9(\Omega(R^{-1}))^{-1}]2t^{-\frac{1}{2}},$$

for $t\geqslant R^{-2}$, we can conclude that

$$\rho_R^{\alpha}(f)(x) \leqslant C(x)\omega(f;R^{-1})\ln\ln[9(\Omega(R^{-1}))^{-1}]2\frac{\Gamma\left(\alpha_0+\frac{1}{2}\right)\Gamma(\alpha_0+\delta+1)}{\Gamma(\alpha_0+1)\Gamma\left(\alpha_0+\delta+\frac{1}{2}\right)}.$$

This completes the proof of Lemma 6.

Proof of theorem 7 By Theorem 5 we have

$$^{l}\rho_R^{\frac{1}{2}k-\frac{1}{2}}(f)(x) \leqslant \sum_{j=0}^{2q+1}|b_j|\rho_{R^0}^{\alpha+j}(f)(x)+\theta_R^l(f)(x).$$

Hence the desired result follows from Lemma 5 and Lemma 6. □

Remark The proof of Theorem 7 does not hold in the case of $l=1$. For this case another argument should be made.

References

[1] Cheng Minde, Chen Yonghe. On the approximation of functions of several variables by trigonometrical polynomials. Acta Scientiarum Naturalium Universitatis Pekinensis. 1956, 2: 411—428.

[2] Stein E M, Weiss G. Introduction to Fourier analysis on Euclidean spaces. Princeton Univ. Press. 1971.

[3] Stein, E M. Localization and Summablility of multiple Fourier series. Acta Mathematica, 1958, 100: 93—147.

[4] Bateman H. Tables of Integral Transforms, Ⅱ. New York, 1954.

[5] Magnus W, Oberhettinger F, Soni R P, Formulas and Theorems for the Special Functions of Mathematical Physics. Springer-Verlag, 1966.

[6] Wang Kunyang. Approximation for continuous periodic function of several variables and its conjugate function by Rieans means on set of total measure. Approximation Theory and Its Applications, 1985, 1: 19—56.

[7] Guo Zhurui. Approximation for continuous function by the Cesàro means of its Fourier series. Acta Mathematica, 1962, 12: 320—329.

[8] Calderon A P, Zygmund A. Singular integrals and periodic functions. Studia Math., 1954, 14: 249—271.

Approx. Theory and Its Appl., 1989, 5(4): 69—77.

L^2-函数用其球型 Fourier 和在全测度集上逼近

On The Approximation of L^2-Functions by Their Spherical Fourier Sums on Set of Full Measure[①]

Abstract In this note, we study the approximation of L^2-functions by their spherical Fourier sums on a set of full measure, and get main three theorems.

Let $Q^n=\{(x_1, x_2, \cdots, x_n)\mid -\pi\leqslant x_j<\pi, j=1,2,\cdots,n\}$, and $\mu>0$. For a function $f\in L^2(Q^n)$ we always denote its Fourier coefficients by $c_m(f)$, $(m\in\mathbf{Z}^n)$. Define two subclasses of $L^2(Q^n)$:

$$A_\mu=A_\mu(Q^n)=\Big\{f\in L^2(Q^n)\mid \|f\|_{A_\mu}\overset{\text{def}}{=\!=}\Big[\sum_{m\in\mathbf{Z}^n}|c_m(f)|^2\ln^{2\mu}(|m|+\mathrm{e})\Big]^{\frac{1}{2}}<+\infty\Big\},$$

$$W_\mu=W_\mu(Q^n)=\Big\{f\in L^2(Q^n)\mid \|f\|_{W_\mu}\overset{\text{def}}{=\!=}\Big[|c_0(f)|^2+\sum_{m\in Z'}|c_m(f)|^2|m|^{2\mu}\Big]^{\frac{1}{2}}<+\infty\Big\}.$$

with tha norms $\|\cdot\|_{A_\mu}$ and $\|\cdot\|_{W_\mu}$, A_μ and W_μ compose respectively Banach spaces. And obviously,

$L^2(Q^n)\supset A_\mu(Q^n)\supset A_\lambda(Q^n)\supset W_\mu(Q^n)\supset W_\lambda(Q^n)$, as $\lambda>\mu>0$.

The Bochner-Riesz means of order α of $f\in L^2(Q^n)$ are defined by

① Supported by NSFC.
Received: 1988-01-11; Revised: 1989-02-06.

$$S_R^\alpha(f)(x)=\sum_{|m|<R}\left(1-\frac{|m|^2}{R^2}\right)^\alpha c_m(f)\mathrm{e}^{imx},\quad R>0.$$

The means $S_R^0(f)(R>0)$ are just the spherical Fourier sums of f and are simply written as $S_R(f)$.

It is known from the general theory of orthogonal series (see[1], p. 190) that for $n=1$ the Fourier sums of A_μ-function converge almost everywhere when $\mu\geqslant 1$. The aim of the present paper is to discuss the order of $|S_R(f)-f|$ on set of full measure for $f\in A_\mu$ and $\mu\geqslant 1$.

In the case of approximation by Bochner-Riesz means of positive order, Wang Shiming[2] proved the following

Theorem A If $f\in W_\mu(Q^n)$ and $0<\mu\leqslant 2$, $\alpha>0$, then

$$S_R^\alpha(f)(x)-f(x)=\begin{cases}O(R^{-2}), & \text{when } \mu=2,\\ o(R^{-\mu}), & \text{when } 0<\mu<2,\end{cases}\quad \text{a. e.},\ R\to+\infty.$$

Now our results are the following three theorems.

Theorem 1 Let $R_k>0$ and $R_{k+1}R_k^{-1}\geqslant q>1$ for all $k\in\mathbf{N}$ and some fixed q. If $f\in A_\mu(Q^n)$ and $\mu\geqslant\frac{1}{2}$ then

$$S_{R_k}(f)(x)-f(x)=o(\ln^{\frac{1}{2}-\mu}R_k)\quad \text{a. e.},\quad k\to+\infty.$$

Theorem 2 If $f\in A_\mu(Q^n)$ and $\mu\geqslant 1$, then

$$S_R(f)(x)-f(x)=o(\ln^{1-\mu}R)\quad \text{a. e.},\quad R\to+\infty.$$

Theorem 3 If $f\in W_\mu(Q^n)$ and $0<\mu\leqslant 2$, then

$$S_R(f)(x)-f(x)=o(R^{-\mu}\ln R)\quad \text{a. e.},\quad R\to+\infty.$$

There is an open problem concerning Theorem 1: Can we improve this theorem to get the conclusion

$$S_{R_k}(f)(x)-f(x)=o(\ln^{-\mu}R_k)\quad \text{a. e.},\quad k\to+\infty$$

with the condition $\mu>0$ in place of $\mu\geqslant\frac{1}{2}$?

Lemma 1 Let $\mu\geqslant\frac{1}{2}$ and $f\in A_\mu$. Define

$$M_{\mu-\frac{1}{2}}(f)(x)=\sup_{\delta>0}\frac{1}{\delta}\ln^{\frac{-1}{2}}\left(\frac{1}{\delta}+\mathrm{e}\right)\int_0^\delta|f_x(t)-f(x)|\,\mathrm{d}t,$$

where

$$f_x(t)=\frac{1}{|S^{n-1}|}\int_{S^{n-1}} f(x+ty)\mathrm{d}\sigma(y), S^{n-1}=\{y\in\mathbf{R}^n: |y|=1\}.$$

Then

$$\|M_{\mu-\frac{1}{2}}(f)\|_2\leqslant c_{n,\mu}\|f\|_{A_\mu}.$$

Proof By HL we denote the Hardy-Littlewood maximal operator. Then we have

$$\begin{aligned}M_{\mu-\frac{1}{2}}(f)(x)&\leqslant\sup_{1>\delta>0}\frac{1}{\delta}\ln^{-\frac{1}{2}}\left(\frac{1}{\delta}+\mathrm{e}\right)\int_0^\delta|f_x(t)-f(x)|\mathrm{d}t+\\&\quad c_{n,\mu}(HL(f)(x)+|f(x)|)\\&\leqslant\left\{\int_0^1\frac{1}{t}\ln^{2\mu-1}\left(\frac{1}{t}+\mathrm{e}\right)|f_x(t)-f(x)|\mathrm{d}t\right\}^{\frac{1}{2}}+\\&\quad c_{n,\mu}(HL(f)(x)+|f(x)|).\end{aligned}$$

Set $N_{\mu-\frac{1}{2}}(f)(x)=\left\{\int_0^1\frac{1}{t}\ln^{2\mu-1}\left(\frac{1}{t}+\mathrm{e}\right)|f_x(t)-f(x)|\mathrm{d}t\right\}^{\frac{1}{2}}$. By the type (2, 2) of HL we see that it is sufficient to prove

$$\|N_{\mu-\frac{1}{2}}(f)\|_2\leqslant c_{n,\mu}\|f\|_{A_\mu}.$$

Since the Fourier coefficients of $f_x(t)$, as a function of x, are

$$c_m(f)\frac{1}{|S^{n-1}|}\int_{S^{n-1}}\mathrm{e}^{-\mathrm{i}tmy}\mathrm{d}\sigma(y)=c_m(f)\frac{2^{\frac{1}{2}n-1}\Gamma\left(\frac{1}{2}n\right)}{(|m|t)^{\frac{1}{2}n-1}}J_{\frac{1}{2}n-1}(|m|t),$$

(see[3]), we have

$$\|f_x(t)-f(x)\|_2^2=(2\pi)^n\sum_{m\neq 0}|c_m(f)|^2\left[\frac{2^{\frac{1}{2}n-1}\Gamma\left(\frac{1}{2}n\right)}{(|m|t)^{\frac{1}{2}n-1}}J_{\frac{1}{2}n-1}(|m|t)-1\right]^2.$$

By the properties of Bessel functions of first kind, we know that

$$\left|\frac{2^{\frac{1}{2}n-1}\Gamma\left(\frac{1}{2}n\right)}{(|m|t)^{\frac{1}{2}n-1}}J_{\frac{1}{2}n-1}(|m|t)-1\right|\leqslant c_n|m|^2t^2$$

and

$$\int_0^{\frac{1}{|m|}}|m|^2t\ln^{2\mu-1}\left(\frac{1}{t}+\mathrm{e}\right)\mathrm{d}t\leqslant\ln^{2\mu-1}(|m|+\mathrm{e}),\quad |m|\neq 0.$$

On the other hand, when $t\geqslant|m|^{-1}$, we have

$$\frac{2^{\frac{1}{2}n-1}\Gamma\left(\frac{1}{2}n\right)}{(|m|t)^{\frac{1}{2}n-1}}J_{\frac{1}{2}n-1}(|m|t)-1|\leqslant c_n$$

and

$$\int_{\frac{1}{|m|}}^{1} \frac{1}{t}\ln^{2\mu-1}\left(\frac{1}{t}+\mathrm{e}\right)\mathrm{d}t \leqslant c\ln^{2\mu}(|m|+\mathrm{e}).$$

Therefore

$$\|N_{\mu-\frac{1}{2}}(f)\|_2^2 \leqslant c_{n,\mu}\|f\|_{A_\mu}^2.$$

Remark The author does not know whether $\|M_\mu(f)\|_2 \leqslant c_{n,\mu}\|f\|_{A_\mu}^2$ for all $\mu>0$.

Lemma 2 Let $R_1\geqslant \mathrm{e}$, $\frac{R_{k+1}}{R_k}\geqslant q>1\,(k\in\mathbf{N})$, $\mu\geqslant 0$, and $f\in A_\mu$. Define

$$U_\mu(f)(x)=\sup_{k\in\mathbf{N}}\ln^\mu R_k\,|S_{R_k}(f)(x)-S_{R_k}^n(f)(x)|.$$

Then

$$\|U_\mu(f)\|_2 \leqslant c_{n,q,\mu}\|f\|_{A_\mu}.$$

The symbol $S_R^\alpha(f)$ denotes the Bochner-Riesz means of order α of f, i. e.,

$$S_R^\alpha(f)(x)=\sum_{|m|<R} c_m(f)\left(1-\frac{|m|^2}{R_2}\right)^\alpha \mathrm{e}^{imx},\quad R>0.$$

Proof For simplicity we assume $R_k=\mathrm{e}^k$ and write

$$\Delta_k(f)=k^\mu\,|S_{\mathrm{e}^k}(f)-s_{\mathrm{e}^k}^n(f)|.$$

Then

$$\|\Delta_k(f)\|_2^2 \leqslant c_n k^{2\mu}\sum_{0<|m|<\mathrm{e}^k}|c_m(f)|^2|m|^4\mathrm{e}^{-4k},$$

$$\|U_\mu(f)\|_2^2 \leqslant \sum_{k=1}^{+\infty}\|\Delta_k(f)\|_2^2 \leqslant c_n\sum_{|m|>0}\Big(\sum_{k>\ln|m|}^{+\infty}k^{2\mu}\mathrm{e}^{-4k}\Big)|m|^4|c_m(f)|^2$$
$$\leqslant c_{n,\mu}^2\|f\|_{A_\mu}^2.$$

Lemma 3 Let $\mu\geqslant\frac{1}{2}$ and $f\in A_\mu$. Define

$$V_{\mu-\frac{1}{2}}(f)(x)=\sup_{R>\mathrm{e}}\ln^{\mu-\frac{1}{2}}R\,|S_R^n(f)(x)-f(x)|.$$

Then

$$\|V_{\mu-\frac{1}{2}}(f)\|_2 \leqslant c_{n,\mu}\|f\|_{A_\mu}.$$

Proof By Bochner's formula (see[3]) we have

$$S_R^n(f)(x)-f(x)=c_n\int_0^{+\infty}(f_x(t)-f(x))t^{n-1}R^nJ_{\frac{3}{2}n}(Rt)(Rt)^{-\frac{3}{2}n}\mathrm{d}t.$$

Using the estimation

$$|J_v(u)u^{-v}|\leqslant c_v(u+1)^{-\frac{1}{2}-v},\qquad v>0,\ u>0,$$

we get

$$|S_R^n(f)(x)-f(x)|\leqslant c_n\int_0^{+\infty}|f_x(t)-f(x)|R(Rt+1)^{-\frac{n+3}{2}}\mathrm{d}t.$$

Writing $\varphi_x(t)=\int_0^l |f_x(s)-f(x)|\mathrm{d}s$ and integrating by parts we calculate

$$|S_R^n(f)(x)-f(x)|\leqslant c_n\int_0^{+\infty}\varphi_a(t)R^2(Rt+1)^{-\frac{n+5}{2}}\mathrm{d}t$$

$$\leqslant c_nM_{\mu-\frac{1}{2}}(f)(x)\int_0^{+\infty}t(Rt+1)^{-\frac{n+5}{2}}\ln^{\frac{1}{2}-\mu}\left(\frac{1}{t}+\mathrm{e}\right)\mathrm{d}tR^2$$

$$\leqslant c_{n,\mu}M_{\mu-\frac{1}{2}}(f)(x)\ln^{\frac{1}{2}-\mu}R,\qquad R\geqslant \mathrm{e}.$$

Hence an application of Lemma 1 completes the proof.

Proof of Theorem 1　Set

$$T^\mu(f)(x)=\sup_{R_k\geqslant \mathrm{e}}\ln^\mu R_k|S_{R_k}(f)(x)-f(x)|,\qquad \mu\geqslant 0.$$

We use the notations of Lemma 2 and 3 and get

$$T^\mu(f)\leqslant U_\mu(f)+V_\mu(f),\qquad \mu\geqslant 0.$$

Hence by these lemmas we get $\qquad \|T^{\mu-\frac{1}{2}}(f)\|_2\leqslant c_{n,\mu}\|f\|_{A_\mu}.$

From this inequality and the density of trigonometric polynomials in A_μ the conclusion of Theorem 1 is deduced immediately.

For the proof of Theorem 2 we introduce the following notations:

$$J=\{R\,|\,R=|m|\quad \text{for some}\quad m\in\mathbf{N}^n\},$$

$$J_k=J\cap(\mathrm{e}^k,\ \mathrm{e}^{k+1}),\qquad k\geqslant 4.$$

J is a subsequence of $\mathbf{N}$. So that for any $R\in J_k$, ($k\geqslant 4$) there must be integers in J which are strictly less than R. Denote by $s=s(R)$ the greatest of them.

For a function $f\in L^2(Q^n)$, define

$$f_k(x)=\sum_{\mathrm{e}^k\leqslant|m|<\mathrm{e}^{k+1}}c_m(f)\mathrm{e}^{\mathrm{i}mx},\qquad k\geqslant 4.$$

If $R\in J_k$ then obviously

$$S_R(f)-S_{\mathrm{e}^k}(f)=S_R(f_k).$$

Assume $\mu\geqslant 0$ and $f\in A_\mu$. Define

$$W_k^\mu(f)(x)=\sup_{R\in J_k}\ln^\mu R\,|S_R(f_k)(x)|,$$

$$H_{kj}^\mu(f)(x)=\sup_{R\in J_k}\ln^\mu R\,|S_R^{j-1}(f_k)(x)-s_R^j(f_k)(x)|,\qquad j=1,2,\cdots,n.$$

Lemma 4 Let $\mu\geqslant 0$. If $f\in A_{\mu+1}$ then

$$\|H_{k,j}^\mu(f)\|_2\leqslant c_{n,\mu}\|f_k\|_{A_{\mu+1}},\qquad j=1,2,\cdots,n.$$

Proof For $k\geqslant 4$, $j\in\mathbf{N}$, $r>0$ and $R>r$, we have

$$S_r^{-\frac{3}{2}+\frac{1}{k}+j}(f;\ x)-S_r^{-\frac{1}{2}+\frac{1}{k}+j}(f;\ x)$$

$$=\sum_{|m|<r}\left(-\frac{|m|^2}{r^2}\right)^{-\frac{3}{2}+\frac{1}{k}+j}\frac{|m|^2}{r^2}\mathrm{e}^{\mathrm{i}mx}c_m(f)$$

$$=\sum_{|m|<R}\left(1-\frac{|m|^2}{r^2}\right)^{-\frac{3}{2}+\frac{1}{k}+j}\frac{|m|^2}{r^2}\chi_{(|m|,R)}(R)\mathrm{e}^{\mathrm{i}mx}c_m(f).$$

So we get

$$R^{-2j}\int_0^R(R^2-r^2)^{-\frac{1}{2}-\frac{1}{k}}r^{2(\frac{1}{k}+j)}[S_r^{-\frac{3}{2}+\frac{1}{k}+j}(f)-S^{-\frac{1}{2}+\frac{1}{k}+j}(f)]\mathrm{d}r$$

$$=R^{-2j}\sum_{|m|<R}\int_0^R(R^2-r^2)^{-\frac{1}{2}-\frac{1}{k}}r^{2(\frac{1}{k}+j)}\frac{|m|^2}{r^2}\left(1-\frac{|m|^2}{r^2}\right)^{-\frac{3}{2}+\frac{1}{k}+j}\cdot$$

$$\chi_{(|m|,R)}(r)\mathrm{d}rc_m(f)\mathrm{e}^{\mathrm{i}mx}$$

$$=R^{-2j}\sum_{|m|<R}\int_{|m|/R}^1(1-r^2)^{-\frac{1}{2}-\frac{1}{k}}r^{2(\frac{1}{k}+j)}\frac{|m|^2}{R^2r^2}\cdot$$

$$\left(1-\frac{|m|^2}{R^2r^2}\right)^{-\frac{3}{2}+\frac{1}{k}+j}R^{-1-\frac{2}{k}+2j+1}\mathrm{d}rc_m(f)\mathrm{e}^{\mathrm{i}mx}$$

$$=\sum_{|m|<R}\int_{\frac{|m|^2}{R^2}}^1(1-t)^{-\frac{1}{2}-\frac{1}{k}}t^{\frac{1}{k}+j}\left(1-\frac{|m|^2}{R^2t}\right)^{-\frac{3}{2}+\frac{1}{k}+j}\frac{|m|^2}{R^2}\frac{1}{t}\frac{1}{2\sqrt{t}}\mathrm{d}tc_m(f)\mathrm{e}^{\mathrm{i}mx}$$

$$=\sum_{|m|<R}\frac{|m|^2}{R^2}c_m(f)\mathrm{e}^{\mathrm{i}mx}\int_{\frac{|m|^2}{R^2}}^1(1-t)^{-\frac{1}{2}-\frac{1}{k}}\left(t-\frac{|m|^2}{R^2}\right)^{-\frac{3}{2}+\frac{1}{k}+j}\frac{1}{2}\mathrm{d}t$$

$$=\sum_{|m|<R}\frac{|m|^2}{R^2}c_m(f)\mathrm{e}^{\mathrm{i}mx}\frac{1}{2}\int_0^{1-\frac{m^2}{R^2}}S^{-\frac{1}{2}-\frac{1}{k}}\left(1-\frac{|m|^2}{R^2}-s\right)^{-\frac{3}{2}+\frac{1}{k}+j}\mathrm{d}s$$

$$=\frac{1}{2}\int_0^1 s^{-\frac{1}{2}-\frac{1}{k}}(1-s)^{-\frac{3}{2}+\frac{1}{k}+j}\mathrm{d}s\sum_{|m|<R}\left(1-\frac{|m|^2}{R^2}\right)^{-\frac{1}{2}-\frac{1}{k}-\frac{3}{2}+\frac{1}{k}+j+1}c_m(f)\mathrm{e}^{\mathrm{i}mx}\frac{|m|^2}{R^2}$$

$$=\frac{1}{2}B\left(\frac{1}{2}-\frac{1}{k},-\frac{1}{2}+\frac{1}{k}+j\right)\sum_{|m|<R}\left(1-\frac{|m|^2}{R^2}\right)^{j-1}\frac{|m|^2}{R^2}c_m(f)\mathrm{e}^{\mathrm{i}mx}$$

$$=\frac{1}{c_{kj}}[S_R^{j-1}(f)-S_R^j(f)].$$

Thus, we have

$$S_R^{j-1}(f)-S_R^j(f)=$$

$$C_{k,j}R^{-2j}\int_0^R(R^2-r^2)^{-\frac{1}{2}-\frac{1}{k}}r^{2(\frac{1}{k}+j)}\left[S_r^{-\frac{3}{2}+\frac{1}{k}+j}(f)-S_r^{-\frac{1}{2}+\frac{1}{k}+j}(f)\right]\mathrm{d}r,$$

where $$C_{k,j}=2B\left(\frac{1}{2}-\frac{1}{k},\ -\frac{1}{2}+\frac{1}{k}+j\right)^{-1}.$$

Let $R\in J_k$, $k\geqslant 4$. It is easy to see that

$$\frac{3}{2}\geqslant R-s>\frac{1}{2R}>\frac{8}{R^{n+1}},\qquad R\leqslant 2S\leqslant 2R.$$

Set $$E_j=\sup_{R\in J_k}\frac{\ln^{\mu}R}{R^{2j}}\int_0^{R-\frac{1}{R^{n+1}}}(R^2-r^2)^{-\frac{1}{2}-\frac{1}{k}}r^{2(\frac{1}{k}+j)}\cdot$$

$$\left|S_r^{-\frac{3}{2}+\frac{1}{k}+j}(f_k)-S_r^{-\frac{1}{2}+\frac{1}{k}+j}(f_k)\right|\mathrm{d}r.$$

Since $$R^{-ej}\left\{\int_0^{R-\frac{1}{R^{n+1}}}(R^2-r^2)^{-1-\frac{2}{k}}r^{4(\frac{1}{k}+j)}\mathrm{d}r\right\}^{\frac{1}{2}}\leqslant\sqrt{\frac{k}{R}}R^{\frac{n+2}{k}}$$

$$\leqslant c_n(R^{-1}\ln R)^{\frac{1}{2}},\quad R\in J_n,$$

we get

$$E_j\leqslant c_n\sup_{R\in J_k}\left\{\frac{\ln^{2\mu+1}R}{R}\int_{\mathrm{e}^k}^R\left|S_r^{-\frac{3}{2}+\frac{1}{k}+j}(f_k)-S_r^{-\frac{1}{2}+\frac{1}{k}+j}(f_k)\right|^2\mathrm{d}r\right\}^{\frac{1}{2}}$$

$$\leqslant c_n\left\{\int_{\mathrm{e}^k}^{\mathrm{e}^{k+1}}\frac{\ln^{2\mu+1}r}{r}\left|S_r^{-\frac{3}{2}+\frac{1}{k}+j}(f_k)-S_r^{-\frac{1}{2}+\frac{1}{k}+j}(f_k)\right|^2\mathrm{d}r\right\}^{\frac{1}{2}},$$

and

$$\|E_j\|_2^2\leqslant c_n\sum_{\mathrm{e}^k\leqslant|m|<\mathrm{e}^{k+1}}|c_m(f)|^2\int_{|m|}^{\mathrm{e}^{k+1}}\left(1-\frac{|m|^2}{r^2}\right)^{-1+\frac{2}{k}}\frac{|m|^4}{r^5}\ln^{2\mu+1}r\mathrm{d}r$$

$$\leqslant c_{n,\mu}\|f_k\|_{A_{\mu+1}}^2,\qquad j=1,2,\cdots,n.$$

On the other hand

$$\sup_{R\in J_k}\frac{\ln^{\mu}R}{R^{2j}}\int_{R-\frac{1}{R^{n+1}}}^R(R^2-r^2)^{-\frac{1}{2}-\frac{1}{k}}r^{2(\frac{1}{k}+j)}\left|S_r^{-\frac{3}{2}+\frac{1}{k}+j}(f_k)-S_r^{-\frac{1}{2}+\frac{1}{k}+j}(f_k)\right|\mathrm{d}r$$

$$\leqslant\sup_{R\in J_k}R^{-\frac{1}{2}+\frac{1}{k}}\ln^{\mu}R\int_{R-\frac{1}{R^{n+1}}}^R(R-r)^{-\frac{1}{2}-\frac{1}{k}}\sum_{|m|<r}\left(1-\frac{|m|^2}{r^2}\right)^{-\frac{1}{2}+\frac{1}{k}}|c_m(f_k)|\mathrm{d}r.$$

Observing $r-|m|\geqslant R-R^{-(n+1)}-s>(4r)^{-1}$ and $R^{\frac{1}{k}}\leqslant\mathrm{e}^{(k+1)\frac{1}{k}}\leqslant\mathrm{e}^2$ we know that the right-hand side of the above inequality does not exceed

$$c_{\mu} \sum_{e^k \leqslant |m| < e^{k+1}} |c_m(f)| k^{\mu} e^{-k^{\frac{n+1}{2}}} \leqslant c_{n,\mu} \| f_k \|_{A_{\mu+1}}.$$

Combining all of the above results we complete the proof of Lemma 4.

Lemma 5 Let $\mu \geqslant 0$. If $f \in A_{\mu+1}$ then $\| W_k^{\mu}(f) \|_2 \leqslant c_{n,\mu} \| f_k \|_{A_{\mu+1}}$.

Proof From the equation

$$S_R(f_k) = \sum_{j=1}^{n} [S_R^{j-1}(f_k) - S_R^{j}(f_k)] + s_R^{n}(f_k) - f_k + f_k,$$

we deduce that

$$W_k^{\mu}(f) \leqslant \sum_{j=1}^{n} H_{k,j}^{\mu}(f) + V_{\mu}(f_k) + (k+1)^{\mu} |f_k|.$$

We have

$$\begin{aligned} \| (k+1)^{\mu} |f_k| \|_2^2 &= \int (k+1)^{2\mu} f_k \cdot \bar{f}_k \mathrm{d}x \\ &= (k+1)^{2\mu} \int \sum_{e^k \leqslant |m| < e^{k+1}} c_m(f) e^{imx} \sum_{e^k \leqslant |m| < e^{k+1}} \bar{c}_m(f) e^{-imx} \mathrm{d}x \\ &= (k+1)^{2\mu} \sum_{e^k \leqslant |m| < e^{k+1}} |c_m(f)|^2 \\ &\leqslant c_{\mu}^2 \sum_{e^k \leqslant |m| < e^{k+1}} |c_m(f)|^2 \ln^{2\mu}(|m| + e) = c_{\mu}^2 |f|_{A_{\mu}}^2 \end{aligned}$$

Hence an application of Lemma 3 and 4 yields the conclusion.

Proof of Theorem 2 By Lemma 5 we get

$$\int_{Q^n} \sum_{k=4}^{+\infty} |W_k^{\mu-1}(f)|^2 \mathrm{d}x \leqslant c_{n,\mu} \| f \|_{A_{\mu}}^2 < +\infty.$$

Hence $\lim_{k \to +\infty} W_k^{\mu-1}(f)(x) = 0$ a. e.

Observing that $W_k^{\mu-1}(f) = \sup_{R \in J_k} \ln^{\mu-1} R |S_R(f) - S_{e^k}(f)|$

and using Theorem 1 we complete the proof.

The proof of Theorem 3 is similar to that of Theorem 2 and is easier. So we omit it.

References

[1] Качмаж С, Штейнгауз Г. Теорня ортогональных рядов. ГИХ-МЛ, Москва, 1958.

[2] Wang Shi-ming. On approximation by Riesz means of multiple Fourier series. Thesis of Master degree at Zhejiang University, 1987.

[3] Bochner S. Trans. AMS. 1936, 40(2): 175—207.

Lehrstuhl für Mathematik I
University of Siegen, D-5900 Siegen (West Germany)

用 Bochber-Riesz 平均强一致逼近

Strong Uniform Approximation by Bochner-Riesz Means

§ 1. Introduction

All those real-valued continuous functions defined on $\mathbf{R}^n$ which are periodic in each variable with period 2π compose a Banach space with max-norm. We denote this space by $C(\mathbf{Q}^n)$, where $\mathbf{Q}^n=\{(x_1, x_2, \cdots, x_n): -\pi\leqslant x_j<\pi, j=1,2,\cdots,n\}$. Suppose $f\in C(\mathbf{Q}^n)$ then the norm of f is

$$\|f\|=\max\{|f(x)|\,|\,x\in\mathbf{Q}^n\}.$$

The Bochner-Riesz means of f is defined by

$$S_R^\alpha(f;x)=\sum_{|m|<R}c_m(f)\mathrm{e}^{\mathrm{i}mx}(1-|m|^2R^{-2})^\alpha.$$

where m denotes n-dimensional integers and $mx=m_1x_1+m_2x_2+\cdots+m_nx_n$ for $x\in\mathbf{R}^n$, $|m|=(mm)^{\frac{1}{2}}$. The index α can be any complex number. The symbol $c_m(f)$ denotes the Fourier coefficients of f, i. e.,

$$c_m(f)=|Q^n|^{-1}\int_{Q^n}f(y)\mathrm{e}^{-\mathrm{i}my}\mathrm{d}y,$$

where $|Q^n|$ denotes the Lebesgue measure of Q^n, that is $(2\pi)^n$. The special value $\alpha_0=\frac{1}{2}n-\frac{1}{2}$ of the index is called the critical index. When $n=1$, $S_R^0(f)$ is just the R-th Fourier sum of f. And for $n>1$ $S_R^{\alpha_0}$ have many properties similar to that of univariate Fourier sums. So we reasonably look upon the B-R means of critical index $S_R^{\alpha_0}$ as the analogue of the univariate Fourier sums. By this viewpoint we consider the problem of the convergence of the quantity

$$\left\| R^{-1}\int_0^R |S_r^{\alpha_0}(f)-f|^q \mathrm{d}r \right\|, \quad q>0$$

as $R\to+\infty$, and call it to be the problem of strong uniform approximation.

For a function $f\in C(Q^n)$ we denote by $\omega_2(f;\ t)(t>0)$ its modulus of continuity of order 2 which is defined by

$$\omega_2(f;\ t)=\sup\{\|f(\cdot+h)+f(\cdot-h)-2f(\cdot)\| \mid |h|<t\}, \qquad t>0.$$

In the present paper we will prove the following

Theorem If $\frac{1}{2}n-1<\alpha\leqslant\frac{1}{2}n-\frac{1}{2}$ then there exists a constant $c(n,\ \alpha)$ depending only on n and α, such that for any $f\in C(Q^n)$ and any $R>0$,

$$\left\| R^{-1}\int_0^R |S_r^{\alpha}(f)-f|^2 \mathrm{d}r \right\| \leqslant c(n,\alpha)R^{-1}\int_0^R \omega_2(f;r^{-1})^2\mathrm{d}r.$$

§ 2. The integral representation of $R^{-1}\int_0^R \{S_r^\alpha(f;x)\}^2\mathrm{d}r$

Suppose $f\in C(Q^n)$, $x\in Q^n$ and $R>0$ are all fixed. For every $\varepsilon\in\left[0,\ \frac{3}{4}\right)$ we define a region

$$S_\varepsilon=\left\{\alpha=u+\mathrm{i}v:\ \varepsilon+\frac{1}{2}n-1<u<\frac{1}{2}n-\frac{1}{4},\ -1<v<1\right\}.$$

For a complex number $\alpha\in S_\varepsilon$ we write

$$\alpha=\frac{1}{2}n-1+\beta, \qquad \delta=\mathrm{Re}\,\beta\in\left(\varepsilon,\ \frac{3}{4}\right).$$

We know, if $\alpha\in S_{\frac{1}{2}}$ then the following Bochner formula ([1]) is valid:

$$S_R^\alpha(f;x) = 2^{\alpha+1-\frac{1}{2}n}\Gamma\left(\alpha+1 \quad \Gamma\left(\frac{1}{2}n\right)\right\}^{-1}\int_0^{+\infty} f_x(t)t^{n-1}r^nV_{\frac{1}{2}n+\alpha}(rt)\mathrm{d}t,$$

where

$$f_x(t) = 2^{-1-\frac{1}{2}n}\Gamma\left(\frac{1}{2}n\right)\int_{|y|=1} f(x+ty)\mathrm{d}\sigma(y), \quad V_\upsilon(z) = z^{-\upsilon}J_\upsilon(z)$$

and J_υ denotes the Bessel function of first kind. From this formula we derive that

$$R^{-1}\int_0^R \{S_r(f;x)\}^2 \mathrm{d}r =$$

$$A(n,\alpha)\int_0^{+\infty}\int_0^{+\infty} f_x(s)f_x(t)(st)^{n-1}\cdot K_R(s,t,\alpha)\mathrm{d}s\mathrm{d}t,$$

where

$$A(n,\ \alpha)=4^{\alpha+1-\frac{1}{2}n}\{\Gamma(\alpha+1)\}^2\left\{\Gamma\left(\frac{1}{2}n\right)\right\}^{-2},$$

$$K_R(s,t,\alpha) = R^{-1}\int_0^R r^{2n}V_{\frac{1}{2}n+\alpha}(sr)V_{\frac{1}{2}n+\alpha}(tr)\mathrm{d}r.$$

We define a function of α

$$F(\alpha) = A(n,\alpha)\int_0^{+\infty}\int_0^{+\infty} f_x(s)f_x(t)(st)^{n-1}K_R(s,t,\alpha)\mathrm{d}s\mathrm{d}t. \quad (2.1)$$

Then we see that the equation

$$R^{-1}\int_0^R \{S_r^\alpha(f;x)\}^2 \mathrm{d}r = F(\alpha) \quad (2.2)$$

is valid when $\alpha\in S_{\frac{1}{2}}$ and F is analytic in $S_{\frac{1}{2}}$. If we are able to prove that F is also well-defined and analytic is S_ε for all $\varepsilon\in\left(0,\ \frac{1}{2}\right)$, then we can immediately conclude that this equation holds for all $\alpha\in S_0$.

We define four sets as follows:

$$E_1=\{(s,\ t)\,|\,0<t<s<R^{-1}\},$$
$$E_2=\{(s,\ t)\,|\,0<t<R^{-1}<s\},$$
$$E_3=\{(s,\ t)\,|\,R^{-1}<t<s<t+R^{-1}\},$$
$$E_4=\{(s,\ t)\,|\,R^{-1}<t<s-R^{-1}\}.$$

Then we estimate $K_R(s,\ t,\ \alpha)$ on E_K for $k=1,2,3,4$, respectively. By a careful calculation and applying some properties of Bessel function of first kind, which can be found in [2], we get

$$|K_R(s,\ t,\ \alpha)|<3R^{2n},\ \text{if } (s,\ t)\in E_1,$$
$$|K_R(s,\ t,\ \alpha)|<c_nR^{n-1-\delta}s^{-n-1-\delta},\ \text{if } (s,\ t)\in E_2,$$
$$|K_R(s,\ t,\ \alpha)|<c_nR^{1-2\delta}s^{-2n+1-2\delta},\ \text{if } (s,\ t)\in E_3,$$

$$|K_R(s,t,\alpha)|<c_n\varepsilon^{-1}R^{-1}(R^{1-2\delta}+t^{2\delta-1})(st)^{-n+\frac{1}{2}-\delta}(s-t)^{-1},\text{if } (s,t)\in E_4.$$

where c_n denotes a constant depending only on n. From these estimates we derive the following lemma.

Lemma If $\varepsilon \in \left(0, \frac{1}{2}\right)$ and $\alpha \in S_\varepsilon$. then

$$\int_0^{+\infty}\int_0^{+\infty}(st)^{n-1}|K_R(s,t,\alpha)|\mathrm{d}s\mathrm{d}t < c_n\varepsilon^{-2}, \quad R>0.$$

From this lemma we derive the following proposition.

Proposition If $\alpha > \frac{1}{2}n-1$, then the expression (2. 2) holds for all $f \in C(Q^n)$ at all x, and when $\frac{1}{2}n-1<\alpha<\frac{1}{2}n-\frac{1}{4}$,

$$\sup\left\{R^{-1}\int_0^R |S_r^\alpha(f;x)|^2\mathrm{d}r \,\middle|\, R>0, x\in Q^n\right\} \leqslant c_n\left(\alpha-\frac{1}{2}n+1\right)^{-2}\|f\|^2. \tag{2.3}$$

Proof Let $\varepsilon \in (0, 2^{-1})$, $\alpha \in S_\varepsilon$ and $f\in C(Q^n)$. We set

$$I_R^\alpha f(x) = |A(n,\alpha)|\int_0^{+\infty}\int_0^{+\infty}|f_x(s)f_x(t)|(st)^{n-1}|K_R(s,t,\alpha)|\mathrm{d}s\mathrm{d}t.$$

Using lemma we get

$$I_R^\alpha f(x) \leqslant c_n\varepsilon^{-2}\|f\|^2. \tag{2.4}$$

Hence we conclude that the function $F(\alpha)$, in (1), for $\alpha\in S_\varepsilon$ is well-defined and is analytic. And we have proved that the equation (2) holds for all $\alpha \in S_0$. Now for $\alpha \in \left(\frac{1}{2}n-1, \frac{1}{2}n-\frac{1}{4}\right)$, we take ε in (4) such that $0<\varepsilon<\min\left(\alpha-\frac{1}{2}n+1, \frac{1}{2}\right)$ and then let ε tend to $\min\left(\alpha-\frac{1}{2}n+1, \frac{1}{2}\right)$. Then we get (3). □

§ 3. The proof of the theorem

Let $\frac{1}{2}n-1<\alpha<\frac{1}{2}(n-1)$, $f\in C(\mathbf{Q}^n)$ and $R>0$. Suppose $T_R(x)$ is the R-th trigonometric polynomial of the best approximation of f. Then obviously

$$\int_0^R |S_r^\alpha(f)-f|^2\mathrm{d}r \leqslant 4\int_0^R\{|S_r^\alpha(f-T_R)|^2+|S_r^\alpha(T_R)-S_r^{\alpha+1}(T_R)|^2+ |S_r^{\alpha+1}(T_R)-T_R|^2\}\mathrm{d}r+4|T_R-f|^2. \tag{2.5}$$

$$\|f-T_R\|\leqslant c_n\omega_2(f;\ R^{-1}). \tag{2.6}$$

Applying our Proposition we get

$$\left\|\frac{1}{R}\int_0^R |S_r^\alpha(f-T_R)|^2\mathrm{d}r\right\|\leqslant c(n,\alpha)|f-T_R|^2\leqslant c(n,\alpha)\omega_2\left(f;\frac{1}{R}\right)^2. \tag{2.7}$$

Noticing $\alpha+1>\frac{1}{2}n>\frac{n-1}{2}$ and applying a result of approximation by B-R means with the index bigger than $\frac{n-1}{2}$ (see [3]) we get for $0<r<R$,

$$|S_r^{\alpha+1}(T_R)-T_R\|\leqslant c(n,\alpha)\omega_2\left(T_R;\frac{1}{r}\right)\leqslant c(n,\alpha)\omega_2\left(f;\frac{1}{r}\right). \tag{2.8}$$

We denote by ∇^2 the Laplace operator. Then we have

$$S_r^\alpha(T_R)-S_r^{\alpha+1}(T_R)=S_r^\alpha\left(-\frac{1}{R^2}\nabla^2 T_R\right).$$

Noticing that

$$\|R^{-2}\nabla^2 T_R\|\leqslant c_n\omega_2(f;\ R^{-1})$$

and applying the Proposition once again we get

$$\left\|R^{-1}\int_0^R |S_r^\alpha(T_R)-S_r^{\alpha-1}(T_R)|^2\mathrm{d}r\right\|\leqslant c(n,\alpha)\omega_2(f;R^{-1})^2. \tag{2.9}$$

Substituting (2.6)~(2.9) into (2.5) we complete the proof.

Acknowledgment

This paper was completed while the author was a visiting scholar at University of Siegen. The author thanks Professor Schempp very much for his kind help.

References

[1] Bochner S. Summation of multiple Fourier series by spherical means. Trans. AMS, 1936, 40: 175—207.

[2] Watson G N. Theory of Bessel functions. Cambridge, 1952.

[3] Wang Kunyang. Approximation for continuous periodic function of several variables and its conjugate function by Riesz means on set of total measure. Approximation theory and its applications, 1985, 1(4): 19—56.

Arkiv för Matematik，1991，29(2)：261—276.

L^2-函数的 Fourier 积分的一类求和法的收敛速度

On the Rate of Convergence of Certain Summability Methods for Fourier Integrals of L^2-Functions[①]

§1. Introduction

Suppose B is a bounded convex symmetric body in $\mathbf{R}^n$ and let $|\cdot|$ be the Minkowski norm associated with B，i. e.

$$|x|=\inf\{t>0 \mid t^{-1}x\in B\},\ x\in\mathbf{R}^n.$$

Let $m\in L^\infty(0,+\infty)$. Denote by m_t the function

$$m_t(s)=m(ts),\quad t>0,\ s>0.$$

We define operator $T_{m_t}(t>0)$ on $L^2(\boldsymbol{R}^n)$ by

$$(T_{m_t}f)\hat{}(\xi)=m_t(|\xi|)\hat{f}(\xi).$$

If $m\in L^\infty(0,+\infty)$ and $0<\alpha\leqslant 1$，the fractional integral of order α of m is defined as in [5] (see also [6]). That is，we set

$$I_\omega^\alpha(m)(t)=\begin{cases}\dfrac{1}{\Gamma(\alpha)}\displaystyle\int_t^\omega(s-t)^{\alpha-1}m(s)\mathrm{d}s, & \text{if}\quad 0<t<\omega,\\ 0, & \text{if}\quad t\geqslant\omega;\end{cases}\tag{1.1}$$

and，if $0<\alpha<1$ and $I_\omega^{1-\alpha}(m)$ is locally absolutely continuous for every $\omega>0$，we define the fractional derivative $m^{(\alpha)}$ by

① 本文与 Müller D 合作.

Received：1990-02-09；Revised：1990-07-07.

$$m^{(\alpha)}(t)=\lim_{\omega\to+\infty}\left(-\frac{\mathrm{d}}{\mathrm{d}t}I_{\omega}^{1-\alpha}m(t)\right). \tag{1.2}$$

Moreover, by induction over the integer part $[\alpha]$ of α, we define for arbitrary $\alpha>0$,

$$m^{(\alpha)}(t)=-\frac{\mathrm{d}}{\mathrm{d}t}m^{(\alpha-1)}(t), \tag{1.2'}$$

Provided this makes sense, i. e. that $I_{\omega}^{1-\delta}$, $m^{(\delta)}$, $\cdots$, $m^{(\alpha-1)}$ are absolutely continuous, where $\delta=\alpha-[\alpha]$.

Notice that for m with compact support in $\mathbf{R}^{+}$,

$$(m^{(\alpha)})^{\wedge}(\tau)=(\mathrm{i}\tau)^{\alpha}\hat{m}(\tau), \tag{1.3}$$

where $(-\mathrm{i}\tau)^{\alpha}$ is defined by the principal branch.

We will consider the localized Riemann-Liouville spaces $RL(2, \alpha)$ which are defined (cf. [3]) by

$$RL(2, \alpha)=\{m\in L^{\infty}(0, +\infty)\mid \|m\|_{RL(2,\alpha)}<+\infty\}, \text{ if } \alpha>\frac{1}{2},$$

where

$$\|m\|_{RL(2,\alpha)}=\sup_{t>0}\|(\chi m_{t})^{(\alpha)}\|_{2}.$$

Here $\chi\in C_0^{\infty}(0, +\infty)$ is an arbitrary fixed non-negative and non-trivial bump function. It is know [3] that the space $RL(2, \alpha)$ does not depend on the choice of χ. For convenience we will choose χ such that

$$\chi\in C_0^{\infty}(0, +\infty),\ \chi(t)\geqslant 0,\ \operatorname{supp}\chi\subset\left[\frac{1}{2}, 1\right], \text{ and } \chi(t)=1 \quad \text{for } \frac{5}{8}<t<\frac{7}{8}. \tag{1.4}$$

We will also consider the space of functions of weak bounded variation $WBV_{q,\alpha}$ in the case $q=2$ and $\alpha>0$. By definition (see [7]) $WBV_{2,\alpha}$ is the space of all $m\in L^{\infty}\cap C(0, +\infty)$ for which $m^{(\alpha)}$ exists in the sense of (2′) and whose norm

$$\|m\|_{2,\alpha}=\|m\|_{\infty}+\sup_{k\in\mathbf{Z}}\left\{\int_{2^{k-1}}^{2^{k}}|t^{\alpha}m^{(\alpha)}(t)|^{2}\frac{\mathrm{d}t}{t}\right\}^{\frac{1}{2}} \tag{1.5}$$

is finite. From [3], Theorem 2, we know that for $\alpha>\frac{1}{2}$,

$$RL(2, \alpha)=WBV_{2,\alpha}, \tag{1.6}$$

with equivalent norms.

Remark If m is supported in a compact interval $[a, b]$, $0<a<b<+\infty$, then for $x<\frac{a}{2}$,

$$m^{\alpha}(x)=\frac{\alpha}{\Gamma(1-\alpha)}\int_{a}^{b}(t-x)^{-\alpha-1}m(t)\mathrm{d}t,$$

hence

$$|m^{(\alpha)}(x)|\leqslant C_{\alpha}\|m\|_{\infty}|(b-x)^{-\alpha}-(a-x)^{-\alpha}|$$

and

$$\left(\int_{-\infty}^{\frac{a}{2}}|m^{(\alpha)}(x)|^{2}\mathrm{d}x\right)^{\frac{1}{2}}\leqslant C_{\alpha}'\|m\|_{\infty}.$$

These estimates easily imply that for $0<\alpha<1$, there exist constants c, $C>0$, depending only on α and a, b, such that

$$c\|m\|_{2,\alpha}\leqslant\|m\|_{\infty}+\|m^{(\alpha)}\|_{2}\leqslant C\|m\|_{2,\alpha}. \tag{1.7}$$

Moreover, if $\alpha>\frac{1}{2}$, by (1.6) we also have

$$c\|m\|_{RL(2,\alpha)}\leqslant\|m^{(\alpha)}\|_{2}\leqslant C\|m\|_{RL(2,\alpha)}. \tag{1.8}$$

We denote by φ a function on $[0, +\infty]$ which is non-decreasing and satisfies the following condition:

$$1\leqslant\varphi(2t)\leqslant\lambda\varphi(t)\quad\text{for some}\quad\lambda\geqslant 1. \tag{1.9}$$

We write throughout this paper

$$\mu=\log_{2}\lambda, \tag{1.10}$$

Where λ is smallest possible to satisfy (1.9).

It is easy to see that

$$\begin{cases}\varphi(t)\leqslant\varphi(2)t^{\mu}, & \text{if}\quad t\geqslant 1,\\ \varphi(st)\geqslant\dfrac{1}{\lambda}s^{\mu}\varphi(t), & \text{if}\quad t\geqslant 0\quad\text{and}\quad 0\leqslant s\leqslant 1.\end{cases} \tag{1.11}$$

Corresponding to φ we define

$$\psi(t)=\varphi(1)+\left\{\int_{1}^{t+1}\frac{\varphi^{2}(s)}{s}\mathrm{d}s\right\}^{\frac{1}{2}}.$$

Then ψ is also non-decreasing and satisfies (1.9), possibly with a different λ, and

$$\psi(t)\geqslant\frac{1}{\lambda}\sqrt{\ln 2}\,\varphi(t).$$

Given φ, we define the space L^2_φ by

$$L^2_\varphi=\{f\in L^2(\mathbf{R}^n)\mid \|f\|_{L^2_\varphi}<+\infty\},$$

where

$$\|f\|_{L^2_\varphi}=\left\{\int_{\mathbf{R}^n}|\hat{f}(\xi)|^2|\varphi(|\xi|)|^2\mathrm{d}\xi\right\}^{\frac{1}{2}}.$$

We shall prove the following results:

Theorem 1 Suppose $\alpha>\frac{1}{2}$, $m\in RL(2,\alpha)$. If for some $\beta>\mu$ and $\gamma>1$,

$$\|(\chi m_t)^{(\alpha)}\|_2=O(t^\beta) \quad \text{as} \quad t\to 0^+, \tag{1.12}$$

$$\|(\chi m_t)^{(\alpha)}\|_2=O((\ln t)^{-\gamma}) \quad \text{as } t\to+\infty, \tag{1.13}$$

where χ is a bump function as in (1.4), then

$$\left\|\sup_{t>0}|T_{m_t}f|\varphi\left(\frac{1}{t}\right)\right\|_2\leqslant c\|f\|_{L^2_\varphi}.$$

Theorem 2 Suppose $\alpha>\frac{1}{2}$, $m\in RL(2,\alpha)$. If for some $\beta>\mu$ (1.2) holds, then

$$\left\|\sup_{0<t<1}|T_{m_t}f|\varphi\left(\frac{1}{t}\right)\right\|_2\leqslant c\|f\|_{L^2\psi}.$$

Theorem 3 Suppose $m\in WBV_{2,\alpha}$ for all $0<\alpha<\frac{1}{2}$ and sup $m\subset\left[\frac{1}{2},1\right]$. If

$$\|m\|^2_{2,\frac{1}{2}-\varepsilon}=O\left(\frac{1}{\varepsilon}\right) \quad \text{as} \quad \varepsilon\to 0^+, \tag{1.14}$$

which holds in particular if m is of bounded variation, then

$$\left\|\sup_{0<t<1}\frac{1}{\ln\left(\frac{1}{t}+1\right)}\varphi\left(\frac{1}{t}\right)|T_{m_t}f|\right\|_{L^2+L^\infty}\leqslant c\|f\|_{L^2_\varphi},$$

where

$$\|f\|_{L^2+L^\infty}=\inf\{\|g\|_2+\|h\|_\infty\mid f=g+h\}.$$

A corollary of Theorem 2 includes the result of Chen Tian-ping [4] on generalized bochner—Riesz means of positive order.

We remark that Theorem 1 has some overlapping with the results in [6], in particular with Theorem 1 and Theorem 4 in that paper. However, in [6] Dappa and Trebels are concerned with L^p-estimates for maximal operators under no smoothness condition whatsoever on the function f (but in the more general context of quasi-radial multipliers), whereas we want to concentrate in this article on the rate of convergence of $T_{m_t}f$ as $t\to 0^+$, given f has a certain degree of smoothness, measured by some L_φ^2-norm of f. Our main result is in fact Theorem 3, which deals with the critical index of smoothness $\alpha=\frac{1}{2}$ for m.

As to the L^p-case, let us also mention some results due to Carbery in order to give a slightly more complete picture of what is known on the subject.

Define D^s by

$$(D^s f)\hat{}(\xi)=\|\xi\|^s\hat{f}(\xi),$$

where $\|\cdot\|$ is the Euclidean norm on $\mathbf{R}^n$. We introduce the global Bessel potential space $L_\alpha^2=L_\alpha^2(\mathbf{R}^+)$ as in [1]: L_α^2 is the completion of the C^∞ functions of compact support in $(0, +\infty)$ under the norm

$$\|m\|_{L_\alpha^2}=\left\{\int_0^{+\infty}\left|s^{\alpha+1}\left(\frac{\mathrm{d}}{\mathrm{d}s}\right)^\alpha\left(\frac{m(s)}{s}\right)\right|^2\frac{\mathrm{d}s}{s}\right\}^{\frac{1}{2}}.$$

Theorem(Carbery) Let $|\cdot|=\|\cdot\|$ be the Euclidean norm. If $\alpha>n\left(\frac{1}{p}-\frac{1}{2}\right)+\frac{1}{2}$ for $1<p\leqslant 2$, or $\alpha>n\left(\frac{1}{2}-\frac{1}{p}\right)+\frac{1}{p}$ for $2\leqslant p<+\infty$, then

$$\|\sup_{t>0}t^{-s}|T_{m_t}f|\|_{L^p(\mathbf{R}^n)}\leqslant c_\alpha\||\cdot|^{-s}m(\cdot)\|_{L_\alpha^2}\|D^s f\|_{L^P(\mathbf{R}^n)}.$$

Furthermore, if $n=1$ or $n=2$, the above estimate even holds if $2\leqslant p<+\infty$ and $\alpha>\max\left(\frac{1}{2}, n\left(\frac{1}{2}-\frac{1}{p}\right)\right)$.

Of course, Carbery's theorem implies results on the pointwise convergence of Bochner-Riesz means of L^p functions, which will be stated later as a remark.

Throughout this paper c will denote a constant which can take different values from statement to statement.

§ 2. Auxiliary results

Lemma 1 Let $0<\alpha<1$. If $m\in WBV_{2,\alpha}$, then there exists a set $E\subset(0, +\infty)$ of one-dimensional measure zero such that for any $\beta>1$ and every $u\in(0, +\infty)\setminus E$,

$$m(u)=\frac{1}{\Gamma(\alpha)}\int_u^{\beta u}(s-u)^{\alpha-1}m^{(\alpha)}(s)\,\mathrm{d}s+\frac{[(\beta-1)u]^{\alpha}}{\Gamma(\alpha)\Gamma(1-\alpha)}\int_{\beta u}^{+\infty}\frac{m(s)}{(s-\beta u)^{\alpha}(s-u)}\mathrm{d}s. \tag{2.1}$$

Proof We first assume that m vanishes on $(a, +\infty)$ for some $0<a<+\infty$. Then we have for $u\in(0, a)$,

$$\int_u^a\left[\int_t^a(s-t)^{\alpha-1}m^{(\alpha)}(s)\,\mathrm{d}s\right]\mathrm{d}t=\int_u^a\left[\int_u^s(s-t)^{\alpha-1}\,\mathrm{d}t\right]m^{(\alpha)}(s)\,\mathrm{d}s$$
$$=\frac{1}{\alpha}\int_u^a(s-u)^{\alpha}m^{(\alpha)}(s)\,\mathrm{d}s.$$

On the other hand, we have

$$m^{(\alpha)}(s)=-\frac{\mathrm{d}}{\mathrm{d}s}I_a^{1-\alpha}(m)(s),\quad s>0, \tag{2.2}$$

and $I_a^{1-\alpha}(m)$ is absolutely continuous on $[\varepsilon, a]$ for every $0<\varepsilon<a$. So, by plugging (2.2) into $\int_u^a(s-u)^{\alpha}m^{(\alpha)}(s)\,\mathrm{d}s$ and integrating by parts, one obtains after some routine calculations

$$\frac{1}{\alpha}\int_u^a(s-u)^{\alpha}m^{(\alpha)}(s)\,\mathrm{d}s=\Gamma(\alpha)\int_u^a m(t)\,\mathrm{d}t.$$

By comparison with the previous formula, we see that for almost every $t>0$,

$$m(t)=\frac{1}{\Gamma(\alpha)}\int_t^{+\infty}(s-t)^{\alpha-1}m^{(\alpha)}(s)\,\mathrm{d}s. \tag{2.3}$$

(Compare also [1], [7]).

Moreover, by partial integration we get from (2.2),

$$\int_{\beta t}^{+\infty}(s-t)^{\alpha-1}m^{(\alpha)}(s)\,\mathrm{d}s=[(\beta-1)t]^{\alpha-1}I_a^{1-\alpha}(m)(\beta t)+$$
$$\frac{-1}{\Gamma(1-\alpha)}\int_{\beta t}^{+\infty}(s-t)^{\alpha-2}\int_s^{+\infty}(u-s)^{-\alpha}m(u)\,\mathrm{d}u\mathrm{d}s$$

$$= \frac{[(\beta-1)t]^{\alpha}}{\Gamma(1-\alpha)}\int_{\beta t}^{+\infty}\frac{m(s)\mathrm{d}s}{(s-\beta t)^{\alpha}(s-t)}.$$

So we have proved (2. 1) if $m(t)$ vanishes for t sufficiently large.

For general m we define m_N, $n\in\mathbf{N}$, by

$$m_N(t)=\begin{cases} m(t), & \text{if} \quad 0\leqslant t\leqslant N, \\ 0, & \text{if} \quad t>N, \end{cases}$$

and let $m_\infty=m$. Define

$$a_N(t) = \frac{1}{\Gamma(\alpha)}\int_t^{\beta t}(s-t)^{\alpha-1}m_N^{(\alpha)}(s)\mathrm{d}s,$$

$$b_N(t) = \frac{[(\beta-1)t]^{\alpha}}{\Gamma(\alpha)\Gamma(1-\alpha)}\int_{\beta t}^{+\infty}\frac{m_N(s)\mathrm{d}s}{(s-\beta t)^{\alpha}(s-t)},$$

with $N\in\mathbf{N}$ or $N=+\infty$. Then we have

$$a_\infty(t)-a_N(t) = \frac{1}{\Gamma(\alpha)}\int_t^{\beta t}(s-t)^{\alpha-1}(m-m_N)^{(\alpha)}(s)\mathrm{d}s.$$

By definition

$$(m-m_N)^{(\alpha)}(s) = \lim_{\omega\to+\infty} -\frac{\mathrm{d}}{\mathrm{d}s}\frac{1}{\Gamma(1-\alpha)}\int_s^{\omega}(u-s)^{-\alpha}(m-m_N)(u)\mathrm{d}u.$$

For any fixed $s>0$ with $N>s$ this implies

$$(m-m_N)^{(\alpha)}(s) = \frac{\alpha}{\Gamma(1-\alpha)}\int_N^{+\infty}(u-s)^{-\alpha-1}m(u)\mathrm{d}u,$$

so that

$$|(m-m_N)^{(\alpha)}(s)|\leqslant c\|m\|_\infty\frac{1}{(N-s)^{\alpha}}, \qquad \text{if} \quad N>s.$$

Therefore

$$|a_\infty(t)-a_N(t)|\leqslant c\|m\|_\infty\int_t^{\beta t}(s-t)^{\alpha-1}\frac{\mathrm{d}s}{(N-s)^{\alpha}}$$

$$\leqslant c\|m\|_\infty\frac{[(\beta-1)t]^{\alpha}}{(N-\beta t)^{\alpha}}, \quad \text{if} \quad N>\beta t.$$

On the other hand, if $N>\beta t$, we also have

$$|b_\infty(t)-b_N(t)|\leqslant c[(\beta-1)t]^{\alpha}\int_N^{+\infty}\frac{|m(s)|}{(s-\beta t)(s-t)}\mathrm{d}s$$

$$\leqslant c\|m\|_\infty\frac{[(\beta-1)t]^{\alpha}}{(N-\beta t)^{\alpha}},$$

since $s-t>s-\beta t$. We conclude that for every $t>0$,

$$a_\infty(t)+b_\infty(t)=\lim_{N\to+\infty}(a_N(t)+b_N(t)).$$

But we have proved that

$$m_N(t)=a_N(t)+b_N(t),\ t\in(0,\ +\infty)\setminus E_N,$$

where E_N is a set of one-dimensional measure zero. Let $E=\bigcup_{N=1}^{+\infty}E_N$. We get

$$a_\infty(t)+b_\infty(t)=m(t),\ \text{if}\quad t\in(0,\ +\infty)\setminus E.\quad \square$$

Lemma 2　If $\alpha\in\left(\frac{1}{2},\ 1\right)$ and $m\in RL(2,\ \alpha)$, then

$$\|\sup_{1<s<2}|T_{m_s}f|\|_2\leqslant c\|m\|_{RL(2,\alpha)}\|f\|_2.$$

Proof　Define operators P_s on $L^2(\mathbf{R}^n)$, with $s\in(1,\ 3)$, by

$$(P_sf)\hat{}(\xi)=\left(\frac{\mathrm{d}}{\mathrm{d}s}\right)^\alpha m(s|\xi|)\hat{f}(\xi).$$

By Plancherel's theorem we have

$$\begin{aligned}\int_{\mathbf{R}^n}\int_1^3|(P_sf)(x)|^2\mathrm{d}s\mathrm{d}x&=\int_1^3\int_{\mathbf{R}^n}|(P_sf)\hat{}(\xi)|^2\mathrm{d}\xi\mathrm{d}s\\&=\int_{\mathbf{R}^n}|\hat{f}(\xi)|^2\int_1^3|m_{|\xi|}^{(\alpha)}(s)|^2\mathrm{d}s\mathrm{d}\xi.\end{aligned}$$

From (1.5) and (1.6) we see that for $t>0$,

$$\begin{aligned}\int_1^3|m_t^{(\alpha)}(s)|^2\mathrm{d}s&=\int_t^{3t}|t^\alpha m^{(\alpha)}(s)|^2\frac{\mathrm{d}s}{t}\leqslant c\int_t^{3t}|s^\alpha m^{(\alpha)}(s)|^2\frac{\mathrm{d}s}{s}\\&\leqslant c\|m\|_{2,\alpha}^2\leqslant c\|m\|_{RL(2,\alpha)}^2.\end{aligned}\tag{2.4}$$

This shows that P_s is well-defined for a. e. $s\in(1,\ 3)$. Now for $s\in(1,\ 2)$ we define two operators as follows:

$$A_sf=\frac{1}{\Gamma(\alpha)}\int_s^3(u-s)^{\alpha-1}P_sf\,\mathrm{d}u,$$

$$B_sf=\frac{(3-s)^\alpha}{\Gamma(\alpha)\Gamma(1-\alpha)}\int_3^{+\infty}\frac{T_{m_u}f}{(u-3)^\alpha(u-s)}\mathrm{d}u.$$

Since $\alpha>\frac{1}{2}$ we have

$$\sup_{1<s<2}|A_sf|\leqslant c\sup_{1<s<2}\left\{\int_s^3(u-s)^{2\alpha-2}\mathrm{d}u\right\}^{\frac{1}{2}}\left\{\int_1^3|P_uf|^2\mathrm{d}u\right\}^{\frac{1}{2}}.$$

So, using (2.4), we get

$$\|\sup_{1<s<2}|A_sf|\|_2\leqslant c\|m\|_{RL(2,\alpha)}\|f\|_2.\tag{2.5}$$

On the other hand we have

$$\| \sup_{1<s<2} | B_s f | \|_2 \leqslant c \int_3^{+\infty} \frac{\| T_{m_u} f \|_2}{(u-3)^{\alpha}(u-2)} \mathrm{d}u \leqslant c \| m \|_{\infty} \| f \|_2. \quad (2.6)$$

For $\xi \neq 0$ and $s \in (1, 2)$ one has

$$\widehat{A_s f}(\xi) = \frac{1}{\Gamma(\delta)} \int_{s|\xi|}^{3|\xi|} (t - s|\xi|)^{\alpha-1} m^{(\alpha)}(t) \mathrm{d}t \hat{f}(\xi),$$

$$\widehat{B_s f}(\xi) = \frac{(3-s)^{\alpha} |\xi|^{\alpha}}{\Gamma(\alpha)\Gamma(1-\alpha)} \int_{3|\xi|}^{+\infty} \frac{m(t)\mathrm{d}t}{(t-3|\xi|)^{\delta}(t-s|\xi|)} \hat{f}(\xi).$$

If we write $u = s|\xi|$ and $\beta = \frac{3}{s} > 1$, then by Lemma 1, for every $u \in (0, +\infty) \setminus E$, i. e. for each $|\xi| \in \frac{1}{s}[(0, +\infty) \setminus E]$ and $s \in (1, 2)$,

$$(T_{m_s} f)\hat{}(\xi) = \widehat{A_s f}(\xi) + \widehat{B_s f}(\xi).$$

This shows that

$$T_{m_s} = A_s + B_s,$$

And so the estimate in Lemma 2 follows from (2.5) and (2.6). □

Lemma 3 Suppose $\alpha \in \left(\frac{1}{2}, 1\right)$ and $m \in RL(2, \alpha)$. If supp $m \subset [1, 2]$, then

$$\left\| \sup_{t>0} | T_{m_t} f | \varphi\left(\frac{1}{t}\right) \right\|_2 \leqslant c \| m^{(\alpha)} \|_2 \| f \|_{L^2_{\varphi}}.$$

Proof We have

$$\sup_{t>0} | T_{m_t} f | \varphi\left(\frac{1}{t}\right) \leqslant c \left\{ \sum_{j=-\infty}^{+\infty} [\sup_{1<t<2} | T_{m_{t2^j}} f | \varphi(2^{-j})]^2 \right\}^{\frac{1}{2}}.$$

For $j \in \mathbf{Z}$ define f_j by

$$\hat{f}(\xi) = \begin{cases} \hat{f}(\xi), & \text{if} \quad 2^{-j-1} < |\xi| < 2^{-j+1}, \\ 0, & \text{otherwise.} \end{cases} \quad (2.7)$$

Since supp $m \subset [1, 2]$ we see that

$$T_{m_{t2^j}} f = T_{m_{t2^j}} f_j \quad \text{for} \quad 1 < t < 2.$$

So, by Lemma 2 and (1.8) we get

$$\left\| \sup_{t>0} | T_{m_t} f | \varphi\left(\frac{1}{t}\right) \right\|_2^2 \leqslant c \sum_{j=-\infty}^{+\infty} [\| m_{2^j} \|_{RL(2,\alpha)} \| f_j \|_2 \varphi(2^{-j})]^2$$

$$\leqslant c \| m^{(\alpha)} \|_2^2 \| f \|_{L^2_{\varphi}}^2. \quad □$$

§ 3. Proof of the theorems

Since $RL(2, \beta)$ is continuously embedded in $RL(2, \alpha)$, if $\beta>\alpha$ (see [3]), we may assume without restriction in the proofs of Theorem 1 and Theorem 2 that $\frac{1}{2}<\alpha<1$.

Proof of Theorem 1 Choose $h\in C_0^\infty(\mathbf{R})$ such that

$$\operatorname{supp} h\subset\left[\frac{1}{2},2\right],\quad \sum_{j=-\infty}^{+\infty}h(2^jt)=1,\quad \forall t>0.$$

Define
$$m_j(t)=m(t)h(2^jt)$$
and
$$(T_t^jf)\hat{}(\xi)=m_j(t|\xi|)\hat{f}(\xi),\ f\in L^2(\mathbf{R}^n).$$
Then we have
$$m=\sum_{j=-\infty}^{+\infty}m_j,T_{m_t}=\sum_{j=-\infty}^{+\infty}T_t^j,$$
and
$$\left\|\sup_{t>0}|T_{m_t}f|\varphi\left(\frac{1}{t}\right)\right\|_2\leqslant\sum_{j=-\infty}^{+\infty}\left\|\sup_{t>0}|T_t^jf|\varphi\left(\frac{1}{t}\right)\right\|_2. \tag{3.1}$$
We write
$$\widetilde{m}_j(t)=m_{2^{-j}}(t)h(t)=m_j(2^{-j}t),$$
and
$$(\widetilde{T}_t^jf)\hat{}(\xi)=\widetilde{m}_j(t|\xi|)\hat{f}(\xi).$$
Then we see that $\sup_{t>0}|T_t^jf|\varphi\left(\frac{1}{t}\right)=\sup_{t>0}|\widetilde{T}_t^jf|\varphi\left(\frac{2^j}{t}\right)$.

Moreover, by (1.9) and (1.10), $\varphi\left(\frac{2^j}{t}\right)\leqslant 2^{\mu j}\varphi\left(\frac{1}{t}\right)$, $j\geqslant 0$.

So, by applying Lemma 3 to $\widetilde{m}_j$ we get

$$\left\|\sup_{t>0}|T_t^jf|\varphi\left(\frac{1}{t}\right)\right\|_2\leqslant c\,2^{\mu j}\|\widetilde{m}_j^{(\alpha)}\|_2\|f\|_{L_\varphi^2},\quad j\geqslant 0.$$

By condition (1.2), $\|\widetilde{m}_j^{(\alpha)}\|_2\leqslant c2^{-\beta j}$, $j\geqslant 0$.

Hence

$$\sum_{j=0}^{+\infty}\left\|\sup_{t>0}|T_t^jf|\varphi\left(\frac{1}{t}\right)\right\|_2\leqslant c\sum_{j=0}^{+\infty}2^{-(\beta-\mu)j}\|f\|_{L_\varphi^2}\leqslant c\|f\|_{L_\varphi^2}. \tag{3.2}$$

On the other hand, if $j<0$, then

$$\left\|\sup_{t>0}|T_t^jf|\varphi\left(\frac{1}{t}\right)\right\|_2=\left\|\sup_{t>0}|\widetilde{T}_t^jf|\varphi\left(\frac{2^j}{t}\right)\right\|_2$$
$$\leqslant\left\|\sup_{t>0}|\widetilde{T}_t^jf|\varphi\left(\frac{1}{t}\right)\right\|_2\leqslant c\|\widetilde{m}_j^{(\alpha)}\|_2\|f\|_{L_\varphi^2},$$

once again by Lemma 3. Condition (1.3) implies

$$\|\widetilde{m}_j^{(\alpha)}\|_2 \leqslant c(-j)^{-\gamma}, \quad j<0,$$

hence

$$\sum_{j=-\infty}^{-1}\left\|\sup_{t>0}|T_t^j f|\varphi\left(\frac{1}{t}\right)\right\|_2 \leqslant c\sum_{j=1}^{+\infty} j^{-\gamma}\|f\|_{L_\varphi^2} \leqslant c\|f\|_{L_\varphi^2}. \tag{3.3}$$

The theorem now is an immediate consequence of (3.1) (3.2) and (3.3). □

Proof of theorem 2 By Theorem 1 we can assume supp $m\subset\left[\frac{1}{2}, +\infty\right]$. We have

$$\left|\sup_{0<t<1}|T_{m_t}f|\varphi\left(\frac{1}{t}\right)\right|^2 \leqslant c\sum_{k=0}^{+\infty}\sup_{1<t<2}|T_{m_{t2^{-k-1}}}f|^2\varphi^2(2^k).$$

Define f_k by $\quad \hat{f}_k(\xi)=\begin{cases}\hat{f}(\xi), & \text{if} \quad 2^{k-1}<|\xi|<2^k,\\ 0, & \text{otherwise.}\end{cases}$

Then for $t\in(1, 2)$, $\quad (T_{m_{t2^{-k-1}}}f)\hat{\ }(\xi)=m(t2^{-k-1}|\xi|)\sum_{j=k}^{+\infty}\hat{f}_j(\xi)$,

i. e.

$$T_{m_{t2^{-k-1}}}f=T_{m_{t2^{-k-1}}}\left(\sum_{j=k}^{+\infty}f_j\right).$$

By Lemma 2 we get $\quad \|\sup_{1<t<2}|T_{m_{t2^{-k-1}}}f|\|_2^2\leqslant c\|m\|_{RL(2,\alpha)}^2\left\|\sum_{j=k}^{+\infty}f_j\right\|_2^2$,

hence

$$\left\|\sup_{0<t<1}|T_{m_t}f|\varphi\left(\frac{1}{t}\right)\right\|_2^2 \leqslant c\sum_{k=0}^{+\infty}\|m\|_{RL(2,\alpha)}^2\left\|\sum_{j=k}^{+\infty}f_j\right\|_2^2\varphi^2(2^k).$$

Since

$$\left\|\sum_{j=k}^{+\infty}f_j\right\|_2^2=\sum_{j=k}^{+\infty}\|\hat{f}_j\|_2^2,$$

we obtain

$$\left\|\sup_{0<t<1}|T_{m_t}f|\varphi\left(\frac{1}{t}\right)\right\|_2^2 \leqslant c\|m\|_{RL(2,\alpha)}^2\sum_{j=0}^{+\infty}\|\hat{f}_j\|_2^2\cdot\sum_{k=0}^{j}\varphi^2(2^k).$$

Noticing that

$$\sum_{k=0}^{j}\varphi^2(2^k)\leqslant c\psi^2(2^j),$$

the proof follows by another application of Plancherel's theorem. □

Proof of theorem 3 Let us first notice that if m is of bounded variation, then $\hat{m}(\tau)=O(|\tau|^{-1})$ as $|\tau|\to+\infty$, which implies

$$\int_{-\infty}^{+\infty}(|\hat{m}(\tau)||\tau|^{\frac{1}{2}-\varepsilon})^2\mathrm{d}t=O\left(\frac{1}{\varepsilon}\right).$$

So, by (1.3)(1.7), m satisfies condition (1.14).

Now assume $m \in WBV_{2,\alpha}$ is such that (1.14) holds. Define operators $P_u = P_u^{k,\alpha}$ by $(P_u^{k,\alpha} f)\hat{}(\xi) = \left(\frac{\mathrm{d}}{\mathrm{d}u}\right)^{\alpha} m\left(u \frac{|\xi|}{2^{k-2}}\right)\hat{f}(\xi)$, $u \in (1,3)$.

By an argument similar to that in the proof of Lemma 2 we get

$$\left\| \left\{ \int_1^3 |P_u f|^2 \mathrm{d}u \right\}^{\frac{1}{2}} \right\|_2 \leqslant c \|m\|_{2,\alpha} \|f\|_2. \tag{3.4}$$

And similarly we get for $s \in (1, 2)$,

$$T_{m_t^k} f = A_t^{k,\alpha} f + B_t^{k,\alpha} f \text{ for every } \alpha \in \left(0, \frac{1}{2}\right), \tag{3.5}$$

Where $\quad m_t^k(s) = m(st2^{-k+2})$,

$$A_t^{k,\alpha} f = \frac{1}{\Gamma(\alpha)} \int_t^3 (u-t)^{\alpha-1} P_u^{k,\alpha} f \mathrm{d}u,$$

$$B_t^{k,\alpha} f = \frac{(3-t)^{\alpha}}{\Gamma(\alpha)\Gamma(1-\alpha)} \int_3^{+\infty} \frac{T_{m_u^k} f}{(u-3)^{\alpha}(u-t)} \mathrm{d}u.$$

Define f_j by $\hat{f}_j(\xi) = \hat{f}(\xi) \chi_{[2^{k-4}, 2^{k-2}]}(|\xi|)$ and

$$M(f)(x) = \sup_{0<t<1} \frac{1}{\ln\left(\frac{1}{t}+1\right)} \varphi\left(\frac{1}{t}\right) |T_{m_t} f(x)|.$$

We have

$$M(f)^2 \leqslant c \sum_{k=3}^{+\infty} \left(\frac{1}{k} \varphi(2^k) \sup_{1<t<2} |T_{m_t^k} f_k| \right)^2. \tag{3.6}$$

Take $\alpha = \alpha_k = \frac{1}{2} - \frac{1}{k}$, $k \geqslant 3$, and split the integral of $A_t^{k,\alpha_k} f$ into the following two parts:

$$D_t^k f_k = \frac{1}{\Gamma(\alpha_k)} \int_t^{t+2^{-k(n+1)}} (u-t)^{\alpha_k - 1} P_u^{k,\alpha_k} f_k \mathrm{d}u,$$

$$E_t^k f_k = \frac{1}{\Gamma(\alpha_k)} \int_{t+2^{-k(n+1)}}^{3^3} (u-t)^{\alpha_k - 1} P_u^{k,\alpha_k} f_k \mathrm{d}u.$$

From (3.5) (3.6) we get

$$M(f)^2 \leqslant c \sum_{k=3}^{+\infty} \left(\frac{1}{k} \varphi(2^k) \right)^2 \sup_{1<t<2} (|B_t^{k,\alpha_k} f_k|^2 + |D_t^k f_k|^2 + |E_t^k f_k|^2),$$

hence

$$M(f) \leqslant c \left\{ \sum_{k=3}^{+\infty} \left(\frac{1}{k} \varphi(2^k) \right)^2 \sup_{1<t<2} (|B_t^{k,\alpha_k} f_k|^2 + |E_t^k f_k|^2) \right\}^{\frac{1}{2}} +$$

$$c\left\{\sum_{k=3}^{+\infty}\left(\frac{1}{k}\varphi(2^k)\right)^2\sup_{1<t<2}|D_t^k f_k|^2\right\}^{\frac{1}{2}}. \tag{3.7}$$

It is easy to see that

$$\|\sup_{1<t<2}B_t^{k,\alpha_k}f_k\|_2\leqslant c\|m\|_\infty\|f_k\|_2. \tag{3.8}$$

Next, for a. e. $u\in(1, 3)$,

$$P_u^{k,\alpha_k}f_k(x)=\int_{2^{k-4}<|\xi|<2^{k-2}}\left(\frac{\mathrm{d}}{\mathrm{d}u}\right)^{\alpha_k}m\left(u\frac{|\xi|}{2^{k-2}}\right)\hat{f}_k(\xi)\mathrm{e}^{\mathrm{i}x\xi}\mathrm{d}\xi.$$

So, by applying Cauchy-Schwarz' estimate and (1. 14), we get the following uniform estimate:

$$|P_u^{k,\alpha_k}f_k(x)|\leqslant c2^{\frac{kn}{2}}\left\{\int_{\frac{1}{4}}^3|m^{(\alpha_k)}(s)|^2\mathrm{d}s\right\}^{\frac{1}{2}}\|f_k\|_2\leqslant c\sqrt{k}2^{\frac{kn}{2}}\|f_k\|_2,$$

which implies

$$\|\sup_{1<t<2}|D_t^k f_k|\|_\infty\leqslant c\sqrt{k}2^{-\frac{k}{2}}\|f_k\|_2. \tag{3.9}$$

Finally, estimating the integral in u defining $E_t^k f_k$ again by Cauchy—Schwarz, we obtain

$$|E_t^k f_k|\leqslant c\sqrt{k}\left\{\int_1^3|P_u^{k,\alpha_k}f_k|^2\mathrm{d}u\right\}^{\frac{1}{2}}, (1<t<2),$$

which, by (3. 4) and (1. 14), implies

$$\|\sup_{1<t<2}|E_t^k f_k|\|_2\leqslant ck\|f_k\|_2. \tag{3.10}$$

From (3. 7)～(3. 10) we conclude

$$\|M(f)\|_{L^2+L^\infty}\leqslant c\left(\sum_{k=3}^{+\infty}\left(\frac{1}{k}\varphi(2^k)k\|f_k\|_2\right)^2\right)^{\frac{1}{2}}\leqslant c\|f\|_{L_\varphi^2}. \quad \square$$

§ 4. Applications

Now we can use the above estimates of maximal functions to get some results on almost everywhere convergence.

Theorem 4 Assume m is a continuous function on $\mathbf{R}^+$, contained in $RL(2, \alpha)$ for some $\alpha>\frac{1}{2}$. If condition (1. 12) holds for the multiplier $m-1$ with $\beta>\mu$, then for every $f\in L_\psi^2$,

$$T_{m_t}f(x)-f(x)=o\left[\frac{1}{\varphi\left(\frac{1}{t}\right)}\right] \quad \text{a. e.} \quad \text{as} \quad t\to 0^+.$$

Proof　Since　$\|\chi(m_t-1)\|_\infty \leqslant c\|[\chi(m_t-1)]^{(\alpha)}\|_2 = O(t^\beta)$,

we have　$m(t)-1=O(t^\beta)$　as　$t\to 0^+$.

We know from (1.11) that　$t^\mu \leqslant c\left[\varphi\left(\frac{1}{t}\right)\right]^{-1}$, $0<t<1$.

Therefore　$|m(t)-1| = O\left(\left[\varphi\left(\frac{1}{t}\right)\right]^{-1}\right)$　as　$t\to 0^+$.

From this we conclude that the theorem is valid for those functions whose Fourier transforms belong to $C_0^\infty(\mathbf{R}^n)$. Since such functions are dense in L_ψ^2, the theorem will be proved provided

$$\left\|\sup_{0<t<1}|T_{(m-1)_t}f|\varphi\left(\frac{1}{t}\right)\right\|_2 \leqslant c\|f\|_{L_\psi^2}.$$

But this is a direct consequence of Theorem 2 applied to the multiplier $m-1$.　□

As a corollary of Theorem 4 we get the following result on Bochner—Riesz means of positive order, which includes the result of Chen Tian-ping [4], who only deals with the case where $|\cdot|$ is the Euclidean norm.

Corollary 1　Let $\alpha>0$, $l>0$, $m(t)=(1-t^l)_+^\alpha$. If $f\in L^2(\mathbf{R}^n)$ satisfies the condition

$$\int_{\mathbf{R}^n}|\hat{f}(x)|^2|x|^{2\mu}\mathrm{d}x < +\infty,\quad \mu>0, \tag{4.1}$$

then for a. e. $x\in\mathbf{R}^n$,

$$T_{m_t}f(x)-f(x)=\begin{cases} o(t^\mu), & \text{if } l>\mu,\\ O(t^\mu), & \text{if } l=\mu,\end{cases} \tag{4.2}$$

as $t\to 0^+$.

Proof　First, one easily estimates

$$\|[\chi(m_t-1)]'\|_2 = O(t^l)\quad\text{as}\quad t\to 0^+.$$

Since $RL(2,1)\subset RL(2,\delta)$, if $\frac{1}{2}<\delta<1$, we get

$$\|[\chi(m_t-1)]^{(\delta)}\|_2 = O(t^l)\quad\text{as}\quad t\to 0^+,$$

for every $\frac{1}{2}<\delta<1$. Moreover, one checks easily that $m\in RL(2,\delta)$

whenever $\frac{1}{2}<\delta<\frac{1}{2}+\alpha$. Notice that the condition $\delta<\frac{1}{2}+\alpha$ is forced by the singularity of m at $t=1$.

So, if $l>\mu$, the required result is a direct consequence of Theorem 4 with $\varphi(t)=1+t^{\mu}$.

Now assume $l>\mu$, $\varphi(t)=1+t^{\mu}$. Choose a function $h\in C^{\infty}(0, +\infty)$ such that

$$\begin{cases} h(t)=1, & \text{if} \quad 0\leqslant t\leqslant \frac{1}{2}, \\ h(t)=0, & \text{if} \quad t>\frac{3}{4}. \end{cases} \tag{4.3}$$

If we define $$\widetilde{m}(t)=m(t)+\alpha t^{l}h(t),$$

then $\widetilde{m}\in RL(2, \delta)$ for $\delta\in\left(\frac{1}{2}, \frac{1}{2}+\alpha\right)$, and

$$\|[\chi(\widetilde{m}_t-1)]^{(\delta)}\|_2=O(t^{2l}),$$

by a similar argument as before. So, by Theorem 4, we have

$$T_{\widetilde{m}_t}f(x)-f(x)=o(t^{\mu}) \quad \text{a. e.} \quad \text{as} \quad t\to 0^+$$

for every $f\in L^2_{\varphi}$. We write $\Delta(t)=\alpha t^{l}h(t)$. Then

$$(T_{\Delta_t}f)\hat{}(\xi)=\alpha t^{l}h(t|\xi|)|\xi|^{l}\hat{f}(\xi).$$

For $f\in L^2_{\varphi}$ let $\tilde{f}$ be defined by $\quad \hat{\tilde{f}}(\xi)=|\xi|^{l}\hat{f}(\xi)$.

We see that $$t^{-l}T_{\Delta_t}f=\alpha T_{h_t}\tilde{f},$$

and so there only remains to prove that

$$\|\sup_{0<t<1}T_{h_t}g\|_2\leqslant c\|g\|_2, \quad g\in L^2(\mathbf{R}^n). \tag{4.4}$$

To this end, write $h(|\xi|)=v(\|\xi\|)+w(\xi)$, where v is smooth, $v=1$ on $\left[0, \frac{1}{4}\right]$ and supp $v\subset\left[0, \frac{1}{2}\right]$.

Clearly, the maximal operator $g\mapsto\sup_{t>0}|T_{v_t(\|\cdot\|)}g|$ is dominated by the Hardy—Littlewood maximal operator, hence bounded on $L^2(\mathbf{R}^n)$. Moreover, one checks easily that w satisfies the condition (1.1) of the proposition in Section 3 of Carbery [2], and so also the maximal operator associated with w is bounded on $L^2(\mathbf{R}^n)$. Together this implies (4.4). □

Corollary 2 Let $m(t)=\chi_{(0,1)}(t)$. Then for every $f\in L_{\varphi}^{2}$ with $\varphi(t)=1+t^{\mu}$ or $\varphi(t)=\ln^{\mu+1}(\mathrm{e}+t)(\mu>0)$ the following estimates hold, respectively:

$$T_{m_t}f(x)-f(x)=o\left(t^{\mu}\ln\frac{1}{t}\right)\quad \text{a. e.}\quad \text{as}\quad t\to 0^{+},\tag{4.5}$$

$$T_{m_t}f(x)-f(x)=o\left(\frac{1}{\ln^{\mu}\frac{1}{t}}\right)\quad \text{a. e.}\quad \text{as}\quad t\to 0^{+}.\tag{4.6}$$

Proof We take $h\in C^{\infty}(0,+\infty)$ satisfying (4.3). Define $\widetilde{m}=m-h$. Then supp $\widetilde{m}\subset\left[\frac{1}{2},1\right]$, and $\widetilde{m}$ is of bounded variation. By Theorem 3 we conclude that for $f\in L_{\varphi}^{2}$,

$$T_{\widetilde{m}_t}f(x)=o\left(\ln\frac{1}{t}\frac{1}{\varphi\left(\frac{1}{t}\right)}\right)\quad \text{a. e.}\quad \text{as}\quad t\to 0^{+}.\tag{4.7}$$

On the other hand the multiplier h satisfies the condition of Theorem 4. So for $f\in L_{\psi}^{2}$,

$$T_{h_t}f(x)-f(x)=o\left(\frac{1}{\varphi\left(\frac{1}{t}\right)}\right)\quad \text{a. e.}\quad \text{as}\quad t\to 0^{+}.\tag{4.8}$$

If $\varphi(t)=1+t^{\mu}(\mu>0)$, then we have $\varphi(t)\sim\psi(t)$. Hence, the combination of (4.7) and (4.8) yields (4.5).

If $\varphi(t)=\ln^{\mu}(\mathrm{e}+t)$, then $\psi(t)\leqslant c\ln^{\mu+1}(\mathrm{e}+t)$. Hence for $f\in L^{2}_{\ln^{\mu+1}(\mathrm{e}+t)}$ (4.7) and (4.8) imply (4.6). □

§ 5. Remarks

(1) Since we only consider convergence of $T_{m_t}f$ as $t\to 0^{+}$, it is clear that we could even replace λ in (1.9) by $\lambda'=\overline{\lim_{t\to+\infty}}\frac{\varphi(2t)}{\varphi(t)}$.

(2) We do not know whether the weight function ψ in Theorem 2 could even be replaced by φ.

(3) If one choues $m(t)=(1+t^{l})h(t)$, h has in (4.3), then obviously for every $f\in L^{2}$ with supp $\hat{f}\subset\{\xi:|\xi|\leqslant\frac{1}{4}\}$,

$$T_{m_t}f-f=t^{l}\widetilde{f},\quad 0<t<1,$$

where $(\widetilde{f})\check{}(\xi)=|\xi|^{t}\hat{f}(\xi)$.

This example shows that the condition $\beta>\mu$ in Theorem 4 is necessary for such a theorem.

(4) By Carbery's theorem, in the case of the Euclidean norm (4. 2) is also valid for those $f\in L^{p}(\mathbf{R}^{n})$ for which $\|D^{\mu}f\|_{L^{p}(\mathbf{R}^{n})}<+\infty$ for the range of p's described in Carbery's theorem.

References

[1] Carbery A. Radial Fourier multipliers and associated maximal functions, in: Recent progress in Fourier analysis, edited by Peral I, Rubio de Francia, T-L. 49—56, North-Holland, Amsterdam, 1985.

[2] Carbery A. An almost-orthogonality principle with applications to maximal functions associated to convex bodies. Bull. Am. Math. Soc., 1986, 14: 269—273.

[3] Carbery A, Gasper G, Trebels W. On localized potential spaces. J. Approximation Theory, 1986, 48: 251—261.

[4] Chen Tian-ping. Generalized Bochner—Riesz means of Fourier integrals. in: Multivariate approximation theory IV: Proceedings of the Mathematical Research Institute at Oberwolfach, Febr. 12—18, 1989, edited by Chui, C K. Schempp W, Zeller K. Birkhäuser, Basel—Boston, 1989.

[5] Cossar J. A theorem on Cesàro summability. J. London Math. Soc., 1941, 16: 56—68.

[6] Dappa H, Trebels W. On maximal functions generated by Fourier multipliers. Ark. Mat., 1985, 23: 241—259.

[7] Gasper G, Trebels W. A Characterization of localized Bessel potential spaces and applications to Jacobi and Hankel multipliers. Studia Math., 1979, 94: 243—278.

北京师范大学学报(自然科学版)，1994，30(2)：163－169.

用 Bochner-Riesz 平均逼近及 Hardy 求和

Approximation by Bochner-Riesz Means and Hardy Summability①

Abstract Let $f\in C(Q^n)$, $n\in\mathbf{N}$ and $S_R^{\frac{n-1}{2}}(f)$ be the Bochner-Riesz means of critical order of f. Get the estimate for the degree of (H, q) approximation:

$$\left\|\frac{1}{R}\int_0^R |f-S_r^{\frac{n-1}{2}}(f)|^q\mathrm{d}r\right\|_\infty \leqslant c\frac{1}{R}\int_0^R \left|\omega_2\left(f;\frac{1}{r}\right)\right|^q\mathrm{d}r,\quad R>0,$$

where ω_2 denotes the second modulus of continuity, $q>0$ and c is a constant. The saturation problem of this kind of approximation is also investigated.

Keywords strong approximation; Bochner-Riesz means; saturation.

§ 0. Introduction

Let $Q^n=\{(x_1, x_2, \cdots, x_n): -\pi\leqslant x_j<\pi, j=1,2,\cdots,n\}$, and let $L(Q^n)$ be the Banach space of all functions which are integrable on Q^n and have period 2π with respect to each variable, with the norm $\|f\|_1=\int_{Q^n}|f(x)|\mathrm{d}x$. All continuous functions in $L(Q^n)$ constitute another Banach space $C(Q^n)$ with max norm: $\|f\|_\infty=\max\{|f(x)|\,|\,x\in Q^n\}$.

① Supported by NSFC and the Australian Research Council.

本文与 Brown G 合作.

Received: 1993-11-10.

The Bochner-Riesz means of $f\in L(Q^n)$ are defined by (see [1])

$$S_R^\alpha(f,x) = \sum_{|m|<R} c_m(f)^{e^{imx}}(1-|m|^2R^{-2})^\alpha,$$

where $mx=m_1x_1+m_2x_2+\cdots+m_nx_n(m\in\mathbf{Z}_n)$, $|m|=(m_1^2+m_2^2+\cdots+m_n^2)^{\frac{1}{2}}$. The index α can be any complex number and $c_m(f)$ denotes the m-th Fourier coefficient of f, i. e.

$$c_m(f) = \frac{1}{(2\pi)^n}\int_{Q^n} f(x)\mathrm{e}^{-imx}\mathrm{d}x.$$

The special value $\alpha_0=\frac{n-1}{2}$ of the index is called critical. When $n=1$, S_R^0 is just the R-th Fourier sum. In the case $n>1$, many results show that $S_R^{\frac{n-1}{2}}$ can be regarded as an analogue of one dimensional Fourier sums (see [1] and [2] chapter 7). From this point of view we are concerned with convergence of

$$\frac{1}{R}\int_0^R |S_r^{\frac{n-1}{2}}(f,x)-f(x)|^q\mathrm{d}r,\quad R\to+\infty. \tag{0.1}$$

In other words, we consider Hardy summability with power $q>0$, briefly $(H,\ q)$ summability (see [3] for the one dimensional case), and we want to estimate the degree of (1) in norm $\|\cdot\|_\infty$. We call the max norm of (0.1), the $(H,\ q)$ approximation of f.

For a function $f\in C(Q^n)$ wc denote by $\omega_2(f,\ t)(t>0)$ its second modulus of continuity, that is

$$\omega_2(f,\ t)=\sup\{\|f(\cdot+h)+f(\cdot-h)-2f(\cdot)\|_\infty\}$$
$$h\in\mathbf{R}^n,\ |h|\leqslant t,\ t>0.$$

Our main result is the following.

Theorem 1 Let $q>0$. Then for any $f\in C(Q^n)$,

$$\left\|\frac{1}{R}\int_0^R|f-S_r^{\frac{n-1}{2}}(f)|^q\mathrm{d}r\right\|_\infty\leqslant c\frac{1}{R}\int_0^R\left|\omega_2\left(f,\frac{1}{r}\right)\right|^q\mathrm{d}r,\quad R>0,$$

where, and throughout this paper, c denotes a constant depending only on n and q.

In the particular case $q=2$, this was proved by Wang[4] using an analytic argument. For general $q>2$, that analytic argument does not work and we will apply the Hausdorff-Young inequality.

We define an operator "sequence" $A_q=\{A_R^q\}_{R>0}$ by

$$A_R^q(f)(x)=\frac{1}{R}\int_0^R |S_r^{\frac{n-1}{2}}(f,x)-f(x)|^q\mathrm{d}r. \tag{0.2}$$

Then we prove that the degree of $\|A_R^q(f)\|_\infty$ is in the best case $o\left(\frac{1}{R}\right)$, and, if for some $f\in C(Q^n)$ $\|A_R^q(f)\|_\infty=o\left(\frac{1}{R}\right)$ then f must be constant. This means that A_q is saturable with saturation $\left\{\frac{1}{R}\right\}$. This is our theorem 2. When $q\geqslant 1$, we will give a description of the functions for which A_q achieves the saturation degree. (theorem 3).

§ 1. Lemma

Let $0<q<+\infty$. There exists a constant $c=c(q, n)$ such that for and $f\in C(Q^n)$ and $R>0$,

$$\left\|\int_R^{2R}|S_r^{\frac{n-1}{2}}(f)|^q\mathrm{d}r\right\|_\infty\leqslant cR\|f\|_\infty^q. \tag{1.1}$$

Proof We quote Bochner's formula (see [1]):

$$S_r^\alpha(f;x)=2^{\alpha+1-\frac{n}{2}}\frac{\Gamma(\alpha+1)}{\Gamma\left(\frac{n}{2}\right)}\int_0^{+\infty}f_x(t)t^{n-1}r^n\frac{J_{\frac{n}{2}+\alpha}(rt)}{(rt)^{\frac{n}{2}+\alpha}}\mathrm{d}t,\quad r>0, \tag{1.2}$$

where J_v denotes the Bessel function of the first kind,

$$f_x(t)=\frac{\Gamma\left(\frac{n}{2}\right)}{2\pi^{\frac{n}{2}}}\int_{|\xi|=1}f(x+t\xi)\mathrm{d}\sigma(\xi),\quad x\in Q^n,\quad t>0,$$

and $\alpha>\frac{n-1}{2}$ (or generally Re $\alpha>\frac{n-1}{2}$). This formula holds only for indices strictly bigger than the critical one. Accordingly we assume $\beta\in(0, 1)$, $\alpha=\frac{n-1}{2}+\beta$, and treat $S_r^{\frac{n-1}{2}+\beta}$ first. We then let $\beta\to 0^+$ to get the result at the critical index.

From well - known results on Bessel functions (see [5]) we claim that there is a constant c, independent of $\beta\in(0, 1)$, such that for any

$r>0$,

$$\left|J_{n-\frac{1}{2}+\beta}(rt)-\sqrt{\frac{2}{\pi rt}}\cos\left(rt-\frac{n+\beta}{2}\pi\right)\right|\leqslant\frac{c}{(rt)^{\frac{3}{2}}},\ \text{if } t\geqslant\frac{1}{r} \tag{1.3}$$

and
$$|J_{n-\frac{1}{2}+\beta}(rt)|\leqslant c(rt)^{n-\frac{1}{2}+\beta},\ \text{if } 0<t<\frac{1}{r}. \tag{1.4}$$

Hence we get (for $\alpha=\frac{n-1}{2}+\beta$)

$$\begin{aligned}&|S_r^\alpha(f;x)|\\ &\leqslant c\int_0^{\frac{1}{r}}\|f\|_\infty t^{n-1}r^n\mathrm{d}t+c\left|\int_{\frac{1}{r}}^{+\infty}f_x(t)t^{n-1}r^n\frac{\cos\frac{rt-(n+\beta)\pi}{2}}{(rt)^{n+\beta}}\mathrm{d}t\right|+\\ &\quad c\left|\int_{\frac{1}{r}}^{+\infty}f_x(t)t^{n-1}r^n\frac{1}{(rt)^{n+1+\beta}}\mathrm{d}t\right|.\end{aligned} \tag{1.5}$$

It is obvious that the third term of right hand side of (1.5) does not exceed

$$c\|f\|_\infty\int_{\frac{1}{r}}^{+\infty}\frac{\mathrm{d}t}{r^{1+\beta}t^{2+\beta}}=\frac{c}{1+\beta}\|f\|_\infty\leqslant c\|f\|_\infty. \tag{1.6}$$

Hence we have

$$|S_r^\alpha(f;x)|\leqslant c\|f\|_\infty+\frac{c}{r^\beta}\left|\int_{\frac{1}{r}}^{+\infty}f_x(t)\frac{1}{t^{1+\beta}}\cos\left(rt-\frac{n+\beta}{2}+\pi\right)\mathrm{d}t\right|. \tag{1.7}$$

We temporarily fix $x\in Q^n$, $R>0$, and $\beta\in(0,1)$ and define

$$Q(t)=\begin{cases}f_x(t)t^{-1-\beta}, & \text{if } \frac{1}{2R}\leqslant t<+\infty,\\ 0, & \text{if } 0\leqslant t<\frac{1}{2R}.\end{cases}$$

Then we see that $Q\in L^p(0,+\infty)$, ⊗$p\geqslant1$.

Now we assume $q\geqslant2$ and $p=\frac{q}{q-1}\in(1,2]$. For $r\in[R,2R]$ $(R>0)$, we have

$$\begin{aligned}&\int_{\frac{1}{r}}^{+\infty}Q(t)\cos(rt-\frac{n+\beta}{2}\pi)\mathrm{d}t\\ &=\int_{\frac{1}{2R}}^{\frac{1}{r}}Q(t)\cos(rt-\frac{n+\beta}{2}\pi)\mathrm{d}t+\int_0^{+\infty}Q(t)\cos(rt-\frac{n+\beta}{2}\pi)\mathrm{d}t.\end{aligned} \tag{1.8}$$

Obviously, $$\left|\int_{\frac{1}{2R}}^{\frac{1}{r}} Q(t)\cos(rt-\frac{n+\beta}{2}\pi)\,\mathrm{d}t\right|\leqslant c\|f\|_{\infty}R^{\beta}. \tag{1.9}$$

So, from (1.7)~(1.9) we obtain, for $r\in[R, 2R]$,

$$|S_r^{\alpha}(f;x)|\leqslant c\|f\|_{\infty}+c\frac{1}{r^{\beta}}\left|\int_0^{+\infty} Q(t)\cos(rt-\frac{n+\beta}{2}\pi)\,\mathrm{d}t\right|. \tag{1.10}$$

Therefore we have

$$\int_R^{2R}|S_r^{\alpha}(f;x)|^q\mathrm{d}r$$
$$\leqslant cR\|f\|_{\infty}^q+\frac{c}{R^{\beta q}}\int_R^{2R}\left|\int_0^{+\infty} Q(t)\cos\left(rt-\frac{n+\beta}{2}\pi\right)\mathrm{d}t\right|^q\mathrm{d}r, \tag{1.11}$$

where the constant c depends on q, n but not on $\beta\in(0, 1)$. Since we assume $q\geqslant 2$, $p=\frac{q}{q-1}$, we can apply the Hausdorff - Young inequality (see [2]) on $Q\in L^p(0, +\infty)$. Then we obtain

$$\int_R^{2R}\left|\int_0^{+\infty} Q(t)\cos\left(rt-\frac{n+\beta}{2}\pi\right)\mathrm{d}t\right|^q\mathrm{d}r$$
$$\leqslant c\left(\int_0^{+\infty}|Q(t)|^q\mathrm{d}t\right)^{\frac{q}{p}}$$
$$=c\|f\|_{\infty}^q\left[\frac{1}{(1+\beta)p-1}(2R)^{(1+\beta)p-1}\right]^{\frac{q}{p}}$$
$$\leqslant c\|f\|_{\infty}^q R^{1+\beta q}, \tag{1.12}$$

where c is independent of $\beta\in(0, 1)$. A combination of (1.11) and (1.12) yields

$$\int_R^{2R}|S_r^{\alpha}(f)|^q\mathrm{d}r\leqslant cR\|f\|_{\infty}^q,\quad q\geqslant 2, \tag{1.13}$$

with $\alpha=\frac{n-1}{2}+\beta$, $\beta\in(0, 1)$, and a constant c independent of β. Let $\beta\to 0^+$ in (1.13). As a limit we get (1.1) for $q\geqslant 2$.

If $0<q<2$, we apply Hölder's inequality go get

$$\int_R^{2R}|S_r^{\alpha}(f)|^q\mathrm{d}r\leqslant\left(\int_R^{2R}|S_r^{\alpha}(f)|^2\mathrm{d}r\right)^{\frac{q}{2}}R^{1-\frac{q}{2}}. \tag{1.14}$$

Let $\alpha=\frac{n-1}{2}$. Applying (1.1) for $q=2$ (part of the case just proved) we get

$$\int_R^{2R} | S_r^{\frac{n-1}{2}}(f) |^2 \mathrm{d}r |^2 \leqslant cR^{\frac{q}{2}} \| f \|_\infty^q.$$

Substituting into (1.14) we get (1.1) for $q \in (0, 2)$ and this completes the proof.

§ 2. Proof of theorem 1

We say that a trigonometric polynomial is of order $R(R \geqslant 0)$, if it has the following form:

$$\sum_{|m| \leqslant R} a_m \mathrm{e}^{\mathrm{i}mx} (x \in \mathbf{R}^n).$$

Let $f \in C(Q^n)$. We define

$E_R(f) = \inf\{ \| f - T_R \|_\infty :$ T_R is trigonometric polynomial of order $R\}$, $(R \geqslant 0)$.

We call $E_R(f)$ the best approximation of f by trigonometric polynomials. It is easy to see that for any $R > 0$, there exists a trigonometric polynomial of order R, say T_R^*, such that

$$E_R(f) = \| f - T_R^* \|_\infty.$$

This T_R^* is called the trigonometric polynomial of best approximation of order R of f, or briefly the TPBA of order R of f.

Let $k \in \mathbf{N}^*$. Assume $R \in [2^k, 2^{k+1})$. We have

$$\int_1^R S_r^{\frac{n-1}{2}}(f) f |^q \mathrm{d}r \leqslant \sum_{j=0}^k \int_{2^j}^{2^{j+1}} | S_r^{\frac{n-1}{2}}(f) - f |^q \mathrm{d}r. \tag{2.1}$$

For each $j \in \mathbf{N}$, Let T_j be the TPBA of order 2^j of j. Then for $r \in [2^j, 2^{j+1})$,

$$| S_r^{\frac{n-1}{2}}(f) - f |^q \leqslant c\{ | S_r^{\frac{n-1}{2}}(f - T_j) |^q + | S_r^{\frac{n-1}{2}}(T_j) - S_r^{\frac{n+1}{2}}(T_j) |^q + | S^{\frac{n+1}{2}}(T_j) - T_j |^q + | T_j - f |^q \}. \tag{2.2}$$

According to the lemma we have

$$\int_{2^j}^{2^{j+1}} | S_r^{\frac{n-1}{2}}(f - T_j) |^q \mathrm{d}r \leqslant c 2^j \| f - T_j \|_\infty^q. \tag{2.3}$$

Noting

$$S_r^{\frac{n-1}{2}}(T_j) - S_r^{\frac{n+1}{2}}(T_j) = -\left(\frac{1}{r^2}\right) S_r^{\frac{n-1}{2}}(\nabla^2 T_j),$$

where ∇^2 denotes the Laplace operator: $\nabla^2=\frac{\partial^2}{\partial x_1^2}+\frac{\partial^2}{\partial x_2^2}+\cdots+\frac{\partial^2}{\partial x_n^2}$, and applying the lemma to the function $\nabla^2 T_j$ we obtain

$$\int_{2^j}^{2^{j+1}}|S_r^{\frac{n-1}{2}}(T_j)-S_r^{\frac{n+1}{2}}(T_j)|^q\mathrm{d}r\leqslant c\,2^{j-2jq}\|\nabla^2 T_j\|_\infty^q. \tag{2.4}$$

On the other hand, by the n-dimensional extension of a theorem of [8] (or see [7]) we have

$$\|\nabla^2 T_j\|_\infty\leqslant c\,2^{2j}\omega_2(f;\ 2^{-j}). \tag{2.5}$$

Applying a theorem of Wang [6], we have

$$\|S_r^{\frac{n+1}{2}}(T_j)-T_j\|_\infty\leqslant c\omega_2\left(T_j;\ \frac{1}{r}\right). \tag{2.6}$$

Since T_j is the TPBA of order 2^j of f, we have for $r\in[2^j,\ 2^{j+1}]$,

$$\omega_2\left(T_j;\ \frac{1}{r}\right)\leqslant c\omega_2(f;\ 2^{-j}). \tag{2.7}$$

Finally, by Jackson's theorem

$$\|f-T_j\|_\infty=E_{2^j}(f)\leqslant c\omega_2(f;\ 2^{-j}). \tag{2.8}$$

A combination of (2.2)～(2.8) yields

$$\int_{2^j}^{2^{j+1}}|S_r^{\frac{n-1}{2}}(f)-f|^q\mathrm{d}r\leqslant c2^j[\omega_2(f;2^{-j})]^q\leqslant c2^j[\omega_2(f;2^{-j-1})]^q. \tag{2.9}$$

Substituting (2.9) into (2.1) we get

$$\int_1^R|S_r^{\frac{n-1}{2}}(f)-f|^q\mathrm{d}r\leqslant c\sum_{j=0}^k 2^j[\omega_2(f;2^{-j-1})]^q\leqslant c\int_1^R\left[\omega_2\left(f;\frac{1}{r}\right)\right]^q\mathrm{d}r. \tag{2.10}$$

If $0<R<2$ then

$$\begin{aligned}&|S_R^{\frac{n-1}{2}}(f;x)-f(x)|\\=&\frac{1}{(2\pi)^n}\left|\int_{Q^n}[f(x+y)-f(x)]D_R^{\frac{n-1}{2}}(y)\mathrm{d}y\right|\\=&\frac{1}{(2\pi)^n}\left|\int_{0<y_j<\pi,j=1,2,\cdots,n}[f(x+y)+f(x-y)-2f(x)]D_R^{\frac{n-1}{2}}(y)\mathrm{d}y\right|\\\leqslant& c\omega_2(f;1),\end{aligned}$$

where $D_R^{\frac{n-1}{2}}(y)=\sum_{|m|<R}\mathrm{e}^{imy}$, is the n-dimensional spherical Dirichlet kernel. Hence

$$\int_0^2 |S_r^{\frac{n-1}{2}}(f)-f|^q \mathrm{d}r \leqslant c\int_0^2 \left[\omega_2\left(f;\frac{1}{r}\right)\right]^q \mathrm{d}r. \tag{2.11}$$

By the inequalities (2. 10) and (2. 11) we complete the proof.

§ 3. Saturation problem

From theorem 1 we see that if f is smooth enough such that

$$\int_0^{+\infty}\left(\omega_2\left(f;\frac{1}{r}\right)\right)^q \mathrm{d}r<+\infty,$$

then $\|A_R^q(f)\|_\infty \leqslant \frac{c}{R}$. Now we assume for some $f\in C(Q^n)$,

$$\|A_R^q(f)\|_\infty = o\left(\frac{1}{R}\right), \qquad R\to+\infty,$$

i. e.

$$\lim_{R\to+\infty}\left\|\int_0^R |S_r^{\frac{n-1}{2}}(f)-f|^q \mathrm{d}r\right\|_\infty = 0.$$

Then we see $|S_r^{\frac{n-1}{2}}(f)-f|\equiv 0$ $\otimes r>0$, $\otimes x\in Q^n$, which shows that f is constant. we have now demonstrated.

Theorem 2 The operator "sequence" $A_q=\{A_R^q\}_{r>0}$ $(q>0)$ is saturable with saturation degree $\left\{\frac{1}{R}\right\}$ $(R\to+\infty)$.

Now we define the saturation class of A_q to be

$$S(A_q)=\left\{f\in C(Q^n) \;\middle|\; \|A_R^q(f)\|_\infty = o\left(\frac{1}{R}\right)\right\}.$$

We are going to prove the following.

Theorem 3 If $f\in S(A_q)$, $q\geqslant 1$, then

$$\omega_2(f;\ t)=o(t^{\frac{1}{q}}), \qquad t\to 0. \tag{3.1}$$

Proof By the definition of $E_R(f)$, we have

$$\begin{aligned} E_R(f) &\leqslant \left\|\frac{1}{R}\int_0^R S_r^{\frac{n-1}{2}}(f)\mathrm{d}r - f\right\|_\infty \\ &\leqslant \left\|\frac{1}{R}\int_0^R |S_r^{\frac{n-1}{2}}(f)-f|^q \mathrm{d}r\right\|_\infty^{\frac{1}{q}}, \quad q\geqslant 1. \end{aligned} \tag{3.2}$$

Since $\|A_R^q(f)\|_\infty = o\left(\frac{1}{R}\right)$, we get

$$E_R(f)\leqslant c\left(\frac{1}{R}\right)^{\frac{1}{q}}. \tag{3.3}$$

In the one dimensional case, from [7](6.1.2) we know that (3.1) and (3.3) are equivalent. In general, in order to prove that (3.3) implies (3.1), we may use the following inequality:

$$\omega_2(f;t)\leqslant ct^2\int_0^{\frac{1}{t}} rE_r(f)\mathrm{d}r,\quad t>0 \tag{3.4}$$

In fact a more comprehensive result is true:

$$\omega_k(f;t)\leqslant c_k t^k\int_0^{\frac{1}{t}} r^{k-1}E_r(f)\mathrm{d}r, \tag{3.5}$$

where $k\in\mathbf{N}$, and

$$\omega_k(f;\ t)=\sup\{\|\Delta_h^k f(\cdot)\|_\infty \mid h\in\mathbf{R}^n,\ |h|\leqslant t\},$$

with

$$\Delta_k^h f(x)=\sum_{j=0}^{k}(-1)^{k-j}C_k^j f(x+jh),$$

This extends from the one - dimensional case which was discussed in [7], 6.1.1(1) and uses similar arguments. We omit the details.

Remark on theorem 3 Since we can not get (3.2) or a similar estimate for the case $q<1$, we only get the result for $q\geqslant 1$. But if $n=1$, we can still get an estimate similar to (3.2) when $q\in(0,\ 1)$. And then we can by this estimate and (3.4), prove that (3.5) still holds. Hence if $n=1$, and $q\in(0,\ 1)$ then for $f\in S(A_q)$,

$$\omega_2(f;\ t)=\begin{cases} o(t^{\frac{1}{q}}), & \text{if } q>\frac{1}{2},\\ o(t^2\ln t^{-1}), & \text{if } q=\frac{1}{2},\\ o(t^2), & \text{if } 0<q<\frac{1}{2}.\end{cases}$$

We guess that this is also true when $n>1$.

References

[1] Bochner S. Summation of multiple Fourier series by spherical means. Trans Amer Math Soc, 1936, 40: 175.

[2] Stein E M, Weiss G. Introduction to Fourier analysis on Euclidean spaces. Princeton: Princeton University Press, 1971.

[3] Zygmund A. Trigonometric series, Vol. I. Cambridge: [s. n.], 1959.

[4] Wang Kunyang. Strong uniform approximation by Bochner-Riesz means. Multivariate approximation IV. In: Proceedings of conference on multivariate approximation theory. Oberwolfach, West Germany, 1989.

[5] Watson G N. Theory of Bessel functions. Cambridge: [s. n.], 1952.

[6] Wang Kunyang. Approximation for continuous periodic function of several variables and its conjugate functions by Riesz means on set of total measure. Approximation theory and its applications, 1985, 4(1): 19.

[7] Тиман А Ф. Теория прибрижения функцийдействитейлйного переменного. [s. 1.]: физматгиз, 1960.

[8] Стечкин С Б. О порядке найлучших прибрижений непрерывных фкнкций. Изв АН СССР, Сер Матем, 1951, 15: 219.

摘要 设 $f\in C(Q^n)$，$n\in\mathbf{N}$ 且 $S_{R^2}^{\frac{n-1}{2}}(f)$是 f 的临界阶 Bochner-Riesz 平均. 求得了(H, q) 逼近的阶的估计:

$$\left\|\frac{1}{R}\int_0^R |f-S_r^{\frac{n-1}{2}}(f)|^q \mathrm{d}r\right\|_\infty \leqslant c\frac{1}{R}\int_0^R \left|\omega_2\left(f;\frac{1}{r}\right)\right|^q \mathrm{d}r,\quad R>0,$$

其中 ω_2表示二阶连续模，$q>0$ 且 c 是常数. 同时研究了这类逼近的饱和问题.

关键词 强逼近；Bochner-Riesz 平均；饱和.

Proceedings of the American Mathematical Society,
1998, 126(12): 3 527—3 537.

关于 F. Móricz 的一个猜测

On a Conjecture of F. Móricz①

Abstract F. Móricz has investigated the integrability of double lacunary sine series. His result, valid for special lacunary sequences, does not extend in the form originally conjectured, but we establish a suitably modified result.

Keywords Double sine series; lacunarity; integrability.

§ 1. Introduction

Let a_{ij}, i, $j\in\mathbf{N}$, be real numbers satisfying the condition

$$\sigma=\left\{\sum_{i=1}^{+\infty}\sum_{j=1}^{+\infty}a_{ij}^{2}\right\}^{\frac{1}{2}}<+\infty, \tag{1.1}$$

Suppose $q>1$ and m_i, n_j are positive numbers satisfying

$$\frac{m_{i+1}}{m_i}\geqslant q,\quad \frac{n_{j+1}}{n_j}\geqslant q,\quad m_1=n_1=1,\ i,\ j\in\mathbf{N}. \tag{1.2}$$

Define

$$f(x,y)=\sum_{i=1}^{+\infty}\sum_{j=1}^{+\infty}a_{ij}\sin m_i x\ \sin n_j y,$$

① This work was supported by a grant from the Australian Research Council.

本文与 Brown G 合作. Communicated by J. Marshall Ash.

Received: 1993-06-22; Revised: 1997-02-12.

$$g_j(x)=\sum_{i=1}^{+\infty}a_{ij}\sin m_i x,\quad h_j(y)=\sum_{j=1}^{+\infty}a_{ij}\sin n_j y.$$

In the general case these limits are to be understood in the sense of L^2-convergence and, as Lemma 1 shows, there is no inherent ambiguity in the definition.

F. Móricz [3] considered the special case when $m_i=n_i=2^{i-1}$, $i\in \mathbf{N}$. In this case he proved that the condition

$$\sum_{i=1}^{+\infty}\sum_{j=1}^{+\infty}\Big(\sum_{k=i}^{+\infty}\sum_{l=j}^{+\infty}a_{kl}^2\Big)^{\frac{1}{2}}<+\infty \tag{1.3}$$

is equivalent to

$$\frac{f(x,y)}{xy}\in L(0,1)^2,\ \frac{g_i(x)}{x}\in L(0,1),\ \frac{h_i(y)}{y}\in L(0,1),\ i\in \mathbf{N}. \tag{1.4}$$

He proposed that in the general case when m_i, n_j are positive integers satisfying condition (1.2) and the integrability condition, then (1.4) is satisfied if and only if

$$\sum_{i=1}^{+\infty}\sum_{j=1}^{+\infty}\ln\frac{m_{i+1}}{m_i}\ln\frac{n_{j+1}}{n_j}\Big(\sum_{k=i}^{+\infty}\sum_{l=j}^{+\infty}a_{kl}^2\Big)^{\frac{1}{2}}<+\infty. \tag{1.5}$$

Our result is the following

Theorem Let a_{ij}, m_i, n_j satisfy (1.1) and (1.2). Let f, g_i, h_i be as above. Define

$$S=\sum_{i=1}^{+\infty}\sum_{j=1}^{+\infty}|a_{ij}|,\quad T=\sum_{i=1}^{+\infty}\sum_{j=1}^{+\infty}\ln\frac{m_{i+1}}{m_i}\Big(\sum_{k=i+1}^{+\infty}a_{kj}^2\Big)^{\frac{1}{2}},$$

$$U=\sum_{i=1}^{+\infty}\sum_{j=1}^{+\infty}\ln\frac{n_{j+1}}{n_j}\Big(\sum_{l=j+1}^{+\infty}a_{il}^2\Big)^{\frac{1}{2}},$$

$$V=\sum_{i=1}^{+\infty}\sum_{j=1}^{+\infty}\ln\frac{m_{i+1}}{m_j}\ln\frac{n_{j+1}}{n_j}\Big(\sum_{k=i+1}^{+\infty}\sum_{l=j+1}^{+\infty}a_{kl}^2\Big)^{\frac{1}{2}}.$$

Then the condition (1.4) is equivalent to the condition

$$S+T+U+V<+\infty. \tag{1.6}$$

We point out that in our theorem, m_i, m_j need not be integers. If $m_i=n_i=2^{i-1}$, then (1.5) is equivalent to (1.6). But in general, as the following example shows, (1.5) is stronger than (1.6) and they are not equivalent.

First we note that the one-dimensional case is subsumed by the two-dimensional case. If we set $a_{ij}=b_i$ for $j=1$ and $a_{ij}=0$ for $j>1$, then (1. 5) reads

$$\sum_{i=1}^{+\infty}\ln\frac{m_{i+1}}{m_i}\Big(\sum_{j=i}^{+\infty}b_j^2\Big)^{\frac{1}{2}}<+\infty \tag{1.5'}$$

and (1. 6) reads

$$\sum_{i=1}^{+\infty}|b_i|+\sum_{i=1}^{+\infty}\ln\frac{m_{i+1}}{m_i}\Big(\sum_{j=i+1}^{+\infty}b_j^2\Big)^{\frac{1}{2}}<+\infty. \tag{1.6'}$$

Let $b_i=\{\mathrm{e}^{-2^{i+1}}-\mathrm{e}^{-2^{i+2}}\}^{\frac{1}{2}}$ and $m_i=\prod_{l=1}^{i}\mathrm{e}^{\mathrm{e}^{2^{l-1}}}$. Then we see that (1. 6′) holds but (1. 5′) does not.

We present the proof of the Theorem in two parts:

Theorem 1 Let $d=1+\frac{4}{(q-1)^2}+\frac{2}{(q-1)}\sqrt{\frac{4}{(q-1)^2}+8}$. Then

$$\int_0^d|g_i(x)|\frac{\mathrm{d}x}{x}\leqslant c_q\Big\{\sum_{i=1}^{+\infty}|a_{ij}|+\sum_{i=1}^{+\infty}\ln\frac{m_{i+1}}{m_i}\Big(\sum_{k=i+1}^{+\infty}a_{kj}^2\Big)^{\frac{1}{2}}\Big\}, \tag{1.7}$$

$$\int_0^d|h_i(y)|\frac{\mathrm{d}y}{y}\leqslant c_q\Big\{\sum_{j=1}^{+\infty}|a_{ij}|+\sum_{j=1}^{+\infty}\ln\frac{n_{j+1}}{n_j}\Big(\sum_{l=j+1}^{+\infty}a_{il}^2\Big)^{\frac{1}{2}}\Big\}, \tag{1.8}$$

$$\int_0^d\int_0^d|f(x,y)|\frac{\mathrm{d}x\mathrm{d}y}{xy}\leqslant c_q(S+T+U+V). \tag{1.9}$$

This theorem demonstrates that (1. 6) implies (1. 4).

Theorem 2 (1. 4) implies (1. 6)

We point out that the number $d=d(q)$ in Theorem 1 is just the positive root of the equation:

$$\frac{1}{4}(d-1)^2-\frac{2}{(q-1)^2}(d+1)-\frac{4}{(q-1)^2}=0.$$

This particular definition of d will be retained throughout the paper.

§ 2. Proof of theorem 1

Lemma 1 Let a, b be arbitrary real numbers and $Q=(a,\ a+d)\times(b,\ b+d)$. Then
$$0.001\,\sigma\leqslant\Big\{\frac{1}{d^2}\int_Q f^2(x,y)\mathrm{d}x\mathrm{d}y\Big\}^{\frac{1}{2}}\leqslant\sigma.$$

Proof Let $I_{ijkl}=\int_Q\sin m_ix\sin m_kx\sin n_j\sin n_ly\mathrm{d}x\mathrm{d}y$. We have

$\int_Q f^2(x,y)\mathrm{d}x\mathrm{d}y = \sum a_{ij}a_{kl}I_{ijkl}$, where the sum $\sum$ is taken over $\mathbf{N}^4$. If $i \neq k$ and $j \neq l$, then

$$|I_{ijkl}| \leqslant \left(\frac{1}{|m_i - m_k|} + \frac{1}{m_i + m_k}\right)\left(\frac{1}{|n_j - n_l|} + \frac{1}{n_j + n_l}\right).$$

Let S_1 denote the subset of $\mathbf{N}^4$ defined by $S_1 = \{(i,j,k,l) \in \mathbf{N}^4 : i \neq k, j \neq l\}$ and let $\sum_1$ denote the sum taken over S_1. By schwarz's inequality $\left|\sum_1 S_{ij}a_{kl}I_{ijkl}\right| \leqslant \left\{\sum_1 |a_{ij}a_{kl}|^2\right\}^{\frac{1}{2}}\left\{\sum_1 I_{ijkl}^2\right\}^{\frac{1}{2}}$. Applying condition (1.2) we find $\sum_1 I_{ijkl}^2 \leqslant 16\left(\frac{1}{1-q^{-2}}\right)^2\left(\frac{1}{q^2-1}\right)^2$ and hence

$$\left|\sum_1 a_{ij}a_{kl}I_{ijkl}\right| \leqslant \left\{\frac{4}{(q-1)^2} - a_q\right\}\sigma^2,$$

where $a_q = \frac{4}{(q+1)(q-1)^2}$.

If $i=k$ and $j \neq l$, then $|I_{ijkl}| \leqslant \frac{d+1}{2}\left(\frac{1}{|n_j - n_l|} + \frac{1}{n_j + n_l}\right)$. Applying condition (1.2) again we find $\sum_2 |a_{ij}a_{kl}I_{ijkl}| \leqslant \left\{\frac{2(d+1)}{(q-1)^2} - b_q\right\}\sigma^2$, where $\sum_2$ denotes the sum over the set $\{i \neq k,\ j=l\} \cup \{i=k,\ j \neq l\}$ and $b_q = \frac{2(d+1)}{(q-1)^2(q+1)}$.

Finally, if $(i, j)=(k, l)$, then by condition (1.2) we have

$$\frac{1}{4}\left\{d - \max\left(|\sin d|,\ \frac{1}{q}\right)\right\}^2 \leqslant I_{ijkl} \leqslant \frac{1}{4}(d+1)^2,$$

$$\frac{1}{4}(d-1)^2\sigma^2 \leqslant \sum_{i=1}^{+\infty}\sum_{j=1}^{+\infty}\sigma_{ij}^2 I_{ijij} \leqslant \frac{1}{4}(d+1)^2\sigma^2. \tag{2.1}$$

Combining the estimates for $\sum_1$, $\sum_2$ and (2.1), noticing

$$\frac{1}{4}(d-1)^2 - \frac{2(d+1)}{(q-1)^2} - \frac{4}{(q-1)^2} = 0$$

we get $\sqrt{a_q + b_q}\sigma \leqslant \left\{\int_Q f^2(x,y)\mathrm{d}x\mathrm{d}y\right\}^{\frac{1}{2}} \leqslant d\sigma$. If $1 < q \leqslant 31$, then $\sqrt{a_q + b_q} > \frac{d}{850}$ and hence $\left\{\frac{1}{|Q|}\int_Q f^2(x,y)\mathrm{d}x\mathrm{d}y\right\}^{\frac{1}{2}} \geqslant \frac{\sigma}{850}$. If $q \geqslant 31$, then $d < 1.2$ and hence $I_{ijij} \geqslant \frac{1}{4}(d - \sin 1.2)^2$. So we modify (2.1) to get

$$\int_Q f^2(x,y)\mathrm{d}x\mathrm{d}y \geqslant \{a_q + b_q + 0.034^2\}\sigma^2$$

and hence for $q>31$ we have $\left\{\frac{1}{|Q|}\int_Q f^2(x,y)\mathrm{d}x\mathrm{d}y\right\}^{\frac{1}{2}} > 0.028\sigma$. The combination of these estimates completes the proof.

The following lemma is a direct corollary of Lemma 1.

Lemma 2 Let a_j be real numbers satisfying $A = \left(\sum_{j=1}^{+\infty} a_j^2\right)^{\frac{1}{2}} < +\infty$ and let m_j be numbers satisfying the condition:

$$m_1 = 1, \quad \frac{m_{i+1}}{m_i} \geqslant q > 1. \tag{2.2}$$

Define $\psi(x) = \sum_{j=1}^{+\infty} a_j \sin m_j x$. Then for any $a \in \mathbf{R}$,

$$\frac{1}{200}A \leqslant \left\{\frac{1}{d}\int_a^{a+d} \psi^2(x)\mathrm{d}x\right\}^{\frac{1}{2}} \leqslant 2A.$$

Proof of theorem 1 We first prove (1.7). For fixed $j \in \mathbf{N}$, we omit the subscript j, and write $g = g_j$, $a_i = a_{ij}$ for simplicity. Then we have

$$\int_0^d x^{-1}|g(x)|\mathrm{d}x \leqslant \sum_{i=1}^{+\infty}\int_{\frac{d}{m_{i+1}}}^{\frac{d}{m_i}} x^{-1}\{|S_i(x)| + |T_i(x)|\}\mathrm{d}x \tag{2.3}$$

where $S_i(x) = \sum_{k=1}^{i} a_k \sin m_k x$, $T_i(x) = \sum_{k=i+1}^{+\infty} a_k \sin m_k x$. Since $|\sin m_k x| \leqslant m_k|x|$ we see $\int_{\frac{d}{m_{i+1}}}^{\frac{d}{m_i}} |S_i(x)|\frac{\mathrm{d}x}{x} \leqslant d\sum_{k=1}^{i}|a_k|\frac{1}{q^{i-k}}$, and

$$\sum_{i=1}^{+\infty}\int_{\frac{d}{m_{i+1}}}^{\frac{d}{m_i}} x^{-1}|S_i(x)|\mathrm{d}x \leqslant c_q\sum_{i=1}^{+\infty}|a_{ij}|. \tag{2.4}$$

Simultaneously, writing $c_k = a_{i+k,j}$, $u_k = \frac{m_{i+k}}{m_{i+1}}$ we get

$$\int_{\frac{d}{m_{i+1}}}^{\frac{d}{m_i}} |T_i(x)|\frac{\mathrm{d}x}{x} = \int_d^{\frac{m_{i+1}}{m_i}d}\left|\sum_{k=1}^{+\infty} c_k \sin u_k x\right|\frac{\mathrm{d}x}{x}.$$

Let $m = \left[\frac{m_{i+1}}{m_i}\right]$ and $\psi(x) = \sum_{k=1}^{+\infty} c_k \sin u_k x$. Then

$$\int_d^{\frac{m_{i+1}}{m_i}d} |\psi(x)|\frac{\mathrm{d}x}{x} \leqslant \sum_{\mu=1}^{m}\int_{\mu d}^{(\mu+1)d}|\psi(x)|\frac{\mathrm{d}x}{x}.$$

By Lemma 2 we have $\int_{\mu d}^{(\mu+1)d} |\psi(x)| \frac{\mathrm{d}x}{x} \leqslant \frac{2}{\mu} (\sum_{k=i+1}^{+\infty} a_{kj}^2)^{\frac{1}{2}}$ and hence

$$\int_{\frac{d}{m_{i+1}}}^{\frac{d}{m_i}} |T_i(x)| \frac{\mathrm{d}x}{x} \leqslant c_q \ln \frac{m_{i+1}}{m_i} \{ \sum_{k=i+1}^{+\infty} a_{kj}^2 \}^{\frac{1}{2}}, \tag{2.5}$$

$$\sum_{i=1}^{+\infty} \int_{\frac{d}{m_{i+1}}}^{\frac{d}{m_i}} x^{-1} |T_i(x)| \mathrm{d}x \leqslant c_q \sum_{i=1}^{+\infty} \ln \frac{m_{i+1}}{m_i} (\sum_{k=i+1}^{+\infty} a_{kj}^2)^{\frac{1}{2}}. \tag{2.6}$$

Combining (2.3)(2.4) and (2.6) we get (1.7). By symmetry (1.8) follows.

Now we write $I_{ij} = \int_{\frac{d}{m_{i+1}}}^{\frac{d}{m_i}} \int_{\frac{d}{n_{j+1}}}^{\frac{d}{n_j}} |f(x,y)| \frac{\mathrm{d}x\mathrm{d}y}{xy}$ and define

$$\phi_{kl}(x, y) = a_{kl} \sin m_k x \sin n_l y,$$

$$f_1(x,y) = \sum_{k=1}^{i} \sum_{l=1}^{j} \phi_{kl}(x,y), \qquad f_2(x,y) = \sum_{k=i+1}^{+\infty} \sum_{l=1}^{j} \phi_{kl}(x,y),$$

$$f_3(x,y) = \sum_{k=1}^{i} \sum_{l=j+1}^{+\infty} \phi_{kl}(x,y), \qquad f_4(x,y) = \sum_{k=i+1}^{+\infty} \sum_{l=j+1}^{+\infty} \phi_{kl}(x,y).$$

Then define $I_{ij}^{(\nu)} = \int_{\frac{d}{m_{i+1}}}^{\frac{d}{m_i}} \int_{\frac{d}{n_{j+1}}}^{\frac{d}{n_j}} |f_\nu(x,y)| \frac{\mathrm{d}x\mathrm{d}y}{xy}, \nu = 1,2,3,4$. We see $I_{ij}^{(1)} \leqslant d^2 \sum_{k=1}^{i} \sum_{l=1}^{j} |a_{kl}| \frac{1}{q^{i-k}} \cdot \frac{1}{q^{j-1}}$ and hence

$$\sum_{i=1}^{+\infty} \sum_{j=1}^{+\infty} I_{ij}^{(1)} \leqslant c_q \sum_{i=1}^{+\infty} \sum_{j=1}^{+\infty} |a_{ij}| = c_q S. \tag{2.7}$$

Secondly, $I_{ij}^{(2)} \leqslant \sum_{l=1}^{j} \frac{d}{q^{j-l}} \int_{\frac{d}{m_{i+1}}}^{\frac{d}{m_i}} \left| \sum_{k=i+1}^{+\infty} a_{kl} \sin m_k x \right| \frac{\mathrm{d}x}{x}$. So by (2.5)

$$\int_{\frac{d}{m_{i+1}}}^{\frac{d}{m_i}} \left| \sum_{k=i+1}^{+\infty} a_{kl} \sin m_k x \right| \frac{\mathrm{d}x}{x} \leqslant c_q \ln \frac{m_{i+1}}{m_i} (\sum_{k=i+1}^{+\infty} a_{kl}^2)^{\frac{1}{2}}.$$

Therefore

$$\sum_{i=1}^{+\infty} \sum_{j=1}^{+\infty} I_{ij}^{(2)} \leqslant c_q T. \tag{2.8}$$

Symmetrically we have

$$\sum_{i=1}^{+\infty} \sum_{j=1}^{+\infty} I_{ij}^{(3)} \leqslant c_q U. \tag{2.9}$$

Finally, to estimate $I_{ij}^{(4)}$ we change the integrating variables by writing $x = \frac{s}{m_{i+1}}$, $y = \frac{t}{n_{j+1}}$. Then we get

$$I_{ij}^{(4)}=\int_d^{\frac{m_{i+1}}{m_i}d}\int_d^{\frac{n_{j+1}}{n_j}d}\left|\sum_{k=1}^{+\infty}\sum_{l=1}^{+\infty}b_{kl}\sin u_k s\sin v_l t\right|\frac{\mathrm{d}s\mathrm{d}t}{st},$$

where $b_{kl}=a_{i+k,j+l}$, $u_k=\dfrac{m_{i+k}}{m_{i+1}}$, $v_l=\dfrac{n_{j+l}}{n_{j+1}}$. Let $m=\left[\dfrac{m_{i+1}}{m_i}\right]$ and $n=\left[\dfrac{n_{j+1}}{n_j}\right]$. Define

$$\theta_{\mu\nu}=\int_{\mu d}^{(1+\mu)d}\int_{\nu d}^{(1+\nu)d}\left|\sum_{k=1}^{+\infty}\sum_{l=1}^{+\infty}b_{kl}\sin u_k s\sin v_l t\right|\frac{\mathrm{d}s\mathrm{d}t}{st}.$$

Applying Lemma 1 we get $\theta_{\mu\nu}\leqslant\dfrac{1}{\mu\nu}\left(\sum\limits_{k=1}^{+\infty}\sum\limits_{l=1}^{+\infty}b_{kl}^2\right)^{\frac{1}{2}}$, and hence

$$\sum_{i=1}^{+\infty}\sum_{j=1}^{+\infty}I_{ij}^{(4)}\leqslant c_qV.\tag{2.10}$$

A combination of (2.7)~(2.10) yields (1.9). The proof is complete.

§ 3. Proof of theorem 2

To prove theorem 2 we need more integral estimates.

Lemma 3 Let m_i, c_{ij}, i, $j\in\mathbf{N}$, be real numbers satisfying condition (2.2) and $B=\left(\sum\limits_{i=1}^{+\infty}c_{ii}\right)^2+\sum\limits_{i=1}^{+\infty}\sum\limits_{j=1}^{+\infty}c_{ij}^2<+\infty$. Define

$$h(x)=\sum_{i=1}^{+\infty}\sum_{j=1}^{+\infty}c_{ij}\cos(m_i-m_j)x.$$

Then, for any $a\in\mathbf{R}$, $\int_a^{a+d}h^2(x)\mathrm{d}x\leqslant c_qB$, where, and throughout this paper, c_q denotes a constant depending only on q.

Proof Define

$$h_1(x)=\sum_{i=2}^{+\infty}\sum_{j=1}^{i-1}c_{ij}\cos(m_i-m_j)x,$$

$$h_2(x)=\sum_{j=2}^{+\infty}\sum_{i=1}^{j-1}c_{ij}\cos(m_i-m_j)x.$$

We see $h^2(x)\leqslant 3\left\{h_1^2(x)+h_2^2(x)+\left(\sum\limits_{i=1}^{+\infty}c_{ii}\right)^2\right\}$. We have $h_1^2(x)=\dfrac{1}{2}\{g_1(x)+g_2(x)\}$ where

$$g_1(x)=\sum_{i>j}\sum_{k>l}c_{ij}c_{kl}\cos(m_i-m_j-m_k+m_l)x,$$

$$g_2(x) = \sum_{i>j}\sum_{k>l} c_{ij}c_{kl}\cos(m_i - m_j + m_k - m_l)x.$$

By condition (2.2) for $i>j$, $k>l$, we have $m_i - m_j + m_k - m_l \geqslant 2(q-1)\cdot\sqrt{m_{i-1}\cdot m_{k-1}}$. Hence by Schwarz's inequality we get

$$\int_d^{a+d} g_2(x)\mathrm{d}x \leqslant c_q \sum_{i=1}^{+\infty}\sum_{j=1}^{+\infty} c_{ij}^2. \tag{3.1}$$

Choose $n\in\mathbf{N}$ such that $1-\frac{1}{q}-\frac{1}{q^n}=\delta>0$. Then define subsets J_μ of $(i, j, k, l)\in\mathbf{N}^4$ by

$J_1=\{i>j,\ j\geqslant k+n,\ k>l\}$, $J_2=\{i>j,\ i\leqslant k-n,\ k>l\}$,

$J_3=\{k+n>i>k>l,\ i\geqslant j+n\}$, $J_4=\{k+n>i>k>l,\ i>j>i-n\}$,

$J_5=\{i+n>k>i>j,\ k>l\}$, $J_6=\{i=k>j,\ k>l\}$

and define $g_{1\nu}(x)=\sum_\nu c_{ij}c_{kl}\cos(m_i-m_j-m_k+m_l)x$, $\nu=1,2,\cdots,6$, where the sum $\sum_\nu$ denotes $\sum_{(i,j,k,l)\in J_\nu}$. Then we see $g_1=\sum_{\nu=1}^{6} g_{1\nu}$. First we have $\int_a^{a+d} g_{1l}(x)\mathrm{d}x \leqslant \sum_1 |c_{ij}c_{kl}|\left(1-\frac{1}{q}-\frac{1}{q^n}\right)^{-1}\frac{2}{m_i}$. So by Schwarz's inequality

$$\int_a^{a+d} g_{11}(x)\mathrm{d}x \leqslant c_q \sum_{i=1}^{+\infty}\sum_{j=1}^{+\infty} c_{ij}^2. \tag{3.2}$$

Symmetrically

$$\int_a^{a+d} g_{12}(x)\mathrm{d}x \leqslant c_q \sum_{i=1}^{+\infty}\sum_{j=1}^{+\infty} c_{ij}^2. \tag{3.3}$$

If $(i, j, k, l)\in J_3$, then $m_i-m_j-m_k+m_l\geqslant\delta m_i$. Hence we have

$$\int_a^{a+d} g_{13}(x)\mathrm{d}x \leqslant \frac{2}{\delta}\sum_{k=2}^{+\infty}\sum_{l=1}^{k-1}|c_{kl}|\sum_{i=\max(n+1,k+1)}^{+\infty}\sum_{j=1}^{i-n}\frac{|c_{ij}|}{m_i}.$$

The using Schwarz's inequality we get

$$\int_a^{a+d} g_{13}(x)\mathrm{d}x \leqslant c_q B. \tag{3.4}$$

For $(i, j, k, l)\in J_4$ let $\varepsilon_{ijkl}=\left|\int_a^{a+d}\cos(m_i-m_j-m_k+m_l)x\mathrm{d}x\right|$. Then

$$\int_a^{a+d} g_{14}(x)\mathrm{d}x \leqslant \sum_{k=2}^{+\infty}\sum_{l=1}^{k-1}\sum_{i=k+1}^{k+n}\sum_{j=\max(1,i-n)}^{i-1}|c_{ij}c_{kl}|\varepsilon_{ijkl}.$$

If we define $\varepsilon_{ijkl}=c_{ij}=0$ when $i\leqslant 0$ or $j\leqslant 0$, then we get

$$\int_a^{a+d} g_{14}(x)\mathrm{d}x \leqslant \sum_{s=1}^{n}\sum_{t=0}^{n-1}\sum_{k=2}^{+\infty} |c_{k+s,k+s-n+t}| \sum_{l=1}^{k-1} |c_{kl}| \eta_l(k,s,t),$$

where $\eta_l(k, s, t) = \varepsilon_{k+s,k+s-n+t,k,l}$ will be written as η_l for simplicity. Write $\sigma_k = \sum_{l=1}^{k-1} |c_{k,l}| \eta_l$. Then $\sigma_k \leqslant \left\{ \sum_{l=1}^{k-1} c_{kl}^2 \sum_{l=1}^{k-1} \eta_l^2 \right\}^{\frac{1}{2}}$. Define

$$l_0 = \min\{l \in \{1, 2, \cdots, k-1\} \mid |m_{k+s} - m_{k+s-n+1} - m_k + m_l| < \frac{1}{2}\left(1 - \frac{1}{q}\right) m_l\},$$

$l_0 = 0$ for the case that the minimum does not exist. As a convention we let $m_0 = 0$. Then for all $l \neq l_0$. $|m_{k+s} - m_{k+s-n+t} - m_k + m_l| \geqslant \frac{1}{2}\left(1 - \frac{1}{q}\right) m_l$. Hence we have $\eta_l \leqslant c_q \frac{1}{m_l}$, for $l \neq l_0$, $1 \leqslant l \leqslant k-1$, and $\eta_{l_0} \leqslant d$. Consequently we get $\left\{ \sum_{l=1}^{k-1} \eta_l^2 \right\}^{\frac{1}{2}} \leqslant c_q$, and $\sigma_k \leqslant c_q \left\{ \sum_{l=1}^{k-1} c_{kl}^2 \right\}^{\frac{1}{2}}$. Now we derive

$$\int_a^{a+d} g_{14}(x)\mathrm{d}x \leqslant c_q B. \tag{3.5}$$

Since J_5 is symmetrically related to $J_3 \cup J_4$ we conclude

$$\int_a^{a+d} g_{15}(x)\mathrm{d}x \leqslant c_q B. \tag{3.6}$$

By a direct calculation we get

$$\int_a^{a+d} g_{16}(x)\mathrm{d}x \leqslant c_q B. \tag{3.7}$$

A combination of (3.2)~(3.7) yields

$$\int_a^{a+d} g_1(x)\mathrm{d}x \leqslant c_q B. \tag{3.8}$$

Then we combine (3.1) and (3.8) to get $\int_a^{a+d} h_1^2(x)\mathrm{d}x \leqslant c_q B$. Symmetrically we conclude $\int_a^{a+d} h_2^2(x)\mathrm{d}x \leqslant c_q B$. And at last we derive $\int_a^{a+d} h^2(x)\mathrm{d}x \leqslant c_q B$ as required. □

Lemma 4　Under the assumptions of Lemma 3 define

$$g(x) = \sum_{i=1}^{+\infty}\sum_{j=1}^{+\infty} c_{ij} \cos(m_i + m_j)x.$$

Then

$$\int_a^{a+d} g^2(x)\mathrm{d}x \leqslant c_q B.$$

The proof is completely similar to that of Lemma 3. We omit it.

Lemma 5 Under the conditions of Lemma 1

$$\left\{\iint_Q f^4(x,y)\mathrm{d}x\mathrm{d}y\right\}\leqslant c_q\sigma.$$

Proof For $i\in\mathbf{N}$ and $y\in\mathbf{R}$ fixed we define $b_i=\sum_{j=1}^{+\infty}a_{ij}\sin n_j y$. Then we see $f^2(x, y)=\frac{1}{2}\{h(x)+g(x)\}$ where h, g are defined respectively as in Lemma 3 and Lemma 4 with coefficients $c_{ij}=b_i b_j$. Then by these lemmas we conclude

$$\int_Q f^4(x,y)\mathrm{d}x\mathrm{d}y\leqslant c_q\sum_{i=1}^{+\infty}\sum_{j=1}^{+\infty}\int_b^{b+d}\{b_i b_j\}^2\mathrm{d}y. \tag{3.9}$$

Since $b_i b_j=\frac{1}{2}\sum_{k=1}^{+\infty}\sum_{l=1}^{+\infty}a_{ij}a_{jl}\{\cos(n_k-n_l)y-\cos(n_k+n_l)y\}$ we apply Lemma 3 and Lemma 4 again to get $\int_b^{b+d}b_i^2 b_j^2\mathrm{d}y\leqslant c_q\sum_{k=1}^{+\infty}a_{ik}^2\sum_{i=1}^{+\infty}a_{jl}^2$. Substituting this into (3.9) we complete the proof. □

The following two estimates follow from Lemma 1 and Lemma 5.

Lemma 6 Under the conditions of Lemma 1 $c_q\int_Q|f(x,y)|\mathrm{d}x\mathrm{d}y\geqslant\sigma$.

Lemma 7 Under the assumptions of Lemma 2 $c_q\int_a^{a+d}|\psi(x)|\mathrm{d}x\geqslant A$.

The following lemma is the essence of Móricz's Lemma 2 and Lemma 3 of [3].

Lemma 8 Let $a_i\geqslant 0$, $i\in\mathbf{N}$. Then for any $r\in\mathbf{N}$ and $n>r$,

$$\sum_{i=r+1}^{n}a_i\leqslant\frac{1}{\sqrt{r}}\sum_{i=1}^{n-1}\left(\sum_{j=i+1}^{n}a_j^2\right)^{\frac{1}{2}}. \tag{3.10}$$

Proof of theorem 2 By Lemma 1, (1.4) is equivalent to

$$\frac{f(x, y)}{xy}\in L(0, \lambda d)^2,\quad \frac{g_i(x)}{x}\in L(0, \lambda d),$$

$$\frac{h_i(y)}{y}\in L(0, \lambda d),\quad i\in\mathbf{N}, \tag{1.4'}$$

where $\lambda=\max\left(1, \frac{1}{q-1}\right)\geqslant 1$ and d is the value defined in Theorem 1. Now assume (1.4′) holds; we are going to prove (1.6).

Let $A_{ij}=\ln\frac{m_{i+1}}{m_i}\left(\sum_{k=i+1}^{+\infty}u_{kj}^2\right)^{\frac{1}{2}}$, $B_{ij}=\ln\frac{n_{j+1}}{n_j}\left(\sum_{l=j+1}^{+\infty}a_{il}^2\right)^{\frac{1}{2}}$.

We first prove $\sum_{k=1}^{+\infty}A_{kj}<+\infty, \sum_{l=1}^{+\infty}B_{il}<+\infty, (i,j\in\mathbf{N})$.

For n big enough and j fixed, let $I_n=\int_{\frac{\lambda d}{m_{n+1}}}^{\lambda d}|g_j(x)|\frac{\mathrm{d}x}{x}$. Then $I_n\geqslant J_n-R_n$ where

$$R_n=\sum_{i=1}^{n}\int_{\frac{\lambda d}{m_{i+1}}}^{\frac{\lambda d}{m_i}}\left|\sum_{k=1}^{i}a_{kj}\sin m_kx\right|\frac{\mathrm{d}x}{x}\leqslant c_q\sum_{i=1}^{n}|a_{ij}|$$

and

$$J_n=\sum_{i=1}^{n}\int_{\frac{\lambda d}{m_{i+1}}}^{\frac{\lambda d}{m_i}}\left|\sum_{k=i+1}^{+\infty}a_{kj}\sin m_kx\right|\frac{\mathrm{d}x}{x}.$$

Applying Lemma 7 we get

$$J_n\geqslant\sum_{i=1}^{n}\sum_{\mu=1}^{m}\frac{1}{(\mu+\lambda)}c_q\left(\sum_{k=i+1}^{+\infty}a_{kj}^2\right)^{\frac{1}{2}}$$

Where $m=\left[\left(\frac{m_{i+1}}{m_i}-1\right)\lambda\right]\in\mathbf{N}$. It is now clear why we take λd instead of d. We obtain, in fact, $J_n\geqslant c_q\sum_{i=1}^{n}A_{ij}$ and hence

$$\sum_{i=1}^{n}A_{ij}\leqslant c_q'\sum_{i=1}^{n}|a_{ij}|+c_q''I_n.$$

If $\sum_{i=1}^{+\infty}A_{ij}=+\infty$, noticing $\int_0^{\lambda d}|g_j(x)|\frac{\mathrm{d}x}{x}<+\infty$ we derive

$$1\leqslant\varlimsup_{n\to+\infty}c_q'\left\{\sum_{i=1}^{n}|a_{ij}|\left(\sum_{i=1}^{n}A_{ij}\right)^{-1}\right\}.$$

But by Lemma 8 we conclude the right part of this inequality should be zero. This contradiction shows $\sum_{i=1}^{+\infty}A_{ij}<+\infty$. Symmetrically we know $\sum_{j=1}^{+\infty}B_{ij}<+\infty$.

Next we define f_ν, $\nu=1,2,3,4$, as in the proof of theorem 1 and define

$$E_{ij}^{(\nu)}=\int_{\frac{\lambda d}{m_{i+1}}}^{\frac{\lambda d}{m_i}}\int_{\frac{\lambda d}{n_{j+1}}}^{\frac{\lambda d}{n_j}}|f_\nu(x,y)|\frac{\mathrm{d}x\mathrm{d}y}{xy},\quad \nu=1,2,3,4,\ i,\ j\in\mathbf{N}.$$

For big s, $t\in\mathbf{N}$ let

$$\sigma_{s,t}^{(\nu)}=\sum_{i=1}^{s}\sum_{j=1}^{t}E_{ij}^{(\nu)},\quad \sigma_{s,t}=\sum_{i=1}^{s}\sum_{j=1}^{t}\int_{\frac{\lambda d}{m_{i+1}}}^{\frac{\lambda d}{m_i}}\int_{\frac{\lambda d}{n_{j+1}}}^{\frac{\lambda d}{n_j}}\frac{|f(x,y)|}{xy}\mathrm{d}x\mathrm{d}y.$$

We have $\sigma_{s,t}\geqslant\sigma_{st}^{(4)}-(\sigma_{st}^{(1)}+\sigma_{st}^{(2)}+\sigma_{st}^{(3)})$. By an argument similar to that used in the proof of theorem 1 we get inequalities similar to (2.7)~(2.9), viz.

$$\sigma_{st}^{(1)}\leqslant c_q\sum_{i=1}^{s}\sum_{j=1}^{t}|a_{ij}|,\quad \sigma_{st}^{(2)}\leqslant c_q\sum_{i=1}^{s}\sum_{j=1}^{t}\ln\frac{m_{i+1}}{m_i}\Big(\sum_{k=j+1}^{+\infty}a_{kj}^2\Big)^{\frac{1}{2}},$$

$$\sigma_{st}^{(3)}\leqslant c_q\sum_{i=1}^{s}\sum_{j=1}^{t}\ln\frac{n_{j+1}}{n_j}\Big(\sum_{l=j+1}^{+\infty}a_{il}^2\Big)^{\frac{1}{2}}.$$

We now estimate $\sigma_{st}^{(4)}$ applying Lemma 6. Let $m=\left[\left(\frac{m_{i+1}}{m_i}-1\right)\lambda\right]$ and $n=\left[\left(\frac{n_{j+1}}{n_j}-1\right)\lambda\right]$. Then $m\in\mathbf{N}$, $n\in\mathbf{N}$ and

$$\sigma_{st}^{(4)}\geqslant\sum_{i=1}^{s}\sum_{j=1}^{t}\sum_{\mu=1}^{m}\sum_{\nu=1}^{n}\int_{\lambda d+(\mu-1)d}^{\lambda d+\mu d}\int_{\lambda d+(\nu-1)d}^{\lambda d+\nu d}S_{ij}(x,y)\mathrm{d}x\mathrm{d}y$$

Where

$$S_{ij}(x,y)=x^{-1}y^{-1}\left|\sum_{k=i+1}^{+\infty}\sum_{l=j+1}^{+\infty}a_{kl}\sin\frac{m_k}{m_{i+1}}x\sin\frac{n_l}{n_{j+1}}y\right|.$$

Hence by lemma 6 we get

$$\sigma_{st}^{(4)}\geqslant c_q\sum_{i=1}^{s}\sum_{j=1}^{t}\ln\frac{m_{i+1}}{m_i}\ln\frac{n_{j+1}}{n_j}\Big(\sum_{k=i+1}^{+\infty}\sum_{l=j+1}^{+\infty}a_{kl}^2\Big)^{\frac{1}{2}}.\tag{3.11}$$

Then applying Lemma 8, by an argument similar to that for the proof of $\sum_{i=1}^{+\infty}A_{ij}<+\infty$ we conclude $V<+\infty$. Then noticing

$$T=\sum_{i=1}^{+\infty}A_{i1}+\sum_{i=1}^{+\infty}\sum_{j=2}^{+\infty}A_{ij}\leqslant\sum_{i=1}^{+\infty}A_{i1}+c_qV,$$

$$U=\sum_{j=1}^{+\infty}B_{1j}+\sum_{j=1}^{+\infty}\sum_{i=2}^{+\infty}B_{ij}\leqslant\sum_{j=1}^{+\infty}B_{1j}+c_qV,$$

$$S=\sum_{j=2}^{+\infty}|a_{ij}|+\sum_{i=2}^{+\infty}|a_{i1}|+|a_{il}|+\sum_{i=2}^{+\infty}\sum_{j=2}^{+\infty}|a_{ij}|$$

$$\leqslant c_q\sum_{j=1}^{+\infty}B_{1j}+c_q\sum_{i=1}^{+\infty}A_{i1}+|a_{11}|+c_qV,$$

we derive (1.6). □

Remarks (1) In our argument the series by which we define functions need not be trigonometric series because the coefficients m_i, n_j

need not be integers. We must understand such series in the sense of L^2-convergence. For example, if conditions (1.1) and (1.2) are satisfied, then the "partial sums"

$$S_{\mu\nu}(x,y)=\sum_{i=1}^{\mu}\sum_{j=1}^{\nu}a_{ij}\sin m_i x\sin n_j y$$

converge in $L^2(Q)$ for any compact set $Q\subset\mathbf{R}^2$. This is a consequence of Lemma 1. Meanwhile we can easily demonstrate that the convergence of $S_{\mu\nu}$ in $L^2(Q)$ does not depend on the manner in which μ and ν tend to infinity. For a discussion of different kinds of multiple limits we refer the reader to [4].

(2) Since $S_{\mu\nu}$ can be non-trigonometric sums it does not appear to be a trivial question whether the convergence of $S_{\mu\nu}$ in L^2 implies almost everywhere convergence.

(3) Our result can be extended to higher dimensional cases in a quite straight-forward manner.

(4) Since this paper was submitted in June, 1993, two related papers of interest have appeared, viz. [1][2]. We thank the referee for providing the details.

References

[1] Chen C P. Integrability of multiple trigonometric series and parseval's formula. J. Math. Anal. Appl., 1994, 186: 182—199.

[2] Chen C P. Weighted integrability and l^1-convergence of multiple trigonometric series. Studia Math., 1994, 108: 177—190.

[3] Móricz F. Integrability of double lacunary sine series. Proc. of the AMS, 1990, 110: 355—364.

[4] Zhizhiashvili L V. Certain questions from the theory of simple and multiple trigonometric and orthogonal series. Uspekhi Mat. Nauk, 1973, 28: 65—127.

Ukrainian Mathematical Journal, 1999, 51(11): 1 549—1 561.

实直线上局部可积函数的逼近

Approximation of Locally Integrable Functions on the Real Line①

We introduce the notion of generalized $\bar{\psi}$-derivatives for functions locally integrable on the real axis and investigate problems of approximation of the classes of functions determined by these derivatives with the use of entire functions of exponential type.

§ 1. Introduction

Recently, Stepanets [1] has investigated the problem of approximation for functions that are locally integrable on the real line. In fact, Stepanets has extended his serious investigation of periodic functions (see [2～5]) to the case of locally integrable functions. This extension [1] only concerns the so-called (ψ, β)-derivatives, which are defined via a single function ψ and phase translation $\frac{\pi}{2}\beta$. At the same time, in the recent papers of Stepanets [6, 7], the concept of (ψ, β)-derivative has been extended by introducing the notion of $\bar{\psi}$-derivative defined via a pair of functions ψ_1 and ψ_2.

The aim of the present paper is to extend (ψ, β)-derivatives to $\bar{\psi}$-derivatives for locally integrable functions on the real line and then,

① Supported by NSFC 19771009.

本文与 Stepanets A I 和张希荣合作.

correspondingly, to extend the results of [1] to the case of approximation of locally integrable functions that have $\bar{\psi}$-derivatives on the real line.

First, we recall certain definitions introduced in [1] and [5] with suitable modifications.

For a function f measurable on R, we define

$$\|f\|_{\hat{p}} = \sup\left\{\left(\int_0^{2\pi} |f(x+y)|^p \mathrm{d}y\right)^{\frac{1}{p}} \middle| x \in \mathbf{R}\right\} \quad \text{for} \quad 1 \leqslant p \leqslant +\infty;$$

$$\|f\|_{\hat{\infty}} = \sup\{|f(x)| \mid x \in \mathbf{R}\}.$$

We also define the following spaces of functions for $1 \leqslant p \leqslant +\infty$:

$$\hat{L}_p = \{f \text{ is measurable on } R \mid \|f\|_{\hat{p}} < +\infty\}.$$

Sometimes, we simply write $\hat{L}$ instead of $\hat{L}_1$.

Definition 1 $\mathcal{U}$ denotes the set of functions ψ satisfying the following conditions:

(1) $\psi(u) \geqslant 0$, $\psi(0) = 0$, and ψ is increasing and continuous on $[0, 1]$;

(2) ψ is convex on $[1, +\infty)$ and $\lim_{u \to +\infty} \psi(u) = 0$;

(3) $\psi'(u) := \psi'(u+0)$ is bounded on $[0, +\infty)$.

Definition 2 We set $F = \left\{\psi \in \mathcal{U} \middle| \int_1^{+\infty} \frac{\psi(t)}{t} \mathrm{d}t < +\infty\right\}$.

In [5], for suitable ψ defined on $[0, +\infty)$ and $\beta \in \mathbf{R}$, the transform $\hat{\psi}_\beta$ was defined as

$$\hat{\psi}_\beta(t) = \frac{1}{\pi}\int_0^{+\infty} \psi(v) \cos\left(vt + \frac{\pi}{2}\beta\right) \mathrm{d}v, \tag{1.1}$$

whenever this integral makes sense. Furthermore, if $\hat{\psi}_\beta \in L(R)$ (this is the case where $\psi \in \mathcal{U}$), the (ψ, β)-derivative of a function $f \in \hat{L}$ is defined as a function $\varphi \in \hat{L}_1$ satisfying the condition

$$f(x) = A_0 + \lim_{R \to +\infty} \int_{-R}^{R} \varphi(x+t) \hat{\psi}_\beta(t) \mathrm{d}t, \tag{1.2}$$

where A_0 is a constant independent of x. We write $\varphi = f_\beta^\psi$.

For simplicity, we assume that $\psi \in \mathcal{U}$. Let ψ_+ and ψ_- be the even and odd extensions of ψ, respectively. Then, by (1.1), we have

$$\hat{\psi}_\beta=\hat{\psi}_+\cos\frac{\pi}{2}\beta+\mathrm{i}\hat{\psi}_-\sin\frac{\pi}{2}\beta, \tag{1.3}$$

where $\hat{\psi}_+$ and $\hat{\psi}_-$ denote the Fourier transforms of ψ_+ and ψ_-, respectively, in the original sense, i. e. ,

$$\forall h\in L(R),\quad \hat{h}(t)=\frac{1}{2\pi}\int_{\mathbf{R}}h(x)\mathrm{e}^{-\mathrm{i}xt}\,\mathrm{d}x. \tag{1.4}$$

Now assume that $\psi_1\in\mathcal{U}$ and $\psi_2\in\mathcal{U}$. Then ψ_{1+}, ψ_{2+} and ψ_{1-}, ψ_{2-} are even and odd extensions of ψ_1, ψ_2 respectively. For the pair (ψ_1, ψ_2), we define

$$\psi=\psi_{1+}+\mathrm{i}\psi_{2-}. \tag{1.5}$$

The corresponding Fourier transform has the form

$$\hat{\psi}=\hat{\psi}_{1+}+\mathrm{i}\hat{\psi}_{2-}. \tag{1.6}$$

We should remember definition (1. 5) [and (1. 6)]. It will be used throughout the paper.

Definition 3 Assume that ψ_1, $\psi_2\in\mathcal{U}$ and $\psi=\psi_{1+}+\mathrm{i}\psi_{2-}$. If a function $f\in\hat{L}$ can be represented as

$$f=\frac{1}{2\pi}\int_{-\pi}^{\pi}f(x)\,\mathrm{d}x+\varphi*\hat{\psi}$$

with $\varphi\in\hat{L}$, then we say that φ is the ψ-derivative of f and denote this as $\varphi=f^\psi$ or $f=I^\psi(\varphi)$.

Note that the convolution " $*$ " in Definition 3 is defined as usual, i. e. , for g, $h\in L(R)$, we have

$$(g*h)(x)=\int_{\mathbf{R}}g(x-t)h(t)\,\mathrm{d}t=(h*g)(x).$$

If $h\in\mathcal{U}$, $\beta\in\mathbf{R}$, and

$$h_1:=h\cos\frac{\pi}{2}\beta,\qquad h_2:=-h\sin\frac{\pi}{2}\beta,$$

then it follows from (1. 3) that $\hat{h}_\beta=\hat{h}_{1+}+\mathrm{i}\hat{h}_{2-}$ and, hence, by virtue of (1. 5) and (1. 6), for the pair (h_1, h_2) we have $\hat{h}=\hat{h}_\beta$. Hence, we see that Definition 3 is a generalization of the concept of (ψ, β)-derivative considered in [1].

Let $\mathcal{N}$ be a subclass of $\hat{L}$. We use the notation

$$\hat{L}^{\psi}\mathcal{N}:=\{f\in\hat{L}\mid f^{\psi}\in\mathcal{N}\}$$

as in [1]. For $\mathcal{N}=\hat{L}$, we also simply write $\hat{L}^{\psi}$ instead of $\hat{L}^{\psi}\hat{L}$. Denote (see [5])

$$L_{2\pi}^{0}=\left\{f\in L_{2\pi}\,\middle|\,\int_{-\pi}^{\pi}f(t)\mathrm{d}t=0\right\},\quad L_{2\pi}^{\psi}=\{f\in L_{2\pi}\mid f^{\bar{\psi}}\in L_{2\pi}\},$$

where $f^{\bar{\psi}}$ is the $\bar{\psi}$-derivative of f defined in [5]. In this notation, the following statement is true:

Proposition 1 Suppose that ψ_1, $\psi_2\in\mathcal{U}$ and $\psi=\psi_{1+}+\mathrm{i}\psi_{2-}$. Then

$$\hat{L}^{\psi}L_{2\pi}^{0}=L_{2\pi}^{\psi}.$$

Proof If $f\in\hat{L}^{\psi}L_{2\pi}^{0}$, then

$$f=A_0+\varphi*\hat{\psi}$$

with $A_0\in R$ and $\varphi\in L_{2\pi}^{0}$. We see that f is 2π-periodic, i.e., $f\in L_{2\pi}$. Furthermore, by virtue of the definition of $f^{\bar{\psi}}$ introduced in [6], we see that $f^{\bar{\psi}}$ coincides with f^{ψ} defined here. This means that $\hat{L}^{\psi}L_{2\pi}^{0}=L_{2\pi}^{\psi}$.

Next, for $0\leqslant c<\sigma$, we define

$$\lambda_{\sigma c}(t)=\begin{cases}1, & 0\leqslant|t|\leqslant c,\\ \dfrac{\sigma-|t|}{\sigma-c}, & c<|t|<\sigma,\\ 0, & \sigma\leqslant|t|,\end{cases}$$

and, for ψ_1, $\psi_2\in\mathcal{U}$,

$$\lambda_{\sigma c}^{*}(t)=\begin{cases}\lambda_{\sigma c}(t), & |t|\in[0,c]\cap[\sigma,+\infty),\\ 1-\dfrac{|t|-c}{\sigma-c}\dfrac{\psi(\sigma\mathrm{sign}(t))}{\psi(t)}, & c<|t|<\sigma.\end{cases}$$

We see that these are the even extensions of the corresponding functions $\lambda_{\sigma c}$ and $\lambda_{\sigma c}^{*}$ defined in [1]. Then we construct the operators $\mathrm{F}_{\sigma c}$ and $F_{\sigma c}^{*}$ as in [1], i.e., as follows:

Definition 4 Let $0\leqslant c<\sigma<+\infty$ and ψ_1, $\psi_2\in\mathcal{U}$. For $f\in\hat{L}^{\psi}$, we denote

$$F_{\sigma c}(f)=f^{\psi}*\widehat{\psi\lambda}_{\sigma c}+\frac{1}{2\pi}\int_{-\pi}^{\pi}f(t)\mathrm{d}t,$$

$$F_{\sigma c}^{*}(f)=f^{\psi}*\widehat{\psi\lambda}_{\sigma c}^{*}+\frac{1}{2\pi}\int_{-\pi}^{\pi}f(t)\mathrm{d}t.$$

§ 2. Class of entire functions of exponential type

Denote by $\mathscr{E}_\sigma$ the class of entire functions of exponential type $\sigma(\sigma>0)$. We refer to [8] for the basic knowledge on $\mathscr{E}_\sigma$. Let

$$W_\sigma^2=\left\{\varphi\in\mathscr{E}_\sigma\,\middle|\,\int_{\mathbf{R}}\frac{|\varphi(t)|^2}{1+t^2}\mathrm{d}t<+\infty\right\}.$$

We now give a generalization of Proposition 2 of [1].

Theorem 1 Suppose that $\psi_1, \psi_2\in\mathcal{U}$ and $f\in\hat{L}^\psi$. Then the following assertions are true:

(1) If $0<\tau\leqslant c<\sigma$ and $f^\psi\in W_\tau^2$ then

$$F_{\sigma c}(f)=f,\quad F_{\sigma c}^*(f)=f.$$

(2) If $f\in L_{2\pi}$ and $\sigma>c\geqslant0$, then

$$F_{\sigma c}(f)(x)=\sum_{|k|<\sigma}\lambda_{\sigma c}(k)c_k(f)\mathrm{e}^{\mathrm{i}kx},$$

$$F_{\sigma c}^*(f)(x)=\sum_{|k|<\sigma}\lambda_{\sigma c}^*(k)c_k(f)\mathrm{e}^{\mathrm{i}kx},$$

where

$$c_k(f)=\frac{1}{2\pi}\int_{-\pi}^{\pi}f(x)\mathrm{e}^{-\mathrm{i}kx}\mathrm{d}x,\quad k\in\mathbf{Z}.$$

(3) If $\psi'\in L_2(0, a)$, $\forall a>0$, $\sigma>c\geqslant0$, and

$$\int_{\mathbf{R}}\frac{|f^\psi(t)|^2}{1+|t|^2}\mathrm{d}t<+\infty,$$

then $F_{\sigma c}(f)\in W_\sigma^2$ and $F_{\sigma c}^*(f)\in W_\sigma^2$.

Proof By the basic result in Fourier analysis, we have

$$\widehat{\psi\lambda}_{\sigma c}=\hat{\psi}*\hat{\lambda}_{\sigma c},\quad \widehat{\psi\lambda}_{\sigma c}^*=\hat{\psi}*\hat{\lambda}_{\sigma c}^*.$$

Then we get

$$F_{\sigma c}(f)=A_0+f^\psi*(\hat{\psi}*\hat{\lambda}_{\sigma c})=A_0+(f^\psi*\hat{\lambda}_{\sigma c})*\hat{\psi},$$

$$F_{\sigma c}^*(f)=A_0+f^\psi*(\hat{\psi}*\hat{\lambda}_{\sigma c}^*)=A_0+(f^\psi*\hat{\lambda}_{\sigma c}^*)+\hat{\psi}.$$

Generally, for any $\varphi\in W_\tau^2$ and $0<\tau\leqslant c<\sigma$, we have

$$\varphi*\hat{\lambda}_{\sigma c}=\varphi*\hat{\lambda}_{\sigma c}^*=\varphi,$$

which is a consequence of the Wiener-Paley theorem [8]. Thus, by the definition of ψ-derivative, we get assertion (1).

In the case $f \in L_{2\pi}$, we can easily check that

$$\frac{1}{2\pi}\int_{-\pi}^{\pi}(f^{\psi} * \hat{\lambda}_{\sigma c})(x)\mathrm{e}^{-ikx}\,\mathrm{d}x = \int_{\mathbf{R}}\left(\frac{1}{2\pi}\int_{-\pi}^{\pi} f^{\psi}(x-t)\mathrm{e}^{-ikx}\,\mathrm{d}x\right)\hat{\lambda}_{\sigma c}(t)\,\mathrm{d}t$$

$$= c_k(f^{\psi})\int_{\mathbf{R}}\hat{\lambda}_{\sigma c}(t)\mathrm{e}^{-ixt}\,\mathrm{d}t = \lambda_{\sigma c}(-k)c_k(f^{\psi}).$$

But $\lambda_{\sigma c}$ is even and, therefore,

$$(f^{\psi} * \hat{\lambda}_{\sigma c})(x) = \sum_{|k|<\sigma}\lambda_{\sigma c}(k)c_k(f^{\psi})\mathrm{e}^{ikx}.$$

Hence,

$$(f^{\psi} * \hat{\lambda}_{\sigma c} * \hat{\psi})(x) = \sum_{|k|<\sigma}\lambda_{\sigma c}(k)c_k(f^{\psi})\int_{\mathbf{R}}\mathrm{e}^{ik(x-t)}\hat{\psi}(t)\,\mathrm{d}t$$

$$= \sum_{|k|<\sigma}\lambda_{\sigma c}(k)c_k(f^{\psi})\psi(-k)\mathrm{e}^{ikx}.$$

By Definition 3, for $k \neq 0$, we have

$$c_k(f) = \frac{1}{2\pi}\int_{-\pi}^{\pi}\left(\int_{\mathbf{R}} f^{\psi}(x-t)\hat{\psi}(t)\,\mathrm{d}t\right)\mathrm{e}^{-ikx}\,\mathrm{d}x$$

$$= c_k(f^{\psi})\int_{\mathbf{R}}\hat{\psi}(t)\mathrm{e}^{-ikt}\,\mathrm{d}t = c_k(f^{\psi})\psi(-k).$$

Therefore,

$$(f^{\psi} * \hat{\lambda}_{\sigma c} * \hat{\psi})(x) = \sum_{0<|k|<\sigma}\lambda_{\sigma c}(k)c_k(f)\mathrm{e}^{ikx}$$

which implies that

$$F_{\sigma c}(f)(x) = \sum_{|k|<\sigma}\lambda_{\sigma c}(k)c_k(f)\mathrm{e}^{ikx}.$$

The same argument yields

$$F_{\sigma c}^{*}(f)(x) = \sum_{|k|<\sigma}\lambda_{\sigma c}^{*}(k)c_k(f)\mathrm{e}^{ikx}.$$

Assertion (2) is proved.

In the case $\psi' \in L_2(0, \sigma)$, we write $\gamma = \psi\lambda_{\sigma c}$ or $\gamma = \psi\lambda_{\sigma c}^{*}$. Then supp $\gamma \subset [-\sigma, \sigma]$, γ is absolutely continuous on $[-\sigma, \sigma]$, and $\gamma' \in L_2(-\sigma, \sigma)$. By using the result of [8, p. 228], we get $g * \hat{\gamma} \in W_{\sigma}^{2}$, whenever g satisfies the condition

$$\int_{\mathbf{R}}\frac{|g(t)|^2}{1+|t|^2}\,\mathrm{d}t < +\infty.$$

Thus we get assertion (3) and complete the proof.

§ 3. Approximation by $F_{\sigma c}^{*}$

Assume the ψ_1, $\psi_2 \in \mathcal{U}$, $0 \leqslant c < \sigma < +\infty$, and $f \in \hat{L}^{\psi}$. We define

$$\rho_{\sigma c}^{*}(f) = f - F_{\sigma c}^{*}(f), \quad \rho_{\sigma c}(f) = f - F_{\sigma c}(f). \tag{3.1}$$

We write

$$r_{\sigma c}(t) = (1 - \lambda_{\sigma c}^{*}(t))\psi(t) = \begin{cases} 0, & 0 \leqslant |t| \leqslant c, \\ \dfrac{|t| - c}{\sigma - c}\psi(\sigma \operatorname{sign}(t)), & c < |t| < \sigma, \\ \psi(t), & \sigma \leqslant |t|. \end{cases} \tag{3.2}$$

Then, by Definition 3 and Definition 4, we get

$$\rho_{\sigma c}^{*}(f) = f^{\psi} * \hat{r}_{\sigma c}. \tag{3.3}$$

By Theorem 1, for any $u \in W_{\tau}^{2}$ and $0 < \tau \leqslant c < \sigma$, we have $u * \hat{r}_{\sigma c} = 0$. Hence,

$$\rho_{\sigma c}^{*}(f) = (f^{\psi} - u) * \hat{r}_{\sigma c}.$$

This implies that, for $1 \leqslant p \leqslant +\infty$ and $f \in \hat{L}^{\psi}\hat{L}_{p}$, we get

$$\|\rho_{\sigma c}^{*}(f)\|_{\hat{p}} \leqslant E_{c}(f^{\psi})_{\hat{p}} \|\hat{r}_{\sigma c}\|_{1}, \tag{3.4}$$

where

$$E_{c}(h)_{\hat{p}} = \inf\{\|h - u\|_{\hat{p}} : u \in W_{c}^{2}\}, \quad \|\hat{r}_{\sigma c}\|_{1} = \int_{\mathbf{R}} |\hat{r}_{\sigma c}(t)| \, dt.$$

By the definition of the Fourier transform (1.4), we have

$$\hat{r}_{\sigma c}(t) = \frac{1}{2\pi}\int_{\mathbf{R}} r_{\sigma c}(s) e^{-ist} ds = \frac{1}{2\pi}\int_{c}^{\sigma} \frac{s - c}{\sigma - c}(\psi(\sigma)e^{-ist} + \psi(-\sigma)e^{ist}) ds +$$

$$\frac{1}{2\pi}\int_{\sigma}^{+\infty} (\psi(s)e^{-ist} + \psi(-s)e^{ist}) ds.$$

By (1.5), we have

$$\psi(s)e^{-ist} + \psi(-s)e^{ist} = 2(\psi_1(s)\cos st + \psi_2(s)\sin st).$$

we write

$$R_1(t) = \frac{\psi_1(\sigma)}{\pi}\int_{c}^{\sigma} \frac{s - c}{\sigma - c}\cos st \, ds + \frac{1}{\pi}\int_{\sigma}^{+\infty} \psi_1(s)\cos st \, ds, \tag{3.5}$$

$$R_2(t) = \frac{\psi_2(\sigma)}{\pi}\int_{c}^{\sigma} \frac{s - c}{\sigma - c}\sin st \, ds + \frac{1}{\pi}\int_{\sigma}^{+\infty} \psi_2(s)\sin st \, ds. \tag{3.6}$$

We see that

$$\hat{r}_{\sigma c}(t)=R_1(t)+R_2(t). \tag{3.7}$$

As in [1], we denote by F_0 the following subclass of $\mathcal{U}$:

$$F_0=\{\psi\in\mathcal{U}\mid\sup\{\eta'(\psi, t)\mid t\geqslant1\}<+\infty\},$$

where $\eta(\psi, t)=\psi^{-1}\left(\frac{1}{2}\psi(t)\right)$, $t\geqslant1$, was introduced by Stepanets in [5].

Similarly, we keep the notation $\mathcal{U}_0$, $\mathcal{U}_c$, and $\mathcal{U}_\infty$ introduced in [1]. We know from [1] that

$$\mathcal{U}_c\cup\mathcal{U}_\infty\subset F_0.$$

For ψ_1, $\psi_2\in\mathcal{U}$, by (1.5), we conclude that

$$|\psi(t)|=(\psi_1^2(t)+\psi_2^2(t))^{\frac{1}{2}}$$

is decreasing on $[0, +\infty)$. Hence, we can assume that $|\psi(t)|>0$ for all $t\in[1, +\infty)$.

Further, for simplicity, we assume that $0\leqslant c=\sigma-1$ and write r_σ instead of $r_{\sigma,\sigma-1}$. In this case, by (3.5) and (3.6), we have

$$R_1(t)=\frac{2\sin\frac{t}{2}}{\pi t^2}\psi_1(\sigma)\sin\left(\left(\sigma-\frac{1}{2}\right)t\right)-\frac{1}{\pi t}\int_\sigma^{+\infty}\psi'_1(s)\sin st\mathrm{d}s, \tag{3.5'}$$

$$R_2(t)=\frac{2\sin\frac{t}{2}}{\pi t^2}\psi_2(\sigma)\cos\left(\left(\sigma-\frac{1}{2}\right)t\right)+\frac{1}{\pi t}\int_\sigma^{+\infty}\psi'_2(s)\cos st\mathrm{d}s. \tag{3.6'}$$

3.1 Case ψ_1, $\psi_2\in F_0$

We define

$$\alpha_j(t)=\frac{2\pi}{\eta(\psi_j, t)-t},\quad s=1,2,\quad t\geqslant1,$$

$$\sup\{\eta(\psi_j, t)\mid t\geqslant1\}=K_j,\quad j=1,2.$$

Theorem 2 Suppose that ψ_1, $\psi_2\in F_0$. If there exist constants $A\geqslant B>0$ such that

$$\forall t\geqslant1,\quad A\geqslant\frac{\eta(\psi_1, t)-t}{\eta(\psi_2, t)-t}\geqslant B, \tag{3.8}$$

then

$$\|\hat{r}_\sigma\|_1=\frac{4}{\pi^2}|\psi(\sigma)|(\ln^+(\eta(\sigma)-\sigma)+O(1))$$

as $\sigma \to +\infty$, where

$$\eta(\sigma)=\min(\eta(\psi_1, \sigma), \eta(\psi_2, \sigma)), \quad \sigma \geqslant 1.$$

Proof We write

$$\alpha=\frac{2\pi}{\eta(\sigma)-\sigma}, \quad \sigma \geqslant 1.$$

By virtue of the monotonicity of ψ_1 and periodicity of cos, we have

$$\left|\int_\sigma^{+\infty} \psi'_1(s)\cos st \mathrm{d}s\right| \leqslant \int_\sigma^{\sigma+\frac{2\pi}{|t|}} \psi_1(s)\mathrm{d}s$$

and, hence,

$$\int_{|t|<\alpha}\left|\int_\sigma^{+\infty} \psi_1(s)\cos st \mathrm{d}s\right| \mathrm{d}t \leqslant 2\int_0^\alpha \left(\int_\sigma^{\sigma+\frac{2\pi}{\alpha}} + \int_{\sigma+\frac{2\pi}{\alpha}}^{\sigma+\frac{2\pi}{t}}\right)\psi_1(s)\mathrm{d}s\mathrm{d}t$$

$$\leqslant 4\pi\psi_1(\sigma) + 4\int_{\frac{2\pi}{\alpha}}^{+\infty} \frac{\psi_1(\sigma+s)}{s}\mathrm{d}s.$$

We note that

$$\int_{\frac{2\pi}{\alpha}}^{+\infty} \frac{\psi_1(\sigma+s)}{s}\mathrm{d}s = \left(\int_{\eta(\sigma)-\sigma}^{\eta(\psi_1,\sigma)-\sigma} + \int_{\eta(\psi_1,\sigma)-\sigma}^{+\infty}\right)\frac{\psi_1(\sigma+s)}{s}\mathrm{d}s,$$

where, by assumption (3.8),

$$\int_{\eta(\sigma)-\sigma}^{\eta(\psi_1,\sigma)-\sigma} \frac{\psi_1(\sigma+s)}{s}\mathrm{d}s \leqslant \psi_1(\sigma)\ln(1+A).$$

On the other hand, it is known from [2, p. 134] that, for $\psi_1 \in F_0$, we have

$$\int_{\eta(\psi_1,\sigma)-\sigma}^{+\infty} \frac{\psi_1(\sigma+s)}{s}\mathrm{d}s \leqslant (1+\sup\{\eta'(\psi_1,t) \mid t \geqslant 1\})\psi_1(\sigma).$$

Hence, we get

$$\int_{|t|<\alpha}\left|\int_\sigma^{+\infty} \psi_1(s)\cos st \mathrm{d}s\right| \mathrm{d}t \leqslant C\psi_1(\sigma), \tag{3.9}$$

where $C>0$ is a constant independent of $\sigma \geqslant 1$. By the same argument, we have

$$\int_{|t|<\alpha}\left|\int_\sigma^{+\infty} \psi_2(s)\sin st \mathrm{d}s\right| \mathrm{d}t \leqslant C\psi_2(\sigma). \tag{3.10}$$

We can assume that $\psi_1(\sigma)>0$ and write

$$\theta=\arctan\frac{\psi_2(\sigma)}{\psi_1(\sigma)}.$$

We define

$$R_3(t) = \frac{|\psi(\sigma)|}{\pi}\int_{\sigma-1}^{\sigma}(s-\sigma+1)\cos(st-\theta)\mathrm{d}s. \tag{3.11}$$

Then, by virtue of (3. 5)～(3. 7), we get

$$\hat{r}_\sigma(t) = R_3(t) + \frac{1}{\pi}\int_{\sigma}^{+\infty}(\psi_1(s)\cos st + \psi_2(s)\sin st)\mathrm{d}s. \tag{3.12}$$

It is obvious that

$$|R_3(t)| \leqslant |\psi(\sigma)|.$$

Thus, if $\alpha\leqslant 1$, then, by virtue of (3. 9)～(3. 12), we have

$$\int_{|t|\leqslant\alpha}|\hat{r}_\sigma(t)|\mathrm{d}t \leqslant C|\psi(\sigma)|. \tag{3.13}$$

If $1<\alpha$, then

$$\int_{|t|<\alpha}|\hat{r}_\sigma(t)|\mathrm{d}t \leqslant C|\psi(\sigma)| + \int_{1<|t|<\alpha}|R_3(t)|\mathrm{d}t.$$

By (3. 11), we get

$$R_3(t) = \frac{|\psi(\sigma)|}{\pi t}\sin(\sigma t-\theta) - \frac{2|\psi(\sigma)|}{\pi t^2}\sin\frac{t}{2}\sin\left[\left(\sigma-\frac{1}{2}\right)t-\theta\right].$$

Thus,

$$\int_{1<|t|<\alpha}|R_3(t)|\mathrm{d}t \leqslant \frac{2}{\pi}|\psi(\sigma)| + \frac{|\psi(\sigma)|}{\pi}\int_{1<|t|<\alpha}\left|\frac{\sin(\sigma t-\theta)}{t}\right|\mathrm{d}t$$

$$\leqslant \frac{4}{\pi^2}|\psi(\sigma)|(\ln\alpha + C). \tag{3.14}$$

In order to consider the case $|t|>\alpha$, we use (3. 5′) and (3. 6′) to get

$$\hat{r}_\sigma(t) = -|\psi(\sigma)|\frac{2\sin\frac{t}{2}}{\pi t^2}\sin\left(\left(\sigma-\frac{1}{2}\right)t-\theta\right) +$$

$$\frac{1}{\pi t}\int_{\sigma}^{+\infty}(-\psi'_1(s)\sin st + \psi'_2(s)\cos st)\mathrm{d}s. \tag{3.15}$$

By virtue of the monotonicity of ψ_1' and periodicity of sin, we have

$$\left|\int_{\sigma}^{+\infty}\psi'_1(s)\sin st\,\mathrm{d}s\right| \leqslant \psi_1(\sigma) - \psi_1\left(\sigma + \frac{2\pi}{|t|}\right).$$

Hence,

$$\int_{|t|>\alpha}\frac{1}{|t|}\left|\int_{\sigma}^{+\infty}\psi'_1(s)\sin st\,\mathrm{d}s\right|\mathrm{d}t \leqslant 2\int_0^{\frac{2\pi}{\alpha}}\frac{\psi_1(\sigma)-\psi_1(\sigma+t)}{t}\mathrm{d}t$$

$$\leqslant 2\int_{\sigma}^{\eta(\psi,\sigma)}\frac{\psi_1(\sigma)-\psi_1(t)}{t-\sigma}\mathrm{d}t.$$

It is also known from [2, p. 134] that, for $\psi_1 \in F_0$, we have

$$\int_{\sigma}^{\eta(\psi,\sigma)} \frac{\psi_1(\sigma)-\psi_1(t)}{t}\mathrm{d}t \leqslant \sup\{\eta'(\psi_1,t)\,|\,t\geqslant 1\}\psi_1(\sigma).$$

Therefore,

$$\int_{|t|>\alpha}\frac{1}{|t|}\left|\int_{\sigma}^{+\infty}\psi'_1(s)\sin st\,\mathrm{d}s\right|\mathrm{d}t \leqslant C\psi_1(\sigma).$$

The same argument yields

$$\int_{|t|>\alpha}\frac{1}{|t|}\left|\int_{\sigma}^{+\infty}\psi'_2(s)\cos st\,\mathrm{d}s\right|\mathrm{d}t \leqslant C\psi_2(\sigma).$$

By using these estimates, we derive from (3.15) that

$$\begin{aligned}&\int_{|t|>\alpha}|\hat{r}_\sigma(t)|\,\mathrm{d}t\\ &\leqslant C|\psi(\sigma)|+\frac{|\psi(\sigma)|}{\pi}\int_{|t|>\alpha}\frac{\left|2\sin\frac{t}{2}\sin\left(\left(\sigma-\frac{1}{2}\right)t-\theta\right)\right|}{t^2}\mathrm{d}t.\end{aligned}$$

If $\alpha\geqslant 1$, then

$$\int_{|t|>\alpha}\left|2\sin\frac{t}{2}\sin\left(\left(\sigma-\frac{1}{2}\right)t-\theta\right)\right|t^{-2}\mathrm{d}t\leqslant 4.$$

If $\alpha<1$, then

$$\begin{aligned}&\int_{\alpha<|t|<1}\left|2\sin\frac{t}{2}\sin\left(\left(\sigma-\frac{1}{2}\right)t-\theta\right)\right|t^{-2}\mathrm{d}t\\ &\leqslant\int_{\alpha<|t|<1}\left|\frac{\sin\left(\left(\sigma-\frac{1}{2}\right)t-\theta\right)}{t}\right|\mathrm{d}t\\ &\leqslant\frac{4}{\pi}\ln\frac{1}{\alpha}+C.\end{aligned}$$

This and (3.15) yield

$$\int_{|t|>\alpha}|\hat{r}_\sigma(t)|\,\mathrm{d}t\leqslant\begin{cases}\left(\frac{4}{\pi^2}\ln\frac{1}{\alpha}+C\right)|\psi(\sigma)|, & \text{if}\quad \alpha<1,\\ C|\psi(\sigma)|, & \text{if}\quad \alpha>1.\end{cases}\tag{3.16}$$

Combining (3.13)(3.14), and (3.15), we complete the proof. □

Theorem 3 If there exist constants $\varepsilon>0$ and $t_0\geqslant 1$ such that

$$\eta(\psi_1,\ t)-t\geqslant 2(K_1+\varepsilon)(\eta(\psi_2,\ t)-t)\tag{3.17}$$

for all $t\geqslant t_0$, then, for $\sigma\geqslant 1$, we have

$$\|\hat{r}_\sigma\|_1\leqslant\frac{4}{\pi^2}\psi_1(\sigma)(\ln^+(\eta(\psi_1,\ \sigma)-\sigma)+C).\tag{3.18}$$

Proof Using the method from the proof of Theorem 2, we can derive

$$\|R_j\|_1\leqslant\frac{4}{\pi^2}\psi_j(\sigma)(\ln^+(\eta(\psi_j, \sigma)-\sigma)+C), \quad j=1,2.$$

Under assumption (3.17), we use the properties of the functions in F_0 and get

$$\psi_2(t)\leqslant A_0t^{-\varepsilon\delta}\psi_1(t), \quad \forall t\geqslant t_0,$$

where

$$\delta:=\inf\left\{\frac{\tau}{\eta(\psi_1, \tau)-\tau}\middle|\tau\geqslant 1\right\}>0, \quad A_0=\frac{\psi_2(t_0)}{\psi_1(t_0)}t_0^{\varepsilon\delta}>0.$$

Thus, we have

$$\psi_2(\sigma)\ln^+(\eta(\psi_2, \sigma)-\sigma)\leqslant A_0\sigma^{-\varepsilon\delta}\psi_1(\sigma)\ln\left(\frac{\eta(\psi_2, \sigma)-\sigma}{\sigma}\sigma\right)$$
$$\leqslant C\psi_1(\sigma), \quad \sigma\geqslant t_0.$$

Theorem 3 is proved.

3.2 Case where $\psi_1\leqslant\mathcal{U}_0$ and $\psi_2\in\mathcal{U}_0$

We recall that $\psi_j\in\mathcal{U}_0$ means that

$$\sup\{\mu(\psi_j, t)\,|\,t\geqslant 1\}=M_j<+\infty,$$

$$\mu(\psi_j, t)=\frac{t}{\psi_j^{-1}\left(\frac{1}{2}\psi_j(t)\right)-t}, \quad j=1,2.$$

Furthermore, we know that [5, p.118]

$$\forall\sigma\geqslant 1, \quad \sigma|\psi_j'(\sigma)|\leqslant(1+M_j)\psi_j(\sigma).$$

Theorem 4 If ψ_1, $\psi_2\in\mathcal{U}_0$, then, for $\sigma\geqslant 1$, we have

$$\|\hat{r}_\sigma\|_1\leqslant|\psi(\sigma)|\left(\frac{4}{\pi^2}\ln\sigma+C\right)+\frac{2}{\pi}\int_\sigma^{+\infty}\frac{\psi_2(t)}{t}\mathrm{d}t.$$

Proof Following the technique of [1], we first prove that, for $t\in\left(0, \frac{\pi}{2\sigma}\right)$, we have

$$\frac{1}{\pi}\int_\sigma^{+\infty}\psi_2(s)\sin st\,\mathrm{d}s\geqslant\psi_2\left(\frac{\pi}{2t}\right)\frac{\cos(\sigma t)}{\pi t}\geqslant 0. \tag{3.19}$$

Indeed, for $t\in\left(0, \frac{\pi}{2\sigma}\right)$, we get

$$\frac{1}{\pi}\int_{\sigma}^{+\infty}\psi_2(s)\sin st\,\mathrm{d}s=\frac{1}{\pi t}\left(\psi_2(\sigma)\cos(\sigma t)+\frac{1}{t}\left(\int_{\sigma t}^{\frac{\pi}{2}}+\int_{\frac{\pi}{2}}^{+\infty}\right)\psi'_2\left(\frac{s}{t}\right)\cos s\mathrm{d}s\right)$$

$$\geqslant\frac{1}{\pi t}\left(\psi_2(\sigma)\cos(\sigma t)+\frac{1}{t}\int_{\sigma t}^{\frac{\pi}{2}}\psi'_2\left(\frac{s}{t}\right)\cos s\mathrm{d}s\right)$$

$$\geqslant\frac{1}{\pi t}\left(\psi_2(\sigma)\cos(\sigma t)+\cos(\sigma t)\frac{1}{t}\int_{\sigma t}^{\frac{\pi}{2}}\psi'_2\left(\frac{s}{t}\right)\mathrm{d}s\right)$$

$$=\psi_2\left(\frac{\pi}{2t}\right)\frac{\cos(\sigma t)}{\pi t}.$$

In order to estimate $\|\hat{r}_\sigma\|$, we begin with (3.12).

It now follows from (3.12) and (3.19) that

$$\int_{|t|<\frac{\pi}{2\sigma}}|\hat{r}_\sigma(t)|\,\mathrm{d}t\leqslant\frac{\pi}{\sigma}|\psi(\sigma)|+\int_{|t|<\frac{\pi}{2\sigma}}\frac{1}{\pi}\left|\int_{\sigma}^{+\infty}\psi_1(s)\cos(st)\mathrm{d}s\right|\mathrm{d}t+$$

$$\frac{2}{\pi}\int_{\sigma}^{+\infty}\psi_2(s)\left(\int_0^{\frac{\pi}{2\sigma}}\sin(st)\mathrm{d}t\right)\mathrm{d}s$$

$$=\frac{2}{\pi}\int_{\sigma}^{+\infty}\frac{\psi_2(s)}{s}\mathrm{d}s-\frac{2}{\pi}\int_{\sigma}^{+\infty}\psi_2(s)\frac{\cos\left(\frac{\pi}{2\sigma}s\right)}{s}\mathrm{d}s+$$

$$\frac{\pi}{\sigma}|\psi(\sigma)|+\int_{|t|<\frac{\pi}{2\sigma}}\frac{1}{\pi}\left|\int_{\sigma}^{+\infty}\psi_1(s)\cos(st)\mathrm{d}s\right|\mathrm{d}t.$$

It is obvious that

$$\int_{\sigma}^{+\infty}\psi_2(s)\frac{\cos\left(\frac{\pi}{2\sigma}s\right)}{s}\mathrm{d}s=O(\psi_2(\sigma)).$$

Integrating by parts, we get

$$\int_{\sigma}^{+\infty}\psi_1(s)\cos(st)\mathrm{d}s=-\frac{1}{t}\psi_1(\sigma)\sin(\sigma t)-\frac{1}{t}\int_{\sigma}^{+\infty}\psi'_1(s)\sin(st)\mathrm{d}s.$$

Since $-\psi'_1$ is decreasing and positive, for $t>0$ we have

$$-\int_0^{\sigma}\psi'_1(\sigma)\sin(st)\mathrm{d}s-\int_{\sigma}^{+\infty}\psi'_1(s)\sin(st)\mathrm{d}s\geqslant 0.$$

Hence,

$$\int_{\sigma}^{+\infty}\psi'_1(s)\sin(st)\mathrm{d}s\leqslant-\frac{1-\cos(\sigma t)}{t}\psi'_1(\sigma).$$

Let

$$E=\left\{t\in(0,+\infty)\,\middle|\,\int_{\sigma}^{+\infty}\psi'_1(s)\sin(st)\mathrm{d}s>0\right\}.$$

We see that

$$\int_0^{+\infty}\left|\frac{1}{t}\int_\sigma^{+\infty}\psi'_1(s)\sin(st)\mathrm{d}s\right|\mathrm{d}t$$

$$=-\int_0^{+\infty}\frac{1}{t}\int_\sigma^{+\infty}\psi'_1(s)\sin(st)\mathrm{d}s\mathrm{d}t+2\int_E\frac{1}{t}\left(\int_\sigma^{+\infty}\sin(st)\mathrm{d}s\right)\mathrm{d}t$$

$$\leqslant 2\int_E|\psi'_1(\sigma)|\frac{1-\cos(\sigma t)}{t^2}\mathrm{d}t+\int_\sigma^{+\infty}(-\psi'_1(s))\int_0^{+\infty}\frac{\sin(st)}{t}\mathrm{d}t\mathrm{d}s.$$

Noting that

$$\sigma|\psi'_1(\sigma)|\leqslant(1+M_1)\psi_1(\sigma),\quad \int_0^{+\infty}\frac{\sin(st)}{t}\mathrm{d}t=\int_0^{+\infty}\frac{\sin t}{t}\mathrm{d}t\in\mathbf{R},$$

we get

$$\int_0^{+\infty}\frac{1}{t}\left|\int_0^{+\infty}\psi'_1(s)\sin(st)\mathrm{d}s\right|\mathrm{d}t\leqslant C\psi_1(\sigma).$$

Then

$$\int_{|t|<\frac{\pi}{2\sigma}}\left|\int_\sigma^{+\infty}\psi_1(s)\cos(st)\mathrm{d}s\right|\mathrm{d}t=O(\psi_1(\sigma)).$$

Therefore,

$$\int_{|t|<\frac{\pi}{2\sigma}}|\hat{r}_\sigma(t)|\mathrm{d}t\leqslant C|\psi(\sigma)|+\frac{2}{\pi}\int_\sigma^{+\infty}\frac{\psi_2(s)}{s}\mathrm{d}s.$$

By virtue of (3. 5′) and (3. 6′), we have

$$\int_{\frac{\pi}{2\sigma}<|t|}|\hat{r}_\sigma(t)|\mathrm{d}t\leqslant\int_{\frac{\pi}{2\sigma}<|t|<\frac{\pi}{2}}\frac{|\psi(\sigma)|}{\pi t^2}\left|2\sin\frac{t}{2}\sin\left(\left(\sigma-\frac{1}{2}\right)t-\theta\right)\right|\mathrm{d}t+$$

$$|\psi(\sigma)|+\int_{\frac{\pi}{2\sigma}<|t|}\left|\frac{1}{\pi t}\int_\sigma^{+\infty}(-\psi'_1(s)\sin(st)+\psi'_2(s)\cos(st))\mathrm{d}s\right|\mathrm{d}t.$$

It is easy to see that

$$\int_{\frac{\pi}{2\sigma}<|t|<\frac{\pi}{2}}\frac{1}{\pi t^2}\left|2\sin\frac{t}{2}\sin\left(\left(\sigma-\frac{1}{2}\right)t-\theta\right)\right|\mathrm{d}t\leqslant\frac{4}{\pi^2}(\ln\sigma+C).$$

By virtue of the monotonicity of ψ'_1, $j=1,2$, and periodicity of trigonometric functions, we have

$$\left|\int_\sigma^{+\infty}\psi'_1(s)\sin(st)\mathrm{d}s\right|\leqslant\int_\sigma^{\sigma+\frac{2\pi}{|t|}}|\psi'_1(\sigma)|\mathrm{d}s=\frac{2\pi}{|t|}|\psi'_1(\sigma)|,$$

$$\left|\int_\sigma^{+\infty}\psi'_2(s)\cos(st)\mathrm{d}s\right|\leqslant\int_\sigma^{\sigma+\frac{2\pi}{|t|}}|\psi'_2(\sigma)|\mathrm{d}s=\frac{2\pi}{|t|}|\psi'_2(\sigma)|.$$

Then
$$\int_{|t|>\frac{\pi}{2\sigma}}\frac{1}{|t|}\left|\int_\sigma^{+\infty}(-\psi'_1(s)\sin(st)+\psi'_2(s)\cos(st))\mathrm{d}s\right|\mathrm{d}t$$

$$\leqslant C(|\psi'_1(\sigma)|+|\psi'_2(\sigma)|)\sigma\leqslant C|\psi(\sigma)|.$$

Combining the above estimates, we complete the proof of Theorem 4.

References

[1] Stepanets A I. Approximation in spaces of locally integrable functions. Ukr. Mat. Zh., 1994, 46(5): 638—670.

[2] Stepanets A I. Classification of periodic functions and the rate of convergence of their Fourier series. Izv. Akad. Nauk SSSR, Ser. Mat., 1986, 50(1): 101—136.

[3] Stepanets A I. On the Lebesgue inequality on the classes of (ψ, β)-differentiable functions, Ukr. Mar. Zh., 1989, 41(5): 449—510.

[4] Stepanets A I. Deviations of Fourier sums on classes of entire functions, Ukr. Mat. Zh., 1989, 41(6): 783—789.

[5] Stepanets A I. Classification and Approximation of Periodic Functions. Kluwer, Dordrecht, 1995.

[6] Stepanets A I. Convergence rate of Fourier series on the classes of $\bar{\psi}$-integrals, Ukr. Mat. Zh., 1997, 49(8): 1 069—1 114.

[7] Stepanets A I. Approximation of $\bar{\psi}$-integrals of periodic functions by Fourier sums (low smoothness), Ukr. Mat. Zh., 1998, 50(2): 274—291.

[8] Akhiezer N I. Lectures in Approximation Theory [in Russian], Nauka, Moscow, 1965.

Journal of Approximation Theory，2002，118：202－224.

关于 Ditzian 和 Runovskii 的一个猜测

On a Conjecture of Ditzian and Runovskii①

Abstract Let M_θ be the mean operator on the unit sphere in $\mathbf{R}^n$, $n\geqslant 3$, which is an analogue of the Steklov operator for functions of single variable. Denote by D the Laplace-Beltrami operator on the sphere which is an analogue of second derivative for functions of single variable. Ditzian and Runovskii have a conjecture on the norm of the operator $\theta^2 D(M_\theta)^m$, $m\geqslant 2$ from $X=L^p(1\leqslant p\leqslant +\infty)$ to itself which can be expressed as

$$\lim_{m\to+\infty}\sup\{\|\theta^2 D(M_\theta)^m\|_{(X,X)}\mid \theta\in(0,\pi)\}=0.$$

We give a proof of this conjecture.

Keywords ultraspherical polynomials; spherical harmonics; mean operator.

§ 1. Introduction

Suppose $n\in\mathbf{N}$ and $n\geqslant 3$. Let $\mathrm{S}^{n-1}=\{(x_1, x_2, \cdots, x_n)\in\mathbf{R}^n: x_1^2+x_2^2+\cdots+x_n^2=1\}$. For a function $f\in L(\mathrm{S}^{n-1})$ define

$$M_\theta(f)(x)=\frac{1}{D(\theta)}\int_{D(x,\theta)} f(y)\mathrm{d}y,\quad x\in \mathrm{S}^{n-1},$$

① Supported by NSFC 10071007.

本文与戴峰和余纯武合作.

Received：2001-07-09；Revised：2002-06-26.

where $D(x, \theta)=\{y\in S^{n-1} \mid x\cdot y>\cos\theta\}$, $\theta\in(0, \pi)$ and

$$D(\theta) = |D(x,\theta)| = |S^{n-2}|\int_0^\theta \sin^{n-2}t\,dt.$$

We call M_θ the mean operator on the sphere.

For a function $f\in L(S^{n-1})$, its Fourier-Laplace expansion is

$$\sum_{k=0}^{+\infty} Y_k(f),$$

Where $Y_k(f)$ is the "projection" of f on the space $\mathcal{H}_k^n$ of spherical harmonics of degree k. Actually,

$$Y_k(f)(x) = c_{nk}(f * P_k^n)(x) = c_{nk}\int_{S^{n-1}} f(y)P_k^n(xy)\mathrm{d}y,$$

where

$$C_{nk}=\frac{\Gamma\left(\frac{n}{2}\right)}{2\pi^{\frac{n}{2}}}\frac{(n+2k-2)(n+k-2)!}{(n+k-2)k!\ (n-2)!}$$

and $P_k^n(t)$ are the normalized ultraspherical polynomials

$$P_k^n(t)=\frac{P_k^{\left(\frac{n-3}{2},\frac{n-3}{2}\right)}(t)}{P_k^{\left(\frac{n-3}{2},\frac{n-3}{2}\right)}(1)}=\frac{\Gamma\left(\frac{n-1}{2}\right)\Gamma(k+1)}{\Gamma\left(k+\frac{n-1}{2}\right)}P_k^{\left(\frac{n-3}{2},\frac{n-3}{2}\right)}(t), \quad -1\leqslant t\leqslant 1,$$

where $P_k^{(\alpha,\beta)}$ are the Jacobi polynomials.

If $f, g\in L(S^{n-1})$ and g has the expansion

$$\sum_{k=1}^{+\infty} -k(k+n-2)Y_k(f),$$

then we call g the second derivative of f and denote it by $D(f)$, where D means the Laplace-Beltrami operator.

For these basic concepts we refer to [WL]. The concept of spherical harmonics can also be found in [SW, Chap. 4, Sect. 2]. The Laplace-Beltrami operator is called the "spherical Laplacean" in [St, Chap. 3, Sect. 3]. For Jacobi polynomials (including ultraspherical polynomials) very detailed materials can be found in [Sz, Chaps. 4, 7, and Sect. 7. 32, 8].

Ditzian and Runovskii [DR] proposed a conjecture which can be stated as follows: Let $m\in\mathbf{N}$, $m>2$, then

$$\lim_{m\to+\infty} \sup\{ \| D(M_\theta)^m \|_{(X,X)} \theta^2 : \theta \in (0, \pi)\} = 0, \tag{1.1}$$

where $\| \cdot \|_{(X,X)}$ denotes the operator norm from $X=L^p(S^{n-1})(1\leqslant p\leqslant +\infty)$ to itself.

Our purpose is to prove this conjecture. Define

$$T_m(\theta)=\theta^2 D(M_\theta)^m, \qquad m>2.$$

We rewrite (1.1) as the following theorem.

Theorem For $X=L^p(S^{n-1})$, $1\leqslant p\leqslant +\infty$,

$$\lim_{m\to+\infty} \sup\{ \| T_m(\theta) \|_{(X,X)} \mid \theta \in (0, \pi)\} = 0.$$

§ 2. Preliminaries

Let

$$h_\theta(t)=\frac{1}{D(\theta)}\chi_{(0,\theta)}(\arccos(t)), \qquad t\in(-1, 1).$$

We see that M_θ is a convolution operator with kernel h_θ, i. e.

$$M_\theta(f)(x) = f * h_\theta(x) := \int_{S^{n-1}} f(y) h_\theta(xy) \mathrm{d}y, \quad x \in S^{n-1}.$$

Hence M_θ is a multiplier operator with the multipliers

$$\hat{h}_\theta := \{\hat{h}_\theta(k)\}_{k=0}^{+\infty},$$

where

$$\hat{h}_\theta(k) = \frac{1}{\int_0^\theta \sin^{n-2} t \,\mathrm{d}t} \int_0^\theta P_k^n(\cos t) \sin^{n-2} t \,\mathrm{d}t.$$

We see that $\hat{h}_\theta(0)=1$. Applying Rodrigues' formula (see [WL, p. 23, Theorem 1. 2. 1], or [Sz, p. 67]), we have for $k>0$,

$$\hat{h}_\theta(k) = \frac{\sin^{n-1}\theta}{(n-1)\int_0^\theta \sin^{n-2} t \,\mathrm{d}t} P_{k-1}^{n+2}(\cos\theta). \tag{2.1}$$

Hence for $f\in L(S^{n-1})$ and $m>2$, we have

$$Y_k(T_m(\theta)(f)) = -k(k+n-2)\theta^2(\hat{h}_\theta(k))^m, \qquad k\in\mathbf{N}. \tag{2.2}$$

For a function $f\in L(S^{n-1})$, we denote its Cesàro means of order $\delta>-1$ by

$$\sigma_N^\delta(f) = \frac{1}{A_N^\delta}\sum_{k=0}^{N} A_{N-k}^\delta Y_k(f),$$

where A_k^δ denote Cesàro numbers which are given by

$$A_k^\delta=\frac{\Gamma(k+\delta+1)}{\Gamma(\delta+1)\Gamma(k+1)},\qquad \delta>-1.$$

It is known that when $\delta>\frac{n-2}{2}$ and $X=L^P(S^{n-1})(1\leqslant p\leqslant+\infty)$ we have

$$\sup\{\|\sigma_N^\delta\|_{(X,X)}\mid N\in\mathbf{N}^*\}\leqslant B(\delta)<+\infty,$$

where $B(\delta)$ denotes a constant depending only on δ and X (see[WL, p. 50], or [So, p. 47]). We will make use of this result.

Let$\{u_k\}_{k=0}^{+\infty}$ be a sequence of numbers. Define $\triangle^0 u_k=u_k$, $\triangle u_k=\triangle^1 u_k=u_k-u_{k+1}$ and $\triangle^{j+1}u_k=\triangle(\triangle^j u_k)$, $j\in\mathbf{N}^*$.

Lemma 1 Suppose $1\leqslant p\leqslant+\infty$ and $f\in L^p(S^{n-1})$. Let$\{u_k\}_{k=0}^{+\infty}$ be a sequence of real numbers such that

$$\lim_{k\to+\infty}u_k=0$$

and

$$\sum_{k=0}^{+\infty}|\triangle^{l+1}u_k|A_k^l=M<+\infty \tag{2.3}$$

with $l\in\mathbf{N}$, $n+1\geqslant l>\lambda:=\frac{n-2}{2}$. Define

$$g(x):=\sum_{k=0}^{+\infty}(\triangle^{l+1}u_k)A_k^l\sigma_k^l(f)(x).$$

Then

$$\|g\|_p\leqslant C_{n,p}M\|f\|_p,$$

where $C_{n,p}$ is a constant depending only on n and p, and

$$Y_k(g)(x)=u_kY_k(f)(x),\qquad k\in\mathbf{N}.$$

Proof Since $\sup_k\|\sigma_k^l(f)\|_p\leqslant C_{np}\|f\|_p$ for $n+1\geqslant l>\frac{n-2}{2}$, we know, by (2.3), that the series

$$\sum_{k=0}^{+\infty}(\triangle^{l+1}u_k)A_k^l\sigma_k^l(f)(x)$$

converges absolutely in $L^p(S^{n-1})$ and $\|g\|_p\leqslant C_{n,p}M\|f\|_p$.

Fix $k\in\mathbf{N}^*$. Since the projection Y_k is a continuous operator from $L^p(S^{n-1})$ to $L^p(S^{n-1})$, we have

$$Y_k(g)(x)=\sum_{j=0}^{+\infty}(\triangle^{l+1}u_j)A_j^lY_k(\sigma_j^l(f))(x).$$

By the definition of Cesàro means we know that when $j<k$, $Y_k(\sigma_j^l(f))=0$ and for $j\geqslant k$,

$$Y_k(\sigma_j^l(f))=\frac{A_{j-k}^l}{A_j^l}Y_k(f).$$

Then we get

$$Y_k(g)(x)=\Big(\sum_{j=k}^{+\infty}\triangle^{l+1}u_jA_{j-k}^l\Big)Y_k(f)(x).$$

Therefore, it is sufficient to prove

$$\sum_{j=k}^{+\infty}\triangle^{l+1}u_jA_{j-k}^l=u_k. \tag{2.4}$$

Since $\lim\limits_{k\to+\infty}u_k=0$, we know that for and $i\in\mathbf{N}$, $\lim\limits_{k\to+\infty}|\triangle^iu_k|=0$ and

$$\triangle^iu_k=\sum_{j=k}^{+\infty}\triangle^{i+1}u_j.$$

Consequently, noticing $\sum\limits_{k=0}^{j}A_k^{l-1}=A_j^l$ we get

$$\sum_{k=0}^{+\infty}|\triangle^lu_k|A_k^{l-1}\leqslant\sum_{k=0}^{+\infty}\sum_{j=k}^{+\infty}|\triangle^{l+1}u_j|A_k^{l-1}=\sum_{j=0}^{+\infty}|\triangle^{l+1}u_j|\sum_{k=0}^{j}A_k^{l-1}$$
$$=\sum_{k=0}^{+\infty}|\triangle^{l+1}u_j|A_j^l<+\infty.$$

Inductively, one can prove that for all $0\leqslant i\leqslant l$,

$$\sum_{k=0}^{+\infty}|\triangle^{i+1}u_k|A_k^i<+\infty.$$

This implies

$$\lim_{k\to+\infty}|\triangle^{i+1}u_k|A_k^i=0,\qquad 0\leqslant i\leqslant l.$$

For any two sequences of numbers $\{u_k\}_{k=0}^{+\infty}$, $\{v_k\}_{k=0}^{+\infty}$ we have the following Abel transformation formula:

$$\sum_{k=0}^{m+1}u_kv_k=\sum_{k=0}^{m}\triangle u_k\sum_{j=0}^{k}v_j+u_{m+1}\sum_{k=0}^{m+1}v_k.$$

If we know $\sum\limits_{k=0}^{+\infty}u_kv_k\in\mathbf{R}$ and $\lim\limits_{m\to+\infty}u_{m+1}\sum\limits_{k=0}^{m+1}v_k=0$, then passing to limit, we obtain

$$\sum_{k=0}^{+\infty}u_kv_k=\sum_{k=0}^{+\infty}\triangle u_k\sum_{j=0}^{k}v_j,$$

which will be the Abel transformation formula for our use.

Now using the Abel transformation once for our sequence $\{u_k\}_{k=0}^{+\infty}$ in the lemma and a special sequence $\{v_k=r^k\}_{k=0}^{+\infty}$ with $0<r<1$, noticing $\sum_{j=0}^{k}v_j=\sum_{k=0}^{j}A_{j-k}^{0}r^k$, we get

$$\sum_{j=0}^{+\infty}u_jr^j=\sum_{j=0}^{+\infty}(\triangle^1u_j)(\sum_{k=0}^{j}A_{j-k}^{0}r^k).$$

Writing $v_j^1=\sum_{k=0}^{j}A_{j-k}^{0}r^k$ and applying the Abel transformation once again, we get

$$\sum_{j=0}^{+\infty}u_jr^j=\sum_{j=0}^{+\infty}(\triangle^2u_j)(\sum_{k=0}^{j}v_k^1).$$

Note that

$$\sum_{k=0}^{j}v_k^1=\sum_{k=0}^{j}\sum_{\mu=0}^{k}A_{k-\mu}^{0}r^\mu=\sum_{\mu=0}^{j}\sum_{k=\mu}^{j}A_{k-\mu}^{0}r^\mu=\sum_{\mu=0}^{j}A_{j-\mu}^{1}r^\mu.$$

Then we get

$$\sum_{j=0}^{+\infty}u_jr^j=\sum_{j=0}^{+\infty}(\triangle^2u_j)(\sum_{\mu=0}^{j}A_{j-\mu}^{1}r^u).$$

So, using the Abel transformation inductively $l+1$ times, we get

$$\sum_{j=0}^{+\infty}u_jr^j=\sum_{j=0}^{+\infty}(\triangle^{l+1}u_j)(\sum_{k=0}^{j}A_{j-k}^{l}r^k)=\sum_{k=0}^{+\infty}(\sum_{j=k}^{+\infty}(\triangle^{l+1}u_j)A_{j-k}^{l})r^k,\quad 0<r<1.$$

Comparing the coefficients of r^k, we get (5) and complete the proof. □

We know (see [Sz, (4.1.1)])

$$P_k^{(\alpha,\beta)}(1)=\frac{\Gamma(k+\alpha+1)}{\Gamma(\alpha+1)\Gamma(k+1)}.$$

Now we define

$$Q_k^{(\alpha,\beta)}(t)=\frac{P_k^{(\alpha,\beta)}(t)}{P_k^{(\alpha,\beta)}(1)}.$$

Lemma 2 If $\alpha\geqslant\beta\geqslant-\frac{1}{2}$, then there is a constant $B(\alpha,\beta)$ depending only on α and β such that for $k\geqslant1$,

$$|Q_k^{(\alpha,\beta)}(\cos\theta)|\leqslant\begin{cases}1 & 0\leqslant\theta\leqslant\pi,\\ \dfrac{B(\alpha,\beta)}{(k\theta)^{\alpha+\frac{1}{2}}}, & 0<\theta\leqslant\dfrac{\pi}{2},\\ \dfrac{B(\alpha,\beta)}{(k(\pi-\theta))^{\beta+\frac{1}{2}}}, & \dfrac{\pi}{2}\leqslant\theta\leqslant\pi,\end{cases}\qquad(2.5)$$

and, in particular,

$$|P_k^{n+2}(\cos\theta)| \leqslant \begin{cases} 1, & 0 \leqslant \theta \leqslant \pi, \\ \dfrac{B(n)}{(k\sin\theta)^{\frac{n}{2}}}, & \theta \in (0, \pi), \end{cases} \tag{2.6}$$

where $B(n)$ denotes a constant depending only on n.

Proof By formula (7.32.2) in [Sz, p.168]

$$\max\{|P_k^{(\alpha,\beta)}(t)| \mid -1 \leqslant t \leqslant 1\} = C_{k+q}^k \sim k^q, \text{ if } q = \max(\alpha, \beta) \geqslant -\frac{1}{2}$$

and formulas (4.1.1) and (4.1.4) in [Sz, pp.58, 59]

$$P_k^{(\alpha,\beta)}(1) = C_{k+\alpha}^k, \quad P_k^{(\alpha,\beta)}(t) = (-1)^k P_k^{(\beta,\alpha)}(-t),$$

we see that, under the condition $\alpha \geqslant \beta \geqslant \frac{n-3}{2}$, we get the first estimate in (2.5).

We apply formula (8.21.18) in [Sz]: for $\frac{1}{k} \leqslant \theta \leqslant \pi - \frac{1}{k}$,

$$P_k^{(\alpha,\beta)}(\cos\theta)$$

$$= \frac{1}{\sqrt{\pi k}\left(\sin\frac{\theta}{2}\right)^{\alpha+\frac{1}{2}}\left(\cos\frac{\theta}{2}\right)^{\beta+\frac{1}{2}}}\left\{\cos\left[\left(k + \frac{\alpha+\beta+1}{2}\right)\theta - \frac{\alpha+\frac{1}{2}}{2}\pi\right] + O\left(\frac{1}{k\sin\theta}\right)\right\}.$$

The term $O\left(\frac{1}{k\sin\theta}\right)$ can be written as $\frac{r(k, \alpha, \beta, \theta)}{k\sin\theta}$ where $|r(k, \alpha, \beta, \theta)| \leqslant B(\alpha, \beta)$ for $B(\alpha, \beta)$ being a constant depending only on α and β. Then we see that the second and third estimates in (2.5) are valid for $\theta \in \left(\frac{1}{k}, \pi - \frac{1}{k}\right)$. Hence by the first estimate they are also valid for $0 < \theta < \frac{1}{k}$ and $\pi - \frac{1}{k} \leqslant \theta < \pi$.

Applying (2.5) to the case $\alpha = \beta = \frac{n-1}{2}$, we get (2.6). □

§ 3. Further lemmas

Define

$$u_k(m,\ \theta)=-k(k+n-2)\theta^2(\hat{h}_\theta(k))^m. \tag{3.1}$$

For simplicity we write

$$\phi(m,\theta)=-\theta^2\left(\frac{\sin^{n-1}\theta}{(n-1)\int_0^\theta \sin^{n-2}t\,\mathrm{d}t}\right)^m,$$

$$\psi_k(m,\ \theta)=(P_{k-1}^{\frac{n+2}{2}}(\cos\theta))^m,$$

and hence, using (2.1), we have

$$u_k(m,\ \theta)=k(k+n-2)\phi(m,\ \theta)\psi_k(m,\ \theta),\quad 0<\theta<\pi. \tag{3.1$'$}$$

Lemma 3 Let $m\in\mathbf{N}$, $m>10+n$. Then for any $f\in L(S^{n-1})$,

$$T_m(\theta)(f)=\sum_{k=1}^{+\infty}u_k(m,\theta)Y_k(f)=\sum_{k=1}^{+\infty}\triangle^{n+1}u_k(m,\theta)A_k^n\sigma_k^n(f).$$

Proof By (2.6) we have roughly (for $\theta\in(0,\ \pi)$),

$$|u_k(m,\ \theta)|\leqslant(B(n,\ \theta))^m k^{-\frac{mn}{2}}\leqslant(B(n,\ \theta))^m k^{-5n},$$

where here and in what follows $B(n,\ \theta)$ denote constants depending only on n and θ which may have different values in different occurrences. Of course, for $j=1,2,\cdots,n+1$ the estimates

$$|\triangle^j u_k(m,\ \theta)|\leqslant(B(n,\ \theta))^m k^{-\frac{mn}{2}}\leqslant(B(n,\ \theta))^m k^{-5n}$$

hold. So, the conditions of Lemma 1 are satisfied and hence Lemma 3 is valid. □

Lemma 4 When $m>10$,

$$\sup\left\{\|T_m(\theta)\|_{(X,X)}\,\middle|\,\theta\in\left[\frac{\pi}{2},\ \pi\right)\right\}\leqslant C(n)\pi^{-\frac{m}{2}}.$$

Proof Throughout this paper we use $C(n)$ to denote constants depending only on n which may have different values in different occurrences.

By Lemma 3 we have

$$\|T_m(\theta)\|_{(X,X)}\leqslant C(n)\sum_{k=1}^{+\infty}k^n|\triangle^{n+1}u_k(m,\theta)|.$$

Since $\theta\in\left[\frac{\pi}{2},\ \pi\right)$, we have

$$(n-1)\int_0^\theta \sin^{n-2}t\,\mathrm{d}t \geqslant (n-1)\int_0^{\frac{\pi}{2}} \sin^{n-2}t\,\mathrm{d}t = \frac{\Gamma\left(\frac{n+1}{2}\right)\Gamma\left(\frac{1}{2}\right)}{\Gamma\left(\frac{\pi}{2}\right)} > \sqrt{\pi}.$$

Write $\xi=\pi-\theta$. Then we get

$$|\phi(m,\ \theta)| \leqslant \pi^{-\frac{m}{2}+2}(\sin\xi)^{(n-1)m}. \tag{3.2}$$

We fix the constant $B(n)$ in (2.6) of Lemma 2 as $b=B(n)>1$. Define

$$I_1=\left\{k\in\mathbf{N}\,\middle|\,k\leqslant\frac{b}{\sin(\pi-\theta)}\right\},\qquad I_2=\left\{k\in\mathbf{N}\,\middle|\,k>\frac{b}{\sin(\pi-\theta)}\right\}.$$

When $k\in I_1$ we use the estimate $|\psi_k(m,\ \theta)|\leqslant 1$ ($\forall k\in\mathbf{N}$) and hence by (3.5),

$$|u_k(m,\ \theta)| \leqslant (n-1)k^2(\sin\xi)^{(n-1)m}\pi^{-\frac{m}{2}-2}. \tag{3.3}$$

Since

$$\triangle^{n+1}u_k(m,\theta)=\sum_{\nu=0}^{n+1}\mathrm{C}_{n+1}^{\nu}(-1)^{\nu}u_{k+\nu}(m,\theta),$$

we get, by (3.3),

$$|\triangle^{n+1}u_k(m,\ \theta)| \leqslant C(n)k^2(\sin\xi)^{(n-1)m}\pi^{-\frac{m}{2}}. \tag{3.4}$$

Therefore, we get (for $m>10$)

$$\sum_{k\in I_1}k^n|\triangle^{n+1}u_k(m,\theta)| \leqslant C(n)\xi^{(n-1)m}\pi^{-\frac{m}{2}}\xi^{-n-3} \leqslant C(n)\pi^{-\frac{m}{2}}. \tag{3.5}$$

When $k\in I_2$, i. e. $k>\frac{b}{\sin\xi}$, we apply (2.6) of Lemma 2. Then we get

$$|\psi_k(m,\ \theta)| \leqslant \frac{b^m}{(k\sin\xi)^{\frac{mn}{2}}}$$

and hence by (3.2),

$$|u_k(m,\ \theta)| \leqslant (\sin\xi)^{(n-1)m}\pi^{-\frac{m}{2}+2}\frac{(n-1)k^2b^m}{(k\sin\xi)^{\frac{mn}{2}}}. \tag{3.6}$$

Therefore

$$|\triangle^{n+1}u_k(m,\ \theta)| \leqslant C(n)(\sin\xi)^{(n-1)m}\pi^{-\frac{m}{2}}\frac{k^2b^m}{(k\sin\xi)^{\frac{mn}{2}}}. \tag{3.7}$$

So we get

$$\sum_{k\in I_2}k^n|\triangle^{n+1}u_k(m,\theta)| \leqslant C(n)\pi^{-\frac{m}{2}}\sum_{k\in I_2}k^{n+2-\frac{mn}{2}}b^m \leqslant C(n)\pi^{-\frac{m}{2}}. \tag{3.8}$$

A combination of (3.5) and (3.8) yields the conclusion of the lemma. □

§ 4. The case $0<\theta<\frac{\pi}{2}$

In what follows we assume $\theta\in\left(0, \frac{\pi}{2}\right)$. We may assume $m>10+n$.

In order to find a suitable estimate for $|\triangle^{n+1}u_k(m, \theta)|$ we first establish the following lemma.

Lemma 5 Let $m\in\mathbf{N}$, $m>10+n$. Assume $\{h_k(\mu)\}_{k=1}^{+\infty}$, $\mu=1, 2,\cdots,m$, are sequences of numbers. Then

$$\triangle^j\Big(\prod_{\mu=1}^m h_k(\mu)\Big)=\sum_{l_1=1}^m\cdots\sum_{l_j=1}^m\Big(\prod_{\mu=1}^m h_k(l_1,l_2,\cdots,l_j,\mu)\Big),\quad j\in\mathbf{N},\tag{4.1}$$

where the numbers $h_k(l_1, l_2, \cdots, l_j, \mu)$ are defined inductively as follows:

$$h_k(l_1, \mu)=\begin{cases}h_k(\mu), & \text{if } 1\leqslant\mu\leqslant l_1,\\ \triangle h_k(\mu), & \text{if } \mu=l_1,\\ h_{k+1}(\mu), & \text{if } l_1<\mu\leqslant m,\end{cases}\tag{4.2}$$

$$h_k(l_1, l_2, \cdots, l_{j+1}, \mu)=\begin{cases}h_k(l_1, l_2, \cdots, l_j, \mu), & \text{if } 1\leqslant\mu<l_{j+1},\\ \triangle h_k(l_1, l_2, \cdots, l_j, \mu), & \text{if } \mu=l_{j+1},\\ h_{k+1}(l_1, l_2, \cdots, l_j, \mu), & \text{if } l_{j+1}<\mu\leqslant m.\end{cases}\tag{4.3}$$

Proof It is easy to verify (4.1) for $j=1$, i.e.

$$\triangle\Big(\prod_{\mu=1}^m h_k(\mu)\Big)=\sum_{l=1}^m\Big(\prod_{\mu=1}^m h_k(l,\mu)\Big),$$

where $h_k(l, \mu)$ is defined as in (4.2). Suppose (4.1) is valid for j. Then we have

$$\triangle^{j+1}\Big(\prod_{\mu=1}^m h_k(\mu)\Big)=\sum_{l_1=1}^m\cdots\sum_{l_j=1}^m\triangle\Big(\prod_{\mu=1}^m h_k(l_1,l_2,\cdots,l_j,\mu)\Big).$$

But by the result for $j=1$ we have

$$\triangle\Big(\prod_{\mu=1}^m h_k(l_1,l_2,\cdots,l_j,\mu)\Big)=\sum_{l_{j+1}=1}^m\Big(\prod_{\mu=1}^m h_k(l_1,l_2,\cdots,l_j,l_{j+1},\mu)\Big),$$

where $h_k(l_1, l_2, \cdots, l_j, l_{j+1}, \mu)$ is defined in (4.3). Then we have proved that (4.1) is valid for all $j\in\mathbf{N}$ and finish the proof. □

Let us now estimate $|\triangle^j P_{k-1}^{n+2}(\cos\theta)|$, $j=1,2,\cdots,n+1$. We apply formula (4.5.4) in [Sz, p. 71]:

$$P_k^{(\alpha+1,\beta)}(x)=\frac{2}{2k+\alpha+\beta+2}\frac{(k+\alpha+1)P_k^{(\alpha,\beta)}(x)-(k+1)P_{k+1}^{(\alpha,\beta)}(x)}{1-x}$$

and get

$$Q_k^{(\alpha,\beta)}(x)-Q_{k+1}^{(\alpha,\beta)}(x)=(1-x)\frac{2k+\alpha+\beta+2}{2(\alpha+1)}Q_k^{(\alpha+1,\beta)}(x). \qquad (4.4)$$

In particular, with $\alpha=\beta=\frac{n-1}{2}$, we have

$$\begin{aligned}&P_k^{n+2}(\cos\theta)-P_{k+1}^{n+2}(\cos\theta)\\ &=\frac{2k+n+1}{n+1}(1-\cos\theta)Q_k^{(\frac{n+1}{2},\frac{n-1}{2})}(\cos\theta).\end{aligned} \qquad (4.5)$$

Applying (4.4) (4.5) inductively and making use of Lemma 2, we get

$$|\triangle^j P_{k-1}^{n+2}(\cos\theta)|\leqslant\begin{cases}\dfrac{B\theta^j}{(k\theta)^{\frac{n}{2}}}, & \text{if } k\theta\geqslant 1,\\ B\theta^j, & \text{if } k\theta<1,\end{cases}$$
$$j=0,1,\cdots,n+1, \qquad (4.6)$$

where $B>1$ denotes a constant depending only on n.

Using the constant B in (4.6), we will treat the cases $k\theta>2B$ and $k\theta\leqslant 2B$ separately.

Lemma 6 Let $m\in\mathbf{N}$, $m>10n$ and $0<\theta<\frac{\pi}{2}$. If $0<\theta<\frac{\pi}{2}$ and $k\theta\geqslant 2B$ with B the constant in (4.6), then

$$|\triangle^{n+1}u_k(m,\theta)|\leqslant C(n)m^{n+1}\theta^{n+1}\left(\frac{B}{(k\theta)^{\frac{n}{2}}}\right)^{m-n}. \qquad (4.7)$$

where $C(n)$ denotes a constant depending only on n .

Proof Recall (see the proof of Lemma 4)

$$u_k(m,\theta)=\phi(m,\theta)(k(k+n-2))\psi_k(m,\theta).$$

For any sequence $u_k=a_kb_k$ we can easily verify by induction that

$$\triangle^{n+1}u_k=\sum_{j=0}^{n+1}C_{n+1}^j\triangle^j a_{k+n+1-j}\triangle^{n+1-j}b_k. \qquad (4.8)$$

Applying (4.8) with $a_k=k(k+n-2)$, $b_k=\psi_k(m,\theta)$, we get

$$\begin{aligned}&\triangle^{n+1}u_k(m,\theta)\\&=\phi(m,\theta)\Big(\triangle^{n+1}\psi_k(m,\theta)a_{k+n+1}+(n+1)\triangle^{n}\psi_k(m,\theta)\triangle a_{k+n}+\\&\quad\frac{n(n+1)}{2}\triangle^{n-1}\psi_k(m,\theta)\triangle^2 a_{k+n-1}\Big)\\&=\phi(m,\theta)((k+n+1)(k+2n-1)\triangle^{n+1}\psi_k(m,\theta)+\\&\quad(n+1)(-2k-3n+1)\times\triangle^{n}\psi_k(m,\theta)+n(n+1)\triangle^{n-1}\psi_k(m,\theta)).\end{aligned}$$

Hence

$$\begin{aligned}|\triangle^{n+1}u_k(m,\theta)|\leqslant 3(n+1)^2|\phi(m,\theta)|(k^2|\triangle^{n+1}\psi_k(m,\theta)|+\\k|\triangle^{n}\psi_k(m,\theta)|+|\triangle^{n-1}\psi_k(m,\theta)|).\end{aligned}\tag{4.9}$$

We apply Lemma 5 by writing $h_k=h_k(\mu)=P_{k-1}^{\frac{n+2}{2}}(\cos\theta)$, $\mu=1,2,\cdots,m$. Then by (4.1) we have

$$\triangle^j\psi_k(m,\theta)=\sum_{l_1=1}^{m}\cdots\sum_{l_j=1}^{m}\Big(\prod_{\mu=1}^{m}h_k(l_1,l_2,\cdots,l_j,\mu)\Big),\quad (j=n-1,\ n,\ n+1).\tag{4.10}$$

In our case, by (4.2) we have

$$h_k(l_1,\mu)=\begin{cases}h_k(\mu)=h_k, & \text{if } 1\leqslant\mu<l_1,\\ \triangle h_k(\mu)=\triangle h_k, & \text{if } \mu=l_1,\\ h_{k+1}(\mu)=h_{k+1}, & \text{if } l_1<\mu\leqslant m,\end{cases}\tag{4.11}$$

and then by (4.3)

$$h_k(l_1,l_2,\mu)=\begin{cases}h_k(l_1,\mu)=h_k, & \text{if } 1\leqslant\mu<\min(l_1,l_2),\\ \triangle h_k(l_1,\mu)=\triangle h_k, & \text{if } \mu=l_1<l_2,\\ \triangle h_k(l_1,\mu)=h_{k+1}, & \text{if } l_1<\mu<l_2,\\ \triangle h_k(l_1,\mu)=\triangle h_k, & \text{if } l_2=\mu<l_1,\\ \triangle h_k(l_1,\mu)=\triangle^2 h_k, & \text{if } \mu=l_1=l_2,\\ \triangle h_k(l_1,\mu)=\triangle h_{k+1}, & \text{if } l_1<\mu=l_2,\\ h_{k+1}(l_1,\mu)=h_{k+1}, & \text{if } l_2<\mu<l_1,\\ h_{k+1}(l_1,\mu)=\triangle h_{k+1}, & \text{if } l_2<\mu=l_1,\\ h_{k+1}(l_1,\mu)=h_{k+2}, & \text{if } l_1<l_2<\mu.\end{cases}\tag{4.12}$$

From (4.11) and (4.12) by using induction we conclude that for all $j\in\mathbf{N}$,

$$h_k(l_1, l_2, \cdots, l_j, \mu)\in\{h_s, \triangle^t h_{k+j-t} \mid s=k, k+1, \cdots, k+j, t=1,2,\cdots,j\}.$$

Furthermore, by induction we see that in each product

$$\prod_{\mu=1}^{m} h_k(l_1,l_2,\cdots,l_j,\mu)$$

the factors having form "h_s" appear totally $m-j$ times and the sum of all degrees "t" over all factors having the form $\triangle^t h_{k+j-t}$ is exactly j. In effect, for any $(l_1, l_2, \cdots, l_j)(j\leqslant n+1)$ we define

$$I_1(l_1, l_2, \cdots, l_j)=\{\mu \mid h_k(l_1, l_2, \cdots, l_j, \mu)\in \{h_k, h_{k+1}, \cdots, h_{k+j}\}\},$$

$$I_2(l_1, l_2, \cdots, l_j)=\{\mu \mid h_k(l_1, l_2, \cdots, l_j, \mu)\in \{\triangle^t h_{k+j-t} \mid t=1,2,\cdots,j\}\}.$$

Then the cardinality of I_1 is exactly $m-j$. Hence by (4.6) we have

$$\Big|\prod_{\mu\in I_1} h_k(l_1,l_2,\cdots,l_j,\mu)\Big|\leqslant\left(\frac{B}{(k\theta)^{\frac{n}{2}}}\right)^{m-j}.$$

Meanwhile, by (4.6)

$$\Big|\prod_{\mu\in I_2} h_k(l_1,l_2,\cdots,l_j,\mu)\Big|\leqslant\frac{(B\theta)^j}{(k\theta)^{\frac{n}{2}}}.$$

Then we get

$$\Big|\prod_{\mu=1}^{m} h_k(l_1,l_2,\cdots,l_j,\mu)\Big|\leqslant\frac{B^m\theta^j}{(k\theta)^{(m-j+1)\frac{n}{2}}}.$$

Then by (4.10) we get

$$|\triangle^j\psi_k(m, \theta)|\leqslant\frac{B^m m^j\theta^j}{(k\theta)^{(m-j+1)\frac{n}{2}}}, \quad j=n-1, n, n+1, (k\theta)\geqslant 2B. \tag{4.13}$$

Substituting (4.13) into (4.9) and observing $|\phi(m, \theta)|\leqslant\theta^2$, we obtain

$$|\triangle^{n+1}u_k(m, \theta)|\leqslant 9(n+1)^2\frac{B^m m^{n+1}\theta^{n+1}}{(k\theta)^{\frac{(m-n)n}{2}}},$$

and complete the proof. □

Now we consider the case $0<k\theta\leqslant 2(n+1)B$.

Lemma 7 Let $m\in\mathbf{N}$, $m>10n$ and $0<k\theta\leqslant 2(n+1)B$, then

$$|\hat{h}_\theta(k)|\leqslant\begin{cases}1-\frac{1}{10(n+2)}(k\theta)^2, & \text{if } 0\leqslant k\theta\leqslant\frac{1}{2},\\ \delta(n), & \text{if } \frac{1}{2(n+1)}\leqslant k\theta\leqslant 2(n+1)B,\end{cases}$$

where $\delta(n)=1-\frac{1}{(2(n+1)^2\pi B)^{n+2}}$.

Proof From the formula

$$\frac{\mathrm{d}}{\mathrm{d}t}P_k^n(t)=\frac{k(k+n-2)}{n-1}P_{k-1}^{n+2}(t)$$

(see [WL, p. 31, Corollary 1.2.8] or [Sz, p. 81, (4.7.14)]) and the Lagrange mean-value theorem we know that there is a value $\xi\in(0,\theta)$ such that

$$1-P_k^n(\cos\theta)=\frac{k(k+n-2)}{n-1}P_{k-1}^{n+2}(\cos\xi)(1-\cos\theta). \tag{4.14}$$

Hence

$$0\leqslant 1-P_k^n(\cos\theta)\leqslant(k\theta)^2.$$

So, if $0<k\theta\leqslant\frac{1}{2}$, then

$$P_k^n(\cos\theta)\geqslant\frac{3}{4}. \tag{4.15}$$

Of course (4.15) is also valid for $P_{k-1}^{n+2}(\cos\theta)$. From (4.14) applying (4.15) to $P_{k-1}^{n+2}(\cos\xi)$ we derive

$$\frac{3}{4}\leqslant P_k^n(\cos\theta)\leqslant 1-\frac{1}{10n}(k\theta)^2,\quad \text{if } 0\leqslant k\theta\leqslant\frac{1}{2}.$$

Then we get

$$\frac{3}{4}\leqslant P_{k-1}^{n+2}(\cos\theta)\leqslant 1-\frac{1}{10(n+2)}(k\theta)^2,\quad \text{if } 0\leqslant k\theta\leqslant\frac{1}{2}. \tag{4.16}$$

By (2.1) and (4.16) we have

$$|\hat{h}_\theta(k)|\leqslant 1-\frac{1}{10(n+2)}(k\theta)^2,\quad \text{if } 0\leqslant k\theta\leqslant\frac{1}{2}. \tag{4.17}$$

When $\frac{1}{2(n+1)k}\leqslant\theta\leqslant 2(n+1)B_{\frac{1}{k}}$, i.e. $\frac{1}{2(n+1)}\leqslant k\theta\leqslant 2(n+1)B$,

we have

$$\hat{h}_\theta(k) = 1 - \frac{1}{\int_0^\theta \sin^{n-2} t \, dt} \int_0^\theta (1 - P_k^n(\cos t)) \sin^{n-2} t \, dt$$

$$\leqslant 1 - \frac{1}{\int_0^\theta \sin^{n-2} t \, dt} \int_0^{\frac{1}{2(n+1)k}} (1 - P_k^n(\cos t)) \sin^{n-2} t \, dt$$

$$= 1 - \frac{1}{\int_0^\theta \sin^{n-2} t \, dt} \int_0^{\frac{1}{2(n+1)k}} \frac{k(k+n-2)}{n-2} \times$$

$$\left(\int_0^t P_{k-1}^{n+2}(\cos u) \sin u \, du\right) \sin^{n-2} t \, dt.$$

Applying (4.15) we get

$$\frac{1}{\int_0^\theta \sin^{n-2} t \, dt} \int_0^{\frac{1}{2(n+1)k}} \frac{k(k+n-2)}{n-2} \left(\int_0^t P_{k-1}^{n+2}(\cos u) \sin u \, du\right) \sin^{n-2} t \, dt$$

$$\geqslant \frac{n-1}{\theta^{n-1}} \int_0^{\frac{1}{2(n+1)k}} \frac{k(k+n-2)}{n-2} \left(\int_0^t \frac{3}{4} \sin u \, du\right) \sin^{n-2} t \, dt$$

$$\geqslant \frac{k(k+n-2)}{n-2} \frac{3}{4} \frac{n-1}{\theta^{n-1}} \int_0^{\frac{1}{2(n+1)k}} 2\sin^2 \frac{t}{2} \sin^{n-2} t \, dt$$

$$\geqslant \frac{3(n-1)k(k+n-2)}{8(n-2)\theta^{n-1}} \left(\frac{2}{\pi}\right)^n \int_0^{\frac{1}{2(n+1)k}} t^n \, dt \geqslant \frac{1}{(2(n+1)^2 \pi B)^{n+2}}.$$

Therefore, when $\frac{1}{2(n+1)} \leqslant k\theta \leqslant 2(n+1)B$,

$$\hat{h}_\theta(k) \leqslant 1 - \frac{1}{(2(n+1)^2 \pi B)^{n+2}}.$$

Thus

$$-\hat{h}_\theta(k) = 1 - \frac{1}{\int_0^\theta \sin^{n-2} t \, dt} \int_0^\theta (1 + P_k^n(\cos t)) \sin^{n-2} t \, dt$$

$$\leqslant 1 - \frac{1}{\int_0^\theta t^{n-2} \, dt} \int_0^{\frac{1}{2k}} \frac{7}{4} \left(\frac{2}{\pi}\right)^{n-2} t^{n-2} dt = 1 - \frac{7}{8\pi^{n-2} (k\theta)^{n-1}}$$

$$\leqslant 1 - \frac{7\pi^2 B}{8(\pi B)^n} < 1 - \frac{1}{(2(n+1)^2 \pi B)^{n+2}}.$$

Then we get for $\delta(n) = 1 - \frac{1}{(2(n+1)^2 \pi B)^{n+2}}$,

$$|\hat{h}_\theta(k)|\leqslant\delta(n)<1,\quad \text{when } \frac{1}{2(n+1)}\leqslant k\theta\leqslant 2(n+1)B. \quad (4.18)$$

A combination of (4.17) and (4.18) completes the proof. □

Lemma 8 Let $m\in\mathbf{N}$, $m>10n$ and let B be the constant in (4.6). Then

$$|\triangle^{n+1}u_k(m,\theta)|\leqslant$$

$$\begin{cases} C(n)m^{n+1}\theta^{n+1}\left(1-\frac{1}{10(n+2)}(k\theta)^2\right)^m, & \text{if } 0\leqslant k\theta\leqslant\frac{1}{2(n+1)}, \\ C(n)m^{n+1}\theta^{n+1}(\delta(n))^m, & \text{if } \frac{1}{2(n+1)}\leqslant k\theta\leqslant 2B, \end{cases}$$

where $\delta(n)=1-\frac{1}{(2(n+1)^2\pi B)^{n+2}}$.

Proof Applying (4.8) with $a_k=k(k+n-2)$, $b_k=(\hat{h}_\theta(k))^m$ we get

$$\triangle^{n+1}u_k(m,\theta)$$

$$=-\theta^2\left(\triangle^{n+1}b_k a_{k+n+1}+(n+1)\triangle^n b_k\triangle a_{k+n}+\frac{n(n+1)}{2}\triangle^{n-1}b_k\triangle^2 a_{k+n-1}\right)$$

$$=-\theta^2((k+n+1)(k+2n-1)\triangle^{n+1}(\hat{h}_\theta(k))^m+$$

$$(n+1)(-2k-3n+1)\triangle^n(\hat{h}_\theta(k))^m+n(n+1)\triangle^{n-1}(\hat{h}_\theta(k))^m).$$

Hence

$$|\triangle^{n+1}u_k(m,\theta)|\leqslant 3(n+1)^2\theta^2(k^2|\triangle^{n+1}(\hat{h}_\theta(k))^m|+$$

$$k|\triangle^n(\hat{h}_\theta(k))^m|+|\triangle^{n-1}(\hat{h}_\theta(k))^m|). \quad (4.19)$$

Now we repeat the same argument as in the proof of Lemma 6. This time we apply Lemma 5 to $\hat{h}_\theta(k)$. We take $\hat{h}_\theta(k)$ in the place of $h_k(\mu)=\hat{h}_\theta(k)$, $\mu=1,2,\cdots,m$ in Lemma 5. Then by (4.1) we have

$$\triangle^j(\hat{h}_\theta(k))^m=\sum_{l_1=1}^m\cdots\sum_{l_j=1}^m\left(\prod_{\mu=1}^m h_k(l_1,l_2,\cdots,l_j,\mu)\right),$$

$$j=n-1,\ n,\ n+1, \quad (4.20)$$

where, by (4.2),

$$h_k(l_1,\ \mu)=\begin{cases}\hat{h}_\theta(k), & \text{if } 1\leqslant\mu<l_1,\\ \triangle(\hat{h}_\theta(k)), & \text{if } \mu=l_1,\\ \hat{h}_\theta(k+1), & \text{if } l_1<\mu\leqslant m,\end{cases} \tag{4.21}$$

and then by (4.3)

$$h_k(l_1,\ l_2,\ \mu)=\begin{cases}h_k(l_1,\ \mu)=\hat{h}_\theta(k), & \text{if } 1\leqslant\mu<\min(l_1,\ l_2),\\ \triangle h_k(l_1,\ \mu)=\triangle\hat{h}_\theta(k), & \text{if } \mu=l_1<l_2,\\ \triangle h_k(l_1,\ \mu)=\hat{h}_\theta(k+1), & \text{if } l_1<\mu<l_2,\\ \triangle h_k(l_1,\ \mu)=\triangle\hat{h}_\theta(k), & \text{if } l_2=\mu<l_1,\\ \triangle h_k(l_1,\ \mu)=\triangle^2\hat{h}_\theta(k), & \text{if } \mu=l_1=l_2,\\ \triangle h_k(l_1,\ \mu)=\triangle\hat{h}_\theta(k+1), & \text{if } l_1<\mu=l_2,\\ h_{k+1}(l_1,\ \mu)=\hat{h}_\theta(k+1), & \text{if } l_2<\mu<l_1,\\ h_{k+1}(l_1,\ \mu)=\triangle\hat{h}_\theta(k+1), & \text{if } l_2<\mu=l_1,\\ h_{k+1}(l_1,\ \mu)=\hat{h}_\theta(k+2), & \text{if } l_1<l_2<\mu.\end{cases} \tag{4.22}$$

From (4.21) and (4.22) by using induction, we conclude that for all $j\in\mathbf{N}$,

$$h_k(l_1,\ l_2,\ \cdots,\ l_j,\ \mu)\in\{\hat{h}_\theta(s),\ \triangle^t\hat{h}_\theta(k+j-t)\mid,\ s=k,\ k+1,\ \cdots,\ k+j,\ t=1,2,\cdots,j\}.$$

Furthermore, by induction we see that in each product

$$\prod_{\mu=1}^{m}h_k(l_1,l_2,\cdots,l_j,\mu)$$

the factors having form "$\hat{h}_\theta(s)$" appear totally $m-j$ times and the sum of all degrees "t" over all factors having the form $\triangle^t\hat{h}_\theta(k+j-t)$ is exactly j. In effect, for any $(l_1,\ l_2,\ \cdots,\ l_j)(j\leqslant n+1)$ we define

$$I_1(l_1,\ l_2,\ \cdots,\ l_j)=\{\mu\mid h_k(l_1,\ l_2,\ \cdots,\ l_j,\ \mu)\in\{\hat{h}_\theta(k),\ \hat{h}_\theta(k+1),\ \cdots,\ \hat{h}_\theta(k+j)\}\},$$

$$I_2(l_1,\ l_2,\ \cdots,\ l_j)=\{\mu\mid h_k(l_1,\ l_2,\ \cdots,\ l_j,\ \mu)\in\{\triangle^t\hat{h}_\theta(k+j-t)\mid t=1,2,\cdots,j\}\}.$$

Then the cardinality of I_1 is exactly $m-j$. Hence by Lemma 7 we get

$$\left|\prod_{\mu\in I_1} h_k(l_1,l_2,\cdots,l_j,\mu)\right| \leqslant \begin{cases}(1-\frac{1}{10(n+2)}(k\theta)^2)^{m-j}, & \text{if } 0\leqslant k\theta\leqslant\frac{1}{2(n+1)},\\ (\delta(n))^{m-j}, & \text{if } \frac{1}{2(n+1)}\leqslant k\theta\leqslant 2B.\end{cases} \tag{4.23}$$

On the other hand, we write

$$h(\theta)=\frac{\sin^{n-1}\theta}{(n-1)\int_0^\theta \sin^{n-2}t\,\mathrm{d}t}.$$

It is obvious that $|h(\theta)|\leqslant 1$. By (2.1) we see

$$\hat{h}_\theta(k)=h(\theta)P_{k-1}^{\frac{n+2}{2}}(\cos\theta).$$

Hence

$$\triangle^t h_s=h(\theta)\triangle^t P_{s-1}^{\frac{n+2}{2}}(\cos\theta),\qquad s\in\{k,\ k+1,\ \cdots,\ k+n\}.$$

Then by (4.6) we have

$$\left|\prod_{\mu\in I_2} h_k(l_1,\ l_2,\ \cdots,\ l_j,\ \mu)\right|\leqslant(B\theta)^j. \tag{4.24}$$

Combining (4.23) and (4.24) we get

$$\left|\prod_{\mu=1}^{m} h_k(l_1,l_2,\cdots,l_j,\mu)\right| \leqslant \begin{cases}(1-\frac{1}{10(n+2)}(k\theta)^2)^{m-j}(B\theta)^j, & \text{if } 0\leqslant k\theta\leqslant\frac{1}{2(n+1)},\\ (\delta(n))^{m-j}(B\theta)^j, & \text{if } \frac{1}{2(n+1)}\leqslant k\theta\leqslant 2B.\end{cases}$$

Then by (4.20) we get

$$|\triangle^j(\hat{h}_\theta(k))^m| \leqslant \begin{cases}m^{n+1}\left(1-\frac{1}{10(n+2)}(k\theta)^2\right)^{m-j}(B\theta)^j, & \text{if } 0\leqslant k\theta\leqslant\frac{1}{2(n+1)},\\ m^{n+1}(\delta(n))^{m-j}(B\theta)^j, & \text{if } \frac{1}{2(n+1)}\leqslant k\theta\leqslant 2B.\end{cases} \tag{4.25}$$

Substituting (4.25) into (4.19), we complete the proof. □

§ 5. Proof of the theorem

For the constant B in (4.6), $m>10n$ and $\theta\in\left(0, \frac{\pi}{2}\right)$, we define

$$J_1=J_1(m, \theta)=\{k\in\mathbf{N} \mid k\theta\leqslant m^{-\frac{1}{4}}\},$$

$$J_2=J_2(m, \theta)=\left\{k\in\mathbf{N} \mid m^{-\frac{1}{4}}<k\theta\leqslant\frac{1}{2(n+1)}\right\},$$

$$J_3=J_3(m, \theta)=\left\{k\in\mathbf{N} \,\middle|\, \frac{1}{2(n+1)}<k\theta<2B\right\},$$

$$J_4=J_4(m, \theta)=\{k\in\mathbf{N} \mid k\theta\geqslant 2B\}.$$

Assume $J_1\neq\varnothing$.

Take a function $\eta\in C^\infty[0, +\infty)$ such that $\chi_{[0,1]}\leqslant\eta\leqslant\chi_{[0,2]}$. Write $N=\left[\frac{1}{\theta m^{\frac{1}{4}}}\right]$ and define

$$\eta_N(f)=\sum_{k=0}^{+\infty}\eta\left(\frac{k}{N}\right)Y_k(f), \quad f\in L(S^{n-1}).$$

Write $\|\ \|$ instead of $\|\ \|_X$ (or $\|\ \|_{(X,X)}$) for simplicity. We have (see [WL, Theorem 4.6.3, pp. 191, 192], or [R])

$$\|D(\eta_N f)\|\leqslant C(n)N^2\|f\|,$$

where here and in what follows we use $C(n)$ to denote constants depending only on n and the choice of η which may have different values in different occurrences. Hence for any $f\in X$,

$$\|T_m(\theta)(\eta_N f)\|=\theta^2\|(M_\theta)^m D(\eta_N f)\|\leqslant\frac{C(n)}{\sqrt{m}}\|f\|. \tag{5.1}$$

Now we have for any $f\in X$,

$$T_m(\theta)(f)=T_m(\theta)(f-\eta_N(f))+T_m(\theta)(\eta_N f).$$

Then

$$\|T_m(\theta)(f)\|\leqslant\|T_m(\theta)(f-\eta_N(f))\|+\frac{C(n)}{\sqrt{m}}\|f\|. \tag{5.2}$$

Write $g=f-\eta_N(f)$. We know $\|g\|\leqslant C\|f\|$ with constant C depending only on the choice of η (see [WL, p. 162], or [R]). Applying Lemma 1, we get

$$T_m(\theta)(g)=\sum_{k=0}^{+\infty}\triangle^{n+1}u_k(m,\theta)A_k^n\sigma_k^n(g).$$

Note that $Y_k(g)=Y_k(f)-\eta\left(\frac{k}{N}\right)Y_k(f)=0$ when $k\leqslant N$. We get

$$T_m(\theta)(g)=\left(\sum_{k\in J_2}+\sum_{k\in J_3}+\sum_{k\in J_4}\right)\triangle^{n+1}u_k(m,\theta)A_k^n\sigma_k^n(g). \tag{5.3}$$

By Lemma 6 we have for $m>10n$,

$$\begin{aligned}\left\|\sum_{k\in J_4}\triangle^{n+1}u_k(m,\theta)A_k^n\sigma_k^n(g)\right\|&\leqslant C(n)m^{n+1}\|g\|\sum_{k\in J_4}k^n\theta^{n+1}\left(\frac{B}{(k\theta)^{\frac{n}{2}}}\right)^{m-n}\\&\leqslant C(n)\frac{m^{n+1}}{(2B)^m}\|f\|.\end{aligned} \tag{5.4}$$

From (5.2)～(5.4) we derive

$$\|T_m(\theta)\|\leqslant C(n)\left(\sum_{k\in(J_2\cup J_3)}|\triangle^{n+1}u_k(m,\theta)|k^n+\left(\frac{1}{\sqrt{m}}+\frac{m^{n+1}}{(2B)^m}\right)\right),$$
$$0<\theta<\frac{\pi}{2}. \tag{5.5}$$

Now we apply Lemma 8 for $k\in(J_2\cup J_3)$. By Lemma 8, when $k\in J_2$,

$$\begin{aligned}|\triangle^{n+1}u_k(m,\theta)|&\leqslant C(n)m^{n+1}\theta^{n+1}\left(1-\frac{1}{10(n+2)}(k\theta)^2\right)^m\\&\leqslant C(n)m^{n+1}\theta^{n+1}\left(1-\frac{1}{10(n+2)\sqrt{m}}\right)^m.\end{aligned}$$

Define

$$\gamma(n)=\sup\left\{\left(1-\frac{1}{10(n+2)\sqrt{m}}\right)^{\sqrt{m}}\ \Big|\ m\geqslant n\right\}.$$

Then we see $0<\gamma(n)<1$. Hence

$$\sum_{k\in J_2}|\triangle^{n+1}u_k(m,\theta)|k^n\leqslant C(n)m^{n+1}(\gamma(n))^{\sqrt{m}}. \tag{5.6}$$

When $k\in J_3$, by Lemma 8,

$$|\triangle^{n+1}u_k(m,\theta)|\leqslant C(n)m^{n+1}\theta^{n+1}(\delta(n))^m,$$

where $0<\delta(n)=1-\frac{1}{(2(n+1)^2\pi B)^{n+2}}<1$. Then we get

$$\sum_{k\in J_3}|\triangle^{n+1}u_k(m,\theta)|k^n\leqslant C(n)m^{n+1}(\delta(n))^m. \tag{5.7}$$

Substituting (5.6) and (5.7) into (5.5), we obtain

$$\| T_m(\theta) \| \leqslant C(n)\left(m^{n+1}(\gamma(n))^{\sqrt{m}}+m^{n+1}(\delta(n))^m+\frac{1}{\sqrt{m}}+\frac{m^{n+1}}{(2B)^m}\right),$$

$$0<\theta<\frac{\pi}{2}. \tag{5.8}$$

Lemma 4 tells that

$$\sup\left\{ \| T_m(\theta) \|_{(X,X)} \mid \theta\in\left[\frac{\pi}{2}, \pi\right)\right\}\leqslant C(n)\pi^{-\frac{m}{2}}. \tag{5.9}$$

Combining (5.8) and (5.9), we obtain (for $m>10n$)

$$\sup\{ \| T_m(\theta) \|_{(X,X)} : \theta\in(0, \pi)\}$$

$$\leqslant C(n)\left(m^{n+1}(\gamma(n))^{\sqrt{m}}+m^{n+1}(\delta(n))^m+\frac{1}{\sqrt{m}}+\frac{m^{n+1}}{(2B)^m}+\pi^{-\frac{m}{2}}\right), \tag{5.10}$$

which completes the proof of the Theorem. □

Remark The theorem can be extended by the same argument without any difficulty to the case when taking the fractional derivatives of the Laplace-Beltrami operator instead of the Laplace-Beltrami operator itself. Precisely, let $r>0$ and $D^{\frac{r}{2}}$ denote the derivative operator of degree r which is defined (see [WL, p. 171, Definition 4. 3. 1]) by

$$D^{\frac{r}{2}}(f) = \sum_{k=1}^{+\infty} e^{ir\frac{\pi}{2}}(k(k+n-2))^{\frac{r}{2}}Y_k(f),$$

where $i^2=-1$. Then

$$\lim_{m\to+\infty} \sup\{ \| \theta^r D^{\frac{r}{2}(m_\theta)^m} \|_{(X,X)} \mid 0<\theta<\pi\}=0.$$

For this extension we have only to replace

$$k(k+n-2)\theta^2 \text{ by } [k(k+n-2)\theta^2]^{\frac{r}{2}}.$$

In [D] a different definition is given for which the result follows from the results for powers of the Laplace-Beltrami operators and a Kolmogorovtype inequality in [D].

Acknowledgments

The authors thank the referee very much for his valuable suggestions and pointing out misprints.

References

[D] Ditzian Z. Fractional derivatives and best approximation. Acta Math. Hungar，1998，81：323—348.

[DR] Ditzian Z. and Runovskii K. Averages on caps of S^{d-1}. J. Math. Anal. Appl.，2000，248：260—274.

[R] Rustamov Kh P. On the approximation of functions on sphere. Izv Akad. Nauk SSSR Ser. Mat. 1993，59：127—148.

[So] Sogge C D. Oscillatory integrals and spherical harmonics. Duke Math J. 1986，53：43—65.

[St] Stein E M. Singular Integrals and Differentiability Properties of Functions. Princeton Univ. Press，Princeton，NJ，1979 (third printing，with corrections).

[SW] Stein E M，Weiss G. Introduction to Fourier Analysis on Euclidean Spaces. Princeton Univ. Press，Princeton，NJ，1971.

[Sz] Szegö G. Orthogonal Polynomials. 4th Ed.，American Mathematical Society Colloquium Publications 23，Amer. Math. Soc. Providence，RI，1975.

[WL] Wang Kunyang，Li Luoqing. Harmonic Analysis and Approximation on the Unit Sphere. Science Press，Beijing，2000.

Journal of Approximation Theory，2003，123：300－304.

有界变差函数的绝对(C, α)收敛性

On the Absolute (C, α) Convergence for Functions of Bounded Variation[①]

Let f be a 2π-periodic function integrable on $[-\pi, \pi]$. For $\alpha > -1$, the Cesàro means of order α of the Fourier series of f are defined by

$$\sigma_n^\alpha(f)(x) = \frac{1}{\pi}\int_{-\pi}^{\pi} f(x+t)K_n^\alpha(t)\mathrm{d}t,$$

where

$$K_n^\alpha(t) = \frac{1}{A_n^\alpha}\sum_{k=0}^{n} A_{n-k}^{\alpha-1}D_k(t), \quad A_n^\alpha = \frac{\Gamma(n+\alpha+1)}{\Gamma(n+1)\Gamma(\alpha+1)},$$

$$D_k(t) = \frac{\sin\left(k+\frac{1}{2}\right)t}{2\sin\frac{1}{2}t}.$$

Define

$$R_n^\alpha(f, x) = \sum_{k=n+1}^{+\infty} \left|\sigma_k^\alpha(f)(x) - \sigma_{k-1}^\alpha(f)(x)\right|.$$

The following result was presented in [1]:

Theorem HB Let $x \in [0, \pi]$ and f a 2π-periodic function of bounded variation on $[-\pi, \pi]$. Then, for $\alpha > 0$ and $n \geqslant 2$, we have

$$R_n^\alpha(f, x) \leqslant \frac{4\alpha}{n\pi}\sum_{k=1}^{n} V_0^{\frac{\pi}{k}}(\varphi_x),$$

---

① Supported by NSFC 10071007.

本文与余冉合作.

Received：2002-12-21；Revised：2003-05-07.

where

$$\varphi_x(t):=f(x+t)+f(x-t)-f(x+0)-f(x-0)$$

and $V_a^b(f)$ denotes the total variation of f on $[a, b]$.

But this is incorrect when $0<\alpha<1$. Our result is the following theorem.

Theorem Suppose $0<\alpha<1$, $x\in[0, \pi]$ and f is a 2π-periodic function of bounded variation on $[-\pi, \pi]$. Then for $n\geqslant 2$,

$$R_n^\alpha(f,x)\leqslant\frac{100}{\alpha^2 n^\alpha}\sum_{k=1}^{n}k^{\alpha-1}V_0^{\frac{\pi}{k}}(\varphi_x),$$

and there exists a 2π-periodic function f^* of bounded variation on $[-\pi, \pi]$ and a point $x\in[0, \pi]$ such that

$$R_n^\alpha(f^*,x)>\frac{1}{20\ 000\alpha n^\alpha}\sum_{k=1}^{n}k^{\alpha-1}V_0^{\frac{\pi}{k}}(\varphi_x),(n\geqslant 8).$$

We need two lemmas.

Lemma 1 Let $0<\alpha<1$. Define

$$\gamma_n^\alpha(t)=K_n^\alpha(t)-K_{n-1}^\alpha(t)-\frac{\cos\left[\left(n+\frac{\alpha}{2}\right)t-\frac{\pi}{2}\alpha\right]}{A_n^\alpha\left(2\sin\frac{t}{2}\right)^\alpha},\quad 0<t\leqslant\pi.$$

Then for $\frac{\pi}{n}\leqslant t\leqslant\pi$, $n\geqslant 2$,

$$|\gamma_n^\alpha(t)|\leqslant\frac{16\alpha}{(nt)^{\alpha+1}}.\tag{1}$$

Proof It follows from formula (5.15) of [2, p. 95] that

$$K_n^\alpha(t)=D_n^\alpha(t)+E_n^\alpha(t),\qquad 0<t\leqslant\pi,$$

where

$$D_n^\alpha(t):=\frac{1}{A_n^\alpha}\frac{\sin\left[\left(n+\frac{1+\alpha}{2}\right)t-\frac{\pi}{2}\alpha\right]}{\left(2\sin\frac{t}{2}\right)^{\alpha+1}},$$

$$E_n^\alpha(t):=-\Re\left\{\frac{e^{i\left(n+\frac{1}{2}\right)t}}{A_n^\alpha\left(2\sin\frac{t}{2}\right)}\sum_{v=n+1}^{+\infty}A_{v+1}^{\alpha-2}\frac{e^{-i(v+1)t}}{1-e^{-it}}\right\}+\frac{\alpha}{n+1}\frac{1}{\left(2\sin\frac{t}{2}\right)^2}.$$

Since

$$D_n^\alpha(t)-D_{n-1}^\alpha(t)=\frac{\cos\left[\left(n+\frac{\alpha}{2}\right)t-\frac{\pi\alpha}{2}\right]}{A_n^\alpha\left(2\sin\frac{t}{2}\right)^\alpha}-\frac{\alpha\sin\left[\left(n+\frac{\alpha-1}{2}\right)t-\frac{\pi\alpha}{2}\right]}{nA_n^\alpha\left(2\sin\frac{t}{2}\right)^{\alpha+1}},$$

we have

$$\gamma_n^\alpha(t)=-\frac{\alpha\sin\left[\left(n+\frac{\alpha-1}{2}\right)t-\frac{\pi\alpha}{2}\right]}{nA_n^\alpha\left(2\sin\frac{t}{2}\right)^{\alpha+1}}+E_n^\alpha(t)-E_{n-1}^\alpha(t).$$

Hence

$$|\gamma_n^\alpha(t)|\leqslant\frac{\alpha}{nA_n^\alpha\left(2\sin\frac{t}{2}\right)^{\alpha+1}}+|E_n^\alpha(t)-E_{n-1}^\alpha(t)|. \tag{2}$$

A straightforward calculation gives

$$E_n^\alpha(t)-E_{n-1}^\alpha(t)=$$

$$\Re\left\{\frac{\mathrm{e}^{-\frac{\mathrm{i}t}{2}}}{2\sin\frac{t}{2}(1-\mathrm{e}^{-\mathrm{i}t})}\sum_{v=1}^{+\infty}\left(\frac{A_{v+n}^{\alpha-2}}{A_{n-1}^\alpha}-\frac{A_{v+1+n}^{\alpha-2}}{A_n^\alpha}\right)\mathrm{e}^{-\mathrm{i}vt}\right\}-\frac{\alpha}{n(n+1)}\frac{1}{\left(2\sin\frac{t}{2}\right)^2}.$$

Then we obtain

$$|E_n^\alpha(t)-E_{n-1}^\alpha(t)|\leqslant\frac{1}{\left(2\sin\frac{t}{2}\right)^2}\left[\frac{\alpha}{n^2}+\frac{2\alpha(1-\alpha)}{n^3\sin\frac{t}{2}}\right]$$

$$<\frac{3\alpha}{2\left(2n\sin\frac{t}{2}\right)^{\alpha+1}},\qquad \frac{\pi}{n}\leqslant t\leqslant\pi. \tag{3}$$

Substituting (3) into (2) and noticing $\frac{1}{A_n^\alpha}\leqslant n^{-\alpha}$ (when $0<\alpha<1$) we get (1). □

Lemma 2 Let $0<\alpha<1$. Define

$$g_n^\alpha(t):=\int_0^t[K_n^\alpha(\theta)-K_{n-1}^\alpha(\theta)]\mathrm{d}\theta.$$

Then

$$|g_n^\alpha(t)|\leqslant\begin{cases}54t, & 0<t<\pi,\\ 40\alpha^{-1}n^{-\alpha-1}t^{-\alpha}, & \frac{\pi}{n}<t<\pi.\end{cases}$$

Proof By the definition of $K_n^\alpha(t)$ we easily obtain

$$|K_n^\alpha(t)-K_{n-1}^\alpha(t)|\leqslant 1+\frac{\alpha\Gamma(n+1)}{\Gamma(\alpha+n+1)}\sum_{k=1}^{n-1}\frac{(n-k)\Gamma(\alpha+k)}{\Gamma(k+1)}\leqslant 54.$$

This shows

$$|g_n^\alpha(t)|\leqslant\int_0^t|K_n^\alpha(\theta)-K_{n-1}^\alpha(\theta)|\,\mathrm{d}\theta\leqslant 54t,\quad 0<t<\pi.$$

We have

$$g_n^\alpha(t)=-\int_t^\pi[K_n^\alpha(\theta)-K_{n-1}^\alpha(\theta)]\mathrm{d}\theta=-I_n(t)-\int_t^\pi\gamma_n^\alpha(\theta)\mathrm{d}\theta,\tag{4}$$

where γ_n^α is defined in Lemma 1 and

$$I_n(t)=\frac{1}{A_n^\alpha}\int_t^\pi\frac{\cos\left[\left(n+\frac{\alpha}{2}\right)\theta-\frac{\pi}{2}\alpha\right]}{\left(2\sin\frac{\theta}{2}\right)^\alpha}\mathrm{d}\theta.$$

Integrating by parts we derive

$$|I_n(t)|\leqslant\frac{1}{\left(n+\frac{\alpha}{2}\right)A_n^\alpha}\left(\frac{2}{\left(2\sin\frac{t}{2}\right)^\alpha}+\int_t^\pi\frac{\cos\frac{\theta}{2}}{\left(2\sin\frac{\theta}{2}\right)^{\alpha+1}}\mathrm{d}\theta\right)$$

$$<\frac{1}{\left(n+\frac{\alpha}{2}\right)A_n^\alpha}\frac{2+\frac{1}{\alpha}}{\left(2\sin\frac{t}{2}\right)^\alpha}.\tag{5}$$

By (4) and (5), applying Lemma 1 we get

$$|g_n^\alpha(t)|\leqslant\frac{40}{\alpha n^{\alpha+1}t^\alpha}\text{ when }\frac{\pi}{n}<t<\pi.\qquad\square$$

Proof of the Theorem We have

$$R_n^\alpha(f,x)=\frac{1}{\pi}\sum_{j=n+1}^{+\infty}\left|\int_0^\pi\varphi_x(t)\mathrm{d}g_j^\alpha(t)\right|=\frac{1}{\pi}\sum_{j=n+1}^{+\infty}\left|\int_0^\pi g_j^\alpha(t)\mathrm{d}\varphi_x(t)\right|$$

$$\leqslant\frac{1}{\pi}\sum_{j=n+1}^{+\infty}\int_0^\pi|g_j^\alpha(t)|\,\mathrm{d}V_0^t(\varphi_x)=:\frac{1}{\pi}(R_{n1}^\alpha+R_{n2}^\alpha),$$

where

$$R_{n1}^\alpha=\sum_{k=n+1}^{+\infty}\int_0^{\frac{\pi}{k}}|g_k^\alpha(t)|\,\mathrm{d}V_0^t(\varphi_x),\quad R_{n2}^\alpha\sum_{k=n+1}^{+\infty}\int_{\frac{\pi}{k}}^\pi|g_k^\alpha(t)|\,\mathrm{d}V_0^t(\varphi_x).$$

By Lemma 2 we have

$$R_{n1}^\alpha\leqslant 54\int_0^{\frac{\pi}{n+1}}\sum_{k=n+1}^{+\infty}t\chi_{(0,\frac{\pi}{k})}(t)\mathrm{d}V_0^t(\varphi_x)\leqslant 54\pi\int_0^{\frac{\pi}{n+1}}\mathrm{d}V_0^t(\varphi_x)=54\pi V_0^{\frac{\pi}{n+1}}(\varphi_x),$$

where $\chi_{(a,b)}$ denotes the characteristic function of (a, b). Also by Lemma 2 we have

$$R_{n2}^{\alpha} \leqslant \frac{40}{\alpha} \sum_{k=n+1}^{+\infty} \int_{\frac{\pi}{k}}^{\pi} \frac{1}{k^{\alpha+1} t^{\alpha}} \mathrm{d}V_0^t(\varphi_x) = \frac{40}{\alpha} \sum_{k=n+1}^{+\infty} \left(\int_{\frac{\pi}{k}}^{\frac{\pi}{n+1}} + \int_{\frac{\pi}{n+1}}^{\pi} \right) \frac{1}{k^{\alpha+1} t^{\alpha}} \mathrm{d}V_0^t(\varphi_x)$$

$$= \frac{40}{\alpha} \left(\int_0^{\frac{\pi}{n+1}} \sum_{k>\frac{\pi}{t}} \frac{1}{k^{\alpha+1} t^{\alpha}} \mathrm{d}V_0^t(\varphi_x) + \int_{\frac{\pi}{n+1}}^{\pi} \sum_{k=n+1}^{+\infty} \frac{1}{k^{\alpha+1} t^{\alpha}} \mathrm{d}V_0^t(\varphi_x) \right)$$

$$\leqslant \frac{40}{\alpha^2} \left(V_0^{\frac{\pi}{n+1}}(\varphi_x) + \frac{1}{n^{\alpha}} \int_{\frac{\pi}{n+1}}^{\pi} \frac{1}{t^{\alpha}} \mathrm{d}V_0^t(\varphi_x) \right)$$

$$= \frac{40}{\alpha^2} \left(V_0^{\frac{\pi}{n+1}}(\varphi_x) + \frac{1}{n^{\alpha}} \sum_{k=1}^{n} \int_{\frac{\pi}{k+1}}^{\frac{\pi}{k}} \frac{1}{t^{\alpha}} \mathrm{d}V_0^t(\varphi_x) \right)$$

$$\leqslant \frac{120}{\alpha^2 n^{\alpha}} \sum_{k=1}^{n} k^{\alpha-1} V_0^{\frac{\pi}{k}}(\varphi_x).$$

Combining the estimates for R_{n1}^{α} and R_{n2}^{α} we obtain

$$R_n^{\alpha}(f,x) \leqslant \frac{100}{a^2 n^{\alpha}} \sum_{k=1}^{n} k^{\alpha-1} V_0^{\frac{\pi}{k}}(\varphi_x), \ (n \geqslant 2).$$

On the other hand, the function

$$f^*(x) = \sum_{k=1}^{+\infty} \frac{\sin kx}{k} = \begin{cases} \frac{\pi - x}{2}, \text{if } 0 < x < 2\pi, \\ 0, \quad \text{if } x = 0 \end{cases}$$

gives the second conclusion of the Theorem at point $x=\frac{\pi}{2}$. □

Acknowledgments

The authors thank Dr. Dai Feng for the valuable discussion.

References

[1] Humphreys N, Bojanic R., Rate of convergence for the absolutely (C, α) summable Fourier series of functions of bounded variation. J. Approx. Theory, 1999, 101: 212—220.

[2] Zygmund A., Trigomometric Series, Vol. I. Cambridge University Press, Cambridge, 1959.

二、球面上的 Fourier-Laplace 分析

Ⅱ. Fourier-Laplace Analysis on the Sphere

北京师范大学学报(自然科学版)，2000，36(6)：718－721.

Fourier-Laplace 级数强可和点的刻画[①]

Characterization of the Strong Summability Points of Fourier-Laplace Series

摘要 引入了球面上函数 H_s-点的概念，推广了 Lebesgue 点的定义．得到了 Fourier-Laplace 级数强可和点的一个刻画，并讨论了该刻画在线性求和中的一系列应用.

关键词 Fourier-Laplace 级数；强可和；H_s-点.

设 $n\geqslant 3$，$S^{n-1}:=\{(x_1, x_2, \cdots, x_n)\,|\,x_1^2+x_2^2+\cdots+x_n^2=1\}$ 为 $\mathbf{R}^n$ 中的单位球面．S^{n-1}上步长为 θ 的平移 S_θ 定义为

$$S_\theta(f)(x)=\frac{1}{l(\theta)}\int_{\{y\in S^{n-1}\,|\,x\cdot y=\cos\theta\}} f(y)\mathrm{d}l_x(y),$$

其中 $x\in S^{n-1}$，$f\in L(S^{n-1})$，$\mathrm{d}l_x(\xi)$为$\{\xi\in S^{n-1}\,|\,\xi\cdot x=0\}$的 Lebesgue 测度元.

设 $f\in L(S^{n-1})$，$f\sim\sigma(f)(x):=\sum\limits_{k=0}^{+\infty}Y_k(f)(x)$为其 Fourier-Laplace 展式，其中 Y_k 是 $L(S^{n-1})$到 k 次球调和空间 H_k^n 上的投影算子．展式$\sigma(f)$的 δ 阶 Cesàro 平均定义为

$$\sigma_N^\delta(f):=(A_N^\delta)^{-1}\sum_{k=0}^{N}A_{N-k}^\delta Y_k(f),$$

① 国家自然科学基金资助项目(19771009).

本文与戴峰合作.

收稿日期：2000-04-19.

其中 $A_N^\delta=\dfrac{\Gamma(N+\delta+1)}{\Gamma(\delta+1)\Gamma(N+1)}$，$\delta>-1$. 显然，$\delta=\lambda=:\dfrac{n-2}{2}$是 Cesàro 平均 σ_N^δ 的临界阶[1].

记
$$H_{N,q}(f;x)=\frac{1}{N+1}\sum_{k=0}^{N}|\sigma_k^\lambda(f)(x)-f(x)|^q,q>0.$$
当 $n\geqslant 3$ 时，$H_{N,q}(f;x)$的 a.e. 收敛问题至今尚未解决. 设 $f\in L(S^{n-1})$，同周期的情形(即 $n=2$ 的情形)相类似，$H_{N,q}(f;x)$的收敛点(强可和点)未必是 f 的 Lebesgue 点，反之亦然(见[2]). 因此，一个自然的问题是：如何刻画 $H_{N,q}(f,x)$的收敛点？本文运用文献[2，3]的方法，给出强可和点的刻画及该刻画的一系列应用.

§1. 强可和点的刻画

定义 设 $x\in S^{n-1}$，$f\in L^1(S^{n-1})$，$s\in(0,+\infty)$. 记
$$h_{N,s}(f,x,\delta)=\sum_{k=0}^{N}\left(\left(\frac{N}{k}\right)^{n-1}\int_{\frac{(k-1)\delta}{N}}^{\frac{k\delta}{N}}|\varphi_x(\theta)|\sin^{n-2}\theta d\theta\right)^s,$$
其中
$$\varphi_x(\theta)=S_\theta(f)(x)-f(x),\ 0<\delta<\pi.$$
若点 $x\in S^{n-1}$满足如下的反极条件
$$\int_0^\theta|S_t(f)(-x)|t^\lambda dt=o(1),\quad \theta\to 0^+,\tag{1.1}$$
并且$\lim\limits_{N\to+\infty}h_{N,s}(f,x,\delta)=0$ 成立，则称 x 为 f 的 H_s-点. f 的 H_s-点的全体记为 $H_{s,f}$或 H_s.

运用文献[2]的方法，我们容易证明 H_s-点的定义与$\delta\in(0,\pi)$的选取无关. $n=2$ 时，Gabisoniya[4]证明了 $f\in L(S^{n-1})$的 H_s-点的全体是全测度集.

定理 1 设 $2\leqslant q<+\infty$，$p=q':=\dfrac{q}{q-1}$，$f\in L^1(S^{n-1})$，则
$$\lim_{N\to+\infty}\frac{1}{N}\sum_{k=0}^{N}|\sigma_k^\lambda(f)(x)-f(x)|^q=0$$
在 f 的任一 H_p-点处成立.

定理 1 的意义在于将强求和的 a.e. 收敛问题转化成了对函数 H_s-点的研究. 其证明需要以下 3 个引理.

引理 1[2] 设 $T_N(x)=\sum\limits_{k=0}^{N}(\alpha_k\cos kx+\beta_k\sin kx)$为 N 阶的三角多项

式，$1\leqslant p<+\infty$. 若$\left[\sum_{k=0}^{N}(|\alpha_k|^p+|\beta_k|^p)\right]^{\frac{1}{p}}\leqslant K\cdot\left(\frac{1}{N}\right)^{\frac{1}{p'}}$，其中$\frac{1}{p'}=1-\frac{1}{p}$，则对任何$\delta>0$，有

$$\Big(\sum_{k=1}^{N}\sup_{x\in\left[\frac{(k-1)\delta}{N},\frac{k\delta}{N}\right]}|T_N(x)|^{p'}\Big)^{\frac{1}{p'}}\leqslant K.$$

引理 2 设$f\in L^p(S^{n-1})$，$x\in S^{n-1}$为f的满足反极条件(1)的 Lebesgue 点，则

$$\sigma_k^{\lambda}(f)(x)-f(x)=O(1)\int_{\frac{1}{N}}^{\frac{\pi}{2}}\frac{\varphi_x(\theta)}{\theta}M(\theta)\cos(k\theta-\gamma)\mathrm{d}\theta+o(1),\ N\rightarrow+\infty,$$

关于$\frac{N}{2}\leqslant k\leqslant N$一致成立，其中$\gamma=\frac{(n-1)\pi}{2}$，$\varphi_x(\theta)=S_{\theta}(f)(x)-f(x)$，$M(\theta)$为有界函数.

证 运用估计式[5] $|P_k^{(\alpha,\beta)}(\cos\theta)|\leqslant Ck^{\alpha}$，$0<\theta<\frac{\pi}{2}$，及渐近表达式[5]

$$P_k^{(\alpha,\beta)}(\cos\theta)=k^{-\frac{1}{2}}M(\theta)\left(\cos(\tilde{k}\theta+\gamma)+O\left(\frac{1}{k\cdot\sin\theta}\right)\right),$$
$$c\cdot k^{-1}\leqslant\theta\leqslant\pi-c\cdot k^{-1},$$

其中$\tilde{k}=k+\frac{\alpha+\beta+1}{2}$，$M(\theta)=\pi^{-\frac{1}{2}}\left(\sin\frac{\theta}{2}\right)^{-\frac{\alpha+1}{2}}\left(\cos\frac{\theta}{2}\right)^{-\frac{\beta+1}{2}}$，$\gamma=-\left(\alpha+\frac{1}{2}\right)\cdot\frac{\pi}{2}$，$c>1$为常数，并借助文献[6]中的等收敛算子，我们可得引理的证明.

引理 3 设n，f，p，q同定理1，$x\in H_p$，则

$$\lim_{N\rightarrow+\infty}\frac{1}{N}\sum_{k=\frac{N}{2}}^{N}|\sigma_k^{\lambda}(f)(x)-f(x)|^q=0.$$

证 由引理2，我们只要证明

$$I_N^{(1)}=\left(\frac{1}{N}\sum_{k=\frac{N}{2}}^{N}\left|\int_{\frac{\delta}{N}}^{\delta}\frac{\varphi_x(t)}{t}\cdot M(t)\cdot\sin kt\,\mathrm{d}t\right|^q\right)^{\frac{1}{q}}=o(1),$$

$$I_N^{(2)}=\left(\frac{1}{N}\sum_{k=\frac{N}{2}}^{N}\left|\int_{\frac{\delta}{N}}^{\delta}\frac{\varphi_x(t)}{t}\cdot M(t)\cdot\cos kt\,\mathrm{d}t\right|^q\right)^{\frac{1}{q}}=o(1),$$

其中$\delta>0$为一固定常数，$M(t)$为$\left[0,\frac{\pi}{2}\right]$上有界函数.

我们下面只证 $I_N^{(2)}=o(1)$，$N\to+\infty$. $I_N^{(1)}$ 的情形可类似证明. 由 l^p 和 l^q 的对偶性，取(C_i)，$\left(\sum_{i=\frac{N}{2}}^{N}|C_i|^p\right)^{\frac{1}{p}}=\left(\frac{1}{N}\right)^{\frac{1}{q}}$，使

$$I_N^{(2)}=\sum_{k=\frac{N}{2}}^{N}C_k\int_{\frac{\delta}{N}}^{\delta}\frac{\varphi_x(t)}{t}\cdot M(t)\cdot\cos kt\,\mathrm{d}t=\int_{\frac{\delta}{N}}^{\delta}\frac{\varphi_x(t)}{t}M(t)T_N(t)\mathrm{d}t,$$

其中 $T_N(t)=\sum_{k=\frac{N}{2}}^{N}C_k\cos kt$. 于是，由引理 1，有

$$I_N^{(2)}\leqslant\int_{\frac{\delta}{N}}^{\delta}\frac{|\varphi_x(t)|}{t}|T_N(t)|\mathrm{d}t\leqslant\delta^{1-n}\sum_{j=1}^{N-1}\Big(\sup_{t\in\left[\frac{j\delta}{N},\frac{(j+1)\delta}{N}\right]}|T_N(t)|\Big).$$

$$\left(\frac{N}{j}\right)^{n-1}\left(\int_{\frac{j\delta}{N}}^{\frac{(j+1)\delta}{N}}|\varphi_x(t)|\cdot\sin^{n-2}t\,\mathrm{d}t\right)\leqslant\delta^{1-n}\Big(\sum_{j=1}^{N-1}\sup_{t\in\left[\frac{j\delta}{N},\frac{(j+1)\delta}{N}\right]}|T_N(t)|^{p'}\Big)^{\frac{1}{p'}}.$$

$$\left(\sum_{j=1}^{N-1}\left(\left(\frac{N}{j}\right)^{n-1}\int_{\frac{j\delta}{N}}^{\frac{(j+1)\delta}{N}}|\varphi_x(t)|\cdot\sin^{n-2}t\,\mathrm{d}t\right)^p\right)^{\frac{1}{p}}$$

$$\leqslant\delta^{1-n}\left(\sum_{j=1}^{N-1}\left(\left(\frac{N}{j}\right)^{n-1}\int_{\frac{j\delta}{N}}^{\frac{(j+1)\delta}{N}}|\varphi_x(t)|\cdot\sin^{n-2}t\,\mathrm{d}t\right)^p\right)^{\frac{1}{p}}$$

$$=o(1),N\to+\infty.$$

定理 1 的证明可由引理 3 直接得到.

§2. 定理 1 的应用

设矩阵 $\boldsymbol{\Lambda}=\{\lambda_k^{(N)}\}$满足条件

(1) $\lambda_k^{(N)}\geqslant0$ 且 $\lim\limits_{N\to+\infty}\lambda_k^{(N)}=0$；

(2) $\lim\limits_{N\to+\infty}\sum\limits_{k=0}^{+\infty}\lambda_k^{(N)}=1$；

(3) (Gogoladze) $\left(\frac{1}{m}\sum\limits_{k=m+1}^{2m}|\lambda_k^{(N)}|^{\gamma}\right)^{\frac{1}{\gamma}}\leqslant\frac{1}{m}\sum\limits_{k=\left[\frac{m+1}{2}\right]}^{m}|\lambda_k^{(N)}|$，

$\gamma>1$ 固定.

我们将所有这样的矩阵 $\boldsymbol{\Lambda}$ 记为 Λ_γ，将只满足条件(1)(2)的所有矩阵记为 Λ_γ'.

定理 2 设 $\gamma>1$，$\gamma'=\frac{\gamma}{\gamma-1}$，$\boldsymbol{\Lambda}=\{\lambda_k^{(N)}\}\in\Lambda_\gamma$，$q\geqslant\frac{2}{\gamma}$，$s=(\gamma'\cdot q)'$. 若 $f\in L(\mathrm{S}^{n-1})$，$x\in H_s$，则

$$R_{N,q}(f;x;\boldsymbol{\Lambda}):=\sum_{k=0}^{+\infty}\lambda_k^{(N)}\left|\sigma_k^{\lambda}(f)(x)-f(x)\right|^q\to 0, N\to+\infty.$$

证 设 $x\in H_s$，则

$$\sum_{j=1}^{+\infty}\sum_{k=2^j+1}^{2^{j+1}}\lambda_k^{(N)}\left|f(x)-\sigma_k^{\lambda}(f)(x)\right|^q$$

$$\leqslant\sum_{j=1}^{+\infty}\Big(\sum_{k=2^j+1}^{2^{j+1}}\left|\lambda_k^{(N)}\right|^{\gamma}\Big)^{\frac{1}{\gamma}}\cdot\Big(\sum_{k=2^j+1}^{2^{j+1}}\left|f(x)-\sigma_k^{\lambda}(f)(x)\right|^{\gamma' q}\Big)^{\frac{1}{\gamma'}}$$

$$\leqslant\sum_{j=1}^{+\infty}\Big(\sum_{k=2^j+1}^{2^{j+1}}\left|\lambda_k^{(N)}\right|\Big)\cdot\Big(2^{-j}\sum_{k=2^j+1}^{2^{j+1}}\left|f(x)-\sigma_k^{\lambda}(f)(x)\right|^{\gamma' q}\Big)^{\frac{1}{\gamma'}}$$

$$\leqslant\sum_{k=1}^{+\infty}\left|\lambda_k^{(N)}\right|\cdot\sup_{m\geqslant k}\Big(\frac{1}{m}\sum_{k=m+1}^{2m}\left|f(x)-\sigma_k^{\lambda}(f)(x)\right|^{\gamma' q}\Big)^{\frac{1}{\gamma'}}\to 0, N\to+\infty.$$

定理 3 设 $\boldsymbol{\Lambda}\in\Lambda_{\gamma}$，$\gamma\in(1,\ 2]$. 若 $f\in L(\mathrm{S}^{n-1})$，$x\in H_{\gamma}$，则

$$U_N(f,x;\boldsymbol{\Lambda}):=\sum_{k=0}^{+\infty}\lambda_k^{(N)}\sigma_k^{\lambda}(f,x)\to f(x), N\to+\infty.$$

证 设 $x\in H_{\gamma,f}$，则由定理 2(取 $q=1$)，有

$$\left|f(x)-U_N(f,x;\boldsymbol{\Lambda})\right|\leqslant\sum_{k=0}^{+\infty}\lambda_k^{(N)}\left|f(x)-\sigma_k^{\lambda}(f)(x)\right|\to 0, N\to+\infty.$$

设 $r>0$，则矩阵 $\boldsymbol{\Lambda}=\{\alpha_k^{(N)}\}$的 r 阶差分定义为

$$\Delta^r\alpha_k^{(N)}=\sum_{j=0}^{+\infty}(-1)^j\mathrm{C}_r^j\alpha_{k+j}^{(N)}.$$

设 $\boldsymbol{\Lambda}=\{\alpha_k^{(N)}\}$为一复下三角矩阵，满足条件

$$(1)\ \alpha_0^{(N)}=1,\qquad (2)\ \Big(N^{p-1}\sum_{k=0}^{+\infty}\big(k^{\lambda}(\left|\Delta^{\lambda+1}\alpha_k^{(N)}\right|)\big)^p\Big)^{\frac{1}{p}}<K,$$

其中 $p\in(1,\ 2]$，K 为与 N 无关的常数. 我们记

$$A_N(f;x)=A_N(f;x;\boldsymbol{\Lambda})=\sum_{k=0}^{+\infty}\alpha_k^{(N)}Y_k(f)(x), f\in L(\mathrm{S}^{n-1}).$$

定理 4 设 $x\in H_{p,f}$，$f\in L(\mathrm{S}^{n-1})$，则 $\lim\limits_{N\to+\infty}A_N(f;\ x;\ \boldsymbol{\Lambda})=f(x)$.

证 运用 Abel 变换，有

$$f(x)-A_N(f,\ x)=\sum_{k=0}^{+\infty}(\Delta^{\lambda+1}\alpha_k^{(N)})\cdot A_k^{\lambda}\cdot(f(x)-\sigma_k^{\lambda}(f)(x)).$$

所以，由 Hölder 不等式及定理 2，

$$\left|f(x)-A_N(f,x)\right|\leqslant\Big(\sum_{k=0}^{N}k^{\lambda p}\left|\Delta^{\lambda+1}\alpha_k^{(N)}\right|^p\Big)^{\frac{1}{p}}.$$

$$\left(\sum_{k=0}^{N}|f(x)-\sigma_k^\lambda(f)(x)|^q\right)^{\frac{1}{q}}$$
$$\ll\left(\frac{1}{N}\sum_{k=0}^{N}|f(x)-\sigma_k^\lambda(f)(x)|^q\right)^{\frac{1}{q}}\to 0,\quad N\to+\infty.$$

参考文献

[1] Bonami A, Clerc J L. Sommes de Cesàro etmultiplicateurs des dèveloppments en harmonique sphériques. Trans AMS, 1973, 183: 223.

[2] Novikov I Ya, Podin V A. Characterization of the points of p-strong summability of trigonometric Fourier series. Izv Vyssh Uchebn Zaved, 1988, 9: 58.

[3] Stepanets A I, Lasuriya R A. Strong summability of orthogonal expansions of summable functions Ⅱ. Ukrainian Math J, 1996, 48: 531.

[4] Gabisoniya O D. On the points of strong summability of Fourier series. Math Zametki, 1973, 14: 615.

[5] SzegöG. Orthogonal Polynomials, Vol 23. Colloquium: AMS Colloquium Publication, 1939.

[6] Wang Kunyang. Equiconvergent operator of Cesàro means on sphere and its application. J BJ Nor Univ (Nat Sci) (北京师范大学学报(自然科学版)), 1993, 29: 143.

Abstract The concept of H_s-point which generalizes the definition of Lebesgue point is introduced. The characterization of the strong summability points is given, and the application of this characterization is discussed.

Keywords Fourier-Laplace series; strong summability; H_s-point.

北京师范大学学报(自然科学版)，2002，38(4)：440—444.

Fourier-Laplace 分析中用连续模给出的几乎处处收敛条件[①]

Almost Every where Convergence Conditions Given by Modulus of Continuity in Fourier-Laplace Analysis

摘要 讨论了用连续模给出的 Fourier-Laplace 级数的临界指标的 Cesàro 平均的几乎处处收敛条件.

关键词 几乎处处收敛；Fourier-Laplace 级数；连续模.

§1. 预备知识和主要结果

先叙述一些必要的基本知识，这些知识可在文献[1]和[2]中找到.

设 $S^{n-1}=\{(x_1, x_2, \cdots, x_n) \mid x_1^2+x_2^2+\cdots+x_n^2=1\}$ 是 $\mathbf{R}^n$ 中的单位球面，并设 S_θ：$L^p \to L^p$ 是**步长为 $\theta \in \mathbf{R}$ 的平移**，其定义为

$$S_\theta(f)(x) := \frac{1}{|S^{n-2}|}\int_{\{y\in S^{n-1} \mid xy=0\}} f(x\cos\theta + y\sin\theta)\,dl(y),$$

其中 $x \in S^{n-1}$，而 $dl(y)$ 表示 $\{y\in S^{n-1} \mid xy=0\}$ 上的 Lebesgue 测度元.

用 I 代表恒等算子，并设 r 是一个正数. 令 $\psi_r(u) := (1-u)^{\frac{r}{2}}$. 我们定义算子

$$\Delta_u^r := (I-S_u)^{\frac{r}{2}} = \sum_{k=0}^{+\infty} \frac{\psi_r^{(k)}(0)}{k!} S_u^k$$

为 r 阶**差算子**，并定义函数 $f \in L^p(S^{n-1})$，$1 \leqslant p < +\infty$ 的 r 阶**连续模**为

$$\omega_r(f, t)_p := \sup\{\|\Delta_u^r f\|_p \mid 0 < u \leqslant t\}.$$

设 $f \in L(S^{n-1})$，它的 Fourier-Laplace 级数为

---

① 国家自然科学基金资助项目(10071007).

本文与戴峰合作.

收稿日期：2002-01-21.

$$\sigma(f)(x):=\sum_{k=0}^{+\infty}Y_k(f)(x),$$

其中 Y_k 代表从 $L(\mathrm{S}^{n-1})$ 到 k 阶球调和函数空间 $\mathscr{H}_k^n$ 上的"投影"算子．函数 f 的 Fourier-Laplace 级数的 δ 阶 Cesàro 平均定义为

$$\sigma_N^\delta(f):=(A_N^\delta)^{-1}\sum_{k=0}^{N}A_{N-k}^\delta Y_k(f),$$

其中
$$A_N^\delta=\frac{\Gamma(N+\delta+1)}{\Gamma(\delta+1)\Gamma(N+1)},\ \delta>-1.$$

在 $L(\mathrm{S}^{n-1})=L^1(\mathrm{S}^{n-1})$ 空间中，Cesàro 平均的阶数 δ 有一个"临界"值 $\lambda:=\frac{n-2}{2}$．我们知道，当 $\delta>\lambda$ 时，对于每个 $f\in L(\mathrm{S}^{n-1})$，

$$\lim_{N\to+\infty}\sigma_N^\delta(f)(x)=f(x)\tag{1.1}$$

几乎处处成立．临界阶的情形是重要的．本文的主要结果是

定理 1 设 $r>0$，$f\in L(\mathrm{S}^{n-1})$．如果

$$\int_0^1\frac{\omega_r(f,t)_1}{t}\mathrm{d}t<+\infty,\tag{1.2}$$

那么 $\lim\limits_{N\to+\infty}\sigma_N^\lambda(f)(x)=f(x)$ 在 S^{n-1} 上几乎处处成立．

使用文献[3]引入的**等收敛算子**．δ 阶 Cesàro 算子的等收敛算子的定义是
$$E_N^\delta(f)(x)=\gamma_N^\delta\left(f*P_N^{\left(\frac{n-1}{2}+\delta,\frac{n-3}{2}\right)}\right)(x),$$

其中 $\gamma_N^\delta=\dfrac{\Gamma(\delta+1)}{(4\pi)^{\frac{n-1}{2}}}\ \dfrac{\Gamma(N+1)\Gamma(N+n-1)}{\Gamma(N+\delta+1)\Gamma\left(N+\frac{n-1}{2}\right)}$，$P_N^{(\alpha,\beta)}$ 代表 Jacobi 多项式，卷积运算"$*$"的定义是 $(f*h)(x)=\displaystyle\int_{\mathrm{S}^{n-1}}f(y)h(xy)\mathrm{d}y.$

极大等收敛算子的定义是 $E_*^\lambda(f)(x)=\sup\{|E_N^\lambda(f)(x)|\mid N\in\mathbf{N}\}$．

我们将证明下述关于极大算子的 L^1 估计．

定理 2 在定理 1 的条件下，存在只与 n 有关的常数 C_n 使得

$$\|E_*^\lambda(f)\|_1\leqslant\left|\int_{\mathrm{S}^{n-1}}f(x)\mathrm{d}x\right|+C_n\int_0^\pi\frac{\omega_r(f,t)_1}{t}\mathrm{d}t.$$

§ 2. 定理的证明

引理 1 设 $x\in\mathrm{S}^{n-1}$，$f\in L(\mathrm{S}^{n-1})$．如果 $-x$ 是 f 的 Lebesgue 点并且作为 θ 的函数
$$\frac{S_\theta(f)(x)-f(x)}{\theta}\in L(0,\pi),$$

那么(1.1)在点 x 处成立.

引理 2 设 $0<\alpha<\beta$, $f\in L(\mathrm{S}^{n-1})$, 那么

$$\omega_\alpha(f,t)_1\leqslant Ct^\alpha\int_t^1\frac{\omega_\beta(f,u)_1}{u^{\alpha+1}}\mathrm{d}u.$$

证 引理 2 可由 K 泛函与连续模的等价性及文献[4]中的关于 K 泛函的一般性结论推得.

定理 1 的证 在定理 1 的条件下

$$+\infty>\int_0^1\frac{\omega_2(f,t)_1}{t}\mathrm{d}t\geqslant\int_0^1\frac{\|S_t(f)(x)-f(x)\|_1}{t}\mathrm{d}t$$

$$=\int_{\mathrm{S}^{n-1}}\int_0^1\frac{|S_\theta(f)(x)-f(x)|}{\theta}\mathrm{d}\theta\mathrm{d}x.$$

使用 Fubini 定理，我们得到

$$\int_0^1\frac{|S_\theta(f)(x)-f(x)|}{\theta}\mathrm{d}\theta<+\infty,\quad \text{a. e.}\quad x\in\mathrm{S}^{n-1}.$$

由此，根据引理 1，我们证得定理 1 的结论.

引理 3 存在正数 C_n 使得

$$\sup\{|\gamma_N^\lambda P_N^{(\frac{n-3}{2},\frac{n-3}{2})}(\cos\theta)\theta\sin^{n-2}\theta|\,|N\in\mathbf{N},\ \theta\in(0,\pi)\}\leqslant C_n.$$

证 引用文献[5]的(7.32.5)和(4.1.3)，我们有

$$|P_N^{(n-\frac{3}{2},\frac{n-3}{2})}(\cos\theta)|\leqslant\begin{cases}C_nN^{\frac{n-3}{2}}, & 0\leqslant\theta\leqslant\frac{1}{N},\\ C_nN^{-\frac{1}{2}}\theta^{-n+1}(\pi-\theta)^{\frac{n-2}{2}}, & \frac{1}{N}\leqslant\theta\leqslant\pi-\frac{1}{N},\\ C_nN^{\frac{n-3}{2}}, & \pi-\frac{1}{N}\leqslant\theta\leqslant\pi.\end{cases}$$

注意到 $\gamma_N^\lambda\leqslant C_n\sqrt{N}$，于是

$$|\gamma_N^\lambda P_N^{(n-\frac{3}{2},\frac{n-3}{2})}(\cos\theta)\theta\sin^{n-2}\theta|\leqslant\begin{cases}C_n(N\theta)^{n-1}, & 0\leqslant\theta\leqslant\frac{1}{N},\\ C_n(\pi-\theta)^{\frac{n}{2}}, & \frac{1}{N}\leqslant\theta\leqslant\pi-\frac{1}{N},\\ C_nN^{\frac{n-2}{2}}(\pi-\theta)^{n-2}, & \pi-\frac{1}{N}\leqslant\theta\leqslant\pi.\end{cases}$$

由此得到所需的结论.

定理 2 的证 我们先证 $r=2$ 的情形. 由等收敛算子和平移算子的定义知

$$E_N^\lambda(f)(x)$$

$$=f(x)+\gamma_N^\lambda|S^{n-2}|\int_0^\pi\frac{S_\theta(f)(x)-f(x)}{\theta}P_N^{\left(n-\frac{3}{2},\frac{n-3}{2}\right)}(\cos\theta)\theta\sin^{n-2}\theta\mathrm{d}\theta.$$

于是，根据引理 3，$E_*^\lambda(f)(x)\leqslant|f(x)|+C_n\int_0^\pi\left|\frac{S_\theta(f)(x)-f(x)}{\theta}\right|\mathrm{d}\theta.$

那么，积分并换序得 $\|E_*^\lambda(f)\|_1\leqslant\|f\|_1+C_n\int_0^\pi\frac{\omega_2(f;\theta)_1}{\theta}\mathrm{d}\theta.$

注意到 $$\|f\|_1\leqslant\left|\int_{S^{n-1}}f(y)\mathrm{d}y\right|+C_n\int_{\frac{\pi}{2}}^\pi\frac{\omega_2(f,t)_1}{t}\mathrm{d}t,$$

我们得 $r=2$ 情形的定理 2 结论.

对于一般情形的 $r>0$，定理 2 可由 $r=2$ 的情形、引理 2 及不等式

$$\omega_r(f,\ t)_p\leqslant C\omega_2(f,\ t)_p,\quad r>2$$

推得.

注 定理 1 可从定理 2 推出. 事实上，我们可取一个函数 $\eta\in C^\infty[0,+\infty)$满足 $\chi_{[0,1]}\leqslant\eta\leqslant\chi_{[0,2]}$，由 η 定义一列**最佳逼近算子** η_k(见文献[1]). 对于满足定理 1 条件的函数 f，记 $f_k=\eta_k(f)$，那么，

$$\Delta_\theta^2(f_k)=\eta_k(\Delta_\theta^2(f)),\ \omega_2(f_k;\ \theta)_1\leqslant C_n\omega_2(f;\ \theta)_1.$$

记 $\mu(x)=\limsup\limits_{N\to+\infty}|E_N^\lambda(f)(x)-f(x)|$. 因为

$$|E_N^\lambda(f)(x)-f(x)|$$

$$=|E_N^\lambda(f-f_k)(x)+E_N^\lambda(f_k)(x)-f_k(x)+f_h(x)-f(x)|,$$

所以 $$\mu(x)\leqslant E_*^\lambda(f-f_k)(x)+|f_k(x)-f(x)|.$$

于是根据定理 2，$$\|\mu\|_1\leqslant C_n\left(\|f-f_k\|_1+\int_0^\pi\frac{\omega_2(f-f_k;\theta)_1}{\theta}\mathrm{d}\theta\right).$$

然而 $$\frac{\omega_2(f-f_k;\ \theta)_1}{\theta}\leqslant C_n\frac{\omega_2(f;\ \theta)_1}{\theta}\in L(0,\ \pi).$$

那么，令 $k\to+\infty$，使用控制收敛定理就得到$\|\mu\|_1=0$，从而

$$\mu(x)=0,\ \text{a. e.}$$

§3. 评注与猜测

关于 Cesàro 平均的几乎处处收敛性，由文献[6]，有以下结果：

定理 A 设 $1\leqslant p\leqslant2$，$\delta>\delta(p):=\lambda\left(\frac{2}{p}-1\right)$且 $f\in L^p(S^{n-1})$，则

$$\lim_{N\to+\infty}\sigma_N^{\delta}(f)(x)=f(x),\ \text{a. e.}\quad x\in S^{n-1}.$$

一个自然的问题是：当$\delta=\delta(p):=\lambda\left(\frac{2}{p}-1\right)$时，以上结论是否正确？同多元周期函数的 Bochner-Riesz 平均一样，这是一个长期未解决问题. 与此问题相关的，是以下 2 个稍简单的问题.

问题 1 若$1\leqslant p\leqslant 2$，$\delta=\delta(p)$，f取自一个与L^p相接近但含于L^p的函数空间，则定理 A 的结论是否成立？

问题 2 若$1\leqslant p\leqslant 2$，$\delta=\delta(p)$，$f\in L^p$且f具有一定的光滑性，则定理 A 的结论是否成立？

对于问题 1，有以下结论：

定理 B[7~8] 设$1\leqslant p\leqslant 2$，$\delta=\delta(p):=\lambda\left(\frac{2}{p}-1\right)$且$\alpha>1$，则当

$$f\in L^p(\ln^+L)^2(\ln^+\ln^+L)^{\alpha(p-1}(S^{n-1})$$

时，
$$\lim_{N\to+\infty}\sigma_N^{\delta}(f)(x)=f(x),\ \text{a. e.}\ x\in S^{n-1}.$$

据我们所知，这是到目前为止最好的结果.

对于问题 2，本文给出了一个关于$L(S^{n-1})$的结果. 文献[9]得到了一个关于$L^2(S^{n-1})$空间的结果，即

定理 C[9] 设$f\in L^2(S^{n-1})$，$r>0$. 如果

$$\int_0^1\frac{\omega_r(f;t)_2^2}{t}\ln\frac{1}{t}\mathrm{d}t<+\infty,\tag{3.1}$$

那么$\lim\limits_{N\to+\infty}\sigma_N^0(f)(x)=f(x)$在$S^{n-1}$上几乎处处成立.

文献[7]中证明了一个关于$L^p(1<p<2)$的结果：

定理 D 设$1<p<2$，$\delta=\delta(p):=\lambda\left(\frac{2}{p}-1\right)$且$r>0$. 若$f\in L^p$且

$$\int_0^1\frac{\omega_r(f;t)_p^p}{t^2}\mathrm{d}t<+\infty,\tag{3.2}$$

则$\lim\limits_{N\to+\infty}\sigma_N^{\delta}(f)(x)=f(x)$在$S^{n-1}$上几乎处处成立.

与定理C 和定理 1 中的条件(1.2)和(3.1)相比，定理 D 的条件(3.2)显然太强. 作为本文的结尾，我们给出一个关于L^p空间的猜测：

猜测 设$1<p\leqslant 2$，$\delta=\delta(p):=\lambda\left(\frac{2}{p}-1\right)$且$r>0$. 若$f\in L^p$且

$$\int_0^1\frac{\omega_r(f;t)_p^p}{t}\mathrm{d}t<+\infty,$$

则 $\lim\limits_{N\to+\infty}\sigma_N^{\delta}(f)(x)=f(x)$在 S^{n-1} 上几乎处处成立.

值得指出的是，即使在 L^2 的情形，以上猜测也未完全解决.

参考文献

[1] Wang Kunyang，Li Luoqing. Harmonic analysis and approximation on the unit sphere[M]. Beijing：Science Press，2000：161.

[2] Wang Kunyang. Some recent progress in Fourier-Laplace analysis on the sphere[J]. J BJ Nor Univ (Nat Sci)（北京师范大学学报(自然科学版)），1999，35(1)：23.

[3] Wang Kunyang. Equiconvergent operator of Cesàro means on sphere and its application[J]. J BJ Nor Univ (Nat Sci)（北京师范大学学报(自然科学版)），1993，29(1)：143.

[4] Ditzian Z. Fractional derivatives and best approximation[J]. Acta Math Hungar，1998，82(2)：122.

[5] Szegöo G. Orthogonal Polynomials[M]. New York：AMS，1939.

[6] Bonami A，Clerc J L. Sommes de Cesàro etmultiplicateurs des dèveloppments en harmonique sphériques[J]. Trans AMS，1973，183(1)：223.

[7] 戴峰. 球面上 Fourier-Laplace 级数的收敛和求和[D]. 北京：北京师范大学数学系，1999.

[8] Brown G，Wang Kunyang. Jacobi polynomial estimates and Fourier-Laplace convergence[J]. J Fourier Analysis and Appl，1997，3：705.

[9] Dai Feng. A note on a. e. convergence of Fourier-Laplace series in L^2[J]. J BJ Nor Univ (Nat Sci)（北京师范大学学报(自然科学版)），1999，35(1)：6.

Abstract The almost everywhere convergence conditions of Cesàro means with critical index of Fourier-Laplace series in terms of modulus of continuity are discussed.

Keywords a. e. convergence；Fourier-Laplace series；modulus of continuity.

数学学报，2003，46(3)：1—4.

Fourier-Laplace 级数的缺项算术平均在 Lebesgue 点处的收敛性①

Summability of Fourier-Laplace Series with the Method of Lacunary Arithmetical Means at Lebesgue Points

摘要 设 $f(x)$为定义于 n-维欧氏空间 $\mathbf{R}^n$ 中的单位球面 S^{n-1}上的 Lebesgue 可积函数，$\sigma_N^\delta(f)$表示 f 的 Fourier-Laplace 级数的 Cesàro 平均. 众所周知，$\lambda:=\frac{n-2}{2}$是 Cesàro 平均的临界阶. 本文就 n 是偶数的情形证明了，使得 $\frac{1}{N}\sum_{k=1}^{N}\sigma_{n_k}^{\lambda}(f)(x)\to f(x), N\to+\infty$ 成立的具有一定"缺项程度"的数列$\{n_k\}$的存在性.

关键词 收敛性；Fourier-Laplace 级数；球调和；Lebesgue 点.

§1. 引言和主要结果

设 $\mathrm{S}^{n-1}:=\{(x_1, x_2, \cdots, x_n)\mid x_1^2+x_2^2+\cdots+x_n^2=1\}$是 n 维欧氏空间 $\mathbf{R}^n$ 中的单位球面. 在 S^{n-1} 上赋以常规的 Lebesgue 测度. 设 S_θ：$L^p(\mathrm{S}^{n-1})\to L^p(\mathrm{S}^{n-1})$ 为球面上步长为 θ 的平移算子，其定义为 $S_\theta(f)(x):=\frac{1}{|\mathrm{S}^{n-2}|}\int_{\{y\in\mathrm{S}^{n-1}\mid xy=0\}}f(x\cos\theta+y\sin\theta)\mathrm{d}\ell(y)$，其中 $x\in\mathrm{S}^{n-1}$，$\mathrm{d}\ell(y)$表示纬线 $\ell_x:=\{y\in\mathrm{S}^{n-1}\mid xy=0\}$上的 Lebesgue 测度.

---

① 本文翻译并压缩于本刊英文版，2001，17(3)：489—496.

国家自然科学基金资助项目(19771009).

本文与戴峰合作.

收稿日期：1999-06-28；修改日期：1999-08-11；

接受日期：2000-02-16.

对于 $f\in L^p(S^{n-1})$，$p\geqslant 1$，我们称满足条件 $\Phi_{x,p}(f,\theta)=\int_0^\theta |S_t(f)(x)-f(x)|^p\sin^{n-2}t\mathrm{d}t=o(\theta^{n-1})$，$\theta\to 0$ 的点 x 为 f 的 L_{p-} 点. 众所周知 $\Phi_{x,p}(f,\theta)=o(1)$，$\theta\to 0$，a. e. $x\in S^{n-1}$. 设 $f\in L(S^{n-1})$，记 $\sigma(f)(x):=\sum\limits_{k=0}^{+\infty}Y_k(f)(x)$ 为 f 的 Fourier-Laplace 展式，其中 Y_k 是 $L(S^{n-1})$ 到 k 次球调和空间上的投影算子. $\sigma(f)$ 的 δ 阶 Cesàro 平均定义为 $\sigma_N^\delta(f):=(A_N^\delta)^{-1}\cdot\sum\limits_{k=0}^{N}A_{N-k}^\delta Y_k(f)$，其中 $A_N^\delta=\dfrac{\Gamma(N+\delta+1)}{\Gamma(\delta+1)\Gamma(N+1)}$，$\delta>-1$. 我们知道，Cesàro 平均 σ_N^δ 的临界阶是 $\lambda:=\dfrac{n-2}{2}$，且当 $n=2$ 时，临界阶的 Cesàro 平均 σ_N^λ 恰好为 Fourier 级数的部分和算子.

我们从文[2]中引入自然数列 $\{n_k\}$ 稠度的概念 $\mathrm{dens}(x)=\max\{k\,|\,n_k\leqslant x\}$. 显然，稠度的概念刻画了 $\{n_k\}$ 的“缺项程度 ”.

在球面上 Fourier-Laplace 级数的理论中；一个重要的问题是，使 $\dfrac{1}{N}\sum\limits_{k=1}^{N}\sigma_{n_k}^\lambda(f)(x)\to f(x)$，$N\to+\infty$ 成立的数列 $\{n_k\}$ 能够有多大的“缺项程度”?

当 $n=2$ 时，这一问题首先由 Zalcwasser[1] 于 1936 年提出，随后许多学者对此作出了深入的研究.

本文将证明：

定理　设 $n\geqslant 2$ 为偶数，则存在一自然数列 $\{n_k\}$，其稠度 $\mathrm{dens}(x)\simeq(\ln x)^3$，使得对于任何 $f\in L^1$，$\dfrac{1}{N+1}\sum\limits_{k=0}^{N}\sigma_{n_k}^\lambda(f;x)\to f(x)$，$N\to+\infty$ 在 f 的每一满足对极条件

$$\int_0^\theta |S_t(f)(-x)|\,t^\lambda\mathrm{d}t=o(1),\quad \theta\to 0 \tag{1.1}$$

的 Lebesgue 点 x 处成立.

$n=2$ 时，定理的证明首先是由文[2]给出的. 球面上的对极条件(1)是王昆扬在研究局部化问题时引入的.

§2. 定理的证明

引理 1 令

$$h(x)=\frac{\Gamma\left(x+\frac{3}{2}n-2\right)\Gamma\left(2x+\frac{3}{2}n-1\right)\Gamma(x+1)}{x^{\frac{1}{2}}\Gamma(2x+n-2)\Gamma\left(x+\frac{n}{2}\right)\Gamma\left(x+\frac{n-1}{2}\right)},\quad x>0.\tag{2.1}$$

则 $\lim\limits_{x\to+\infty}h(x)=h(+\infty)$ 存在，并且 $h(x)-h(+\infty)=O\left(\frac{1}{x}\right)$.

证 运用以下两个公式(见文[5])：$\Gamma(x)=\sqrt{2\pi}x^{x-\frac{1}{2}}\mathrm{e}^{-x+\mu(x)}$，

$$0<\mu(x)<\frac{1}{12x},\qquad \frac{\mathrm{dln}\,\Gamma(x)}{\mathrm{d}x}=-\gamma-\frac{1}{x}+\sum_{k=1}^{+\infty}\left(\frac{1}{k}-\frac{1}{k+x}\right),$$

其中 γ 是 Euler-常数，我们可得引理 1 的证明.

引理 2 设 $n\geqslant3$ 为偶数，$f\in L(S^{n-1})$，且 $x\in S^{n-1}$ 为 f 的满足对极条件(1.1)的 Lebesgue 点，则

$$\sigma_k^{\lambda}(f)(x)-f(x)=c_n\int_0^{\frac{\pi}{2}}\varphi_x(t)\frac{\sin^{n-2}t}{t^{n-1}}M(t)\sin kt\,\mathrm{d}t+o(1),k\to+\infty,$$

其中 $\varphi_x(t)=S_t(f)(x)-f(x)$，$M(t)$为$\left[0,\ \frac{\pi}{2}\right]$上的连续函数.

证 我们运用公式(见文[4])

$$\begin{aligned}\sigma_k^{\lambda}(f)(x)-f(x)=&[E_k^{\lambda}(f)(x)-E_k^{\lambda}(1)f(x)]+\\&[T_k^{\lambda}(f)(x)-T_k^{\lambda}(1)f(x)],\end{aligned}\tag{2.2}$$

其中 1 为取值为 1 的常函数.

$$\begin{aligned}&T_k^{\lambda}(f)(x):=\sum_{v=1}^{+\infty}O(v^{-\frac{3}{2}n})\sigma_k^{\lambda+v}(f)(x),\\&E_k^{\lambda}(f)(x):=\gamma_k^{\lambda}\int_{S^{n-1}}f(y)P_k^{\left(n-\frac{3}{2},\frac{n-3}{2}\right)}(xy)\mathrm{d}y\end{aligned}\tag{2.3}$$

为 $\sigma_k^{\lambda}(f)$的等收敛算子，这里

$$\gamma_k^{\lambda}=\frac{\Gamma\left(\frac{n}{4}\right)\Gamma\left(k+\frac{3}{2}n-2\right)\Gamma\left(2k+\frac{3}{2}n-1\right)\Gamma(k+1)}{(4\pi)^{\frac{n-1}{2}}\Gamma(2k+n-2)\Gamma\left(k+\frac{n}{2}\right)\Gamma\left(k+\frac{n-1}{2}\right)},$$

$P_k^{(\alpha,\beta)}(t)$为 Jacobi 多项式.

易验证

$$T_k^\lambda(f)(x)-T_k^\lambda(1)f(x)=o(1),\quad k\to+\infty. \tag{2.4}$$

为了处理$E_k^\lambda(f)(x)\to E_k^\lambda(1)f(x)$这一项，写成

$$E_k^\lambda(f)(x)-E_k^\lambda(1)f(x)=\gamma_k^\lambda|S^{n-2}|\int_0^\pi\varphi_x(\theta)\cdot\sin^{n-2}\theta P_k^{\left(n-\frac{3}{2},\frac{n-3}{2}\right)}(\cos\theta)\mathrm{d}\theta.$$

在区间$0<\theta<k^{-1}$，我们运用估计式$|P_k^{(\alpha,\beta)}(\cos\theta)|\leqslant C_{\alpha,\beta}k^\alpha$，而在区间$k^{-1}<\theta<\frac{\pi}{2}$上，运用渐进表达式(见文[6])：$P_k^{(\alpha,\beta)}(\cos\theta)=(k\pi)^{-\frac{1}{2}}\left(\sin\frac{\theta}{2}\right)^{-\left(\alpha+\frac{1}{2}\right)}\left(\cos\frac{\theta}{2}\right)^{-\left(\beta+\frac{1}{2}\right)}\left(\cos(\tilde{k}\theta+\gamma)+O\left(\frac{1}{k\sin\theta}\right)\right)$，其中$\tilde{k}=k+\frac{\alpha+\beta+1}{2}$，$\gamma=-\left(\alpha+\frac{1}{2}\right)\frac{\pi}{2}$. 于是，我们有

$$E_k^\lambda(f)(x)-E_k^\lambda(1)f(x)=\sum_{i=1}^4 I_k^{(i)}+R_k, \tag{2.5}$$

其中

$$|I_k^{(1)}|\leqslant O(k^{n-1})\int_0^{k^{-1}}|\varphi_x(\theta)|\sin^{n-2}\theta\mathrm{d}\theta=o(1), \tag{2.6}$$

$$I_k^{(2)}:=O(1)\int_{k^{-1}}^{\frac{\pi}{2}}\varphi_x(\theta)\frac{\sin^{n-2}\theta}{\theta^{n-1}}M_1(\theta)\sin\left(\frac{3}{4}n-1\right)\theta\cos k\theta\mathrm{d}\theta,$$

$$I_k^{(3)}:=|S^{n-2}|\gamma_k^\lambda k^{-\frac{1}{2}}\int_{k^{-1}}^{\frac{\pi}{2}}\varphi_x(\theta)\frac{\sin^{n-2}\theta}{\theta^{n-1}}M_1(\theta)\cos\left(\frac{3}{4}n-1\right)\theta\sin k\theta\mathrm{d}\theta,$$

$$I_k^{(4)}:=O(1)\int_{k^{-1}}^{\frac{\pi}{2}}\varphi_x(\theta)\frac{\sin^{n-2}\theta}{\theta^{n-1}}M_1(\theta)O\left(\frac{1}{k\theta}\right)\mathrm{d}\theta,$$

这里$M_1(\theta)=\left(\frac{\theta}{\sin\frac{\theta}{2}}\right)^{n-1}\left(\cos\frac{\theta}{2}\right)^{-\frac{n-2}{2}}$，且由文[4]的结果，我们知(2.5)式中余项

$$R_k=\gamma_k^\lambda\int_{\frac{\pi}{2}}^\pi\varphi_x(\theta)P_k^{\left(n-\frac{3}{2},\frac{n-3}{2}\right)}(\cos\theta)\sin^{n-2}\theta\mathrm{d}\theta=o(1), \tag{2.7}$$

$k\to+\infty$. 而由 Riemman-Lebesgue 定理，易得

$$I_k^{(2)}=o(1),\quad k\to+\infty. \tag{2.8}$$

对于第三项$I_k^{(3)}$和第四项$I_k^{(4)}$，运用引理1和分步积分，我们有

$$I_k^{(3)}:=c_n\int_0^{\frac{\pi}{2}}\varphi_x(\theta)\frac{\sin^{n-2}\theta}{\theta^{n-1}}M(\theta)\sin k\theta\mathrm{d}\theta+o(1),\quad k\to+\infty, \tag{2.9}$$

$$I_k^{(4)}=o(1),\quad k\to+\infty, \tag{2.10}$$

其中$M(\theta)=\left(\frac{\theta}{\sin\frac{\theta}{2}}\right)^{n-1}\left(\cos\frac{\theta}{2}\right)^{-\frac{n-2}{2}}\cos\left(\frac{3n}{4}-1\right)\theta.$

联立(2.2)～(2.10)式，完成证明. □

引理 3[7] 对于每一$M>1$，存在自然数n_1，n_2，…，n_k，使得$M<n_j<M^2$，$j=1,2,\cdots,k$，$k\simeq(\ln M)^3$ 且

$$\sup_{1\leqslant s\leqslant k}\left\|\sum_{j=1}^{s}\sin n_jx\right\|_\infty\ll(\ln M)^2. \tag{2.11}$$

定理的证 我们考虑区间$[2^{2^k}, 2^{2^{k+1}})$. 在每一区间之中，选取引理 3 中给出的数列. 稠度估计是显然的. 下面证明在L_1-点处的收敛性.

由引理 2，只要证明

$$I_N=\frac{1}{N}\sum_{k=1}^{N}\int_0^{\frac{\pi}{2}}\varphi_x(\theta)\frac{\sin^{n-2}\theta}{\theta^{n-1}}\sin(n_k\theta)\mathrm{d}\theta=o(1),\quad N\to+\infty.$$

设$2^{3m}\leqslant N<2^{3(m+1)}$，则

$$|I_N|\leqslant$$

$$\frac{1}{N}\sum_{k=1}^{N}\int_0^{\delta}|\varphi_x(\theta)|\frac{\sin^{n-2}\theta}{\theta^{n-2}}n_k\mathrm{d}\theta+\frac{1}{N}\left|\sum_{k=1}^{N}\int_\delta^{\frac{\pi}{2}}|\varphi_x(\theta)|\frac{\sin^{n-2}\theta}{\theta^{n-1}}M(\theta)\sin n_k\theta\mathrm{d}\theta\right|$$

$$\overset{\text{def}}{=\!=\!=}I_{N,1}+I_{N,2}.$$

对于第一项$I_{N,1}$，运用稠度估计式，并分步积分有

$$\begin{aligned}I_{N,1}&=\frac{1}{N}\Big(\sum_{j=1}^{m+1}\sum_{2^{2^j}\leqslant n_k<2^{2^j+1}}n_k\Big)\left[\frac{\Phi_x(\theta)}{\theta^{n-2}}\bigg|_0^\delta+\int_0^\delta\frac{\Phi_x(\theta)}{\theta^{n-1}}\mathrm{d}\theta\right]\\&\ll o(\delta)2^{2^{m+2}},\quad\delta\to0^+.\end{aligned}$$

对于第二项，我们应用(12)式可得

$$\begin{aligned}I_{N,2}&\ll\frac{1}{N}\left(\int_\delta^{\frac{\pi}{2}}|\varphi_x(\theta)|\theta^{-1}\mathrm{d}\theta\right)\left\|\sum_{k=1}^{N}\sin n_k\theta\right\|_\infty\\&\ll\frac{1}{N}\Big(\sum_{j=1}^{m+1}\Big\|\sum_{2^{2^j}<n_k<2^{2^{j+1}}}\sin n_k\theta\Big\|_\infty\Big)\left(\left|\frac{\Phi_x(\theta)}{\theta^{n-1}}\right|_\delta^{\frac{\pi}{2}}+c_n\int_\delta^{\frac{\pi}{2}}\frac{\Phi_x(\theta)}{\theta^n}\mathrm{d}\theta\right)\\&\ll\frac{1}{2m}o\left(\ln\frac{1}{\delta}\right),\quad\delta\to0^+.\end{aligned}$$

取$\delta=2^{-2^{m+2}}$，我们即得证明. □

注 1 在定理的证明中，引理 3(Bourgain 的结果)起了关键性的作用. 只有当n为偶数的时候，我们才可应用这一结果. 这就是定理中我们要求n为偶数的原因. 对于n为奇数的情形，这一问题尚未解决.

注 2 当$n>2$时，定理中的对极条件是必要的. 事实上，我们可证明对任意的$x\in S^{n-1}$，存在$f\in L(S^{n-1})$，使得x为f的 Lebesgue 点，

但即使对于奇数或偶数列$\{n_k\}$，都有

$$\frac{1}{N}\sum_{k=0}^{N}\sigma_{n_k}^{\lambda}\cdot(f)(x)\to\infty, N\to+\infty.$$

参考文献

[1] Zalcwasser Z. Sur la sommability des series de Fourier. Studia Mathematica，1936，6(1)：82—88.

[2] Belinsky E L. Summability of Fourier series with the method of lacunary arithmetical means at the Lebesgue points. Proc. Amer. Math. Soc.，1997，125：3 689—3 693.

[3] Bonami A，Clerc J L. Sommes de Cesàro et multiplicateurs des dèveloppments en harmonique sphériques. Trans. AMS，1973，183：223—263.

[4] Wang K Y. Equiconvergent operator of Cesàro means on sphere and its application. J. Beijing Normal Univ. 1993，29：143—154.

[5] Artin E. The gamma function. Holt：Rinehart and Winston，1964.

[6] Szegö G. Orthogonal polynomials. New York：Amer. Math. Soc.，1959.

[7] Bourgain J.，Sur les sommes de sinus，Harmonic analysis：study group on translation-invariant Banach spaces. Publ. Math. Orsay，83，Univ. Paris XI，Orsay，1983.

[8] Watson G. Theory of Bessel functions，Cambridge：Ar the University Prcss，1944.

Abstract Let S^{n-1} be the unit sphere in n-dimensional Euclidean space $\mathbf{R}^n$. For a function $f\in L(S^{n-1})$ denote by $\sigma_N^{\delta}(f)$ the Cesàro means of order δ of the Fourier-Laplace series of f. The special value $\lambda := \frac{n-2}{2}$ of δ is known as the critical index. In the case of n is even，this paper proves the existence of the "rare" sequence $\{n_k\}$ such that the summability $\frac{1}{N}\sum_{k=1}^{N}\sigma_{n_k}^{\lambda}(f)(x)\to f(x), N\to+\infty$ takes place at each Lebesgue point satisfying some antipole conditions.

Keywords Convergence；Fourier-Laplace series；Spherical harmonics；Lebesgue point.

北京师范大学学报(自然科学版)，1993，29(2)：143－153.

球面上 Cesàro 平均的等收敛算子及其应用

Equiconvergent Operator of Cesàro Means on Sphere and Its Applications①

Abstract An operator S_N^δ which is equiconvergent with Cesàro means on sphere is given. Its kernel has the form $(N+1)^\gamma P_N^{(\alpha,\beta)}$ ($\gamma=\frac{n-1}{2}-\delta$, $\alpha=\frac{n-1}{2}+\delta$, $\beta=\frac{n-3}{2}$, n is the number of variables) where $P_N^{(\alpha,\beta)}$ are Jacobi polynomials. By investigating S_N^b the antipole conditions for convergence at a point x, i. e. the necessary conditions required at the point $-x$, are obtained. Localization theorems are established. This gives a convenient way to investigate the convergence of Cesàro means on sphere.

Keywords spherical harmonics; Cesàro means; Jacobi polynomials; Iocalization.

§ 0. Introduction

Let $\mathbf{R}^n$ be n-dimensional Euclidean space and $n\geqslant 3$. Denote by S^{n-1} the unit sphere of $\mathbf{R}^n$, i. e. $S^{n-1}=\{x\in\mathbf{R}^n \mid |x|=(x_1^2+x_2^2+\cdots+x_n^2)^{\frac{1}{2}}=1\}$. If $f\in L(S^{n-1})$ then f has Fourier-Laplace expansion $\sigma(f)(x)=\sum_{k=0}^{+\infty} f_k$, where f_k is the projection of f on H_k^n, the space of spherical

① Supported by NSFC.
Received: 1992-07-16.

harmonics of degree k. f_k can be expressed as a spherical convolution

$$f_k(x) = c_{n,k}(f * P_k^n)(x) = c_{n,k}\int_{S^{n-1}} f(y)P_k^n(xy)\mathrm{d}y, \qquad (0.1)$$

where $c_{n,k} = \dfrac{\Gamma\left(\frac{n}{2}\right)}{2\pi^{\frac{n}{2}}} \dfrac{(n+2k-2)(n+k-3)!}{k!(n-2)!}$ and P_k^n is Gegenbeuer polynominal normalized by the condition $P_k^n(1) = 1$, which is generated by

$$\frac{1}{(1-2rt+r^2)^{\frac{n-2}{2}}} = \sum_{k=0}^{+\infty} \frac{\Gamma(n-2+k)}{k!\Gamma(n+2)} P_k^n(t) r^k, \quad 0 \leqslant r < 1.$$

For these basic concepts we refer C. Müller[1].

The Cesàro means of $\sigma(f)$ are defined by

$$\sigma_N^\delta(f)(x) = \frac{1}{A_N^\delta}\sum_{k=0}^{N} A_{N-k}^\delta f_k(x), \qquad (0.2)$$

where δ is the order assumed to be strictly bigger than -1 in our discussion. According to (0.1) we can write (0.2) as a convolution, i.e. $\sigma_N^\delta(f)(x) = (f * K_N^\delta)(x)$ where the kernel $K_N^\delta(t) = \frac{1}{A_N^\delta}\sum_{k=0}^{N} A_{N-k}^\delta c_{n,k} P_k^n(t)$. By the formula (9.41.13) of Szegö's book [2] we have

$$K_N^\delta(t) = a_N^\delta (N+1)^{\frac{n-1}{2}-\delta} P_N^{\left(\frac{n-1}{2}+\delta, \frac{n-3}{2}\right)}(t) + \sum_{v=1}^{+\infty} b_v(N,\delta) K_N^{\delta+v}(t). \qquad (0.3)$$

where $P_N^{(\alpha,\beta)}$ denote Jacobi polynomials and

$$a_N^\delta = \frac{(N+1)^{\delta-\frac{n-1}{2}}}{2^{n-1}\Gamma\left(\frac{n-1}{2}\right)} \frac{\Gamma(N+\delta+n-1)}{A_N^\delta \Gamma\left(N+\frac{n-1}{2}\right)} \frac{\Gamma(2N+\delta+n)}{\Gamma(2N+2\delta+n)},$$

$$b_v(N, \delta) = \mu_v(\delta) \frac{(N+\delta+1)\cdots(N+\delta+v)}{(2N+\delta+n)\cdots(2N+\delta+n+v-1)}, \qquad (0.4)$$

$$\mu_v(\delta) = (-1)^{v+1}\frac{\delta(\delta-1)\cdots(\delta-v+1)}{v!}.$$

We see that a_N^δ is bounded as $N \to +\infty$. According to (0.3) we define operators S_N^δ and T_N^δ by

$$S_N^\delta = (N+1)^{\frac{n-1}{2}-\delta} f * P_N^{\left(\frac{n-1}{2}+\delta, \frac{n-3}{2}\right)}, \quad T_N^\delta = \sum_{v=1}^{+\infty} b_v(N,\delta)\sigma_N^{\delta+v}(f),$$

for $f \in L(S^{n-1})$. then we have

$$\sigma_N^\delta = a_N^\delta S_N^\delta(f) + T_N^\delta(f). \qquad (0.5)$$

Our main purpose is to prove that for the convergence of $\sigma_N^\delta(f)(x)$ sufficient and necessary is the convergence of $S_N^\delta(f)(x)$. This result will provide a great convenience for the investigation of convergence problem. A lot of convergence theorems will follow some of them will be established in our next paper (Pointwise convergence of Cesàro means on sphere) which will appear in the same issue of this journal. Although the theory of F-L series has already a not short history (see, for example, [6][7] and [8]) we have not found similar results up to now.

Theorem 1 Assume $\delta>-1$, D is a non-empty subset of S^{n-1}, $f\in L(S^{n-1})$ and $\sup\{|f(x)|\,|\,x\in D\}<+\infty$. For uniform convergence on D the following two relations an equivalent:

$$\lim_{N\to+\infty} S_N^\delta(g_x)(x)=0, \tag{0.6}$$

$$\lim_{N\to+\infty} \sigma_N^\delta(g_x)(x)=0, \tag{0.7}$$

where the function g_x is defined for each $x\in D$ by $g_x(y)=f(y)-f(x)$.

§ 1. Proof of theorem 1

Lemma 1 There is a constant $c(n, \delta)$ depending only on n and δ such that

$$|b_v(N, \delta)|\leqslant c(n, \delta)v^{-(1+n+\delta)}, \quad v\in \mathbf{N}. \tag{1.1}$$

Proof By (0.4), if $v\leqslant[\delta]+n$ then $|b_v(N, \delta)|\leqslant c(n, \delta)$ obviously. If $v\geqslant[\delta]+n$ then we have

$$|b_v(N,\delta)|\leqslant c(\delta)\frac{\Gamma(v-\delta)}{\Gamma(v+1)}\prod_{i=1}^{v}\frac{N+\delta+i}{2N+\delta+n-1+i}.$$

Since for $N\in\mathbf{N}$,

$$\frac{N+\delta+i}{2N+\delta+n-1+i}=\frac{1}{2}\left\{1+\frac{\delta-(n-1)+i}{2N+\delta+(n-1)+i}\right\}$$

$$\leqslant\frac{1}{2}\left\{1+\frac{|\delta-(n-1)+i|}{\delta+n+1+i}\right\},$$

we have

$$|b_v(N,\delta)|\leqslant c(\delta)\frac{\Gamma(v-\delta)}{\Gamma(v+1)}\prod_{i=1}^{n}\frac{n}{\delta+n+1+i}\prod_{i=n+1}^{v}\frac{\delta+1+i}{\delta+n+1+i}$$

$$\leqslant c(n,\delta)\frac{\Gamma(v-\delta)}{\Gamma(v+1)}\frac{\Gamma(v+\delta+2)}{\Gamma(v+\delta+n+2)}.$$

For $s\geqslant 2$ and $a>0$ we have

$$\frac{\Gamma(s)}{\Gamma(s+a)}=\frac{\Gamma(s-[s]+1)}{\Gamma(s-[s]+1+a)}\frac{(s-[s]+1)\cdots(s-1)}{(s-[s]+1+a)\cdots(s-1+a)}$$

$$\leqslant\left\{\lim_{x>0}\Gamma(x)\left(1+\frac{a}{s-[s]+1}\right)\cdots\left(1+\frac{a}{s-1}\right)\right\}^{-1}\leqslant\frac{c(a)}{(s+a)^{a}}.$$

Hence we get (1. 1). □

Lemma 2 Under the assumption of Theorem 1 if the relation

$$\lim_{k\to+\infty}|\sigma_k^{\delta+1}(f)(x)-f(x)|=0$$

holds uniformly on D then $\lim\limits_{k\to+\infty}T_k^{\delta}(g_x)(x)=0$ also, where the function g_x is defined by $g_x(y)=f(y)-f(x)$, $y\in S^{n-1}$.

Proof For and $\varepsilon>0$ there is a number $N\in\mathbf{N}$ such that $|\sigma_k^{\delta+1}(f)(x)-f(x)|=|\sigma_k^{\delta+1}(g_x)(x)|<\varepsilon$ when $k>N$ and $x\in D$. Define $M=\sup\{|\sigma_k^{\delta+1}(f)(x)-f(x)|\,|\,x\in D,\ k=0,1,2,\cdots,N\}$. Then we see $M\in[0,\ +\infty)$. since for

$$v\geqslant 2,\ \sigma_k^{\delta+v}(f)=\frac{1}{A_k^{\delta+v}}\sum_{i=0}^{k}A_{k-i}^{v-2}A_i^{\delta+1}\sigma_i^{\delta+1}(f)$$

we get for $k>N$ and $x\in D$,

$$|\sigma_k^{\delta+v}(f)(x)-f(x)|\leqslant\frac{1}{A_k^{\delta+v}}\sum_{i-0}^{N}A_{k-i}^{v-2}A_i^{\delta+1}|\sigma_i^{\delta+1}(g_x)(x)|+$$

$$\frac{1}{A_k^{\delta+v}}\sum_{i=N+1}^{k}A_{k-i}^{v-2}A_i^{\delta+1}\varepsilon\leqslant\varepsilon+\frac{M}{A_k^{\delta+v}}\sum_{i=0}^{N}A_{k-i}^{v-2}A_i^{\delta+1}.$$

For $\delta>-1$ we have $A_i^{\delta+1}\leqslant A_N^{\delta+1}\ (0\leqslant i\leqslant N)$, and for $v\geqslant 2$ we have $A_j^{v-2}\leqslant A_k^{v-2}\ (0\leqslant j\leqslant k)$. So we get $|\sigma_k^{\delta+v}(f)(x)-f(x)|\leqslant\varepsilon+M(N+1)\cdot A_N^{\delta+1}\cdot\frac{A_k^{v-2}}{A_k^{\delta+v}}$. From the equality $\frac{A_k^{v-2}}{A_k^{\delta+v}}=\prod\limits_{i=1}^{k}\left(1+\frac{\delta+2}{v-2+i}\right)^{-1}$ we see $\frac{A_k^{v-2}}{A_k^{\delta+v}}$ is increasing with v. And we have $\ln\left(1+\frac{\delta+2}{v-2+i}\right)\geqslant\frac{\delta+2}{v-2+i}-\frac{1}{2}\left(\frac{\delta+2}{v-2+i}\right)^2$. Then we see that $\sum\limits_{i=0}^{k}\ln\left(1+\frac{\delta+2}{v-2+i}\right)\geqslant(\delta+2)\cdot\int_{v-1}^{k+v-1}\frac{\mathrm{d}t}{t}-\frac{(\delta+2)^2}{2}\left(1+\int_{v-1}^{k+v-2}\frac{\mathrm{d}t}{t^2}\right)\geqslant(\delta+2)\ln\left(1+\frac{k}{v-1}\right)-(\delta+2)^2$ for $v\geqslant 2$. If we restrict $2\leqslant v\leqslant\lambda^{-1}k$ for k big enough and $\lambda\in\mathbf{N}$ then we derive from this inequality that $\sum_{i=1}^{k}\ln\left(1+\frac{\delta+2}{v-2+i}\right)\geqslant\ln\frac{(1+\lambda)k-\lambda}{k-\lambda}-(\delta+2)^2>\ln(1+\lambda)^{\delta+2}-(\delta+2)^2$. Therefore for $\lambda\in\mathbf{N}$ and $2\leqslant v\leqslant\lambda^{-1}k$ (k big enough) we have $\frac{A_k^{v-2}}{A_k^{\delta+v}}\leqslant(1+\lambda)^{-(\delta+2)}e^{(\delta+2)^2}$. Now we choose $\lambda\in\mathbf{N}$ such that $M(N+1)A_N^{\delta+1}e^{(\delta+2)^2}\cdot(1+\lambda)^{-(\delta+2)}<\varepsilon$ then for $2\leqslant v\leqslant\lambda^{-1}k$ ($k>N$),

$$|\sigma_k^{\delta+v}(g_x)(x)|\leqslant 2\varepsilon \tag{1.2}$$

holds on D uniformly. On the other hand for $v\geqslant[\lambda^{-1}k]+1>\lambda^{-1}k$ and $x\in D$,

$$|\sigma_k^{\delta+v}(g_x)(x)|\leqslant\varepsilon+M(N+1)A_N^{\delta+1}, \tag{1.3}$$

since $\frac{A_k^{v-2}}{A_k^{\delta+v}}<1(\delta>-1)$.

We assume now $k>2\lambda+N$. Then we have

$$T_k^{\delta}(g_x)(x)=\left(\sum_{v=1}^{[\lambda^{-1}k]}+\sum_{v=[\lambda^{-1}k]+1}^{+\infty}\right)b_v(k,\delta)\sigma_k^{\delta+v}(g_x)(x)=\Sigma_1+\Sigma_2. \tag{1.4}$$

By (1.2) and Lemma 1 we have

$$|\Sigma_1|\leqslant c(n,\delta)\sum_{v=1}^{[\lambda^{-1}k]}v^{-\delta-n-1}|\sigma_k^{\delta+v}(g_x)(x)|\leqslant c(n,\delta)\varepsilon. \tag{1.5}$$

By (1.3) and Lemma 1 we have

$$\begin{aligned}|\Sigma_2|&\leqslant c(n,\delta)\{\varepsilon+M(N+1)A_N^{\delta+1}\}\sum_{v=[\lambda^{-1}k]+1}^{+\infty}v^{-\delta-n-1}\\&\leqslant c(n,\delta)\{\varepsilon+M(N+1)A_N^{\delta+1}\}\left(\frac{\lambda}{k}\right)^{n+\delta}.\end{aligned}$$

Obviously, there is a number $k_0>2\lambda+N$ such that $\left(\frac{\lambda}{k}\right)^{n+\delta}\{\varepsilon+M(N+1)A_N^{\delta+1}\}<\varepsilon$ when $k\geqslant k_0$. Hence

$$|\Sigma_2|<\varepsilon. \tag{1.6}$$

if $k\geqslant k_0$. Combining (1.4)~(1.6) we get $|T_k^{\delta}(g_x)(x)|<c(n,\delta)\varepsilon$ for $k\geqslant k_0$ and $x\in D$, and complete the proof. □

Lemma 3 Under the assumption of Theorem 1, if (0.6) holds uniformly on D then (0.7) also.

Proof Let k be the smallest integer bigger than δ and let $\varepsilon=k-\delta$. Then $0<\varepsilon\leqslant 1$ and $k\leqslant\delta+1$. We apply formula (9.4.3) of Szegö's book [2] to get

$$P_{\mu}^{\left(\frac{n-1}{2}+k,\frac{n-3}{2}\right)}(t)=\sum_{v=0}^{\mu}d_{v}(\mu,\delta)P_{v}^{\left(\frac{n-1}{2}+\delta,\frac{n-3}{2}\right)}(t). \tag{1.7}$$

where

$$d_{v}(\mu,\ \delta)=\frac{A_{\mu-v}^{\varepsilon-1}(2v+n+\delta)\Gamma(\mu+v+n+k-1)\Gamma(v+n+\delta-1)\Gamma\left(\mu+\frac{n-1}{2}\right)}{\Gamma(\mu+v+n+\delta)\Gamma\left(v+\frac{n-1}{2}\right)\Gamma(\mu+n-1+k)}.$$

Then we get $\qquad S_{\mu}^{k}=\sum\limits_{v=0}^{\mu}\dfrac{(\mu+1)^{\frac{n-2}{2}-k}}{(v+1)^{\frac{n-1}{2}-\delta}}d_{v}(\mu,\delta)S_{v}^{\delta}(f).$

If $\delta\geqslant-\dfrac{1}{2}$ then we have $0\leqslant\left[\dfrac{(\mu+1)^{\frac{n-1}{2}-k}}{(v+1)^{\frac{n-1}{2}-\delta}}\right]d_{v}(\mu,\ \delta)\leqslant c(n,\ \delta)\left(\dfrac{A_{\mu-v}^{\varepsilon-1}}{A_{\mu}^{\varepsilon}}\right)$

and hence $\qquad |S_{\mu}^{k}(f)|\leqslant\dfrac{c(n,\delta)}{A_{\mu}^{\varepsilon}}\sum\limits_{v=0}^{\mu}A_{\mu-v}^{\varepsilon-1}|S_{v}^{\delta}(f)|. \qquad (1.8)$

If $-1<\delta<-\dfrac{1}{2}$ then $k=0$, $\varepsilon=-\delta\in\left(\dfrac{1}{2},\ 1\right)$ and

$$0\leqslant\left[\frac{(\mu+1)^{\frac{n-1}{2}-k}}{(v+1)^{\frac{n-1}{2}-\delta}}\right]\cdot d_{v}(\mu,\ \delta)\leqslant c(n,\ \delta)A_{\mu-v}^{\varepsilon-1}\frac{A_{v}^{1-2\varepsilon}}{A_{\mu}^{1-\varepsilon}}.$$

Hence in this case

$$|S_{\mu}^{k}(f)|=|S_{\mu}^{0}(f)|\leqslant\frac{c(n,\delta)}{A_{\mu}^{1-\varepsilon}}\sum_{v=0}^{\mu}A_{\mu-v}^{\varepsilon-1}A_{v}^{1-2\varepsilon}|S_{v}^{\delta}(f)|. \tag{1.9}$$

So we conclude in both cases the relation

$$\lim_{\mu\to+\infty}|S_{\mu}^{k}(g_{x})(x)|=0 \tag{1.10}$$

holds uniformly on D since (0. 6) is assumed and f is bounded on D.

Now we apply formula (9. 4. 5) of [2] for the indexes $\alpha=\beta=\dfrac{n-3}{2}$.

Then we get

$$K_{N}^{k}(t)=\frac{1}{A_{N}^{k}}\sum_{v=0}^{N}G_{v}(N,k)H_{v}^{k}P_{v}^{\left(\frac{n-1}{2}+k,\frac{n-3}{2}\right)}(t),$$

where $G_{v}(N,K)=\sum\limits_{j=0}^{N-v}A_{N-v-j}^{k}A_{j}^{-(k+2)}(2v+n-2+2j)\dfrac{\Gamma(2v+n-2+j)}{\Gamma(2v+n+k+j)}$,

and $H_{v}^{k}=\dfrac{\Gamma(2v+2k+n)\Gamma(v+k+1)}{\Gamma(2v+k+n-1)\Gamma(v+1)\Gamma(k+1)}\cdot a_{v}^{k}(v+1)^{\frac{n-1}{2}-k}$. For $k=0$

we have

$$G_v(N,0)=\sum_{j=0}^{n-v}A_j^{-2}(2v+n-2+2j)\frac{\Gamma(2v+n-2+j)}{\Gamma(2v+n+j)}$$

$$=\begin{cases}\dfrac{1}{2N+n-1}, & \text{if } v=N,\\ 0, & \text{if } 0\leqslant v<N.\end{cases}$$

Hence $K_N^0(t)=a_N^0(N+1)^{\frac{n-1}{2}}P_N^{\left(\frac{n-1}{2},\frac{n-3}{2}\right)}(t)$, or, equivalently,

$$a_N^0S_N^0=\sigma_N^0. \tag{1.11}$$

If $k\in\mathbf{N}$ then by (9.41.7) of [2] $G_v(N,k)=O\left(\sum_{\rho=0}^{k-1}(N-v)^\rho v^{-k-\rho-2}\right)$. So in this case we have

$$|\sigma_N^k(g_x)(x)|\leqslant c(n,k)\frac{1}{N+1}\sum_{v=0}^{N}|S_v^k(g_x)(x)|. \tag{1.12}$$

A combination of (1.10)~(1.12) yields the relation

$$\lim_{N\to+\infty}\sigma_N^k(g_x)(x)=0 \tag{1.13}$$

which holds on D uniformly.

Since $k\leqslant\delta+1$ and $\sigma_N^{\delta+1}=\dfrac{1}{A_N^{\delta+1}}\sum_{v=0}^{N}A_{N-v}^{\delta-k}A_v^k\sigma_v^k$ we see that (1.13) implies

$$\lim_{N\to+\infty}\sigma_N^{\delta+1}(g_x)(x)=0 \tag{1.14}$$

uniformly on D. Then an application of Lemma 2 yields

$$\lim_{N\to+\infty}T_N^\delta(g_x)(x)=0 \tag{1.15}$$

uniformly on D. Then the relation (0.7) follows.

If (0.7) is assumed uniformly on D, then the assumption of Lemma 2 is valid obviously. Hence by this lemma, (1.15) holds. Then the relation (0.6) follows. □

§ 2. Localization of σ_N^δ when $\delta\geqslant n-2$

Theorem 2 Suppose $f\in L(S^{n-1})$ and f vanishes in $D=D(p;\ \gamma)=\{y\in S^{n-1}\mid py>\cos\gamma\}$, $p\in S^{n-1}$, $\gamma\in\left(0,\ \frac{\pi}{2}\right)$. If $\delta\geqslant n-2$ then

$$\lim_{N\to+\infty}\sigma_N^\delta(f)(x)=0 \tag{2.1}$$

uniformly in any closed subset of D.

Proof Suppose E is a closed subset of D. Then there exists a positive number $\varphi<\frac{\pi}{2}$ such that $D(x;\varphi)\subset D$ for any $x\in E$. By S_θ we denote the spherical translation operator (see [3]), i. e. for $g\in L(S^{n-1})$ and $p\in S^{n-1}$, $S_\theta(g)(p)=\frac{1}{l_\theta}\int_{l_{p,\theta}}g(y)\,\mathrm{d}\,l_{p,\theta}(y)\ (0<\theta\leqslant\pi)$, where $l_{p,\theta}=\{y\in S^{n-1}\mid yp=\cos\theta\}$ and $l_\theta=|S^{n-2}|\sin^{n-2}\theta$, $|S^{n-2}|=\frac{2\pi^{\frac{n-1}{2}}}{\Gamma\left(\frac{n-1}{2}\right)}$. Then for $x\in E$,

$$S_N^\delta(f)(x)=|S^{n-2}|N^{\frac{n-1}{2}-\delta}\int_\varphi^\pi S_\theta(f)(x)\sin^{n-2}\theta P_N^{\left(\frac{n-1}{2}+\delta,\frac{n-3}{2}\right)}(\cos\theta)\mathrm{d}\theta \quad (2.2)$$

since f vanishes on $D(x;\varphi)$. By Theorem 1 the relation (2. 1) on E is equivalent to $\lim\limits_{N\to\infty}S_N^\delta(f)(x)=0$ (uniformly) on E.

We apply Theorem 8. 21. 12 of [2]:

$$P_N^{\alpha,\beta}(\cos\theta)=$$

$$b_N(\alpha,\beta)M_{\alpha,\beta}(\theta)\frac{J_\alpha(\widetilde{N}\theta)}{\theta^\alpha}+\begin{cases}O(N^{-\frac{3}{2}}\theta^{\frac{1}{2}-\alpha}), & \text{if } N^{-1}\leqslant\theta\leqslant\frac{\pi}{2},\\ O(N^\alpha\theta^2), & \text{if } 0<\theta\leqslant N^{-1},\end{cases} \quad (2.3)$$

where $\widetilde{N}=N+\frac{\alpha+\beta+1}{2}$, $b_N(\alpha,\beta)=\left(\frac{2}{\widetilde{N}}\right)^\alpha\frac{\Gamma(N+\alpha+1)}{\Gamma(N+1)}$, $M_{\alpha,\beta}(\theta)=\left(\frac{\theta}{2\sin\frac{\theta}{2}}\right)^{\alpha+\frac{1}{2}}\times\left(\frac{1}{\cos\frac{\theta}{2}}\right)^{\beta+\frac{1}{2}}$ and J_α denotes Bessel function of first kind. Also we make use the asymptotic formula of Bessel functions (see [4], p. 199):

$$J_v(t)=\sqrt{\frac{2}{\pi t}}\cos\left(t-\frac{v\pi}{2}-\frac{\pi}{4}\right)+O\left(\frac{1}{t^{\frac{3}{2}}}\right),\quad t\geqslant1,\ J_v(t)=O(t^v),\quad 0<t\leqslant1.$$

Then we have

$$P_N^{(\alpha,\beta)}(\cos\theta)$$

$$=\begin{cases}\sqrt{\frac{2}{\pi}}b_N(\alpha,\beta)M_{\alpha,\beta}(\theta)\frac{\cos(\widetilde{N}\theta-\gamma)}{\sqrt{\widetilde{N}}\theta^{\frac{1}{2}+\alpha}}+O\left(\frac{1}{N^{\frac{3}{2}}\theta^{\frac{3}{2}+\alpha}}\right), & \text{if } N^{-1}\leqslant\theta\leqslant\frac{\pi}{2},\\ O(N^\alpha), & \text{if } 0<\theta\leqslant N^{-1},\end{cases} \quad (2.4)$$

where $\gamma=\left(\frac{\alpha}{2}+\frac{1}{4}\right)\pi$. By (2. 4) if $\theta\in\left(\varphi,\frac{\pi}{2}\right)$ and $N>\frac{1}{\varphi}$ then

$|P_N^{(\alpha,\beta)}(\cos\theta)| \leqslant \frac{C}{\sqrt{N}}$ with a constant depending on φ, α, β only. So

$$S_N^{\delta}(f)(x) =$$

$$|S^{n-2}|(N+1)^{\frac{n-1}{2}-\delta}\int_{\frac{\pi}{2}}^{\pi} S_\theta(f)(x)\sin^{n-2}\theta P_N^{(\frac{n-1}{2}+\delta,\frac{n-3}{2})}(\cos\theta)\mathrm{d}\theta + o(1) \quad (2.5)$$

uniformly on E since $\delta \geqslant n-2$ (actually this holds for $\delta > \frac{n-2}{2}$). Now we apply the formula $P_N^{(\alpha,\beta)} = (-1)^N P_N^{(\alpha,\beta)}(-t)$ which can be found in [2]. Noticing $S_\theta(f)(x) = S_{\pi-\theta}(f)(-x)$ we can rewrite (2.5) as

$$S_N^{\delta}(f)(x) = (-1)^N |S^{n-2}| \cdot I_N + o(1) \quad (2.6)$$

where $I_N = \int_0^{\frac{\pi}{2}} (N+1)^{\frac{n-1}{2}-\delta} S_\theta(f)(x)\sin^{n-2}\theta P_N^{(\frac{n-3}{2},\frac{n-1}{2}+\delta)}(\cos\theta)\mathrm{d}\theta$. We break the integral $\int_0^{\frac{\pi}{2}}$ into two parts $\int_0^{N^{-1}}$ and $\int_{N^{-1}}^{\frac{\pi}{2}}$ and apply (2.4). then we get

$$I_N = O(N^{n-2-\delta})\int_0^{N^{-1}} |S_\theta(f)(-x)|\sin^{n-2}\theta\mathrm{d}\theta +$$

$$O(N^{\frac{n-2}{2}-\delta})\int_0^{\frac{\pi}{2}} |S_\theta(f)(-x)|\sin^{n-2}\theta\mathrm{d}\theta.$$

By the absolute continuity of Lebesgue integral we conclude $I_N = o(1)$ uniformly on E when $\delta \geqslant n-2$. Hence from (2.6) we derive (2.2) and establish the theorem.

From the proof of Theorem 2 we can find that when $\delta < n-2$ the localization of σ_N^{δ} will concern the property of the kernel $P_N^{(\frac{n-1}{2}+\delta,\frac{n-3}{2})} \cdot (\cos\theta)$ at $\theta = \pi$. This causes the discussion on "antipole conditions", which is the task of next two sections.

§ 3. The necessity of "antipole conditions" when $-1 < \delta < n-2$

We now give examples to show the if $-1 < \delta < n-2$ then for the convergence of $S_N^{\delta}(f)$ at a point $x \in S^{n-1}$, the property of f at the neighbourhood of the symmetric point $-x$ is not negligible.

We first deal with the case $\frac{n-2}{2} - 1 < \delta < n-2$. Fix $x \in S^{n-1}$ and define function $g=g_\delta$ by

$$g(y)=\begin{cases}0, & \text{if } xy\geqslant 0,\\ \{\arccos(-xy)\}^{-(1+\delta)}, & \text{if } xy<0.\end{cases}$$

Then we have

$$S_N^\delta(g)(x)$$

$$=(-1)^N|S^{n-2}|(N+1)^{\frac{n-1}{2}-\delta}\int_0^{\frac{\pi}{2}}S_\theta(g)(-x)\sin^{n-2}\theta P_N^{\left(\frac{n-3}{2},\frac{n-1}{2}+\delta\right)}(\cos\theta)\mathrm{d}\theta.$$

Now $S_\theta(g)(-x)=\theta^{-(1+\delta)}$, $\left(0<\theta<\frac{\pi}{2}\right)$. So by (2.3) we get

$$S_N^\delta(g)(x)=c_N^\delta(N+1)^{\frac{n-1}{2}-\delta}\int_0^{\frac{\pi}{2}}\theta^{-(1+\delta)-\frac{n-3}{2}}\sin^{-2}\theta M^\delta(\theta)J_{\frac{n-3}{2}}(\widetilde{N}\theta)\mathrm{d}\theta+r_N$$

where $\qquad c_N^\delta=(-1)^N|S^{n-2}|b_N\left(\frac{n-3}{2},\ \frac{n-1}{2}+\delta\right),$

$M^\delta(\theta)=\left(\dfrac{\theta}{2\sin\dfrac{\theta}{2}}\right)^{\frac{n-2}{2}}\left(\dfrac{1}{\cos\dfrac{\theta}{2}}\right)^{\frac{n}{2}+\delta}$, $\widetilde{N}=N+\dfrac{n-1+\delta}{2}$ and the remainder

$$r_N=\int_0^{\frac{1}{N}}O(N^{n-2-\delta}\theta^{n-1-\delta})\mathrm{d}\theta+\int_{\frac{1}{N}}^{\frac{\pi}{2}}O(N\theta)^{\frac{n-4}{2}-\delta}\theta\mathrm{d}\theta=o(1)$$

since $\frac{n-4}{2}-\delta<0$. Write $\varphi(\theta)=\theta^{-2}\left\{\left(\frac{\sin\theta}{\theta}\right)^{n-2}M^\delta(\theta)-1\right\}$. It is easy to verify $\varphi\in C^\infty(\mathbf{R})$. Then we get

$$S_N^\delta(g)(x)=$$

$$d_N^\delta\left\{\int_0^{\frac{\pi}{2}}(\widetilde{N}\theta)^{\frac{n-1}{2}-\delta}\theta^{-1}J_{\frac{n-3}{2}}(\widetilde{N}\theta)\mathrm{d}\theta+\int_0^{\frac{\pi}{2}}(\widetilde{N}\theta)^{\frac{n-1}{2}-\delta}\theta\varphi(\theta)J_{\frac{n-3}{2}}(\widetilde{N}\theta)\mathrm{d}\theta\right\}+o(1),$$

where $d_N^\delta=c_N^\delta\left(\frac{N+1}{\widetilde{N}}\right)^{\frac{n-1}{2}-\delta}$. We write the first and second integral in right hand side by I_N and J_N respectively. Write $v=\frac{n-3}{2}$, $\mu=n-2-\delta$. Then we have $J_N=\frac{1}{\widetilde{N}^2}\int_0^{\frac{\pi\widetilde{N}}{2}}t^{v+2-\delta}\varphi\left(\frac{t}{\widetilde{N}}\right)J_v(t)\mathrm{d}t$. Using the formula $(t^{v+1}J_{v+1}(t))'=t^{v+1}J_v(t)$ and integrating by parts we get $J_N=o(1)$. Hence we get $S_N^\theta(g)(x)=d_N^\delta I_N+o(1)$, where $I_N=\int_0^{\frac{\pi\widetilde{N}}{2}}\frac{J_v(t)}{t^{v-\mu+1}}\mathrm{d}t$. Therefore

$$\lim_{N\to+\infty}|S_N^\delta(g)(x)|=\lim_{N\to+\infty}|d_N^\delta|\int_0^{+\infty}\frac{J_v(t)}{t^{v-\mu+1}}\mathrm{d}t.\tag{3.1}$$

where $\lim\limits_{N\to+\infty}|d_N^\delta|=|S^{n-2}|\lim\limits_{N\to+\infty}\dfrac{\Gamma\left(N+\frac{n-1}{2}\right)}{N^{\frac{n-3}{2}}\Gamma(N+1)}>0$, and by Weber-Schafheitlin formula ([4], p. 391) $\displaystyle\int_0^{+\infty}\frac{J_v(t)}{t^{v-\mu+1}}\mathrm{d}t=\frac{\Gamma\left(\frac{\mu}{2}\right)}{2^{v-\mu+1}\Gamma\left(v-\frac{\mu}{2}+1\right)}>0$ when $0<\mu=n-2-\delta<v+\frac{3}{2}=\frac{n}{2}$.

From (3.1) we see that when $\delta\in\left(\frac{n-2}{2}-1,\ n-2\right)$ for function $g=g_\delta\in L(S^{n-1})$ the localization of $\sigma_N^\delta(g)$ at x fails.

If we define for a given $x\in S^{n-1}$ and $\delta\in(-1,\ n-2)$,

$$h(y)=h_\delta(y)=\begin{cases}0, & \text{when } xy\geqslant 0,\\ \{\arccos(-xy)\}^{-\frac{n+\delta}{2}}, & \text{when } xy<0.\end{cases}$$

then $h\in L(S^{n-1})$ and by a similar argument we can prove that

$$\lim_{N\to+\infty}|S_N^\delta(h)(x)|=+\infty.$$

This shows the localization of $\sigma_N^\delta(h)$ at x fails for $\delta\in(-1,\ n-2)$. Of course it covers the case $-1<\delta\leqslant\frac{n-2}{2}-1$.

We have seen that for the localization of $S_N^\delta(f)$ (hence of $\sigma_N^\delta(f)$) at a point $x\in S^{n-1}$ some conditions on the property of f at $-x$ are necessary.

We use the terminology "antipole conditions" which has appeared in [2](p. 239) to call this kind of localization condition.

§ 4. Antipole conditions

(when $\frac{n-2}{2}-1<\delta<n-2$)

For $f\in L(S^{n-1})$ we define $f_x(\theta)=\dfrac{1}{|S^{n-2}|}\displaystyle\int_{D(x;\theta)}f(y)\mathrm{d}y$ $(0<\theta\leqslant\pi)$ where $D(x;\ \theta)=\{y\in S^{n-1}\mid xy\geqslant\cos\theta\}$. Then $f_x(\theta)=\displaystyle\int_0^\theta S_t(f)(x)\sin^{n-2}t\mathrm{d}t$. Also we define $M(f)(x;\theta)=\displaystyle\int_0^\theta|S_t(f)(x)|t^{n-2}\mathrm{d}t$.

Then obviously $|f_x(\theta)|\leqslant M(f)(x;\ \theta)$.

For the function g defined in §4 we have

$$M(g)(-x;\theta)=\int_0^\theta t^{-(1+\delta)}\sin^{n-2}t\mathrm{d}t\geqslant A(n,\delta)\theta^{n-2-\delta}.$$

so we see that it is natural to require (for $\frac{n-2}{2}-1<\delta<n-2$)

$$M(f)(-x;\ \theta)=o(\theta^{n-2-\delta})\quad \text{as } \theta\to 0^+ \tag{4.1}$$

as a prerequisite for the localization of $\sigma_N^\delta(f)$ at x.

Lemma 4 Let $-1<\delta<n-2$, $x\in S^{n-1}$ and $f\in L(S^{n-1})$. If the condition (4.1) holds for x in $D\subset S^{n-1}$ uniformly then also the following relation

$$\lim_{N\to+\infty}N^{\frac{n-1}{2}-\delta}\int_{\pi-N^{-1}}^{\pi}S_\theta(f)(x)\sin^{n-2}\theta P_N^{(\frac{n-1}{2}+\delta,\frac{n-3}{2})}(\cos\theta)\mathrm{d}\theta=0. \tag{4.2}$$

Proof We write the integral in (4.2) as I_N. Then we have

$$I_N=(-1)^N\int_0^{N^{-1}}S_\theta(f)(-x)\sin^{n-2}\theta P_N^{(\frac{n-3}{2},\frac{n-1}{2}+\delta)}(\cos\theta)\mathrm{d}\theta.$$

Hence, accounting the estimate $P_N^{(\alpha,\beta)}(t)=O(N^\alpha)$, we get

$$|I_N|\leqslant c_nN^{\frac{n-3}{2}}M(f)(-x;\ N^{-1})=o(N^{-(n-2-\delta)+\frac{n-3}{2}})$$

by the condition (4.1). Then (4.2) follows.

We remark that if we replace (4.1) by a weaker condition

$$f_{-x}(\theta)=o(\theta^{n-2-\delta})$$

then (4.2) still holds. We write

$$V_N^\delta(f)(x)=(N+1)^{\frac{n-1}{2}-\delta}\int_{\frac{\pi}{2}}^{\pi-N^{-1}}S_\theta(f)(x)\sin^{n-2}\theta P_N^{(\frac{n-1}{2}+\delta,\frac{n-3}{2})}(\cos\theta)\mathrm{d}\theta.$$

Lemma 5 Let $f\in L(S^{n-1})$, D be a non-empty subset of S^{n-1}. If $\delta>\frac{n-2}{2}$ and (4.1) holds for $x\in D$ uniformly the $\lim\limits_{N\to+\infty}V_N^\delta(f)(x)=0$ for $x\in D$ uniformly.

Proof Write $d=\delta-\frac{n-2}{2}>0$. By (2.4) and $P_N^{(\alpha,\beta)}(-t)=(-1)^NP_N^{(\beta,\alpha)}(t)$ we get $|V_N^\delta(f)(x)|=O(N^{-d})\int_{N^{-1}}^{\frac{\pi}{2}}|S_\theta(f)(-x)|\theta^{n-2}\theta^{-\frac{n-2}{2}}\mathrm{d}\theta$.

Integrating by parts we have

$$\int_{N^{-1}}^{\frac{\pi}{2}}|S_\theta(f)(-x)|\theta^{n-2}\theta^{-\frac{n-2}{2}}\mathrm{d}\theta$$

$$= M(f)(-x;\theta)\theta^{-\frac{n-2}{2}}\Big|_{N^{-1}}^{\frac{\pi}{2}} + \frac{n-2}{2}\int_{N^{-1}}^{\frac{\pi}{2}} M(f)(-x;\theta)\theta^{-\frac{n}{2}}\mathrm{d}\theta$$

$$= O(1) + \frac{n-2}{2}\int_{N^{-1}}^{\frac{\pi}{2}} o(\theta^{\frac{n-2}{2}-1-\delta})\mathrm{d}\theta = o(N^d)$$

uniformly on D by condition (4.1). Hence $V_N^{\delta}(f)(x)=o(1)$ uniformly for $x\in D$.

Lemma 6 Let $f\in L(S^{n-1})$, D be a non-empty subset of S^{n-1} and $\delta=\frac{n-2}{2}$. If for $x\in D$.

$$\int_0^{\theta} |S_t(f)(-x)|t^{\frac{n-2}{2}}\mathrm{d}t = o(1) \quad \text{as } \theta\to 0^+ \tag{4.3}$$

uniformly, then $\lim\limits_{N\to\infty} V_N^{\frac{n-2}{2}}(f)(x)=0$ uniformly on D.

Proof By (2.4) and $P_N^{(\alpha,\beta)}(-t)=(-1)^N P_N^{(\beta,\alpha)}(t)$ we have

$$V_N^{\frac{n-2}{2}}(f)(x) = O(1)\int_{N^{-1}}^{\frac{\pi}{2}} S_\theta(f)(-x)\sin^{n-2}\theta M^{\frac{n-2}{2}}(\theta)\frac{\cos(\widetilde{N}\theta-\gamma_n)}{\theta^{\frac{n-2}{2}}}\mathrm{d}\theta + r_N$$

where $M^{\frac{n-2}{2}}(\theta)=\left(\frac{\theta}{2\sin\frac{\theta}{2}}\right)^{\frac{n}{2}}\left(\frac{1}{\cos\frac{\theta}{2}}\right)^{n-1}$, $\gamma_n=\frac{n-2}{4}\pi$, $\widetilde{N}=N+\frac{3n}{4}-1$

and $$r_N = O\Big(\frac{1}{N}\Big)\int_{N^{-1}}^{\frac{\pi}{2}} |S_\theta(f)(-x)|\theta^{\frac{n-2}{2}}\frac{\mathrm{d}\theta}{\theta}.$$

Write $\Phi_x(\theta)=\int_0^{\theta}|S_t(f)(-x)t^{\frac{n-2}{2}}\mathrm{d}t$. then condition (4.3) reads $\Phi_x(\theta)=o(1)$ uniformly on D. Integrating by parts we get

$$r_N = O\Big(\frac{1}{N}\Big)\left\{\frac{\Phi_x(\theta)}{\theta}\Big|_{\frac{1}{N}}^{\frac{\pi}{2}} + \int_{\frac{1}{N}}^{\frac{\pi}{2}}\frac{\Phi_x(\theta)}{\theta^2}\mathrm{d}\theta\right\}$$

$$= O\Big(\frac{1}{N}\Big)\left\{1+o\Big(\int_{\frac{1}{N}}^{\frac{\pi}{2}}\frac{\mathrm{d}\theta}{\theta^2}\Big)\right\} = o(1)$$

uniformly on D.

For any $\varepsilon>0$, by condition (4.3) there exists a number $\eta\in\left(0,\frac{\pi}{2}\right)$ such that $\int_0^{\eta}|S_t(f)(-x)|t^{\frac{n-2}{2}}\mathrm{d}t<\varepsilon$ uniformly on D. Then we get

$$|V_N^{\frac{n-2}{2}}(f)(x)|\leqslant A_n\{\varepsilon+|I_N(f)(x)|\}+o(1)$$

uniformly on D, where

$$I_N(f)(x) = \int_{\eta}^{\frac{\pi}{2}} S_\theta(f)(-x)\sin^{n-2}\theta M^{\frac{n-2}{2}}(\theta)\frac{\cos(\widetilde{N}\theta-\gamma_n)}{\theta^{\frac{n-2}{2}}}\mathrm{d}\theta.$$

We choose a function $g\in C^{\infty}(S^{n-1})$ such that $\| f-g \|_1<\varepsilon\eta^{\frac{n-2}{2}}$. Then we have $|I_N(f)(x)|\leqslant |I_N(f-g)(x)|+|I_N(g)(x)|$. It is easy to see $|I_N(f-g)(x)|\leqslant\frac{A_n}{\eta^{\frac{n-2}{2}}}\times \| f-g \|_1<A_n\varepsilon$. Since g is smooth we can easily verify that $I_N(g)(x)=o(1)$ uniformly on S^{n-1}. Hence $|V_N^{\frac{n-2}{2}}(f)(x)|\leqslant A_n\varepsilon+o(1)$ on D. This establishes the lemma.

A combination of Lemma 4～6 yields the following two localization theorems.

Theorem 3 Suppose $\frac{n-2}{2}<\delta<n-2$, $f\in L(S^{n-1})$, $D\subset S^{n-1}$ (D nonempty). If the antipole condition (4.1) is satisfied on D uniformly, then

$$S_N^{\delta}(f)(x)=(N+1)^{\frac{n-1}{2}-\delta}\int_0^{\frac{\pi}{2}}S_\theta(f)(x)\sin^{n-2}\theta P_N^{\left(\frac{n-1}{2}+\delta,\frac{n-3}{2}\right)}(\cos\theta)\mathrm{d}\theta+o(1)$$

uniformly on D.

Theorem 4 Let $\delta=\frac{n-2}{2}$, $f\in L(S^{n-1})$, $D\subset S^{n-1}$. If the antipole condition (4.3) is satisfied on D uniformly then

$$S_N^{\frac{n-2}{2}}(f)(x)=\sqrt{N+1}\int_0^{\frac{\pi}{2}}S_\theta(f)(x)\sin^{n-2}\theta P_N^{\left(n-\frac{3}{2},\frac{n-3}{2}\right)}(\cos\theta)\mathrm{d}\theta+o(1)$$

uniformly on D.

It is easy to verify that the following condition at $x\in S^{n-1}$,

$$S_\theta(f)(-x)\theta^{\delta}\in L\left(0,\frac{\pi}{2}\right) \tag{4.4}$$

implies (4.1) when $\delta<n-2$.

Finally, we give an improvement of a result of Bonami and Clerc-Theorem 3.2 of [6] which reads:

If $\varepsilon>0$, $(\varepsilon<\pi)$, $\delta\geqslant\frac{n-2}{2}$, $f\in L^p(S^{n-1})$, $p\geqslant 1$ and $p>\frac{n-1}{\delta+1}$ and f vanishes on $D(x,\varepsilon)=\{y\in S^{n-1}\mid xy>\cos\varepsilon\}$, then $\lim\limits_{N\to+\infty}\sigma_N^{\delta}(f)(x)=0$.

We point out that this result coincides with Theorem 2 when $\delta\geqslant n-2$, but is implied by the following assertion when $\frac{n-2}{2}\leqslant\delta<n-2$.

Corollary of Theorem 3 and 4 Suppose $\frac{n-2}{2}\leqslant\delta<n-2$, $f\in$

$L(S^{n-1})$, $x \in S^{n-1}$. If the antipole condition (4.4) is satisfied and f vanishes on $D(x;\ \varepsilon)\left(\varepsilon \in \left(0,\ \frac{\pi}{2}\right)\right)$ then $\lim\limits_{N \to +\infty} \sigma_N^\delta(f)(x)=0$.

We verify that if $f \in L^p(S^{n-1})$, $p>\frac{n-1}{\delta+1}$ then for any $x \in L(S^{n-1})$, $S_\theta(f)(x)\theta^\delta \in L(0,\ \pi)$. We have

$$\int_0^\pi |S_\theta(f)(x)|\theta^\delta \mathrm{d}\theta \leqslant \left\{\int_0^\pi |S_\theta(f)(x)|^p \theta^{n-2} \mathrm{d}\theta\right\}^{\frac{1}{p}} \left\{\int_0^\pi \theta^{\left(\delta-\frac{n-2}{p}\right)p'} \mathrm{d}\theta\right\}^{\frac{1}{p'}}$$

by Hölder inequality with $p'=\frac{p}{p-1}$. Since $\left(\delta-\frac{n-2}{p}\right)p' > \left(\frac{n-1}{p}-1-\frac{n-2}{p}\right)p'=-1$ and $|S_\theta(f)(x)|^p \leqslant S_\theta(|f|^p)(x)$ we get

$$\int_0^\pi |S_\theta(f)(x)|\delta^\delta \mathrm{d}\theta \leqslant c\int_0^\pi \{S_\theta(|f|^p)(x)\theta^{n-2}\mathrm{d}\theta\}^{\frac{1}{p}} \leqslant c\,\|f\|_p <+\infty,$$

when $f \in L^p(S^{n-1})$.

References

[1] Müller C. Spherical harmonics. In: Lecture Notes in Mathematics [s. 1.]: [s. n.], 1966. 17.

[2] Szegö G. Orthogonal polynomials Vol 23. Colloquium: AMS Colloquium Publication, 1939.

[3] Berens H, Butzer P L, Pawelke S. Limitierungsverfahren von reihen mehrdimensionaler kugelfunktionen und deren saturationsverhalten, Vol 4, Kyoto: Publ Rims Kyoto Univ Ser A, 1968. 201—268.

[4] Watson G N. Theory of Bessel functions. 2nd ed. Cambridge: Cambridge University Press, 1952.

[5] Zygmund A. Trigonometric series, Vol 1, 2. Cambridge: Cambeidge University Press, 1959.

[6] Bonami A, Clerc J L. Sommes de Cesàro et multiplicateues des développements et harmonique sphériques. Trans AMS, 1973, 183: 223.

[7] Colzani L, Taibleson M H, Weiss G. Maximal estimates for Cesàro and Riesz means on spheres. Indiana University Mathematics Journal, 1984, 33 (6): 873.

[8] Meaney C. Localization of spherical harmonic expansions. Montsh Math, 1984, 98: 65.

北京师范大学学报(自然科学版)，1993，29(2)：158－164.

球面上 Cesàro 平均的点态收敛

Pointwise Convergence of Cesàro Means on Sphere①

Abstract The convergence problem of the Cesàro means of critical order of Fourier-Laplace series on sphere is investigated. The convergence tests of Dini type，Dini-Lipschitz type，Lebesgue type and Salem type are established.

Keywords spherical harmonics；Cesàro means；pointwise convergence.

§ 1. Preliminary

This paper is a continuation of [1]. We keep all notations used there. By σ_N^δ we denote Cesàro means of order δ of Fourier-Laplace series on sphere $S^{n-1}=\{(x_1, x_2, \cdots, x_n) \mid x_1^2+x_2^2+\cdots+x_n^2=1\}$，$n\geqslant 3$. We know the order δ has a critical value $\frac{n-2}{2}$. Because the Cesàro means of critical order，$\sigma_N^{\frac{n-2}{2}}$，plays a role as like as the partial sum of single Fourier series in the classical theory，the investigation on it has obvious significance. For simplicity we will delete the index to write σ_N instead of $\sigma_N^{\frac{n-2}{2}}$，S_N instead of $S_N^{\frac{n-2}{2}}$ and so on. By definitions (see [1])

---

① Supported by NSFC.
Received：1992-07-16.

we have for $f \in L(\mathrm{S}^{n-1})$

$$\sigma_N(f) = f * K_N, \quad K_N = \frac{1}{A_N^{\frac{n-2}{2}}} \sum_{k=0}^{N} A_{N-k}^{\frac{n-2}{2}} c_{n,k} P_k^n \tag{1.1}$$

and

$$S_N(f) = \sqrt{N+1} f * P_N^{\left(n-\frac{3}{2}, \frac{n-3}{2}\right)}, \tag{1.2}$$

where $A_N^{\alpha} = \dfrac{(\alpha+1)(\alpha+2)\cdots(\alpha+N)}{N!}$, P_k^n denote Gegenbeuer polynomials generated by

$$(1-2rt+t^2)^{-\frac{n-2}{2}} = \sum_{k=0}^{+\infty} \frac{\Gamma(n-2+k)}{k!(n+1)!} P_k^n(t) r^k, \quad 0 \leqslant r < 1,$$

and $P_N^{(\alpha,\beta)}$ denote Jacobi polynomials.

Our main purpose is to search conditions under which the Cesàro means $\sigma_N(f)$ of function f converge at a point.

Our starting point is Theorem 1 and Theorem 4 of [1]. By Theorem 1 of [1], for $f \in L(\mathrm{S}^{n-1})$ and $D \subset \mathrm{S}^{n-1}$, if f is bounded on D then for $\lim\limits_{N\to+\infty} \sigma_N(f)(x) = f(x)$ uniformly on D the necessary and sufficient condition is that $\lim\limits_{N\to+\infty} [S_N(f)(x) - f(x) S_N(1)(x)] = 0$ uniformly on D, where $1(x) \equiv 1$. By Theorem 4 of [1], for $f \in L(\mathrm{S}^{n-1})$ and $D \subset \mathrm{S}^{n-1}$, if the antipole condition

$$\int_0^{\theta} |S_t(f)(-x)| t^{\frac{n-2}{2}} \mathrm{d}t = o(1) \quad \text{as} \quad \theta \to 0^+ \tag{1.3}$$

is uniformly satisfied on D then

$$S_N(f)(x) = \int_0^{\frac{\pi}{2}} S_\theta(f)(x) \sin^{n-2}\theta u_N(\theta) \mathrm{d}\theta + o(1) \tag{1.4}$$

uniformly on D, where

$$S_\theta(f)(x) = \frac{1}{l_\theta} \int_{l_{x,\theta}} f(y) \mathrm{d}l_{x,\theta}(y), \quad \theta \in (0, \pi],$$

with $I_{x,\theta} = \{ y \in \mathrm{S}^{n-1} \mid xy = \cos\theta \}$, $l_\theta = |l_{x,\theta}| = |\mathrm{S}^{n-2}| \sin^{n-2}\theta = \left(\dfrac{2\pi^{\frac{n-1}{2}}}{\Gamma\left(\frac{n-1}{2}\right)}\right) \sin^{n-2}\theta$, and $u_N(\theta) = \sqrt{N+1} |\mathrm{S}^{n-2}| P_N^{\left(n-\frac{3}{2}, \frac{n-3}{2}\right)}(\cos\theta)$. We define operator U_N on $L(\mathrm{S}^{n-1})$ by

$$U_N(f)(x) = \int_0^{\frac{\pi}{2}} S_\theta(f)(x) \sin^{n-2}\theta u_N(\theta) \mathrm{d}\theta, \quad f \in L(\mathrm{S}^{n-1}). \tag{1.5}$$

Then we see that for the constant function 1 the condition (1.3) is uniformly satisfied on S^{n-1}. Hence by (1.4) $S_N(1)(x)=U_N(1)(x)+o(1)$ is uniformly on S^{n-1}. So we get

Proposition 1 Let $f\in L(S^{n-1})$, $D\subset S^{n-1}$. If f is bounded on D and the antipole condition (1.3) is uniformly satisfied on D then for

$$\lim_{N\to+\infty}\sigma_N(f)(x)=f(x) \text{ uniformly on } D \tag{1.6}$$

the necessary and sufficient condition is

$$\lim_{N\to+\infty}[U_N(f)(x)-f(x)U_N(1)(x)]=0 \quad \text{uniformly on } D. \tag{1.7}$$

Now assume $f\in L(S^{n-1})$ and $x\in S^{n-1}$ is a Lebesgue point of f. Then

$$\int_0^\theta |S_t(f)(x)-f(x)|t^{n-2}\mathrm{d}t=o(\theta^{n-1}) \text{ as } \theta\to 0^+. \tag{1.8}$$

Applying asymptotic formula for Jacobi polynomial (see (2.4) for [1] for details) we have

$$u_N(\theta)=\begin{cases} c_N M(\theta)\theta^{1-n}\cos(\tilde{N}\theta-\gamma_n)+O\left(\dfrac{1}{N\theta^n}\right), & \text{if } N^{-1}\leqslant\theta\leqslant\dfrac{\pi}{2}, \\ O(N^{n-1}), & \text{if } 0<\theta\leqslant N^{-1}, \end{cases} \tag{1.9}$$

where

$$c_N=\sqrt{\frac{N+1}{\pi}}\left(\frac{2}{\tilde{N}}\right)^{n-1}\frac{\Gamma\left(N+n-\frac{1}{2}\right)}{\Gamma(N+1)},\quad M(\theta)=\left[\frac{\theta}{2\sin\frac{\theta}{2}}\right]^{n-1}\left[\frac{1}{\cos\frac{\theta}{2}}\right]^{\frac{n-2}{2}},$$

$\tilde{N}=N+\dfrac{3n}{4}-1$ and $\gamma_n=\dfrac{(n-1)\pi}{2}$. So if condition (1.8) holds uniformly on D then $U_N(f)(x)-f(x)U_N(1)(x)=\int_{N^{-1}}^{\frac{\pi}{2}}(S_\theta(f)(x)-f(x))\sin^{n-2}\theta u_N(\theta)\mathrm{d}\theta+o(1)$ uniformly on D. Noticing $\sin^{n-2}\theta u_N(\theta)=\left(\dfrac{c_N}{\theta}\right)(1+O(\theta^2))$ and defining V_N by

$$V_N(f)(x)=\int_{N^{-1}}^{\frac{\pi}{2}}S_\theta(f)(x)\theta^{-1}\cos(N\theta-\gamma_n)\mathrm{d}\theta,$$

we derive from (1.7) the following

Proposition 2 Let $f\in L(S^{n-1})$, $D\subset S^{n-1}$. Assume f is bounded on D. If condition (1.3) and (1.8) are all uniformly satisfied on D then for (1.6) the necessary and sufficient condition is

$$\lim_{N\to+\infty}[V_N(f)(x)-f(x)V_N(\mathbf{1})(x)]=0 \quad \text{uniformly on } D. \tag{1.10}$$

§ 2. Tests for convergence of $\sigma_N(f)$

We first give a test of Dini type (Compare with Theorem (6.1) of [2], p. 52).

Theorem 1 Let $f\in L(S^{n-1})$, $x\in S^{n-1}$. Assume condition (1.3) is satisfied. If

$$\theta^{-1}\{S_\theta(f)(x)-f(x)\}\in L\left(0,\ \frac{\pi}{2}\right), \tag{2.1}$$

then (1.6) holds for $D=\{x\}$.

Proof Define

$$\varphi_x(\theta)=S_\theta(f)(x)-f(x). \tag{2.2}$$

By (2.1) we have $\int_0^l |\varphi_x(\theta)|\theta^{n-2}\mathrm{d}\theta\leqslant t^{n-1}\int_0^l \theta^{-1}|\varphi_x(\theta)|\mathrm{d}\theta=o(t^{n-1})$. So we see the condition (1.8) holds and hence the conditions of Proposition 2 are all satisfied at x. Then by Riemann-Lebesgue theorem and (2.1) we see (1.10) holds at x. Hence by Proposition 2 (1.6) holds at x.

Remark If condition (1.3) and instead of (2.1) the relation

$$\int_0^\theta t^{-1}|S_t(f)(x)-f(x)|\mathrm{d}t=o(1) \tag{2.3}$$

are all uniformly satisfied on $D\subset S^{n-1}$ then the conclusion is that (1.6) holds uniformly on D.

Next two tests are of Dini-Lipschitz type (compare with the results in [2] p. 63).

Theorem 2 Let $f\in L(S^{n-1})$, $x\in S^{n-1}$. Assume (1.3) holds. If $\varphi_x(\theta)$, defined by (2.2), is continuous on $[0,\ \varepsilon]$ for some $\varepsilon\in\left(0,\ \frac{\pi}{2}\right)$ and its modulus of continuity $\omega(u)=\sup\{|\varphi_x(s)-\varphi_x(t)|\mid s,\ t\in[0,\ \varepsilon],\ |s-t|\leqslant u\}(u>0)$ satisfies the condition

$$\omega(u)=o\left(\frac{1}{\ln u}\right)\text{ as }u\to 0^+, \tag{2.4}$$

then (1.6) holds.

Proof Write $I_N=\int_{N^{-1}}^{\varepsilon}\varphi_x(\theta)\theta^{-1}\cos(N\theta-\gamma_n)\mathrm{d}\theta$. By Proposition 2 and by Riemann-Lebesgue theorem we only need to prove $I_N \to 0$. We have

$$I_N = -\int_{\frac{1-N}{N}}^{\varepsilon-\frac{\pi}{N}} \varphi_x\left(\theta+\frac{\pi}{N}\right)\left(\theta+\frac{\pi}{N}\right)^{-1}\cos(N\theta-\gamma_n)\mathrm{d}\theta = r_N + s_N + J_N + L_N$$

where

$$r_N = \frac{1}{2}\int_{\varepsilon-\frac{\pi}{N}}^{\varepsilon} \varphi_x(\theta)\theta^{-1}\cos(N\theta-\gamma_n)\mathrm{d}\theta = O\left(\frac{1}{N}\right),$$

$$s_N = \frac{1}{2}\int_{\frac{1}{N}}^{\frac{1+\pi}{N}} \varphi_x(\theta)\theta^{-1}\cos(N\theta-\gamma_n)\mathrm{d}\theta = O\left(\omega\left(\frac{1+\pi}{N}\right)\right),$$

and by condition (2. 4)

$$J_N = \frac{1}{2}\int_{\frac{1}{N}}^{\varepsilon-\frac{\pi}{N}} \left(\varphi_x(\theta)-\varphi_x\left(\theta+\frac{\pi}{N}\right)\right)\theta^{-1}\cos(N\theta-\gamma_n)\mathrm{d}\theta$$

$$= \int_{\frac{1}{N}}^{\varepsilon} \frac{O\left(\omega\left(\frac{1}{N}\right)\right)}{\theta}\mathrm{d}\theta = o(1),$$

$$L_N = \frac{1}{2}\int_{\frac{1}{N}}^{\varepsilon-\frac{\pi}{N}} \varphi_x\left(\theta+\frac{\pi}{N}\right)\left(\frac{1}{\theta}-\frac{1}{\theta+\frac{\pi}{N}}\right)\cos(N\theta-\gamma_n)\mathrm{d}\theta$$

$$= \int_{\frac{1}{N}}^{\varepsilon} \frac{O(\omega(\theta))}{N\theta^2}\mathrm{d}\theta = o(1).$$

So we get $I_N = o(1)$ and establish the theorem.

Denote by $D(p;\ h)$ ($p\in S^{n-1}$, $h\in(0,\ \pi]$) the spherical crown with center P and spherical radius h, i. e. $D(p;\ h)=\{y\in S^{n-1}\mid yp\geqslant \cos h\}$. Define for $0<t\leqslant h$, $f\in L(S^{n-1})$,

$$\omega(f;\ t)_{C(D(p;t))} = \sup\{|f(x)-f(y)|\mid x,\ y\in D(p;\ h),\ xy\geqslant \cos t\}.$$

Obviously, the condition $\omega(f;\ t)_{C(D(x;t))} = o\left(\frac{1}{|\ln t|}\right)(t\to 0^+)$ implies (2. 4) and hence (1. 6) when (1. 3) is assumed.

Theorem 3 Let $f\in L(S^{n-1})$, $p\in S^{n-1}$, $0<h\leqslant\pi$. Assume f is bounded on $D(p;\ h)$ and the antipole condition (1. 3) holds uniformly on $D(p;\ h)$. If

$$\omega(f;\ t)_{C(D(p;h))} = o\left(\frac{1}{|\ln t|}\right) \quad \text{as } t\to 0^+, \tag{2. 5}$$

then for any $r\in(0,\ h)$ (1. 6) holds uniformly on $D(p;\ r)$.

Proof Since f is continuous on $D(p;\ h)$, noticing $0<r<h$, we

conclude that (1.8) holds uniformly on $D(p; r)$. Then we see that the conditions of Proposition 2 are all satisfied on $D(p; r)$. So by this proposition we know that we only need to verify (1.10) for $D=D(p; r)$. Noticing $h-r=\varepsilon>0$ we see that an argument similar to the proof of theorem 2 leads to (1.10) and completes the proof.

The next test is of Lebesgue type (compare with [2], Chapter Ⅱ § 11).

Theorem 4 Let $f\in L(S^{n-1})$, $D\subset S^{n-1}$. Assume the conditions of proposition 2 are satisfied. If

$$\int_{\frac{\pi}{N}}^{\pi}\theta^{-1}\left|S_\theta(f)(x)-S_{\theta+\frac{\pi}{N}}(f)(x)\right|\mathrm{d}\theta$$
$$=o(1)\,(N\to+\infty)\text{ uniformly on } D,\tag{2.6}$$

then (1.6) also.

Proof Let us verify (1.10). we have (as like in the proof of Theorem 2) $V_n(f)(x)-f(x)V_N(1)(x)=\int_{N^{-1}}^{\frac{\pi}{2}}\varphi_x(\theta)\theta^{-1}\cos(N\theta-\gamma_n)\mathrm{d}\theta$

$$=r_N+s_N+J_N+L_N,$$

where

$$r_N=\frac{1}{2}\int_{\frac{\pi}{2}-\frac{\pi}{N}}^{\frac{\pi}{2}}\varphi_x(\theta)\theta^{-1}\cos(N\theta-\gamma_n)\mathrm{d}\theta,$$

$$s_N=\frac{1}{2}\int_{\frac{1}{N}}^{\frac{1+\pi}{N}}\varphi_x(\theta)\theta^{-1}\cos(N\theta-\gamma_n)\mathrm{d}\theta,$$

$$J_N=\frac{1}{2}\int_{\frac{1}{N}}^{\frac{\pi}{2}-\frac{\pi}{N}}\left(\varphi_x(\theta)-\varphi_x\left(\theta+\frac{\pi}{N}\right)\right)\theta^{-1}\cos(N\theta-\gamma_n)\mathrm{d}\theta,$$

$$L_N=\frac{1}{2}\int_{\frac{1}{N}}^{\frac{\pi}{2}-\frac{\pi}{N}}\varphi_x\left(\theta+\frac{\pi}{N}\right)\left(\frac{1}{\theta}-\frac{1}{\theta+\frac{\pi}{N}}\right)\cos(N\theta-\gamma_n)\mathrm{d}\theta.$$

Since $\varphi_x(\theta)\in L\left(\frac{\pi}{4}, \frac{\pi}{2}\right)$ we see by the absolute continuity of Lebesgue integral that $|r_N|=O\left(\int_{\frac{\pi}{2}-\frac{\pi}{N}}^{\frac{\pi}{2}}|\varphi_x(\theta)|\,\mathrm{d}\theta\right)=o(1)$ as $N\to+\infty$. By condition (1.8) $|s_N|\leqslant\int_{\frac{1}{N}}^{\frac{5}{N}}|\varphi_x(\theta)|\theta^{-1}\mathrm{d}\theta=\int_{\frac{1}{N}}^{\frac{5}{N}}|\varphi_x(\theta)|\theta^{n-2}\theta^{1-n}\mathrm{d}\theta$

$$=o(1)+\int_{\frac{1}{N}}^{\frac{5}{N}}o(\theta^{-1})\mathrm{d}\theta=o(1).$$

Also by condition (1.8)

$$L_N \leqslant \frac{\pi}{N}\int_{\frac{1}{N}}^{\frac{\pi}{2}} |\varphi_x(\theta)|\theta^{-2}\mathrm{d}\theta = \frac{\pi}{N}\int_{\frac{1}{N}}^{\frac{\pi}{2}} |\varphi_x(\theta)|\theta^{n-2}\theta^{-n}\mathrm{d}\theta$$

$$= o(1) + \frac{1}{N}\int_{\frac{1}{N}}^{\frac{\pi}{2}} \frac{o(\theta^{n-1})}{\theta^{n+1}}\mathrm{d}\theta = o(1).$$

Finally,

$$|J_N| \leqslant \int_{\frac{1}{N}}^{\frac{\pi}{2}} \left|\varphi_x(\theta) - \varphi_x\left(\theta + \frac{\pi}{N}\right)\right|\theta^{-1}\mathrm{d}\theta$$

$$= \int_{\frac{1}{N}}^{\frac{\pi}{2}} |S_\theta(f)(x) - S_{\theta+\frac{\pi}{N}}(f)(x)|\theta^{-1}\mathrm{d}\theta = o(1)$$

by condition (2.6). Hence we establish the theorem.

§ 3. A test of Salem type and almost everywhere convergence

Classical Salem theorem (see [5], p. 379) says: for $f \in L_{2\pi}$, if

$$\int_0^h \{f(x+t) - f(x-t)\}\mathrm{d}t = o\left(\frac{h}{|\ln h|}\right)(h \to 0^+) \text{ uniformly} \tag{3.1}$$

then Fourier sums of f convergence to f almost everywhere. Now we extend this theorem to sphere. Let $f \in L(S^{n-1})$. Define

$$\psi_x(\theta) = \int_{l_{x,\theta}} \{f(y) - f(x)\}\mathrm{d}l_{x,\theta}(y)(0 < \theta \leqslant \pi), \text{ i. e.}$$

$$\psi_x(\theta) = |S^{n-2}|\sin^{n-2}\theta\,\varphi_x(\theta).$$

For convenience we call a point $x \in S^{n-1}$ double-Lebesgue point of f if both x and $-x$ are all Lebesgue points of f.

Theorem 5 Let $f \in L(S^{n-1})$, $x \in S^{n-1}$. Assume x is a double-Lebesgue point of f. If as $h \to 0^+$,

$$\frac{1}{u^{n-2}}\int_0^h \{\psi_x(u+t) - \psi_x(u-t)\}\mathrm{d}t$$

$$= o\left(\frac{h}{|\ln h|}\right) \text{ uniformly for } u \in \left[2h, \frac{\pi}{2}\right], \tag{3.2}$$

then (1.6) holds at x. Hence if (3.2) is satisfied at every double Lebesgue point of f then $\sigma_N(f)$ tends to f almost everywhere.

Before the proof we make an observation on the conditions (3.1) and (3.2). If $n=2$ the $\psi_x(t) = f(x+t) + f(x-t) - 2f(x)$. Hence

(3.2) becomes for $n=2$,

$$\int_0^h\{f(x+u+t)-f(x+u-t)\}\mathrm{d}t-\int_0^h\{f(x-u+t)-f(x-u-t)\}\mathrm{d}t$$
$$=o\left(\frac{h}{|\ln h|}\right). \tag{3.2'}$$

But (3.1) implies (3.2′) obviously. So we can say that Theorem 5 is an improvement of Salem's result.

For multiple Fourier series there is an extension of Salem theorem which was obtained by Lu[3] and can be found in [4](p.155).

We next prove the Theorem 5 for $n\geqslant 3$.

Proof of Theorem 5 Since x is double-Lebesgue point the conditions (1.3) and (1.8) are all satisfied. Hence by proposition 2 we only need to prove (1.10). By condition (1.8) we know $\varphi_x(\theta)\in L\left(0,\ \frac{\pi}{2}\right)$. Hence by this condition, by Riemann-Lebesgue theorem on Fourier coefficients, and by the absolute continuity of Lebesgue integral we get

$$V_N(f)(x)-f(x)V_N(1)(x)=\int_{\frac{x+\gamma_n}{N}}^{\frac{mx+\gamma_n}{N}}\psi_x(\theta)\theta^{1-n}\cos(N\theta-\gamma_n)\mathrm{d}\theta+o(1),$$

where $m=\left[\frac{N}{4}\right](N>4)$. We write the integral on the right hand side as I_N and write $u_k=N^{-1}\left\{\left(k+\frac{1}{2}\right)\pi+\gamma_n\right\}$ for simplicity. By an elementary calculation we get

$$I_N=\sum_{k=1}^{m-1}(-1)^{k+1}\int_0^{\frac{\pi}{2N}}\{\psi_x(u_k+t)(u_k+t)^{1-n}-\psi_x(u_k-t)(u_k-t)^{1-n}\}\sin Nt\mathrm{d}t.$$

Write $\alpha_k=\int_0^{\frac{\pi}{2N}}\psi_x(u_k+t)\{(u_k+t)^{1-n}-(u_k-t)^{1-n}\}\sin Nt\mathrm{d}t$,

$$\beta_k=\int_0^{\frac{\pi}{2N}}\{\psi_x(u_k+t)-\psi_x(u_k-t)\}\{(u_k-t)^{1-n}-u_k^{1-n}\}\sin Nt\mathrm{d}t,$$

$$\theta_k=\int_0^{\frac{\pi}{2N}}u_k^{1-n}\{\psi_x(u_k+t)-\psi_x(u_k-t)\}\sin Nt\mathrm{d}t,$$

Then $I_N=\sum_{k=1}^{m-1}(-1)^{k+1}(\alpha_k+\beta_k+\theta_k)$. We have

$$|\alpha_k|\leqslant c_n\frac{N^{n-1}}{k^n}\int_0^{\frac{\pi}{2N}}|\psi_x(u_x+k)|\mathrm{d}t\leqslant c_nN^{n-1}k^{-n}\int_{u_k}^{u_{k+1}}|\psi_x(t)|\mathrm{d}t.$$

Hence $\sum_{k=1}^{m-1}|\alpha_k|\leqslant c_n N^{n-1}\left\{\sum_{k=1}^{m-2}\frac{1}{k^{n+1}}\int_0^{u_{k+1}}|\psi_x(t)|\,\mathrm{d}t+\frac{1}{m^n}\int_0^{u_m}|\psi_x(t)|\,\mathrm{d}t\right\}.$

Then by (1.8) we get

$$\sum_{k=1}^{m-1}|\alpha_k|\leqslant N^{n-1}\left\{\sum_{k=1}^{m-2}o\left(\frac{1}{k^2N^{n-1}}\right)+O\left(\frac{1}{N^n}\right)\right\}=o(1)\quad(\text{as } N\to+\infty).$$

By the same reason, $\sum_{k=1}^{m-1}|\beta_k|=o(1)$. Finally, integrating by parts we get $\theta_k=u_k^{1-n}\int_0^{\frac{\pi}{2N}}\left\{\int_0^{\theta}(\psi_x(u_k+t)-\psi_x(u_k-t))\mathrm{d}t\cdot N\cos N\theta\right\}\mathrm{d}\theta.$

Hence by condition (3.2) $\theta_k=Nu_k^{1-n}\int_0^{\frac{\pi}{2N}}o\left(\frac{u_k^{n-2}\theta}{|\ln\theta|}\right)\mathrm{d}\theta=o\left(\frac{1}{k\ln N}\right).$

Therefore $\sum_{k=1}^{m-1}|\theta_k|=o\left(\frac{1}{\ln N}\right)\sum_{k=1}^{m-1}\frac{1}{k}=o(1)$. Combining there estimates we get $I_N=o(1)(\text{as } N\to+\infty)$, and complete the proof. □

References

[1] Wang kunyang. Equiconvergent operator of Cesàro means on sphere and its applications. J BJ Nor Univ (Nat Sci), 1993, 29(2): 143.

[2] Zygmund A. Trigonometric Series, Ⅰ～Ⅱ. Cambridge: Cambridge University Press, 1959.

[3] Lu Shanzhen. Spherical integral and convergence of Riesz means. Acta Mathematica, 1980, 23(4): 609.

[4] Lu Shanzhen, Wang Kunyang. Bochner-Riesz means. Beijing: Publishing House of Beijing Normal University, 1988.

[5] Bari N K. Trigonometric series (Russian). Moscow: National Press of Physical and Mathematicl Literatures, 1961.

摘要 研究球面上 Fourier-Laplace 级数的临界阶 Cesàro 平均的收敛问题. 建立了 Dini 型、Dini-Lipschta 型、Lebesgue 型及 Salem 型的收敛判别法.

关键词 球调和；Cesàro 平均；收敛.

数学进展，1995，24(2)：184—186.

球面上的 Jackson 型逼近定理

Approximation Theorems of Jackson Type on the Sphere①

Let $n \geqslant 3$, $n \in \mathbf{N}$. Denote by S^{n-1} the unit sphere of $\mathbf{R}^n$. We consider the function spaces $L^p(S^{n-1})$, $1 \leqslant p < +\infty$ and $C(S^{n-1})$ which are generally expressed by X. A function $f \in X$ has a Fourier-Laplace series expansion

$$f \sim \sum_{k=0}^{+\infty} Y_k(f), \tag{1}$$

where $Y_k(f)$ denotes the projection of f onto the space $\mathcal{H}_k^n$ of spherical harmonics of degree k. We denote by D the Bertrami-Laplace operator which has eigenvalues $-k(k+2\lambda)$ corresponding to the eigenspaces $\mathcal{H}_k^n$ with $\lambda = \frac{n-2}{2}$ and $k \in \mathbf{N}^*$. We refer to [1] and [2] for all basic concepts. The generalized translation of f with spherical distance t, $S_t(f)$, is defined by

$$S_t(f)(x) := \frac{1}{|S^{n-2}|} \int_{S^{n-2,x}} f(x\cos t + y\sin t) \mathrm{d}S^{n-2,x}(y),$$

where $S^{n-2,x} = \{y \in S^{n-1} \mid xy = 0\}$, $\mathrm{d}S^{n-2,x}(y)$ is the measure element on

---

① This research is partially supported by NSERC Canada under grant No. A7687. The second author thanks for the support of the University of Alberta and also for the support of NSFC No. 19471007.

本文与 Riemenschneider S 合作.

Received：1994-12-22.

$S^{n-2,x}$ and $|S^{n-2}|=\dfrac{2\pi\dfrac{n-1}{2}}{\Gamma\left(\dfrac{n-1}{2}\right)}$ is the measure of S^{n-2}. The concept of generalized translation can be found in the 1969 paper of Löfström and Peetre[3]. A finite difference operator Δ_t^r and a corresponding modulus of continuity based on translation operator S_t are defined for $r>0$, $t>0$ by

$$\Delta_t^r=\sum_{k=0}^{+\infty}(-1)^k C_{\frac{r}{2}}^k (S_t)^k,$$

$$w_r(f;\ t)=\sup\{\|\Delta_s^r(f)\|\mid 0<s<t\}.$$

Here and in what follows the norm is always understood in the space X and is written without subscript since no confusion can occur. For these definitions and relative references see Rustamov's paper[4].

We define $H_k=\oplus\sum_{i=0}^{k}\mathcal{H}_i^n$ to be the space of all spherical polynomials of degree k. The best approximation fo degree k of f from this space is defined by

$$E_k(f)=\inf\{\|f-P\|\mid P\in H_k\}.$$

The concept of derivatives on the sphere based on the Beltrami-Laplace operator is also needed: For $f\in X$ having expansion (1) and $r>0$, if there exists a function $g\in X$ having expansion

$$g\sim\sum_{k=1}^{+\infty}(k(k+2\lambda))^{\frac{r}{2}}e^{ir\frac{\pi}{2}}Y_k(f),\tag{2}$$

then g is called the derivative of degree r of f and is written as

$$g=D^{\frac{r}{2}}f=f^{(r)},$$

where $(-1)^{\frac{r}{2}}$ is defined as $e^{ir\frac{\pi}{2}}$ (the definition in [4] was modified by using the factor $e^{ir\frac{\pi}{2}}$ here). Correspondingly, there is concept of fractional integrals based on the results of Askey and Wainger[5]: The integral of degree $r>0$ of $f\in X$ is defined as the function $I^{(r)}(f)$ which has the expansion

$$I^{(r)}(f)\sim\sum_{k=1}^{+\infty}(k(k+2\lambda))^{-\frac{r}{2}}Y_k(f)e^{-ir\frac{\pi}{2}}.$$

This fractional integral can be expressed as a convolution, $I^{(r)}(f)=\mathrm{e}^{-\mathrm{i}r\frac{r}{2}}f*\xi_r$, with the function

$$\xi_r(t)=\sum_{k=1}^{+\infty}(k(k+2\lambda))^{-\frac{r}{2}}c_{n,k}P_k^n(t),\quad t\in\mathbf{R},$$

where $c_{n,k}=\frac{1}{|\mathrm{S}^{n-1}|}d_k^n$, $d_k^n=\frac{2(k+\lambda)\Gamma(k+2\lambda)}{\Gamma(2\lambda)\Gamma(k+1)}$ is the dimension of $\mathcal{H}_k^n$, and P_k^n denotes ultraspherical (or Gegenbauer) polynomials normalized by the condition $P_k^n(1)=1$, i. e. $P_k^n(t)=\frac{P_k^{(\lambda)}(t)}{P_k^{(\lambda)}(1)}$ (see [6] or [7] for $P_k^{(\lambda)}$). By the results of [5], for any $r>0$ and $y\in\mathrm{S}^{n-1}$, $\xi_r(xy)$ as a function of x belongs to $L(\mathrm{S}^{n-1})$. So $I^{(r)}$ is a bounded operator from X to X.

Obviously, for $f\in X$,

$$D^{\frac{r}{2}}\circ I^{(r)}(f)=f-Y_0(f)$$

and if $f^{(r)}\in X$ then

$$I^{(r)}\circ D^{\frac{r}{2}}(f)=f-Y_0(f).$$

To measure the smoothness of functions we also make use of the K-functional defined for $r>0$, $t>0$, $f\in X$ by

$$K_t(f;\ t):=\inf\{\|f-g\|+t^r\|g\|\ |g^{(r)}\in X\}.$$

Over several years, many mathematicians have contributed to the establishment of the following approximation theorem of Jackson type:

Theorem 1 For any $f\in X$, and $N\in\mathbf{N}$, and any $r>0$, there holds

$$E_N(f)\leqslant B_{n,r}\omega^r\left(f;\ \frac{1}{N}\right).\tag{3}$$

But this effort was only partly successful. Up to 1987, Kalyabin[8] established the equivalence between moduli of continuity and K-functionals in spaces $L^p(\mathrm{S}^{n-1})$ for $1<p<+\infty$ which can be applied to prove (3) for $X=L^p(\mathrm{S}^{n-1})$, $1<p<+\infty$. In Rustamov's recent paper[4], he employed a functional argument in his Lemma 3. 9 to give (3) a general proof which he thought would be applicable to general orthogonal series. But we could not follow his method.

The purpose of the present paper is to give a complete proof for

Theorem 1. Related to (3), we will establish the equivalence between moduli of continuity and K-functionals of high degree in all spaces X. That is,

Theorem 2 For any $f\in X$, any $t>0$, and any $r>0$, there holds

$$w_r(f;\ t)\leqslant B_{n,r}K_r(f;\ t)\leqslant B'_{n,r}\omega_r(f;\ t), \tag{4}$$

where $B_{n,r}$, $B'_{n,r}$ are positive constants.

References

[1] Müller C. Spherical Harmonics. Lectures Notes in Mathematics 17, Springer-Verlag, 1966.

[2] Stein E M and Weiss G. Introduction of Fourier Analysis on Euclidean Spaces. Princeton University Press, 1971.

[3] Löfström J and Peetre J. Approximation theorems connected with generalized translation. Math. Ann., 1969, 181: 255—268.

[4] Rustamov Kh P. On the approximation of functions on spheres. Izv Akad. Nauk SSSR ser. Mat., 1993, 59(5): 127—148.

[5] Askey R and Wainger S. On the behavior of special classes of ultraspherical expansions-I. Journal D'Analyse Mathématique, 1965, 15: 192—220.

[6] Szegö G. Orthogonal Polynomials. American Mathematics Society Colloquium Publication, Vol 23, 1939.

[7] Erdélyi, Magnus, Oberhettinger and Tricomi. Higher Transcendental Functions, Vol. 2, McGraw-Hill Book Company, INC, 1953.

[8] Kalyabin G A. On moduli of smoothness of functions given on the sphere. Soviet Math. Dokl., 1987, 35: 619—622.

The Journal of Fourier Analysis and Applications, 1997, 3(6): 706—714.

Jacobi 多项式的估计及 Fourier-Laplace 收敛

Jacobi Polynomial Estimates and Fourier-Laplace Convergence①

§ 1. Introduction

Wang has shown in [5] how to reduce convergence problems for Fourier-Laplace series on the sphere to the consideration of a certain equiconvergent operator defined by convolution with certain Jacobi polynomials. Here we derive new estimates for Jacobi polynomials with complex indices and use Stein's interpolation theorem to obtain L^p estimates for the corresponding maximal equiconvergent operator. This leads, in turn, to new results for almost everywhere convergence of Fourier-Laplace series.

The Jacobi polynomials, $P_k^{(\alpha,\beta)}(x)$ with $\alpha>-1$, $\beta>-1$, can be defined by (see [1, p. 62], formula (4.21.2))

$$P_k^{(\alpha,\beta)}(x) = \frac{\Gamma(\alpha+k+1)}{\Gamma(\alpha+1)\Gamma(k+1)}\sum_{j=0}^{k} \mathrm{C}_k^j \frac{\Gamma(\alpha+1)\Gamma(\alpha+\beta+k+j+1)}{\Gamma(\alpha+j+1)\Gamma(\alpha+\beta+k+1)}\left(\frac{x-1}{2}\right)^j.$$

① The authors thank the Australian Research Council for support of their collaboration.

Wang K Y is also supported by NSFC 19471007.

本文与 Brown G 合作.

Received: 1996-01-23.

This formula obviously also gives the definition for Jacobi polynomials with complex indices α, β such that $\Re\alpha>-1$, $\Re\beta>-1$. These are the polynomials we must estimate but our purposes require us to consider only certain types of complex index. In fact, we will prove the following.

Theorem 1 Let $\alpha\in[0,\ 2n]$, $\beta\in[0,\ n]$, $n\in\mathbf{N}$, $\mu=\frac{1}{2}+\mathrm{i}\tau$, $\tau\in\mathbf{R}$. Then for $k\in\mathbf{N}$,

$$|P_k^{(\alpha+\mu,\beta)}(\cos\theta)|\leqslant B_n\mathrm{e}^{3|\tau|}k^{\alpha+\frac{1}{2}},\qquad 0<\theta<2k^{-1},\tag{1.1}$$

$$|P_k^{(\alpha+\mu,\beta)}(\cos\theta)|\leqslant B_n\mathrm{e}^{3|\tau|}k^{-\frac{1}{2}}\theta^{-\alpha-1}(\pi-\theta)^{-\beta-1},\quad 2k^{-1}<\theta<\pi-k^{-1},\tag{1.2}$$

$$|P_k^{(\alpha+\mu,\beta)}(\cos\theta)|\leqslant B_n\mathrm{e}^{3|\tau|}k^{\beta+\frac{1}{2}},\qquad \pi-k^{-1}<\theta<\pi.\tag{1.3}$$

Using this theorem and applying Stein's interpolation theorem (see [2]) we will prove the following.

Theorem 2 Let $n\geqslant 3$, $\mathrm{S}^{n-1}=\{(x_1,\ x_2,\ \cdots,\ x_n)\in\mathbf{R}^n\mid x_1^2+x_2^2+\cdots+x_n^2=1\}$. If $f\in L\ \ln^2 L(\mathrm{S}^{n-1})$, i. e.,

$$\int_{\mathrm{S}^{n-1}}|f(x)|(1+\ln_+^2|f(x)|)\mathrm{d}x<+\infty,$$

Then

$$\lim_{N\to+\infty}S_N(f)(x)=f(x),\qquad \text{a. e.}\tag{1.4}$$

where $S_N(f)$ denotes the Cesàro means at critical index of the Fourier-Laplace series of f.

We will prove Theorem 1 in Section 2. To prove Theorem 2 we will make use of the equicovergent operator of Cesàro means which was discussed in detail in [5]. This operator is convenient for convergence problems. We will investigate the corresponding maximal operators in Section 3. Then, by the results in Section 3 we complete the proof of Theorem 2 in Section 4.

§ 2. The proof of theorem 1

The following formula is obtained by analytic extension from the formula (3.9) of [3, p. 20].

$$P_k^{(\alpha+\mu,\beta)}(x)=\frac{2^{\mu}\Gamma(\alpha+\mu+1)(1+x)^{\alpha+k+1}P_k^{(\alpha+\mu,\beta)}(1)}{\Gamma(\alpha+1)\Gamma(\mu)(1-x)^{\alpha+\mu}P_k^{(\alpha,\beta)}(1)}\cdot$$

$$\int_x^1\frac{(1-y)^{\alpha}}{(1+y)^{\alpha+\mu+1+k}}P_k^{(\alpha,\beta)}(y)(y-x)^{\mu-1}\mathrm{d}y. \qquad (2.1)$$

For simplicity, we write

$$f(\theta,\ t)=\left(\frac{1}{\sin\frac{\theta}{2}\cos\frac{t}{2}}\right)^{\mathrm{i}2\tau}\left(\frac{\sin\frac{t}{2}}{\sin\frac{\theta}{2}}\right)^{2\alpha+1}\left(\frac{\cos\frac{\theta}{2}}{\cos\frac{t}{2}}\right)^{2\alpha+2k+2}, \qquad (2.2)$$

Then, by (2.1) we have for $0<\theta<\pi$ and $\mu=\frac{1}{2}+\mathrm{i}\tau$ $(\tau\in\mathbf{R})$,

$$P_k^{(\alpha+\mu,\beta)}(\cos\theta)$$

$$=\frac{\Gamma(\alpha+\mu+k+1)}{2^{\mu}\Gamma(\mu)\Gamma(\alpha+k+1)}\int_0^{\theta}\frac{f(\theta,t)}{(\cos t-\cos\theta)^{\frac{1}{2}-\mathrm{i}\tau}}P_k^{(\alpha,\beta)}(\cos t)\mathrm{d}t. \qquad (2.3)$$

Now we apply the formulae 1.3.(2) and 1.3.(3) of [4], which state

$$\frac{\Gamma(u+v)}{\Gamma(u)}=\mathrm{e}^{-\gamma v}\prod_{j=0}^{+\infty}\frac{1}{1+\frac{v}{v+j}}\mathrm{e}^{\frac{v}{j+1}},\quad u>0,v>0, \qquad (2.4)$$

$$\frac{1}{|\Gamma(u+\mathrm{i}v)|}=\frac{1}{\Gamma(u)}\prod_{j=0}^{+\infty}\sqrt{1+\frac{v^2}{(j+u)^2}},\quad u>0,v\in\mathbf{R}. \qquad (2.5)$$

By (2.4) we have

$$\ln\frac{\Gamma(u+v)}{\Gamma(u)}=-\gamma v+\sum_{j=0}^{+\infty}\left(\frac{v}{j+1}+\ln\frac{j+u}{j+u+v}\right).$$

Write

$$f(x)=\frac{v}{x+1}+\ln\frac{x+u}{x+u+v},\qquad x\geqslant 0.$$

Noticing $u\geqslant 1$, $v>0$ we have

$$\frac{\mathrm{d}}{\mathrm{d}x}f(x)=-\frac{v}{(x+1)^2}+\frac{v}{(x+u)(x+u+v)}<0.$$

Hence, we get

$$\int_0^{+\infty} f(x)\mathrm{d}x < \sum_{j=0}^{+\infty} f(j) < f(0) + \int_0^{+\infty} f(x)\mathrm{d}x.$$

Integrating by parts we find

$$\int_0^{+\infty} f(x)\mathrm{d}x = -\int_0^{+\infty} x\frac{\mathrm{d}}{\mathrm{d}x}f(x)\mathrm{d}x = -v + u\ln\frac{u+v}{u} + v\ln(u+v).$$

From these we derive that

$$\mathrm{e}^{-1-\gamma v}(u+v)^v \leqslant \frac{\Gamma(u+v)}{\Gamma(u)} \leqslant \mathrm{e}^{(1-\gamma)v}(u+v)^v,\quad u\geqslant 1, v>0. \tag{2.6}$$

Similarly, by (2.5) we have

$$\ln\frac{1}{|\Gamma(u+\mathrm{i}v)|} = \ln\frac{1}{\Gamma(u)} + \sum_{j=0}^{+\infty}\ln\sqrt{1+\frac{v^2}{(j+u)^2}}.$$

Since

$$\begin{aligned}\sum_{j=0}^{+\infty}\ln\sqrt{1+\frac{v^2}{(j+u)^2}} &< \ln\sqrt{1+\frac{v^2}{u^2}} + \frac{1}{2}\int_0^{+\infty}\ln\left(1+\frac{v^2}{(x+u)^2}\right)\mathrm{d}x \\ &< \ln\sqrt{1+\frac{v^2}{u^2}} + \int_0^{+\infty}\frac{v^2}{(x+u)^2+v^2}\mathrm{d}x,\end{aligned}$$

we get

$$\sqrt{1+\frac{v^2}{u^2}}\frac{1}{\Gamma(u)} \leqslant \frac{1}{|\Gamma(u+\mathrm{i}v)|} \leqslant \sqrt{1+\frac{v^2}{u^2}}\frac{1}{\Gamma(u)}\mathrm{e}^{\frac{\pi}{2}|v|},\quad u>0, v\in\mathbf{R}. \tag{2.7}$$

From (2.3) (2.6) (2.7), we derive

$$|P_k^{(\alpha+\mu,\beta)}(\cos\theta)| \leqslant B_n k^{\frac{1}{2}}\mathrm{e}^{2|\tau|}\left|\int_0^\theta \frac{f(\theta,t)}{(\cos t-\cos\theta)^{\frac{1}{2}-\mathrm{i}\tau}}P_k^{(\alpha,\beta)}(\cos t)\mathrm{d}t\right|. \tag{2.8}$$

Then from the well-known estimate (see [1, p. 197])

$$|P_k^{(\alpha,\beta)}(\cos t)| \leqslant B_n k^\alpha,\quad 0<t<\frac{2}{k},$$

we get (1.1).

When $\frac{2}{k}<\theta<\pi$, we break the integral on the right side of (2.8) into two parts, $\int_0^{\frac{1}{k}}$ and $\int_{\frac{1}{k}}^\theta$, and write them as I_k^1 and I_k^2, respectively.

For the first part we note that

$$\cos t-\cos\theta>\cos\frac{\theta}{2}-\cos\theta>B\theta^2,\quad 0<t<2k^{-1}<\theta<\pi.$$

Then we have

$$|I_k^1|=\left|\int_0^{k^{-1}}\frac{f(\theta,t)}{(\cos t-\cos\theta)^{\frac{1}{2}-\mathrm{i}\tau}}P_k^{(\alpha,\beta)}(\cos t)\mathrm{d}t\right|$$

$$\leqslant B_n k^{\alpha}\int_0^{k^{-1}}\frac{t^{2\alpha+1}}{\theta^{2\alpha+\frac{3}{2}}}\frac{\mathrm{d}t}{(\theta-t)^{\frac{1}{2}}}\leqslant B_n\frac{1}{k\theta^{\alpha+1}}.\tag{2.9}$$

In order to estimate I_k^2, we apply the following asymptotic estimate for Jacobi polynomial with real indices, which is the formula (8. 21. 18) in [1, p. 192].

$$P_k^{(\alpha,\beta)}(\cos\theta)=$$

$$\sqrt{\frac{1}{\pi k}}\frac{\cos(N\theta-\gamma)+r(\alpha,\beta,k,\theta)(k\sin\theta)^{-1}}{\left(\sin\frac{\theta}{2}\right)^{\alpha+\frac{1}{2}}\left(\cos\frac{\theta}{2}\right)^{\beta+\frac{1}{2}}},\quad\frac{1}{k}<\theta<\pi-\frac{1}{k},\tag{2.10}$$

where $N=k+\dfrac{\alpha+\beta+1}{2}$, $\gamma=\left(\alpha+\dfrac{1}{2}\right)\dfrac{\pi}{2}$ and $r(\alpha,\beta,k,\theta)$ satisfies

$$|r(\alpha,\beta,k,\theta)|\leqslant B_n,\quad\alpha,\beta\in[0,2n],\quad\frac{1}{k}<\theta<\pi-\frac{1}{k},\tag{2.11}$$

We see that

$$I_k^2=$$

$$(\pi k)^{\frac{1}{2}}\int_{k^{1}}^{\theta}\frac{f(\theta,t)\cos(Nt-\gamma)}{(\cos t-\cos\theta)^{\frac{1}{2}-\mathrm{i}\tau}\left(\sin\frac{t}{2}\right)^{\alpha+\frac{1}{2}}\left(\cos\frac{t}{2}\right)^{\beta+\frac{1}{2}}}\mathrm{d}t+R(\alpha,\beta,k,\theta)\tag{2.12}$$

where

$$R(\alpha,\beta,k,\theta)=$$

$$(\pi k)^{-\frac{1}{2}}\int_{k^{-1}}^{\theta}\frac{f(\theta,t)}{(\cos t-\cos\theta)^{\frac{1}{2}-\mathrm{i}\tau}}\frac{r(\alpha,\beta,k,t)}{k\sin t\left(\sin\frac{t}{2}\right)^{\alpha+\frac{1}{2}}\left(\cos\frac{t}{2}\right)^{\beta+\frac{1}{2}}}\mathrm{d}t.\tag{2.13}$$

By (2. 2) and (2. 11), we get for $\dfrac{2}{k}<\theta<\pi-\dfrac{2}{k}$,

$$|R(\alpha,\beta,k,\theta)|\leqslant\frac{B_n}{k\left(\sin\frac{\theta}{2}\right)^{\alpha+1}\left(\cos\frac{\theta}{2}\right)^{\beta}}\int_0^{\theta}\frac{\mathrm{d}t}{\cos\frac{t}{2}(\cos t-\cos\theta)^{\frac{1}{2}}}.\tag{2.14}$$

By an easy calculation we know

$$\int_0^\theta \frac{\mathrm{d}t}{\cos\frac{t}{2}(\cos t-\cos\theta)^{\frac{1}{2}}}=\frac{\pi}{\sqrt{2}\cos\frac{\theta}{2}}.$$

Hence, we obtain for $\frac{2}{k}<\theta<\pi-\frac{1}{k}$,

$$|R(\alpha,\beta,k,\theta)|\leqslant\frac{B_n}{k\left(\sin\frac{\theta}{2}\right)^{\alpha+1}\left(\cos\frac{\theta}{2}\right)^{\beta+1}}. \tag{2.15}$$

Write

$$h(t)=\left(\sin\frac{t}{2}\right)^{\alpha+\frac{1}{2}}\left(\cos\frac{t}{2}\right)^{-\left(2\alpha+\beta+2k+\frac{5}{2}+\mathrm{i}2\tau\right)}$$

and write the integral on the right side of (2.12) as F_k. Then

$$F_k=\left(\cos\frac{\theta}{2}\right)^{2\alpha+2k+2}\left(\sin\frac{\theta}{2}\right)^{-(2\alpha+1+\mathrm{i}2\tau)}\int_{k^{-1}}^{\theta}\frac{h(t)\cos(Nt-\gamma)}{(\cos t-\cos\theta)^{\frac{1}{2}-\mathrm{i}\tau}}\mathrm{d}t.$$

If we write $a=\alpha+\frac{1}{2}$, $b=2\alpha+\beta+2k+\frac{5}{2}$, then we have

$$\frac{\mathrm{d}}{\mathrm{d}t}h(2t)=a\frac{(\sin t)^{a-1}}{(\cos t)^{b-1+\mathrm{i}2\tau}}+(b+\mathrm{i}2\tau)\frac{(\sin t)^{a+1}}{(\cos t)^{b+1+\mathrm{i}2\tau}}.$$

Because $b>2$, we have $|b+\mathrm{i}2\tau|<b\mathrm{e}^{|\tau|}$ and hence

$$\left|\frac{\mathrm{d}}{\mathrm{d}t}h(t)\right|\leqslant\mathrm{e}^{|\tau|}\frac{\mathrm{d}}{\mathrm{d}t}|h(t)|. \tag{2.16}$$

Write

$$u_k(\theta)=\int_{k^{-1}}^{\theta-k^{-1}}\frac{h(t)\cos(Nt-\gamma)}{(\cos t-\cos\theta)^{\frac{1}{2}-\mathrm{i}\tau}}\mathrm{d}t,$$

$$v_k(\theta)=\int_{\theta-k^{-1}}^{\theta}\frac{h(t)\cos(Nt-\gamma)}{(\cos t-\cos\theta)^{\frac{1}{2}-\mathrm{i}\tau}}\mathrm{d}t.$$

Integrating by parts we have

$$u_k(\theta)=$$

$$N^{-1}\frac{h(t)\sin(Nt-\gamma)}{(\cos t-\cos\theta)^{\frac{1}{2}-\mathrm{i}\tau}}\Bigg|_{k^{-1}}^{\theta-k^{-1}}+$$

$$N^{-1}\int_{k^{-1}}^{\theta-k^{-1}}\sin(Nt-\gamma)\left(\frac{h'(t)}{(\cos t-\cos\theta)^{\frac{1}{2}-\mathrm{i}\tau}}+\frac{\left(\frac{1}{2}-\mathrm{i}\tau\right)h(t)\sin t}{(\cos t-\cos\theta)^{\frac{3}{2}-\mathrm{i}\tau}}\right)\mathrm{d}t. \tag{2.17}$$

By (2.16) we have

$$\left|\int_{k^{-1}}^{\theta-\frac{1}{k}}\sin(Nt-\gamma)\left[\frac{h'(t)}{(\cos t-\cos\theta)^{\frac{1}{2}-\mathrm{i}\tau}}+\frac{\left(\frac{1}{2}-\mathrm{i}\tau\right)h(t)\sin t}{(\cos t-\cos\theta)^{\frac{3}{2}-\mathrm{i}\tau}}\right]\mathrm{d}t\right|$$

$$\leqslant\int_{k^{-1}}^{\theta-\frac{1}{k}}2\mathrm{e}^{|\tau|}\frac{\mathrm{d}}{\mathrm{d}t}\frac{|h(t)|}{(\cos t-\cos\theta)^{\frac{1}{2}}}\mathrm{d}t=2\mathrm{e}^{|\tau|}\left.\frac{|h(t)|}{(\cos t-\cos\theta)^{\frac{1}{2}}}\right|_{\frac{1}{k}}^{\theta-\frac{1}{k}}. \tag{2.18}$$

Meanwhile, we have

$$|v_k(\theta)|\leqslant\int_{\theta-\frac{1}{k}}^{\theta}\frac{|h(t)|}{(\cos t-\cos\theta)^{\frac{1}{2}}}\mathrm{d}t\leqslant B\frac{|h(\theta)|}{\sqrt{k\sin\theta}}. \tag{2.19}$$

Combining (2.17)~(2.19) we get

$$|F_k|\leqslant B_n\mathrm{e}^{|\tau|}\left(\sqrt{k\sin\theta}\left(\sin\frac{\theta}{2}\right)^{\alpha+\frac{1}{2}}\left(\cos\frac{\theta}{2}\right)^{\beta+\frac{1}{2}}\right)^{-1} \tag{2.20}$$

Substituting (2.15) and (2.20) into (2.12), we get for $\frac{2}{k}<\theta<\pi-\frac{1}{k}$,

$$|I_k^2|\leqslant B_n\mathrm{e}^{|\tau|}\frac{1}{k\left(\sin\frac{\theta}{2}\right)^{\alpha+1}\left(\cos\frac{\theta}{2}\right)^{\beta+1}}. \tag{2.21}$$

A combination of (2.8) (2.9) and (2.21) yields (1.2).

Next we assume $\pi-\frac{1}{k}<\theta<\pi$. By (2.8) we have

$$|P_k^{\alpha+\mu,\beta}(\cos\theta)|\leqslant B_nk^{\frac{1}{2}}\mathrm{e}^{2|\tau|}\int_0^{\theta}\frac{|f(\theta,t)|}{\sqrt{\cos t-\cos\theta}}|P_k^{(\alpha,\beta)}(\cos t)|\,\mathrm{d}t.$$

We have for $\pi-k^{-1}<\theta<\pi$,

$$\frac{|f(\theta,t)|}{\sqrt{\cos t-\cos\theta}}|P_k^{(\alpha,\beta)}(\cos t)|$$

$$\leqslant B_n\begin{cases}t^{\alpha+1}, & \text{if } 0<t<\frac{1}{k},\\ \dfrac{1}{\sqrt{k(\cos t-\cos\theta)}(\pi-t)^{\beta+\frac{1}{2}}}, & \text{if } \frac{1}{k}<t<\pi-\frac{1}{k},\\ \dfrac{k^{\beta}}{\sqrt{\cos t-\cos\theta}}, & \text{if } \pi-\frac{1}{k}<t<\theta.\end{cases}$$

From this derive (1.3). □

Remark If we take $\mu=\varepsilon+\mathrm{i}\tau$, $\varepsilon\in(0,1)$ in Theorem 1, then the same argument will yield the following estimate:

$$|P_k^{(\alpha+\varepsilon+i\tau,\beta)}(\cos\theta)| \leqslant B_n e^{3|\tau|}\begin{cases} k^{\alpha+\varepsilon}, & \text{if } 0<\theta<\frac{1}{k}, \\ \varepsilon^{-1}k^{-\frac{1}{2}}\theta^{-\alpha-\varepsilon-\frac{1}{2}}(\pi-\theta)^{-\beta-\varepsilon-\frac{1}{2}}, & \text{if } \frac{1}{k}<\theta<\pi-\frac{1}{k}, \\ \varepsilon^{-1}k^{\beta+\varepsilon}, & \text{if } \pi-\frac{1}{k}<\theta<\pi. \end{cases}$$

This estimate is better than Theorem 1, which is good enough for our use but is definitely not the best possible. We guess that the factor $\frac{1}{\varepsilon}$ in the above estimate is removable and hence it should yield an estimate for $\varepsilon=0$.

§ 3. Equiconvergent operators

Denote by σ_N^δ the Cesàro means of the Fourier-Laplace series on the unit sphere S^{n-1}. We will apply the main result of [5]:

Proposition 1 Assume $\delta>-1$, D is a non-empty subset of S^{n-1}, $f\in L(S^{n-1})$ and $\sup\{|f(x)|\,|\,x\in D\}<+\infty$. For uniform convergence on D, the following two relations are equivalent:

$$\lim_{N\to+\infty} S_N^\delta(g_x)(x)=0, \tag{3.1}$$

$$\lim_{N\to+\infty} \sigma_N^\delta(g_x)(x)=0 \tag{3.2}$$

where the function g_x is defined for each $x\in D$ by $g_x(y)=f(y)-f(x)$ and the operator S_N^δ is defined by

$$S_N^\delta(f)(x)=a_N^\delta\left(f*P_N^{\left(\frac{n-1}{2}+\delta,\frac{n-3}{2}\right)}\right)(x),\quad f\in L(S^{n-1}),x\in S^{n-1},\delta>-1 \tag{3.3}$$

where

$$a_N^\delta=|S^{n-2}|^{-1}\frac{\Gamma(N+\delta+n-1)\Gamma(2N+\delta+n)}{2^{n-2}\Gamma\left(\frac{n-1}{2}\right)A_N^\delta\Gamma\left(N+\frac{n-1}{2}\right)\Gamma(2N+2\delta+n)}, \tag{3.4}$$

$$A_k^\delta=\frac{\Gamma(\delta+k+1)}{\Gamma(\delta+1)\Gamma(k+1)}. \tag{3.5}$$

Remark The expression (3.4) is a correction of the formula (4) in [5] by inserting a factor

$$2|S^{n-2}|^{-1}=\frac{\Gamma\left(\frac{n-1}{2}\right)}{\pi^{\frac{n-1}{2}}}.$$

Now we normalize the operator S_N^δ to obtain the following.

Definition 1 We call the following operator

$$E_N^\delta(f)=S_N^\delta(f)(S_N^\delta(\mathbf{1}))^{-1}$$

where $\mathbf{1}$ denotes the constant function of value 1, the equiconvergent operator of σ_N^δ.

Next we give a multiplier representation for this operator. Denote by Y_k the projection operator on the subspace $\mathcal{H}_k^n$ of spherical harmonics of degree k. Write $P_N^{\left(\frac{n-1}{2}+\delta,\frac{n-3}{2}\right)}$ as g_N, $c_{n,k}P_k^n$ as h_k temporarily for simplicity. Then we have

$$\begin{aligned}Y_k(S_N^\delta(f))(x)&=a_N^\delta((f*g_N)*h_k)\\&=a_N^\delta\int_{S^{n-1}}\left(\int_{S^{n-1}}f(u)g_N(uv)\mathrm{d}u\right)h_k(xv)\mathrm{d}v\\&=a_N^\delta\int_{S^{n-1}}f(u)\left(\int_{S^{n-1}}g_N(uv)h_k(xv)\mathrm{d}v\right)\mathrm{d}u.\end{aligned}$$

By the Funk-Hecke formula (see [6]) we have

$$\int_{S^{n-1}}g_N(uv)h_k(xv)\mathrm{d}v=\alpha_{N,k}^\delta h_k(xu),$$

where

$$\alpha_{N,k}^\delta=\frac{2\pi^{\frac{n-1}{2}}}{\Gamma\left(\frac{n-1}{2}\right)}\int_{-1}^{1}g_N(t)P_k^n(t)(1-t^2)^{\frac{n-3}{2}}\mathrm{d}t. \tag{3.6}$$

Hence, we get

$$Y_k(S_N^\delta(f))(x)=a_N^\delta\alpha_{N,k}^\delta\int_{S^{n-1}}f(u)h_k(xu)\mathrm{d}u=a_N^\delta\alpha_{N,k}^\delta Y_k(f)(x). \tag{3.7}$$

To calculate the value of $\alpha_{N,k}^\delta$, we apply Rodrigues' formula (see [6]) and get $\alpha_{N,k}^\delta=\dfrac{2\pi^{\frac{n-1}{2}}}{2^k\Gamma\left(k+\frac{n-1}{2}\right)}\displaystyle\int_{-1}^{1}(1-t^2)^{k+\frac{n-3}{2}}\left(\frac{\mathrm{d}}{\mathrm{d}t}\right)^k g_N(t)\mathrm{d}t.$

We see that $\alpha_{N,k}^\delta=0$, if $k>N$. When $0\leqslant k\leqslant N$ we make use of the formula (see [1, p. 63], (4.21.7))

$$\left(\frac{\mathrm{d}}{\mathrm{d}t}\right)^k g_N(t)=2^{-k}\frac{\Gamma(N+\delta+n+k-1)}{\Gamma(N+\delta+n-1)}P_{N-k}^{\left(\frac{n-1}{2}+\delta+k,\frac{n-3}{2}+k\right)}(t)$$

and get by partial integration that

$$\alpha_{N,k}^{\delta}=$$

$$\frac{2\pi^{\frac{n-1}{2}}\Gamma(N+\delta+n+k-1)}{4^k\Gamma\left(k+\frac{n-1}{2}\right)\Gamma(N+\delta+n-1)}\int_{-1}^{1}P_{N-k}^{\left(\frac{n-1}{2}+\delta+k,\frac{n-3}{2}+k\right)}(t)(1-t^2)^{k+\frac{n-3}{2}}\mathrm{d}t$$

$$=2^{n-1}\pi^{\frac{n-1}{2}}A_{N-k}^{\delta}\frac{\Gamma(N+\delta+n+k-1)\Gamma\left(N+\frac{n-1}{2}\right)}{\Gamma(N+\delta+n-1)\Gamma(N+k+n-1)}. \tag{3.8}$$

Then, by (3.7) we get $S_N^{\delta}(\mathbf{1})=Y_0(S_N^{\delta}(\mathbf{1}))=a_N^{\delta}\alpha_{N,0}^{\delta}$.

We write

$$\gamma_N^{\delta}=(\alpha_{N,0}^{\delta})^{-1}=\frac{\Gamma(\alpha+1)\Gamma(N+1)\Gamma(N+n-1)}{(4\pi)^{\frac{n-1}{2}}\Gamma(N+\delta+1)\Gamma\left(N+\frac{n-1}{2}\right)}. \tag{3.9}$$

$$b_{N,k}^{\delta}=\frac{\alpha_{N,k}^{\delta}}{\alpha_{N,0}^{\delta}}=\frac{A_{N-k}^{\delta}}{A_N^{\delta}}\frac{\Gamma(N+\delta+n+k-1)\Gamma(N+n-1)}{\Gamma(N+\delta+n-1)\Gamma(N+n+k-1)}. \tag{3.10}$$

Then, by an analytic extension for the indexes we get the following.

Proposition 2 Let δ be a complex number with $\Re\delta>-1$ and $N\in\mathbf{N}^*$. Then

$$E_N^{\delta}(f)(x)=\gamma_N^{\delta}\left(f*P_N^{\left(\frac{n-1}{2}+\delta,\frac{n-3}{2}\right)}\right)(x)$$

$$=\sum_{k=0}^{N}b_{N,k}^{\delta}Y_k(f)(x),\quad f\in L(\mathrm{S}^{n-1}),x\in\mathrm{S}^{n-1},$$

$E_N^{\delta}(\mathbf{1})(x)=1$, for constant function $\mathbf{1}(x)=1$, $x\in\mathrm{S}^{n-1}$.

Definition 2 Let δ be complex number such that $\Re\delta>-1$. The maximal equiconvergent operator E_*^{δ} of Cesàro mean is defined by

$$E_*^{\delta}(f)(x)=\sup\{|E_N^{\delta}(f)(x)|\mid N\in\mathbf{N}\}.$$

Theorem 3 Let $n\geqslant 3$, $\lambda=\frac{n-2}{2}\delta=\lambda+\varepsilon+\mathrm{i}\tau$, $\tau\in\mathbf{R}$. If $\varepsilon\in(0,n)$, then

$$E_*^{\delta}(f)(x)\leqslant B_n\varepsilon^{-1}\mathrm{e}^{3|\tau|}(HL(f)(x)+HL(f)(-x))$$

where HL denotes the Hardy-Littlewood maximal operator on sphere.

Proof $\delta=\lambda+\varepsilon+\mathrm{i}\tau$, $\varepsilon\in(0,n)$, $\tau\in\mathbf{R}$. First we note that by (3.9)

$$|\gamma_N^{\delta}| \leqslant B_n e^{2|\tau|} N^{\frac{1}{2}-\varepsilon}, \quad N \in \mathbf{N}.$$

Then $$|E_N^{\delta}(f)(x)| \leqslant B_n e^{2|\tau|} N^{\frac{1}{2}-\varepsilon} |f * P_N^{(\alpha,\beta)}(x)|$$

where $\alpha = n - \frac{3}{2} + \varepsilon + i\tau$, $\beta = \frac{n-3}{2}$. write

$$F(x,t) = \int_{xy=\cos t} f(y) d\sigma_t(y), \quad 0 < t < \pi,$$

where $\sigma_t(y)$ denotes the measure element on the surface $\{y \in S^{n-1} | xy = \cos t\}$. We have

$$\int_0^{\theta} |F(x,t)| dt \leqslant \theta^{n-1} HL(f)(x), \quad \theta \in (0, \pi) \tag{3.11}$$

and $$F(x,t) = F(-x, \pi - t), \quad t \in (0, \pi). \tag{3.12}$$

Then we get

$$|E_N^{\delta}(f)(x)| \leqslant B_n e^{2|\tau|} N^{\frac{1}{2}-\varepsilon} \int_0^{\pi} |F(x,t) P_N^{(\alpha,\beta)}(\cos t)| dt. \tag{3.13}$$

Now we break the integral in (3.13) into four parts:

$$I_1 = N^{\frac{1}{2}-\varepsilon} \int_0^{N^{-1}}, \quad I_2 = N^{\frac{1}{2}-\varepsilon} \int_{N^{-1}}^{\frac{\pi}{2}}, \quad I_3 = N^{\frac{1}{2}-\varepsilon} \int_{\frac{\pi}{2}}^{\pi - N^{-1}}, \quad I_4 = N^{\frac{1}{2}-\varepsilon} \int_{\pi - N^{-1}}^{\pi},$$

and apply Theorem 1 to estimate these integrals. Accounting (3.11) and (3.12) we get

$$|I_1| \leqslant B_n e^{3|\tau|} N^{n-1} \int_0^{N^{-1}} |F(x,t)| dt \leqslant B_n e^{3|\tau|} HL(f)(x), \tag{3.14}$$

$$|I_2| \leqslant B_n e^{3|\tau|} \frac{1}{N^{\varepsilon}} \int_{N^{-1}}^{\frac{\pi}{2}} \frac{|F(x,t)|}{t^{n-1+\varepsilon}} dt \leqslant B_n e^{3|\tau|} \varepsilon^{-1} HL(f)(x), \tag{3.15}$$

$$|I_3| \leqslant B_n e^{3|\tau|} \frac{1}{N^{\varepsilon}} \int_{N^{-1}}^{\frac{\pi}{2}} \frac{|F(-x,t)|}{t^{\frac{n-1}{2}}} dt \leqslant B_n e^{3|\tau|} N^{-\varepsilon} HL(f)(-x), \tag{3.16}$$

$$|I_4| \leqslant B_n e^{3|\tau|} N^{\frac{n-1}{2}-\varepsilon} \int_0^{N^{-1}} |F(-x,t)| dt \leqslant B_n e^{3|\tau|} N^{-\frac{n-1}{2}-\varepsilon} HL(f)(-x). \tag{3.17}$$

A combination of (3.13) through (3.17) yields the desired inequality. □

Corollary of theorem 3 If $\delta = \lambda + \varepsilon + i\tau$, $\varepsilon \in (0, n)$, $\tau \in \mathbf{R}$ and $1 < p < +\infty$, then

$$\| E_*^{\delta}(f) \|_p \leqslant B_n e^{3|\tau|} \frac{p}{\varepsilon(p-1)} \| f \|_p. \tag{3.18}$$

Remark For the maximal Cesàro operator $\sigma_*^\delta(f)=\sup\{|\sigma_N^\delta(f)|\mid N\in \mathbf{N}\}$ it is well known (see [7, p. 239]) that

$$\|\sigma_*^\delta(f)\|_p\leqslant B_{\varepsilon,p}e^{c\tau^2}\|f\|_p,\ 1<p<+\infty.$$

But no estimate for the bound $B_{\varepsilon,p}$ exists. By the method in [7], which is based on the estimates for the Cesàro coefficients with complex indices, we would be able to obtain only $B_{\varepsilon,p}\leqslant B_n\varepsilon^{-2}\frac{p}{p-1}$ ($\varepsilon\in(0, n)$) where the power of ε is -2. Our theorem gives -1. This is important because this power decides the power of $\ln_+$ appearing in the conclusion of Theorem 2. This is the reason that we first give an appropriate estimate for Jacobi polynomials with complex indices and then make use of the equiconvergent operator which is a convolution with Jacobi polynomial as kernel.

Now we turn to estimate the L^2 bound of E_*^δ.

Theorem 4 Let $\delta=\varepsilon+i\tau$, $\tau\in\mathbf{R}$. If $\varepsilon\in(0, n)$, then

$$\|E_*^\delta(f)\|_2\leqslant B_ne^{4|\tau|}\varepsilon^{-1}\|f\|_2. \tag{3.19}$$

Proof It is also known from [7, p. 239] that

$$\|\sigma_*^\delta(f)\|_2\leqslant Be^{c\tau^2}\|f\|_2. \tag{3.20}$$

In [7] the dependence of the constant B on the parameters n, ε was not discussed (and the factor) $e^{c\tau^2}$ is not exact. But by the same argument we may refine (3.20) and get

$$\|\sigma_*^\delta(f)\|_2\leqslant B_n\varepsilon^{-1}e^{3|\tau|}\|f\|_2,\ \ \varepsilon>0. \tag{3.21}$$

Then we apply the formula (5) of [5]:

$$\sigma_N^\delta(f)=S_N^\delta(f)+|S^{n-2}|^{-1}T_N^\delta(f)$$

where the factor $|S^{n-2}|^{-1}$ in front of T_N^δ was missed in [5]. Then we get

$$T_N^\delta(f)=\sum_{k=1}^{+\infty}b_k(N,\delta)\sigma_N^{\delta+k}(f)$$

with $|b_k(N,\delta)|\leqslant B_ne^{|\tau|}k^{-1-n-\varepsilon}$ (see [5], Lemma 1) and hence

$$E_N^\delta(f)=(a_N^\delta\alpha_{N,0}^\delta)^{-1}\{\sigma_N^\delta(f)-|S^{n-2}|^{-1}T_N^\delta(f)\}$$

So, by (3.21) we derive (3.19). □

§ 4. Proof of theorem 2

It is a routine application of Stein's interpolation theorem (see [2] or [8] for more details) to derive the $L^p(1<p<2)$ bound from Theorem 3 and Theorem 4. We omit the details and state the result:

Theorem 5 If $1<p<2$, Then

$$\|E_*^\lambda(f)\|_p \leqslant B_n(p-1)^{-2}\|f\|_p. \tag{4.1}$$

Also, it can be regarded as routine to derive the $L\ \ln_+^r L$ bound $(r>0)$ from the known L^p bound $(p-1)^{-r}$ following the argument stated in [8] (see [8]: Lemma 2 of Section 9 and the proof of Theorem D^*). Then we have the following.

Theorem 6

$$\|E_*^\lambda(f)\|_1 \leqslant B_n\left(\int_{S^{n-1}}|f(x)|(1+\ln_+^2|f(x)|)\mathrm{d}x\right). \tag{4.2}$$

As a consequence of Theorem 6 we establish Theorem 2.

References

[1] Szegö G. Orthogonal Polynomials. AMS Colloquium Publications, 23, 1957.

[2] Stein EM, Weiss G. Introduction to Fourier Analysis on Euclidean Spaces. Princeton University Press, Princeton, NJ, 1971.

[3] Askey R. Orthogonal polynomials and special functions. Regional Conf. Lect. Appl. Math., 21, SIAM. Philadelphia, PA, 1975.

[4] Erdélyi A et al. Higher Transcedental Functions, Vol. I. McGraw-Hill Book Company, 1953.

[5] Wang K-Y. Equiconvergent operator of Cesàro means on sphere and its applications. J. Beijing Normal University (NS), 1993, 29(2): 143—154.

[6] Müller. C. Spherical harmonics, Lecture Notes in Mathematics, 17, 1966.

[7] Bonamn A, Clerc J-L. Sommes de Cesàro et multiplicateuars des developpements en harmoniques sphériques. Trans, AMS, 1973, 183; 223—263.

[8] Stein E M. Localization and summability of multiple Fourier series. Acta Mathematica, 1958, 100: 93—147.

Approx，Theory and its Appl.，1999，15(4)：50－59.

球面上函数的逼近

On Approximation of Functions on Sphere[①]

Abstract Let f be an integrable function on the unit sphere S^{n-1} of $\mathbf{R}^n(n\geqslant 3)$ and let $\sigma_N^\delta(f)$ be the Cesàro means of order δ of the Fourier-Laplace series of f. The special value $\lambda := \frac{n-2}{2}$ of δ is known as the critical index. This paper proves that

$$\| f-\sigma_N^\delta(f) \|_p \leqslant C(\delta)\omega\left(f, \frac{1}{N}\right)_p, \quad \text{when } 1\leqslant p\leqslant +\infty \text{ and } \delta>\lambda$$

and

$$\left\| \left(\frac{1}{N}\sum_{k=1}^{N} |\sigma_k^\lambda(f)-f|^q\right)^{\frac{1}{q}} \right\|_c \leqslant C_n \frac{q^2}{q-1}\omega\left(f, \left(\frac{\ln N}{N}\right)^{\frac{1}{q}}\right)_c,$$

when $1<q<+\infty$,

where $\omega(f, t)_p$ is the 1st-order modulus of continuity of f in L^p-metric which is defined in a way different than in the classical case of $n=2$.

§ 1. Introduction

Let f be an integrable function on the unit sphere

$$S^{n-1} := \{(x_1, x_2, \cdots, x_n) \mid x_1^2+x_2^2+\cdots+x_n^2=1\}$$

of $\mathbf{R}^n(n\geqslant 3)$ and let

① Supported by NSFC 19771009.

本文与戴峰合作.

$$f \sim \sigma(f)(x):=\sum_{k=0}^{+\infty} Y_k(f)(x)$$

be the Fourier-Laplace expansion of f, where Y_k is the projection operator from $L(S^{n-1})$ to the space of all spherical harmonics of degree k. The Cesàro means of order δ of $\sigma(f)$ are defined (as usual) by

$$\sigma_N^{\delta}(f):=(A_N^{\delta})^{-1}\sum_{k=0}^{N} A_{N-k}^{\delta} Y_k(f),$$

where

$$A_N^{\delta}=\frac{\Gamma(N+\delta+1)}{\Gamma(\delta+1)\Gamma(N+1)}, \quad \delta>-1.$$

The special value $\lambda:=\frac{n-2}{2}$ of δ is known as the "critical index" of Cesàro means $\sigma_N^{\delta}(f)(x)$. (See [1]).

We will say that f has derivative of order $\alpha>0$ in $L^p(S^{n-1})$ if there exists a function $h_\alpha \in L^p(S^{n-1})$ such that

$$h_\alpha \sim \sum_{k=0}^{+\infty}(-k(k+2\lambda))^{\frac{\alpha}{2}} Y_k(f)$$

and we then denote h_α by $D^\alpha f$ or $f^{(\alpha)}$.

The translation operator is defined by

$$S_\theta(f)(x):=\frac{1}{|S^{n-2}|}\int_{\{y\in S^{n-1}\mid xy=0\}} f(x\cos\theta + y\sin\theta)\,dl(y), \quad x\in S^{n-1},$$

where $dl(y)$ denotes the usual Lebesgue measure elements on $\{y\in S^{n-1} \mid xy=0\}$ (see [2], for example).

Let I be the identity operator and let r be a positive number. We set $\psi_r(u):=(1-u)^{\frac{r}{2}}$. Following [4], we call the operator

$$\Delta_u^r:=(I-S_u)^{\frac{r}{2}}=\sum_{k=0}^{+\infty}\frac{\psi_r^{(k)}(0)}{k!}S_u^k$$

the rth-order difference and defined the rth-order modulus of continuity in L^p-metric of a function $f\in L^p(S^{n-1})$ by

$$\omega_r(f,t)_p:=\sup\{\|\Delta_u^r f\|_p \mid 0<u\leqslant t\}.$$

Throughout this paper, we will use $L^\infty(S^{n-1})$ to denote $C(S^{n-1})$ and $\omega(f,t)_p$ is the 1st-order modulus of continuity in L^p-metric of f.

Our goal in this paper is to relate the strong uniform approximation

$$\left\| \left(\frac{1}{N} \sum_{k=0}^{N} | f - \sigma_k^{\lambda}(f) |^q \right)^{\frac{1}{q}} \right\|_c$$

to 1st-order modulus of continuity $\omega(f, t)_c$. We organize this paper as follows: In § 2, we consider approximation by Cesàro means of order equal to or greater than the critical index $\lambda = \frac{n-2}{2}$. The results of § 2 are Theorem 2.1 and Theorem 2.4. Using Theorem 2.1 we establish our main result Theorem 3.1 in § 3. Because the modulus we use here is essentially a maximal multiplier, which is different from that of [8], the proof of Theorem 3.1 seems to be a little difficult.

§ 2. Approximation by Cesàro means

Theorem 2.1　Let $\delta_0 > \lambda$. Then there exists a constant $C(\delta_0) > 0$ such that

$$\| f - \sigma_N^{\delta}(f) \|_p \leqslant C(\delta_0) \delta \omega\left(f, \frac{1}{N}\right)_p, \quad 1 \leqslant p \leqslant +\infty$$

for all $\delta \geqslant \delta_0$ and $f \in L^p(S^{n-1})$.

To prove Theorem 2.1, we need several lemmas.

Lemma 2.2　Let $\delta_0 > \lambda$ and $1 \leqslant p \leqslant +\infty$. Then for all $\delta \geqslant \delta_0$ and all N

$$\| \sigma_N^{\delta}(f) \|_p \leqslant C(\delta_0) \| f \|_p.$$

Proof　For $\delta > \delta_0$, by Abel transform, we have

$$\sigma_N^{\delta}(f) = \frac{1}{A_N^{\delta}} \sum_{k=0}^{N} A_{N-k}^{\delta-\delta_0-1} A_k^{\delta_0} \sigma_k^{\delta_0}(f).$$

On the other hand, it follows from [1]that

$$\| \sigma_k^{\delta_0}(f) \|_p \leqslant C(\delta_0) \| f \|_p$$

for $\delta_0 > \lambda$ and $k \in \mathbf{N}$. Putting the above two inequalities together and noticing that $A_N^{\delta} = \sum_{k=0}^{N} A_{N-k}^{\sigma-\delta_0-1} A_k^{\delta_0}$, we then derive Lemma 2.2.

Lemma 2.3[4]　Suppose $\eta \in C^{\infty}[0, +\infty)$ and $\chi_{[0,1]} \leqslant \eta \leqslant \chi_{[0,2]}$. Define

$$\eta_N(f)=\sum_{k=0}^{+\infty}\eta\left(\frac{k}{N}\right)Y_k(f),\quad N>0.$$

Then for $f\in L^p(S^{n-1})$ and $r>0$,

$$\frac{1}{N^r}\|(\eta_N f)^{(r)}\|_p\leqslant C_{n,r}\omega_r\left(f,\ \frac{1}{N}\right)_p,\ 1\leqslant p\leqslant+\infty.$$

Proof of theorem 2.1 Take $\eta\in C[0,\ +\infty)$ such that $\chi_{[0,1]}\leqslant\eta\leqslant\chi_{[0,2]}$. Let $\xi=\eta^2$. Then it follows from [3] that

$$\|f-\xi_N(f)\|_p\leqslant C(\xi)\omega\left(f,\ \frac{1}{N}\right)_p,\quad 1\leqslant p\leqslant+\infty,\qquad (2.1)$$

from which we get

$$\begin{aligned}\|\sigma_N^\delta(f)-f\|_p &\leqslant\|\sigma_N^\delta(f)-\sigma_N^\delta(\xi_{\frac{N}{4}}f)\|_p+\|\sigma_N^\delta(\xi_{\frac{N}{4}}f)-\xi_{\frac{N}{4}}(f)\|_p+\\ &\qquad\|\xi_{\frac{N}{4}}(f)-f\|_p\\ &\leqslant C(\delta_0)\omega\left(f,\ \frac{1}{N}\right)_p+\|\sigma_N^\delta(\xi_{\frac{N}{4}}f)-\xi_{\frac{N}{4}}(f)\|_p\qquad (2.2)\end{aligned}$$

by Lemma 2.2.

On the other hand, it follows from Lemma 2.2 that

$$\|\sigma_k^\delta(\xi_{\frac{N}{4}}(f))-\xi_{\frac{N}{4}}(f)\|_p=o(1),\quad k\to+\infty,$$

which implies that in L^p-metric

$$\begin{aligned}&\sigma_N^\delta(\xi_{\frac{N}{4}}f)-\xi_{\frac{N}{4}}(f)\\ &=\sum_{k=N}^{+\infty}(\sigma_k^\delta(\xi_{\frac{N}{4}}f)-\sigma_{k+1}^\delta(\xi_{\frac{N}{4}}f))\\ &=\sum_{k=N}^{+\infty}\sum_{j=0}^{\left[\frac{N}{2}\right]}\frac{A_{k-j}^\delta}{A_k^\delta}\left[1-\frac{A_{k+1-j}^\delta A_k^\delta}{A_{k+j}^\delta A_{k-j}^\delta}\right]Y_j(\xi_{\frac{N}{4}}f)\\ &=-\delta\sum_{k=N}^{+\infty}\frac{1}{k(k+\delta+1)}\sum_{j=1}^{\left[\frac{N}{2}\right]}\frac{A_{k-j}^\delta}{A_k^\delta}\frac{k}{k-j+1}\left(\frac{j}{j+2\lambda}\right)^{\frac{1}{2}}\eta\left(\frac{4j}{N}\right)Y_j((\eta_{\frac{N}{4}}f)'),\end{aligned}$$

where $(\eta_{\frac{N}{4}}(f))'=D^{\frac{1}{2}}(\eta_{\frac{N}{4}}(f))$. So, applying Lemma 2.2 again, we get

$$\begin{aligned}&\|\sigma_N^\delta(\xi_{\frac{N}{4}}f-\xi_{\frac{N}{4}}(f))\|_p\\ &\leqslant C(\delta_0)\delta\sum_{k=N}^{+\infty}\frac{1}{k(k+\delta+1)}\left\|\sum_{j=1}^{\frac{N}{2}}\frac{k}{k-j+1}\left(\frac{j}{j+2\lambda}\right)^{\frac{1}{2}}\eta\left(\frac{4j}{N}\right)Y_j((\eta_{\frac{N}{4}}f)')\right\|_p.\end{aligned}\qquad (2.3)$$

Write

$$a(k,\ j)=\frac{k}{k-j+1}\left(\frac{j}{j+2\lambda}\right)^{\frac{1}{2}}\eta\left(\frac{4j}{N}\right),\ j=1,2,\cdots,\frac{N}{2},\quad k\geqslant N.$$

Now noticing that $\eta\left(\frac{4j}{N}\right)=0$ whenever $j>\frac{N}{2}$, we get

$$|\Delta^{n+1}a(k,\ j)|\leqslant C_n\left(\left(\frac{1}{j+1}\right)^{n+2}+\frac{1}{N^{n+1}}\right),\quad 1\leqslant j\leqslant N\leqslant k.$$

Therefore, by Abel transform and Lemma 2.2,

$$\begin{aligned}&\left\|\sum_{j=1}^{\frac{N}{2}}a(k,j)Y_j((\eta_{\frac{N}{4}}f)')\right\|_p\\ \leqslant\ & C_n\sum_{j=1}^{+\infty}|\Delta^{n+1}a(k,j)|j^n\|\sigma_j^n((\eta_{\frac{N}{4}}f)')\|_p\leqslant C_n\|(\eta_{\frac{N}{4}}f)'\|_p.\end{aligned}\qquad(2.4)$$

Combining (2.3) and (2.4), we get

$$\|\sigma_N^\delta(\xi_{\frac{N}{4}}f)-\xi_{\frac{N}{4}}(f)f\|_p\leqslant C(\delta_0)\frac{\delta}{N}\|D^{\frac{1}{2}}\eta_{\frac{N}{4}}(f)\|_p\leqslant C(\delta_0)\delta\omega\left(f,\ \frac{1}{N}\right)_p\qquad(2.5)$$

on account of Lemma 2.3.

Now Theorem 2.1 follows from (2.2) and (2.5).

Theorem 2.4 Let $1\leqslant p\leqslant+\infty$ and $f\in L^p(\mathrm{S}^{n-1})$. Then (for $N\geqslant 2$)

$$\|f-\sigma_N^\lambda(f)\|_p\leqslant\begin{cases}C_n\ln N\omega\left(f,\ \frac{1}{N}\right)_p, & \text{if } p=1 \text{ or } p=+\infty,\\ C_n\omega\left(f,\ \frac{1}{N}\right)_p, & \text{if } 1<p<+\infty.\end{cases}$$

To prove Theorem 2.4, we need the following lemma.

Lemma 2.5[1] Let $f\in L^p(\mathrm{S}^{n-1})$ with $1\leqslant p\leqslant+\infty$. Then

$$\|\sigma_N^\lambda(f)\|_p\leqslant\begin{cases}C_n\ln N\|f\|_p, & \text{if } p=1 \text{ or } p=+\infty,\\ C_n\|f\|_p, & \text{if } 1<p<+\infty.\end{cases}$$

Proof of theorem 2.4 Write

$$A_{pN}=\begin{cases}\ln N, & \text{if } p=1 \text{ or } p=+\infty,\\ 1, & \text{if } 1<p<+\infty.\end{cases}$$

Take $\eta(t)\in C^\infty(\mathbf{R})$ such that $\chi_{[0,1]}\leqslant\eta\leqslant\chi_{[0,2]}$. Then by (2.1), Theorem 2.1 and Lemma 2.5, we get

$$\| \sigma_N^\lambda(f) - f \|_p$$

$$\leqslant \| \sigma_N^\lambda(\eta_{\frac{N}{2}} f) - \sigma_N^\lambda(f) \|_p + \| \sigma_N^\lambda(\eta_{\frac{N}{2}} f) - \sigma_N^{\lambda+1}(\eta_{\frac{N}{2}} f) \|_p +$$

$$\| \sigma_N^{\lambda+1}(\eta_{\frac{N}{2}} f) - \sigma_N^{\lambda+1}(f) \|_p + \| \sigma_N^{\lambda+1}(f) - f \|_p$$

$$\leqslant C_n A_{pN} \| f - \eta_{\frac{N}{2}}(f) \|_p + \| \sigma_N^\lambda(\eta_{\frac{N}{2}} f) - \sigma_N^{\lambda+1}(\eta_{\frac{N}{2}} f) \|_p + C_n \omega\left(f, \frac{1}{N}\right)_p$$

$$\leqslant C_n A_{pN} \omega\left(f, \frac{1}{N}\right)_p + \| \sigma_N^\lambda(\eta_{\frac{N}{2}} f) - \sigma_N^{\lambda+1}(\eta_{\frac{N}{2}} f) \|_p. \tag{2.6}$$

Write $g = \eta_{\frac{N}{2}}(f)$ simply. We have

$$\sigma_N^\lambda(g) - \sigma_N^{\lambda+1}(g) = \sum_{k=0}^{N} \frac{A_{N-k}^\lambda}{A_N^\lambda}\left(1 - A_{N-k}^{\lambda+1} \frac{A_N^\lambda}{A_N^{\lambda+1} A_{N-k}^\lambda}\right) Y_k(g)$$

$$= \frac{1}{N+\lambda+1} \sum_{k=0}^{N} \frac{A_{N-k}^\lambda}{A_N^\lambda} k Y_k(g).$$

By the definition of derivative we have

$$Y_k(g') = (-k(k+2\lambda))^{\frac{1}{2}} Y_k(g).$$

We write $a_k = -\left(\frac{-k}{k+2\lambda}\right)^{\frac{1}{2}}$ and define

$$h = \sum_{k=0}^{+\infty} a_k Y_k(g').$$

Noticing that $\eta\left(\frac{2k}{N}\right) = 0$ whenever $k > \frac{N}{2}$, we have

$$\sigma_N^\lambda(g) - \sigma_N^{\lambda+1}(g) = \frac{1}{N+\lambda+1} \sigma_N^\lambda(h).$$

Applying Lemma 2.5 we get

$$\| \sigma_N^\lambda(g) - \sigma_N^{\lambda+1}(g) \|_p \leqslant C_n \frac{A_{pN}}{N+\lambda+1} \| h \|_p. \tag{2.7}$$

Using Abel transform we get

$$\| h \|_p \leqslant \left\| \sum_{k=0}^{+\infty} | \Delta^{n+1} a_k | A_k^n \| \sigma_k^n(g') \| \right\|_p.$$

Noticing that

$$| \Delta^{n+1} a_k | \leqslant C_n \frac{1}{k^{n+2}}$$

and applying Lemma 2.3 we get

$$\| h \|_p \leqslant C_n \| g' \|_p \leqslant C_n N \omega\left(f, \frac{1}{N}\right)_p. \tag{2.8}$$

So, combining (2.6)～(2.8), we have

$$\| f-\sigma_N^\lambda(f) \|_p \leqslant C_n A_{pN}\omega\left(f, \frac{1}{N}\right)_p$$

which completes the proof. □

§ 3. Strong uniform approximation

Theorem 3.1 Let $1<q<+\infty$. Then for $f\in C(S^{n-1})$,

$$\left\| \left(\frac{1}{N}\sum_{k=1}^{N} | \sigma_k^\lambda(f)(x) - f(x) |^q\right)^{\frac{1}{q}} \right\|_c \leqslant C_n \frac{q^2}{q-1}\omega\left(f, \left(\frac{\ln N}{N}\right)^{\frac{1}{q}}\right)_c.$$

Theorem 3.1 will be derived fom the following theorem.

Theorem 3.2 Suppose $1<q<+\infty$ and $f\in C(S^{n-1})$. Then

$$\left\| \left(\frac{1}{N}\sum_{k=\frac{N}{2}}^{N} | \sigma_N^\lambda(f)(x) - f(x) |^q\right)^{\frac{1}{q}} \right\|_c \leqslant C_n \frac{q^2}{q-1}\omega(f, N^{-\frac{1}{q}})_c.$$

To prove Theorem 3.2, we make use of the following formula which can be found in [7]:

$$\begin{aligned} &\sigma_k^\lambda(f)(x)-f(x) \\ &=[E_k^\lambda(f)(x)-E_k^\lambda(\mathbf{1})f(x)]+[T_k^\lambda(f)(x)-T_k^\lambda(\mathbf{1})f(x)], \end{aligned} \tag{3.1}$$

where **1** denotes the constant function with value 1 and

$$E_k^\lambda(f)(x) := O(k^{\frac{1}{2}})\int_{S^{n-1}} f(y) P_k^{\left(n-\frac{3}{2}, \frac{n-3}{2}\right)}(xy)\mathrm{d}y$$

is called equiconvergent operator of $\sigma_k^\lambda(f)(x)$ in [7], $P_k^{(\alpha,\beta)}(t)$ denotes Jacobi polynomial and

$$T_k^\lambda(f)(x) := \sum_{v=1}^{+\infty} O(v^{-\frac{3}{2}n})\sigma_k^{\lambda+1}(f)(x).$$

It follows easily from Theorem 2.1 that

$$\left\| \left(\frac{1}{N}\sum_{k=\frac{N}{2}}^{N} | T_k^\lambda(f)(x) - T_k^\lambda(\mathbf{1})f(x) |^q\right)^{\frac{1}{q}} \right\|_c \leqslant C_n\omega\left(f, \frac{1}{N}\right)_c. \tag{3.2}$$

To deal with the term $E_k^\lambda(f)(x)-E_k^\lambda(\mathbf{1})f(x)$, in the interval $k^{-1}\leqslant\theta\leqslant \pi-k^{-1}$ we use the formula

$$P_k^{(\alpha,\beta)}(\cos\theta)=\frac{1}{(\pi k)^{\frac{1}{2}}}\left(\frac{1}{\sin\frac{\theta}{2}}\right)^{\alpha+\frac{1}{2}}\left(\frac{1}{\cos\frac{\theta}{2}}\right)^{\beta+\frac{1}{2}}\left[\cos(\tilde{k}\theta+\gamma)+O\left(\frac{1}{k\sin\theta}\right)\right]$$

and elsewhere the estimate

$$|P_k^{(\alpha,\beta)}(\cos\theta)|\leqslant\begin{cases}C_{\alpha\beta}k^{\alpha}, & 0\leqslant\theta\leqslant\frac{\pi}{2},\\ C_{\alpha\beta}k^{\beta}, & \pi-k^{-1}\leqslant\theta\leqslant\pi,\end{cases}$$

where $\tilde{k}=k+\frac{\alpha+\beta+1}{2}$, $\gamma=-\left(\alpha+\frac{1}{2}\right)\frac{\pi}{2}$, (see [6]). We then have

$$|E_k^{\lambda}(f)(x)-E_k^{\lambda}(\mathbf{1})f(x)|\leqslant C_n\sum_{i=1}^{4}|I_k^{(i)}(x)|,\tag{3.3}$$

where $\quad I_k^{(1)}(x)=k^{n-1}\int_0^{\frac{\pi}{2k}}|S_\theta(f)(x)-f(x)|\sin^{n-2}\theta\,\mathrm{d}\theta,$

$$I_k^{(2)}(x)=\int_{\frac{\pi}{2k}}^{\frac{\pi}{2}}(S_\theta(f)(x)-f(x))\frac{M_n(\theta)}{\theta^{n-1}}\sin^{n-2}\theta\cos(\tilde{k}\theta+\gamma)\,\mathrm{d}\theta,$$

$$I_k^{(3)}(x)=\frac{1}{k}\int_{\frac{\pi}{2k}}^{\frac{\pi}{2}}|S_\theta(f)(x)-f(x)|\theta^{-2}\,\mathrm{d}\theta,$$

$$I_k^{(4)}(x)=O(k^{\frac{1}{2}})\int_{\frac{\pi}{2}}^{\pi}[S_\theta(f)(x)-f(x)]P_k^{\left(n-\frac{3}{2},\frac{n-3}{2}\right)}(\cos\theta)\sin^{n-2}\theta\,\mathrm{d}\theta,$$

With $\tilde{k}=k+\frac{3}{4}n-1$, $\gamma=-\frac{n-1}{2}\pi$ and $M_n(\theta)=\left(\frac{\theta}{\sin\frac{\theta}{2}}\right)^{n-1}\left(\frac{1}{\cos\frac{\theta}{2}}\right)^{\frac{n-2}{2}}$.

We see $M_n(\theta)\in C\left[0,\ \frac{\pi}{2}\right]$.

It is obvious that

$$\left\|\left(\frac{1}{N}\sum_{k=\frac{N}{2}}^{N}|I_k^{(1)}(x)|^q\right)^{\frac{1}{q}}\right\|_c\leqslant C_n\omega\left(f,\frac{1}{N}\right)_c.\tag{3.4}$$

In order to estimate $I_k^{(3)}(x)$ we use the following inequality (see [3])

$$\omega_r(f,\ lt)\leqslant C_{n,r}l^r\omega_r(f,\ t),\ r>0,\ l>1.\tag{3.5}$$

Then we get

$$\begin{aligned}\|I_k^{(3)}(x)\|_c&\leqslant C_nk^{-1}\int_{k^{-1}}^{\frac{\pi}{2}}\frac{\omega(f,\theta)}{\theta^2}\mathrm{d}\theta\leqslant C_nk^{-1}\sum_{j=1}^{k}\omega\left(f,\frac{1}{j}\right)\\&\leqslant C_nk^{-1}\sum_{1\leqslant j\leqslant k^{\frac{1}{q}}}\frac{k^{\frac{1}{q}}}{j}\omega(f,k^{-\frac{1}{q}})+C_nk^{-1}\sum_{k^{\frac{1}{q}}<j\leqslant k}\omega(f,k^{-\frac{1}{q}})\\&\leqslant C_n\frac{q^2}{q-1}\omega(f,k^{-\frac{1}{q}}),\end{aligned}$$

Which implies

$$\left\| \left(\frac{1}{N} \sum_{k=\frac{N}{2}}^{N} | I_k^{(3)}(x) |^q \right)^{\frac{1}{q}} \right\|_c \leqslant C_n \frac{q^2}{q-1} \omega(f, N^{-\frac{1}{q}})_c. \tag{3.6}$$

To estimate the term $I_k^{(2)}(x)$, we need the following two lemmas:

Lemma 3.3[5] Let $1 \leqslant p \leqslant +\infty$ and let

$$T_N(x) = \sum_{k=0}^{N} (\alpha_k \cos kx + \beta_k \sin kx)$$

be a trigonometric polynomial of degree not more than N. If

$$\left(\sum_{k=0}^{N} (|\alpha_k|^p + |\beta_k|^p) \right)^{\frac{1}{p}} \leqslant K \left(\frac{1}{N} \right)^{\frac{1}{p'}}, \quad \frac{1}{p} + \frac{1}{p'} = 1,$$

then for any $\delta > 0$, $\quad \left(\sum_{k=1}^{N} \sup_{x \in \left[\frac{k-1}{N}\delta, \frac{k}{N}\delta \right]} | T_N(x) |^{p'} \right)^{\frac{1}{p'}} \leqslant K,$

where K is a constant independent of N and T_N.

Lemma 3.4 Let $\sigma > 0$, $\tau > 0$ and $M(\theta)$ be a bounded function on $\left[0, \frac{\pi}{2} \right]$. Define $A_k^{(N)}(g) := \int_{\frac{\pi}{2N}}^{\frac{\pi}{2}} \frac{g(\theta) M(\theta)}{\theta^\sigma} (\alpha \cos k\theta + \beta \sin k\theta) \sin^\tau \theta \, d\theta.$

Then for $1 < q < +\infty$ and $\frac{1}{q} + \frac{1}{q'} = 1$,

$$\left(\frac{1}{N} \sum_{k=1}^{N} | A_k^{(N)}(g) |^q \right)^{\frac{1}{q}} \leqslant (|\alpha| + |\beta|) \left(\sum_{j=1}^{N-1} \left(\left(\frac{N}{j} \right)^\sigma \int_{\frac{j}{N}\frac{\pi}{2}}^{\frac{j+1}{N}\frac{\pi}{2}} | g(\theta) | \sin^\tau \theta \, d\theta \right)^{q'} \right)^{\frac{1}{q'}}.$$

Proof Choose $\{c_i\}$ such that

$$\left(\sum_{k=1}^{N} | c_k |^{q'} \right)^{\frac{1}{q'}} \leqslant \left(\frac{1}{N} \right)^{\frac{1}{q}} \tag{3.7}$$

and

$$\left(\frac{1}{N} \sum_{k=1}^{N} | A_k^{(N)}(g) |^q \right)^{\frac{1}{q}} = \sum_{k=1}^{N} c_k A_k^{(N)}(g) = \int_{\frac{\pi}{2N}}^{\frac{\pi}{2}} \frac{g(t)}{t^\sigma} M(t) T_N(t) \sin^\tau t \, dt,$$

where

$$T_N(t) = \sum_{k=1}^{N} c_k (\alpha \cos k\theta + \beta \sin k\theta).$$

Then by Hölder inequality, we get

$$\begin{aligned} & \left(\frac{1}{N} \sum_{k=1}^{N} | A_k^{(N)}(g) |^q \right)^{\frac{1}{q}} \\ \leqslant & \sum_{j=1}^{N-1} \sup_{t \in \left[\frac{j}{N}\frac{\pi}{2}, \frac{j+1}{N}\frac{\pi}{2} \right]} | T_N(t) | \left(\frac{N}{j} \right)^\sigma \int_{\frac{j}{N}\frac{\pi}{2}}^{\frac{j+1}{N}\frac{\pi}{2}} | g(\theta) | \sin^\tau \theta \, d\theta \end{aligned}$$

$$\leqslant \Big(\sum_{j=1}^{N-1} \sup_{t\in\left[\frac{j}{N}\frac{\pi}{2},\frac{j+1}{N}\frac{\pi}{2}\right]} |T_N(t)|^q\Big)^{\frac{1}{q}} \left(\sum_{j=1}^{N-1}\left(\left(\frac{N}{j}\right)^{\sigma}\int_{\frac{j}{N}\frac{\pi}{2}}^{\frac{j+1}{N}\frac{\pi}{2}} |g(\theta)|\sin^{\tau}\theta d\theta\right)^{q'}\right)^{\frac{1}{q'}},$$

which implies Lemma 3.4 on account of (3.7) and Lemma 3.3.

For $\frac{N}{2}\leqslant k\leqslant N$, we write $I_k^{(2)}(x)=I_k^{(2,1)}(x)+I_k^{(2,2)}(x)$,

where

$$I_k^{(2,1)}(x)=\int_{\frac{\pi}{2N}}^{\frac{\pi}{2}}(S_\theta(f)(x)-f(x))\frac{M_n(\theta)}{\theta^{n-1}}\sin^{n-2}\theta\cos(\tilde{k}\theta+\gamma)d\theta,$$

$$I_k^{(2,2)}(x)=-\int_{\frac{\pi}{2N}}^{\frac{\pi}{2k}}(S_\theta(f)(x)-f(x))\frac{M_n(\theta)}{\theta^{n-1}}\sin^{n-2}\theta\cos(\tilde{k}\theta+\gamma)d\theta.$$

It is obvious that

$$\left\|\left(\frac{1}{N}\sum_{k=\left[\frac{N}{2}\right]}^{N} I_k^{(2,2)}(x)\ |^q\right)^{\frac{1}{q}}\right\|_c\leqslant C_n\omega\left(f,\frac{1}{N}\right)_c. \tag{3.8}$$

On the other hand, it follows from Lemma 3.4 that

$$\left\|\left(\frac{1}{N}\sum_{k=\frac{N}{2}}^{N}|I_k^{(2,1)}|^q\right)^{\frac{1}{q}}\right\|_c\leqslant C_n\left(\sum_{j=1}^{N-1}\left\|\left(\frac{N}{j}\right)^{n-1}\int_{\frac{j\pi}{2N}}^{\frac{(j+1)\pi}{2N}}|S_\theta(f)-f|\sin^{n-2}\theta d\theta\right\|_c^{q'}\right)^{\frac{1}{q'}}.$$

Now using the inequality (3.5), we have

$$\left\|\left(\frac{1}{N}\sum_{k=\left[\frac{N}{2}\right]}^{N}|I_k^{(2,1)}|^q\right)^{\frac{1}{q}}\right\|_c\leqslant C_n\left(\sum_{j=1}^{N}\left[\frac{\omega\left(f,\frac{j+1}{N}\right)_c}{j}\right]^{q'}\right)^{\frac{1}{q'}}$$

$$\leqslant C_n\Big(\sum_{1\leqslant j\leqslant N^{\frac{1}{q}}} j^{-q'}\Big)^{\frac{1}{q'}}\omega(f,N^{-\frac{1}{q}})_c+C_n\left(\sum_{N^{\frac{1}{q}}<j\leqslant N-1}\left[\frac{1}{j}\frac{j}{N^{\frac{1}{q}}}\omega(f,N^{-\frac{1}{q}})_c\right]^{q'}\right)^{\frac{1}{q'}}$$

$$\leqslant C_n\frac{q^2}{q-1}\omega(f,N^{-\frac{1}{q}})_c. \tag{3.9}$$

Combining (3.8) and (3.9), we get

$$\left\|\left(\frac{1}{N}\sum_{k=\frac{N}{2}}^{N}|I_k^{(2)}|^q\right)^{\frac{1}{q}}\right\|_c\leqslant C_n\frac{q^2}{q-1}\omega(f,N^{-\frac{1}{q}})_c. \tag{3.10}$$

By a similar argument and using the formula $P_k^{(\alpha,\beta)}(-t)=(-1)^kP_k^{(\beta,\alpha)}(t)$, we can derive

$$\left\|\left(\frac{1}{N}\sum_{k=\frac{N}{2}}^{N}|I_k^{(4)}|^q\right)^{\frac{1}{q}}\right\|_c\leqslant C_n\frac{q^2}{q-1}\omega\left(f,\frac{1}{N}\right)_c. \tag{3.11}$$

Now Theorem 3.2 follows form (3.1)~(3.4)(3.6)(3.10)(3.11).

Proof of Theorem 3.1　Suppose $2^{k_0-1}\leqslant N<2^{k_0}$. Then by Theorem 3.2, we have

$$\left\|\left(\frac{1}{N}\sum_{k=1}^{N}|\sigma_k^\lambda(f)-f|^q\right)^{\frac{1}{q}}\right\|_c\leqslant\left(\frac{1}{N}\sum_{j=1}^{k_0}\left\|\left(\sum_{2^{j-1}\leqslant k<2^j}|\sigma_k^\lambda(f)-f|^q\right)^{\frac{1}{q}}\right\|_c^q\right)^{\frac{1}{q}}$$

$$\leqslant C_n\frac{q^2}{q-1}\left(\frac{1}{N}\sum_{j=1}^{k_0}2^j\omega^q(f,2^{-\frac{j}{q}})\right)^{\frac{1}{q}}$$

$$\leqslant C_n\frac{q^2}{q-1}\left(\frac{1}{N}\sum_{\substack{1\leqslant j\leqslant k_0\\ 2^j\geqslant\frac{N}{\ln N}}}2^j\omega^q\left(f,\left(\frac{\ln N}{N}\right)^{\frac{1}{q}}\right)\right)^{\frac{1}{q}}+$$

$$C_n\frac{q^2}{q-1}\left(\frac{1}{N}\sum_{\substack{1\leqslant j\leqslant k_0\\ 2^j<\frac{N}{\ln N}}}2^j\frac{N}{2^j\ln N}\omega^q\left(f,\left(\frac{\ln N}{N}\right)^{\frac{1}{q}}\right)\right)^{\frac{1}{q}}$$

$$\leqslant C_n\frac{q^2}{q-1}\omega\left(f,\left(\frac{\ln N}{N}\right)^{\frac{1}{q}}\right).$$

Acknowledgments

The authors are grateful to the referees for their valuable comments about this paper.

References

[1] Bonami A, Clerc J L. Sommes de Cesàro et multiplicateures des developments en harmonique sphériques, Trans. AMS., 1973, 183—223.

[2] Löfsröm J, Peetre J. Approximation theorems connected with generanized translations, Math. Ann., 1969, 181—225.

[3] Riemenschneider S, Wang Kun-yang. Approximation theorems of Jackson type on the sphere, Adv. Math., 1995, 24—184.

[4] Rustamov. On the approximation of functions on sphere. Izv. Akad. Nauk SSSR Ser. Math., 1993, 59—127.

[5] Stepanets A I, Lasuriya R L. Strong summability of orthogonal expansions of summable functions I. Ukrainian Math. J., 1996, 48: 121.

[6] Szegö. Orthogonal Polynomials. Amer. Math. Soc. Colloq. Publ., 1975.

[7] Wang Kun-yang. Equiconvergent operator of Cesàro means on sphere and its application. J. Beijing Normal Univ., 1993, 29—143.

[8] Wang Kun-yang, Zhang Pu. Strong uniform approximation by Cesàro means of Fourier-Laplace series. J. Beijing Normal Univ., 1994, 30—321.

Approx. Theory and its Appl., 2000, 16(1): 26-35.

定义在球面上的函数的一些构造性质

Some Constructive Properties of Functions Defined on the Sphere[①]

Abstract The norm of difference operators for functions defined on the sphere is investigated. A mistake in Rustamov's results is pointed out by a counterexample. And correct result is given.

Results of convergence in norm of difference operators acting on the differentiable functions on the sphere are obtained. The highest possible degree tending to zero for the moduli of continuity of fractional order for non-constant valued functions are discussed.

§1. Notations and results

Let $n \in \mathbf{N}$, $n \geqslant 2$. Define

$$S^{n-1} = \{\boldsymbol{x} = (x_1, x_2, \cdots, x_n) \in \mathbf{R}^n \,|\, x_1^2 + x_2^2 + \cdots + x_n^2 = 1\}.$$

For $x \in S^{n-1}$ $(n \geqslant 3)$, and $\gamma \in \mathbf{R}$ define

$$l_{x,\gamma} = \{y \in S^{n-1} \,|\, xy = \cos\gamma\}, \quad l(\gamma) = |S^{n-2}|\,|\sin\gamma|^{n-2},$$

where $|S^{n-2}|$ denotes the surface measure of S^{n-2}.

In what follows we always assume $n \geqslant 3$. On the space $L(S^{n-1})$, the translation operator S_γ is defined by

① Supported by NSFC 19771009.

本文与王晟合作.

Received: 1998-05-07; Revised: 1999-05-30.

$$S_\gamma(f)(x)=\frac{1}{l(\gamma)}\int_{l(x,\gamma)}f(y)\mathrm{d}l_{x,\gamma}(y),\quad f\in L(\mathrm{S}^{n-1}).$$

This definition may be regarded as motivated by the work [1] of Löfström and Peetre. Based on this operator the difference operator of degree $r>0$ is defined (see [2]) as

$$\Delta_t^r=(I-S_t)^{\frac{r}{2}}=\sum_{k=0}^{+\infty}(-1)^k\mathrm{C}_{\frac{r}{2}}^k(S_t)^k,$$

where I denotes the identity operator and

$$\mathrm{C}_{\frac{r}{2}}^k=\frac{1}{k!}\frac{r}{2}\left(\frac{r}{2}-1\right)\cdots\left(\frac{r}{2}-k+1\right).$$

Let X denote one of the function spaces $L^p(\mathrm{S}^{n-1})$, $1\leqslant p<+\infty$ or $C(\mathrm{S}^{n-1})$, generally. Then the mudulus of continuity of degree $r>0$ in the space X is defined by

$$\omega_r(f;\ t)_X=\sup\{\|\Delta_s^r(f)\|_X\mid 0<s<t\},\quad 0<t\leqslant\pi.$$

Concerning the relation between the moduli of different degrees Rustamov (see [2] and [3]) gives the following inequality

$$\omega_r(f;\ t)\leqslant 2^{\frac{r-q}{2}}\omega_q(f;\ t),\quad 0<q<r.$$

But, as we will show, this is incorrect in some cases. Our first result is the following

Theorem 1 If $0<q<r$, then for all $f\in X$ and $t>0$,

$$\omega_r(f;\ t)_X\leqslant 2^{\frac{r-q}{2}}\omega_q(f;\ t)_X,$$

where $\{r\}$ is the smallest integer not less than r. The factor $2^{\frac{r-q}{2}}$ in the right side is best possible when $X=C$ and can be replaced exactly by $2^{\frac{r-q}{2}}$ when $X=L^2$.

As a very basic fact, it is known that for any periodic continuous function f the condition

$$\lim_{t\to 0}t^{-1}\sup\{|f(x)-f(y)|\mid |x-y|<t\}=0$$

implies $f=$constant.

We will establish a generalization of this kind of conclusion on the sphere. This is our second result, i.e.,

Theorem 2 Let $f\in X$, $r>0$. If

$$\omega_r(f;\ t)_X = o(t^r) \quad \text{as } t \to 0,$$

then $f=$constant.

We use $\mathscr{H}_k^n$, $k \in \mathbf{N}^*$ to denote the function space of spherical harmonics of n variables of degree k. If $f \in L(S^{n-1})$, for a function $f \in L(S^{n-1})$ its projection to the space $\mathscr{H}_k^n$ is defined by a convolution

$$Y_k(f) = f * c_{nk} P_k^n,$$

where

$$c_{nk} = \frac{\Gamma\left(\frac{n}{2}\right)(n+2k-2)\Gamma(n+k-2)}{2\pi^{\frac{n}{2}}\Gamma(k+1)\Gamma(n-1)},$$

and $P_k^n(t)$ denote the ultraspherical polynomials which can be expressed as

$$P_k^n(t) = \frac{\Gamma(k+1)\Gamma\left(\frac{n-1}{2}\right)}{\Gamma\left(k+\frac{n-1}{2}\right)} P_k^{\left(\frac{n-3}{2},\frac{n-3}{2}\right)}(t)$$

with $P_k^{(\alpha,\beta)}$ denoting Jacobi polynomials. For Jacobi polynomials we refer the reader to [4]. For the convolution of functions defined on the sphere we refer to [5]. The fractional derivatives of functions on the sphere are defined as follows.

Definition Let $r>0$, f, $g \in L(S^{n-1})$. If g has Fourier-Laplace series

$$e^{ir\frac{\pi}{2}} \sum_{k=1}^{+\infty} (k(k+n-2))^{\frac{r}{2}} Y_k(f),$$

then g is called the derivative of degree r of f and is written as $f^{(r)}$ or $D^{\frac{r}{2}}(f)$ with D denoting the Laplace-Beltrami operator (see [6]).

By a result of Askey and Wainger in [7] we can prove easily that for any $g \in X$ and $r>0$, there exists a unique $f \in L(S^{n-1})$ such that

$$f \in X, \quad Y_0(f)=0, \quad f^{(r)} = g - Y_0(g).$$

In fact this f has the expansion

$$e^{-ir\frac{\pi}{2}} \sum_{k=1}^{+\infty} (k(k+n-2))^{-\frac{r}{2}} Y_k(g).$$

We call this function f the integral of degree r of $g-Y_0(g)$ and is written as $I^{(r)}(g-Y_0(g))$. Define

$$W_X^{2r}=\{f\in X\mid f^{(2r)}\in X\}=\{f\in L(S^{n-1})\mid f^{(2r)}\in X\}.$$

Our third result is

Theorem 3 Let $r>0$. If $f\in W_X^{(2r)}$, then

$$\lim_{t\to 0^+}\|(-1)^r(2n-2)^r t^{-2r}\Delta_t^{(2r)}(f)-f^{(2r)}\|_X=0.$$

§ 2. Proof of Theorem 1

By the definition we have for $0<q<r$,

$$\Delta_t^r=\Delta_t^{r-q}\circ\Delta_t^q.$$

Noticing that S_γ is even and 2π-periodic with respect to γ we get

$$\omega_r(f;\ t)_X\leqslant\sup_{0<h<\pi}\|\Delta_h^{r-q}\|_{(X,X)}\omega_q(f;\ t)_X.$$

We now estimate the norm from X to X of the operator Δ_t^r. Since the space X is fixed in the discussion we omit the index X in the notation. If $r\in\mathbf{N}$ then by definition the operator $\Delta_t^{(2r)}$ is exactly a r-time composition of $(I-S_t)$. Hence

$$\|\Delta_t^{(2r)}\|\leqslant\|I-S_t\|^r\leqslant(\|I\|+\|S_t\|)^r=2^r.$$

Here we have applied the basic fact $\|S_t\|=1$.

We assume now $0<r<1$. By the definition we have

$$\Delta_t^{(2r)}=(I-S_t)^r=I-r\sum_{k=1}^{+\infty}\frac{\Gamma(k-r)}{\Gamma(1-r)\Gamma(k+1)}(S_t)^k.$$

Then we see

$$\|\Delta_t^{(2r)}\|\leqslant\|I\|+r\sum_{k=1}^{+\infty}\frac{\Gamma(k-r)}{\Gamma(1-r)\Gamma(k+1)}\|(S_t)^k\|\leqslant 2=2^{(r)}.$$

To show the bound 2 is best possible in the case, when $X=C$ we take $\varepsilon\in(0,\ 1)$ and define

$$h(t)=\begin{cases}1-\dfrac{2}{\varepsilon}t, & 0\leqslant t\leqslant\varepsilon,\\ -1, & \varepsilon<t\leqslant\pi,\end{cases}$$

and

$$f(x)=h(\arccos xe),\qquad \boldsymbol{e}=(1,\ 0,\ \cdots,\ 0)\in S^{n-1},\qquad x\in S^{n-1}.$$

Let $t=\dfrac{\pi}{2}$. Then for any $x\in S^{n-1}$,

$$S_t(f)(x)\leqslant -1+\frac{\varepsilon}{\pi}.$$

In order to prove this inequality we define

$$l(e,\ x)=\{y\in S^{n-1}\mid ye\geqslant\cos\varepsilon\}$$

which is the intersection of the set $l_{x,x}$ (which can be called a parallel of the sphere S^{n-1}) with the set $\{y\in S^{n-1}\mid ye\geqslant\cos\varepsilon\}$ (which can be called a crown on S^{n-1}). Then we have for $t=\frac{\pi}{2}$,

$$\begin{aligned}S_t(f)(x) &= \frac{1}{l(t)}\int_{l(x,t)} f(y)\mathrm{d}l_{x,t}(y)\\ &=-1+\frac{1}{l(t)}\int_{l(x,t)}(f(y)+1)\mathrm{d}l_{x,t}(y)\\ &=-1+\frac{1}{l(t)}\int_{l(e,x)}(f(y)+1)\mathrm{d}l_{x,t}(y).\end{aligned}$$

By a geometrical observation we know that the integral

$$\frac{1}{l(t)}\int_{l(e,x)}(f(y)+1)\mathrm{d}l_{x,t}(y)$$

reaches its maximum as $\boldsymbol{ex}=0$. In this case, without loss of generality we may assume $\boldsymbol{x}=(0,\ \cdots,\ 0,\ 1)$. Noticing $\boldsymbol{e}=(1,\ 0,\ \cdots,\ 0)$ we have

$$l(\boldsymbol{e},\ \boldsymbol{x})=\{y=(y_1,\ y_2,\ \cdots,\ y_{n-1},\ 0)\in S^{n-1}\mid y_1\geqslant\cos\varepsilon\}.$$

Then we see

$$|l(\boldsymbol{e},\boldsymbol{x})|=\int_{\{y\mid y_1^2+y_2^2+\cdots+y_n^2=1,\,y_1\geqslant\cos\varepsilon\}}\left(2-\frac{2\arccos y_1}{\varepsilon}\right)\mathrm{d}S^{n-2}(y).$$

We make use of the following coordinate transform of the sphere S^{n-2}:

$$\begin{cases}y_1=\cos\theta_1,\\ y_2=\sin\theta_1\cos\theta_2,\\ \cdots\cdots\\ y_{n-2}=\sin\theta_1\cdots\sin\theta_{n-3}\cos\theta_{n-2},\\ y_{n-1}=\sin\theta_1\cdots\sin\theta_{n-3}\sin\theta_{n-2},\end{cases}$$

where $0<\theta_j<\pi$, $j=1,2,\cdots,n-3$, $0<\theta_{n-2}<2\pi$.

Then by the basic fact of the calculus we have for $n=3$,

$$|l(e,x)|=\int_0^{\varepsilon}\left(2-\frac{2\theta_1}{\varepsilon}\right)\mathrm{d}\theta_1=\varepsilon.$$

Meanwhile

$$l\left(\frac{\pi}{2}\right)=|S^1|=\int_0^{2\pi}\mathrm{d}\theta_1=2\pi.$$

Then we get exactly

$$\frac{2|l(e,x)|}{l\left(\frac{\pi}{2}\right)}=\frac{\varepsilon}{\pi}.$$

When $n\geqslant 4$ we have

$$|l(e,x)|=\int_0^{\varepsilon}\left(2-\frac{2\theta_1}{\varepsilon}\right)(\sin\theta_1)^{n-3}\mathrm{d}\theta_1\int_{(0,\pi)^{n-4}}(\sin\theta_2)^{n-4}\cdots$$
$$(\sin\theta_{n-3})^1\mathrm{d}\theta_2\cdots\mathrm{d}\theta_{n-3}\int_0^{2\pi}\mathrm{d}\theta_{n-2}.$$

Meanwhile

$$l\left(\frac{\pi}{2}\right)=|S^{n-2}|=\int_{(0,\pi)^{n-3}}(\sin\theta_1)^{n-3}\cdots(\sin\theta_{n-3})^1\mathrm{d}\theta_1\cdots\mathrm{d}\theta_{n-3}\int_0^{2\pi}\mathrm{d}\theta_{n-2}.$$

From these we find

$$\frac{2|l(e,x)|}{l\left(\frac{\pi}{2}\right)}=\frac{2\int_0^{\varepsilon}\left(2-\frac{2\theta}{\varepsilon}\right)(\sin\theta)^{n-3}\mathrm{d}\theta}{\int_0^{\pi}(\sin\theta)^{n-3}\mathrm{d}\theta}.$$

If $n=4$ the above equality yields

$$\frac{2|l(e,x)|}{l\left(\frac{\pi}{2}\right)}<\frac{1}{\pi}\varepsilon^2.$$

When $n\geqslant 5$ we make use of the formula

$$\int_0^{\pi}(\sin\theta)^{n-3}\mathrm{d}\theta=\frac{\Gamma\left(\frac{n-2}{2}\right)\Gamma\left(\frac{1}{2}\right)}{\Gamma\left(\frac{n-1}{2}\right)},$$

where Γ denotes the γ-function. Then noticing the inequality $\sin\theta<\theta$ for $0<\theta<1$ we get

$$\frac{2|l(e,x)|}{l\left(\frac{\pi}{2}\right)}\leqslant\frac{4B(2,n-2)\Gamma\left(\frac{n-1}{2}\right)}{\Gamma\left(\frac{n-1}{2}\right)\Gamma\left(\frac{1}{2}\right)}\varepsilon^{n-2},$$

where $B(u,v)$ denotes the β-function. Then we obtain for $n>4$,

$$\frac{2|l(e,x)|}{l\left(\frac{\pi}{2}\right)}\leqslant\frac{2(n-3)}{(n-1)(n-2)\sqrt{\pi}}\varepsilon^{n-2}<\frac{\varepsilon}{\pi}.$$

Therefore we get in all cases when $n\geqslant 3$,

$$S_t(f)(x)\leqslant -1+\frac{\varepsilon}{\pi}.$$

Since for all $x\in S^{n-1}$ and $t=\frac{\pi}{2}$,

$$S_t(f)(x)\leqslant -1+\frac{\varepsilon}{\pi},$$

we have

$$S_t^2(f)(x)=\frac{1}{l(t)}\int_{l(x,t)}S_t(f)(y)\mathrm{d}l_{x,t}(y)\leqslant -1+\frac{\varepsilon}{\pi}.$$

And furthermore for all $k\in\mathbf{N}$,

$$(S_t)^k(f)(x)\leqslant -1+\frac{\varepsilon}{\pi}.$$

We get

$$\Delta_t^{2r}(f)(e)=f(e)-r\sum_{k=1}^{+\infty}\frac{\Gamma(k-r)}{\Gamma(1-r)\Gamma(k+1)}(S_t)^k(f)(e)\geqslant 2-\frac{\varepsilon}{\pi}.$$

Since $\|f\|=1$ we see

$$\|\Delta_t^{2r}\|\geqslant 2-\frac{\varepsilon}{\pi}.$$

Taking limit $\varepsilon\to 0$ we get $\|\Delta_t^{2r}\|=2$.

We have proved the inequality in the theorem for all spaces X and have shown the constant $2^{\frac{r-q}{2}}$ in the right side is best possible in the particular case when $X=C$.

Now we consider the case of $X=L^2(S^{n-1})$. For $f\in L^2(S^{n-1})$ we have

$$Y_l(\Delta_h^{2r}(f))=\sum_{k=0}^{+\infty}(-1)^k C_r^k Y_l(S_h^k(f)).$$

According to the fact

$$Y_l(S_h(f))=P_l^n(\cos h)Y_l(f),$$

we get

$$Y_l(\Delta_h^{2r}(f))=\sum_{k=0}^{+\infty}(-1)^h C_r^k(P_l^n(\cos h))^h Y_l(f)=(1-P_l^n(\cos h))^r Y_l(f).$$

Hence

$$\Delta_h^{2r}(f)=\sum_{k=1}^{+\infty}(1-P_k^n(\cos h))^r Y_k(f).$$

By Parseval equality we have

$$\begin{aligned}\|\Delta_h^{2r}(f)\|^2&=\sum_{k=1}^{+\infty}(1-P_h^n(\cos h))^{2r}\|Y_k(f)\|^2\\&\leqslant\sum_{k=1}^{+\infty}2^{2r}\|Y_k(f)\|^2.\end{aligned}$$

Hence

$$\|\Delta_h^{2r}(f)\|\leqslant 2^r\|f\|.$$

For odd k we take $f\in\mathscr{H}_k^n$, i. e., f is a spherical harmonic of degree $k(\|f\|>0)$. Then for $h=\pi$ we have

$$\Delta_h^{2r}(f)=(1-P_k^n(-1))^r Y_h(f)=2^r f,$$

which shows the best possibility of the factor 2^r. □

§ 3. Proof of theorem 2

We still omit the index X in the notation since it is fixed in the discussion.

Take a function $\eta\in C^\infty[0, +\infty)$ such that $\chi_{[0,1]}\leqslant\eta\leqslant\chi_{[0,2]}$. For any $m\in\mathbf{N}$, we define an operator

$$\eta_m(f):=\sum_{k=0}^{+\infty}\eta\left(\frac{k}{m}\right)Y_k(f),\quad f\in X.$$

It is obvious that $\eta_m(f)\in H_{2m-1}$, where

$$H_k:=\oplus\sum_{j=0}^{k}\mathscr{H}_j^n,\quad k\in\mathbf{N}.$$

By a result of [2]

$$\|(\eta_m(f))^{(r)}\|\leqslant c(n, r, \eta)t^{-r}\|\Delta_t^r(f)\|,$$

where $c(n, r, \eta)$ is a constant depending only on the choice of n, r and η. We get

$$\|(\eta_m(f))^{(r)}\|\leqslant c(n, r, \eta)^{-r}\omega_r(f; t).$$

Then

$$\|(\eta_m(f))^{(r)}\|\leqslant\varliminf_{t\to 0}\|(\eta_m(f))^{(r)}\|=0.$$

Since

$$(\eta_m(f))^{(r)} = e^{ir\frac{\pi}{2}}\sum_{k=1}^{2m}\eta\left(\frac{k}{m}\right)(k(k+n-2))^{\frac{r}{2}}Y_k(f)$$

$$=\sum_{k=1}^{m}(-k(k+n-2))^{\frac{r}{2}}Y_k(f)+\sum_{k=m+1}^{2m}\eta\left(\frac{k}{m}\right)(-k(k+n-2))^{\frac{r}{2}}Y_k(f),$$

we conclude

$$Y_k(f)=0,\qquad k=1,2,\cdots,m.$$

Then by the arbitrariness of $m\in\mathbf{N}$ we see

$$f=Y_0(f).\qquad\square$$

Remark Wehrens [8] defined in a different manner the difference operator and modulus of continuity only for even degree and stated a result similar to Theorem 2 for $r\in\mathbf{N}$ (corresponding to the even degree). Earlier on the result for $r=1$ (corresponding to the degree 2) was proved in [9].

§ 4. Proof of Theorem 3

Write

$$a(k,\ t)=\frac{n-1}{k(k+n-2)}\ \frac{1-P_k^n(\cos t)}{1-\cos t}.$$

By the formula

$$\frac{d}{dx}P_k^n(x)=\frac{k(k+n-2)}{n-1}P_{k-1}^{n+2}(x),$$

(see [4], (4.21.7)) we get

$$a(k,t)=\frac{1}{1-\cos t}\int_{\cos t}^{1}P_{k-1}^{n+2}(x)dx=\frac{1}{1-\cos t}\int_0^t\sin uP_{k-1}^{n+2}(\cos u)du.$$

Therefore

$$1-P_k^n(\cos t)=\frac{k(k+n-2)}{n-1}\int_0^t\sin uP_{k-1}^{n+2}(\cos u)du.$$

Hence we get

$$(2-2n)^r t^{-2r}\Delta_t^{2r}(f)=(2-2n)^r t^{-2r}\sum_{k=1}^{+\infty}(1-P_k^n(\cos t))^r Y_k(f)$$

$$=\sum_{k=1}^{+\infty}(-k(k+n-2))^r Y_k(f)\left(2t^{-2}\int_0^t\sin uP_{k-1}^{n+2}(\cos u)du\right)^r.$$

We define

$$G := \{g \in X \mid Y_0(g) = 0\}.$$

It is easy to see that G is a Banach space with the norm of X. Now we define the linear operator T_s, $s>0$ on G by

$$T_s(g) = (-1)^r (2n-2)^r s^{-2r} \Delta_s^{2r}(I^{(2r)}(g)), \qquad g \in G.$$

Then we find for every $g \in G$,

$$T_s(g) = \sum_{k=1}^{+\infty} (-k(k+n-2))^r Y_k(I^{(2r)}(g)) \left(2t^{-2} \int_0^t \sin u P_{k-1}^{\frac{n+2}{2}}(\cos u) \mathrm{d}u\right)^r.$$

By a result of Rustamov (see [2], Lemma 3. 8) we have

$$\omega_r(f; t)_X \leqslant B_{nr} t^r \| f^{(r)} \|_X, \qquad r>0.$$

Then we derive that for every $g \in G$, $s>0$,

$$\begin{aligned} \| T_s(g) \| &= \| (-1)^r (2n-2)^r s^{-2r} \Delta_t^{2r}(I^{(2r)}(g)) \|_X \\ &\leqslant (2n-2)^r B_{nr} \| g \|_X. \end{aligned}$$

This shows that T_s is a uniformly bounded linear operator on G with respect to $s>0$.

On the other hand, for any spherical polynomial $P \in \oplus \sum_{k=1}^{N} \mathscr{H}_k^n \subset G$, $N \in \mathbf{Z}$, we have

$$P = \sum_{k=1}^{N} Y_k(P),$$

hence uniformly in Ω_n.

$$\begin{aligned} & \lim_{s \to 0^+} T_s(P) \\ &= \lim_{s \to 0^+} \sum_{k=1}^{N} (-k(k+n-2))^r Y_k(I^{(2r)})(P) \left(2s^{-2} \int_0^s \sin u P_{k-1}^{\frac{n+2}{2}}(\cos u) du\right)^r \\ &= \sum_{k=1}^{m} (-k(k+n-2))^r Y_k(I^{(2r)})(P) \lim_{s \to 0^+} \left(2s^{-2} \int_0^s \sin u P_{k-1}^{\frac{n+2}{2}}(\cos u) \mathrm{d}u\right)^r \\ &= D^r(I^{(2r)}(P)) = P. \end{aligned}$$

It is obvious that the set of spherical polynomials is dense in G, so by Banach-Steinhaus Theorem we conclude

$$\lim_{s \to 0^+} \| T_s(g) - g \|_X = 0, \qquad \forall g \in G.$$

For any $f \in W_X^{2r}$, write $g = f^{(2r)}$. Then $g \in G$ and it is easy to verify that

$$T_s(g) = (-1)^r (2n-2)^r s^{-2r} \Delta_s^{2r}(f).$$

Then we derive

$$\lim_{s\to 0^+}\|(-1)^r(2n-2)^r s^{-2r}\Delta_s^{2r}(f)-f^{(2r)}\|_X=0,\qquad \forall f\in W_X^{2r}. \quad \square$$

Remark When $r=1$, the conclusion of Theorem 3 was proved in [8](and earlier on in [9]) by applying an integral representation for $\Delta_t(f)$. Such kind of representation can be easily extended to the case of $r\in\mathbf{N}$(i. e., the case of even degree) but can not be established in the case when r is not an integer including the case of odd degree and the fractional (non-integral) degree.

References

[1] Löfström J, Peetre J. Approximation theorems connected with generalized translation. Math. Ann., 1969, 181: 255—268.

[2] Rustamov Kh P. On the approximation of functions on sphere. Izv. Akad. Nauk SSSR Ser. Math., 1993, 59(5): 127—148.

[3] Rustamov Kh P. On the equivalence of some moduli of smoothness on the sphere. Dokl. Akad. Nauk SSSR, 1991, 321(1).

[4] Szegö G. Orthogonal polynominals. American Mathematics Society Colloquium Publication. Vol 23, 1939.

[5] Dunkl C. Operators and harmonic analysis on the sphere. Trans. AMS., 1966, 125: 250—263.

[6] Müller C. Sphereical harmonics. Lecture Notes in Math., 17, 1966.

[7] Askey R, Wainger S. On the behavior of special classes of ultraspherical-I. Journal D'Analyse Mathématique, 1965, 15: 193—220.

[8] Wehrens M. Best approximation on the unit sphere in $\mathbf{R}^k$. Functional Analysis and Approximation, (Proceeding of Conference at Oberwolfach, 1980), Birkhäuser, 1981, 233—245.

[9] Berens H, Butzer P L, Pawelke S. Lmmtierungserfahren von reihen mehrdimensionaler kungelfunktionen und deren saturationsverhalten. Publ. RIMS, Kyoto Univ. Ser. A, 1968, 4: 201—268.

Journal of Approximation Theory，2004，128：103－114.

L^2-函数的 Fourier-Laplace 级数的收敛速度

Convergence Rate of Fourier-Laplace Series of L^2-Functions①②

Abstract　The almost everywhere convergence rates of Fourier-Laplace series are given for functions in certain subclasses of $L^2(S^{n-1})$ defined in terms of moduli of continuity.

Keywords　Almost everywhere convergence；Fourier-Laplace series；Modulus of continuity.

§1. Introduction and main results

Let $n \geqslant 3$ and let $S^{n-1} = \{(x_1, x_2, \cdots, x_n) \mid x_1^2 + x_2^2 + \cdots + x_n^2 = 1\}$ be the unit sphere of $\mathbf{R}^n$ equipped with the normal Lebesgue measure. Let $f \in L^2(S^{n-1})$ and let

$$f \sim \sigma(f)(x) := \sum_{k=0}^{+\infty} Y_k(f)(x) \tag{1.1}$$

be the Fourier-Laplace series of f，where Y_k is the projection operator from $L^2(S^{n-1})$ to the space of all spherical harmonics of degree k. The Cesàro means of order δ of $\sigma(f)$ are defined as usual by

$$\sigma_N^\delta(f) := (A_N^\delta)^{-1} \sum_{k=0}^{N} A_{N-k}^\delta Y_k(f),$$

① Supported in part by Science Council of Taipei in China.

② Supported in part by NSFC.

本文与 Lin Chin-Cheng 合作.

Received：2002-07-09；Revised：2004-04-27.

where

$$A_N^\delta=\frac{\Gamma(N+\delta+1)}{\Gamma(\delta+1)\Gamma(N+1)},\qquad \delta>-1, \tag{1.2}$$

are the coefficients of the power series of the function $(1-x)^{-\delta-1}$, $|x|<1$; i. e.

$$(1-x)^{-\delta-1}=\sum_{k=0}^{+\infty}A_k^\delta x^k. \tag{1.3}$$

It is obvious that $\sigma_N^0(f)=\sum_{k=0}^{N}Y_k(f)$ is just the Nth partial sum of the Fourier-Laplace series of f.

Our main purpose is to find the convergence rate of $\sigma_N^0(f)$ on a set of full measure in S^{n-1} for any f in certain subclasses of $L^2(\mathrm{S}^{n-1})$ defined in terms of modulus of continuity.

In order to define modulus of continuity on the sphere, we first introduce the translation operator S_θ with step $\theta\in\mathbf{R}$. As in [7, p. 58], we define

$$S_\theta(f)(x):=\frac{1}{|\mathrm{S}^{n-2}|}\int_{\{y\in \mathrm{S}^{n-1}\mid xy=0\}}f(x\cos\theta+y\sin\theta)\mathrm{d}\ell(y),$$

here $x\in\mathrm{S}^{n-1}$ and $\mathrm{d}\ell(y)$ denotes the usual Lebesgue measure elements on the $n-2$ dimensional manifold $\{y\in\mathrm{S}^{n-1}\mid xy=0\}$. We know (see [7, p. 61, (2. 4. 6)]) that, for every $k\in\mathbf{N}$.

$$Y_k(S_\theta(f))=P_k^n(\cos\theta)Y_k(f),$$

where P_k^n is Gegenbauer polynomial defined by

$$P_k^n(t)=\frac{P_k^{(\frac{n-3}{2},\frac{n-3}{2})}(t)}{P_k^{(\frac{n-3}{2},\frac{n-3}{2})}(t)},\qquad |t|\leqslant 1, \tag{1.4}$$

with $P_k^{\alpha,\beta}$ being Jacobi polynomial. By the formulas [6, p. 58, (4. 1. 1)] and [p. 168, (7. 32. 2)], we have

$$|P_k^n(\cos\theta)|\leqslant\begin{cases}1,\\ \dfrac{\gamma}{(k\theta(\pi-\theta))^{\frac{n-2}{2}}},\end{cases}\qquad \text{for all }\theta\in(0,\pi), \tag{1.5}$$

where $\gamma>1$ is a constant depending only on n. Then we get

$$\|S_\theta(f)\|_2=\Big(\sum_{k=0}^{+\infty}|P_k^n(\cos\theta)|^2\|Y_k(f)\|_2^2\Big)^{\frac{1}{2}}\leqslant\|f\|_2.$$

Thus, we conclude that as an operator from $L^2(\mathrm{S}^{n-1})$ to $L^2(\mathrm{S}^{n-1})$, S_θ has norm 1; that is, $\|S_\theta\|_{(L^2(\mathrm{S}^{n-1}),L^2(\mathrm{S}^{n-1}))}=1$. Let I be the identity operator and let s be a positive number. We set $\psi_s(u):=(1-u)^{\frac{S}{2}}$, Following [5], we call the operator

$$\Delta_\theta^s := (I-S_\theta)^{\frac{S}{2}} = \sum_{k=0}^{+\infty}\frac{\psi_s^{(k)}(0)}{k!}S_\theta^k$$

an sth order difference operator. It is obvious that

$$\|\Delta_\theta^s\|_{(L^2(\mathrm{S}^{n-1}),L^2(\mathrm{S}^{n-1}))} \leqslant \sum_{k=0}^{+\infty}\frac{|\psi_s^{(k)}(0)|}{k!} < +\infty.$$

We define (as in [5]) the sth order modulus of continuity of a function $f\in L^2(\mathrm{S}^{n-1})$ by

$$\omega_s(f,\ t)_2 := \sup\{\|\Delta_u^s f\|_2 \mid 0<u\leqslant t\}.$$

Since all of our discussion are in $L^2(\mathrm{S}^{n-1})$, in what follows, we will omit the subscription "2" in the norm and in the moduli.

It is well known (see [5]) that, for $0<\alpha<\beta$ and $f\in L^2(\mathrm{S}^{n-1})$,

$$\omega_\beta(f,\ t)\leqslant C(\alpha,\ \beta)\omega_\alpha(f,\ t), \tag{1.6}$$

where $C(\alpha,\ \beta)$ is a constant depending only on α and β.

Definition 1 Let $s>0$, $r\in\mathbf{R}$, and $f\in L^2(\mathrm{S}^{n-1})$. If

$$\int_0^1 \frac{\omega_s(f,t)^2}{t}\ln^r\left(\frac{2}{t}\right)\mathrm{d}t < +\infty,$$

then we say that f satisfies the condition $(\{s,\ r\})$.

Our first result is the following theorem:

Theorem 1 Let $r\geqslant 1$. If there exists $s>0$ such that f satisfies the condition $(\{s,\ r\})$, then

$$\sigma_N^0(f)(x)-f(x)=o\left(\frac{1}{\ln^{\frac{r-1}{2}}N}\right) \quad \text{as } N\to+\infty$$

holds on S^{n-1} almost everywhere.

Our second theorem is about the case $0\leqslant r<1$.

Theorem 2 Let $0\leqslant r<1$. If there exists $s>0$ such that f satisfies the condition $(\{s,\ r\})$, then

$$\lim_{N\to+\infty}\ln^{\frac{r-1}{2}}N(\sigma_N^0(f)(x)-f(x))=0$$

holds on S^{n-1} almost everywhere.

Remark 1 Since the modulus of continuity is defined in square integrable terms, the convergence rate in L^2-norm can be obtained easily. For example, if f satisfies the condition $(\{2, r\})$, i. e.

$$f_{\frac{r+1}{2}} \sim \sum_{k=0}^{+\infty} \ln^{\frac{r+1}{2}}(k+2) Y_k(f) \in L^2(S^{n-1}),$$

then we can easily get

$$\begin{aligned}\|\sigma_N^0(f)-f\|_2 &= \Big\{\sum_{k=N+1}^{+\infty} \ln^{-(r+1)}(k+2)\|Y_k(f_{\frac{r+1}{2}})\|_2^2\Big\}^{\frac{1}{2}} \\ &= o\Big(\frac{1}{\ln^{\frac{r+1}{2}}(N+2)}\Big).\end{aligned}$$

This order $\ln^{-\frac{r+1}{2}} N$ is better than the almost everywhere convergence rate $\ln^{-\frac{r-1}{2}} N$.

Remark 2 Both Theorems 1 and 2 have the same form. However, when $r \geqslant 1$, we really get the convergence rate like Theorem 1; when $0 \leqslant r < 1$, we cannot conclude whether $\sigma_N^0(f)$ converges almost everywhere but get only the almost everywhere convergence of $\ln^{\frac{r-1}{2}} N(\sigma_N^0(f)-f)$.

Before proving the theorems, we will introduce some function classes related to the condition $(\{s, r\})$. Given a function $f \in L^2(S^{n-1})$ and a positive number r, if

$$\sum_{k=0}^{+\infty} \ln^{2r}(k+2)\|Y_k(f)\|_2^2 < +\infty,$$

then we say $f \in L_r^2(S^{n-1})$ and write

$$f_r = \sum_{k=0}^{+\infty} \ln^r(k+2) Y_k(f) \quad \text{in } L^2(S^{n-1}) \text{ sense.} \tag{1.7}$$

In Section 2, we give a characterization for the function class $L_r^2(S^{n-1})$ in terms of the modulus of continuity. In Section 3, we give a domination for maximal partial sum with "ln" factor of Fourier-Laplace series which plays the key role for proving the main theorems. In Section 4 we complete the proofs of the theorems.

§ 2. Characterization of the class $L_r^2(S^{n-1})$

Theorem 3 Let $r > -1$. The necessary and sufficient condition for $f \in L_{\frac{r+1}{2}}^2(S^{n-1})$ is that f satisfies $(\{2, r\})$.

Proof　Assume $f\in L^2_{\frac{r+1}{2}}(\mathrm{S}^{n-1})$ first. Then

$$f_{\frac{r+1}{2}}(x)\sim\sigma(f_{\frac{r+1}{2}})(x):=\sum_{k=0}^{+\infty}\ln^{\frac{r+1}{2}}(k+2)Y_k(f)(x).$$

We have

$$\Delta_t^2(f)\sim\sum_{k=1}^{+\infty}(1-P_k^n(\cos t))Y_K(f)$$

and

$$\|\Delta_t^2(f)\|^2=\sum_{k=1}^{+\infty}(1-P_k^n(\cos t))^2\|Y_k(f)\|^2. \tag{2.1}$$

Applying the formula (see [6, p. 81], [7, p. 31])

$$\frac{\mathrm{d}}{\mathrm{d}t}P_k^n(t)=\frac{k(k+n-2)}{n-1}P_{k-1}^{n+2}(t),$$

we have

$$1-P_k^n(\cos\theta)=\frac{k(k+n-2)}{n-1}P_{k-1}^{n+2}(\cos\xi)(1-\cos\theta),$$

where $\xi\in(0,\theta)$. Taking (1.5) into account, we hence have

$$0\leqslant 1-P_k^n(\cos\theta)\leqslant(k\theta)^2. \tag{2.2}$$

It follows from (1.5) and $\frac{n-2}{2}\geqslant\frac{1}{2}$ that

$$|P_k^n(\cos\theta)|\leqslant\frac{1}{2}\quad\text{for}\quad\theta\in(0,1)\text{ and }k\theta\geqslant 4\gamma^2. \tag{2.3}$$

Write

$$\delta_k(t)=\sup\{|1-P_k^n(\cos\theta)|^2:0\leqslant\theta\leqslant t\},\qquad t\in(0,1).$$

By (2.1) we get

$$\int_0^1\frac{\omega_2(f,t)^2}{t}\ln^r\left(\frac{2}{t}\right)\mathrm{d}t\leqslant\sum_{k=1}^{+\infty}\int_0^1\frac{\delta_k(t)}{t}\ln^r\left(\frac{2}{t}\right)\mathrm{d}t\|Y_k(f)\|^2.$$

Applying (2.2) and (2.3), we have

$$\delta_k(t)\leqslant\begin{cases}1, & 4\gamma^2k^{-1}<t<1, & k>4\gamma^2,\\(kt)^4, & 0<t\leqslant 4\gamma^2k^{-1}, & k>4\gamma^2,\\(kt)^4\leqslant B_nt^4, & 0<t<1, & k\leqslant 4\gamma^2,\end{cases} \tag{2.4}$$

where B_n is a constant depending only on n. From (2.4) we derive that

$$\int_0^1\frac{\delta_k(t)}{t}\ln^r\left(\frac{2}{t}\right)\mathrm{d}t\leqslant B_n\ln^{r+1}(k+2).$$

Since $f\in L^2_{\frac{r+1}{2}}(\mathrm{S}^{n-1})$ as an assumption, we see that f satisfies the condition $(\{2, r\})$.

We now assume that f satisfies the condition $(\{2, r\})$. We start from (2.1). Then

$$\sum_{k=1}^{+\infty}\int_0^1 (1-P_k^n(\cos t))^2 t^{-1}\ln^r\left(\frac{2}{t}\right)dt\|Y_k(f)\|^2$$
$$=\int_0^1\|\Delta_t^2(f)\|^2 t^{-1}\ln^r\left(\frac{2}{t}\right)dt<+\infty.$$

Hence

$$\sum_{k>4\gamma^2}^{+\infty}\int_{4\gamma^2k^{-1}}^1 (1-P_k^n(\cos t))^2 t^{-1}\ln^r\left(\frac{2}{t}\right)dt\|Y_k(f)\|^2<+\infty.$$

So, by (2.3) we obtain

$$\sum_{k>4\gamma^2}^{+\infty}\int_{4\gamma^2k^{-1}}^1 t^{-1}\ln^r\left(\frac{2}{t}\right)dt\|Y_k(f)\|^2<+\infty,$$

which implies $f\in L^2_{\frac{r+1}{2}}(S^{n-1})$. □

§ 3. Domination for Fourier-Laplace partial sums with ln factors

From (1.2) we derive the well-known formula for Cesàro numbers:

$$A_k^{\alpha+\beta}=\sum_{j=0}^k A_{k-j}^{\alpha-1}A_j^\beta,\quad \alpha>0,\ \beta>-1.$$

By this formula and (1.3) we get, for $f\in L^2(S^{n-1})$,

$$\sigma_N^0(f)=\sum_{k=0}^N A_{N-k}^{-\frac{1}{2}}A_k^{-\frac{1}{2}}\sigma_k^{-\frac{1}{2}}(f). \tag{3.1}$$

Denote by σ_*^α the maximal Cesàro operator of order α; that is,

$$\sigma_*^\alpha(f)(x):=\sup\{|\sigma_k^\alpha(f)(x)|\,|\,k\in\mathbf{N}\},\quad \alpha>-1,\quad x\in S^{n-1}.$$

It is well known (see [1]) that, for $\alpha>0$, σ_*^α is bounded from $L^2(S^{n-1})$ to $L^2(S^{n-1})$.

We introduce an operator on $L^2(S^{n-1})$ as follows:

Definition 2 For $f\in L^2(S^{n-1})$, define

$$\delta(f):=\left(\sum_{k=0}^{+\infty}|A_k^{-\frac{1}{2}}(\sigma_k^{-\frac{1}{2}}(f)-\sigma_k^{\frac{1}{2}}(f))|^2\right)^{\frac{1}{2}}.$$

By (3.1), applying Schwarz inequality, we have

$$\begin{aligned}|\sigma_N^0(f)| &\leqslant \sum_{k=0}^{N} A_{N-k}^{-\frac{1}{2}} A_k^{-\frac{1}{2}} |\sigma_k^{-\frac{1}{2}}(f)-\sigma_k^{\frac{1}{2}}(f)| + \sum_{k=0}^{N} A_{N-k}^{-\frac{1}{2}} A_k^{-\frac{1}{2}} |\sigma_k^{\frac{1}{2}}(f)| \\ &\leqslant \Big(\sum_{k=0}^{N} (A_{N-k}^{-\frac{1}{2}})^2\Big)^{\frac{1}{2}} \delta(f) + \sigma_*^{\frac{1}{2}}(f) \\ &\leqslant C\ln^{\frac{1}{2}}(N+2)\delta(f) + \sigma_*^{\frac{1}{2}}(f),\end{aligned} \tag{3.2}$$

where C is a proper constant.

Lemma 1 The operator δ is bounded from $L^2_{\frac{1}{2}}(S^{n-1})$ to $L^2(S^{n-1})$; i. e.

$$\|\delta(f)\| \leqslant C\|f_{\frac{1}{2}}\| \quad \text{for } f\in L^2_{\frac{1}{2}}(S^{n-1}),$$

where C is a proper constant.

Proof We have

$$\begin{aligned}\sigma_k^{-\frac{1}{2}}(f)-\sigma_k^{\frac{1}{2}}(f) &= \frac{1}{A_k^{-\frac{1}{2}}} \sum_{j=0}^{k} A_{k-j}^{-\frac{1}{2}} \left(1-\frac{A_k^{-\frac{1}{2}} A_{k-j}^{\frac{1}{2}}}{A_{k-j}^{-\frac{1}{2}} A_k^{\frac{1}{2}}}\right) Y_j(f) \\ &= \frac{1}{A_k^{-\frac{1}{2}}} \sum_{j=0}^{k} A_{k-j}^{-\frac{1}{2}} \frac{j}{k+\frac{1}{2}} Y_j(f).\end{aligned}$$

So,

$$\begin{aligned}\|\delta(f)\|^2 &= \int_{S^{n-1}} \sum_{k=1}^{+\infty} \left|\sum_{j=1}^{k} A_{k-j}^{-\frac{1}{2}} \frac{j}{k+\frac{1}{2}} Y_j(f)(x)\right|^2 dx \\ &= \sum_{k=1}^{+\infty} \sum_{j=1}^{k} \left|A_{k-j}^{-\frac{1}{2}} \frac{j}{k+\frac{1}{2}}\right|^2 \|Y_j(f)\|^2 \\ &\leqslant C\sum_{j=1}^{+\infty} \sum_{k=j}^{+\infty} \frac{j^2}{(k-j+1)k^2} \|Y_j(f)\|^2 \\ &\leqslant C\sum_{j=1}^{+\infty} \ln(j+1) \|Y_j(f)\|^2 \\ &\leqslant C\|f_{\frac{1}{2}}\|^2. \quad \square\end{aligned}$$

Definition 3 Let $\alpha>-1$. For $f\in L^2(S^{n-1})$, define

$$\rho_\alpha(f)(x) := \sup\{\ln^\alpha N\,|\sigma_N^0(f)(x)|\,|\,N\geqslant 3\}, \tag{3.3}$$

$$h_\alpha(f)(x) := \sup\{\ln^\alpha N\,|f(x)-\sigma_N^0(f)(x)|\,|\,N\geqslant 3\}. \tag{3.4}$$

Lemma 2 For $f\in L^2_{\frac{1}{2}}(S^{n-1})$,

$$\|\rho_{-\frac{1}{2}}(f)\| \leqslant C\|f_{\frac{1}{2}}\|.$$

Proof This is a direct consequence of (3.2), Lemma 1, and the boundedness of $\sigma_*^{\frac{1}{2}}$. $\square$

Corollary 1 If $f\in L^2_{\frac{1}{2}}(\mathrm{S}^{n-1})$, then $\lim\limits_{N\to+\infty} \ln^{-\frac{1}{2}} N|\sigma_N^0(f)(x)|=0$ almost everywhere.

Proof By Definition 3, it is obvious that

$$h_{-\frac{1}{2}}(f)(x)\leqslant\rho_{-\frac{1}{2}}(f)(x)+|f(x)|.$$

Hence $\|h_{-\frac{1}{2}}(f)\|\leqslant C\|f_{\frac{1}{2}}\|$ by Lemma 2. Given $\varepsilon>0$, we choose $m\in\mathbf{N}$ big enough such that

$$\|f-\sigma_m^0(f)\|\leqslant\|f_{\frac{1}{2}}-\sigma_m^0(f_{\frac{1}{2}})\|<\varepsilon$$

and write $g=\sigma_m^0(f)$ for simplicity. Since

$$\begin{aligned}\limsup_{N\to+\infty} \ln^{-\frac{1}{2}} N|\sigma_N^0(f)(x)| &=\limsup_{N\to+\infty} \ln^{-\frac{1}{2}} N|\sigma_N^0(f-g)(x)|\\ &\leqslant h_{-\frac{1}{2}}(f-g)(x),\end{aligned}$$

we get

$$\|\limsup_{N\to+\infty} \ln^{-\frac{1}{2}} N|\sigma_N^0(f)|\,\|\leqslant C\|(f-g)_{\frac{1}{2}}\|<C\varepsilon.$$

So, by the arbitrariness of ε, the left-hand side of the above inequality is zero. □

§ 4. Proof of the theorems

We first prove the following lemma.

Lemma 3 For all $s>0$ and $r>-1$, the condition $(\{s, r\})$ implies the condition $(\{2, r\})$.

We will verify this by using K-functionals concerning the derivatives. Let $f\in L^2(\mathrm{S}^{n-1})$ and $s>0$. If there exists a function $g\in L^2(\mathrm{S}^{n-1})$ such that

$$g\sim\sum_{k=1}^{+\infty}(k(k+n-2))^{\frac{s}{2}}Y_k(f),$$

then g is called the derivative of degree s of f and is written as

$$g=D^s f=f^{(s)}.$$

Following [5], the sth K-functional $K_s(\cdot, t)$ on $L^2(\mathrm{S}^{n-1})$ is defined by

$$K_s(f,t)=\inf\{\|f-g\|+t^s\|g^{(s)}\|\,|\,g^{(s)}\in L^2(\mathrm{S}^{n-1})\}.$$

Lemma 4 (see Ditzian [3]) If $0<\alpha<\beta$, then

$$K_\alpha(f,t) \leqslant C(\alpha,\beta)t^\alpha \int_t^1 K_\beta(f,u)u^{-\alpha-1}\mathrm{d}u.$$

Lemma 5 (see Kalyabin [5]) Suppose $s>0$ and $f\in L^2(S^{n-1})$. Then

$$\omega_s(f, t)\leqslant B_{n,s}K_s(f, t)\leqslant B'_{n,s}\omega_s(f, t) \quad \text{for all } t>0.$$

By Lemmas 2 and 3, we get directly that, for $0<\alpha<\beta<+\infty$,

$$\omega_\alpha(f,t) \leqslant C(\alpha,\beta)t^\alpha \int_t^1 \frac{\omega_\beta(f,u)}{u^{\alpha+1}}\mathrm{d}u. \tag{4. 1}$$

Proof of Lemma 3 Assume $0<s<2$. By (1. 6) we know that if the condition ($\{s, r\}$) holds, then ($\{2, r\}$) holds also. Now we assume that $2<s$ and ($\{s, r\}$) holds. Then by (4. 1),

$$\omega_2(f,t) \leqslant C_s t^2 \int_t^1 \frac{\omega_s(f,u)}{u^3}\mathrm{d}u$$

and hence, by Schwarz inequality,

$$\omega_2(f,t)^2 \leqslant C_s t^4 \int_t^1 \frac{\omega_s(f,u)^2}{u^2}\mathrm{d}u \int_t^1 u^{-4}\mathrm{d}u \leqslant C_s t \int_t^1 \frac{\omega_s(f,u)^2}{u^2}\mathrm{d}u.$$

Then we get

$$\begin{aligned}\int_0^1 \frac{\omega_2(f,t)^2}{t}\ln^r\left(\frac{2}{t}\right)\mathrm{d}t &\leqslant C_s \int_0^1 \ln^r\left(\frac{2}{t}\right)\int_t^1 \frac{\omega_s(f,u)^2}{u^2}\mathrm{d}u\mathrm{d}t \\ &\leqslant C_s \int_0^1 \frac{\omega_s(f,u)^2}{u^2}\left(\int_0^u \ln^r\left(\frac{2}{t}\right)\mathrm{d}t\right)\mathrm{d}u \\ &\leqslant C_s \int_0^1 \frac{\omega_s(f,t)^2}{t}\ln^r\left(\frac{2}{t}\right)\mathrm{d}t. \quad \square\end{aligned}$$

From Lemma 3, to prove Theorems 1 and 2, it suffices for us to prove both for $s=2$.

Proof of Theorem 1 Assume $r\geqslant 1$.

Let $f\in L^2(S^{n-1})$. If f satisfies the condition ($\{2, r\}$), then by Theorem 3,

$$f\in L^2_{\frac{r+1}{2}}(S^{n-1})\subset L^2_1(S^{n-1}).$$

We first consider the case of $r=1$. In this case the result is known (see [2]). In fact, for any function $f\in L^2_1$, the almost everywhere convergence of the orthogonal expansion of f holds by a general theorem for orthogonal series (see [4, p. 190 for Russian translation]). But for the completeness we give a very short proof here. Generally, for $\alpha>0$

and $f\in L^2_{\alpha+\frac{1}{2}}(S^{n-1})$, by using Abel transform twice, we have, for $N\geqslant3$,

$$\sigma_N^0(f)=\sum_{k=0}^{N}Y_k(f)=\sum_{k=0}^{N}\mu_k^\alpha Y_k(f_\alpha)$$
$$=\sum_{k=0}^{N-2}(k+1)\Delta^2\mu_k^\alpha\sigma_k^1(f_\alpha)+(N-1)\Delta\mu_{N-1}^\alpha\sigma_{N-1}^1(f_\alpha)+\mu_N^\alpha\sigma_N^0(f_\alpha),$$

where

$$\mu_k^\alpha:=\frac{1}{\ln^\alpha(k+2)},\quad \Delta\mu_k^\alpha:=\mu_k^\alpha-\mu_{k+1}^\alpha,\quad \Delta^2\mu_k^\alpha:=\Delta\mu_k^\alpha-\Delta\mu_{k+1}^\alpha.$$

Then we get

$$\sigma_*^0(f)\leqslant C_\alpha(\sigma_*^1(f_\alpha)+\rho_{-\alpha}(f_\alpha)).$$

Hence, by Lemma 2 and the boundedness of σ_*^1, we get

$$\|\sigma_*^0(f)\|\leqslant C\|\sigma_*^1(f_{\frac{1}{2}})\|+\|\rho_{-\frac{1}{2}}(f_{\frac{1}{2}})\|\leqslant C\|f_1\|.$$

Thus, for all $f\in L_1^2(S^{n-1})$,

$$\lim_{N\to+\infty}\sigma_N^0(f)(x)=f(x)\quad \text{almost everywhere.}\tag{4.2}$$

Next we assume $r>1$. Fix $N>2$ temporarily and let $m>N$. Then

$$\sigma_m^0(f)-\sigma_{N-1}^0(f)=\sum_{k=N}^{m}Y_k(f)=\sum_{k=N}^{m}\mu_k^{\frac{r}{2}}Y_k(f_{\frac{r}{2}}).$$

Using Abel transform, we get

$$\sigma_m^0(f)-\sigma_{N-1}^0(f)=\sum_{k=N}^{m-1}\Delta\mu_k^{\frac{r}{2}}\sigma_k^0(f_{\frac{r}{2}})+\mu_m^{\frac{r}{2}}(\sigma_m^0(f_{\frac{r}{2}})-\sigma_{N-1}^0(f_{\frac{r}{2}})).$$

Since $f_{\frac{r}{2}}\in L_{\frac{1}{2}}^2(S^{n-1})$, by Corollary 1 we know

$$\lim_{m\to\infty}\mu_m^{\frac{r}{2}}(\sigma_m^0(f_{\frac{r}{2}})-\sigma_{N-1}^0(f_{\frac{r}{2}}))=0\quad \text{almost everywhere }(r\geqslant1).\tag{4.3}$$

Taking the limit $m\to+\infty$ and applying (18) yield

$$f-\sigma_{N-1}^0(f)=\sum_{k=N}^{+\infty}\Delta\mu_k^{\frac{r}{2}}\sigma_k^0(f_{\frac{r}{2}})\quad \text{almost everywhere.}$$

Notice that $|\Delta\mu_k^{\frac{r}{2}}|\leqslant C\frac{1}{k\ln^{1+\frac{r}{2}}k}(k>2)$. We get

$$|f-\sigma_{N-1}^0(f)|\leqslant C\sum_{k=N}^{+\infty}\frac{1}{k\ln^{1+\frac{r-1}{2}}k}\rho_{-\frac{1}{2}}(f_{\frac{r}{2}})\leqslant\frac{C}{\ln^{\frac{r-1}{2}}N}\rho_{-\frac{1}{2}}(f_{\frac{r}{2}})$$

almost everywhere.

Applying Lemma 2, we obtain

$$\|h_{\frac{r-1}{2}}(f)\|\leqslant C\|f_{\frac{r+1}{2}}\|,$$

which implies, by a routine argument, the conclusion of Theorem 1 for

$r>1$. □

Proof of Theorem 2 Assume $0\leqslant r<1$ and f satisfies $(\{2, r\})$. We write $\alpha=\frac{r-1}{2}$ for convenience. Then $-\frac{1}{2}\leqslant\alpha<0$. By Theorem 3, $f\in L^2_{\alpha+1}(S^{n-1})$. We have, for $N>3$,

$$\sigma_N^0(f)=\sum_{k=0}^{N-1}\Delta\mu_k^{\frac{r}{2}}\sigma_k^0(f_{\frac{r}{2}})+\mu_N^{\frac{r}{2}}\sigma_N^0(f_{\frac{r}{2}}).$$

Then

$$\begin{aligned}&\ln^{\alpha}(N+2)\,|\sigma_N^0(f)|\\ \leqslant\ & C\sum_{k=0}^{N-1}\frac{\ln^{\alpha}(N+2)}{(k+2)\ln^{\alpha+\frac{3}{2}}(k+2)}|\sigma_k^0(f_{\frac{r}{2}})|+\frac{1}{\ln^{\frac{1}{2}}(N+2)}|\sigma_N^0(f_{\frac{r}{2}})|\\ \leqslant\ & C_{\alpha}\rho(f_{\alpha+1}).\end{aligned}$$

Therefore,

$$\|h_{\alpha}(f)\|\leqslant C_{\alpha}\|f_{\alpha+1}\|=C_{\alpha}\|f_{\frac{r+1}{2}}\|.$$

By this we finish the proof. □

References

[1] Bonami A., Clerc J L. Sommes de Cesàro et multiplicateurs des dèvelopments en harmonique sphérique. Trans. Amer. Math. Soc., 1973, 183: 223—263.

[2] Dai F. A note on a. e. convergence of Fourier-Laplace series in L^2. J. Beijing Normal Univ. (NS), 1999, 35(1): 6—9.

[3] Ditzian Z. Fractional derivatives and best approximation. Acta Math. Hungar., 1998, 81: 323—348.

[4] Kaczmarz S., Steinhaus H. Theorie der Orthogonalreihen. Chelsea, New York, 1951 (Russian translation with additional material, Teoriya Ortogonalnykh Ryadov [Theory, of Orthogonal Series], GIH-ML, Moscow, 1958).

[5] Kalyabin G A. On moduli of smoothness of functions given on the sphere. Sov. Math. Dokl., 1987, 35: 619—660.

[6] Szegö G. Orthogonal Polynomials, American Mathematical Society Colloquium Publication, Vol. 23, American Mathematical Society, Providence, RI, 1939.

[7] Wang K., Li L. Harmonic Analysis and Approximation on the Unit Sphere. Science Press, Beijing, 2000.

Journal of Approximation Theory, 2004, 130: 38—45.

联系于 Laplace 算子的平均和 *K* 泛函的等价性

A Note on the Equivalences between the Averages and the *K*-Functionals Related to the Laplacian[①]

Abstract For $\mathbf{R}^d$ or $\mathbf{T}^d$, a strong converse inequality of type A (in the terminology of Ditzian and Ivanov(J. Anal. Math. 61(1993)61)) is obtained for the high order averages on balls and the *K*-functionals generated by the high order Laplacian, which answers a problem raised by Ditzian and Runovskii (J. Approx. Theory 97 (1999) 113).

Keywords *K*-functionals; High order averages; Strong converse inequalities of type A.

§ 1. Introduction and main result

Given a function $f \in L(\mathbf{R}^d)$, its Fourier transform is defined by

$$\hat{f}(\xi) = \int_{\mathbf{R}^d} f(x) \mathrm{e}^{-\mathrm{i}x\xi} \mathrm{d}x, \quad \xi \in \mathbf{R}^d.$$

For a positive integer ℓ, the ℓth order Laplacian Δ^ℓ is defined, in a distributional sense, by

$$(\Delta^\ell f)^\wedge(\xi) = (-1)^\ell |\xi|^{2\ell} \hat{f}(\xi).$$

Associated with the operator Δ^ℓ, there is a *K*-functional

① Supported by NSFC 10071007. Dai Feng was also supported by University of Alberta Start-up Fund.

本文与戴峰合作.

Received: 2002-10-29; Revised: 2004-01-01.

$$K_{\Delta,\ell}(f,\ t^{2\ell})_p\ :=\inf\{\|f-g\|_p+t^{2\ell}\|\Delta^{\ell}g\|_p\,|_g,\ \Delta^{\ell}g\in L^p(\mathbf{R}^d)\},\tag{1.1}$$

where $t>0$, $1\leqslant p\leqslant+\infty$ and $\|\cdot\|_p$ denotes the usual L^p-norm on $\mathbf{R}^d$.

Let V_d denote the volume of the unit ball of $\mathbf{R}^d$. For $t>0$ and a locally integrable function f, we define the average $B_t(f)$ by

$$B_t(f)(x)=\frac{1}{t^dV_d}\int_{\{u\in\mathbf{R}^d\,|\,|u|\leqslant t\}}f(x+u)\mathrm{d}u$$

and the ℓth order average $B_{\ell,t}(f)$ (for a given positive integer ℓ) by

$$B_{\ell,t}(f)(x)=\frac{-2}{\mathrm{C}_{2\ell}^{\ell}}\sum_{j=1}^{\ell}(-1)^j\mathrm{C}_{2\ell}^{\ell-j}B_{jt}(f)(x).\tag{1.2}$$

We remark that for $\ell>1$ the operator $B_{\ell,t}$ was first introduced by Ditzian and Runovskii in [DR, p. 117, (2.6)].

For more background information we refer to [Di1, Di2, DR, Di-Iv, To].

Our main goal in this paper is to prove the following strong converse inequality of type A (in the terminology of [Di-Iv]), which was conjectured in [DR, p. 138].

Theorem 1 Let $\ell\in\mathbf{N}$, $1\leqslant p\leqslant+\infty$ and $f\in L^p(\mathbf{R}^d)$. Then

$$\|f-B_{\ell,t}(f)\|_p\approx K_{\Delta,\ell}(f,\ t^{2\ell})_p,$$

where $t>0$ and

$$A(f,\ t)\approx B(f,\ t)$$

means that there is a $C>0$, independent off and t, such that

$$C^{-1}A(f,\ t)\leqslant B(f,\ t)\leqslant CA(f,\ t).$$

Theorem 1 for $\ell=1$ was proved in [DR, p. 133, Theorem 6. 1] and for $d=1$, ℓ small, as it was indicated in [DR, p. 138], can be obtained by following the technique developed in [Di-Iv]. For $\ell\geqslant2$ and $d\geqslant2$, the following strong converse inequality of type B (in the terminology of [Di-Iv]) was obtained in [DR, p. 127, Theorem 4. 8 and p. 131, Theorem 5. 7]:

$$K_{\Delta,\ell}(f,\ t^{2\ell})_p\approx\|f-B_{\ell,t}(f)\|_p+\|f-B_{\ell,t\rho}(f)\|_p,\ 1\leqslant p\leqslant+\infty\tag{1.3}$$

for some $\rho>1$. The proof of our Theorem 1 will be based on this equivalence.

We remark that with a sligh modification of the proof below a similar result for the periodic case can also be obtained.

§ 2. Basic lemmas

The following lemma can be easily obtained by a straightforward computation.

Lemma 1 Let $\chi_{B(0,1)}(x)$ denote the characteristic function of the unit ball

$B(0,1):=\{x=(x_1,x_2,\cdots,x_d)\in\mathbf{R}^d \mid x_1^2+x_2^2+\cdots+x_d^2\leqslant 1\}$,

V_d denote the volume of $B(0,1)$ and let $I(x)=\frac{1}{V_d}\chi_{B(0,1)}(x)$. Then

$$\hat{I}(x)=\gamma_d\int_0^1\cos(u|x|)(1-u^2)^{\frac{d-1}{2}}\mathrm{d}u \tag{2.1}$$

with

$$\gamma_d=\left(\int_0^1(1-u^2)^{\frac{d-1}{2}}\mathrm{d}u\right)^{-1}. \tag{2.2}$$

Lemma 2 Let $B_{\ell,t}$ be defined by (1.2) and $I(x)$ the same as in Lemma 1. Then for $f\in L(\mathbf{R}^d)$,

$$\widehat{B_{\ell,t}(f)}(x)=m_\ell(t|x|)\hat{f}(x), \tag{2.3}$$

where

$$m_\ell(|x|)=\frac{-2}{C_{2\ell}^\ell}\sum_{j=1}^{\ell}(-1)^jC_{2\ell}^{\ell-j}\hat{I}(jx) \tag{2.4}$$

$$=1-A_\ell(|x|), \tag{2.5}$$

$$A_\ell(|x|)=\gamma_d\frac{4^\ell}{C_{2\ell}^\ell}\int_0^1(1-u^2)^{\frac{d-1}{2}}\left(\sin\frac{u|x|}{2}\right)^{2\ell}\mathrm{d}u \tag{2.6}$$

and γ_d is given by (2.2).

Proof For $t>0$, we write

$$I_t(x)=\frac{1}{t^d}I\left(\frac{x}{t}\right).$$

Then from definition (1.2), it follows that

$$B_{\ell,t}(f)(x)=\frac{-2}{C_{2\ell}^\ell}\sum_{j=1}^{\ell}(-1)^jC_{2\ell}^{\ell-j}(f*I_{jt})(x),$$

which implies (2.3) and (2.4). Substituting (2.1) into (2.4) yields

$$m_\ell(|x|)=\frac{-2\gamma_d}{C_{2\ell}^{\ell}}\sum_{j=1}^{\ell}(-1)^j C_{2\ell}^{\ell-j}\int_0^1\cos(ju|x|)(1-u^2)^{\frac{d-1}{2}}\mathrm{d}u, \tag{2.7}$$

which, together with the following identity

$$\left(\sin\frac{x}{2}\right)^{2\ell}=\frac{C_{2\ell}^{\ell}}{4^\ell}+\frac{2}{4^\ell}\sum_{j=1}^{\ell}(-1)^j C_{2\ell}^{\ell-j}\cos jx,$$

gives (2.5) and (2.6). This completes the proof. □

Lemma 3 Let $m_\ell(u)$ be the same as in Lemma 2. Then for $j\in\mathbf{N}^*$ and $u\geqslant 0$,

$$\left|\left(\frac{\mathrm{d}}{\mathrm{d}u}\right)^j m_\ell(u)\right|\leqslant C_{\ell,j}\left(\frac{1}{u+1}\right)^{\frac{d+1}{2}},$$

where $C_{\ell,j}>0$ is independent of u.

Proof By identity (2.7), it suffices to show that for $j\in\mathbf{N}^*$ and $u\geqslant 0$,

$$\left|\left(\frac{\mathrm{d}}{\mathrm{d}u}\right)^j\int_0^1\cos(uv)(1-v^2)^{\frac{d-1}{2}}\mathrm{d}v\right|\leqslant C_j\left(\frac{1}{u+1}\right)^{\frac{d+1}{2}}. \tag{2.8}$$

We use formula (4.7.5) of [An-As-R, p. 204] to obtain that

$$\int_0^1\cos(uv)(1-v^2)^{\frac{d-1}{2}}\mathrm{d}u=2^{\frac{d-2}{2}}\sqrt{\pi}\Gamma\left(\frac{d+1}{2}\right)\frac{J_{\frac{d}{2}}(u)}{u^{\frac{d}{2}}}, \tag{2.9}$$

where $J_\alpha(u)$ denotes the Bessel function of the first kind of order α. Now (2.8) is a consequence of (12) and the following well-known estimates on Bessel functions:

$$\frac{\mathrm{d}}{\mathrm{d}u}u^{-\alpha}J_\alpha(u)=-u^{-\alpha}J_{\alpha+1}(u), \quad \text{[An-As-R, (4.6.2), p. 202]},$$

$$J_\alpha(u)=O\left(\frac{1}{(u+1)^{\frac{1}{2}}}\right)\quad \text{for } u\geqslant 0 \quad \text{[An-As-R, (4.8.5), p. 209]},$$

$$J_\alpha(u)=O(u^\alpha)\text{ as } u\to 0 \quad \text{[An-As-R, (4.7.6), p. 128]}.$$

This concludes the proof. □

Lemma 4 Suppose that a is a C^∞-function defined on $[0,+\infty)$ with the property that for $u\geqslant 0$ and $0\leqslant j\leqslant d+1$,

$$\left|\left(\frac{\mathrm{d}}{\mathrm{d}u}\right)^j a(u)\right|\leqslant C(a)\left(\frac{1}{1+u}\right)^{d+1}. \tag{2.10}$$

For $t>0$, define the operator T_t, in a distributional sense, by

$$(T_t(f))^\wedge(\xi)=a(t|\xi|)\hat{f}(\xi), \quad \xi\in\mathbf{R}^d.$$

Then for $1\leqslant p\leqslant+\infty$ and $f\in L^p(\mathbf{R}^d)$,

$$\sup_{t>0}\|T_t(f)\|_p\leqslant C_{p,a}\|f\|_p.$$

This lemma is well known (see [St]), but for the sake of completeness, we give its proof here.

Proof Let

$$K(x)=\int_{\mathbf{R}^d}e^{ix\xi}a(|\xi|)d\xi. \tag{2.11}$$

Since

$$T_t(f)(x)=f*K_t(x),$$

with

$$K_t(x)=\frac{1}{t^d}K\left(\frac{x}{t}\right),$$

it is sufficient to prove

$$\|K\|_{L^1(\mathbf{R}^d)}<+\infty. \tag{2.12}$$

By (14), we get for $\gamma=(\gamma_1, \gamma_2, \cdots, \gamma_d)\in\mathbf{Z}_+^d$,

$$(-x)^\gamma K(x)=\int_{\mathbf{R}^d}e^{ix\xi}\left(\frac{\partial}{\partial\xi}\right)^\gamma(a(|\xi|))d\xi,$$

which, by (2.10), implies

$$|x^\gamma K(x)|\leqslant C\int_{\mathbf{R}^d}\frac{d\xi}{(1+|\xi|)^{d+1}}<+\infty,$$

with $|\gamma|=\gamma_1+\gamma_2+\cdots+\gamma_d\leqslant d+1$. Now taking the supremum over all γ with $|\gamma|=d+1$ yields

$$|K(x)|\leqslant\frac{C}{|x|^{d+1}},$$

which, together with the fact that $K\in C(\mathbf{R}^d)$, implies (2.12) and so completes the proof. □

§ 3. Proof of Theorem 1

The upper estimate

$$\|f-B_{l,t}(f)\|_p\leqslant C_{l,p}K_{\Delta,l}(f, t^{2l})_p$$

follows directly from (1.3), which, as indicated in Section 1, was proved in [DR]. Hence it remains to prove the lower estimate

$$\| f-B_{\ell,t}(f)\|_p\geqslant C_{\ell,p}K_{\Delta,\ell}(f,\ t^{2\ell})_p.$$

Lemma 3 implies that there is a number $\mu=\mu(\ell,\ d)>1$ such that for $u>\mu$,

$$|m_\ell(u)|\leqslant\frac{1}{2}. \tag{3.1}$$

We will keep this special number μ throughout the proof.

Let η be a C^∞-function on $[0,\ +\infty)$ with the properties that $\eta(x)=0$ for $x>2$, $\eta(x)=1$ for $0\leqslant x\leqslant 1$, and $0\leqslant\eta(x)\leqslant 1$ for all $x\in[0,\ +\infty)$. For $t>0$, we define the operator V_t by

$$(V_t(f))^\wedge(\xi)=\eta(t|\xi|)\hat{f}(\xi), \tag{3.2}$$

where $f\in L^p(\mathbf{R}^d)$ and $\xi\in\mathbf{R}^d$.

According to definition (1.1), the estimates

$$\| f-V_{\frac{t}{2\mu}}(f)\|_p\leqslant C_{\ell,p}\| f-B_{\ell,t}(f)\|_p \tag{3.3}$$

and

$$t^{2\ell}\|\Delta^\ell V_{\frac{t}{2\mu}}(f)\|_p\leqslant C_{\ell,p}\| f-B_{\ell,t}(f)\|_p \tag{3.4}$$

will prove

$$K_{\Delta,\ell}(f,\ t^{2\ell})_p\leqslant\| f-V_{\frac{t}{2\mu}}(f)\|_p+t^{2\ell}\|\Delta^\ell V_{\frac{t}{2\mu}}(f)\|_p\leqslant C_{\ell,p}\| f-B_{\ell,t}(f)\|_p$$

and so complete the proof of Theorem 1. Thus, it has remained to prove (3.3) and (3.4).

Let

$$\phi(u)=\left(1-\eta\left(\frac{u}{2\mu}\right)\right)\frac{(m_\ell(u))^3}{1-m_\ell(u)} \tag{3.5}$$

and

$$\psi(u)=\frac{u^{2\ell}\eta\left(\frac{u}{2\mu}\right)}{A_\ell(u)}, \tag{3.6}$$

with $A_\ell(u)$ and $m_\ell(u)$ the same as in Lemma 2. For $t>0$, we define two operators Φ_t and Ψ_t as follows:

$$(\Phi_t(f))^\wedge(\xi):=\varphi(t|\xi|)\hat{f}(\xi),$$

$$(\Psi_t(f))^\wedge(\xi):=\psi(t|\xi|)\hat{f}(\xi), \tag{3.7}$$

It follows from (3.1)(3.5) and Lemma 3 that for $u\geqslant 0$ and $0\leqslant j\leqslant$

$d+1$,

$$|\varphi^{(j)}(u)|\leqslant C_{l,d}\left(\frac{1}{u+1}\right)^{\frac{3(d+1)}{2}}. \tag{3.8}$$

On the other hand, by (2.6) and a straightforward computation, we obtain that for $u\geqslant\frac{\pi}{2}$,

$$A_l(u)\geqslant C_{l,d}\int_0^{\frac{2}{3}}\left(\sin\frac{uv}{2}\right)^{2l}\mathrm{d}v\geqslant C'_{l,d}>0 \tag{3.9}$$

and for $0<u<\frac{\pi}{2}$,

$$\frac{A_l(u)}{u^{2l}}\geqslant C_{l,d}\frac{1}{u^{2l}}\int_0^1(1-v^2)^{\frac{d-1}{2}}(uv)^{2l}\mathrm{d}v\geqslant C_{l,d}>0, \tag{3.10}$$

which, together with (3.6), implies that

$$\psi\in C^\infty[0,+\infty)\text{ and supp }\psi\subset[0,4\mu]. \tag{3.11}$$

Now invoking Lemma 4 three times, with $a=\eta$, ϕ and ψ, respectively, in view of (3.8), (3.11) and the fact that η is a C^∞-function with compact support, we obtain from (3.2) and (3.7) that for $1\leqslant p\leqslant+\infty$,

$$\sup_{t>0}\|V_t(f)\|_p+\sup_{t>0}\|\Phi_t(f)\|_p+\sup_{t>0}\|\Psi_t(f)\|_p\leqslant C_p\|f\|_p. \tag{3.12}$$

We claim that (3.3) and (3.4) follow from (3.12). In fact, from the identity $\quad (f-V_{\frac{t}{2\mu}}(f))^\wedge(\xi)=W(t\xi)(f-B_{l,t}(f))^\wedge(\xi)$,

where

$$W(\xi):=\left(1-\eta\left(\frac{|\xi|}{2\mu}\right)\right)\left(\frac{m_l(|\xi|)^3}{1-m_l(|\xi|)}+1+m_l(|\xi|)+(m_l(|\xi|))^2\right),$$

it follows that

$$f-V_{\frac{t}{2\mu}}(f)=\Phi_t(f-B_{l,t}(f))+(I-V_{\frac{t}{2\mu}})(1+B_{l,t}+B_{l,t}^2)(f-B_{l,t}(f)),$$

where I denotes the identity operator on $L^p(\mathbf{R}^d)$. This, together with (3.12) and the fact that $\|B_{l,t}\|_{(p,p)}\leqslant C_l$, gives (3.3).

Similarly, from the identities

$$(t^{2l}\Delta^l V_{\frac{t}{2\mu}}(f))^\wedge(\xi)=\frac{(-1)^l t^{2l}|\xi|^{2l}\eta\left(\frac{t|\xi|}{2\mu}\right)}{1-m_l(t|\xi|)}(f-B_{l,t}(f))^\wedge(\xi)$$
$$=(-1)^l\Psi_t(f-B_{l,t}(f))^\wedge(\xi),$$

it follows that $\quad t^{2l}\Delta^l V_{\frac{t}{2\mu}}(f)=(-1)^l\Psi_t(f-B_{l,t}(f))$,

which, again by (3.12), implies (3.4). This completes the proof.

Acknowledgments

The authors would like to thank Professor Z. Ditzian for supplying them with some preprints of his excellent papers on K-functionals, which gave a better perspective on their proof. The authors would also like to thank the anonymous referee for pointing out some misprints of their paper.

References

[An-As-R] Andrews G E., Askey R., Roy R. Special Functions, Encyclopedia of Mathematics and its Applications. Vol 71, Cambridge University Press, Cambridge, 1999.

[Di1] Ditzian Z. Fractional derivatives and best approximation. Acta Math. Hungar. 1998, 81: 323—348.

[Di2] Ditzian Z. Measure of smoothness related to the Laplacian. Trans. Amer. Math. Soc., 1991, 326: 407—422.

[Di-Iv] Ditzian Z., Ivanov K. Strong converse inequalities, J. Anal. Math., 1993, 61: 61—111.

[DR] Ditzian Z., Runovskii K. Averages and K-functionals related to the Laplacian. J. Approx. Theory, 1999, 97: 113—139.

[St] Stein E M. Singular Integrals and Differentiability Properties of Functions. Princeton University Press, Princeton, NJ, 1970.

[To] Totik V. Approximation by Bernstein polynomials. Amer. J. Math. 1994, 116: 995—1 018.

Acta Mathematica Sinica, English Series, 2005, 21(2): 439—448.

Hardy 空间 $H^p(S^{d-1})(0<p\leqslant 1)$ 中临界阶 Cesàro 平均的强逼近

Strong Approximation by Cesàro Means with Critical Index in the Hardy Spaces $H^p(S^{d-1})$ $(0<p\leqslant 1)$ ①

Abstract Let $S^{d-1}=\{x \mid |x|=1\}$ be a unit sphere of the d-dimensional Euclidean space $\mathbf{R}^d$ and let $H^p\equiv H^p(S^{d-1})(0<p\leqslant 1)$ denote the real Hardy space on S^{d-1}. For $0<p\leqslant 1$ and $f\in H^p(S^{d-1})$, let $E_j(f, H^p)(j\in\mathbf{N})$ be the best approximation of f by spherical polynomials of degree less than or equal to j, in the space $H^p(S^{d-1})$. Given a distribution f on S^{d-1}, its Cesàro mean of order $\delta>-1$ is denoted by $\sigma_k^\delta(f)$. For $0<p\leqslant 1$, it is known that $\delta(p):=\frac{d-1}{p}-\frac{d}{2}$ is the critical index for the uniform summability of σ_k^δ in the metric H^p. In this paper, the following result is proved:

Theorem Let $0<p<1$ and $\delta=\delta(p):=\frac{d-1}{p}-\frac{d}{2}$. Then for $f\in H^p(S^{d-1})$,

$$\sum_{j=1}^{N}\frac{1}{j}\|\sigma_j^\delta(f)-f\|_{H^p}^p\approx\sum_{j=1}^{N}\frac{1}{j}E_j^p(f,H^p),$$

where $A_N(f)\approx B_N(f)$ means that there's a positive constant C, independent of N and f, such that

$$C^{-1}A_N(f)\leqslant B_N(f)\leqslant CA_N(f).$$

---

① Supported by NSFC 10071007.

本文与戴峰合作.

Received: 2002-03-20; Revised: 2003-03-04.

In the case $d=2$, this result was proved by Belinskii in 1996.

Keywords H^p spaces; spherical harmonics; Cesàro means; K-functionals; strong approximation.

§ 1. Introduction

Let $S^{d-1}=\{(x_1, x_2, \cdots, x_d) \mid x_1^2+x_2^2+\cdots+x_d^2=1\}$ be a unit sphere of the d-dimensional Euclidean space $\mathbf{R}^d$ and let $H^p \equiv H^p(S^{d-1})$ $(0<p\leqslant 1)$ denote the real Hardy space on S^{d-1}. Given a distribution f on S^{d-1}, its Cesàro mean of order $\delta>-1$ is denoted by $\sigma_k^\delta(f)$. (We shall give the precise definitions of $H^p(S^{d-1})$ and $\sigma_k^\delta(f)$ in Section 2.) The following statements on the uniform boundedness of the Cesàro means in $H^p(S^{d-1})$ were proved in [1]:

(1) For $0<p\leqslant 1$ and $\delta>\delta(p):=\dfrac{d-1}{p}-\dfrac{d}{2}$, $\sup_k \|\sigma_k^\delta\|_{(H^p, H^p)}<+\infty$.

(2) The estimate in (1) is sharp in the sense that $\sup_k \|\sigma_k^\delta\|_{(H^p, H^p)}=+\infty$ whenever $\delta\leqslant\delta(p)$. In fact, more can be stated: For $\delta<\delta(p)$, $\sup_k \|\sigma_k^\delta\|_{w(H^p, L^p)}=+\infty$, while for $\delta=\delta(p)$, $\|\sigma_k^\delta\|_{(H^p, L^p)}\geqslant C(\ln k)^{\frac{1}{p}}$.

(3) For $0<p<1$ and the critical index $\delta=\delta(p)$, $\sup_k \|\sigma_k^\delta\|_{w(H^p, L^p)}<+\infty$. For $p=1$ and $\delta=\lambda$, the above weak type estimate fails.

The main goal of this paper, as suggested in the title, is to consider the strong approximation by Cesàro means with critical index. Before stating the main result, we have to introduce some necessary notations. Given a distribution f on S^{d-1}, let $f^{(r)}$ $(r>0)$ denote the r-th order derivative of f. (See Section 2 for its precise definition.) For $r>0$ and $0<p\leqslant 1$, Let W_p^r denote the function class

$$W_p^r:=\{f\in H^p(S^{d-1}) \mid f^{(r)}\in H^p(S^{d-1})\}.$$

Define the K-functional $K_r(f, t)_{H^p}$ by

$$K_r(f, t)_{H^p}:=\inf\{\|f-g\|_{H^p}+t^r\|g^{(r)}\|_{H^p} \mid g\in W_p^r\}.$$

For $0<p\leqslant 1$ and $f\in H^p$, let $E_j(f, H^p)$ $(j\in\mathbf{N})$ be the best approximation of f by spherical polynomials of degree less than or equal to j, in the space $H^p(S^{d-1})$.

In this paper, we shall prove

Theorem 1.1 Let $0<p<1$ and $\delta=\delta(p):=\frac{d-1}{p}-\frac{d}{2}$. Then for $f\in H^p(\mathbb{S}^{d-1})$,

$$\sum_{j=1}^{N}\frac{1}{j}\|\sigma_j^{\delta}(f)-f\|_{H^p}^p\approx\sum_{j=1}^{N}\frac{1}{j}E_j^p(f,H^p),$$

where $A_N(f)\approx B_N(f)$ means that there's a positive constant C, independent of N and f, such that $C^{-1}A_N(f)\leqslant B_N(f)\leqslant CA_N(f)$.

As a consequence, we have

Corollary 1.2 Suppose $0<p<1$ and $\delta=\delta(p):=\frac{d-1}{p}-\frac{d}{2}$. Then for $f\in H^p(\mathbb{S}^{d-1})$,

$$\frac{1}{\ln N}\sum_{k=1}^{N}\frac{\|\sigma_k^{\delta}(f)-f\|_{H^p}^p}{k}\leqslant CK_1^p\left(f,\left(\frac{1}{\ln N}\right)^{\frac{1}{p}}\right)_{H^p}.$$

We remark that in the case of Riesz means of the periodic functions, Theorem 1.1 is due to [2] while Corollary 1.2 is due to [3] (see also [4]).

The proof of Theorem 1.1 depends on the following

Theorem 1.3 Let $0<p<1$, $\delta=\delta(p):=\frac{d-1}{p}-\frac{d}{2}$ and $f\in H^p$. Then

$$\frac{1}{\ln N}\sum_{k=1}^{N}\frac{\|\sigma_k^{\delta}(f)\|_{H^p}^p}{k}\leqslant C_p\|f\|_{H^p}^p. \tag{1.1}$$

In the case of the Riesz means of multiple periodic functions, Theorem 1.3 is due to [4]. For the Cesàro means of functions on the high dimensional sphere $\mathbb{S}^{d-1}$, to the best of our knowledge, the best previously known result is due to Chen [5], who proved Theorem 1.3 with H^p-norms on the left-hand side of (1.1) being replaced by L^p-norms. We point out that the proof in [4] relies on the atomic decomposition theorem of the Hardy spaces and seemingly does not apply to our case, as demonstrated in paper [5]. A new technique will be used in our proof of Theorem 1.3.

We can also obtain some similar results for the generalized Riesz means $R_k^{\delta,\alpha}(f)$ (We give the precise definition of $R_k^{\delta,\alpha}(f)$ in Section 2 below).

Theorem 1.4 Let $\alpha>0$, $0<p<1$ and $\delta=\delta(p):=\frac{d-1}{p}-\frac{d}{2}$. Then for $f\in H^p(\mathrm{S}^{d-1})$,

$$\sum_{j=1}^{N}\frac{1}{j}\|R_j^{\delta,\alpha}(f)-f\|_{H^p}^p\approx\sum_{j=1}^{N}\frac{1}{j}E_j^p(f,H^p).$$

Corollary 1.5 Let $\alpha>0$, $0<p<1$ and $\delta=\delta(p):=\frac{d-1}{p}-\frac{d}{2}$. Then for $f\in H^p(\mathrm{S}^{d-1})$,

$$\frac{1}{\ln N}\sum_{k=1}^{N}\frac{\|R_k^{\delta,\sigma}(f)-f\|_{H^p}^p}{k}\leqslant CK_1^p\left(f,\left(\frac{1}{\ln N}\right)^{\frac{1}{p}}\right)_{H^p}.$$

The organization of this paper is as follows. Section 2 contains some basic definitions and results on the real Hardy spaces $H^p(\mathrm{S}^{d-1})$. A new characterization of $H^p(\mathrm{S}^{d-1})$ is given via the maximal Cesàro operator in Section 3. In Section 4, we briefly sketch some of the basic approximation results in $H^p(\mathrm{S}^{d-1})$. Finally, we prove our main results in Section 5 by applying these approximation results and the characterization theorem in Section 3.

§ 2. Hardy Spaces on S^{d-1}

Most materials described in this section can be found in [1] and [6].

Let $\mathscr{S}\equiv\mathscr{S}(\mathrm{S}^{d-1})$ denote the set of indefinitely differentiable functions on S^{d-1} endowed with the usual test function topology, and let $\mathscr{S}'\equiv\mathscr{S}'(\mathrm{S}^{d-1})$ be the dual of $\mathscr{S}$. $\mathscr{S}$ is called the space of test functions and $\mathscr{S}'$ the space of distributions. (One may think of a function on S^{d-1} as a function defined on an annulus about S^{d-1} by extending the function to be constant along rays through the origin. This allows us to associate with

$$\gamma=(\gamma_1,\gamma_2,\cdots,\gamma_d),\qquad D^\gamma=\left(\frac{\partial}{\partial x_1}\right)^{\gamma_1}\left(\frac{\partial}{\partial x_2}\right)^{\gamma_2}\cdots\left(\frac{\partial}{\partial x_d}\right)^{\gamma_d},$$

$$|\gamma|=\gamma_1+r_2+\cdots+\gamma_d$$

a differential operator of order $|\gamma|$ by differentiating in $\mathbf{R}^d$ and restricting to S^{d-1}. The topology on $\mathcal{S}$ is that induced by the seminorms $N_m(\varphi)=\sum_{|\gamma|=m}\|D^\gamma\varphi\|_\infty, m\in\mathbf{N}$) The pairing of $f\in\mathcal{S}'$ and $\varphi\in\mathcal{S}$ is given by $\langle f, \varphi\rangle$. If f is an integrable function on S^{d-1}, we set

$$\langle f,\varphi\rangle=\int_{S^{d-1}} f(u)\varphi(u)\mathrm{d}\sigma(u).$$

For $x\in S^{d-1}$ and $z\in B_d:=\{(z_1, z_2, \cdots, z_d)\in\mathbf{R}^d \mid z_1^2+z_2^2+\cdots+z_d^2\leqslant 1\}$, let

$$P_z(x)=c_d\frac{1-|z|^2}{|z-x|^d},$$

where P_z belongs to $\mathscr{S}$ and is called the Poisson kernel, c_d is chosen so that $\int_{S^{d-1}} P_z(x)\mathrm{d}\sigma(x)=1$ for all $z\in B_d$. For $f\in\mathscr{S}'$, we write $f(z)=\langle f, P_z\rangle$. $f(z)$ is called the Poisson integral of f.

For a distribution f we define the radial maximal function, $P^+f(x)$,

$$P^+f(x)=\sup_{0\leqslant r<1}|f(rx)|, \quad x\in S^{d-1}.$$

Definition 2.1([6]) The Hardy space $H^p(S^{d-1})$ is the linear space of distributions f with $\|P^+f\|_p<+\infty$. We set $\|f\|_{H^p}=\|P^+f\|_p$.

Definition 2.2([6]) Let m be a non-negative integer and $x\in S^{d-1}$. We say that $\varphi\in K_m$ if $\varphi\in\mathscr{S}$, and:

1) supp $\varphi\subset B(x, h)$ for some $h>0$; and

2) $\sup\{|h|^{d-1+|\gamma|}|D^\gamma\varphi(z)| \mid z\in B_d, |\gamma|\leqslant m\}\leqslant 1$.

Definition 2.3([6]) For $f\in\mathscr{S}'$, the grand maximal function f^* of f is the function

$$f^*(x)=\sup\{|\langle f, \varphi\rangle| \mid \varphi\in K_m(x)\}, \quad x\in S^{d-1}.$$

Theorem 2.1([6]) If $m>\frac{d-1}{p}$, $0<p<+\infty$, then $A\|P^+f\|_p\leqslant\|f^*\|_p\leqslant B\|P^+f\|_p$, where $0<A<B<+\infty$ are constants that depend on p, m and d.

It is well known that if $p>1$, $\|p^+f\|_p$ is equivalent to $\|f\|_p$. Thus, $H^p(S^{d-1})$ coincides with $L^p(S^{d-1})$ if $p>1$. For the remainder of this paper we assume $0<p\leqslant 1$.

The conclusion of Theorem 2.1 is often described as the "grand maximal function" characterization of Hardy spaces. We now turn to the "atomic characterization".

Definition 2.4 A regular p-atom, $0<p\leqslant 1$, centered at $x\in S^{d-1}$, is a function $a\in L^{\infty}(S^{d-1})$ satisfying:

1) supp $a\subset B(x, s)$ for some $s>0$;

2) $\|a\|_{\infty}\leqslant s^{-\frac{d-1}{p}}$; and

3) $\int_{S^{d-1}} a(u)Y(u)\mathrm{d}\sigma(u)=0$, for every spherical polynomial of degree less than or equal to $\left[(d-1)\left(\frac{1}{p}-1\right)\right]$.

An exceptional atom is a function $a\in L^{\infty}(S^{d-1})$ with $\|a\|_{\infty}\leqslant 1$.

Theorem 2.2([6]) Let $0<p\leqslant 1$. If $\{a_j\}_{j=0}^{+\infty}$ is a sequence of exceptional or regular p-atoms, and $\{c_j\}_{j=0}^{+\infty}$ is a sequence of complex numbers with

$$\left(\sum_{j=0}^{+\infty}|c_j|^p\right)^{\frac{1}{p}}<+\infty,$$

then $\sum_{j=0}^{+\infty} c_j a_j$ converges in H^p and

$$\left\|\sum_j c_j a_j\right\|_{H^p}\leqslant A\left(\sum_j |c_j|^p\right)^{\frac{1}{p}},$$

where $A>0$ depends on p and d.

Conversely, if $f\in H^p(S^{d-1})$, there exists a sequence $\{a_j\}$ of complex numbers such that

$$f=\sum_j c_j a_j \quad \text{and} \quad \left(\sum_j |c_j|^p\right)^{\frac{1}{p}}\leqslant B\|f\|_{H^p},$$

where B depends on p and d.

For $f\in \mathscr{S}'(S^{d-1})$ we associate its expansion in spherical harmonics, $f\sim\sum_{k=0}^{+\infty} Y_k(f)$, where $Y_k(f)(x)=\langle f, Z_x^{(k)}\rangle$ with $Z_x^{(k)}$ the zonal harmonics of degree k with pole x. For $\delta>-1$ and $\alpha>0$, the Cesàro mean σ_k^{δ} and the Riesz mean $R_k^{\delta,\alpha}$ are defined by

$$\sigma_k^{\delta}(f)(x):=\sum_{j=0}^{k}\frac{A_{k-j}^{\delta}}{A_k^{\delta}}Y_j(f)(x)$$

and

$$R_k^{\delta,\alpha}(f)(x) := \sum_{j=0}^{k}\left(1-\left(\frac{j}{k+1}\right)^{\alpha}\right)^{\delta} Y_j(f)(x),$$

respectively. Given $f\in\mathscr{S}'$ and a number $r>0$, we say $g := f^{(r)}$ is the r-th order derivative of f if $g\in\mathscr{S}'$ and for any non-negative integer k,

$$Y_k(g)=(-k(k+d-2))^{\frac{r}{2}}Y_k(f).$$

§ 3. A new characterization of the Hardy space $H^p(S^{d-1})$

The main goal in this section is to establish a "maximal Cesàro operator" characterization of $H^p(S^{d-1})$, which will play a basic role in our later proof of Theorem 1.3.

We first introduce some necessary notations. Let $\{\mu_k\}$ be a sequence of complex numbers. Given a positive integer l, we define $\Delta^l\mu_k$ by $\Delta\mu_k=\mu_k-\mu_{k+1}$, $\Delta^{i+1}\mu_k=\Delta(\Delta^i\mu_k)$, $i=1,2,\cdots,l-1$, and define $\overleftarrow{\Delta}^l\mu_k$ by $\overleftarrow{\Delta}^l\mu_k=(-1)^l\Delta^l\mu_k$. Given $f\in\mathscr{S}'$, its maximal Cesàro mean $\sigma_*^\delta(f)$ of order δ is $\sigma_*^\delta(f)=\sup_k|\sigma_k^\delta(f)|$.

In this section, we shall prove:

Theorem 3.1 Suppose $0<p\leqslant 1$, $\delta>\delta(p) := \frac{d-1}{p}-\frac{d}{2}$ and f is a distribution on S^{d-1}. Then $f\in H^p(S^{d-1})$ if and only if $\sigma_*^\delta(f)\in L^p(S^{d-1})$. Furthermore,

$$\|f\|_{H^p(S^{d-1})}\approx\|\sigma_*^\delta(f)\|_{L^p(S^{d-1})}.$$

To the best of our knowledge, Theorem 3.1 is new even in the case $d=2$ (the periodic case), where only the boundedness of σ_*^δ from $H^p(T)$ to $L^p(T)$ is known (see [7]).

Proof Suppose $\sigma_*^\delta(f)\in L^p(S^{d-1})$. Then for a. e. $x\in S^{d-1}$, $\sigma_*^\delta(f)(x)<+\infty$. Observing

$$\sigma_*^{[\delta]+1}(f)(x)\leqslant\sigma_*^\delta(f)(x),$$

we get

$$|Y_k(f)(x)|\leqslant C_\delta k^{\delta+1}\sigma_*^\delta(f)(x),\qquad \text{a. e. } x\in S^{d-1},$$

since

$$Y_k(f)(x)=\overset{\leftarrow[\delta]+2}{\Delta}\left[A_k^{[\delta]+1}\sigma_k^{[\delta]+1}(f)(x)\right].$$

Thus for every $r\in(0,1)$,

$$\sum_{k=0}^{+\infty} r^k\left|Y_k(f)(x)\right|<+\infty,\quad \text{a. e. } x\in S^{d-1}.$$

Taking into account that

$$(1-r)^{-1-\delta}=\sum_{k=0}^{+\infty} A_k^{\delta} r^k,\tag{3.1}$$

we get for a. e. $x\in S^{d-1}$,

$$\begin{aligned}(1-r)^{-1-\delta}\sum_{k=0}^{+\infty} r^k Y_k(f)(x) &= \Big(\sum_{k=0}^{+\infty} A_k^{\delta} r^k\Big)\Big(\sum_{k=0}^{+\infty} r^k Y_k(f)(x)\Big)\\ &= \sum_{k=0}^{+\infty} A_k^{\delta} r^k \sigma_k^{\delta}(f)(x).\end{aligned}$$

Noticing

$$P_r(f)(x)=\sum_{k=0}^{+\infty} r^k Y_k(f)(x),$$

we obtain that for a. e. $x\in S^{d-1}$,

$$P_r(f)(x)=(1-r)^{1+\delta}\sum_{k=0}^{+\infty} A_k^{\delta} r^k \sigma_k^{\delta}(f)(x),$$

which, together with (3.1), implies

$$P^+(f)(x)\leqslant\sigma_*^{\delta}(f)(x),\qquad \text{a. e. } x\in S^{d-1}.$$

So according to Definition 1.1, we conclude that $f\in H^p(S^{d-1})$ and

$$\|f\|_{H^p}=\|P^+(f)\|_p\leqslant\|\sigma_*^{\delta}(f)\|_p.$$

The proof of the converse part of the theorem was essentially contained in [1]. In fact, with a slight modification of the proof of Lemma 4.2 of [1], we have

$$\sigma_*^{\delta}(a)(x)\leqslant\begin{cases} Cr^{-\frac{d-1}{p}+\frac{d}{2}+\delta}|x-y|^{-(\frac{d}{2}+\delta)}, & 0<|x-y|\leqslant\frac{\pi}{2},\\ Cr^{-\frac{d-1}{p}+\frac{d}{2}+\delta}|x+y|^{-(\frac{d}{2}+\delta)}, & 0<|x+y|\leqslant\frac{\pi}{2},\\ Cr^{-\frac{d-1}{p}}, & x\in S^{d-1},\end{cases}$$

where a is a regular p-atom, supported in $B(y,r)$ and with the property $\|a\|_\infty\leqslant r^{-\frac{d-1}{p}}$. Taking into account $\|\sigma_*^{\delta}\|_{(\infty,\infty)}<+\infty$, we get, by a standard method, that $\|\sigma_*^{\delta}(f)\|_p\leqslant C\|f\|_{H^p}$. This concludes the proof.

As a consequence, we have the following:

Corollary 3.2 Let $\{u_k\}_{k=0}^{+\infty}$ be a sequence of complex numbers, $0<p\leqslant 1$, $\delta(p):=\frac{d-1}{p}-\frac{d}{2}$ and $\ell=[\delta(p)]+1$. Suppose the following conditions are satisfied:

(1) $\sup_k|\mu_k|\leqslant M<\infty$;

(2) $\sum_{k=0}^{+\infty}|\Delta^{\ell+1}\mu_k|(k+1)^{\ell}\leqslant M.$

Then

$$\Big\|\sum_{k=0}^{+\infty}\mu_k Y_k(f)\Big\|_{H^p}\leqslant MC\|f\|_{H^p},$$

where $C>0$ is independent of M, $\{\mu_k\}$ and f.

Proof Let

$$T(f):=\sum_{k=0}^{+\infty}\mu_k Y_k(f).$$

Then by Theorem 3.1, it suffices to prove

$$\sigma_*^{\ell+2}(Tf)\leqslant MC\sigma_*^{\ell}(f). \tag{3.2}$$

Applying Abel's transform $\ell+1$ times gives

$$\sigma_N^{\ell+2}(Tf)=\sum_{k=0}^{N}\Delta^{\ell+1}\left(\frac{A_{N-k}^{\ell+2}}{A_N^{\ell+2}}\mu_k\right)A_k^{\ell}\sigma_k^{\ell}(f)(x), \tag{3.3}$$

where $A_j^{\ell+2}=0$ for $j<0$.

On the other hand, according to conditions (1) and (2), one can easily verify that for all $v=0,1,2,\cdots,\ell$,

$$\sum_{k=0}^{+\infty}|\Delta^{v+1}\mu_k|k^{v}\leqslant MC.$$

Thus

$$\begin{aligned}\sum_{k=0}^{N}\left|\Delta^{\ell+1}\left(\frac{A_{N-k}^{\ell+2}}{A_N^{\ell+2}}\mu_k\right)\right|A_k^{\ell}&\leqslant C\sum_{v=0}^{\ell+1}\sum_{k=0}^{N}\left|\Delta^{\ell+1-v}\left(\frac{A_{N-k-v}^{\ell+2}}{A_N^{\ell+2}}\Delta^{v}\mu_k\right)\right|(k+1)^{\ell}\\&\leqslant MC,\end{aligned} \tag{3.4}$$

where we define $\Delta^0\mu_k=\mu_k$.

Now combining (3.3) with (3.4), we get (3.2) and complete the proof.

§ 4. Basic approximation results in the space $H^p(S^{d-1})(0<p\leqslant 1)$

Lemma 4.1 (Bernstein inequality) Let $r>0$. Then for every spherical polynomial T_N of degree less than or equal to N, $\|T_k^{(r)}\|_{H^p}\leqslant CN^r\|T_N\|_{H^p}$, where $C>0$ is independent of N and T_N.

Lemma 4.2 Suppose $r>0$ and η is a C^∞ function with the properties that $0\leqslant\eta\leqslant 1$ for $x\in\mathbf{R}$, $\eta(x)=1$ for $0\leqslant|x|\leqslant 1$ and $\eta(x)=0$ for $|x|>2$. For $N>0$, let η_N be an operator defined by

$$\eta_N(f):=\sum_{k=0}^{2N}\eta\left(\frac{k}{N}\right)Y_k(f).$$

Then for $f\in H^p(S^{d-1})$,

$$\|f-\eta_N f\|_{H^p}+\frac{1}{N^r}\|(\eta_N f)^{(r)}\|_{H^p}\approx K_r\left(f, \frac{1}{N}\right)_{H^p}.$$

Lemma 4.1 can be obtained by applying Corollary 3.2, while Lemma 4.2 is a consequence of Lemma 4.1 and Corollary 3.2. The proofs of both lemmas follow the standard method (see the proofs of Theorems 3.2 and 5.1 in [8]). We omit the details.

Lemma 4.3 Suppose $0<p\leqslant 1$ and $r>0$. Then for $f\in H^p(S^{d-1})$,

$$K_r^p\left(f,\frac{1}{N}\right)_{H^p}\leqslant CN^{-rp}\sum_{k=1}^{N}k^{rp-1}E_k^p(f,H^p).$$

Lemma 4.3 can be obtained by invoking Lemma 4.2 and the routine method (see the proof of Theorem 6.4 of [8]).

Lemma 4.4 Suppose $0<p\leqslant 1$ and $\delta>\delta(p):=\dfrac{d-1}{p}-\dfrac{d}{2}$. Then for $f\in H^p$,

$$\|\sigma_N^\delta(f)-f\|_{H^p}\approx K_1\left(f, \frac{1}{N}\right)_{H^p}.$$

In the special case $d=2$, Lemma 4.4 for the Riesz means is due to [2], while for the general case $d\geqslant 3$, its proof runs along the same lines as that of Theorem 3.1 in [9]. We omit the details here.

§ 5. Proofs of main results

First, we prove Theorem 1.3. We need the following lemma:

Lemma 5.1([5]) Suppose $0<p<1$, $\delta=\delta(p):=\frac{d-1}{p}-\frac{d}{2}$ and $f\in H^p(\mathrm{S}^{d-1})$. Then

$$\frac{1}{\ln N}\sum_{k=1}^{N}\frac{\|\sigma_k^\delta(f)\|_{L^p}^p}{k}\leqslant C\|f\|_{H^p}^p. \tag{5.1}$$

We remark that Lemma 5.1 was essentially contained in [1] though not explicitly stated there. In fact, it is a direct consequence of the following estimates of $\sigma_L^\delta(a)(x)$, which were obtained in the proof of Lemma 4.2 of [1]:

$$|\sigma_L^\delta(a)(x)|\leqslant\begin{cases}C(rL)^{s-(d-1)\left(\frac{1}{p}-1\right)}|x-y|^{-\frac{d-1}{p}}, & 0<|x-y|\leqslant\frac{\pi}{2},\\ C(rL)^{s-(d-1)\left(\frac{1}{p}-1\right)}|x+y|^{-\frac{d-1}{p}}, & |x+y|\leqslant\frac{\pi}{2},\\ Cr^{-\frac{d-1}{p}}, & x\in\mathrm{S}^{d-1},\end{cases}$$

where a is a regular p-atom, supported in $B(y, r)$, $\|a\|_\infty\leqslant r^{-\frac{1}{p}}$ and

$$s=\left[(d-1)\left(\frac{1}{p}-1\right)\right]+1 \text{ or } 0.$$

Proof of Theorem 1.3 For $j<0$ we define $A_j^\delta=0$ and throughout the proof we will consider A_j^δ as a function for all $j\in\mathbf{Z}$.

According to Theorem 3.1 and Lemma 5.1, it suffices to prove

$$\sigma_*^{\ell+4}(\sigma_L^\delta(f))(x)\leqslant C\sigma_*^\ell(f)(x)+C|\sigma_L^\delta(f)(x)|, \tag{5.2}$$

where $\ell=[\delta]+1$. To this end, we have to estimate $\sigma_N^{\ell+4}(\sigma_L^\delta(f))(x)$. We consider three cases.

Case 1 $0\leqslant N\leqslant L$.

Let $\mu_k=\frac{A_{N-k}^{\ell+4}A_{L-k}^\delta}{A_N^{\ell+4}A_L^\delta}$ for all $k\in\mathbf{N}^*$. It is easy to verify that

$$|\Delta^{\ell+1}\mu_k|\leqslant\begin{cases}\frac{C}{N^{\ell+1}}, & \text{if } 0\leqslant N\leqslant\frac{L}{2},\\ \frac{C(L-k+1)^{\delta+3}}{L^{\delta+\ell+4}}, & \text{if } \frac{L}{2}\leqslant N\leqslant L.\end{cases}$$

So, invoking Abel's transform $\ell+1$ times yields

$$|\sigma_N^{\ell+4}\sigma_L^{\delta}(f)(x)|\leqslant C\sigma_*^{\ell}(f)(x). \tag{5.3}$$

Case 2 $N\geqslant L+1$ and $\delta>0$ is not an integer.

We write

$$\sigma_N^{\ell+4}\sigma_L^{\delta}(f)(x)=\sum_{k=0}^{L}a_kb_kY_k(f)(x)+O(1)\sigma_L^{\delta}(f)(x), \tag{5.4}$$

where

$$a_k=\begin{cases}\dfrac{A_{N-k}^{\ell+4}}{A_N^{\ell+4}}-\dfrac{A_{N-L}^{\ell+4}}{A_N^{\ell+4}}, & 0\leqslant k\leqslant N,\\ 0, & k\geqslant N+1,\end{cases}$$

and

$$b_k=\begin{cases}\dfrac{A_{L-k}^{\delta}}{A_L^{\delta}}, & 0\leqslant k\leqslant L,\\ 0, & \text{otherwise.}\end{cases}$$

The following inequalities can be verified for a non-integer δ by a straightforward computation:

$$|a_k|\leqslant C\frac{L-k+1}{N}, \tag{5.5}$$

$$|\Delta^i a_k|\leqslant C\left(\frac{1}{N}\right)^i,\ i=1,2,\cdots,\ell, \tag{5.6}$$

$$|\Delta^i b_k|\leqslant C\frac{(L-k+1)^{\delta-1}}{(L+1)^{\delta}},\ i\in\mathbf{N}. \tag{5.7}$$

Invoking (5.5)~(5.7), we conclude that for a non-integer δ,

$$\sum_{k=0}^{L}|\Delta^{\ell+1}(a_kb_k)|(k+1)^{\ell}\leqslant C, \tag{5.8}$$

which, by applying Abel's transform $\ell+1$ times to the first sum on the right-hand side of (5.4), implies

$$|\sigma_N^{\ell+4}(\sigma_L^{\delta}(f))(x)|\leqslant C\sigma_*^{\ell}(f)(x)+C|\sigma_L^{\delta}(f)(x)|. \tag{5.9}$$

Case 3 $N\geqslant L+1$, δ is an integer.

We start with the identity (5.4). Observing $\ell=\delta+1$, we can easily verify that

$$\Delta^{\ell}a_k=\begin{cases}O\left(\dfrac{1}{L^{\delta}}\right), & L-\ell\leqslant k\leqslant L,\\ 0, & \text{otherwise,}\end{cases} \tag{5.10}$$

$$\Delta^{\ell+1}b_k=\begin{cases}O\left(\frac{L}{L^{\delta}}\right), & L-\ell\leqslant k\leqslant L,\\ 0, & \text{otherwise.}\end{cases} \tag{5.11}$$

Now invoking (5.5)～(5.7) and (5.10)(5.11) gives (5.8), which again implies (5.9).

Putting the above three cases together, we obtain (5.2) and conclude the proof.

Proof of Theorem 1.1 We follow the proof in [2]. The lower estimate is evident. For the proof of the upper estimate we take $\alpha>\delta(p)$. Then

$$\begin{aligned}&\sum_{j=1}^{N}\frac{1}{j}\|f-\sigma_j^{\delta}(f)\|_{H^p}^p\\ &=\sum_{j=1}^{\ln\ln N}\sum_{m=2^{2^j}+1}^{2^{2^{j+1}}}\frac{1}{m}\|\sigma_m^{\delta}(f)-f\|_{H^p}^p\\ &\leqslant\sum_{j=1}^{\ln\ln N}\sum_{m=2^{2^j}+1}^{2^{2^{j+1}}}\frac{1}{m}\|(\sigma_m^{\delta}(f)-f)-\sigma_m^{\alpha}(\sigma_m^{\delta}(f)-f)\|_{H^p}^p+\\ &\quad\sum_{j=1}^{\ln\ln N}\sum_{m=2^{2^j}+1}^{2^{2^{j+1}}}\frac{1}{m}\|\sigma_m^{\alpha}(\sigma_m^{\delta}(f)-f)\|_{H^p}^p.\end{aligned} \tag{5.12}$$

With a slight modification of the proof of Theorem 3.1 in [9], we can obtain

$$\|\sigma_m^{\alpha}(\sigma_m^{\delta}(f)-f)\|_{H^p}^p\leqslant CK_1^p\left(f,\frac{1}{m}\right)_{H^p}. \tag{5.13}$$

Therefore, the second sum in (5.12) can be estimated to be not greater than

$$C\sum_{j=1}^{N}\frac{1}{j}K_1^p\left(f,\frac{1}{j}\right)_{H^p},$$

which, by the monotonicity of the K-functional $K_1(f,\ \cdot)_{H^p}$, is bounded by

$$C\sum_{j=1}^{\ln N}K_1^p\left(f,\frac{1}{2^j}\right)_{H^p}.$$

According to Lemma 4.4, we have that for $2^{2^j}+1\leqslant m\leqslant 2^{2^{j+1}}$ and every $g\in H^p$,

$$\|g-\sigma_m^{\alpha}(g)\|_{H^p}\leqslant C\|g-\sigma_{2^{2^j}}^{\alpha}(g)\|_{H^p}. \tag{5.14}$$

Now applying (5. 14), with $g=\sigma_m^\delta(f)-f$, to the first sum in (5. 12), we have

$$\begin{aligned}&\|(\sigma_m^\delta(f)-f)-\sigma_m^\alpha(\sigma_m^\delta(f)-f)\|_{H^p}^p\\ \leqslant& C\|(\sigma_m^\delta(f)-f)-\sigma_{2^{2^j}}^\alpha(\sigma_m^\delta(f)-f)\|_{H^p}^p\\ \leqslant& C\|\sigma_m^\delta(f-\sigma_{2^{2^j}}^\alpha(f))\|_{H^p}^p+C\|f-\sigma_{2^{2^j}}^\alpha(f)\|_{H^p}^p.\end{aligned}$$

Applying Theorem 1. 3 to the sum

$$\sum_{m=2^{2^j}}^{2^{2^{j+1}}}\frac{1}{m}\|\sigma_m^\delta(f-\sigma_{2^{2^j}}^\alpha(f))\|_{H^p}^p,$$

and using Lemma 4. 4, we obtain, by the monotonicity of the K-functional, that

$$\sum_{j=1}^{N}\frac{1}{j}\|\sigma_j^\delta(f)-f\|_{H^p}^p\leqslant C\sum_{j=1}^{\ln N}K_1^p\left(f,\frac{1}{2^j}\right)_{H^p}.$$

Lemma 4. 3 gives

$$\sum_{j=1}^{\ln N}K_1^p\left(f,\frac{1}{2^j}\right)_{H^p}\leqslant C\sum_{j=1}^{\ln N}2^{-jp}\sum_{k=1}^{2^j}k^{p-1}E_k^p(f,H^p)\leqslant C\sum_{k=1}^{N}\frac{E_k^p(f,H^p)}{k},$$

which is the required upper estimate. This concludes the proof.

Proof of Corollary 1. 2 By Lemma 4. 4, we have the following Jackson type inequality:

$$E_k(f,\ H^p)\leqslant CK_r\left(f,\ \frac{1}{k}\right)_{H^p},\ r>0. \tag{5. 15}$$

Invoking (5. 15) and by the monotonicity of the K-functional, we conclude that

$$\sum_{j=1}^{N}\frac{1}{j}E_j^p(f,H^p)\leqslant C\sum_{j=1}^{N}\frac{1}{j}K_1^p\left(f,\frac{1}{j}\right)_{H^p}\approx\int_1^{N+1}\frac{K_1^p\left(f,\frac{1}{t}\right)_{H^p}}{t}\mathrm{d}t. \tag{5. 16}$$

We decompose the integral into two parts $\int_1^A+\int_A^{N+1}$, with $A>0$ a constant to be decided later. Then a straightforward computation shows that (5. 16) is estimated to be not greater than

$$CK_1^p\left(f,\ \frac{1}{A}\right)_{H^p}\ln N+CK_1^p\left(f,\ \frac{1}{A}\right)_{H^p}A^p.$$

Taking $A=(\ln N)^{\frac{1}{p}}$ and invoking Theorem 1. 1 gives the corollary.

This concludes the proof.

The proofs of Theorem 1.4 and Corollary 1.5 are analogous to those of Theorem 1.1 and Corollary 1.2. We omit the details.

References

[1] Colzani L, Taibleson M H, Weiss G. Maximal estimates for Cesàro and Riesz means on spheres. Indiana Univ. Math. J., 1984, 33(6): 873—889.

[2] Belinskii E S. Strong summability of Fourier series of the periodic functions from $H^p(0<p\leqslant 1)$. Constr. Approx., 1996, 12: 187—195.

[3] Chen G L, Jiang Y S, Lu S Z. Strong approximation of Riesz means at critical index on $H^p(0<p\leqslant 1)$. Approx. Theory and its Appl., 1989, 5(2): 39—49.

[4] Jiang Y S, Liu H P, Lu S Z. Research Report CMA-R39-87. The Australian National University, 1987.

[5] Chen G L. Boundedness of strong means of Cesàro means on $H^p(\sum_n)(0<p<1)$. J. of Zhejiang Univ. NS, 1990, 24(1): 153—162. (in Chinese)

[6] Colzani L. Hardy spaces on unit spheres. Boll. Un. Mat. Ital. C(6), 1985, 4(1): 219—244.

[7] Móricz F. The maximal Fejér operator is bounded from $H^1(\mathbf{T})$ into $L^1(\mathbf{T})$. Analysis, 1996, 16: 125—135.

[8] Ditzian Z. Fractional derivatives and best approximation. Acta Math. Hungar., 1998, 81(4): 323—348.

[9] Feng D. Some equivalence theorems with K-functionals. J. Approx. Theory., 2003, 121(1): 143—157.

Acta Mathematica Sinica，English Series，2007，23(7)：1 327—1 332.

单位球面的一个覆盖引理及其对于 Fourier-Laplace 级数的收敛的应用

A Covering Lemma on the Unit Sphere and Application to the Fourier-Laplace Convergence①

Abstract　A covering lemma on the unit sphere is established and then is applied to establish an almost everywhere convergence test of Marcinkiewicz type for the Fourier-Laplace series on the unit sphere which can be stated as follows:

Theorem　Suppose $f\in L(\mathrm{S}^{n-1})$，$n\geqslant 3$. If f satisfies the condition

$$\frac{1}{\theta^{n-1}}\int_{D(x,\theta)}|f(y)-f(x)|\,\mathrm{d}y=O\left(\frac{1}{|\ln\theta|}\right),\quad \text{as } \theta\to 0^{+},$$

at every point x in a set E of positive measure in S^{n-1}, then the Cesàro means of critical order $\frac{n-2}{2}$ of the Fourier-Laplace series of f coverage to f at almost every point x in E.

Keywords　sphere; covering lemma; Fourier-Laplace series; a. e. convergence.

§ 1. Introduction

In harmonic analysis, the convergence problem is always a main topic. For Fourier series of a single variable there are already plenty of

---

① Supported by NSFC 10471010 and the Natural Science Foundation of Tianjin Normal University, 52LJ32.

本文与黄蓉合作.

Received: 2005-02-25; Revised: 2005-07-27.

results, including point-wise and almost everywhere convergence tests. Among them a sufficient condition for almost everywhere convergence was found in 1935 by Marcinkiewicz (see [1]). This is the so-called Marcinkiewicz test, which is a deep result and can be stated as follows:

Theorem A Suppose $f\in L[-\pi, \pi]$. If f satisfies, at every point x in a set E of positive measure, the condition

$$\frac{1}{h}\int_0^h |f(x+h)-f(x)|\,\mathrm{d}t = O\left(\frac{1}{|\ln|h||}\right), \quad |h|\to 0^+,$$

then the Fourier series f converges almost everywhere in E.

It has been pointed out in [2, p. 303] that, at an individual point x, even the stronger condition

$$|f(x+h)-f(x)| = o\left(\frac{1}{|\ln|h||}\right), \quad \text{as } |h|\to 0^+$$

does not always imply the convergence of Fourier series of f at x.

This test for almost everywhere convergence was extended to multiple Fourier series in [3], in 1965, in terms of the convergence of Bochner-Riesz means with critical index.

But for the Fourier-Laplace series on the unit sphere, this test has not been established for long. The main difficulty is to find a suitable covering lemma on the unit sphere.

In the present paper we will give such a covering lemma and then apply it to establish a test of Marcinkiewicz type for the almost everywhere convergence of the Fourier-Laplace series on the unit sphere.

We will use the following notations and definitions: $\mathbf{N}$ denotes the set of all non-negative integers, and $\mathbf{N}^*$ is the set of all positive integers. $\mathbf{R}^n$ denotes the Euclidian space of dimension $n\in\mathbf{N}^*$. For $x, y\in\mathbf{R}^n$, xy denotes their inner product, $|x|$ denotes the Euclidian norm of x.

In what follows we will always assume $n\in\mathbf{N}^*$ and $n\geqslant 3$. Other notations we will use are:

$S^{n-1} := \{(x_1, x_2, \cdots, x_n)\in\mathbf{R}^n \mid x_1^2+x_2^2+\cdots+x_n^2=1\}$, the unit sphere in $\mathbf{R}^n$;

$d(x, y) := \arccos(xy)$, $x, y \in S^{n-1}$, the distance between x, y;

$D(x, \theta) := \{y \in S^{n-1} \mid d(x, y) \leqslant \theta\}$, $x \in S^{n-1}$, $\theta \in [0, \pi]$;

$d(A, B) := \inf\{d(x, y) \mid x \in A, y \in B\}$, $\varnothing \neq A \subset S^{n-1}$, $\varnothing \neq B \subset S^{n-1}$, the distance between A, B;

$d(x, A) := D(\{x\}, A)$, $x \in S^{n-1}$, $\varnothing \neq A \subset S^{n-1}$, the distance between the single point x and the set A;

$l_{x,t} := \{y \in S^{n-1} \mid d(x, y) = t\}$, $x \in S^{n-1}$, $t \in (0, \pi)$, the parallel of latitude with pole at x and angle t;

$l(t) := |S^{n-2}| \sin^{n-2} t$,　the length of $l_{x,t}$.

We now verify the triangular inequality on S^{n-1}:

$$d(x,y) \leqslant d(x,z) + d(y,z). \tag{1.1}$$

To do so, we may assume $z = (1, 0, 0, \cdots, 0)$ since rotations preserve inner products in $\mathbf{R}^n$. Then

$$d(x, z) = \arccos(x_1), \ d(y, z) = \arccos(y_1).$$

So,

$$\begin{aligned} \cos(d(x, z) + d(y, z)) &= x_1 y_1 - \sqrt{1 - x_1^2}\sqrt{1 - y_1^2} \\ &= -\Big(\sum_{k=2}^{n} x_k^2\Big)^{\frac{1}{2}} \Big(\sum_{k=2}^{n} y_k^2\Big)^{\frac{1}{2}} + x_1 y_1 \\ &\leqslant \sum_{k=2}^{n} x_k y_k + x_1 y_1 = xy = \cos(d(x,y)), \end{aligned}$$

which shows (1.1) in the case $d(x, z) + d(y, z) \leqslant \pi$. If $d(x, z) + d(y, z) > \pi$ then (1.1) is obviously valid.

Let $f \in L(S^{n-1})$. The (C, α) means of the Fourier-Laplace series of f are denoted by $\sigma_N^\alpha(f)$. For the basic knowledge on the Fourier-Laplace series we refer to [4].

For the Cesàro means, the special index $\alpha = \frac{n-2}{2}$ is critical, and $\sigma_N^{\frac{n-2}{2}}(f)$ takes a role as the partial sums in a single Fourier series. Since only $\sigma_N^{\frac{n-2}{2}}(f)$ with critical index will be considered in our paper we will write it as $\sigma_N(f)$, for short.

In the next section we will prove the following covering lemma on the unit sphere:

Covering lemma Suppose P is a non-empty closed subset of S^{n-1} and $P \neq S^{n-1}$. Then there exist points $x_j \in \Omega_n$, and numbers $d_j \in (0, \pi)$, $j \in \mathbf{N}^*$, such that

$$D(x_i, d_i) \bigcap D(x_j, d_j) = \varnothing \text{ when } i \neq j, \tag{1.2}$$

$$\bigcup_{j=1}^{+\infty} D(x_j, d_j) \subset S^{n-1} \setminus P = \bigcup_{j=1}^{+\infty} D\left(x_j, \frac{7}{2} d_j\right), \tag{1.3}$$

$$4d_j \geqslant d(D(x_j, d_j), P) > d\left(D\left(x_j, \frac{7}{2} d_j\right), P\right) \geqslant \frac{1}{2} d_j, \quad j \in \mathbf{N}^*. \tag{1.4}$$

Then in §3 we will prove our theorem:

Theorem Suppose $f \in L(S^{n-1})$, $n \geqslant 3$, $E \subset S^{n-1}$ and the measure $|E| > 0$. If the condition

$$\frac{|\ln \theta|}{\theta^{n-1}} \int_{D(x,\theta)} |f(y) - f(x)| \, dy = O(1), \quad \text{as } \theta \to 0^+, \tag{1.5}$$

is fulfilled at every $x \in E$ then

$$\lim_{N \to +\infty} \sigma_N(f)(x) = f(x) \quad \text{a. e. } x \in E. \tag{1.6}$$

§2. Proof of the Covering Lemma

Define the function δ: $S^{n-1} \to [0, \pi]$ by $\delta(x) = d(x, P)$. It is obvious that δ is continuous and has maximum $r_0 := \max\{\delta(x) \mid x \in S^{n-1}\} > 0$ and minimum 0. Write $r_k = 2^{-k} r_0$, $k \in \mathbf{N}^*$. Define

$$F_k = \{x \in S^{n-1} \mid r_k \leqslant \delta(x) \leqslant r_{k-1}\}, \quad k \in \mathbf{N}^*.$$

Write the interior of F_k as G_k. By the continuity of the function δ, all F_k are non-empty closed sets and $\bigcup_{k=1}^{+\infty} F_k = S^{n-1} \setminus P$.

Also by the continuity of the function δ, there exists $x \in F_k$ such that $\delta(x) = \frac{3}{2} r_k$ and hence $D\left(x, \frac{1}{4} r_k\right) \subset G_k$. This shows that each G_k contains at least one crown $D\left(x, \frac{1}{4} r_k\right)$, for all $k \in \mathbf{N}^*$. Then, for each k, there exist finitely many points $x_{k1}, x_{k2}, \cdots, x_{kj_k}$ in G_k such that

$$D\left(x_{k\mu}, \frac{1}{4} r_k\right) \subset G_k, \quad D\left(x_{k\mu}, \frac{1}{4} r_k\right) \bigcap D\left(x_{kv}, \frac{1}{4} r_k\right) = \varnothing \tag{2.1}$$

when μ, $\nu\in\{1, 2, \cdots, j_k\}$, $\mu\neq\nu$, and if any $D\left(y, \frac{1}{4}r_k\right)\subset G_k$, then

$$D\left(y,\frac{1}{4}r_k\right)\cap\left(\bigcup_{j=1}^{j_k}D\left(x_{kj},\frac{1}{4}r_k\right)\right)\neq\varnothing. \qquad (2.2)$$

Suppose $z\in F_k$, i. e. $r_k\leqslant\delta(z)\leqslant r_{k-1}$. Then there exists a point $y\in D\left(z, \frac{3}{8}r_k\right)$ such that $\frac{11}{8}r_k\leqslant\delta(y)\leqslant\frac{13}{8}r_k$. Since $D\left(y, \frac{1}{4}r_k\right)\subset G_k$, the relation (2.2) shows that there is some x_{kj} $(1\leqslant j\leqslant j_k)$ such that $d(y, k_{kj})\leqslant\frac{1}{2}r_k$. Then

$$d(z, x_{kj})\leqslant d(z, y)+d(y, x_{kj})\leqslant\frac{3}{8}r_k+\frac{1}{2}r_k=\frac{7}{8}r_k.$$

This shows that $z\in D\left(x_{kj}, \frac{7}{8}r_k\right)\subset S^{n-1}\setminus P$. We conclude that

$$F_k\subset\bigcup_{j=1}^{j_k}D\left(x_{kj},\frac{7}{8}r_k\right)\subset S^{n-1}\setminus P. \qquad (2.3)$$

So, the collection of all $D\left(x_{kj}, \frac{1}{4}r_k\right)$, $1\leqslant j\leqslant k_j$, $k\in\mathbf{N}^*$ (with any arrangement) gives the desired $D(x_k, d_k)$, $k\in\mathbf{N}^*$ with $x_\mu=x_{kj}$, $d_\mu=\frac{1}{4}r_k$, $\frac{7}{2}d_\mu=\frac{7}{8}r_k$ for every $\mu\in\mathbf{N}^*$ and correspondingly some $k\in\mathbf{N}^*$, $j\in\{1, 2, \cdots, j_k\}$.

§ 3. Proof of the Theorem

Denote by l a continuous, increasing and up-convex function on $[0, \pi]$ with $l(0)=0$.

Lemma 1 Suppose $f\in L(S^{n-1})P\subset S^{n-1}$. If there is a constant $M>0$ such that, for every $x\in P$,

$$\int_{D(x,\theta)}|f(y)-f(x)|\,\mathrm{d}y\leqslant M\theta^{n-1}l(\theta),\quad \theta\in[0,\pi], \qquad (3.1)$$

then there is a constant C such that

$$|f(y)-f(x)|\,\mathrm{d}y\leqslant Cl(d(x,y)),\quad x,y\in P. \qquad (3.2)$$

Proof For x, $y\in P$, $x\neq y$ and $d(x, y)=2\theta<\pi$. Let $z=\frac{x+y}{|x+y|}$. Then $d(x, z)=d(y, z)=\theta$. Define two functions ξ, η on

$D\left(z, \frac{1}{2}\theta\right)$ by

$$\xi(u)=|f(u)-f(x)|\frac{1}{l(d(x, u))},$$

$$\eta(u)=|f(u)-f(y)|\frac{1}{l(d(y, u))}.$$

For $\nu>0$ we define

$$H_\nu=\left\{u\in D\left(z, \frac{1}{2}\theta\right)\middle|\xi(u)>\nu\right\}, \quad K_\nu=\left\{u\in D\left(z, \frac{1}{2}\theta\right)\middle|\eta(u)>\nu\right\}.$$

Then, for any point $u\in D\left(z, \frac{1}{2}\theta\right)$, we have

$$d(x, u)\geqslant d(x, z)-d(z, u)\geqslant\frac{1}{2}\theta,$$

$$d(x, u)\leqslant d(x, z)+d(z, u)\leqslant\frac{3}{2}\theta.$$

So $D\left(z, \frac{1}{2}\theta\right)\subset D\left(x, \frac{3}{2}\theta\right)$. Then, by the condition (3.1), we have the measure

$$|H_\nu|\leqslant\frac{1}{\nu}\int_{D\left(z,\frac{1}{2}\theta\right)}\xi(u)\mathrm{d}u\leqslant\frac{1}{\nu l\left(\frac{1}{2}\theta\right)}\int_{D\left(x,\frac{3}{2}\theta\right)}|f(u)-f(x)|\mathrm{d}u$$

$$\leqslant\frac{M}{\nu}\left(\frac{3}{2}\theta\right)^{n-1}.$$

Also,

$$|K_\nu|\leqslant\frac{M}{\nu}\left(\frac{3}{2}\theta\right)^{n-1}.$$

We know $|D(z, t)|=c_nt^{n-1}$, where and in what follows c_n denotes a constant depending on n, which may take different values at different places. Then, for $\nu=\frac{4^n}{c_n}M$,

$$|H_\nu|\leqslant\frac{1}{4}\left|D\left(z, \frac{\theta}{2}\right)\right|, \quad |K_\nu|\leqslant\frac{1}{4}\left|D\left(z, \frac{\theta}{2}\right)\right|,$$

$$|H_\nu|+|K_\nu|\leqslant\frac{1}{2}\left|D\left(z, \frac{1}{2}\theta\right)\right|.$$

The last inequality shows that there exists at least a point $u_0\in D\left(z, \frac{1}{2}\theta\right)\setminus(H_\nu\cup K_\nu)$. Hence $\xi(u_0)\leqslant\nu$, $\eta(u_0)\leqslant\nu$. These yield the following two inequalities:

$$|f(u_0)-f(x)|\leqslant \nu l(d(x,\ u_0))\leqslant \nu l\left(\frac{3}{2}\theta\right),$$

$$|f(u_0)-f(y)|\leqslant \nu l(d(y,\ u_0))\leqslant \nu l\left(\frac{3}{2}\theta\right).$$

Combining them we get

$$|f(x)-f(y)|\leqslant|f(u_0)-f(x)|+|f(u_0)-f(y)|\leqslant 2\nu l(d(x,\ y)),$$

which shows that (3.2) is valid with a constant C depending on n and M.

Lemma 2 Let P be a closed subset of S^{n-1}. If a real-valued function $f\in C(P)$ satisfies the condition

$$|f(x)-f(y)|\leqslant l(d(x,\ y)),\quad x,\ y\in P,$$

then f can be extended to $\varphi\in C(S^{n-1})$ such that

$$|\varphi(x)-\varphi(y)|\leqslant l(d(x,\ y)),\quad x,\ y\in S^{n-1}.$$

Proof We construct the extension φ by defining

$$\varphi(x)=\sup_{y\in P}\{f(y)-l(d(x,\ y))\}=\max_{y\in P}\{f(y)-l(d(x,\ y))\}.$$

It is obvious that $\varphi|_P=f$. For $x\in S^{n-1}$ we have

$$\varphi(x)=f(u)-l(d(u,\ x))\quad \text{for some point } u\in P,$$

and for this point u we have the inequality

$$\varphi(y)\geqslant f(u)-l(d(u,\ y)).$$

By the fact that l is increasing, up-convex and that $|d(u,\ y)-d(u,\ x)|\leqslant d(x,\ y)$, we have

$$\begin{aligned}\varphi(x)-\varphi(y)&\leqslant 2l(d(u,\ y))-l(d(u,\ x))\\&\leqslant l(|d(u,\ y)-d(u,\ x)|)\\&\leqslant l(d(x,\ y)).\end{aligned}$$

Hence $$|\varphi(x)-\varphi(y)|\leqslant l(d(x,\ y)).$$

Proof of theorem 1 Define

$$l(t)=\begin{cases}0, & \text{if } t=0,\\ \dfrac{1}{|\ln t|}, & \text{if } 0<t\leqslant \mathrm{e}^{-2},\\ 1+\dfrac{1}{4}\ln t, & \text{if } \mathrm{e}^{-2}<t\leqslant\pi.\end{cases}$$

Then l is continuous, increasing and up-convex on $[0, \pi]$, which satisfies:

$$l(x+y)\leqslant l(x)+l(y), \quad l(kx)\leqslant kl(x), \ x, \ y\geqslant 0, \ k\geqslant 1.$$

The condition (1.5) of the Theorem can be rewritten as

$$\int_{D(x,\theta)} |f(y)-f(x)|\,\mathrm{d}y\leqslant M_x\theta^{n-1}l(\theta), \quad 0\leqslant\theta\leqslant\pi \qquad (1.5)'$$

where M_x is a constant depending on $x\in E$. Since $(1.5)'$ is assumed to be valid at every $x\in E$ it follows that, for any $\varepsilon>0$, there exist a closed subset $P\subset E$ and a constant $M>0$ such that the measure $|E\setminus P|<\varepsilon$ and

$$\int_{D(x,\theta)} |f(y)-f(x)|\,\mathrm{d}y\leqslant M\theta^{n-1}l(\theta), \quad 0\leqslant\theta\leqslant\pi \qquad (1.5)''$$

holds for every $x\in P$.

Applying Lemma 1, we conclude that there is a constant C such that (3.2) holds for all $x, y\in P$. Then, applying Lemma 2, we define $\varphi\in C(S^{n-1})$ such that

$$\varphi(x)=f(x) \text{ for } x\in P,$$
$$|\varphi(u)-\varphi(v)|\leqslant Cl(d(u, v)), \ u, \ v\in\Omega_{n-1}.$$

Define $\psi=f-\varphi$.

We know (see [4], or [5]) that, if $f\in L\ln^2 L(S^{n-1})$, then $\sigma_N(f)\to f$ almost everywhere. Hence,

$$\lim_{N\to+\infty}\sigma_N(\varphi)(x)=\varphi(x), \quad \text{a. e.} \quad x\in S^{n-1}. \qquad (3.3)$$

We claim that ψ satisfies the Dini convergence condition (see [6] for the Dini convergence test) at almost every point in P. Then

$$\lim_{N\to+\infty}\sigma_N(\psi)(x)=\psi(x), \text{ for almost every } x\in P. \qquad (3.4)$$

A combination of (3.3) and (3.4) then yields the almost every convergence of $\sigma_N(f)$ on P. Since ε is arbitrary, the Theorem will then be established.

In order to verify the Dini convergence condition for ψ at almost every point in P, it is sufficient to prove

$$\int_{D(x,1)}\frac{|\psi(y)-\psi(x)|}{d(y,x)^{n-1}}\mathrm{d}y\leqslant\int_{S^{n-1}\setminus P}\frac{|\psi(y)|}{d(y,x)^{n-1}}\mathrm{d}y<+\infty$$

$$\text{for almost every } x \in P. \tag{3.5}$$

Write the integral on the right side of (3.5) as $I(x)$. Now we apply the Covering Lemma to get

$$I(x) = \int_{S^{n-1} \setminus P} \frac{|\psi(y)|}{d(y,x)^{n-1}} \mathrm{d}y \leqslant \sum_{j=1}^{+\infty} \int_{D(x_j, d'_j)} \frac{|\psi(y)|}{d(y,x)^{n-1}} \mathrm{d}y, \tag{3.6}$$

where $D(x_j, d_j)$ and $d'_j = \frac{7}{2} d_j$ are as in the Covering Lemma. Write the summands on the right side of (15) as I_j, $j \in \mathbf{N}^*$. By (4), for $y \in D(x_j, d'_j)$ and $x \in P$, we have $d(x, y) \geqslant \frac{1}{8} d(x, x_j)$. Hence

$$I_j = \int_{D(x_j, d'_j)} \frac{|\psi(y)|}{d(y,x)^{n-1}} \mathrm{d}y \leqslant \frac{c_n}{d(x_j, x)^{n-1}} \int_{D(x_j, d'_j)} |\psi(y)| \mathrm{d}y.$$

Since $D(x_j, d'_j)$ and P are all compact there exist points $a_j \in P$ and $b_j \in D(x_j, d'_j)$ such that $d(D(x_j, d'_j), P) = d(a_j, b_j)$. Then, for any $y \in D(x_j, d'_j)$,

$$d(y, a_j) \leqslant d(y, b_j) + d(a_j, b_j) \leqslant 7d_j + 4d_j = 11d_j.$$

Noticing that $\psi = f - \varphi$ and $\psi(a_j) = 0$ (for $a_j \in P$) we have, for $y \in D(x_j, d'_j)$,

$$\begin{aligned} |\psi(y)| &\leqslant |f(y) - f(a_j)| + |\varphi(y) - \varphi(a_j)| \\ &\leqslant |f(y) - f(a_j)| + Cl(11d_j). \end{aligned}$$

Hence, by (1.5)″,

$$\begin{aligned} \int_{D(x_j, d'_j)} |\psi(y)| \mathrm{d}y &\leqslant \int_{D(a_j, 11d_j)} |f(y) - f(a_j)| \mathrm{d}y + c_n d_j^{n-1} l(d_j) \\ &\leqslant M(11d_j)^{n-1} l(11d_j) + c_n l(d_j) \\ &\leqslant c_n d_j^{n-1} l(d_j). \end{aligned} \tag{3.7}$$

Therefore,

$$\begin{aligned} I_j &\leqslant \frac{c_n d_j^{n-1} l(d_j)}{d(x_j, x)^{n-1}} = c_n \int_{D(x_j, d_j)} \frac{l(d_j)}{d(x_j, x)^{n-1}} \mathrm{d}y \\ &\leqslant c_n \int_{D(x_j, d_j)} \frac{l(\delta(y))}{d(x, y)^{n-1}} \mathrm{d}y, \end{aligned} \tag{3.8}$$

where $\delta(y) = d(y, P)$. We have

$$\int_P I_j(x) \mathrm{d}x = c_n \int_P \int_{D(x_j, d_j)} \frac{l(\delta(y))}{d(x, y)^{n-1}} \mathrm{d}y \mathrm{d}x$$

$$= c_n \int_{D(x_j, d_j)} l(\delta(y)) \left\{ \int_P \frac{1}{(d(x,y))^{n-1}} \mathrm{d}x \right\} \mathrm{d}y.$$

Hence

$$\int_P I_j(x) \mathrm{d}x \leqslant c_n \int_{D(x_j, d_j)} l(\delta(y)) \left\{ \int_{\{x \mid \delta(y) \leqslant d(x,y) \leqslant \pi\}} \frac{1}{(d(x,y))^{n-1}} \mathrm{d}x \right\} \mathrm{d}y$$

$$= c_n \int_{D(x_j, d_j)} l(\delta(y)) \int_{\delta(y)}^{\pi} \frac{1}{\theta^{n-1}} |\Sigma_{n-2}| \theta^{n-2} \mathrm{d}\theta$$

$$\leqslant c_n \int_{D(x_j, d_j)} l(\delta(y)) |\ln(\delta(y))| \mathrm{d}y$$

$$\leqslant c_n |D(x_j, d_j)|. \tag{3.9}$$

From (3.6) and (3.9) we derive that

$$\int_P I(x) \mathrm{d}x \leqslant c_n \sum_{j=1}^{+\infty} |D(x_j, d_j)| \leqslant c_n,$$

which shows that (3.5) holds and completes the proof.

References

[1] Marcinkiewicz J. On the Convergence of Fourier Series. J. London Math. Soc., 1935, 10: 264—268.

[2] Zygmund A. Trigonometric Series. China Machine Press, Beijing, 2004.

[3] Chang C P. On Bochner-Riesz summability almost everywhere of multiple Fourier Series. Studia Mathematica. 1965, 26: 25—66.

[4] Wang K Y, Li L Q. Harmonic Analysis and Approximation on the Unit Sphere. Science Press, Beijing, 2000.

[5] Brown G, Wang K Y. Jacobi polynomial estimates and Fourier-Laplace convergence. The Journal of Fourier Analysis and Applications, 1997, 3(6): 705—714.

[6] Wang K Y. Pointwise convergence of Cesàro means on sphere. Journal of Beijing Normal University (NS), 1993, 29(2): 158—164.

Journal of Mathematical Analysis and Applications，2008，348：28—33.

光滑函数的球调和展开的收敛速度

Convergence Rate of Spherical Harmonic Expansions of Smooth Functions①

Abstract We extend a well -known result of Bonami and Clerc on the almost everywhere (a. e.) convergence of Cesàro means of spherical harmonic expansions. For smooth functions measured in terms of φ-derivatives on the unit sphere, we obtained the sharp a. e. convergence rate of Cesàro means of their spherical harmonic expansions.

Keywords Almost everywhere convergence; Spherical harmonics; Cesàro means; φ-Derivatives.

§ 1. Introduction

Let $\mathrm{S}^{d-1}=\{x\in\mathbf{R}^d \mid |x|=1\}$ be the unit sphere in d-dimensional Euclidean space $\mathbf{R}^d$ equipped with the usual Lebesgue measure $\mathrm{d}\sigma(x)$, and let $L^p(\mathrm{S}^{d-1})$, $0<p<+\infty$, denote the Lebesgue space on S^{d-1} endowed with the quasi-norm $\|f\|_p=\left(\int_{\mathrm{S}^{d-1}}|f(x)|^p\mathrm{d}\sigma(x)\right)^{\frac{1}{p}}$. For an integrable f on S^{d-1}, we associate its expansion in spherical harmonics:

① Dai F was partially supported by NSERC Canada under grant G121211001. Wang K Y was partially supported by NSFC under the Grant NO. 10471010.

本文与戴峰合作.

Received：2007-11-30；Revised：2008-07-05.

$$f \sim \sum_{k=0}^{+\infty} Y_k(f), \tag{1.1}$$

where and throughout, $Y_k(f)$ denotes the orthogonal projection of f onto the space of spherical harmonics of degree k.

Given a nonnegative function φ on $[0, +\infty)$, we define the φ-derivative $f^{(\varphi)}$ of $f \in L(\mathrm{S}^{d-1})$ in a distributional sense by

$$Y_k(f^{(\varphi)}) = \varphi(k) Y_k(f), \quad k \in \mathbf{N},$$

and we denote by $L^p_\varphi(\mathrm{S}^{d-1})(1 \leqslant p < +\infty)$ the space of all functions in $L^p(\mathrm{S}^{d-1})$ whose φ-derivatives also belong to $L^p(\mathrm{S}^{d-1})$. For background information on φ-derivatives, we refer to [8] and [1].

Given $\delta > -1$, the Cesàro (C, δ)-operators of the spherical harmonic expansion (1.1) are defined by

$$\sigma_N^\delta(f) = \frac{1}{A_N^\delta} \sum_{k=0}^{N} A_{N-k}^\delta Y_k(f), \quad N \in \mathbf{N},$$

where $A_k^\delta = \mathrm{C}_{k+\delta}^k$, $k \in \mathbf{N}$, while the maximal (C, δ)-operator is defined as

$$\sigma_*^\delta(f)(x) := \sup_N |\sigma_N^\delta(f)(x)|, \quad x \in \mathrm{S}^{d-1}.$$

A well-known result of Bonami and Clerc [2] states that the maximal Cesàro (C, δ)-operator σ_*^δ is bounded on $L^p(\mathrm{S}^{d-1})$ whenever $1 < p \leqslant 2$ and $\delta > \delta_p := (d-2)\left(\frac{1}{p} - \frac{1}{2}\right)$; while it is of weak type (1, 1) if $\delta > \delta_1 := \frac{d-2}{2}$. As a matter of fact, we have

$$\lim_{N \to +\infty} \sigma_N^\delta(f)(x) = f(x), \quad \text{for a. e. } x \in \mathrm{S}^{d-1}, \tag{1.2}$$

whenever $f \in L^p(\mathrm{S}^{d-1})$, $1 \leqslant p \leqslant 2$ and $\delta > \delta_p$. In the case when $\delta = \delta_1 := \frac{d-2}{2}$, G. Brown and K. Y. Wang [3] proved that (1.2) remains true if $f \in L\ln_+^2 L(\mathrm{S}^{d-1})$. For more information on spherical harmonic analysis, we refer to [4, 6, 9, 11].

Our main aim in this paper is to study the a. e. convergence rate of the Cesàro means of smooth functions on S^{d-1}. Throughout the paper, the letter c denotes an inessential positive constant which may vary at each occurrence, and $[\cdot]$ denotes the usual integer part function. We write $A \asymp B$ to mean that $c^{-1}A \leqslant B \leqslant cA$ for some constant c. Given two sequences $\{a_n\}_{n=1}^{+\infty}$ and $\{b_n\}_{n=1}^{+\infty}$ of real numbers, we write $a_n = O(b_n)$ for the statement $|a_n| \leqslant c|b_n|$ and write $a_n = o(b_n)$ for $\lim\limits_{n\to+\infty}\frac{a_n}{b_n}=0$.

Our main result can be stated as follows.

Theorem 1 Assume that $\delta>0$, l is a positive integer, and $\varphi \in C^{l+1}[0, +\infty)$ satisfies $\varphi(0)=0$, $\lim\limits_{t\to+\infty}\varphi(t)=+\infty$ and $\varphi(t)>0$ for $t\in(0, +\infty)$. Then for $f\in L_\varphi^1(\mathrm{S}^{d-1})$ and a. e. $x\in \mathrm{S}^{d-1}$, we have

$$|f(x)-\sigma_N^\delta(f)(x)| \leqslant c\left[\int_1^{+\infty} |\mu^{(l+1)}(t)| \frac{t^{l+1}}{t+N}\mathrm{d}t\right]\sigma_*^\alpha(f^{(\varphi)})(x),$$

where $\alpha=\min\{\delta, l\}$, $\mu(t)=\dfrac{1}{\varphi(t)}$, and the constant c is independent of N, x and f.

Using the usual density argument, one can deduce the following:

Corollary 2 Assume that $1\leqslant p\leqslant 2$, $\delta>\delta_p := (d-2)\left(\frac{1}{p}-\frac{1}{2}\right)$, and l is a positive integer greater than δ_p. Assume further that $\varphi\in C^{l+1}[0, +\infty)$ satisfies $\varphi(0)=0$, $\lim\limits_{t\to+\infty}\varphi(t)=+\infty$ and $\varphi(t)>0$ for $t\in(0, +\infty)$. Then for $f\in L_\varphi^p(\mathrm{S}^{d-1})$, we have

$$|f(x)-\sigma_N^\delta(f)(x)| = o\left(\int_1^{+\infty}|\mu^{(l+1)}(t)|\frac{t^{l+1}}{t+N}\mathrm{d}t\right)+o\left(\frac{1}{N}\right),$$

$$\text{a. e. } x\in \mathrm{S}^{d-1}, \text{ as } N\to+\infty.$$

Our work here was motivated by [5, 7, 8, 10]. The main difficulty in our proof comes from the case when δ is not an integer and $\delta< l$, where the usual argument of summation by parts fails.

Here are a few interesting examples of φ for which Corollary 2 is applicable:

Example 1 Let $\varphi(t)=t(t+d-2)$. As is well known, the φ-derivative in this case coincides with the usual Laplace-Beltrami operator $\Delta_{\mathrm{S}^{d-1}}$ on the sphere S^{d-1}, that is, for $f\in C^2(\mathrm{S}^{d-1})$, $\Delta_{\mathrm{S}^{d-1}}f=f^{(\varphi)}$. Note that in this case, $\mu(t)=\dfrac{1}{t(t+d-2)}$ and for any positive integer l,

$$\int_1^{+\infty} |\mu^{(l+1)}(t)| \frac{t^{l+1}}{t+N} dt \asymp \frac{1}{N}.$$

Example 2 Let $\varphi(t)=t^r$ with $r>0$. In this case, $\mu(t)=t^{-r}$ and for any positive integer l,

$$\int_1^{+\infty} |\mu^{(l+1)}(t)| \frac{t^{l+1}}{t+N} dt \asymp \begin{cases} N^{-r}, & \text{if } r \in (0,1), \\ N^{-1}\ln N, & \text{if } r = 1, \\ N^{-1}, & \text{if } r > 1. \end{cases}$$

Example 3 Let $\varphi(t)=\ln^r(t+1)$ with $r>0$. Then $\mu(t)=\ln^{-r}(1+t)$ and for any positive integer l.

$$\int_1^{+\infty} |\mu^{(l+1)}(t)| \frac{t^{l+1}}{t+N} dt \asymp \ln^{-r} N.$$

We remark that in the special case $p=2$, Corollary 2 with $\varphi(t)=\ln^r(t+1)$ with $r>0$ was previously obtained in [5].

Example 4 Let $\varphi(t)=e^t-1$. In this case, we have $\mu(t)=\frac{1}{e^t-1}$ and for any positive integer l.

$$\int_1^{+\infty} |\mu^{(l+1)}(t)| \frac{t^{l+1}}{t+N} dt \asymp \frac{1}{N}.$$

§ 2. Proof of Theorem 1

For a sequence $\{a_k\}$ of complex numbers, we define $\Delta^0 a_k = a_k$, $\Delta a_k = a_k - a_{k+1}$ and $\Delta^{j+1} = \Delta(\Delta^j)$. We start with the following simple lemma.

Lemma 2.1 Suppose j is a nonnegative integer, $g \in L^1(\mathbb{S}^{d-1})$, $x \in \mathbb{S}^{d-1}$ and $\{a_k\}_{k=0}^{+\infty}$ is a sequence of complex numbers satisfying $\lim_{k\to+\infty} a_k = 0$. Then for any $0 \leqslant \beta \leqslant j$,

$$\left| \sum_{k=0}^{+\infty} a_k Y_k(g)(x) \right| \leqslant \left(\sum_{k=0}^{+\infty} |\Delta^{j+1} a_k| A_k^j \right) \sigma_*^\beta(g)(x),$$

where and throughout, it is agreed that the absolute value of a divergent series is $+\infty$.

This lemma is probably well known, but for completeness, we sketch its proof as follows.

Proof Without loss of generality, we may assume that $\sum_{k=0}^{+\infty}|\Delta^{j+1}a_k|A_k^j<+\infty$. Since $\lim_{k\to+\infty}a_k=0$, it follows that $|\Delta^i a_k|\leqslant\sum_{u=k}^{+\infty}|\Delta^{i+1}a_u|$, which implies that for $0\leqslant i\leqslant j$, $\sum_{k=0}^{+\infty}|\Delta^{i+1}a_k|A_k^i\leqslant c\sum_{k=0}^{+\infty}|\Delta^{j+1}a_k|A_k^j<+\infty$. Thus, $\lim_{k\to+\infty}|\Delta^{i+1}a_k|A_k^i=0$ for $0\leqslant i\leqslant j$, and the desired inequality follows by applying the Abel transformation $j+1$ times and using the fact that $\sigma_*^{\alpha_1}(g)\leqslant\sigma_*^{\alpha_2}(g)$ if $\alpha_1\geqslant\alpha_2\geqslant 0$. □

Recall that $\mu(t)=\frac{1}{\varphi(t)}$. We shall use the following simple lemma repeatedly.

Lemma 2.2

(i) For $k\in\mathbf{N}$,

$$|\Delta^{l+1}\mu(k)|\leqslant\int_k^{k+l+1}|\mu^{(l+1)}(t)|\,\mathrm{d}t. \tag{2.1}$$

(ii) For $0\leqslant i\leqslant l$ and $k\in\mathbf{N}^*$,

$$|\Delta^i\mu(t)|\leqslant c\sum_{j=k}^{+\infty}|\Delta^{l+1}\mu(j)|j^{l-i}. \tag{2.2}$$

Proof Inequality (2.1) follows by applying a change of variable to the following well-known formula:

$$\Delta^{l+1}\mu(k)=\int_{[0,1]^{l+1}}\mu^{(l+1)}(k+t_1+\cdots+t_{l+1})\mathrm{d}t_1\cdots\mathrm{d}t_{l+1},$$

while inequality (2.2) follows from the fact that

$$\Delta^v\mu(k)=\sum_{j=k}^{+\infty}\Delta^{v+1}\mu(j),\quad v\in\mathbf{N}.\quad\square \tag{2.3}$$

Let η be a C^∞-function on $[0,+\infty)$ with the property that $\eta(t)=1$ for $0\leqslant t\leqslant\frac{1}{4}$ and $\eta(t)=0$ for $t\geqslant\frac{1}{2}$. Define

$$V_N f=\sum_{k=0}^{+\infty}\eta\left(\frac{k}{N}\right)Y_k(f).$$

Without loss of generality, we may assume, for the rest of the proof, that $f\in L_\varphi^1(\mathrm{S}^{d-1})$ with $Y_0(f)=0$. Since

$$f-\sigma_N^\delta(f)=[f-V_Nf]+[V_Nf-\sigma_N^\delta(V_Nf)]+[\sigma_N^\delta(V_Nf)-\sigma_N^\delta(f)],$$

Theorem 1 is a direct consequence of Lemmas 2.3～2.5 below and the fact that $\sigma_*^{\alpha_1}(f^{(\varphi)})(x)\leqslant\sigma_*^{\alpha_2}(f^{(\varphi)})(x)$ whenever $0\leqslant\alpha_2\leqslant\alpha_1$.

Lemma 2.3 For a. e. $x\in S^{d-1}$, we have

$$|f(x)-V_Nf(x)|\leqslant c\left(\int_{\frac{N}{8}}^{+\infty}|\mu^{(l+1)}(t)|t^l\mathrm{d}t\right)\sigma_*^l(f^{(\varphi)})(x).$$

Proof By definition, we may write

$$f-V_Nf=\sum_{k\geqslant\frac{N}{4}}\left(1-\eta\left(\frac{k}{N}\right)\right)\mu(k)Y_k(f^{(\varphi)}).$$

Thus, an application of Lemma 2.1 gives

$$|f(x)-V_Nf(x)|\leqslant M_1(N)\sigma_*^l(f^{(\varphi)})(x),$$

where

$$M_1(N)=\sum_{k\geqslant\frac{N}{4}-l-1}\left|\Delta^{l+1}\left(\left(1-\eta\left(\frac{k}{N}\right)\right)\mu(k)\right)\right|A_k^l.$$

Since the derivative of η is supported in $\left(\frac{1}{4},\ \frac{1}{2}\right)$, it is easily seen that

$$\begin{aligned}M_1(N)&\leqslant c\max_{0\leqslant j\leqslant l}\sum_{k\geqslant\frac{N}{4}-l-1}k^j|\Delta^{j+1}\mu(k)|+c\frac{1}{N}\sum_{k=\left[\frac{N}{4}-l-1\right]}^{N}|\mu(k)|\\&\leqslant c\max_{0\leqslant j\leqslant l}\sum_{k\geqslant\frac{N}{8}}k^j|\Delta^{j+1}\mu(k)|\\&\leqslant c\sum_{k\geqslant\frac{N}{8}}k^l|\Delta^{l+1}\mu(k)|\\&\leqslant c\int_{\frac{N}{8}}^{+\infty}|\mu^{(l+1)}(t)|t^l\mathrm{d}t,\end{aligned}$$

where we have used the inequality $|\mu(k)|\leqslant\sum_{u=k}^{+\infty}|\Delta\mu(u)|$ in the second step, (2.3) in the third step, and (2.1) in the last step. This proves Lemma 2.3. □

Lemma 2.4 For a. e. $x\in S^{d-1}$,

$$|V_Nf(x)-\sigma_N^\delta(V_Nf)(x)|\leqslant c\sigma_*^l(f^{(\varphi)})(x)\int_1^{+\infty}|\mu^{(l+1)}(t)|\frac{t^{l+1}}{t+N}\mathrm{d}t.$$

Proof Noticing that

$$V_Nf-\sigma_N^\delta(V_Nf)=\sum_{k\leqslant\frac{N}{2}}\left(1-\frac{A_{N-k}^\delta}{A_N^\delta}\right)\eta\left(\frac{k}{N}\right)\mu(k)Y_k(f^{(\varphi)}),$$

we obtain from Lemma 2.1 that

$$|V_N f(x)-\sigma_N^\delta(V_N f)(x)|\leqslant cM_2(N)\sigma_*^l(f^{(\varphi)})(x),$$

where

$$M_2(N)=\sum_{1\leqslant k\leqslant\frac{N}{2}}\left|\Delta^{l+1}\left(\left(1-\frac{A_{N-k}^\delta}{A_N^\delta}\right)\eta\left(\frac{k}{N}\right)\mu(k)\right)\right|k^l.$$

To estimate $M_2(N)$, we observe that $\left|1-\frac{A_{N-k}^\delta}{A_N^\delta}\right|\leqslant c\frac{k}{N}$ and that

$$\frac{1}{A_N^\delta}|\Delta^j(A_{N-k}^\delta)|\leqslant cN^{-j},\quad j\in\mathbf{N},\ 0\leqslant k\leqslant\frac{3N}{4}. \tag{2.4}$$

Since η' is supported in $\left(\frac{1}{4},\ \frac{1}{2}\right)$, it is easily seen that

$$\begin{aligned}M_2(N)&\leqslant\frac{1}{N}\max_{0\leqslant i\leqslant l+1}\sum_{k=1}^{N}|\Delta^i\mu(k)|k^i\\&\leqslant c\frac{1}{N}\sum_{j=1}^{+\infty}|\Delta^{l+1}\mu(j)|j^l\min\{N,j\}\\&\leqslant c\int_1^{+\infty}|\mu^{(l+1)}(t)|\frac{t^{l+1}}{t+N}\mathrm{d}t.\end{aligned}$$

where we have used (2.2) in the second step, and (2.1) in the last step. □

Lemma 2.5 For a. e. $x\in S^{d-1}$,

$$|\sigma_N^\delta(f)(x)-\sigma_N^\delta(V_N f)(x)|\leqslant c\sigma_*^\alpha(f^{(\varphi)})(x)\int_0^{+\infty}|\mu^{(l+1)}(t)|\frac{t^{l+1}}{t+N}\mathrm{d}t,$$

where $\alpha=\min\{\delta,\ l\}$.

Proof We write

$$\sigma_N^\delta(f)-\sigma_N^\delta(V_N f)=\mu(N)\sigma_N^\delta(f^{(\varphi)})-\mu(N)\sigma_N^\delta(V_N(f^{(\varphi)}))+D_N(f),$$

where

$$D_N(f)=\sum_{k=0}^{+\infty}a_kY_k(f^{(\varphi)})$$

and

$$a_k=\begin{cases}\frac{A_{N-k}^\delta}{A_N^\delta}\left(1-\eta\left(\frac{k}{N}\right)\right)[\mu(k)-\mu(N)], & \text{if } 0\leqslant k\leqslant N,\\ 0, & \text{if } k\geqslant N+1.\end{cases} \tag{2.5}$$

Since, by (2.2) and (2.1),

$$|\mu(N)|\leqslant c\sum_{j=N}^{+\infty}|\Delta^{l+1}\mu(j)|j^l\leqslant c\int_N^{+\infty}|\mu^{(l+1)}(t)|t^l\mathrm{d}t, \tag{2.6}$$

it follows that

$$|\mu(N)\sigma_N^\delta(f^{(\varphi)})|\leqslant|\mu(N)|\sigma_*^\delta(f^{(\varphi)})$$
$$\leqslant c\sigma_*^\alpha(f^{(\varphi)})\int_1^{+\infty}|\mu^{(l+1)}(t)|\frac{t^{l+1}}{t+N}\mathrm{d}t.$$

Moreover, since η is supported in $\left[0,\ \frac{1}{2}\right]$, using (2.4), we have

$$\left|\Delta^v\left(\frac{A_{N-k}^\delta}{A_N^\delta}\eta\left(\frac{k}{N}\right)\right)\right|\leqslant cN^{-v},\quad v\in\mathbf{N},\quad k=0,1,2,\cdots,N,$$

therefore, an application of (2.6) and Lemma 2.1 gives

$$|\mu(N)\sigma_N^\delta(V_N(f^{(\varphi)}))|=|\mu(N)|\left|\sum_{k=0}^N\frac{A_{N-k}^\delta}{A_N^\delta}\eta\left(\frac{k}{N}\right)Y_k(f^{(\varphi)})\right|$$
$$\leqslant c\sigma_*^\alpha(f^{(\varphi)})\int_1^{+\infty}|\mu^{(l+1)}(t)|\frac{t^{l+1}}{t+N}\mathrm{d}t.$$

Thus, it remains to prove

$$|D_N(f)|\leqslant c\sigma_*^\alpha(f^{(\varphi)})\int_1^{+\infty}|\mu^{(l+1)}(t)|\frac{t^{l+1}}{t+N}\mathrm{d}t.\tag{2.7}$$

Let l_1 denote the smallest positive integer $\geqslant\alpha\ :=\min\{\delta,\ l\}$. Then applying Lemma 2.1, and taking into account the fact that $a_k=0$ for $k\leqslant\frac{N}{4}$, we obtain

$$|D_N(f)|\leqslant cM_3(N)\sigma_*^{l_1}(f^{(\varphi)})\leqslant cM_3(N)\sigma_*^\alpha(f^{(\varphi)}),$$

where

$$M_3(N)=\sum_{\frac{N}{4}-l_1-1\leqslant k\leqslant N}|\Delta^{l_1+1}a_k|k^{l_1}.\tag{2.8}$$

Since $\Delta^i(A_{N-k}^\delta)=A_{N-k}^{\delta-i}$, it is easily seen that for $\frac{N}{8}\leqslant k\leqslant N$.

$$|\Delta^{l_1+1}a_k|\leqslant cN^{-\delta}\max_{0\leqslant i\leqslant l_1}(N-k+1)^{\delta-i}|\Delta^{l_1+1-i}\mu(k)|+$$
$$cN^{-\delta}(N-k+1)^{\delta-l_1-1}|\mu(k)-\mu(N)|.\tag{2.9}$$

However,

$$|\mu(k)-\mu(N)|\leqslant\sum_{j=k}^N|\Delta\mu(j)|\leqslant c\sum_{j=k}^N\sum_{v=j}^{+\infty}|\Delta^{l+1}\mu(v)|v^{l-1}$$
$$\leqslant c(N-k+1)\sum_{v=k}^{+\infty}|\Delta^{l+1}\mu(v)|v^{l-1},\tag{2.10}$$

where we have used (2.2) in the second step. Since $l_1<\delta+1$, it follows from (2.8)~(2.10) that

$$M_3(N) \leqslant cN^{l_1-\delta} \max_{0\leqslant i\leqslant l_1} \sum_{\frac{N}{8}\leqslant k\leqslant N} (N-k+1)^{\delta-i} | \Delta^{l_1+1-i}\mu(k) | +$$

$$cN^{l_1-\delta} \sum_{\frac{N}{8}\leqslant k\leqslant N} (N-k+1)^{\delta-l_1} \sum_{v=k}^{+\infty} | \Delta^{l+1}\mu(v) | v^{l-1}$$

$$\leqslant c \sum_{k\geqslant\frac{N}{8}} | \Delta^{l+1}\mu(k) | k^l \leqslant c \int_{\frac{N}{8}}^{+\infty} | \mu^{(l+1)}(t) | t^l \mathrm{d}t,$$

where we have used (2.2) in the second step, and (2.1) in the last step. This implies the desired inequality (2.7). □

References

[1] Askey R., Wainger S. On the behavior of special classes of ultraspherical expansions I. J. Anal. Math., 1965, 15: 193—220.

[2] Bonami A., Clerc J L. Sommes de Cesàro et multiplicateurs des développements en harmoniques sphériques. Trans. Amer. Math. Soc., 1973, 183: 223—263.

[3] Brown G., Wang Kunyang. Jacobi polynomial estimates and Fourier-Laplace convergence. J. Fourier Anal. Appl. 1997, 3(6): 705—714.

[4] Dunkl C. F. Xu Yuan Orthogonal Polynomials of Several Variables. Cambridge Univ. Press, 2001.

[5] Lin Chin-Cheng, Wang Kunyang. Convergence rate of Fourier-Laplace series of L^2-functions, J. Approx. Theory. 2004, 128(2): 103—114.

[6] Müller C. Spherical Harmonics, Lecture Notes in Math., Vol. 17. Springer-Verlag, Berlin-New York, 1996.

[7] Oskolkov K I. Approximation properties of integrable functions on sets of full measure. Mat. Sb. (N. S.), 1977, 103(145): 563—589, 631 (in Russian).

[8] Stepanets A I. Classification of periodic functions and rate of convergence of their Fourier series. Math. USSR lzv., 1987, 28(1): 99—132.

[9] Stein F M., Weiss G. Introduction to Fourier Analysis on Euclidean Spaces. Princeton Univ. Press, Princeton, N J, 1971.

[10] Wang Kunyang. On the approximation of L^2-functions by their spherical Fourier sums on set of full measure. in: Approximation Theory and Its Applications, Anal. Theory Appl. 1989, 5: 69—77.

[11] Wang Kunyang. Li Luoqing. Harmonic Analysis and Approximation on the Unit Sphere. Science Press, Beijing, 2000.

三、正交和的正性

Ⅲ.
Positivity of Orthogonal Sums

Mathematical Proceedings of the Cambridge Philosophical Society, 1993, 114(3): 383－391.

一些基本的余弦和的正性

Positivity of Some Basic cosine Sums①

Abstract We show that all partial sums of $1+\sum_{k=1}^{n}k^{-\alpha}\cos k\theta$ are non-negative for $\alpha\geqslant\alpha_0$, where $0.308\,443<\alpha_0<0.308\,444$ and α_0 is the unique root of the equation $\int_0^{\frac{3\pi}{2}}t^{-\alpha}\cos t\mathrm{d}t=0$.

§ 1. Introduction

We consider the cosine sums

$$T_n^\alpha(\theta)=1+\sum_{k=1}^{n}k^{-\alpha}\cos k\theta.$$

where $n\in\mathbf{N}$ and $\alpha>0$, and seek to determine conditions under which the property

$$T_n^\alpha(\theta)\geqslant0 \text{ for all } \theta\in\mathbf{R} \text{ and all } n\geqslant1 \tag{1.1}$$

holds. The classical result of W. H. Young is that (1.1) holds for $\alpha=1$ [7]. If (1.1) holds for α then partial summation shows that (1.1) also holds for any $\beta>\alpha$. Further from the continuity of $T_n^\alpha(\theta)$ as a function of α we deduce that those α for which (1.1) fails form an open subset of (0, 1). Thus there is a critical value $\alpha_1\in(0, 1]$ such that (1.1) holds for every $\alpha\geqslant\alpha_1$ and for no $\alpha<\alpha_1$.

① 本文与 Brown G 和 Wilson D C 合作.

Received: 1993-01-11; Revised: 1993-04-05.

Littlewood and Salem showed that there is $\alpha_0 \in (0, 1)$ such that for $\alpha \geqslant \alpha_0$ the T_n^α are uniformly bounded below, and for $\alpha < \alpha_0$ they are not; Izumi (see [8], V. 2. 29 showed that α_0 is the unique root of the equation $\int_0^{\frac{3\pi}{2}} t^{-\alpha} \cos t\mathrm{d}t = 0$). It follows that $\alpha_0 \leqslant \alpha_1$; we show that in fact $\alpha_0 = \alpha_1$.

Our main result is

Theorem 1 For all $\theta \in \mathbf{R}$ and $n > 1$ we have $T_n^{\alpha_0}(\theta) > 0.0376908$.

An immediate corollary to Theorem 1 is

Theorem 2 Let $\alpha \geqslant \alpha_0$ and suppose $\{\alpha_k\}_{k\geqslant 0}$ is a non-increasing sequence of non-negative real numbers satisfying $k^\alpha a_k \geqslant (k+1)^\alpha a_{k+1}$. Then

$$\sum_{k=0}^{n} a_k \cos k\theta \geqslant 0$$

for all $\theta \in \mathbf{R}$ and $n \geqslant 0$.

Note that the positivity of such cosine sums forces $a_k = O(k^{-\alpha_0})$ ([1], theorem 2).

§ 2. The value of α_0

Our proof requires some detailed numerical estimates: we have used the computer package Mathematica 2. 1 [6] for this purpose, running on a DEC5000/200 workstation under Ultrix 4. 3. Copies of both input and output are available upon request from the last-named author.

We take α_0 as the unique root in $(0, 1)$ of the equation ([8]. V. 2. 29)

$$\int_0^{\frac{3}{2}\pi} \frac{\cos u}{u^\alpha} \mathrm{d}u = 0;$$

numerical integration using Mathematica shows $0.308443 < \alpha_0 < 0.308444$.

§ 3. Proof of Theorem 1

3. 1 The case $2 \leqslant n \leqslant 10$. Note that for $\delta = \beta - \alpha > 0$ and $n \geqslant 2$ we obtain

$$T_n^\beta(\theta) = \sum_{k=1}^{n-1} (k^{-\delta} - (k+1)^{-\delta}) T_k^\alpha(\theta) + n^{-\delta} T_n^\alpha(\theta)$$

by partial summation. In particular, the non-negativity of $T_n^{0.308\,443}(\theta)$ on $\mathbf{R}$, where $1\leqslant n\leqslant 10$, and the fact that $0.308\,443<\alpha_0<0.308\,444$ leads to

$$T_n^{\alpha_0}(\theta)>n^{-0.000\,001}T_n^{0.308\,443}(\theta),\qquad 2\leqslant n\leqslant 10.$$

Writing $x=\cos\theta$ we find that $T_n^{0.308\,443}(\theta)$ is a polynomial of degree n in x, which we denote by $\phi_n(x)$. Since $\phi_n(1)=T_n^{0.308\,443}(0)>0$, the positivity of $T_n^{0.308\,443}(\theta)$ on $\mathbf{R}$ follows from the absence of zeros of $\phi_n(x)$ in $[-1, 1]$. The method of Sturm sequences [3], section 89) reduces the verification of this fact to a straightforward computation. By locating the local minima of $\phi_n(x)$ in $[-1, 1]$ we can also derive a lower bound for $T_n^{0.308\,443}(\theta)$; we find

$$T_n^{\alpha_0}(\theta)\geqslant 0.037\,690\,8,\qquad 2\leqslant n\leqslant 10,\qquad \theta\in\mathbf{R}.$$

3.2 The interval $\left[\frac{1}{2}\pi, \pi\right]$, $n\geqslant 11$. We write $a_0=2$, $a_k=k^{-\alpha}$, $k\geqslant 1$, so that $T_n^{\alpha}(\theta)=\frac{1}{2}a_0+\sum_{k=1}^{n}a_k\cos k\theta$. Summing twice by parts we obtain

$$T_n^{\alpha}(\theta)=\sum_{k=0}^{n-2}\Delta^2 a_k(k+1)\sigma_k(\theta)+\Delta a_{n-1}n\sigma_{n-1}(\theta)+a_nD_n(\theta)$$

where

$$\Delta a_k=a_k-a_{k+1},$$

$$\Delta^2 a_k=\Delta a_k-\Delta a_{k+1},$$

$$D_k(\theta)=\frac{1}{2}+\sum_{j=1}^{k}\cos j\theta=\frac{\sin\left(k+\frac{1}{2}\right)\theta}{2\sin\frac{1}{2}\theta},\tag{3.1}$$

$$(k+1)\sigma_k(\theta)=\sum_{j=0}^{k}D_j(\theta)=\frac{1}{2}\left(\frac{\sin\left[\frac{1}{2}(k+1)\theta\right]}{\sin\frac{1}{2}\theta}\right)^2.$$

In particular, $T_n^{\alpha}(\theta)>0$ on each interval $\left(\frac{2l\pi}{n+\frac{1}{2}}, \frac{(2l+1)\pi}{n+\frac{1}{2}}\right)$. Further,

$$\begin{aligned}T_n^{\alpha_0}(\theta)&\geqslant\Delta a_0^2\sigma_0(\theta)+a_nD_n(\theta)\\&\geqslant 2^{-1-\alpha_0}-\frac{n^{-\alpha_0}}{2\sin\frac{1}{2}\theta}\end{aligned}$$

so that

$$T_n^{\alpha_0}(\theta) \geqslant 2^{-1.308\,444} - \left(2 \times 11^{0.308\,443} \sin \frac{\pi}{4}\right)^{-1} = 0.066\,255\,4$$

for $\frac{1}{2}\pi \leqslant \theta \leqslant \pi$ and $n \geqslant 11$.

3.3 The interval $\left[0, \frac{1}{2}\pi\right]$, $n \geqslant 11$. We begin by considering the sine sums

$$\begin{aligned} U_n^\alpha(\theta, s) &= \sum_{k=1}^{n} \frac{\sin k\theta}{(k+s)^{\alpha+1}} \\ &= \sum_{k=1}^{n-1} \delta_k(s) V_k(\theta) + \frac{n}{(n+s)^{\alpha+1}} V_n(\theta), \end{aligned} \tag{3.2}$$

summing by parts and writing

$$\delta_k(s) = \frac{k}{(k+s)^{\alpha+1}} - \frac{k+1}{(k+1+s)^{\alpha+1}}, \tag{3.3}$$

$$V_k(\theta) = U_n^0(\theta, 0) = \sum_{j=1}^{k} \frac{\sin j\theta}{j}. \tag{3.4}$$

Now since
$$\left(1 + \frac{1}{k+s}\right)^{\alpha+1} \geqslant 1 + \frac{\alpha+1}{k + \frac{1}{2}}$$

for $0 \leqslant s \leqslant \frac{1}{2}$, it follows that $\delta_k(s) > 0$ for $k > \frac{1}{2\alpha}$, and (for $\alpha > \frac{1}{4}$, and in particular for α near α_0) certainly for $k \geqslant 2$. Moreover, for $0 \leqslant u \leqslant \frac{1}{2}$ we have

$$\int_0^u \delta_1(s)\,\mathrm{d}s = 1 - 2^{1-\alpha} - (1+u)^{-\alpha} + 2(2+u)^{-\alpha} \geqslant 0, \tag{3.5}$$

so by the positivity of the Jackson-Gronwall sums $V_k(\theta)$ [4, 5] it follows that

$$\int_0^u U_n^\alpha(\theta, s)\,\mathrm{d}s \geqslant 0, \quad 0 \leqslant u \leqslant \frac{1}{2}.$$

(See [2], especially Lemmas 4 and 13 for closely related results.) From [8], V. 2. 29 et seq. we obtain

$$\begin{aligned} &T_n^\alpha(\theta) \\ &= 1 + \frac{\theta}{2\sin\frac{1}{2}\theta} \int_{\frac{1}{2}}^{n+\frac{1}{2}} \frac{\cos u\theta}{u^\alpha}\,\mathrm{d}u - \frac{\theta}{2\sin\frac{1}{2}\theta} \sum_{k=1}^{n} \int_{k-\frac{1}{2}}^{k+\frac{1}{2}} (u^{-\alpha} - k^{-\alpha}) \cos u\theta\,\mathrm{d}u \end{aligned}$$

$$= 1-\rho\left(\frac{\theta}{2}\right)\left\{f_0\left(\frac{\theta}{2}\right)-f_n\left(\frac{\theta}{2}\right)+\sum_{k=1}^{n}\int_{k-\frac{1}{2}}^{k+\frac{1}{2}}(u^{-\alpha}-k^{-\alpha})\cos u\theta \mathrm{d}u\right\},$$

writing
$$\rho(x)=\frac{x}{\sin x} \tag{3.6}$$

and
$$f_n(x) = (2x)^{\alpha-1}\int_0^{(2n+1)x} t^{-\alpha}\cos t\mathrm{d}t. \tag{3.7}$$

Now

$$\int_{k-\frac{1}{2}}^{k+\frac{1}{2}}(u^{-\alpha}-k^{-\alpha})\cos u\theta \mathrm{d}u$$
$$=\int_{-\frac{1}{2}}^{\frac{1}{2}}((k+u)^{-\alpha}-k^{-\alpha})\cos (k+u)\theta \mathrm{d}u$$
$$=\int_0^{\frac{1}{2}}\Big\{-\cos (k+u)\theta\int_0^u\alpha(k+s)^{-\alpha-1}\mathrm{d}s+$$
$$\cos (k-u)\theta\int_0^u\alpha(k-u+s)^{-\alpha-1}\mathrm{d}s\Big\}\mathrm{d}u$$
$$=\int_0^{\frac{1}{2}}\{\cos (k-u)\theta-\cos (k+u)\theta\}\int_0^u\alpha(k+s)^{-\alpha-1}\mathrm{d}s\mathrm{d}u+$$
$$\int_0^{\frac{1}{2}}\cos (k-u)\theta\int_0^u\alpha\{(k-u+s)^{-\alpha-1}-(k+s)^{-\alpha-1}\}\mathrm{d}s\mathrm{d}u$$
$$=2\int_0^{\frac{1}{2}}\sin u\theta\int_0^u\alpha(k+s)^{-\alpha-1}\sin k\theta\mathrm{d}s\mathrm{d}u+$$
$$\int_0^{\frac{1}{2}}\cos (k-u)\theta\int_0^u\int_0^u\alpha(\alpha+1)(k-u+s+t)^{-\alpha-2}\mathrm{d}t\mathrm{d}s\mathrm{d}u$$

so writing

$$A_k(\theta)=\int_0^{\frac{1}{2}}\cos (k-u)\theta\int_0^u\int_0^u\alpha(\alpha+1)(k-u+s+t)^{-\alpha-2}\mathrm{d}t\mathrm{d}s\mathrm{d}u \tag{3.8}$$

we have

$$T_n^{\alpha}(\theta)\geqslant 1-\rho\left(\frac{\theta}{2}\right)\left\{f_0\left(\frac{\theta}{2}\right)-f_n\left(\frac{\theta}{2}\right)+\sum_{k=1}^{n}A_k(\theta)+2\alpha\theta\int_0^{\frac{1}{2}}u\int_0^u U_n^{\alpha}(\theta,s)\mathrm{d}s\mathrm{d}u\right\}$$
$$=1-\rho\left(\frac{\theta}{2}\right)\left\{f_0\left(\frac{\theta}{2}\right)-f_n\left(\frac{\theta}{2}\right)+A_1(\theta)+A_2(\theta)+A_3(\theta)+\Sigma_n(\theta)+g_n(\theta)\right\},$$

where
$$\Sigma_n(\theta)=\sum_{k=4}^{n}A_k(\theta) \tag{3.9}$$

and
$$g_n(\theta)=2\alpha\theta\int_0^{\frac{1}{2}}u\int_0^u U_n^{\alpha}(\theta,s)\mathrm{d}s\mathrm{d}u. \tag{3.10}$$

We turn first to the estimation of $\Sigma_n(\theta)$. Observe that

$$\left\{(s,\ t,\ u)\in\mathbf{R}^3 \mid 0\leqslant s,\ t\leqslant u\leqslant\frac{1}{2}\right\}=$$

$$\left\{(s,\ t,\ u)\in\mathbf{R}^3 \mid \max(s,\ t)\leqslant u\leqslant\frac{1}{2},\ 0\leqslant s,\ t\leqslant\frac{1}{2}\right\},$$

so that

$$\begin{aligned}\Sigma_n(\theta)&\leqslant\sum_{k=4}^{+\infty}\int_0^{\frac{1}{2}}\int_0^u\int_0^u\frac{\alpha(\alpha+1)}{(k-u+s+t)^{\alpha+2}}\mathrm{d}s\mathrm{d}t\mathrm{d}u\\&=\int_0^{\frac{1}{2}}\int_0^{\frac{1}{2}}\sum_{k=4}^{+\infty}\int_{\max(s,t)}^{\frac{1}{2}}\frac{\alpha(\alpha+1)}{(k-u+s+t)^{\alpha+2}}\mathrm{d}u\mathrm{d}s\mathrm{d}t\\&\leqslant\int_0^{\frac{1}{2}}\int_0^{\frac{1}{2}}\frac{1}{2}\sum_{k=4}^{+\infty}\int_{\max(s,t)}^{1-\max(s,t)}\frac{\alpha(\alpha+1)}{(k-u+s+t)^{\alpha+2}}\mathrm{d}u\mathrm{d}s\mathrm{d}t\\&=\frac{1}{2}\int_0^{\frac{1}{2}}\int_0^{\frac{1}{2}}\sum_{k=4}^{+\infty}\int_{k-1+\max(s,t)}^{k-\max(s,t)}\frac{\alpha(\alpha+1)}{(u+s+t)^{\alpha+2}}\mathrm{d}u\mathrm{d}s\mathrm{d}t\\&<\frac{1}{2}\int_0^{\frac{1}{2}}\int_0^{\frac{1}{2}}\sum_{k=4}^{+\infty}\int_{k-1+\max(s,t)}^{k+\max(s,t)}\frac{\alpha(\alpha+1)}{(u+s+t)^{\alpha+2}}\mathrm{d}u\mathrm{d}s\mathrm{d}t\\&<\frac{1}{2}\int_0^{\frac{1}{2}}\int_0^{\frac{1}{2}}\int_{3+\frac{1}{2}(s+t)}^{+\infty}\frac{\alpha(\alpha+1)}{(u+s+t)^{\alpha+2}}\mathrm{d}u\mathrm{d}s\mathrm{d}t\\&=\frac{1}{2}\int_0^{\frac{1}{2}}\int_0^{\frac{1}{2}}\frac{\alpha}{\left(3+\frac{3}{2}(s+t)\right)^{\alpha+1}}\mathrm{d}s\mathrm{d}t\\&=\frac{2}{9(1-\alpha)}\left(2\left(\frac{15}{4}\right)^{1-\alpha}-\left(\frac{9}{2}\right)^{1-\alpha}-3^{1-\alpha}\right)\\&=C_1(\alpha).\end{aligned}\tag{3.11}$$

Now consider $g_n(\theta)$. Notice that

$$\begin{aligned}V_n(\theta)&=\int_0^\theta D_n(t)\mathrm{d}t-\frac{\theta}{2}\\&=\int_0^{(n+\frac{1}{2})\theta}\frac{\sin t}{t}\mathrm{d}t-\frac{\theta}{2}+\int_0^\theta\left[\frac{1}{2\sin\frac{1}{2}t}-\frac{1}{t}\right]\sin\left(n+\frac{1}{2}\right)t\mathrm{d}t\\&\leqslant C_2-\frac{\theta}{2}+\frac{\theta^2}{48}\rho\left(\frac{\theta}{2}\right).\end{aligned}$$

since
$$0\leqslant\int_0^u\frac{\sin t}{t}\mathrm{d}t\leqslant\int_0^\pi\frac{\sin t}{t}\mathrm{d}t=C_2\tag{3.12}$$

and for $0<t\leqslant\theta$ we have

$$\left|\left(\frac{1}{2\sin\frac{1}{2}t}-\frac{1}{t}\right)\sin\left(n+\frac{1}{2}\right)t\right|$$

$$\leqslant\frac{t-2\sin\frac{1}{2}t}{2t\sin\frac{1}{2}t}\leqslant\frac{t-2\left(\frac{1}{2}t-\frac{1}{6}\left(\frac{1}{2}t\right)^3\right)}{2t\sin\frac{1}{2}t}\leqslant\frac{t}{24}\rho\left(\frac{\theta}{2}\right).$$

Elementary calculus yields

$$V_2(\theta)\leqslant V_2\left(\frac{1}{3}\pi\right)=\frac{3\sqrt{3}}{4},$$

$$V_3(\theta)\leqslant V_3\left(\frac{1}{4}\pi\right)=\frac{4\sqrt{2}+3}{6}.$$

Thus

$$U_n^\alpha(\theta,s)\leqslant\sum_{k=1}^{3}\delta_k(s)V_k(\theta)+\left(C_2-\frac{\theta}{2}+\frac{\theta^2}{48}\rho\left(\frac{\theta}{2}\right)\right)\left(\sum_{k=4}^{n-1}\delta_k(s)+\frac{n}{(n+s)^{\alpha+1}}\right)$$

$$\leqslant\delta_1(s)\sin\theta+\frac{3\sqrt{3}}{4}\delta_2(s)+\frac{4\sqrt{2}+3}{6}\delta_3(s)+$$

$$\left(C_2-\frac{\theta}{2}+\frac{\theta^2}{48}\rho\left(\frac{\theta}{2}\right)\right)\frac{4}{(4+s)^{\alpha+1}},$$

so for $0\leqslant u\leqslant\frac{1}{2}$ we have

$$0\leqslant\int_0^u\alpha U_n^\alpha(\theta,s)\mathrm{d}s$$

$$\leqslant(1-2^{1-\alpha}-(1+u)^{-\alpha}+2(2+u)^{-\alpha})\theta+C_3(\alpha)-\frac{3\sqrt{3}}{2}(2+u)^{-\alpha}-$$

$$\frac{8\sqrt{2}+6-9\sqrt{3}}{4}(3+u)^{-\alpha}+\frac{8\sqrt{2}+6}{3}(4+u)^{-\alpha}+$$

$$\left(C_2-\frac{\theta}{2}+\frac{\theta^2}{48}\rho\left(\frac{\theta}{2}\right)\right)(4^{1-\alpha}-4(4+u)^{-\alpha}),$$

where $$C_3(\alpha)=\frac{3\sqrt{3}}{4}(2^{1-\alpha}-3^{1-\alpha})+\frac{4\sqrt{2}+3}{6}(3^{1-\alpha}-4^{1-\alpha}).\tag{3.13}$$

Writing $$I_k(\alpha)=\int_0^{\frac{1}{2}}u(k+u)^{-\alpha}\mathrm{d}u$$

$$=\frac{1}{(1-\alpha)(2-\alpha)}\left(k^{2-\alpha}+\left(-k+\frac{1}{2}-\frac{1}{2}\alpha\right)\left(k+\frac{1}{2}\right)^{1-\alpha}\right)\tag{3.14}$$

we obtain $\quad g_n(\theta)\leqslant 2\theta\left(C_4(\alpha)+C_5(\alpha)\theta+C_6(\alpha)\theta^2\rho\left(\frac{1}{2}\theta\right)\right)$,

where
$$C_4(\alpha)=\frac{C_3(\alpha)}{8}-\frac{3\sqrt{3}}{2}I_2(\alpha)-\frac{8\sqrt{2}+6-9\sqrt{3}}{4}I_3(\alpha)+\frac{8\sqrt{2}+6}{3}I_4(\alpha)+\frac{C_2 4^{1-\alpha}}{8}-4C_2 I_4(\alpha), \tag{3.15}$$

$$C_5(\alpha)=\frac{1-2^{1-\alpha}}{8}-I_1(\alpha)+2I_2(\alpha)-\frac{4^{1-\alpha}}{16}+2I_4(\alpha), \tag{3.16}$$

$$C_6(\alpha)=\frac{4^{1-\alpha}}{384}-\frac{4I_4(\alpha)}{48}. \tag{3.17}$$

Noting that $\cos x\leqslant 1-\frac{1}{2}x^2+\frac{1}{24}x^4$, and remembering that $0\leqslant\theta\leqslant\frac{1}{2}\pi$, we obtain $f_0\left(\frac{\theta}{2}\right)\leqslant\frac{1}{(1-\alpha)2^{1-\alpha}}-\frac{\theta^2}{2(3-\alpha)2^{3-\alpha}}+\frac{\theta^4}{24(5-\alpha)2^{5-\alpha}}$; also, for $k=1$ and $k=2$ we have $\sin\left(k-\frac{1}{2}\right)\theta\geqslant 0$ on $\left[0,\ \frac{1}{2}\pi\right]$, so $\cos(k-u)\theta\leqslant\cos\left(k-\frac{1}{2}\right)\theta$ for $0\leqslant u\leqslant\frac{1}{2}$, giving

$$A_k(\theta)\leqslant\left(1-\frac{1}{2}\left(k-\frac{1}{2}\right)^2\theta^2+\frac{1}{24}\left(k-\frac{1}{2}\right)^4\theta^4\right)\cdot\left(\left[\frac{1}{1-\alpha}\right]\left(\left(k+\frac{1}{2}\right)^{1-\alpha}-\left(k-\frac{1}{2}\right)^{1-\alpha}\right)-k^{-\alpha}\right),$$

for $k=1,2$. Thus on $\left[0,\ \frac{1}{2}\pi\right]$ and for $n\geqslant 11$ we have

$$T_n^\alpha(\theta)\geqslant 1-\rho\left(\frac{1}{2}\theta\right)\left\{C_7(\alpha)+2C_4(\alpha)\theta+C_8(\alpha)\theta^2+2C_6(\alpha)\rho\left(\frac{1}{2}\theta\right)\theta^3+C_9(\alpha)\theta^4+A_3(\theta)-f_n\left(\frac{1}{2}\theta\right)\right\}, \tag{3.18}$$

where
$$C_7(\alpha)=\frac{5^{1-\alpha}}{(1-\alpha)2^{1-\alpha}}-1-2^{-\alpha}+C_1(\alpha), \tag{3.19}$$

$$C_8(\alpha)=\frac{-1}{2(3-\alpha)2^{3-\alpha}}-\frac{9\times 5^{1-\alpha}-8\times 3^{1-\alpha}-1}{(1-\alpha)2^{4-\alpha}}+\frac{1+9\times 2^{-\alpha}}{8}+2C_5(\alpha), \tag{3.20}$$

and

$$C_9(\alpha)=\frac{1}{24(5-\alpha)2^{5-\alpha}}+\frac{81\times 5^{1-\alpha}-80\times 3^{1-\alpha}-1}{3(1-\alpha)2^{8-\alpha}}-\frac{1+81\times 2^{-\alpha}}{384}. \tag{3.21}$$

3.3.1 The interval $\left[0,\ \frac{5}{23}\pi\right]$, $n\geqslant 11$. For $0\leqslant\theta\leqslant\frac{5}{23}\pi$, we have

$\sin\frac{5}{2}\theta\geqslant 0$ so that $\cos(3-u)\theta\leqslant\cos\left(3-\frac{1}{2}\right)\theta$ for $0\leqslant u\leqslant\frac{1}{2}$, giving

$$A_3(\theta)\leqslant C_{10}(\alpha)-\frac{25C_{10}(\alpha)}{8}\theta^2+\frac{625C_{10}(\alpha)}{384}\theta^4,$$

where
$$C_{10}(\alpha)=\frac{7^{1-\alpha}-5^{1-\alpha}}{(1-\alpha)2^{1-\alpha}}-3^{-\alpha}. \tag{3.22}$$

We have
$$\rho\left(\frac{\theta}{2}\right)\leqslant\frac{\frac{5}{46}\pi}{\sin\frac{5}{46}\pi}=C_{11}, \tag{3.23}$$

and for $\alpha\geqslant\alpha_0$ we have $f_n\left(\frac{1}{2}\theta\right)\geqslant 0$. Using $0.308\,443<\alpha_0<0.308\,444$ we obtain $C_6(\alpha_0)>0.000\,164\,6>0$ and $C_9(\alpha_0)+\frac{625}{384}C_{10}(\alpha_0)>0.003\,304\,6>0$, so that $T_n^{\alpha_0}(\theta)\geqslant 1-\rho\left(\frac{1}{2}\theta\right)((C_7(\alpha_0)+C_{10}(\alpha_0))+2C_4(\alpha_0)\theta+C_{12}(\alpha_0)\theta^2)$ where

$$C_{12}(\alpha)=C_8(\alpha)-\frac{25}{8}C_{10}(\alpha)+2C_6(\alpha)C_{11}\frac{5}{23}\pi+\left(C_9(\alpha)+\frac{625}{384}C_{10}(\alpha)\right)\left(\frac{5}{23}\pi\right)^2. \tag{3.24}$$

Since $C_{12}(\alpha_0)<-0.043\,000\,9<0$, it follows that

$$C_7(\alpha_0)+C_{10}(\alpha_0)+2C_4(\alpha_0)\theta+C_{12}(\alpha_0)\theta^2\leqslant C_7(\alpha_0)+C_{10}(\alpha_0)-\frac{C_4(\alpha_0)^2}{C_{12}(\alpha_0)},$$

and this being positive yields

$$\begin{aligned}T_n^{\alpha_0}(\theta)&\geqslant 1-C_{11}\left(C_7(\alpha_0)+C_{10}(\alpha_0)-\frac{C_4(\alpha_0)^2}{C_{12}(\alpha_0)}\right)\\&\geqslant 0.049\,522\,9.\end{aligned} \tag{3.25}$$

3.3.2 The interval $\left(\frac{5}{23}\pi,\frac{1}{2}\pi\right]$, $n\geqslant 11$. For $\frac{5}{23}\pi<\theta\leqslant\frac{1}{2}\pi$ and $n\geqslant 11$, we have $(2n+1)\frac{1}{2}\theta>\frac{5}{2}\pi$, so that

$$f_n\left(\frac{\theta}{2}\right)\geqslant\left(\frac{\pi}{2}\right)^{\alpha-1}\int_{\frac{3}{2}\pi}^{\frac{7}{2}\pi}t^{-\alpha}\cos t\,dt=C_{13}(\alpha). \tag{3.26}$$

Also, $\cos(3-u)\theta\leqslant 0$ on $\left(\frac{5}{23}\pi,\frac{1}{2}\pi\right]$ for $0\leqslant u\leqslant\frac{1}{2}$, so $A_3(\theta)\leqslant 0$. We have $\rho\left(\frac{1}{2}\theta\right)\leqslant\frac{1}{4}\sqrt{2}\pi$, giving (since $C_9(\alpha_0)>0.001\,125\,1>0$ and, as before, $C_6(\alpha_0)>0$),

$$T_n^{\alpha_0}(\theta)\geqslant 1-\rho\left(\frac{1}{2}\theta\right)\left((C_7(\alpha_0)-C_{13}(\alpha_0))+2C_4(\alpha_0)\theta+C_{14}(\alpha_0)\theta^2\right)$$

where $$C_{14}(\alpha)=C_8(\alpha)+2C_6(\alpha)\frac{1}{4}\sqrt{2}\pi\left(\frac{1}{2}\pi\right)+C_9(\alpha)\left(\frac{1}{2}\pi\right)^2. \tag{3.27}$$

Since $C_{14}(\alpha_0)<-0.0372427<0$, it follows that

$$C_7(\alpha_0)-C_{13}(\alpha_0)+2C_4(\alpha_0)\theta+C_{14}(\alpha_0)\theta^2\leqslant C_7(\alpha_0)-C_{13}(\alpha_0)-\frac{C_4(\alpha_0)^2}{C_{14}(\alpha_0)},$$

and this being positive yields

$$\begin{aligned}T_n^{\alpha_0}(\theta)&\geqslant 1-\frac{1}{4}\sqrt{2}\pi\left(C_7(\alpha_0)-C_{13}(\alpha_0)-\frac{C_4(\alpha_0)^2}{C_{14}(\alpha_0)}\right) \\ &\geqslant 0.0748784. \quad \square\end{aligned} \tag{3.28}$$

§ 4. Sine sums

There is no analogue of Theorem 1 for sine sums. Indeed, for any $\alpha<1$, the sums $$U_n^\alpha(\theta)=\sum_{k=1}^n k^{-\alpha}\sin k\theta$$

satisfy $$\limsup_{n\to+\infty}\min_{\theta\to\mathbf{R}} n^\alpha U_n^\alpha(\theta)\leqslant -\frac{1}{2}. \tag{4.1}$$

To see this we write, for $n\geqslant 3$,

$$U_n^\alpha(\theta)=\sum_{k=1}^{n-2}\Delta^2 a_k(k+1)\tilde{\sigma}_k(\theta)+\Delta a_{n-1}n\tilde{\sigma}_{n-1}(\theta)+n^{-\alpha}\widetilde{D}_n(\theta)$$

where $$\widetilde{D}_k(\theta)=\sum_{j=1}^k\sin j\theta=\frac{\cos\frac{1}{2}\theta-\cos\left(n+\frac{1}{2}\right)\theta}{2\sin\frac{1}{2}\theta}$$

and $$(k+1)\tilde{\sigma}_k(\theta)=\sum_{j=1}^k\widetilde{D}_j(\theta)=\frac{(k+1)\sin\theta-\sin(k+1)\theta}{4\sin^2\frac{1}{2}\theta}.$$

Now let $\theta_n=\left(n+\frac{1}{2}\right)^{-1}\left(1-\frac{1}{2}(-1)^n\right)\frac{1}{2}\pi$: then

$$\widetilde{D}_n(\pi-2\theta_n)=\frac{1}{2}\tan\theta_n-\frac{1}{2\cos\theta_n}.$$

It is easy to see that $\Delta a_n\leqslant cn^{-\alpha-1}$,

$$\Delta^2 a_k\leqslant ck^{-\alpha-2},$$

$$0<(k+1)\tilde{\sigma}_k(\pi-2\theta_n)\leqslant c\,\frac{k}{n},$$

where c is an absolute constant. Hence

$$n^\alpha U_n^\alpha(\pi-2\theta_n)\leqslant c^2\sum_{k=1}^{n}k^{-\alpha-1}n^{\alpha-1}+c^2n^{-1}+\frac{1}{2}\tan\theta_n-\frac{1}{2\cos\theta_n},$$

from which (4. 1) follows.

This research was supported by a grant from the Australian Research Council.

References

[1] Belov. A S. Coefficients of trigonometric series with non-negative partial sums. Mat. Zametki 41: 1987, 152—158.

[2] Brown G, Wang K-Y. An extension of the Fejér-Jackson inequality. preprint, 1991.

[3] Burnside W. S. Panton. A W. Theory of Equations. Hodges, Figgis & Co., Dublin, 1886.

[4] Gronwall. T H. Über die Gibbsche Erscheinung und die trigonometrischen Summen $\sin x+\frac{1}{2}\sin 2x+\cdots+\frac{1}{n}\sin nx$. Math. Ann. 1912, 72: 228—243.

[5] Jackson. D. Über eine trigonometrische Summe. Rend. Circ. Mat. Palermo, 1911, 32: 257—262.

[6] Wolfram. S. Mathematica: a System for Doing Mathematics by Computer, Addison-Wesley, Redwood City, 1988.

[7] Young. W H. On a certain series of Fourier. Proc. London Math. Soc., 1912, 11: 357—366.

[8] Zygmund A. Trigonometric Series, 2nd ed. Cambridge University Press, 1959.

Math. Proc. Camb. Phil. Soc., 1996, 119: 681—694.

再谈 Jacobi 多项式和的正性

Positivity of More Jacobi Polynomial Sums①

§ 1. Introduction

Our main result can be stated as follows:

Theorem

$$\sum_{k=0}^{n}(-1)^k\frac{C_{2k}^{\lambda}(\cos\theta)}{C_{2k}^{\lambda}(1)}\geqslant 0,\quad n\in\mathbf{N}^*,\quad 0\leqslant\theta\leqslant\frac{\pi}{2},\tag{1.1}$$

for all $\lambda\geqslant\lambda_0$, where λ_0 is the unique root in (0, 1) of the equation

$$\int_0^{\frac{3}{2}\pi}\frac{\cos t}{t^{\lambda}}\mathrm{d}t=0,$$

and C_{2k}^{λ} is the Gegenbauer polynomial of degree $2k$ and order λ. The only cases of equality in (1.1) are when $\theta=0$ and n is odd.

This result is best viewed in the context of the Jacobi polynomials $P_n^{(\alpha,\beta)}(x)$ which can be defined by

$$\begin{aligned}P_n^{(\alpha,\beta)}(x)&=\frac{(\alpha+1)_n}{n!}\sum_{k=0}^{n}\frac{(-n)_k(n+\alpha+\beta+1)_k}{k!(\alpha+1)_k}\left(\frac{1-x}{2}\right)^k\\&=\frac{(\alpha+1)_n}{n!}{}_2F_1\left(-n,\ n+\alpha+\beta+1;\ \alpha+1;\ \frac{1-x}{2}\right),\end{aligned}\tag{1.2}$$

where $$(a)_k=a(a+1)\cdots(a+k-1)=\frac{\Gamma(k+a)}{\Gamma(a)}.$$

Over the years, some special cases of the inequality

① 本文与 Brown G, Koumandos S 合作.

Received: 1994-12-12; Revised: 1995-04-24.

$$\sum_{k=0}^{n} \frac{P_k^{(\alpha,\beta)}(x)}{P_k^{(\beta,\alpha)}(1)} \geqslant 0, \quad -1 \leqslant x \leqslant 1, \tag{1.3}$$

have been proved by several authors. In particular, Fejér [11, 12] observed this when $\alpha=\frac{1}{2}$, $\beta=-\frac{1}{2}$ (positivity of the classical Fejér kernel), and when $\alpha=\beta=0$, which is the inequality

$$\sum_{k=0}^{n} P_k(\cos\theta) \geqslant 0, \quad 0 \leqslant \theta \leqslant \pi,$$

where $P_k(x)$ is the Legendre polynomial of degree k. He also conjectured (1.3) when $\alpha=\beta=\frac{1}{2}$ and this was proved by Jackson [16] and Gronwall [15].

Feldheim [13] extended Fejér's results proving (1.3) for $\alpha=\beta\geqslant 0$ by the use of a fractional integral connecting ultraspherical polynomials of different order.

Some special cases of (1.3) have been demonstrated by Askey [1, 2] and Askey and Gasper [4] and some of its interesting applications have also been mentioned there. Gasper in [14] proved (1.3) under the more general assumptions $\alpha+\beta\geqslant 0$, $\beta\geqslant -\frac{1}{2}$ or $\alpha+\beta\geqslant -2$, $\beta\geqslant 0$. He also conjectured that (1.3) may be true for $-1<\alpha<\frac{1}{2}$, $\beta(\alpha)\leqslant\beta<0$, where $\beta(\alpha)$ is the unique solution of the equation

$$\int_0^{j_{\alpha,2}} t^{-\beta(\alpha)} J_\alpha(t)\,\mathrm{d}t = 0, \quad -\frac{1}{2} < \beta(\alpha) < 0,$$

$j_{\alpha,2}$ being the second positive zero of the Bessel function $J_\alpha(t)$. In addition, he pointed out that any cases of this conjecture should be treated by completely different methods than his own.

In a recent paper [5], Askey draws attention to this conjecture, mentioning that it would be interesting to prove it for $-1<\alpha<0$ and some $\beta<0$, predicting, however, that this is probably quite hard.

Recently, the authors [7] established (1.3) for $\alpha=\beta\geqslant\alpha'$, where α' is the unique root of the equation

$$\int_0^{j_{\alpha,2}} t^{-\alpha} J_\alpha(t)\,\mathrm{d}t = 0$$

whose numerical value is $\alpha' = -0.269\ 387\cdots$. This result is the best possible extension of Feldheim's inequality [13, (3)] and partially answers in the affirmative Gasper's conjecture above.

In this paper we extend progress on the same conjecture by considering (1.3) for $-1<\alpha\leqslant-\frac{1}{2}$ and negative β. It is true that if (1.3) holds for some (α, β), then it holds for $(\alpha-\mu, \beta+\mu)$, $\mu>0$ and this can be verified by the application of Bateman's integral [3, (3.4)].

Thus, it is sufficient to concentrate on the case $\alpha=-\frac{1}{2}$. We set $\beta=\lambda-\frac{1}{2}$ and are concerned with showing

$$\sum_{k=0}^{n} \frac{P_k^{\left(-\frac{1}{2},\lambda-\frac{1}{2}\right)}(x)}{P_k^{\left(\lambda-\frac{1}{2},-\frac{1}{2}\right)}(1)} \geqslant 0, \quad -1\leqslant x\leqslant 1, \tag{1.4}$$

for the largest possible range of $\lambda<\frac{1}{2}$.

It should be noted that (1.4) is true for all $\lambda\geqslant\frac{1}{2}$ and this is obtained from the results of Gasper in [14].

Since Jacobi polynomials with $\alpha=-\frac{1}{2}$ can be expressed in terms of Gegenbauer polynomials $C_n^\lambda(x)$ by means of a quadratic transformation (see [17, (4.1.5) and (4.7.1)]), it follows that (1.4) is equivalent to (1.1).

Using the well-known formula (see for example [5, (1.8)]),

$$\lim_{n\to+\infty}\left(\frac{\theta}{n}\right)^{\alpha-\beta+1} \sum_{k=0}^{n} \frac{P_k^{(\alpha,\beta)}\left(\cos\left(\frac{\theta}{n}\right)\right)}{P_k^{(\beta,\alpha)}(1)} = 2^\alpha\Gamma(\beta+1)\int_0^\theta t^{-\beta}J_\alpha(t)\,\mathrm{d}t,$$

$$\beta<\alpha+1$$

and bearing in mind that

$$J_{-\frac{1}{2}}(t)=\left(\frac{2}{\pi t}\right)^{\frac{1}{2}}\cos t,$$

we see that

$$\lim_{n\to+\infty}\left(\frac{\theta}{n}\right)^{1-\lambda}\sum_{k=0}^{n}\frac{P_k^{\left(-\frac{1}{2},\lambda-\frac{1}{2}\right)}\left(\cos\left(\frac{\theta}{n}\right)\right)}{P_k^{\left(\lambda-\frac{1}{2},-\frac{1}{2}\right)}(1)}=\frac{\Gamma\left(\lambda+\frac{1}{2}\right)}{\Gamma\left(\frac{1}{2}\right)}\int_0^{\theta}\frac{\cos t}{t^{\lambda}}\mathrm{d}t.$$

Choosing $\theta=\frac{3}{2}\pi$, we deduce

$$\lim_{n\to+\infty}\sum_{k=0}^{n}\frac{P_k^{\left(-\frac{1}{2},\lambda-\frac{1}{2}\right)}\left(\cos\frac{3\pi}{2n}\right)}{P_k^{\left(\lambda-\frac{1}{2},-\frac{1}{2}\right)}(1)}$$

$$=\lim_{n\to+\infty}\left(\frac{2n}{3\pi}\right)^{1-\lambda}\frac{\Gamma\left(\lambda+\frac{1}{2}\right)}{\Gamma\left(\frac{1}{2}\right)}\int_0^{\frac{3}{2}\pi}\frac{\cos t}{t^{\lambda}}\mathrm{d}t=-\infty,$$

for $\lambda<\lambda_0$, where λ_0 is the unique root in (0, 1) of the equation

$$\int_0^{\frac{3}{2}\pi}\frac{\cos t}{t^{\lambda}}\mathrm{d}t=0,$$

whose numerical valus is $\lambda_0=0.308\,443\,7\cdots$. This is the Littlewood-Salem-Izumi constant, (see [18] and [6]).

It follows from what is above that (1.1) cannot hold for $\lambda<\lambda_0$. However, we shall prove that it does hold for all $\lambda\geqslant\lambda_0$ as our main result states.

§ 2. Proof of the theorem

Let us define

$$S_N(\theta,\lambda)=\sum_{k=0}^{N}(-1)^k\frac{C_{2k}^{\lambda}(\cos\theta)}{C_{2k}^{\lambda}(1)}.$$

In this section, we are concerned with proving

$$S_N(\theta,\lambda)\geqslant 0,\quad 0\leqslant\theta\leqslant\frac{\pi}{2},\ \text{for all}\quad N,\ \text{and}\quad \lambda=\lambda_0. \tag{2.1}$$

($\lambda_0=0.308\,443\,7\cdots$). Then in view of Feldheim's integral [13], this will be true for all $\lambda\geqslant\lambda_0$.

Using the Dirichlet-Mehler formula [10, 10.9, (32)]

$$\frac{C_n^{\lambda}(\cos\theta)}{C_n^{\lambda}(1)}=\frac{2^{\lambda}(\sin\theta)^{1-2\lambda}}{B\left(\lambda,\frac{1}{2}\right)}\int_0^{\theta}\frac{\sin(n+\lambda)t}{(\cos t-\cos\theta)^{1-\lambda}}\mathrm{d}t,$$

it can be easily shown that

$$S_N(\theta, \lambda)=\frac{1}{2}+r(\theta, \lambda)+(-1)^N Q_{2N+1}(\theta, \lambda), \qquad (2.2)$$

where $$r(\theta,\lambda)=\frac{(\sin\theta)^{1-2\lambda}}{2^{1-\lambda}B\left(\lambda,\frac{1}{2}\right)}\int_0^{\theta}\frac{\sin t\cdot\sin\lambda t}{\cos t(\cos t-\cos\theta)^{1-\lambda}}\mathrm{d}t$$

and $$Q_n(\theta,\lambda)=\frac{(\sin\theta)^{1-2\lambda}}{2^{1-\lambda}B\left(\lambda,\frac{1}{2}\right)}\int_0^{\theta}\frac{\cos(n+\lambda)t}{\cos t(\cos t-\cos\theta)^{1-\lambda}}\mathrm{d}t.$$

Thus, we seek to prove the non-negativity of the right-hand side of (2.2) for $0\leqslant\theta\leqslant\frac{1}{2}\pi$ and $\lambda=\lambda_0$.

Throughout this section we shall employ, for simplicity, the notation $\delta=\frac{\pi}{2N+1+\lambda}$. When N is even and $0\leqslant\theta\leqslant\frac{\delta}{2}$, it follows immediately from (2.2) that.

$$S_N(\theta, \lambda)\geqslant\frac{1}{2}+r(\theta, \lambda)>0.$$

Suppose that N is odd and $0\leqslant\theta\leqslant\frac{3}{2}\delta$. Then, using again (2.2), we have

$$\begin{aligned}S_N(\theta,\lambda)&=\frac{(\sin\theta)^{1-2\lambda}}{2^{1-\lambda}B\left(\lambda,\frac{1}{2}\right)}\int_0^{\theta}\frac{\cos(1-\lambda)t-\cos(2N+1+\lambda)t}{\cos t(\cos t-\cos\theta)^{1-\lambda}}\mathrm{d}t\\&=\frac{2^{\lambda}(\sin\theta)^{1-2\lambda}}{B\left(\lambda,\frac{1}{2}\right)}\int_0^{\theta}\frac{\sin(N+1)t\cdot\sin(N+\lambda)t}{\cos t(\cos t-\cos\theta)^{1-\lambda}}\mathrm{d}t\geqslant 0,\end{aligned}$$

with equality only when $\theta=0$.

The interval $(2N+1)\frac{\delta}{2}\leqslant\theta\leqslant\frac{1}{2}\pi$.

We see that (2.2) can also be written

$$S_N(\theta,\lambda)=\frac{(\sin\theta)^{1-2\lambda}}{2^{1-\lambda}B\left(\lambda,\frac{1}{2}\right)}\left[\int_0^{\theta}\frac{\cos(1-\lambda)t}{\cos t(\cos t-\cos\theta)^{1-\lambda}}\mathrm{d}t+(-1)^N\int_0^{\theta}\frac{\cos\frac{\pi}{\delta}t}{\cos t(\cos t-\cos\theta)^{1-\lambda}}\mathrm{d}t\right]. \qquad (2.3)$$

We set $\sigma=\frac{1}{2}\pi-\frac{1}{2}\lambda\delta=(2N+1)\frac{1}{2}\delta$. Then, we can easily show that

$$(-1)^N\int_0^\theta \frac{\cos\frac{\pi}{\delta}t}{\cos t(\cos t-\cos\theta)^{1-\lambda}}dt$$

$$>-\int_0^{\theta-\sigma}\frac{\sin\frac{\pi}{\delta}t}{\cos(t+\sigma)[\cos(t+\sigma)-\cos\theta]^{1-\lambda}}dt. \tag{2. 4}$$

On the other hand,

$$\int_0^\theta \frac{\cos(1-\lambda)t}{\cos t(\cos t-\cos\theta)^{1-\lambda}}dt$$

$$=\int_0^\sigma \frac{\cos(1-\lambda)t}{\cos t(\cos t-\cos\theta)^{1-\lambda}}dt+\int_0^{\theta-\sigma}\frac{\cos[(1-\lambda)(t+\sigma)]}{\cos(t+\sigma)[\cos(t+\sigma)-\cos\theta]^{1-\lambda}}dt.$$

Combining this with (2. 4) we get

$$\int_0^\theta \frac{\cos(1-\lambda)t}{\cos t(\cos t-\cos\theta)^{1-\lambda}}dt+(-1)^N\int_0^\theta \frac{\cos\frac{\pi}{\delta}t}{\cos t(\cos t-\cos\theta)^{1-\lambda}}dt$$

$$>\int_0^\sigma \frac{\cos(1-\lambda)t}{\cos t(\cos t-\cos\theta)^{1-\lambda}}dt+\int_0^{\theta-\sigma}\frac{\cos[(1-\lambda)(t+\sigma)]-\sin\frac{\pi}{\delta}t}{\cos(t+\sigma)[\cos(t+\sigma)-\cos\theta]^{1-\lambda}}dt$$

$$\geqslant\int_0^\sigma \frac{\cos(1-\lambda)t}{\cos t(\cos t-\cos\theta)^{1-\lambda}}dt>0, \tag{2. 5}$$

because $0\leqslant\theta-\sigma\leqslant\frac{1}{2}\lambda\delta$.

In view of (2. 5), it follows directly from (2. 3) that

$$S_N(\theta,\lambda)>0,\quad (2N+1)\frac{\delta}{2}\leqslant\theta\leqslant\frac{\pi}{2}\quad \text{for all}\quad N. \qquad \square$$

Next, we give an appropriate lower bound for the function $\frac{1}{2}+r(\theta,\lambda)$ appearing on the right-hand side of (2. 2).

We have

$$\frac{1}{2}+r(\theta,\lambda)=\frac{(\sin\theta)^{1-2\lambda}}{2^{1-\lambda}B\left(\lambda,\frac{1}{2}\right)}\int_0^\theta \frac{\cos(1-\lambda)t}{\cos t(\cos t-\cos\theta)^{1-\lambda}}dt, \tag{2. 6}$$

and observe that

$$\frac{\cos(1-\lambda)t}{\cos\frac{t}{2}}>\frac{\cos(1-\lambda)\theta}{\cos\frac{\theta}{2}},\quad 0<t<\theta<\frac{\pi}{2},$$

whence

$$\int_0^\theta \frac{\cos(1-\lambda)t}{\cos t(\cos t-\cos\theta)^{1-\lambda}}\mathrm{d}t$$

$$>\frac{\cos[(1-\lambda)\theta]}{\cos\frac{\theta}{2}}\int_0^\theta \frac{\cos\frac{t}{2}}{\cos t(\cos t-\cos\theta)^{1-\lambda}}\mathrm{d}t. \tag{2.7}$$

A simple calculation shows that

$$\int_0^\theta \frac{\cos\frac{t}{2}}{\cos t(\cos t-\cos\theta)^{1-\lambda}}\mathrm{d}t$$

$$=\frac{1}{2^{1-\lambda}\left(\sin\frac{\theta}{2}\right)^{1-2\lambda}}\sum_{k=0}^{+\infty}B\left(k+\frac{1}{2},\lambda\right)\left(2\sin^2\frac{\theta}{2}\right)^k. \tag{2.8}$$

Since

$$B\left(k+\frac{1}{2},\ \lambda\right)=B\left(\frac{1}{2},\ \lambda\right)\frac{\left(\frac{1}{2}\right)_k}{\left(\lambda+\frac{1}{2}\right)_k},\qquad \text{for all } k,$$

and
$$\frac{\left(\frac{1}{2}\right)_k\cdot k!}{\left(\lambda+\frac{1}{2}\right)_k\cdot(1-\lambda)_k}>\frac{1}{2}B\left(\lambda+\frac{1}{2},\ 1-\lambda\right),\qquad k\in\mathbf{N},$$

we infer that

$$B\left(k+\frac{1}{2},\ \lambda\right)>\frac{1}{2}B\left(\lambda+\frac{1}{2},\ 1-\lambda\right)B\left(\lambda,\ \frac{1}{2}\right)\frac{(1-\lambda)_k}{k!}=\frac{\pi}{\sin\pi\lambda}\ \frac{(1-\lambda)_k}{k!}.$$

From this it follows that

$$\sum_{k=0}^{+\infty}B\left(k+\frac{1}{2},\lambda\right)\left(2\sin^2\frac{\theta}{2}\right)^k>\frac{\pi}{\sin\pi\lambda}\ \frac{1}{(\cos\theta)^{1-\lambda}}.$$

Thus by (2.8) we have

$$\int_0^\theta \frac{\cos\frac{t}{2}}{\cos t(\cos t-\cos\theta)^{1-\lambda}}\mathrm{d}t>\frac{\pi}{2^{1-\lambda}\sin\pi\lambda}\cdot\frac{1}{\left(\sin\frac{\theta}{2}\right)^{1-2\lambda}(\cos\theta)^{1-\lambda}}.$$

This in combination with (2.6) and (2.7) yields

$$\frac{1}{2}+r(\theta,\lambda)>\frac{1}{2B\left(\lambda,\frac{1}{2}\right)}\ \frac{\pi}{\sin\pi\lambda}\ \frac{\cos(1-\lambda)\theta}{(\cos\theta)^{1-\lambda}\left(\cos\frac{\theta}{2}\right)^{2\lambda}}. \tag{2.9}$$

Let us define $\theta_\nu=(2\nu+1)\frac{1}{2}\delta$, $\nu=0,1,\cdots,N$, and

$$J_N(\theta,\lambda)=\int_0^\theta \frac{\cos\frac{\pi}{\delta}t}{\cos t(\cos t-\cos\theta)^{1-\lambda}}\mathrm{d}t.$$

Suppose now that N is even. We can easily verify that when $\theta_{2k}\leqslant\theta\leqslant\theta_{2k+1}$, $k=0,1,2,\cdots,\left[\frac{N}{2}\right]-1$, then

$$J_N(\theta,\lambda)>-\int_0^{\theta-\theta_{2k}}\frac{\sin\frac{\pi}{\delta}t}{\cos(t+\theta_{2k})[\cos(t+\theta_{2k})-\cos\theta]^{1-\lambda}}\mathrm{d}t \tag{2.10}$$

and when $\theta_{2k+1}\leqslant\theta\leqslant\theta_{2k+2}$, $k=0,1,2,\cdots,\left[\frac{N}{2}\right]-1$, then

$$J_N(\theta,\lambda)>-\int_0^{\delta}\frac{\sin\frac{\pi}{\delta}t}{\cos(t+\theta_{2k})[\cos(t+\theta_{2k})-\cos\theta]^{1-\lambda}}\mathrm{d}t. \tag{2.11}$$

In a similar manner, we see, when N is odd, $\theta_{2k+1}\leqslant\theta\leqslant\theta_{2k+2}$ and $k=0,1,2,\cdots,\left[\frac{N}{2}\right]-1$, that

$$J_N(\theta,\lambda)<\int_0^{\theta-\theta_{2k+1}}\frac{\sin\frac{\pi}{\delta}t}{\cos(t+\theta_{2k+1})[\cos(t+\theta_{2k+1})-\cos\theta]^{1-\lambda}}\mathrm{d}t \tag{2.12}$$

and for $\theta_{2k+2}\leqslant\theta\leqslant\theta_{2k+3}$ and $k=0,1,2,\cdots,\left[\frac{N}{2}\right]-1$, that

$$J_N(\theta,\lambda)<\int_0^{\delta}\frac{\sin\frac{\pi}{\delta}t}{\cos(t+\theta_{2k+1})[\cos(t+\theta_{2k+1})-\cos\theta]^{1-\lambda}}\mathrm{d}t. \tag{2.13}$$

Suppose now that $\theta_{2k}\leqslant\theta\leqslant\theta_{2k+1}$, $k=0,1,2,\cdots,\left[\frac{N}{2}\right]-1$. We have

$$\begin{aligned}
&\int_0^{\theta-\theta_{2k}}\frac{\sin\frac{\pi}{\delta}t}{\cos(t+\theta_{2k})[\cos(t+\theta_{2k})-\cos\theta]^{1-\lambda}}\mathrm{d}t\\
&\leqslant\frac{1}{\cos\theta}\frac{1}{\left(\sin\frac{\theta+\theta_{2k}}{2}\right)^{1-\lambda}}\int_0^{\xi\delta}\frac{\sin\frac{\pi}{\delta}t}{\left(2\sin\frac{\xi\delta-t}{2}\right)^{1-\lambda}}\mathrm{d}t\\
&\leqslant\frac{1}{\cos\theta}\frac{1}{\left(\sin\frac{\theta+\theta_{2k}}{2}\right)^{1-\lambda}}\left(\frac{\delta}{2\sin\frac{\delta}{2}}\right)^{1-\lambda}\left(\frac{\delta}{\pi}\right)^{\lambda}g(\xi\pi,\lambda),\quad 0\leqslant\xi\leqslant1,
\end{aligned} \tag{2.14}$$

where
$$g(x,\lambda)=\int_0^{x}\frac{\sin(x-t)}{t^{1-\lambda}}\mathrm{d}t.$$

When $\theta_{2k+1}\leqslant\theta\leqslant\theta_{2k+2}$, $k=0,1,2,\cdots,\left[\frac{N}{2}\right]-1$, a similar calculation gives

$$\int_0^{\delta} \frac{\sin \frac{\pi}{\delta} t}{\cos(t+\theta_{2k})[\cos(t+\theta_{2k})-\cos\theta]^{1-\lambda}} \mathrm{d}t$$

$$\leqslant \frac{1}{\cos\theta_{2k+1}} \frac{1}{\left(\sin \frac{\theta+\theta_{2k}}{2}\right)^{1-\lambda}} \int_0^{\delta} \frac{\sin \frac{\pi}{\delta} t}{\left(2\sin \frac{\delta-t}{2}\right)^{1-\lambda}} \mathrm{d}t$$

$$\leqslant \frac{1}{\cos\theta_{2k+1}} \frac{1}{\left(\sin \frac{\theta_{2k}+\theta_{2k+1}}{2}\right)^{1-\lambda}} \left(\frac{\delta}{\sin\delta}\right)^{1-\lambda} \left(\frac{\delta}{\pi}\right)^{\lambda} g(\pi,\lambda). \tag{2.15}$$

Recalling that

$$Q_{2N+1}(\theta, \lambda)=\frac{(\sin\theta)^{1-2\lambda}}{2^{1-\lambda}B\left(\lambda, \frac{1}{2}\right)} J_N(\theta, \lambda),$$

and using (2.10)~(2.15) we deduce, when N is even and $\theta_{2k}\leqslant\theta\leqslant\theta_{2k+1}$ or N is odd and $\theta_{2k+1}\leqslant\theta\leqslant\theta_{2k+2}$, $k=0,1,2,\cdots,\left[\frac{N}{2}\right]-1$, that

$$(-1)^N Q_{2N+1}(\theta, \lambda)>$$

$$-\frac{1}{2B\left(\lambda, \frac{1}{2}\right)} \frac{1}{(\cos\theta)^{1-\lambda}} \cdot \left[\frac{\sin\theta}{\sin\left(\theta-\frac{\xi\delta}{2}\right)}\right]^{1-\lambda} \left(\frac{2\delta}{\sin 2\theta}\right)^{\lambda} \cdot C_1,$$

$$0\leqslant\xi\leqslant 1, \tag{2.16}$$

where $$C_1=C_1(N, \lambda)=\left(\frac{2}{\pi}\right)^{\lambda} \left[\frac{\delta}{2\sin\frac{\delta}{2}}\right]^{1-\lambda} \cdot M$$

and $$M=\max_{0\leqslant x\leqslant\pi} g(x, \lambda).$$

Numerical integration gives $M<3.43863$ for $0.308443<\lambda<0.308444$.

Similarly, when N is even and $\theta_{2k+1}\leqslant\theta\leqslant\theta_{2k+2}$ or N is odd and $\theta_{2k+2}\leqslant\theta\leqslant\theta_{2k+3}$, $k=0,1,2,\cdots,\left[\frac{N}{2}\right]-1$, we have

$$(-1)^N Q_{2N+1}(\theta, \lambda)>$$

$$-\frac{1}{2B\left(\lambda, \frac{1}{2}\right)} \frac{1}{(\cos\theta)^{1-\lambda}} \left[\frac{\sin\theta}{\sin\left(\theta-\frac{\delta}{2}-\xi\delta\right)}\right]^{1-\lambda} \left(\frac{2\delta}{\sin(2\theta-2\xi\delta)}\right)^{\lambda} \cdot C_2,$$

$$0\leqslant\xi\leqslant 1, \tag{2.17}$$

where $$C_2=C_2(N, \lambda)=\left(\frac{2}{\pi}\right)^{\lambda} \left(\frac{\delta}{\sin\delta}\right)^{1-\lambda} g(\pi, \lambda).$$

By numerical integration we find $g(\pi, \lambda)<1.77312$.

Next, we assume that $N\geqslant 12$, so that

$$C_1<2.9929 \quad \text{and} \quad C_2<1.54532$$

and write
$$f(\theta, \lambda)=\frac{\pi}{\sin \pi\lambda} \frac{\cos(1-\lambda)\theta}{\left(\cos \frac{\theta}{2}\right)^{2\lambda}}.$$

In view of (2.2)(2.9)(2.16) and (2.17) we see that in order to prove the positivity of $S_N(\theta, \lambda)$ it is sufficient to establish the inequalities

$$f(\theta, \lambda)-C_1\left[\frac{\sin \theta}{\sin\left(\theta-\frac{\xi\delta}{2}\right)}\right]^{1-\lambda}\left(\frac{2\delta}{\sin 2\theta}\right)^{\lambda}>0 \tag{2.18}$$

and
$$f(\theta, \lambda)-C_2\left[\frac{\sin \theta}{\sin\left(\theta-\frac{\delta}{2}-\xi\delta\right)}\right]^{1-\lambda}\left(\frac{2\delta}{\sin(2\theta-2\xi\delta)}\right)^{\lambda}>0. \tag{2.19}$$

It is not hard to see that $f(\theta, \lambda)$ is a decreasing function of θ in $\left[0, \frac{1}{2}\pi\right]$ so that

$$f(\theta, \lambda)\geqslant f\left(\frac{\pi}{2}, \lambda\right)>2.1982. \tag{2.20}$$

The interval $\theta_{N-1}\leqslant\theta\leqslant\theta_N$.

In this case, we see that we have only to prove (2.19). This follows immediately from (2.20) and the fact that

$$C_2\left[\frac{\sin \theta}{\sin\left(\theta-\frac{\delta}{2}-\xi\delta\right)}\right]^{1-\lambda}\left(\frac{2\delta}{\sin(2\theta-2\xi\delta)}\right)^{\lambda}$$

$$\leqslant C_2\left[\frac{\cos\left(\frac{\lambda}{2}+1\right)\delta}{\cos\left(\frac{\lambda}{2}+\frac{5}{2}\right)\delta}\right]^{1-\lambda}\left(\frac{2\delta}{\sin(\lambda+2)\delta}\right)^{\lambda}\leqslant 1.5317, \quad \text{for} \quad N\geqslant 12. \quad \square$$

The interval $\frac{15}{2}\delta\leqslant\theta\leqslant(2N-7)\frac{1}{2}\delta$, $N\geqslant 12$.

We have

$$C_1\left[\frac{\sin \theta}{\sin\left(\theta-\frac{\xi\delta}{2}\right)}\right]^{1-\lambda}\left(\frac{2\delta}{\sin 2\theta}\right)^{\lambda}\leqslant C_1\left(\frac{\sin \frac{15}{2}\delta}{\sin 7\delta}\right)^{1-\lambda}\left(\frac{2\delta}{\sin(\lambda+8)\delta}\right)^{\lambda}<2.1124$$

and $$C_2\left[\frac{\sin\theta}{\sin\left(\theta-\frac{\delta}{2}-\xi\delta\right)}\right]^{1-\lambda}\left(\frac{2\delta}{\sin(2\theta-2\xi\delta)}\right)^{\lambda}<1.1845.$$

A combination of the above inequalities with the estimate (2.20) gives (2.18) and (2.19). □

The interval $\frac{5}{2}\delta\leqslant\theta\leqslant\frac{15}{2}\delta$.

Now,

$$C_1\left[\frac{\sin\theta}{\sin\left(\theta-\frac{\xi\delta}{2}\right)}\right]^{1-\lambda}\left(\frac{2\delta}{\sin 2\theta}\right)^{\lambda}<C_1\left[\frac{\sin\frac{5}{2}\delta}{\sin 2\delta}\right]^{1-\lambda}\left(\frac{2\delta}{\sin 5\delta}\right)^{\lambda}<2.686$$

and

$$C_2\left[\frac{\sin\theta}{\sin\left(\theta-\frac{\delta}{2}-\xi\delta\right)}\right]^{1-\lambda}\left(\frac{2\delta}{\sin(2\theta-2\xi\delta)}\right)^{\lambda}<C_2\left[\frac{\sin\frac{5}{2}\delta}{\sin\delta}\right]^{1-\lambda}\left(\frac{2\delta}{\sin 3\delta}\right)^{\lambda}$$
$$<2.59.$$

In this case (2.18) and (2.19) follow from the above inequalities and the fact that for $N\geqslant 12$, $\frac{15}{2}\delta\leqslant\frac{15\pi}{50+2\lambda}$, so that $f(\theta,\lambda)>3.267$ in the interval in question. □

The interval $\frac{3}{2}\delta\leqslant\theta\leqslant\frac{5}{2}\delta$.

In this case the validity of (2.18) and (2.19) is obtained by observing that $f(\theta,\lambda)>3.751$,

$$C_1\left[\frac{\sin\theta}{\sin\left(\theta-\frac{\xi\delta}{2}\right)}\right]^{1-\lambda}\left(\frac{2\delta}{\sin 2\theta}\right)^{\lambda}\leqslant C_1\left[\frac{\sin\frac{3}{2}\delta}{\sin 2\delta}\right]^{1-\lambda}\left(\frac{2\delta}{\sin 3\delta}\right)^{\lambda}<3.53$$

and

$$C_2\left[\frac{\sin\theta}{\sin\left(\theta-\frac{\delta}{2}-\xi\delta\right)}\right]^{1-\lambda}\left(\frac{2\delta}{\sin(2\theta-2\xi\delta)}\right)^{\lambda}\leqslant C_2\left[\frac{\sin\frac{5}{2}\delta}{\sin\delta}\right]^{1-\lambda}\left(\frac{2\delta}{\sin 3\delta}\right)^{\lambda}$$
$$<2.59.$$ □

The interval $\frac{1}{2}\delta\leqslant\theta\leqslant\frac{3}{2}\delta$.

Now, we have only to consider the case where N is even, hence we need to establish (2.18). First we observe that $f(\theta,\lambda)>3.789$. Setting $\theta=\left(\frac{1}{2}+\xi\right)\delta$, $0\leqslant\xi\leqslant 1$, we have

$$C_1\left[\frac{\sin\theta}{\sin\left(\theta-\frac{\xi\delta}{2}\right)}\right]^{1-\lambda}\left(\frac{2\delta}{\sin 2\theta}\right)^{\lambda}=C_1\left[\frac{\sin\left(\frac{1}{2}+\xi\right)\delta}{\sin\left(\frac{1}{2}+\frac{\xi}{2}\right)\delta}\right]^{1-\lambda}\left(\frac{2\delta}{\sin(2\xi+1)\delta}\right)^{\lambda}$$

$$\leqslant C_1\left(\frac{2\xi+1}{\xi+1}\right)^{1-\lambda}\left(\frac{2\delta}{\sin(2\xi+1)\delta}\right)^{\lambda}<3.73,$$

as desired. □

It remains to establish the positivity of $S_N(\theta,\lambda)$ for $(2N-7)\frac{1}{2}\delta\leqslant\theta\leqslant(2N-1)\frac{1}{2}\delta$, $N\geqslant 12$.

Keeping in mind that

$$[2(N-k)+3]\frac{\delta}{2}=\frac{\pi}{2}-\left(\frac{\lambda}{2}+k-1\right)\delta,\qquad k=2,3,4,5,$$

we shall use, for this purpose, an argument different from the above. In fact, setting $\phi=\frac{1}{2}\pi-\theta$, it is easy to see that

$$r(\theta,\lambda)+(-1)^N Q_{2N+1}(\theta,\lambda)=\frac{(\cos\phi)^{1-2\lambda}}{2^{1-\lambda}B\left(\lambda,\frac{1}{2}\right)}I_N(\phi,\lambda),$$

where

$$I_N(\phi,\lambda)=\int_{\phi}^{\frac{\pi}{2}}\frac{\sin\left[(2N+1+\lambda)t-\frac{\lambda\pi}{2}\right]+\cos t\sin\left[\lambda\left(\frac{\pi}{2}-t\right)\right]}{\sin t(\sin t-\sin\phi)^{1-\lambda}}\mathrm{d}t.$$

Hence

$$S_N\left(\frac{\pi}{2}-\phi,\lambda\right)=\frac{1}{2}+\frac{(\cos\phi)^{1-2\lambda}}{2^{1-\lambda}B\left(\lambda,\frac{1}{2}\right)}I_N(\phi,\lambda).\qquad(2.21)$$

It is not hard to prove that

$$\int_1^{+\infty}\frac{\sin\left(xt-\frac{\lambda\pi}{2}\right)+\sin\frac{\lambda\pi}{2}}{t(t-1)^{1-\lambda}}\mathrm{d}t=\Gamma(\lambda)\int_0^x\frac{\cos t}{t^{\lambda}}\mathrm{d}t,\quad x>0.$$

Using this identity we get

$$I_N(\phi,\lambda)=\frac{\Gamma(\lambda)}{\phi^{1-\lambda}}\int_0^{(2N+1+\lambda)\phi}\frac{\cos t}{t^{\lambda}}\mathrm{d}t-R_N(\phi,\lambda),\qquad(2.22)$$

where $R_N(\phi,\lambda)=\sum_{i=1}^{3}r_i(\phi,\lambda)-r_4(\phi,\lambda)-r_5(N,\phi,\lambda)+r_6(N,\phi,\lambda)$

and

$$r_1(\phi,\lambda)=\sin\frac{\lambda\pi}{2}\int_{\phi}^{\frac{\pi}{2}}\frac{1-\cos t\cos\lambda t}{\sin t(\sin t-\sin\phi)^{1-\lambda}}\mathrm{d}t,$$

$$r_2(\phi,\lambda)=\sin\frac{\lambda\pi}{2}\int_{\frac{\pi}{2}}^{+\infty}\frac{\mathrm{d}t}{t(t-\phi)^{1-\lambda}},$$

$$r_3(\phi,\lambda)=\cos\frac{\lambda\pi}{2}\int_{\phi}^{\frac{\pi}{2}}\frac{\cos t\sin\lambda t}{\sin t(\sin t-\sin\phi)^{1-\lambda}}\mathrm{d}t,$$

$$r_4(\phi,\lambda)=\sin\frac{\lambda\pi}{2}\int_{\phi}^{\frac{\pi}{2}}\left[\frac{1}{\sin t(\sin t-\sin\phi)^{1-\lambda}}-\frac{1}{t(t-\phi)^{1-\lambda}}\right]\mathrm{d}t,$$

$$r_5(N,\phi,\lambda)=$$

$$\int_{\phi}^{\frac{\pi}{2}}\left[\frac{1}{\sin t(\sin t-\sin\phi)^{1-\lambda}}-\frac{1}{t(t-\phi)^{1-\lambda}}\right]\sin\left[(2N+1+\lambda)t-\frac{\lambda\pi}{2}\right]\mathrm{d}t,$$

$$r_6(N,\phi,\lambda)=\int_{\frac{\pi}{2}}^{+\infty}\frac{\sin\left[(2N+1+\lambda)t-\frac{\lambda\pi}{2}\right]}{t(t-\phi)^{1-\lambda}}\mathrm{d}t.$$

Now we need to find an appropriate upper bound for the functions $R_N(\phi,\lambda)$ defined above.

Choosing y such that $\phi<y<\frac{1}{2}\pi$, we have

$$\int_{\phi}^{y}\frac{1-\cos t\cos\lambda t}{\sin t(\sin t-\sin\phi)^{1-\lambda}}\mathrm{d}t=\int_{\phi}^{y}\frac{2-\cos(1+\lambda)t-\cos(1-\lambda)t}{2\sin t(\sin t-\sin\phi)^{1-\lambda}}\mathrm{d}t<$$
$$\frac{1+\lambda^2}{2}\frac{y}{\sin y}\frac{1}{(\cos y)^{1-\lambda}}\left(\frac{y-\phi}{2\sin\frac{y-\phi}{2}}\right)^{1-\lambda}\int_{\phi}^{y}\frac{t\,\mathrm{d}t}{(t-\phi)^{1-\lambda}}.$$

It follows from this that

$$r_1(\phi,\lambda)<$$
$$\frac{1+\lambda^2}{2}\sin\frac{\lambda\pi}{2}\frac{y}{\sin y}\frac{1}{(\cos y)^{1-\lambda}}\left(\frac{y-\phi}{2\sin\frac{y-\phi}{2}}\right)^{1-\lambda}\left[\frac{(y-\phi)^{\lambda+1}}{\lambda+1}+\frac{\phi(y-\phi)^{\lambda}}{\lambda}\right]+$$
$$\sin\frac{\lambda\pi}{2}\int_{y}^{\frac{\pi}{2}}\frac{1-\cos t\cos\lambda t}{\sin t(\sin t-\sin\phi)^{1-\lambda}}\mathrm{d}t. \tag{2.23}$$

Clearly,

$$r_2(\phi,\lambda)<\sin\frac{\lambda\pi}{2}\cdot\frac{1}{(1-\lambda)\left(\frac{\pi}{2}-\phi\right)^{1-\lambda}}. \tag{2.24}$$

Furthermore, it is easy to see that

$$r_3(\phi, \lambda)<\cos\frac{\lambda\pi}{2}\cdot\frac{\sin\lambda y}{\lambda\sin y}(\sin y-\sin\phi)^{\lambda}+$$

$$\cos\frac{\lambda\pi}{2}\int_y^{\frac{\pi}{2}}\frac{\sin\lambda t\cos t}{\sin t(\sin t-\sin\phi)^{1-\lambda}}\mathrm{d}t. \qquad (2.25)$$

Let

$$w(\phi,y,\lambda)=\int_y^{\frac{\pi}{2}}\frac{\sin\frac{\lambda\pi}{2}-\cos t\cdot\sin\left[\lambda\left(\frac{\pi}{2}-t\right)\right]}{\sin t(\sin t-\sin\phi)^{1-\lambda}}\mathrm{d}t$$

and

$u(\phi, y, \lambda)=$

$$\frac{1+\lambda^2}{2}\sin\frac{\lambda\pi}{2}\frac{y}{\sin y}\frac{1}{(\cos y)^{1-\lambda}}\left(\frac{y-\phi}{2\sin\frac{y-\phi}{2}}\right)^{1-\lambda}\left[\frac{(y-\phi)^{\lambda+1}}{\lambda+1}+\frac{\phi(y-\phi)^{\lambda}}{\lambda}\right]+$$

$$\sin\frac{\lambda\pi}{2}\frac{1}{(1-\lambda)\left(\frac{\pi}{2}-\phi\right)^{1-\lambda}}+\cos\frac{\lambda\pi}{2}\frac{\sin\lambda y}{\lambda\sin y}(\sin y-\sin\phi)^{\lambda}.$$

It follows from (2.23)～(2.25) and the above that

$$\sum_{i=1}^{3}r_i(\phi,\lambda)<u(\phi,y,\lambda)+w(\phi,y,\lambda). \qquad (2.26)$$

Now, let

$$b(t, \phi, \lambda)=\frac{1}{t(t-\phi)^{1-\lambda}}.$$

Integrating twice by parts, we get

$$r_6(N,\phi,\lambda)=\int_{\frac{\pi}{2}}^{+\infty}b(t,\phi,\lambda)\sin\left[(2N+1+\lambda)t-\frac{\lambda\pi}{2}\right]\mathrm{d}t$$

$$=\frac{1}{(2N+1+\lambda)^2}\left\{(-1)^{N+1}\frac{\partial b}{\partial t}\left(\frac{\pi}{2},\phi,\lambda\right)-\int_{\frac{\pi}{2}}^{+\infty}\frac{\partial^2 b}{\partial t^2}\sin\left[(2N+1+\lambda)t-\frac{\lambda\pi}{2}\right]\mathrm{d}t\right\}.$$

Since

$$\frac{\partial b}{\partial t}<0, \quad \frac{\partial^2 b}{\partial t^2}>0, \quad \text{for} \quad t>0,$$

it follows from the above that

$$|r_6(N, \phi, \lambda)|\leqslant\frac{2}{(2N+1+\lambda)^2}\left(\frac{4}{\pi^2\left(\frac{\pi}{2}-\phi\right)^{1-\lambda}}+\frac{2(1-\lambda)}{\pi\left(\frac{\pi}{2}-\phi\right)^{2-\lambda}}\right). \qquad (2.27)$$

Next, we assume that $\left(\frac{1}{2}\lambda+1\right)\delta\leqslant\phi\leqslant\left(\frac{1}{2}\lambda+4\right)\delta$, $N\geqslant12$. We define

$$F(t,\ \phi,\ \lambda)=\frac{1}{\sin t(\sin t-\sin\phi)^{1-\lambda}}-\frac{1}{t(t-\phi)^{1-\lambda}}.$$

Clearly, $F(t,\ \phi,\ \lambda)>0$, $0<\phi<t<\frac{1}{2}\pi$. It is then easy to see that

$$|r_5(N,\phi,\lambda)|\leqslant\int_\phi^{\phi+\delta}F(t,\phi,\lambda)\mathrm{d}t,$$

which, in combination with

$$|r_4(\phi,\lambda)|>\sin\frac{\lambda\pi}{2}\int_\phi^{\phi+\delta}F(t,\phi,\lambda)\mathrm{d}t,$$

gives

$$-[r_4(\phi,\lambda)+r_5(N,\phi,\lambda)]<\left(1-\sin\frac{\lambda\pi}{2}\right)\int_\phi^{\phi+\delta}F(t,\phi,\lambda)\mathrm{d}t.\tag{2.28}$$

Writing

$$F(t,\ \phi,\ \lambda)=\frac{t}{(t-\phi)^{1-\lambda}}\left\{\left(\frac{t-\phi}{\sin t-\sin\phi}\right)^{1-\lambda}\left(\frac{1}{t\sin t}-\frac{1}{t^2}\right)+\frac{1}{t^2}\left[\left(\frac{t-\phi}{\sin t-\sin\phi}\right)^{1-\lambda}-1\right]\right\},$$

it can be easily shown that

$$F(t,\ \phi,\ \lambda)<\left\{\frac{t}{6\sin t}\left[1+(1-\lambda)\frac{t^2}{2-t^2}\right]+(1-\lambda)\frac{1}{2-t^2}\right\}\frac{t}{(t-\phi)^{1-\lambda}},$$

whence

$$\int_\phi^{\phi+\delta}F(t,\phi,\lambda)\mathrm{d}t\leqslant\left\{\frac{\phi+\delta}{6\sin(\phi+\delta)}\left[1+(1-\lambda)\frac{(\phi+\delta)^2}{2-(\phi+\delta)^2}\right]+(1-\lambda)\frac{1}{2-(\phi+\delta)^2}\right\}\left(\phi\cdot\frac{\delta^\lambda}{\lambda}+\frac{\delta^{\lambda+1}}{\lambda+1}\right).\tag{2.29}$$

Let

$$V_N(\phi,\ \lambda)=\left(1-\sin\frac{\lambda\pi}{2}\right)\left\{\frac{\phi+\delta}{6\sin(\phi+\delta)}\left[1+(1-\lambda)\frac{(\phi+\delta)^2}{2-(\phi+\delta)^2}\right]+\frac{1-\lambda}{2-(\phi+\delta)^2}\right\}\times\left(\phi\cdot\frac{\delta^\lambda}{\lambda}+\frac{\delta^{\lambda+1}}{\lambda+1}\right)+2\left(\frac{\delta}{\pi}\right)^2\left[\frac{4}{\pi^2\left(\frac{\pi}{2}-\phi\right)^{1-\lambda}}+\frac{2(1-\lambda)}{\pi\left(\frac{\pi}{2}-\phi\right)^{2-\lambda}}\right].$$

It is evident that $V_N(\phi,\ \lambda)$ is a decreasing sequence of functions, so that $$V_N(\phi, \lambda) \leqslant V_{12}(\phi, \lambda), \quad N \geqslant 12.$$

It follows from (2.26)~(2.29) that

$$R_N(\phi, \lambda) < u(\phi, y, \lambda) + w(\phi, y, \lambda) + V_{12}(\phi, \lambda), \quad N \geqslant 12. \tag{2.30}$$

By the application of (2.21)(2.22) and (2.30) we are now able to prove the positivity of $S_N(\theta, \lambda)$, when

$$(2N-7)\frac{1}{2}\delta \leqslant \theta \leqslant (2N-1)\frac{1}{2}\delta, \quad N \geqslant 12.$$

The interval $\left(\frac{1}{2}\lambda+1\right)\delta \leqslant \phi \leqslant \left(\frac{1}{2}\lambda+2\right)\delta.$

We choose $y=\frac{1}{5}\pi$ and use the formulae above to get

$$u\left(\phi, \frac{\pi}{5}, \lambda\right) < 1.49.$$

Numerical integration shows that

$$w\left(\phi, \frac{\pi}{5}, \lambda\right) < 0.332\,6.$$

Since $V_{12}(\phi, \lambda)$ is an increasing function of ϕ in this interval, we obtain

$$V_{12}(\phi, \lambda) < 0.151\,9.$$

Then by (2.30) we get

$$R_N(\phi, \lambda) < 1.974\,5, \quad N \geqslant 12.$$

In view of (2.22), we now have

$$I_N(\phi, \lambda) \geqslant -1.974\,5.$$

From this and (2.21) we finally obtain

$$S_N\left(\frac{\pi}{2}-\phi, \lambda\right) > 0.225\,8. \quad \square$$

The interval $$\left(\frac{1}{2}\lambda+2\right)\delta \leqslant \phi \leqslant \left(\frac{1}{2}\lambda+4\right)\delta.$$

In this case, we choose $y=\frac{1}{4}\pi$ and by a similar computation we get

$$u\left(\phi, \frac{\pi}{4}, \lambda\right) < 1.72,$$

$$w\left(\phi, \frac{\pi}{4}, \lambda\right) < 0.676\,6,$$

and $$V_{12}(\phi,\ \lambda)<0.3225.$$

Hence, by (2.30),

$$R_N(\phi,\ \lambda)<2.7191.$$

By virtue of (2.22),

$$I_N(\phi,\lambda) > \Gamma(\lambda)\left(\frac{50+2\lambda}{8\pi+\lambda\pi}\right)^{1-\lambda}\int_{\frac{3\pi}{2}}^{\frac{7\pi}{2}}\frac{\cos t}{t^{\lambda}}\mathrm{d}t-2.7191>-2.099,\quad N\geqslant 12,$$

which in combination with (2.21) gives

$$S_N\left(\frac{\pi}{2}-\phi,\ \lambda\right)>0.2085.\qquad\square$$

Finally, we consider the cases $1\leqslant N\leqslant 11$. We write $x=\cos\theta$, so that

$$S_N(x)=\sum_{k=0}^{N}(-1)^k\frac{C_{2k}^{\lambda}(x)}{C_{2k}^{\lambda}(1)},\quad \lambda=0.308443,$$

is a polynomial of $2N$ degree in x. We observe that $S_N(1)=0$, when N is odd, $S_N(1)=1$, when N is even and in any case $S_N(0)>0$. The desired result now follows from the application of Sturm sequences (see [8], section 89), which shows that the polynomials $S_N(x)$, $N=1, 2,\cdots,11$, have no zeros in $[0,\ 1)$. (The calculations in the implementation of Sturmian algorithm can, for example, be facilitated by the use of the package Maple V cf. [9].)

The proof of (2.1) is now complete. $\square$

References

[1] Askey R. Positive Jacobi polynomial sums. Tôhoku Math. Journ., 1972, 24: 109—119.

[2] Askey R. Orthogonal polynomials and special functions. Regional Conf. Lect. Appl. Math., 21, SIAM, Philadelphia, 1975.

[3] Askey R, Fitch J. Integral representations for Jacobi polynomials and some applications. J. Math. Anal. Appl., 1969, 26: 411—437.

[4] Askey R. Gasper G. Positive Jacobi polynomial sums, Ⅱ. Amer. J. Math., 1976, 98: 709—737.

[5] Askey R. Problems which interest and/or annoy me. J. Gomp. Appl. Math., 1993, 48: 3—15.

[6] Brown G, Wang K Y, Wilson D C. Positivity of some basic cosine sums. Math. Proc. Camb. Phil. Soc., 1993, 114: 383—391.

[7] Brown G, Koumandos S, Wang K Y. Positivity of basic sums of ultraspherical polynomials. Analysis, 1998, 18: 313—331.

[8] Burnside W S, Panton A W. Theory of Equations. Dublin University Press, 1886.

[9] Char B W, Geddes K O, Gonnet G H, Leong B L, Monagan M B, Watt S M. Maple V first leaves, A tutorial introduction to Maple V and library reference manual. Springer-Verlag, 1992.

[10] Erdélyi A, Magnus W, Oberhettinger F, Tricomi F G. Higher transcendental functions, Vol. 2. McGraw-Hill, 1953.

[11] Fejér L. Sur les fonctions bornées et intégrables. C. R. Acad. Sci. Paris., 1900, 131: 984—987.

[12] Fejér L. Sur le développement d'une function arbitraire suivant les fonctions de Laplace. C. R. Acad. Sci. Paris, 1908, 224—227; Gesammelte Arbeiten I, 319—322, Birkhäuser Verlag, 1970.

[13] Feldheim E. On the positivity of certain sums of ultraspherical polynomials. J. Analyse Math., 1963, 11: 275—284; reprinted in G. Szegö collected papers Vol. 3 Birkhäuser, 1982, 821—830.

[14] Gasper G. Positive sums of the classical orthogonal polynomials. SIAM J. Math. Anal. 1977, 8: 423—447.

[15] Gronwall T H. Über die Gibbssche Erscheinung und die trigonometrischen Summen $\sin x+\frac{1}{2}\sin 2x+\cdots+\frac{1}{n}\sin nx$. Math. Ann., 1912, 72: 228—243.

[16] Jackson D. Uber eine trigonometrische Summe. Rend. Circ. Math. Palermo, 1911, 32: 257—262.

[17] Szegö G. Orthogonal polynomials. Amer. Math. Soc. Colloq. Publ., 23, 4th edition. Amer. Math. Soc., Providence, R. I., 1975.

[18] Zygmund A. Trigonometric Series. 2nd edition. Cambridge University Press, 1959.

Monatschefte für Mathematik，1996，122(1)：9—19.

再谈 Jacobi 横坐标上的 Cotes 数的正性

Positivity of Cotes Numbers at More Jacobi Abscissas①

Abstract The positivity of certain finite sums of even ultraspherical polynomials has been identified by Askey as a specially interesting case of a more general problem concerning positivity of Cotes numbers at Jacobi abscissas. The authors establish several new inequalities of this type.

Keywords Jacobi polynomials；positive Cotes numbers.

§ 1. Introduction

Let $P_n^{(\alpha,\beta)}(x)$ be the Jacobi polynomials defined by the Rodrigues formula：

$$(1-x)^\alpha(1+x)^\beta P_x^{(\alpha,\beta)}(x)=\frac{(-1)^n}{2^n n!}\frac{\mathrm{d}^n}{\mathrm{d}x^n}[(1-x)^{n+\alpha}(1+x)^{n+\beta}] \qquad (1.1)$$

or，equivalently，by the hypergeometric representation

$$P_n^{(\alpha,\beta)}(x)=\frac{(\alpha+1)_n}{n!}{}_2F_1\left(-n,\ n+\alpha+\beta+1;\ \alpha+1;\ \frac{1-x}{2}\right). \qquad (1.2)$$

The shifted factorial $(a)_k$ is defined by

---

① This research was supported by a grant from the Australian Research Council.

本文与 Brown G 和 Koumandos S 合作.

Received：1994-06-17；Revised：1994-10-09.

$$(a)_k=a(a+1)\cdots(a+k-1)=\frac{\Gamma(k+a)}{\Gamma(a)}.$$

These polynomials are orthogonal on $[-1, 1]$ with respect to $(1-x)^{\alpha}(1+x)^{\beta}$ when α, $\beta>-1$. Expand

$$(1-x)^{-\gamma}(1+x)^{-\delta}\sim\sum_{k=0}^{+\infty}a_kP_k^{(\alpha,\beta)}(x) \tag{1.3}$$

and ask when

$$\sum_{k=0}^{n}a_kP_k^{(\alpha,\beta)}(x)\geqslant 0,\quad -1\leqslant x\leqslant 1,\quad n\in\mathbf{N}. \tag{1.4}$$

The values of the left side of (1.4) at the zeros of $P_n^{(\alpha,\beta)}(x)$ are (positive multiples of) the Cotes numbers for integration with respect to

$$(1-x)^{\alpha-\gamma}(1+x)^{\beta-\delta}\mathrm{d}x$$

(the case $\gamma=\delta=0$ being Gaussian quadrature).

In his SIAM conference Lectures, ASKEY [2], draws attention to these questions noticing that they are more than mere test problems and warrant further work as the source of new and interesting inequalities. He notes moreover that the case $\beta=-\frac{1}{2}$ is especially interesting. This is the case of even ultraspherical polynomials and we concentrate on it here.

We establish several new instances of (1.4) with strict inequality and hence of positive Cotes numbers following in the footsteps of [1]. Our results do indeed correspond to new inequalities. Some of these, concerning even Legendre polynomials $\left(\text{the case } \alpha=0,\ \beta=-\frac{1}{2}\right)$, were established in [5] viz.

$$\frac{1}{2}+\sum_{k=1}^{n}P_{2k}(\cos\theta)\geqslant 0,\quad 0\leqslant\theta\leqslant\pi, \tag{1.5}$$

$$\frac{1}{2}+\sum_{k=1}^{n}P_{4k}(\cos\theta)>0,\quad 0\leqslant\theta\leqslant\pi. \tag{1.6}$$

In [5] we found also an analogous result for sums of odd Legendre polynomials;

$$\frac{\sqrt{15}}{45}+\sum_{k=1}^{n}P_{2k-1}(\cos\theta)\geqslant 0,\quad 0\leqslant\theta\leqslant\frac{\pi}{2}. \tag{1.7}$$

Of course FEJÉR proved in [6] that

$$\sum_{k=0}^{n} P_k(\cos\theta) \geqslant 0, \quad 0 \leqslant \theta \leqslant \pi, \tag{1.8}$$

so it is easy to obtain the weak version of (1.5) in which the leading constant $\frac{1}{2}$ is replaced by 1. Similarly using (1.8) together with (5.8) of GASPER's paper, [10], we can obtain the weak version of (1.6) in which the leading constant $\frac{1}{2}$ is replaced by 1. It is further confirmation of the predictive value of (1.4) that the sharper results are needed here. We note, in passing, that (1.8) itself is the special case of (1.4) in which $\alpha=\beta=\delta=0$, $\gamma=\frac{1}{2}$.

Known cases of (1.4) for $\beta=-\frac{1}{2}$ are $\alpha=\frac{1}{2}$, $\gamma=1$, $\delta=0$ (positivity of the Fejér kernel), $\alpha=-\frac{1}{2}$, $\gamma=\frac{1}{4}$, $\delta=-\frac{1}{4}$ (a cosine sum of VIETORIS [13], set in context and given an elegant proof in [3] then sharpened in [4]), $\alpha=\frac{1}{2}=\gamma$, $\beta=\delta$ (contained in results of FEJÉR [7] and SZEGö [11]) and $\alpha>\frac{1}{2}$, $\delta=0$, $\gamma=\frac{1}{2}\alpha+\frac{3}{4}$ (which follows from work of GASPER in [9]).

Our main results are the following:

Theorem 1 Let $\delta=0$ and $\beta=-\frac{1}{2}$. All the partial sums in (1.4) are strictly positive when

$$\alpha \geqslant 0, \quad 0<\gamma \leqslant \frac{1}{2}.$$

Theorem 2 All the partial sums in (1.4) are strictly positive when

$$\alpha=0, \quad \beta=-\frac{1}{2}, \quad \delta=\frac{1}{4}, \quad \gamma=\frac{1}{2}.$$

Theorem 3 Let $\delta=0$. We suppose that $\alpha+\beta+1>0$. The strict inequality in (1.4) holds for $\alpha \geqslant \beta$, $-1<\gamma<0$. Moreover the partial sums in (1.4) form a decreasing function of x for all $n \geqslant 1$, when

(1) $\alpha \geqslant \beta+1$, $\quad -1<\gamma<0$,

(2) $|\alpha|<\beta+1$, $\quad -1<\gamma \leqslant \frac{1}{2}(\alpha-\beta-1)$.

An immediate consequence of (2) is

Corollary Let $\delta=0$. The strict inequality in (1.4) holds for

$$-\beta-1<\alpha<\beta \text{ and } -1<\gamma \leqslant \frac{1}{2}(\alpha-\beta-1).$$

§ 2. Preliminary reductions

Writing

$$h_k^{(\alpha,\beta)}=\frac{(2k+\alpha+\beta+1)\Gamma(k+1)\Gamma(k+\alpha+\beta+1)}{2^{\alpha+\beta+1}\Gamma(k+\alpha+1)\Gamma(k+\beta+1)}, \tag{2.1}$$

and bearing in mind that

$$\int_{-1}^{1}(P_k^{(\alpha,\beta)}(x))^2(1-x)^\alpha(1+x)^\beta \mathrm{d}x=(h_k^{(\alpha,\beta)})^{-1}, \tag{2.2}$$

it is natural to rewrite a_k in (1.4) as

$$a_k=h_k^{(\alpha,\beta)}\hat{f}_k, \tag{2.3}$$

where

$$\hat{f}_k=\int_{-1}^{1}(1-x)^{\alpha-\gamma}(1+x)^{\beta-\delta}P_k^{(\alpha,\beta)}(x)\mathrm{d}x \tag{2.4}$$

is the Fourier-Jacobi coefficient in the expansion of $f(x)=(1-x)^{-\gamma}\cdot(1+x)^{-\delta}$. In summary

$$f \sim \sum h_k^{(\alpha,\beta)}\hat{f}_k P_k^{(\alpha,\beta)}$$

and we are concerned with proving

$$\sum_{k=0}^{n} h_k^{(\alpha,\beta)}\hat{f}_k P_k^{(\alpha,\beta)}(x)>0, \quad -1 \leqslant x \leqslant 1, n \in \mathbf{N}. \tag{2.5}$$

Of course we can give an explicit formula for the right hand side of (2.4). In fact, using (1.2), we have, for $\alpha-\gamma>-1$, $\beta-\delta>-1$,

$$\hat{f}_k=\frac{2^{\alpha-\gamma+\beta-\delta+1}\Gamma(\alpha-\gamma+1)\Gamma(\beta-\delta+1)}{\Gamma(\alpha-\gamma+\beta-\delta+2)}\cdot\frac{(\alpha+1)_k}{k!}\times$$
$$_3F_2(-k,\ k+\alpha+\beta+1,\ \alpha-\gamma+1;\ \alpha+1,\ \alpha-\gamma+\beta-\delta+2;\ 1). \tag{2.6}$$

Note that, in the special case $\delta=0$, the terminating hypergeometric series is Saalschützian and hence (2.6) simplifies to give

$$\hat{f}_k=\frac{2^{\alpha-\gamma+\beta+1}\Gamma(\alpha-\gamma+1)\Gamma(\beta+1)}{\Gamma(\alpha-\gamma+\beta+2)}\frac{(\beta+1)_k(\gamma)_k}{k!\ (\alpha+\beta-\gamma+2)_k}. \tag{2.7}$$

In the other special case of interest to us $\left(\alpha=0,\ \beta=-\frac{1}{2},\ \gamma=\frac{1}{2},\ \delta=\frac{1}{4}\right)$ observe that the terminating hypergeometric series takes the form

$$s_k={}_3F_2\left(-k,\ k+\frac{1}{2},\ \frac{1}{2};\ 1,\ \frac{3}{4};\ 1\right). \tag{2.8}$$

Using a standard formula which is due to WATSON [14], we find that

$$s_k=\begin{cases}0, & k=2j+1,\\ \dfrac{\left(\frac{1}{2}\right)_j\left(\frac{1}{4}\right)_j}{j!\ \left(\frac{3}{4}\right)_j}, & k=2j,\ j\in\mathbf{N}.\end{cases} \tag{2.9}$$

Let us recall also from [12] the quadratic transformation

$$\frac{P_k^{(\alpha,-\frac{1}{2})}(2x^2-1)}{P_k^{(\alpha,-\frac{1}{2})}(1)}=\frac{C_{2k}^{\alpha+\frac{1}{2}}(x)}{C_{2k}^{\alpha+\frac{1}{2}}(1)}=\frac{P_{2k}^{(\alpha,\alpha)}(x)}{P_{2k}^{(\alpha,\alpha)}(1)}, \tag{2.10}$$

where $C_n^\lambda(x)$ is the Gegenbauer polynomial of degree n, order λ.

In particular, for $\alpha=0$ we have

$$C_{2k}^{\frac{1}{2}}(x)=P_{2k}(x)$$

the Legendre polynomial of degree $2k$.

§ 3. Proof of Theorem 1

We define

$$B_k(\alpha,\ \gamma)=\left(2k+\alpha+\frac{1}{2}\right)\frac{\left(\alpha+\frac{1}{2}\right)_k(\gamma)_k}{k!\ \left(\alpha-\gamma+\frac{3}{2}\right)_k},\qquad k\in\mathbf{N}.$$

Taking into account (2.1)(2.7) and (2.10) inequality (2.5) reduces to

$$\sum_{k=0}^{n}B_k(\alpha,\gamma)\frac{P_{2k}^{(\alpha,\alpha)}(x)}{P_{2k}^{(\alpha,\alpha)}(1)}>0,\quad 0\leqslant x\leqslant 1. \tag{3.1}$$

It can be easily checked that

$$\frac{B_{k+1}(\alpha,\ \gamma)}{B_{k+1}\left(\alpha,\ \frac{1}{2}\right)}<\frac{B_k(\alpha,\ \gamma)}{B_k\left(\alpha,\ \frac{1}{2}\right)},\qquad k\in\mathbf{N}.$$

when $\alpha \geqslant 0$, $0<\gamma<\frac{1}{2}$.

A summation by parts shows that inequality (3.1) is an immediate consequence of

$$\sum_{k=0}^{n} b_k(\alpha) \frac{P_{2k}^{(\alpha,\alpha)}(x)}{P_{2k}^{(\alpha,\alpha)}(1)}>0 \quad \text{for all } n \quad \text{and } 0 \leqslant x \leqslant 1, \tag{3.2}$$

where

$$b_k(\alpha)=B_k\left(\alpha, \frac{1}{2}\right)=\left(2k+\alpha+\frac{1}{2}\right)\frac{\left(\alpha+\frac{1}{2}\right)_k\left(\frac{1}{2}\right)_k}{k!\ (\alpha+1)_k}, \quad k \in \mathbf{N}.$$

In order to prove (3.2), we observe that

$$b_{k+1}(\alpha)-b_k(\alpha)=\frac{1-4\alpha^2}{8(\alpha+1)}\frac{\left(\alpha+\frac{1}{2}\right)_k\left(\frac{1}{2}\right)_k}{(k+1)!\ (\alpha+2)_k}, \quad k \in \mathbf{N}. \tag{3.3}$$

and

$$\lim_{k \to +\infty} b_k(\alpha)=\frac{2\Gamma(\alpha+1)}{\Gamma\left(\frac{1}{2}\right)\Gamma\left(\alpha+\frac{1}{2}\right)}. \tag{3.4}$$

Then we set

$$S_n^\alpha(x)=\sum_{k=0}^{n} \frac{P_{2k}^{(\alpha,\alpha)}(x)}{P_{2k}^{(\alpha,\alpha)}(1)}, \quad n \in \mathbf{N}.$$

It is well-known that

$$S_n^\alpha(x)>0, \quad \text{for all } n, \quad 0 \leqslant x \leqslant 1,\ \alpha \geqslant 0. \tag{3.5}$$

(This is the even part of (3) in [8].)

When $\alpha>\frac{1}{2}$ the sequence $b_k(\alpha)$, $k \in \mathbf{N}$. is strictly decreasing hence (3.5) implies (3.2) by a summation by parts. Note that for $\alpha=\frac{1}{2}$, (3.2) is the classical inequality

$$\sum_{k=0}^{n} \frac{\sin (2k+1)\theta}{(2k+1)\sin \theta}>0, \quad 0 \leqslant \theta \leqslant \frac{\pi}{2}.$$

Next we assume that $0 \leqslant \alpha<\frac{1}{2}$.

It follows from (1.5) and Feldheim's integral [8]

$$\frac{P_n^{(a,a)}(\cos \theta)}{P_n^{(a,a)}(1)}=\frac{2\Gamma(a+1)}{\Gamma(a-b)\Gamma(b+1)} \times$$

$$\int_0^{\frac{\pi}{2}} \sin^{2b+1} t \cos^{2a-2b-1} t (1-\sin^2\theta\cos^2 t)^{\frac{n}{2}} \times$$

$$\frac{P_n^{(b,b)}(\cos\theta(1-\sin^2\theta\cos^2 t)^{-\frac{1}{2}})}{P_n^{(b,b)}(1)} \mathrm{d}t,$$

$$a > b > -1, 0 \leqslant \theta \leqslant \pi, \quad (3.6)$$

that

$$S_n^\alpha(x) \geqslant -\frac{1}{2}, \quad \text{for all } n \quad \text{and } 0 \leqslant x \leqslant 1. \quad (3.7)$$

Now summing by parts the left hand side of (3.2) and taking into consideration (3.3), (3.7) and the standard estimate

$$S_k^\alpha(x) \leqslant k+1, \quad \text{for all } k,$$

we obtain

$$\sum_{k=0}^{n} b_k(\alpha) \frac{P_{2k}^{(\alpha,\alpha)}(x)}{P_{2k}^{(\alpha,\alpha)}(1)}$$

$$= \sum_{k=0}^{n-1} (b_k(\alpha) - b_{k+1}(\alpha)) S_k^\alpha(x) + b_n(\alpha) S_n^\alpha(x)$$

$$\geqslant \sum_{k=0}^{n-1} (k+1)(b_k(\alpha) - b_{k+1}(\alpha)) + \frac{b_n(\alpha)}{2}$$

$$= \frac{b_n(\alpha)}{2} - \frac{1-4\alpha^2}{8(\alpha+1)} \sum_{k=0}^{n-1} \frac{\left(\frac{1}{2}\right)_k \left(\alpha+\frac{1}{2}\right)_k}{(\alpha+2)_k k!}. \quad (3.8)$$

Since the sequence $b_n(\alpha)$ is strictly increasing, it is easy to see that the sequence

$$l_n^\alpha = \frac{b_n(\alpha)}{2} - \sum_{k=0}^{n-1} (k+1)(b_{k+1}(\alpha) - b_k(\alpha))$$

is strictly decreasing.

Now taking into account (3.4) and the fact that

$$\sum_{k=0}^{+\infty} \frac{\left(\frac{1}{2}\right)_k \left(\alpha+\frac{1}{2}\right)_k}{(\alpha+2)_k k!} = {}_2F_1\left(\frac{1}{2}, \alpha+\frac{1}{2}; \alpha+2; 1\right) = \frac{\Gamma(\alpha+2)}{\Gamma\left(\alpha+\frac{3}{2}\right)\Gamma\left(\frac{3}{2}\right)},$$

we get

$$l_n^\alpha \geqslant \lim_{n\to+\infty} l_n^\alpha = \frac{\frac{1}{2}+\alpha}{B\left(\frac{1}{2}, \frac{1}{2}+\alpha\right)}. \quad (3.9)$$

The combination of (3. 8) with (3. 9) gives

$$\sum_{k=0}^{+\infty} b_k(\alpha) \frac{P_{2k}^{(\alpha,\alpha)}(x)}{P_{2k}^{(\alpha,\alpha)}(1)} > \frac{\frac{1}{2}+\alpha}{B\left(\frac{1}{2}, \frac{1}{2}+\alpha\right)}$$

which establishes (3. 2). The proof of Theorem 1 is then complete.

□

§ 4. Proof of Theorem 2

We follow an argument similar to that in the proof of Theorem 1 by making use of (1. 6). More specifically, by virtue of Watson's formula (2. 9), inequality (2. 5) reduces to

$$\sum_{k=0}^{n} b_k P_{4k}(x) > 0, \quad 0 \leqslant x \leqslant 1, \tag{4. 1}$$

where

$$b_k = (8k+1) \frac{\left(\frac{1}{2}\right)_k \left(\frac{1}{4}\right)_k}{k!\ \left(\frac{3}{4}\right)_k}, \quad k \in \mathbf{N}.$$

A simple calculation shows that (b_k) is an increasing sequence, such that

$$b_{k+1} - b_k = \frac{1}{2} \frac{\left(\frac{1}{2}\right)_k \left(\frac{1}{4}\right)_k}{(k+1)!\ \left(\frac{7}{4}\right)_k}, \quad k \in \mathbf{N}. \tag{4. 2}$$

On the other hand it can be easily verified that

$$\lim_{n \to +\infty} b_n = \frac{8\Gamma\left(\frac{3}{4}\right)}{\Gamma\left(\frac{1}{2}\right)\Gamma\left(\frac{1}{4}\right)}. \tag{4. 3}$$

Let

$$\sigma_k(x) = \sum_{j=0}^{k} P_{4j}(x), \quad k \in \mathbf{N}.$$

It follows from (1. 6) that

$$\sigma_n(x) > \frac{1}{2}, \quad 0 \leqslant x \leqslant 1, \ n \in \mathbf{N}. \tag{4. 4}$$

Now summing by parts the left hand side of (4.1) using (4.2) (4.4) and the obvious inequality

$$\sigma_k(x)\leqslant k+1,\quad k\in\mathbf{N}$$

we get

$$\sum_{k=0}^{n} b_k P_{4k}(x) = \sum_{k=0}^{n-1}(b_k - b_{k+1})\sigma_k(x) + b_n\sigma_n(x) > \sum_{k=0}^{n-1}(k+1)(b_k - b_{k+1}) + \frac{b_n}{2}. \tag{4.5}$$

Clearly the sequence

$$q_n = \frac{b_n}{2} - \sum_{k=0}^{n-1}(k+1)(b_{k+1} - b_k) = \frac{b_n}{2} - \frac{1}{2}\sum_{k=0}^{n-1}\frac{\left(\frac{1}{2}\right)_k\left(\frac{1}{4}\right)_k}{\left(\frac{7}{4}\right)_k k!} \tag{4.6}$$

decreases with n.

Since

$$\sum_{k=0}^{+\infty}\frac{\left(\frac{1}{2}\right)_k\left(\frac{1}{4}\right)_k}{\left(\frac{7}{4}\right)_k k!} = {}_2F_1\left(\frac{1}{2},\frac{1}{4};\frac{7}{4};1\right) = 6\,\frac{\Gamma\left(\frac{3}{4}\right)}{\Gamma\left(\frac{1}{2}\right)\Gamma\left(\frac{1}{4}\right)},$$

it follows from (4.3) (4.6) and the above that

$$q_n \geqslant \lim_{n\to+\infty} q_n = \frac{1}{B\left(\frac{1}{2},\ \frac{1}{4}\right)}. \tag{4.7}$$

Finally, combining (4.5) with (4.7) we conclude that

$$\sum_{k=0}^{n} b_k P_{4k}(x) > \frac{1}{B\left(\frac{1}{2},\frac{1}{4}\right)} = 0.190\,6\cdots. \quad \square$$

§ 5. Proof of Theorem 3

Let

$$T_n(x) = \sum_{k=0}^{n}(2k+\alpha+\beta+1)\frac{(\alpha+\beta+1)_k(\gamma)_k}{k!(\alpha+\beta-\gamma+2)_k}\cdot\frac{P_k^{(\alpha,\beta)}(x)}{P_k^{(\alpha,\beta)}(1)}$$

for $-1\leqslant x\leqslant 1$ and $n\in\mathbf{N}$.

By virtue of (2.1) and (2.7), (2.5) reduces to

$$T_n(x)>0 \quad \text{for all } n.$$

The condition $\alpha+\beta+1>0$ implies $\max(\alpha, \beta)\geqslant-\frac{1}{2}$. Then, by [12, 7.32.2.] we have

$$\frac{P_k^{(\alpha,\beta)}(x)}{P_k^{(\alpha,\beta)}(1)}\leqslant 1, \quad k\in\mathbf{N}.$$

From this it follows that

$$T_n(x)\geqslant T_n(1), \quad -1\leqslant x\leqslant 1,$$

for $\alpha+\beta+1>0$ and $-1<\gamma<0$. So it suffices to prove

$$T_n(1)=\sum_{k=0}^{n}(2k+\alpha+\beta+1)\frac{(\alpha+\beta+1)_k(\gamma)_k}{k!(\alpha+\beta-\gamma+2)_k}>0 \quad \text{for all } n\geqslant 1.$$

We have

$$\begin{aligned}
T_n(1)&>(\alpha+\beta+1)+\sum_{k=1}^{+\infty}(2k+\alpha+\beta+1)\frac{(\alpha+\beta+1)_k(\gamma)_k}{k!(\alpha+\beta-\gamma+2)_k}= \\
&=(\alpha+\beta+1)\Big[{}_2F_1(\alpha+\beta+1,\gamma;\alpha+\beta-\gamma+2;1)+ \\
&\quad\frac{2\gamma}{\alpha+\beta-\gamma+2}\,{}_2F_1(\alpha+\beta+2,\gamma+1;\alpha+\beta-\gamma+3;1)\Big] \\
&=(\alpha+\beta+1)\left[\frac{\Gamma(\alpha+\beta-\gamma+2)\Gamma(1-2\gamma)}{\Gamma(\alpha+\beta-2\gamma+2)\Gamma(1-\gamma)}+\right. \\
&\quad\left.\frac{2\gamma}{\alpha+\beta-\gamma+2}\frac{\Gamma(\alpha+\beta-\gamma+3)\Gamma(-2\gamma)}{\Gamma(\alpha+\beta-2\gamma+2)\Gamma(1-\gamma)}\right] \\
&=0.
\end{aligned}$$

Now using the differentiation formula

$$\frac{\mathrm{d}}{\mathrm{d}x}P_k^{(\alpha,\beta)}(x)=\frac{1}{2}(k+\alpha+\beta+1)P_{k-1}^{(\alpha+1,\beta+1)}(x).$$

we see that, the inequality

$$\frac{\mathrm{d}}{\mathrm{d}x}T_n(x)<0, \quad -1<x<1, \quad \text{and} \quad n\in\mathbf{N}^*$$

is equivalent to

$$\sum_{k=0}^{n}c_k\frac{P_k^{(\alpha+1,\beta+1)}(x)}{P_k^{(\beta+1,\alpha+1)}(1)}>0 \tag{5.1}$$

for $-1<x<1$, where

$$c_k=(2k+\alpha+\beta+3)\frac{(\alpha+\beta+3)_k(\gamma+1)_k(\beta+2)_k}{(\alpha+2)_k(\alpha+\beta-\gamma+3)_k k!}.$$

It is known that

$$\sum_{k=0}^{n} \frac{P_k^{(\alpha+1,\beta+1)}(x)}{P_k^{(\beta+1,\alpha+1)}(1)} > 0, \quad -1 < x \leqslant 1, \quad \text{for all } n$$

(see [10, Theorem 4]).

A summation by parts shows that (5.1) is true for all the values of α, β and γ such that

$$c_{k+1} \leqslant c_k, \quad k \in \mathbf{N}. \tag{5.2}$$

A routine calculation proves that the above inequality is equivalent to

$$\gamma \leqslant (\alpha-\beta-1)\frac{k^2+(\alpha+\beta+4)k+\alpha+\beta+3}{2k^2+(2\alpha+2\beta+8)k+(\beta+3)(\alpha+\beta+3)}.$$

From this we conclude that (5.2) is valid for the values of α, β and γ stated in assertions (1) and (2). The proof of Theorem 3 is complete. □

References

[1] Askey R. Positivity of the Cotes Numbers for some Jacobi Abscissas. Numer. Math. 1972, 19: 46 — 48. et seq. J. Inst. Maths. Applics, 1979, 24: 95—98.

[2] Askey R. Orthogonal polynomials and special functions. Regional Conf. Lect. Appl. Math, 21, Philadelphia: SIAM. 1975.

[3] Askey R, Steinig J. Some positive trigonometric sums. Trans. Amer. Math. Soc. 1974, 187: 295—307.

[4] Brown G, Hewitt E. A class of positive trigonometric sums. Math. Ann, 1984, 268: 91—122.

[5] Brown G, Koumandos S, Wang K-Y. On the positivity of some basic Legendre polynomial sums. Journal of the London Mathematical Society, 1999, 59(3): 939—954.

[6] Fejér L. Über die Laplacesche Reihe. Math. Ann., 1909, 67: 76—109. Gesammelte Arbeiten, I, 503—537. Basel: Birkhäuser, 1970.

[7] Fejér L. Mechanische Quadraturen mit positiven Cotesschen Zahlen. Math. Z., 1933, 37: 287—309. Gesammelte Arbeiten Ⅱ, 457—478.

[8] Feldheim E. On the positivity of certain sums of ultraspherical polynomials. J. Analyse Math. 1963, 11: 275—284. reprinted in: G. Szegö, Collected Papers V3, 821—830. Boston: Birkhäuser, 1982.

[9] Gasper G. Positivity and special functions. In: Theory and Application of Special Functions (ed. R. Askey) 375 — 433, New York: Academic Press.

[10] Gasper G. Positive sums of the classical orthogonal polynomials. SIAM J. Math Anal, 1977, 8: 423—447.

[11] Szegö G. Asymptotische Entwicklungen der Jacobischen Polynome. Schriften der Königsberger Gelehrten Gesellschaft. Naturw. Klasse 1933, 10: 35 — 111. Also in: Collected Papers of Gabor Szegö, Vol. Ⅱ, 401—478. Boston: Birkhäuser, 1982.

[12] Szegö G. Orthogonal polynomials. Amer. Math. Soc. Colloq. Publ. 23, 4th edition. Providence, R. I.: Amer. Math. Soc, 1975.

[13] Vietoris L. Über das Vorzeichen gewisser trigonometrischer Summen. S-B. Österr. Akad. Wiss. 1958, 167: 125 — 135. Teil Ⅱ: Anzeiger Österr. Akad. Wiss. 1959, 167: 92—103.

[14] Watson G. N. A note on generalized hypergeometric series. Proc. London Math. Soc., 1925, 23: 13—15.

J. Austral. Math. Soc., Series A, 1997, 62: 1—12.

Fejér-Jackson 不等式的一个推广

An Extension of the Fejér-Jackson Inequality①

Abstract Best-possible results are established for positivity of the partial sums of $\sum \sin k\theta(k+\alpha)^{-1}$. In fact odd sums are positive for $-1\leqslant\alpha\leqslant\alpha_0=2.1\cdots$, while sums with $2k$ terms are positive on the subinterval $[0, \pi-2\mu_0\pi(4k+1)^{-1}]$, $\mu_0=0.812\,8\cdots$. This is analagous to the Gasper extension of the Szegö-Rogosinski-Young inequality for cosine sums.

§ 1. Introduction

It is, of course, well-known that, for every positive integer n,

$$\sum_{k=1}^{n}\frac{\sin k\theta}{k}>0,\quad 0<\theta<\pi. \tag{1.1}$$

This inequality was first mentioned by Fejér in 1910 (see [3]), was proved by Jackson, and subsequently revisited by many mathematicians who have offered different proofs. In parallel, for cosine series, Young [6] proved the following inequality:

$$1+\sum_{k=1}^{n}\frac{\cos k\theta}{k}\geqslant 0,\quad 0\leqslant\theta\leqslant\pi. \tag{1.2}$$

---

① 本文与 Brown G 合作.

Received: 1992-12-01; Revised: 1994-02-09.

Rogosinski and Szegö [5] extended Young's inequality to

$$\frac{1}{1+\alpha}+\sum_{k=1}^{n}\frac{\cos k\theta}{k+\alpha}\geqslant 0,\quad 0\leqslant\theta\leqslant\pi,\quad -1<\alpha\leqslant 1. \tag{1.3}$$

The case $\alpha=1$ is interesting but Gasper showed in [4] that the result admits considerable improvement. In fact he showed that (1.3) holds for $-1<\alpha\leqslant\tilde{\alpha}=4.567\cdots$, and this $\tilde{\alpha}$ is best possible.

In the present paper, we will extend the Fejér-Jackson inequality in a similar way. Thus we shall be concerned with the partial sums

$$T_n^{\alpha}(\theta)=\sum_{k=1}^{n}\frac{\sin k\theta}{k+\alpha},\quad \alpha>-1, n\in\mathbf{N}. \tag{1.4}$$

The odd partial sums are positive for $-1<\alpha\leqslant\alpha_0$, where $\alpha_0=2.1\cdots$ is best possible. The result for even partial sums holds only on a subinterval $[0,\ \pi-2\mu_0\pi(4n+1)^{-1}]$, where $\mu_0=0.812\ 825\ 2\cdots$ is best possible.

To make matters precise we must define three constants λ_0, μ_0, α_0. The first of these is the solution of the equation

$$(1+\lambda)\pi=\tan(\lambda\pi),\quad 0<\lambda<\frac{1}{2},$$

and it is easy to see that $\lambda_0=0.430\ 296\ 7\cdots$ is the point at which the function $\frac{\sin\lambda\pi}{1+\lambda}$ attains its maximum for $0<\lambda<\frac{1}{2}$. We define μ_0 to be the solution of

$$\frac{\sin\mu\pi}{\mu\pi}=\frac{\sin\lambda_0\pi}{(1+\lambda_0)\pi},$$

and α_0 to be the solution of the equation

$$\sum_{k=1}^{+\infty}\frac{2k}{(2k-1+\alpha)(2k+\alpha)(2k+1+\alpha)}=\frac{\sin\lambda_0\pi}{2(1+\lambda_0)\pi}.$$

The main results can now be stated.

Theorem 1 If $-1<\alpha\leqslant\alpha_0$ then

$$T_{2n-1}^{\alpha}(\theta)>0,\quad 0<\theta<\pi,\quad n\in\mathbf{N}.$$

Theorem 2 If $-1<\alpha\leqslant\alpha_0$ then

$$T_{2n}^{\alpha}(\theta)>0,\quad 0<\theta\leqslant\pi-\frac{\mu_0\pi}{2n+0.5}.$$

Theorem 3 If $\alpha>\alpha_0$ then there exists an infinite subset $N\subset\mathbf{N}$

such that

$$T_{2n-1}^{\alpha}\left(\pi-\frac{(1+\lambda_0)\pi}{2n-\frac{1}{2}}\right)<0, \qquad n\in\mathbf{N}.$$

Theorem 4 If $0<\gamma<\mu_0$ then there exists an α near to but strictly smaller than α_0 such that

$$T_{2n}^{\alpha}\left(\frac{\pi-\gamma\pi}{2n+0.5}\right)<0$$

for an infinite number on n.

From Theorem 3 we see that α_0 is best possible in Theorem 1. Theorem 4 shows that μ_0 is best possible in Theorem 2.

The particular case $\alpha=1$ has been considered by Brown and Wilson [2]. They obtained the following conclusion:

$$T_{2n-1}^{1}(\theta)>0, \qquad 0<\theta<\pi; \qquad T_{2n}^{1}(\theta)>0, \qquad 0<\theta<\pi-\frac{\pi}{2n}.$$

§ 2. Basic lemmas

Lemma 1 For $n\geqslant 3$, $\alpha>-1$, double partial summation gives

$$T_n^{\alpha}(\theta)=\sum_{k=1}^{n-2}a_k(\alpha)\sigma_k(\theta)+\frac{\sigma_{n-1}(\theta)}{(n-1+\alpha)(n+\alpha)}+\frac{S_n(\theta)}{n+\alpha} \tag{2.1}$$

where

$$a_k(\alpha)=\frac{2}{(k+\alpha)(k+1+\alpha)(k+2+\alpha)},$$

$$S_k(\theta)=\sum_{j=1}^{k}\sin j\theta, \quad \sigma_k(\theta)=\sum_{j=1}^{k}S_j(\theta).$$

Lemma 2 For $0<\theta<\pi$,

$$\sigma_k(\theta)>0. \tag{2.2}$$

Lemma 3 Let $0<\delta<2$, $t_n=\dfrac{\delta\pi}{n+\frac{1}{2}}$ and $\theta_n=\pi-t_n$. Then

$$\lim_{n\to+\infty}\frac{\sigma_k(\theta_n)}{\sin\theta_n}=\begin{cases}\dfrac{k+1}{2}, & \text{if } k \text{ is odd},\\ 0, & \text{if } k \text{ is even},\end{cases} \tag{2.3}$$

and

$$0<\frac{\sigma_k(\theta_n)}{\sin\theta_n}<k+1,\quad 1\leqslant k\leqslant n,\ n\geqslant 7. \tag{2.4}$$

All these lemmas admit simple direct proofs. We learned of (6) from [1, (1.9) p. 8]. The inequality was found by Lukacs and published in Fejér's paper [3].

§ 3. Proof of Theorems 3 and 4

Proof of Theorem 3 Here we write $\theta_n=\pi-\dfrac{(1+\lambda_0)\pi}{2n-\frac{1}{2}}$. Suppose $T^{\alpha}_{2n-1}(\theta_n)>0$ for all sufficiently large $n\in\mathbf{N}$. Then by (2.1) we get

$$\sum_{k=1}^{2n-3}a_k(\alpha)\frac{\sigma_k(\theta_n)}{\sin\theta_n}+\frac{1}{(2n-2+\alpha)(2n-1+\alpha)}\frac{\sigma_{2n-2}(\theta_n)}{\sin\theta_n}>$$
$$-\frac{1}{2n-1+\alpha}\frac{S_{2n-1}(\theta_n)}{\sin\theta_n}, \tag{3.1}$$

and the right hand side of (3.1) tends to $\dfrac{\sin\lambda_0\pi}{2(1+\lambda_0)\pi}$ as n tends to infinity.

By Lemma 3 and using the Dominated Convergence Theorem we deduce from (3.1) that

$$\sum_{k=1}^{+\infty}ka_{2k-1}(\alpha)\geqslant\frac{\sin\lambda_0\pi}{2(1+\lambda_0)\pi}. \tag{3.2}$$

Noticing that $ka_{2k-1}(\alpha)$ is decreasing when α is increasing, we conclude that (3.2) is equivalent to $\alpha\leqslant\alpha_0$. This completes the proof.

Proof of Theorem 4 Write $u_n=\dfrac{\gamma\pi}{2n+\frac{1}{2}}$, $0<\gamma<\mu_0$. For $n\geqslant 2$, we have

$$\frac{1}{\sin u_n}T^{\alpha}_{2n}(\pi-u_n)$$
$$=\sum_{k=1}^{2n-2}a_k(\alpha)\frac{\sigma_k(\pi-u_n)}{\sin u_n}+\frac{1}{2n-1+\alpha}\frac{\sigma_{2n-1}(\pi-u_n)}{\sin u_n}+\frac{1}{2n+\alpha}\frac{S_{2n}(\pi-u_n)}{\sin u_n}.$$

Let us write $F(\alpha)=\sum_{k=1}^{+\infty}ka_{2k-1}(\alpha)$. Then by Lemma 3 and the Dominated

Convergence Theorem we get

$$\lim_{n\to+\infty}\frac{1}{\sin u_n}T_{2n}^{\alpha}(\pi-u_n)=F(\alpha)-\frac{\sin\gamma\pi}{2\gamma\pi}. \tag{3.3}$$

As we chose $F(\alpha_0)=\dfrac{\sin\mu_0\pi}{2\mu_0\pi}$, it follows that $T_{2n}^{\alpha}(\pi-u_n)$ is negative for α near to but strictly less than α_0 and n big enough.

§ 4. Proof of theorem 1

We achieve the proof by developing a sequence of lemmas.

Lemma 4　Let

$$-\frac{\mathrm{d}}{\mathrm{d}\alpha}T_n^{\alpha}(\theta)=U_n^{\alpha}(\theta)=\sum_{k=1}^{n}\frac{\sin k\theta}{(k+\alpha)^2},\quad \alpha>-1,\quad n\in\mathbf{N}.$$

If $-1<\alpha\leqslant 2.5$ then

$$U_{2n-1}^{\alpha}(\theta)>0,\qquad 0<\theta<\pi,\ n\in\mathbf{N}.$$

Proof　Simple direct estimates for the other cases allow us to assume $n\geqslant 3$ and $1<\alpha\leqslant 2.5$. By using partial summation twice we get

$$U_{2n-1}^{\alpha}(\theta)=\sum_{k=1}^{2n-3}b_k(\alpha)\sigma_k(\theta)+(c_{2n-2}(\alpha)-c_{2n-1}(\alpha))\sigma_{2n-2}(\theta)+ c_{2n-1}(\alpha)S_{2n-1}(\theta),$$

where $b_k(\alpha)=c_k(\alpha)-2c_{k+1}(\alpha)+c_{k+2}(\alpha)$, $c_k(\alpha)=(k+\alpha)^{-2}$.

Noticing

$$S_{2n-1}(\theta)=\frac{\cos\dfrac{\theta}{2}-\cos\left(2n-\dfrac{1}{2}\right)\theta}{2\sin\dfrac{\theta}{2}}\geqslant-\frac{1}{2}\tan\frac{\theta}{4},$$

we get

$$2\cot\frac{\theta}{4}\cdot U_{2n-1}^{\alpha}(\theta)>(b_1(\alpha)+b_3(\alpha))8\cos^2\frac{\theta}{4}\cos\frac{\theta}{2}-\frac{1}{(2n-1+\alpha)^2},\qquad n\geqslant 3.$$

If $0<\theta\leqslant 0.85\pi$ and $1\leqslant\alpha\leqslant 2$ then

$$2\cot\frac{\theta}{4}U_{2n-1}^{\alpha}(\theta)>(b_1(2)+b_3(2))\times 1.151\,77-\frac{1}{36}>0,\qquad n\geqslant 3.$$

When $2<\alpha\leqslant 2.5$ and $0<\theta\leqslant 0.85\pi$, since

$$b_1(\alpha)+b_3(\alpha)\geqslant b_1(2.5)+b_3(2.5)=0.019\,42$$

we have

$$2\cot\frac{\theta}{4}U_{2n-1}^{\alpha}(\theta)>0.019\,42\times 1.151\,77-\frac{1}{49}>0,\qquad n\geqslant 3.$$

On the other hand, for $0.85\pi<\theta<\pi$, we set $t=\pi-\theta$. Then

$$U_{2n-1}^{\alpha}(\theta)\geqslant(b_1(\alpha)+b_3(\alpha))\sin t+\frac{1}{(2n-1+\alpha)^2}\frac{\sin\frac{t}{2}+\sin\left(2n-\frac{1}{2}\right)t}{2\cos\frac{t}{2}}.$$

We see, if $t\leqslant\dfrac{\pi}{2n-\frac{1}{2}}$, the above sum is strictly positive. Assume $\dfrac{\pi}{2n-\frac{1}{2}}<t<0.15\pi$. This occurs only when $n\geqslant 4$. Then we have

$$\frac{1}{t}U_{2n-1}^{\alpha}(\pi-t)>(b_1(\alpha)+b_3(\alpha))\frac{\sin 0.15\pi}{0.15\pi}-\frac{0.514\,21\left(2n-\frac{1}{2}\right)}{(2n-1+\alpha)^2\pi}$$

and we estimate for $1\leqslant\alpha\leqslant 2$, $2\leqslant\alpha\leqslant 2.5$ as before.

Lemma 5 Let

$$f(\theta)=a_1(2.11)\sigma_1(\theta)+a_2(2.11)\sigma_2(\theta)+a_3(2.11)\sigma_3(\theta),$$

$$g_n(\theta)=\frac{\sigma_{2n-2}(\theta)}{2n+0.11}+S_{2n-1}(\theta),\qquad n\geqslant 3.$$

If $0<\theta\leqslant 0.75\pi$ and $n\geqslant 3$ then

$$T_{2n-1}^{\alpha_0}(\theta)>f(\theta)+\frac{1}{2n+1.11}g_n(\theta)>0.$$

Proof By Lemma 4 we have $T_{2n-1}^{\alpha_0}(\theta)>T_{2n-1}^{2.11}(\theta)$. Then using Lemma 1 we get for $n\geqslant 3$,

$$T_{2n-1}^{\alpha_0}(\theta)>T_{2n-1}^{2.11}(\theta)\geqslant f(\theta)+\frac{1}{2n+1.11}g_n(\theta).$$

It is easy to establish that $0.050\,3\sin\theta$ is a lower bound for $f(\theta)$, for $0<\theta\leqslant 0.75\pi$. For $0<\theta<0.5\pi$, this can be combined with the lower bound $-0.5\tan\dfrac{\theta}{4}$ for $g_n(\theta)$ to establish the result. For $0.5\pi<\theta\leqslant 0.75\pi$, we use the estimate

$$g_n(\theta)>\left[\frac{2n-1-\operatorname{cosec}\theta}{(2n+0.11)4\sin^2\frac{\theta}{2}}-\frac{\sin^2\frac{\theta}{4}}{\sin\frac{\theta}{2}-\sin\theta}\right]\sin\theta>-0.300\,6\sin\theta.$$

The next two lemmas are straightforward estimates based on Lemma 1.

Lemma 6 Let $t=\frac{2}{11}(1+\delta)\pi$, $0<\delta\leqslant 0.375$. Then $T_5^{\alpha_0}(\pi-t)>0$.

Lemma 7 Let $t=\frac{(2l-1+\delta)\pi}{2n-0.5}$, $l\in\mathbf{N}$, $0<\delta<1$, $n\geqslant 4$. Define

$$U_n(t)=\frac{\cos^2\frac{t}{2}\cdot T_{2n-1}^{\alpha_0}(\pi-t)}{\sin t},$$

$$D_n(t)=\sum_{k=1}^{+\infty}ka_{2k-1}(\alpha_0)-\frac{\sin\delta\pi}{2(2l-1+\delta)\pi}u_n(t)-$$

$$\frac{1}{2}\sum_{k=1}^{n-2}a_{2k+1}\frac{\sin^2\left(k+\frac{3}{4}\right)t}{\cos\frac{t}{2}}-v_n(t)$$

with

$$u_n(t)=\left(1-\frac{\alpha_0-0.5}{2n-1+\alpha_0}\right)\frac{t}{2\sin\frac{t}{2}},$$

$$v_n(t)=\frac{(\alpha_0-1)\sin t+\sin(2n-1)t}{(2n-2+\alpha_0)(2n-1+\alpha_0)4\sin t},$$

and

$$\Delta(t)=\frac{\alpha_0}{2}\sum_{k=n}^{+\infty}a_{2k-1}+\frac{\sin^2\frac{t}{4}}{2\cos\frac{t}{2}}\sum_{k=1}^{n-2}a_{2k+1}-\sin^2\frac{t}{2}a_1,$$

where $a_k=a_k(\alpha_0)$. Then

$$U_n(t)>\Delta_n(t)+D_n(t).$$

Lemma 8 If $n\geqslant 4$, $0<\delta\leqslant 0.55$ and $t=\frac{(1+\delta)\pi}{2n-0.5}$ then

$$D_n(t)>0.$$

Proof By definition of D_n and the fact that $F(\alpha_0)\geqslant\frac{\sin\delta\pi}{2(1+\delta)\pi}$ we get

$$D_n(t)\cos\frac{t}{2}>F(\alpha_0)(1-u_n(t))\cos\frac{t}{2}-v_n(t)-\frac{1}{2}\sum_{k=1}^{n-2}a_{2k+1}\sin^2\left(k+\frac{3}{4}\right)t.$$

Simple power series estimates yield

$$u_n(t)<1-\frac{1.6}{2n+1.1}+\frac{0.9880}{(2n-0.05)^2-0.9880}, \tag{4.1}$$

$$v_n(t)<\frac{0.4053}{(2n+0.1)(2n+1.1)}, \quad 0<\delta<1, \tag{4.2}$$

$$\cos\frac{t}{2}\geqslant 1-\frac{2.9640}{(2n-0.5)^2}. \tag{4.3}$$

Combining these inequalities with (crude) direct estimates of the upper bound of $0.5\sum_{k=1}^{n-2}a_{2k+1}\sin^2\left(k+\frac{3}{4}\right)t$ for each n from 4 to 12 allows us to verify that $D_n(t)>0$ for $4\leqslant n\leqslant 12$.

Now we assume $n\geqslant 13$ and observe that

$$\begin{aligned}&(2n-0.5)^2D_n(t)\cos\frac{t}{2}\\&>F(\alpha_0)\left(3.2n-4.3640-\frac{0.5950}{2n+1.11}-\frac{0.9761}{(2n-0.5)^2-0.9880}\right)-\\&\frac{(1.55\pi)^2}{8}\sum_{k=1}^{n-2}\frac{1}{k+4.65}-0.4053\\&>0.3476n-0.8821-2.9640\ln\frac{n+2.65}{4.65}.\end{aligned}$$

If we write $\varphi(n)=0.3476n-0.8821-2.9640\ln\frac{n+2.65}{4.65}$,

then we find $\phi(13)=0.0396>0$, $\frac{\mathrm{d}\phi}{\mathrm{d}n}>0$ for $n\geqslant 13$.

Hence we conclude $D_n(t)>0$ for $n\geqslant 13$ and complete the proof.

Lemma 9 If $n\geqslant 4$, $0.55<\delta<1$ and $t=\frac{(1+\delta)\pi}{2n-0.5}$ then $D_n(t)>0$.

Proof It is easy to verify that

$$\sin\left(\delta\pi-\frac{0.5(1+\delta)\pi}{2n-0.5}\right)>(\alpha_0-1)\sin\frac{(1+\delta)\pi}{2n-0.5}$$

for $0.55<\delta\leqslant 0.8$ and $n\geqslant 4$. Hence by (4.2), $-v_n(t)>0$ $(0.55<\delta\leqslant 0.8,\ n\geqslant 4)$.

On the other hand, $\frac{\sin\delta\pi}{2(1+\delta)\pi}\leqslant\frac{\sin 0.55\pi}{2(1.55\pi)}=0.10142$, $\delta>0.55$.

Therefore

$$D_n(t)\geqslant 0.10862-0.10142u_n(t)-\frac{1}{2\cos\frac{t}{2}}\sum_{k=1}^{n-2}a_{2k+1}\sin^2\left(k+\frac{3}{4}\right)t \tag{4.4}$$

when $0.55<\delta\leqslant 0.8$ and $n\geqslant 4$. In this case,

$$u_n(t)<\left(1-\frac{1.6}{2n+1.11}\right)\left(1+\frac{(1.8\pi)^2}{24(2n-0.5)^2-(1.8\pi)^2}\right) \tag{4.5}$$

$$<1-\frac{1.6}{2n+1.11}+\frac{1.3324}{(2n-0.5)^2-1.3324}.$$

According to (4.4) and (4.5), calculating directly we get

$$D_4(t)\geqslant 0.00720+\frac{0.16227}{2n+1.11}-\frac{0.13513}{(2n-0.5)^2-1.3324}-$$

$$0.53776\left(\alpha_3\sin^2\frac{7}{4}t+a_5\right)>0.01592,$$

$$D_5(t)\geqslant 0.02029-0.52999\left(a_3\sin^2\frac{7}{4}t+a_5+a_7\right)>0.01374,$$

$$D_6(t)\geqslant 0.01854-0.51550\left(a_3\sin^2\left(\frac{7}{4}\frac{1.8\pi}{11.5}\right)+a_5+a_7+a_9\right)>0.01229.$$

For $n\geqslant 7$ we have

$$(2n-0.5)^2D_n(t)>0.00720(2n-0.5)^2+0.15099(2n-2.11)-$$

$$4.08649\ln\frac{n+2.65}{4.65}.$$

It is easy to see this is positive by an argument similar to that in the proof of Lemma 8.

Assume now $0.8<\delta<1$. In this case

$$\frac{\sin\delta\pi}{2(1+\delta)\pi}<0.05197,\qquad u_n<1,\qquad v_n<0.00207.$$

Hence

$$D_n(t)>0.05458-5.40181\frac{1}{(2n-0.5)^2}\ln\frac{n+2.65}{4.65}>0.$$

Lemma 10 If $t=\dfrac{(1+\delta)\pi}{2n-0.5}$, $0<\delta<1$ and $n\geqslant 4$, then $\Delta_n(t)>0$.

Proof Since $2.10<\alpha_0<2.11$, we have

$$\frac{1}{2}\alpha_0\sum_{k=n}^{+\infty}a_{2k-1}>2.10\sum_{k=n}^{+\infty}(2k+2.11)^{-3}.$$

It turns out that the ensuing estimate

$$\Delta_n(t)>0.13125(n+1.055)^{-3}+\sin^2\frac{t}{4}\frac{a_3+a_5}{2}-a_2\sin^2\frac{t}{2}$$

leads quickly to the required result.

Lemmas 8～10 treat the case $l=1$. Now we assume $l\geqslant 2$, $t=\frac{(2l-1+\delta)\pi}{2n-0.5}$, $0<\delta<1$. And we keep $t<0.25\pi$. This restriction requires $n\geqslant 7$.

Lemma 11 If $t=\frac{(2l-1+\delta)\pi}{2n-0.5}<0.25\pi$, $0<\delta<1$, and $l\geqslant 2$ then $T_{2n-1}^{\alpha_0}(\pi-t)>0$.

Proof In this case $u(t)<\pi\left(8\sin\frac{\pi}{8}\right)^{-1}$, $\frac{\sin\delta\pi}{2(2l-1+\delta)\pi}<\frac{1}{6\pi}$, $v_n(t)<0.001\,97$ ($n\geqslant 7$) and

$$\frac{1}{2}\sec\frac{t}{2}\sum_{k=1}^{n-2}a_{2k+1}\sin^2\left(k+\frac{3}{4}\right)t<\frac{1}{2}\sec\frac{\pi}{8}\sum_{2}^{+\infty}a_{2k-1}.$$

These estimates combined show that $D_n(t)>a_1$, and since $\Delta_n(t)>-a_1$, we achieve the proof.

Lemmas 5～11 now demonstrate that $T_{2n-1}^{\alpha_0}(\theta)>0$, $0<\theta<\pi$, for $n\geqslant 3$. We complete the proof of Theorem 1 by checking the case $n=2$ directly.

§ 5. Proof of Theorem 2

Because of the many parallels with the proof of Theorem 1 we merely sketch the salient steps. A fuller version of this (and the preceding proofs) is available from the authors on request.

Lemma 12 For U_n^{α} as in Lemma 4, if $-1<\alpha\leqslant 2.3$ then $U_{2n}^{\alpha}(\theta)>0$, $0<\theta\leqslant\pi-\frac{\mu_0\pi}{2n+0.5}$ (recall $\mu\approx 0.812\,825\,2$).

Proof We have, in a way similar to the proof of Lemma 4,

$$2\cot\frac{\theta}{4}U_{2n}^{\alpha}(\theta)>(b_1(\alpha)+b_3(\alpha))8\cos^2\frac{\theta}{4}\cos\frac{\theta}{2}-\frac{1}{(2n+\alpha)^2},\qquad n\geqslant 3.$$

So, we get $U_{2n}^{\alpha}(\theta)>0$ for $0<\theta\leqslant 0.85\pi$ and $-1<\alpha\leqslant 2.3$, $n\geqslant 3$. On the other hand, when $0.85\pi<\theta\leqslant\pi-\frac{\mu_0\pi}{2n+0.5}$ by writing $t=\pi-\theta$ we have

$$U_{2n}^{\alpha}(\theta)>(b_1(\alpha)+b_3(\alpha))\sin t+\frac{1}{(2n+\alpha)^2}\frac{\sin\frac{t}{2}-\sin(2n+0.5)t}{2\cos\frac{t}{2}},$$

and this leads quickly to a proof. The cases $n=1,2$ are handled separately. Now to establish Theorem 4 we need only prove

$$T_{2n}^{\alpha_0}(\theta)>0,\quad 0<\theta\leqslant\left(1-\frac{\mu_0}{2n+0.5}\right)\pi,\quad n\in\mathbf{N}. \tag{5.1}$$

By an argument similar to Lemma 5, we get

Lemma 13 If $0<\theta\leqslant 0.75\pi$ and $n\geqslant 3$ then $T_{2n}^{\alpha_0}(\theta)>0$.

Now we assume $\theta=\pi-t$, $\frac{\mu_0\pi}{2n+0.5}\leqslant t<0.25\pi$ and write $V_n(t)=\cos^2\frac{t}{2}(\sin t)^{-1}T_{2n}^{\alpha_0}(\pi-t)$. By a calculation similar to that in the proof of Lemma 6 we obtain

$$\begin{aligned}V_n(t)>&-a_1\sin^2\frac{t}{2}+\sum_{k=1}^{+\infty}ka_{2k-1}-\sum_{k=n+1}^{+\infty}ka_{2k-1}+\\&\frac{1}{2\cos\frac{t}{2}}\sum_{k=1}^{n-1}a_{2k+1}\left(\sin^2\frac{t}{4}-\sin^2\left(k+\frac{3}{4}\right)\right)+\\&\frac{n}{2(2n-1+\alpha_0)(2n+\alpha_0)}\left(1-\frac{2}{2n+1+\alpha_0}\right)+\\&\frac{\sin 2nt}{4(2n-1+\alpha_0)(2n+\alpha_0)\sin t}\left(1-\frac{2}{2n+1+\alpha_0}\right)+\\&\frac{1}{4(2n+\alpha_0)}-\frac{\sin(2n+0.5)t}{2(2n+\alpha_0)2\sin\frac{t}{2}}.\end{aligned} \tag{5.2}$$

We write

$$\delta_n=\frac{n}{2(2n-1+\alpha_0)(2n+\alpha_0)}\left(1-\frac{2}{2n+1+\alpha_0}\right)-\frac{1}{2}\frac{1}{2n+1+\alpha_0}+\frac{1}{4(2n+\alpha_0)}, \tag{5.3}$$

and note that

$$\delta_n>\frac{0.2225-0.7775(n+0.555)^{-1}}{(2n+\alpha_n)(2n+1+\alpha_0)}. \tag{5.4}$$

We still use the notation Δ_n defined in Lemma 7, and write also

$$G_n(t)=F(\alpha_0)-\frac{\sin(2n+0.5)t}{2(2n+\alpha_0)2\sin\frac{t}{2}}-$$

$$\frac{1}{2\cos\frac{t}{2}}\sum_{k=1}^{n-1}a_{2k+1}\sin^2\left(k+\frac{3}{4}\right)t+\delta_n+$$

$$\left(1-\frac{2}{2n+1+\alpha_0}\right)\frac{\sin 2nt}{4(2n-1+\alpha_0)(2n+\alpha_0)\sin t}.$$

Then we get $$V_n(t)>G_n(t)+\Delta_{n+1}(t).$$

Lemma 14 If $\frac{\mu_0 t}{2n+0.5}\leqslant t<\frac{\pi}{2n+0.5}$, $n\geqslant 3$ then $G_n(t)>0$ and $\Delta_{n+t}(t)>0$.

Lemma 15 If $t=\frac{2l\pi+\delta\pi}{2n+0.5}<0.25\pi$, $l\geqslant 1$ and $0<\delta<1$, then $V_n(t)>0$. In this case, $n\geqslant 4$ and we have $G_n(t)>0.025\,98$ and $\Delta_{n+1}(t)>-0.004\,52$.

A combination of Lemmas 13～15 together with a direct check of low order cases completes the proof of Theorem 2.

Acknowledgment

This work was supported by a prant from the Australian Research Council.

References

[1] Askey R, Gasper G. Inequalities for Polynomials. in: The Bieberbach conjecture. Math. Surveys Monographs 21 (Amer. Math. Soc., Providence), 1986: 7-32.

[2] Brown G, Wilson D C. A class of positive trigonometric sums Ⅱ. Math. Ann., 1989, 285: 57-74.

[3] Fejér L. Einige Sätze, die sich auf das Vorzeichen einer ganzen rationalen Funktion beziehen. Monats. Math. Phys., 1928, 35: 305-344.

[4] Gasper G. Nonnegative sums of cosine, ultraspherical and Jacobi polynomials. J. Math. Anal. Appl. 1969, 26: 60-68.

[5] Rogosinski W, Szegö G. Über die Abschnitte von Potenzreihen, die in einem Kreise beschränkt bleiben. Math. Z. 1928, 28: 73-94.

[6] Young W H. On a certain series of Fourier. Proc. London Math. Soc. 1912, 11: 357-366.

Analysis，1998，18：313－331.

基本超球多项式和的正性

Positivity of Basic Sums of Ultraspherical Polynomials[①]

Abstract We determine the maximum range of λ for the positivity of sums $\sum_{k=0}^{n}\frac{C_k^\lambda(x)}{C_k^\lambda(1)}$, where $C_n^\lambda(x)$ is the ultraspherical polynomial of degree n and order λ. This result resolves a problem posed by Szegö thirty five years ago.

§ 1. Introduction

Consider the sums

$$S_n^\lambda(x)=\sum_{k=0}^{n}\frac{C_k^\lambda(x)}{C_k^\lambda(1)},\quad \text{for}\quad -1<x<1,\quad n\in\mathbf{N},$$

where $C_n^\lambda(x)$ is the ultraspherical polynomial of degree n and order λ.

It has long been known that, for $\lambda\geqslant\frac{1}{2}$, all the sums are positive (cf [2][4][12]) but determination of the best lower bound for λ has been a recurring source of speculation, [3][13][11].

Here we give a complete solution, showing that the sums are positive whenever $\lambda\geqslant\lambda'=\alpha'+\frac{1}{2}$, where α' is the solution of

① 本文与 Brown G 和 Koumandos S 合作.

Received：1995-03-30.

$$\int_0^{j_{\alpha,2}} t^{-\alpha} J_\alpha(t)\,\mathrm{d}t = 0, \tag{1.1}$$

and J_α is the Bessel function of the first kind of order α with $j_{\alpha,2}$ being its second positive root. Moreover for every $\lambda<\lambda'$ the sums take arbitrarily large negative values. Numerical integration shows that $\lambda' = 0.230\,612\,97\cdots$.

The special case $\lambda=\frac{1}{2}$ corresponds to a famous result of Fejér, [10], viz.,

$$\sum_{k=0}^{n} P_k(x) > 0, \quad \text{for} \quad -1<x<1, \quad n\in\mathbf{N}, \tag{1.2}$$

where P_k is the Legendre polynomial of degree k. In work published posthumously on his behalf by Szegö, [11], Feldheim showed how to extend Fejér's result to larger values of λ using the formula

$$\frac{C_n^\nu(\cos\theta)}{C_n^\nu(1)} = \frac{2\Gamma\left(\nu+\frac{1}{2}\right)}{\Gamma\left(\lambda+\frac{1}{2}\right)\Gamma(\nu-\lambda)} \times$$

$$\int_0^{\frac{\pi}{2}} \sin^{2\lambda}\varphi\cos^{2\nu-2\lambda-1}\varphi(1-\sin^2\theta\cos^2\varphi)^{\frac{n}{2}} \frac{C_n^\lambda(\cos\theta(1-\sin^2\theta\cos^2\varphi)^{-\frac{1}{2}})}{C_n^\lambda(1)}\mathrm{d}\varphi,$$

$$\nu>\lambda>-\frac{1}{2}, \quad 0<\theta<\pi.$$

See Feldheim [11] or Vilenkin [15].

Szegö commented in [11] that there exists a critical value λ', $0<\lambda'\leqslant\frac{1}{2}$, such that $S_n^\lambda>0$ for $\lambda>\lambda'$ but $S_n^\lambda(x)$ takes negative values for appropriate x and n when $\lambda<\lambda'$. He gave $\alpha'+\frac{1}{2}$ (α' defined as in (1.1)) as a lower bound for λ' but noted that "evaluation of this number [viz. λ'] seems to be difficult".

We resolve the problem as follows:

Theorem Let $\lambda'=\alpha'+\frac{1}{2}$, where α' is the solution of the equation (1.1). Then

$$\inf\{S_n^\lambda(x): -1\leqslant x\leqslant 1, n\in\mathbf{N}\}=-\infty \text{ for } \lambda<\lambda'' \tag{1.3}$$

and

$$S_n^\lambda(x)>0 \text{ for } \lambda\geqslant\lambda', -1<x<1. \tag{1.4}$$

Note that $S_n^\lambda(-1)=0$ for n odd.

Although the theorem answers the Szegö problem, it may be regarded as only partial progress towards the understanding of something more general, namely what can be said, for α, β negative, concerning the sum of Jacobi polynomials

$$\sum_{k=0}^{n}\frac{P_k^{(\alpha,\beta)}(x)}{P_k^{(\beta,\alpha)}(1)}\geqslant 0,\quad -1\leqslant x\leqslant 1?$$

Some special cases of this inequality have been obtained by Askey in [1][2] and Askey and Gasper [5] and some of its applications have also been mentioned there. Gasper in [13] proved the inequality in question under the more general assumptions $\alpha+\beta\geqslant 0$, $\beta\geqslant-\frac{1}{2}$ or $\alpha+\beta\geqslant-2$, $\beta\geqslant 0$. He also conjectured that this may be true for $-1<\alpha<\frac{1}{2}$, $\beta(\alpha)\leqslant\beta<0$, where $\beta(\alpha)$ is the unique solution of the equation

$$\int_0^{j_{\alpha,2}} t^{-\beta(\alpha)}J_\alpha(t)\,\mathrm{d}t=0,\quad -\frac{1}{2}<\beta(\alpha)<0,$$

$j_{\alpha,2}$ being the second positive zero of the Bessel function $J_\alpha(t)$. In [3], Askey discusses this conjecture, mentioning that it would be interesting to consider it for $-1<\alpha<0$ and some $\beta<0$, predicting, however that this is a quite hard problem.

In [6], the authors of this paper completely solved the Gasper conjecture for $\alpha=-\frac{1}{2}$ by translating the Jacobi polynomial problem into an alternating sum of even ultraspherical polynomials by a quadratic transformation.

In this paper we prove Gasper's conjecture for $\alpha=\beta$.

§ 2. Proof of the theorem

It is well-known that [11]

$$\lim_{n\to+\infty}\frac{\theta}{n}\sum_{k=0}^{n}\frac{C_k^\lambda\left(\cos\frac{\theta}{n}\right)}{C_k^\lambda(1)}=2^\alpha\Gamma(\alpha+1)\int_0^\theta t^{-\alpha}J_\alpha(t)\,dt,\quad \alpha=\lambda-\frac{1}{2}.$$

It follows from this that

$$\lim_{n\to+\infty}\sum_{k=0}^{n}\frac{C_k^\lambda\left(\cos\frac{j_{\alpha,2}}{n}\right)}{C_k^\lambda(1)}$$
$$=\left\{2^\alpha\Gamma(\alpha+1)\int_0^{j_{\alpha,2}}t^{-\alpha}J_\alpha(t)\,dt\right\}\cdot\lim_{n\to+\infty}\frac{n}{j_{\alpha,2}}=-\infty$$

for all $\alpha<\alpha'$. Setting $\lambda'=\alpha'+\frac{1}{2}$, we obtain (1.3).

We shall prove that

$$\sum_{k=0}^{n}\frac{C_k^\lambda(\cos\theta)}{C_k^\lambda(1)}>0\quad\text{for}\quad 0<\theta<\pi,\ n\in\mathbf{N},\ \lambda\geqslant\lambda'.$$

It is clear that it suffices to prove this inequality for $\lambda=\lambda'$. Then, an application of the Feldheim-Vilenkin formula establishes the inequality for all $\lambda\geqslant\lambda'$.

At this stage we distinguish two cases:

The case $0\leqslant\theta\leqslant\frac{\pi}{2}$.

Let

$$S_n^\lambda(\theta)=\sum_{k=0}^{n}\frac{C_k^\lambda(\cos\theta)}{C_k^\lambda(1)}.$$

By applying the Dirichlet-Mehler formula for the ultraspherical polynomials [9]

$$\frac{C_k^\lambda(\cos\theta)}{C_k^\lambda(1)}=\frac{2^\lambda(\sin\theta)^{1-2\lambda}}{B\left(\lambda,\frac{1}{2}\right)}\int_0^\theta\frac{\cos(k+\lambda)t}{(\cos t-\cos\theta)^{1-\lambda}}\,dt \tag{2.1}$$

we obtain

$$S_n^\lambda(\theta)=1-Q_0(\theta,\lambda)+Q_n(\theta,\lambda), \tag{2.2}$$

where

$$Q_n(\theta,\lambda)=\frac{(\sin\theta)^{1-2\lambda}}{2^{1-\lambda}B\left(\lambda,\frac{1}{2}\right)}\int_0^\theta\frac{\sin\left(n+\lambda+\frac{1}{2}\right)t}{\sin\frac{t}{2}(\cos t-\cos\theta)^{1-\lambda}}\,dt$$

for $n\in\mathbf{N}$.

We first give an estimate of the term in (2.2) which is independent of n.

Lemma 1

$$1-Q_0(\theta,\lambda)\geqslant\frac{1}{2}-\lambda,\ 0\leqslant\theta\leqslant\frac{\pi}{2}.$$

Proof Clearly,

$$Q_0(\theta,\lambda)=\frac{1}{2}+\frac{(\sin\theta)^{1-2\lambda}}{2^{1-\lambda}B\left(\lambda,\frac{1}{2}\right)}\int_0^\theta\frac{\sin\lambda t}{(\cos t-\cos\theta)^{1-\lambda}}\cot\frac{t}{2}\mathrm{d}t.$$

Since

$$\int_0^\theta\frac{\cos\frac{t}{2}}{(\cos t-\cos\theta)^{1-\lambda}}\mathrm{d}t=\frac{B\left(\lambda,\frac{1}{2}\right)}{2^{1-\lambda}\left(\sin\frac{\theta}{2}\right)^{1-2\lambda}},\tag{2.3}$$

and the function $\dfrac{\sin\lambda t}{\sin\frac{t}{2}}$ is increasing in $\left[0,\ \frac{\pi}{2}\right]$, it follows that

$$1-Q_0(\theta,\lambda)\geqslant\frac{1}{2}-\frac{\sin\lambda\theta\left(\cos\frac{\theta}{2}\right)^{1-2\lambda}}{2\sin\frac{\theta}{2}}\geqslant\frac{1}{2}-\lambda.\quad\square$$

We see that, in order to establish the positivity of $S_n^\lambda(\theta)$, we need to find appropriate estimates of the functions $Q_n(\theta,\ \lambda)$. It is convenient to give first the proof of two Lemmas.

Lemma 2 For $0<t<\theta\leqslant\frac{\pi}{2}$, we have

$$\frac{1}{(\cos t-\cos\theta)^{1-\lambda}}=\left(\frac{2}{\theta^2-t^2}\right)^{1-\lambda}\cdot\{1+\theta^2p(\theta,\lambda)+t^2q(\theta,t,\lambda)\},$$

where

$$p(\theta,\lambda)=\sum_{k=0}^{+\infty}\frac{(1-\lambda)_{k+1}}{(k+1)!}\theta^{2k}a(\theta)^{k+1},$$

$$q(\theta,t,\lambda)=\sum_{k=1}^{+\infty}\frac{(1-\lambda)_k}{k!}\sum_{j=0}^{k-1}\mathrm{C}_k^{j+1}\theta^{2(k-1-j)}t^{2j}a(\theta)^{k-1-j}b(\theta,t)^{j+1}$$

and

$$a(\theta)=2\sum_{k=0}^{+\infty}\frac{(-1)^k}{(2k+4)!}\theta^{2k}$$

$$b(\theta) = 2\sum_{k=0}^{+\infty} \frac{(-1)^k}{(2k+4)!} \sum_{j=0}^{k} \theta^{2(k-j)} t^{2j}$$

As usual, $(c)_k = c(c+1)\cdots(c+k-1)$.

Proof By Taylor's expansion, we have

$$\cos t - \cos\theta = \frac{1}{2}(\theta^2 - t^2)(1 - u(\theta,t)), \quad 0 < t < \theta \leqslant \frac{\pi}{2},$$

where

$$u(\theta,t) = 2\sum_{k=0}^{+\infty} \frac{(-1)^k}{(2k+4)!} \sum_{j=0}^{k+1} \theta^{2(k+1-j)} t^{2j}.$$

Now, it is easy to see that

$$u(\theta,t) = \theta^2 a(\theta) + t^2 b(\theta,t).$$

Since for all $k \in \mathbf{N}$,

$$u(\theta,t)^k = \theta^{2k} a(\theta)^k + t^2 \sum_{j=0}^{k-1} C_k^{j+1} \theta^{2(k-1-j)} t^{2j} a(\theta)^{k-1-j} b(\theta,t)^{j+1},$$

the identity of the Lemma follows. □

Lemma 3 The function $q(\theta, \theta, \lambda)$ is bounded in $\left[0, \frac{\pi}{2}\right]$, so that

$$q(\theta,\theta,\lambda) \leqslant \beta(\lambda) = 0.103\ 25\cdots, \quad 0 \leqslant \theta \leqslant \frac{\pi}{2}, (\lambda = 0.230\ 612\cdots).$$

Proof We observe that

$$q(\theta,\theta,\lambda) = (1-\lambda) b(\theta,\theta) + \sum_{k=2}^{+\infty} \frac{(1-\lambda)_k}{k!} \theta^{2k-2} \{[a(\theta) + b(\theta,\theta)]^k - a(\theta)^k\}$$

and

$$b(\theta,\theta) = 2\sum_{k=0}^{+\infty} (-1)^k \frac{k+1}{(2k+4)!} \theta^{2k}.$$

It is easy to see that

$$a(\theta) \leqslant a_1(\theta) \quad \text{and} \quad b(\theta,\theta) \leqslant b_1(\theta),$$

where

$$a_1(\theta) = \frac{1}{12} - 0.002\ 65\theta^2, \quad b_1(\theta) = \frac{1}{12} - 0.005\ 18\theta^2, \quad 0 \leqslant \theta \leqslant \frac{\pi}{2}.$$

Let $c_1(\theta) = a_1(\theta) + b_1(\theta)$. Then, plainly,

$$q(\theta,\theta,\lambda) \leqslant (1-\lambda) b_1(\theta) + \sum_{k=2}^{+\infty} \frac{(1-\lambda)_k}{k!} \theta^{2k-2} \{c_1(\theta)^k - a_1(\theta)^k\}.$$

From this we can easily get the estimate

$$q(\theta,\theta,\lambda) < 0.10325\cdots, 0 \leqslant \theta \leqslant \frac{\pi}{2}. \quad \square$$

In what follows, in order to simplify the notation, we write $\gamma=\gamma_n(\lambda)=\dfrac{\pi}{n+\lambda+\frac{1}{2}}$ and $\lambda=\lambda'$, $\alpha=\alpha'$. As (2.2) in combination with the inequality of Lemma 1 gives

$$S_n^{\lambda}(\theta) \geqslant \frac{1}{2}-\lambda+Q_n(\theta,\lambda), \tag{2.4}$$

it suffices to prove the positivity of $S_n^{\lambda}(\theta)$ for $\gamma\leqslant\theta\leqslant\frac{\pi}{2}$, by finding estimates of $Q_n(\theta, \lambda)$ in this interval.

Let

$$r(t)=\frac{1}{\sin\frac{t}{2}}-\frac{2}{t}.$$

Using the identity of Lemma 2 we have

$$\frac{1}{\sin\frac{t}{2}(\cos t-\cos\theta)^{1-\lambda}}$$
$$=\frac{2^{2-\lambda}(1-\theta^2 p(\theta,\lambda))}{t(\theta^2-t^2)^{1-\lambda}}+\frac{2^{2-\lambda}tq(\theta,t,\lambda)}{(\theta^2-t^2)^{1-\lambda}}+\frac{r(t)}{(\cos t-\cos\theta)^{1-\lambda}}.$$

It follows from this that

$$Q_n(\theta,\lambda)=A_n(\theta,\lambda)+B_n(\theta,\lambda)+C_n(\theta,\lambda), \tag{2.5}$$

where

$$A_n(\theta,\lambda)=\frac{2}{B\left(\lambda,\frac{1}{2}\right)}\left(\frac{\sin\theta}{\theta}\right)^{1-2\lambda}\frac{1+\theta^2 p(\theta,\lambda)}{\theta}\int_0^1\frac{\sin\frac{\pi\theta}{\gamma}t}{t(1-t^2)^{1-\lambda}}\mathrm{d}t$$

$$B_n(\theta,\lambda)=\frac{2(\sin\theta)^{1-2\lambda}}{B\left(\lambda,\frac{1}{2}\right)}\int_0^{\theta}\frac{tq(\theta,t,\lambda)\sin\frac{\pi}{\gamma}t}{(\theta^2-t^2)^{1-\lambda}}\mathrm{d}t$$

$$C_n(\theta,\lambda)=\frac{(\sin\theta)^{1-2\lambda}}{2^{1-\lambda}B\left(\lambda,\frac{1}{2}\right)}\int_0^{\theta}\frac{r(t)\sin\frac{\pi}{\gamma}t}{(\cos t-\cos\theta)^{1-\lambda}}\mathrm{d}t.$$

We define

$$\Phi(z,\alpha)=\int_0^z t^{-\alpha}J_\alpha(t)\mathrm{d}t,\quad z>0$$

where $J_\alpha(t)$ is the Bessel function of the first kind and order α.

It is well-known that [11]

$$\Phi(z,\alpha)\geqslant 0,\quad z>0,\quad \alpha=\lambda-\frac{1}{2}.$$

This, and the fact that

$$\int_0^1\frac{\sin(zt)}{t(1-t^2)^{\frac{1}{2}-\alpha}}\mathrm{d}t=\frac{\Gamma\left(\alpha+\frac{1}{2}\right)\Gamma\left(\frac{1}{2}\right)}{2^{1-\alpha}}\Phi(z,\alpha),$$

(which can be easily proved using the Poisson's integral formula for Bessel functions, see [16, p. 48]), yield the inequality

$$A_n(\theta,\lambda)\geqslant 2^\alpha\Gamma(\alpha+1)\left(\frac{\sin\theta}{\theta}\right)^{-2\alpha}\frac{1}{\theta}\Phi\left(\frac{\pi}{\gamma}\theta,\alpha\right),\tag{2.6}$$

for all n, $0\leqslant\theta\leqslant\frac{\pi}{2}$, $\alpha=\lambda-\frac{1}{2}$.

In order to give an estimate of the term $B_n(\theta,\lambda)+C_n(\theta,\lambda)$ appearing on the right hand side of (2.5), we write

$$\theta=(2k-1)\gamma+\xi\gamma,\quad 0\leqslant\xi\leqslant 2,\quad k\in\mathbf{N}^*.$$

As it can be easily checked, the function $q(\theta,t,\lambda)$, defined in Lemma 2, is an increasing function of $t\in(0,\theta)$. Then, we have

$$\begin{aligned}
&\int_0^\theta\frac{q(\theta,t,\lambda)\sin\frac{\pi}{\gamma}t}{(\theta^2-t^2)^{1-\lambda}}\mathrm{d}t\\
&>-\int_0^{\xi\gamma}\frac{[t+(2k-1)\gamma]q(\theta,t+(2k-1)\gamma,\lambda)\sin\frac{\pi}{\gamma}t}{[\theta+t+(2k-1)\gamma]^{1-\lambda}(\xi\gamma-t)^{1-\lambda}}\mathrm{d}t\\
&>-\frac{1}{2^{1-\lambda}}\theta^\lambda q(\theta,\theta,\lambda)\int_0^{\eta\gamma}\frac{\sin\frac{\pi}{\gamma}t}{(\xi\gamma-t)^{1-\lambda}}\mathrm{d}t\\
&\geqslant-\frac{1}{2^{1-\lambda}}\theta^\lambda q(\theta,\theta,\lambda)\left(\frac{\gamma}{\pi}\right)^\lambda g(\eta\pi,\lambda),
\end{aligned}\tag{2.7}$$

where

$$g(x,\lambda)=\int_0^x\frac{\sin(x-t)}{t^{1-\lambda}}\mathrm{d}t\tag{2.8}$$

and

$$\eta = \min(1, \xi).$$

It is easy to verify that $\dfrac{r(t)}{\left(\sin\dfrac{\theta+t}{2}\right)^{1-\lambda}}$ is an increasing function of $t \in (0, \theta)$. Then, by repeating the previous argument, we get

$$\int_0^\theta \frac{r(t)\sin\frac{\pi}{\gamma}t}{(\cos t - \cos\theta)^{1-\lambda}}\mathrm{d}t > -\frac{r(\theta)}{(\sin\theta)^{1-\lambda}}\int_0^{\eta\gamma}\frac{\sin\frac{\pi}{\gamma}t}{\left(2\sin\frac{\xi\gamma - t}{2}\right)^{1-\lambda}}\mathrm{d}t$$

$$> -\left[\frac{\gamma}{2\sin\frac{\gamma}{2}}\right]^{1-\lambda}\frac{r(\theta)}{(\sin\theta)^{1-\lambda}}\left(\frac{\gamma}{\pi}\right)^{\lambda} g(\eta\pi, \lambda). \tag{2.9}$$

Let

$$\psi_n(\lambda) = \left[\frac{\gamma}{2\sin\frac{\gamma}{2}}\right]^{1-\lambda}. \tag{2.10}$$

It follows from (2.7) and (2.9) that

$$B_n(\theta, \lambda) + C_n(\theta, \lambda) \geqslant$$

$$-\frac{(\sin\theta)^{1-2\lambda}\theta^\lambda}{2^{1-\lambda}B\left(\lambda, \frac{1}{2}\right)}\left(\frac{\gamma}{\pi}\right)^{\lambda} g(\eta\pi, \lambda)\left\{2q(\theta, \theta, \lambda) + \psi_n(\lambda)\frac{r(\theta)}{\theta^\lambda(\sin\theta)^{1-\lambda}}\right\}.$$

Now, the estimate in Lemma 3 and the fact that the function $\dfrac{r(\theta)}{\theta^\lambda(\sin\theta)^{1-\lambda}}$ is increasing for $\theta \in \left[0, \frac{\pi}{2}\right]$ guarantee the boundedness

$$2q(\theta, \theta, \lambda) + \psi_n(\lambda)\frac{r(\theta)}{\theta^\lambda(\sin\theta)^{1-\lambda}} \leqslant c(\lambda). \tag{2.11}$$

We shall determine the precise value of the constant $c(\lambda)$ later.

Now, we have

$$B_n(\theta, \lambda) + C_n(\theta, \lambda)$$

$$\geqslant -\frac{c(\lambda)g(\eta\pi, \lambda)}{2^{1-\lambda}B\left(\lambda, \frac{1}{2}\right)}\left(\frac{\gamma}{\pi}\right)^{\lambda}\theta^{1-\lambda}\left(\frac{\sin\theta}{\theta}\right)^{1-2\lambda}$$

$$= -2^\alpha\Gamma(\alpha+1)\left(\frac{\sin\theta}{\theta}\right)^{-2\alpha}\left\{\theta^{\frac{1}{2}-\alpha}\frac{c(\lambda)g(\eta\pi, \lambda)}{2^{\frac{1}{2}}\pi^{\frac{1}{2}}\Gamma\left(\alpha+\frac{1}{2}\right)}\left(\frac{\gamma}{\pi}\right)^{\alpha+\frac{1}{2}}\right\}$$

for $\alpha=\lambda-\frac{1}{2}$. A combination of this with (2.6) gives

$$Q_n(\theta,\lambda) > 2^\alpha\Gamma(\alpha+1)\left(\frac{\sin\theta}{\theta}\right)^{-2\alpha}\cdot V_n(\theta,\alpha), \qquad (2.12)$$

where

$$V_n(\theta,\alpha) = \frac{1}{\theta}\Phi\left(\frac{\pi}{\gamma}\theta,\alpha\right)-\theta^{\frac{1}{2}-\alpha}\frac{c(\lambda)g(\eta\pi,\lambda)}{2^{\frac{1}{2}}\pi^{\frac{1}{2}}\Gamma\left(\alpha+\frac{1}{2}\right)}\left(\frac{\gamma}{\pi}\right)^{\alpha+\frac{1}{2}}.$$

Let $j_{\alpha,1}$, $j_{\alpha,2}$, $\cdots$ be the positive roots of the Bessel function $J_\alpha(t)$, in increasing order. It is well-known that the function $\Phi(z, \alpha)$ attains its local minima at $j_{\alpha,2k}$ and $\Phi(j_{\alpha,2k}, \alpha)>0$, $k=2,3,\cdots$ and $\Phi(j_{\alpha,2}, \alpha)=0$ (cf [11] and [14]). Obviously, its local maxima are $\Phi(j_{\alpha,2k-1}, \alpha)$, $k\in \mathbf{N}^*$. On the other hand, since the zeros of J_α are increasing with α[16, p. 508], $-\frac{1}{2}<\alpha<\frac{1}{2}$, and

$$J_{-\frac{1}{2}}(t) = \left(\frac{2}{\pi t}\right)^{\frac{1}{2}}\cos t,\quad J_{\frac{1}{2}}(t) = \left(\frac{2}{\pi t}\right)^{\frac{1}{2}}\sin t$$

we have

$$\left(\nu-\frac{1}{2}\right)\pi < j_{\alpha,\nu} < \nu\pi,\quad \nu\in\mathbf{N}^*. \qquad (2.13)$$

Then we easily get

$$V_n(\theta,\alpha) > \begin{cases} \dfrac{\Phi(j_{\alpha,2k},\alpha)}{2k\gamma}-c_1(\alpha)k^{\frac{1}{2}-\alpha}\gamma, & (2k-1)\gamma\leqslant\theta\leqslant 2k\gamma,\ k\in\mathbf{N}^*, \\ \dfrac{\Phi(2k\pi,\alpha)}{(2k+1)\gamma}-c_2(\alpha)(2k+1)^{\frac{1}{2}-\alpha}\cdot\gamma, & 2k\gamma\leqslant\theta\leqslant(2k+1)\gamma,\ k\in\mathbf{N}^*, \end{cases} \qquad (2.14)$$

where

$$c_1(\alpha) = \frac{c(\lambda)M(\lambda)}{2^\alpha\pi^{1+\alpha}\Gamma\left(\alpha+\frac{1}{2}\right)},$$

$$c_2(\alpha) = \frac{c(\lambda)g(\pi,\lambda)}{2^{\frac{1}{2}}\pi^{\alpha+1}\Gamma\left(\alpha+\frac{1}{2}\right)}$$

and

$$M(\lambda) = \max_{0\leqslant x\leqslant \pi} g(x,\lambda).$$

Notice also that the function $g(x, \lambda)$ is decreasing for $x\in[\pi, 2\pi]$.

We are now able to prove the following:

Proposition 1 We suppose that $n\geqslant 18$ and $\gamma\leqslant\theta\leqslant 9\gamma$. Then,

$$S_n^\lambda(\theta) > 0.235\,1.$$

Proof It follows from our assumptions and the estimate in Lemma 3 that $c(\lambda)=0.329\,27\cdots$. Moreover, $M(\lambda)=4.459\,687\,7\cdots$ and $g(\pi, \lambda)=1.778\,95\cdots$, as it can be easily checked by numerical integration. Hence $c_1(\alpha)=0.196\,41\cdots$ and $c_2(\alpha)=0.045\,964\cdots$. (The numerical value of $\Phi(j_{\alpha,2k}, \alpha)$ and $\Phi(2k\pi, \alpha)$ can also be determined for $k=1,2,3,4$, and this computation can be facilitated by the use of Maple V [8].) Then, by (2.14) we get

$$V_n(\theta,\alpha) > -0.033 \quad \text{for} \quad \gamma\leqslant\theta\leqslant 9\gamma \quad \text{and} \quad n\geqslant 18.$$

A combination of this with (2.12) and (2.14) gives

$$S_n^\lambda(\theta) \geqslant \frac{1}{2} - \lambda - 0.033\cdot 2^\alpha\cdot\Gamma(\alpha+1)\cdot\left(\frac{\sin\theta}{\theta}\right)^{-2\alpha}$$

$$\geqslant -\alpha - 0.033\cdot 2^\alpha\cdot\Gamma(\alpha+1) = 0.235\,1\cdots.$$

Note that $\alpha=-0.269\,387\cdots$. □

Next we shall prove the following:

Proposition 2 Let $n\geqslant 18$ and $9\gamma\leqslant\theta\leqslant\frac{\pi}{2}$. Then

$$S_n^\lambda(\theta) > 0.079.$$

Proof We shall use a completely different argument in estimating the functions $Q_n(\theta, \lambda)$. In particular, write

$$U(t,x,\lambda) = \left(\sin\frac{t}{2}\right)^{\frac{1}{1-\lambda}}(\cos t - \cos\theta)$$

and observe that $U(t, x, \lambda)$ as a function of t in $(0, \theta)$, $0\leqslant\theta\leqslant\frac{\pi}{2}$, attains its absolute maximum at

$$t_0(\theta,\lambda) = \arccos\left(\frac{2-2\lambda+\cos\theta}{3-2\lambda}\right).$$

Thus, $U(t, x, \lambda)$ is increasing in $(0, t_0)$ and decreasing in (t_0, θ).

Hence, for $\theta=(2k-1)\gamma+\xi\gamma$, $0\leqslant\xi\leqslant 2$, $k\geqslant 5$, $\left(\theta\leqslant\frac{\pi}{2}\right)$ we have

$$Q_n(\theta,\lambda)>\sum_{\mu=0}^{3}(-1)^{\mu}I_{\mu}(\theta,\lambda)-J_0(\theta,\lambda), \tag{2.15}$$

where

$$I_{\mu}(\theta,\lambda)=\frac{(\sin\theta)^{1-2\lambda}}{2^{1-\lambda}B\left(\lambda,\frac{1}{2}\right)}\int_0^{\gamma}\frac{\sin\frac{\pi}{\gamma}t}{\sin\left(\frac{\mu\gamma+t}{2}\right)[\cos(\mu\gamma+t)-\cos\theta]^{1-\lambda}}\mathrm{d}t$$

and

$$J_0(\theta,\lambda)=$$

$$\frac{(\sin\theta)^{1-2\lambda}}{2^{1-\lambda}B\left(\lambda,\frac{1}{2}\right)}\int_0^{\eta\gamma}\frac{\sin\frac{\pi}{\gamma}t}{\sin\left[\frac{t+(2k-1)\gamma}{2}\right][\cos(t+(2k-1)\gamma)-\cos\theta]^{1-\lambda}}\mathrm{d}t,$$

where $\eta=\min(1,\ \xi)$. Then we write

$$I_{\mu}(\theta,\lambda)=\frac{\left(\cos\frac{\theta}{2}\right)^{2-2\lambda}}{B\left(\lambda,\frac{1}{2}\right)\sin\theta}\tilde{I}_{\mu}(\theta,\lambda), \tag{2.16}$$

where

$$\tilde{I}_{\mu}(\theta,\lambda)=\int_0^{\gamma}\frac{\sin\frac{\pi}{\gamma}t}{\sin\left(\frac{\mu\gamma+t}{2}\right)\left[1-\frac{\sin^2\left(\frac{\mu\gamma+t}{2}\right)}{\sin^2\frac{\theta}{2}}\right]^{1-\lambda}}\mathrm{d}t.$$

It is easy to see that

$$2s(\mu)<\tilde{I}_{\mu}(\theta,\lambda)<2s(\mu)\cdot K_n(\theta,\lambda,\mu), \tag{2.17}$$

where

$$s(\mu)=\int_0^{\pi}\frac{\sin t}{\mu\pi+t}\mathrm{d}t$$

and

$$K_n(\theta,\lambda,\mu)=\frac{(\mu+1)\frac{\gamma}{2}}{\sin\left[(\mu+1)\frac{\gamma}{2}\right]\left(1-\frac{\sin^2(\mu+1)\frac{\gamma}{2}}{\sin^2\frac{\theta}{2}}\right)^{1-\lambda}}.$$

Numerical integration yields

$$s(0)=1.851\,937\cdots,\qquad s(1)=0.433\,79\cdots,$$

$$s(2)=0.256\,61\cdots,\qquad s(3)=0.182\,6\cdots.$$

On the other hand, for $n\geqslant 18$ and $9\gamma\leqslant\theta\leqslant\frac{\pi}{2}$,

$$K_n(\theta,\lambda,1)<1.053\,151\cdots,\quad K_n(\theta,\lambda,3)<1.247\,062\,2\cdots.$$

It follows from (2.17) and the above that

$$\sum_{\mu=0}^{3}(-1)^{\mu}\tilde{I}_{\mu}(\theta,\lambda)>0.847\,97,$$

Which in combination with (2.16) gives

$$\sum_{\mu=0}^{3}(-1)^{\mu}I_{\mu}(\theta,\lambda)>\frac{0.509\,501}{\sin\theta}\left(\cos\frac{\theta}{2}\right)^{2-2\lambda}\tag{2.18}$$

since $B\left(\lambda,\ \frac{1}{2}\right)=5.589\,7\cdots$.

In order to estimate $J_0(\theta,\ \lambda)$ appearing in (2.15) we distinguish two cases:

(1) $0\leqslant\xi\leqslant 1$.

Then, we can easily obtain

$$J_0(\theta,\lambda)\leqslant\frac{(\sin\theta)^{1-2\lambda}}{2^{1-\lambda}B\left(\lambda,\frac{1}{2}\right)}\frac{1}{\sin\left[(2k-1)\frac{\gamma}{2}\right]}\frac{1}{\left[\sin\frac{\theta+(2k-1)\gamma}{2}\right]^{1-\lambda}}\times$$

$$\int_0^{\xi\gamma}\frac{\sin\frac{\pi}{\gamma}t}{\left(2\sin\frac{\xi\gamma-t}{2}\right)^{1-\lambda}}\mathrm{d}t.$$

Since

$$\int_0^{\xi\gamma}\frac{\sin\frac{\pi}{\gamma}t}{\left(2\sin\frac{\xi\gamma-t}{2}\right)^{1-\lambda}}\mathrm{d}t\leqslant\left(\frac{\gamma}{\pi}\right)^{\lambda}\psi_n(\lambda)\cdot M(\lambda),$$

($\psi_n(\lambda)$, and $M(\lambda)$ as above), writing

$$F_n(\theta,k,\lambda)=\frac{1}{\sin\left[(2k-1)\frac{\gamma}{2}\right]}\frac{(\sin\theta)^{1-\lambda}}{\left[2\sin\frac{\theta+(2k-1)\gamma}{2}\right]^{1-\lambda}}\left(\frac{\gamma}{\pi}\right)^{\lambda}$$

we have

$$J_0(\theta,\lambda) \leqslant \frac{1}{B\left(\lambda,\frac{1}{2}\right)} \cdot \psi_n(\lambda) \cdot M(\lambda) \cdot F_n(\theta,k,\lambda). \qquad (2.19)$$

Suppose that $n \geqslant 20$. Then,

$$F_n(\theta,k,\lambda)\sin\theta$$

$$= \left\{\frac{\sin^2\theta}{2\sin\frac{\theta+(2k-1)\gamma}{2}}\right\}^{1-\lambda} \frac{1}{\sin\left[(2k-1)\frac{\gamma}{2}\right]} \cdot \left(\frac{\gamma}{\pi}\right)^{\lambda}$$

$$\leqslant \left\{\frac{\sin^2 2k\gamma}{2\sin\left[(2k-1)\gamma+\frac{\gamma}{2}\right]\sin\left[(2k-1)\frac{\gamma}{2}\right]}\right\}^{1-\lambda} \left\{\frac{\gamma}{\pi\sin\left[(2k-1)\frac{\gamma}{2}\right]}\right\}^{\lambda}$$

$$\leqslant \left\{\frac{\sin^2 10\gamma}{2\sin\left(\frac{19\gamma}{2}\right)\sin\left(\frac{9\gamma}{2}\right)}\right\}^{1-\lambda} \left[\frac{\gamma}{\pi\sin\frac{9\gamma}{2}}\right]^{\lambda}$$

$$\leqslant \frac{200}{171} \cdot \left(\frac{19}{100\pi}\right)^{\lambda} = 0.612\,43\cdots.$$

For $n=18$, it follows by a straightforward calculation that

$$F_{18}(\theta,5,\lambda)\sin\theta \leqslant 0.435\,8\cdots, \quad 9\gamma_{18} \leqslant \theta \leqslant \frac{\pi}{2}$$

and similarly, for $n=19$,

$$F_{19}(\theta,5,\lambda)\sin\theta \leqslant 0.450\,007\cdots, \quad 9\gamma_{19} \leqslant \theta \leqslant \frac{\pi}{2}.$$

Therefore

$$F_n(\theta,k,\lambda) \leqslant \frac{0.612\,43}{\sin\theta} \text{ for all } n \geqslant 18.$$

It follows from this and (2.19) that for $(2k-1)\gamma \leqslant \theta \leqslant 2k\gamma$, $k \geqslant 5$,

$$J_0(\theta,\lambda) < \frac{0.489\,06}{\sin\theta}. \qquad (2.20)$$

(2) $1 \leqslant \xi \leqslant 2$. This occurs only when $n \geqslant 20$.

In this case (2.19) takes the form

$$J_0(\theta,\lambda) \leqslant \frac{1}{B\left(\lambda,\frac{1}{2}\right)} \psi_n(\lambda) g(\pi,\lambda) F_n(\theta,k,\lambda), \qquad (2.21)$$

and with the assumption $n \geqslant 22$,

$$F_n(\theta, k, \lambda)\sin\theta$$

$$\leqslant\left\{\frac{\sin^2(2k+1)\gamma}{2\sin 2k\gamma\cdot\sin\left[(2k-1)\frac{\gamma}{2}\right]}\right\}^{1-\lambda}\left\{\frac{\gamma}{\pi\sin\left[(2k-1)\frac{\gamma}{2}\right]}\right\}^{\lambda}$$

$$\leqslant\left[\frac{\sin^2 11\gamma}{2\sin 10\gamma\sin\frac{9\gamma}{2}}\right]^{1-\lambda}\left(\frac{\gamma}{\pi\sin\frac{9\gamma}{2}}\right)^{\lambda}$$

$$\leqslant\frac{121}{90}\left(\frac{20}{121\pi}\right)^{\lambda}=0.681\ 73\cdots$$

but

$$F_{20}(\theta,5,\lambda)\sin\theta\leqslant 0.464\ 799,\quad 10\gamma_{20}\leqslant\theta\leqslant\frac{\pi}{2}$$

and

$$F_{21}(\theta,5,\lambda)\sin\theta\leqslant 0.480\ 093,\quad 10\gamma_{21}\leqslant\theta\leqslant\frac{\pi}{2}.$$

Therefore, for $2k\gamma\leqslant\theta\leqslant(2k+1)\gamma$, $k\geqslant 5$, $n\geqslant 20$,

$$F_n(\theta,k,\lambda)\leqslant\frac{0.681\ 73}{\sin\theta}.$$

From this and (2.21) we get

$$J_0(\theta,\lambda)\leqslant\frac{0.217\ 2}{\sin\theta}.$$

Finally, taking into accout (2.20), (2.18) and (2.15), we conclude that

$$Q_n(\theta,\lambda)>\frac{1}{\sin\theta}\left\{0.509\ 501\left(\cos\frac{\theta}{2}\right)^{2-2\lambda}-0.489\ 06\right\}.$$

The right hand side of this inequality defines a decreasing function of θ in $\left(0,\ \frac{\pi}{2}\right]$, therefore

$$Q_n(\theta,\lambda)\geqslant -0.190\ 155$$

and by (2.4) we obtain $S_n^{\lambda}(\theta)>0.079$, and complete the proof of the proposition. □

In order to prove the positivity of $S_n^{\lambda}(\theta)$, $n=1,2,\cdots,17$ one can apply the methods in the proof of Proposition 1 and Proposition 2, with some obvious modifications. However, an elegant way of establishing this polynomial inequality is the application of Sturm's Theorem (see [7], Section 89) and the use of Maple V [8]. In fact, the polynomials

$S_n^\lambda(x)$, $n=1,2,\cdots,17$ have no roots in $(-1, 1]$ and their positivity follows from the fact $S_n^\lambda(0)>0$. It should be noted that, when n is odd we have $S_n^\lambda(-1)=0$. □

The case $\frac{\pi}{2}<\theta\leqslant\pi$.

We are now concerned with proving

$$\sum_{k=0}^{n}\frac{C_k^\lambda(x)}{C_k^\lambda(1)}\geqslant 0,\quad -1\leqslant x<0 \tag{2.22}$$

for $\lambda=\lambda'$, $n\geqslant 18$.

For this purpose, we set

$$S_n^\lambda(\theta)=\sum_{k=0}^{n}(-1)^k\frac{C_k^\lambda(\cos\theta)}{C_k^\lambda(1)},$$

and we see that (2.22) is equivalent to

$$S_n^\lambda(\theta)\geqslant 0,\quad 0\leqslant\theta<\frac{\pi}{2}. \tag{2.23}$$

The application of the Dirichlet-Mehler formula (2.1) now gives

$$S_n^\lambda(\theta)=1-R_0(\theta,\lambda)+(-1)^nR_n(\theta,\lambda), \tag{2.24}$$

where

$$R_n(\theta,\lambda)=\frac{(\sin\theta)^{1-2\lambda}}{2^{1-\lambda}B\left(\lambda,\frac{1}{2}\right)}\int_0^\theta\frac{\cos\left(n+\lambda+\frac{1}{2}\right)t}{\cos\frac{t}{2}(\cos t-\cos\theta)^{1-\lambda}}\mathrm{d}t,\quad n\in\mathbf{N}.$$

Throughout this section we maintain the notation $\gamma=\dfrac{\pi}{n+\lambda+\frac{1}{2}}$.

We observe that

$$1-R_0(\theta,\lambda)=\frac{(\sin\theta)^{1-2\lambda}}{2^{1-\lambda}B\left(\lambda,\frac{1}{2}\right)}\int_0^\theta\frac{\cos\left(\frac{1}{2}-\lambda\right)t}{\cos\frac{t}{2}(\cos t-\cos\theta)^{1-\lambda}}\mathrm{d}t. \tag{2.25}$$

It follows from this and (2.24) that when n is even and $0\leqslant\theta\leqslant\frac{\gamma}{2}$,

$$S_n^\lambda(\theta)\geqslant 1-R_0(\theta,\lambda)>0.$$

Suppose that n is odd and $0\leqslant\theta\leqslant\frac{3}{2}\gamma$. A combination of (2.25) with (2.24) gives

$$S_n^\lambda(\theta)=\frac{2^\lambda(\sin\theta)^{1-2\lambda}}{B\left(\lambda,\frac{1}{2}\right)}\int_0^\theta\frac{\sin(n+1)\frac{t}{2}\cdot\sin(n+2\lambda)\frac{t}{2}}{\cos\frac{t}{2}(\cos t-\cos\theta)^{1-\lambda}}\mathrm{d}t\geqslant 0,$$

with equality only when $\theta=0$.

Using (2.3) (2.25) and the fact that

$$\frac{\cos\left(\frac{1}{2}-\lambda\right)t}{\cos^2\frac{t}{2}}\geqslant 1,\quad 0\leqslant t\leqslant\theta<\frac{\pi}{2},$$

we get

$$1-R_0(\theta,\lambda)\geqslant\frac{1}{2}\left(\cos\frac{\theta}{2}\right)^{1-2\lambda}. \tag{2.26}$$

From this and (2.24) we deduce

$$S_n^\lambda(\theta)\geqslant\frac{1}{2}\left(\cos\frac{\theta}{2}\right)^{1-2\lambda}+(-1)^nR_n(\theta,\lambda). \tag{2.27}$$

Let

$$H_n(\theta,\lambda)=\int_0^\theta\frac{\cos\frac{\pi}{\gamma}t}{\cos\frac{t}{2}(\cos t-\cos\theta)^{1-\lambda}}\mathrm{d}t.$$

We assume that $\frac{\gamma}{2}\leqslant\theta<\frac{\pi}{2}$. It is then easy to see that when

$$(4k-3)\frac{\gamma}{2}\leqslant\theta\leqslant(4k-1)\frac{\gamma}{2},\quad k\in\mathbf{N}^*.$$

$$\begin{aligned}
&H_n(\theta,\lambda)\\
&>\int_{(4k-3)\frac{\gamma}{2}}^\theta\frac{\cos\frac{\pi}{\gamma}t}{\cos\frac{t}{2}(\cos t-\cos\theta)^{1-\lambda}}\mathrm{d}t\\
&\geqslant-\frac{1}{\cos\frac{\theta}{2}\left[\sin\dfrac{\theta+\frac{(4k-3)\gamma}{2}}{2}\right]^{1-\lambda}}\left[\frac{\gamma}{2\sin\frac{\gamma}{2}}\right]^{1-\lambda}\left(\frac{\gamma}{\pi}\right)^\lambda M(\lambda)\\
&\geqslant-\frac{1}{\cos\frac{\theta}{2}}\frac{1}{\left[\sin\left(\theta-\frac{\xi\gamma}{2}\right)\right]^{1-\lambda}}\left[\frac{\gamma}{2\sin\frac{\gamma}{2}}\right]^{1-\lambda}\left(\frac{\gamma}{\pi}\right)^\lambda M(\lambda),\ 0\leqslant\xi\leqslant 1,
\end{aligned} \tag{2.28}$$

where $M(\lambda)$ as defined in the previous section. Also, when $(4k-1)\frac{\gamma}{2}\leqslant\theta\leqslant(4k+1)\frac{\gamma}{2}$, $k\in\mathbf{N}^*$. we have

$$\begin{aligned}
&H_n(\theta,\lambda)\\
&>\int_{(4k-3)\frac{\gamma}{2}}^{(4k-1)\frac{\gamma}{2}}\frac{\cos\frac{\pi}{\gamma}t}{\cos\frac{t}{2}(\cos t-\cos\theta)^{1-\lambda}}\mathrm{d}t\\
&\geqslant-\frac{1}{\cos\left[(4k-1)\frac{\gamma}{4}\right]}\frac{1}{[\sin(2k-1)\gamma]^{1-\lambda}}\left(\frac{\gamma}{\sin\gamma}\right)^{1-\lambda}\left(\frac{\gamma}{\pi}\right)^{\lambda}g(\pi,\lambda)\\
&\geqslant-\frac{1}{\cos\frac{\theta}{2}}\frac{1}{\left[\sin\left(\theta-\frac{\gamma}{2}-\xi\gamma\right)\right]^{1-\lambda}}\left(\frac{\gamma}{\sin\gamma}\right)^{1-\lambda}\left(\frac{\gamma}{\pi}\right)^{\lambda}g(\pi,\lambda),\ 0\leqslant\xi\leqslant1,
\end{aligned}\tag{2.29}$$

Let us suppose now that $\frac{3}{2}\gamma\leqslant\theta<\frac{\pi}{2}$. In the case where $(4k-1)\frac{\gamma}{2}\leqslant\theta\leqslant(4k+1)\frac{\gamma}{2}$, $k\in\mathbf{N}^*$. we have

$$\begin{aligned}
&-H_n(\theta,\lambda)\\
&>-\int_{(4k-1)\frac{\gamma}{2}}^{\theta}\frac{\cos\frac{\pi}{\gamma}t}{\cos\frac{t}{2}(\cos t-\cos\theta)^{1-\lambda}}\mathrm{d}t\\
&>-\frac{1}{\cos\frac{\theta}{2}}\frac{1}{\sin\left[\frac{\theta+(4k-1)\frac{\gamma}{2}}{2}\right]^{1-\lambda}}\left[\frac{\gamma}{2\sin\frac{\gamma}{2}}\right]^{1-\lambda}\left(\frac{\gamma}{\pi}\right)^{\lambda}M(\lambda)\\
&=-\frac{1}{\cos\frac{\theta}{2}}\frac{1}{\left[\sin\left(\theta-\frac{\xi\gamma}{2}\right)\right]^{1-\lambda}}\left[\frac{\gamma}{2\sin\frac{\gamma}{2}}\right]^{1-\lambda}\left(\frac{\gamma}{\pi}\right)^{\lambda}M(\lambda),\quad 0\leqslant\xi\leqslant1,
\end{aligned}\tag{2.30}$$

Similarly, for $(4k+1)\frac{\gamma}{2}\leqslant\theta\leqslant(4k+3)\frac{\gamma}{2}$, $k\in\mathbf{N}^*$. we have

$$\begin{aligned}
&-H_n(\theta,\lambda)\\
&>-\int_{(4k-1)\frac{\gamma}{2}}^{(4k+1)\frac{\gamma}{2}}\frac{\cos\frac{\pi}{\gamma}t}{\cos\frac{t}{2}(\cos t-\cos\theta)^{1-\lambda}}\mathrm{d}t
\end{aligned}$$

$$>-\frac{1}{\cos\left[(4k+1)\frac{\gamma}{4}\right]}\frac{1}{(\sin 2k\gamma)^{1-\lambda}}\left(\frac{\gamma}{\sin\gamma}\right)^{1-\lambda}\left(\frac{\gamma}{\pi}\right)^{\lambda}g(\pi,\lambda)$$

$$\geqslant-\frac{1}{\cos\frac{\theta}{2}}\frac{1}{\left[\sin\left(\theta-\frac{\gamma}{2}-\xi\gamma\right)\right]^{1-\lambda}}\left(\frac{\gamma}{\sin\gamma}\right)^{1-\lambda}\left(\frac{\gamma}{\pi}\right)^{\lambda}g(\pi,\lambda),\quad 0\leqslant\xi\leqslant 1, \tag{2.31}$$

We recall that

$$M(\lambda)=4.4596877\cdots,\quad g(\pi,\lambda)=1.77895\cdots$$

so that, for $n\geqslant 18$,

$$\frac{1}{B\left(\lambda,\frac{1}{2}\right)}\left(\frac{2}{\pi}\right)^{\lambda}\left[\frac{\gamma}{2\sin\frac{\gamma}{2}}\right]^{1-\lambda}M(\lambda)<0.7196$$

and

$$\frac{1}{B\left(\lambda,\frac{1}{2}\right)}\left(\frac{2}{\pi}\right)^{\lambda}\left(\frac{\gamma}{\sin\gamma}\right)^{1-\lambda}g(\pi,\lambda)<0.2879.$$

For $0\leqslant\xi\leqslant 1$, we let

$$U(\theta,\xi,\gamma)=\left[\frac{\sin\theta}{\sin\left(\theta-\frac{\xi\gamma}{2}\right)}\right]^{1-\lambda}\left(\frac{\gamma}{\sin\theta}\right)^{\lambda},$$

and

$$V(\theta,\xi,\gamma)=\left[\frac{\sin\theta}{\sin\left(\theta-\frac{\gamma}{2}-\xi\gamma\right)}\right]^{1-\lambda}\left(\frac{\gamma}{\sin\theta}\right)^{\lambda}.$$

Using (2.27)～(2.31) and the above, we get

$$2\cos\frac{\theta}{2}S_n^{\lambda}(\theta)\geqslant\left(\cos\frac{\theta}{2}\right)^{2-2\lambda}-0.7196\,U(\theta,\xi,\gamma), \tag{2.32}$$

when n is even and $(4k-3)\frac{\gamma}{2}\leqslant\theta\leqslant(4k-1)\frac{\gamma}{2}$ or n is odd and

$$(4k-1)\frac{\gamma}{2}\leqslant\theta\leqslant(4k+1)\frac{\gamma}{2},\quad k\in\mathbf{N}^*\ (n\geqslant 18)$$

and

$$2\cos\frac{\theta}{2}S_n^{\lambda}(\theta)\geqslant\left(\cos\frac{\theta}{2}\right)^{2-2\lambda}-0.2879\,V(\theta,\xi,\gamma), \tag{2.33}$$

when n is even and $(4k-1)\frac{\gamma}{2}\leqslant\theta\leqslant(4k+1)\frac{\gamma}{2}$ or n is odd and

$$(4k+1)\frac{\gamma}{2}\leqslant\theta\leqslant(4k+3)\frac{\gamma}{2},\ k\in\mathbf{N}^*\ (n\geqslant 18).$$

It is now clear that in order to establish the positivity of $S_n^\lambda(\theta)$, we need appropriate upper bounds for the functions $U(\theta, \xi, \gamma)$ and $V(\theta, \xi, \gamma)$ defined above. These are given in the following lemma.

Lemma 4

$$U(\theta,\xi,\gamma)<\begin{cases}1.26, & \frac{\gamma}{2}\leqslant\theta\leqslant\frac{5\gamma}{2},\\ 0.967\,8, & \frac{5\gamma}{2}\leqslant\theta\leqslant\frac{9\gamma}{2},\\ 0.791\,5, & \frac{9\gamma}{2}\leqslant\theta\leqslant\frac{\pi}{2}\end{cases}$$

and

$$V(\theta,\xi,\gamma)<1.847\,7,\ \frac{3\gamma}{2}\leqslant\theta\leqslant\frac{\pi}{2}.$$

Proof When $\frac{\gamma}{2}\leqslant\theta\leqslant\frac{3\gamma}{2}$, we set $\theta=\frac{\gamma}{2}+\xi\gamma$, $0\leqslant\xi\leqslant 1$, so that

$$\begin{aligned}U(\theta,\xi,\gamma)&=\left[\frac{\sin(2\xi+1)\frac{\gamma}{2}}{\sin(\xi+1)\frac{\gamma}{2}}\right]^{1-\lambda}\left[\frac{\gamma}{\sin(2\xi+1)\frac{\gamma}{2}}\right]^{\lambda}\\ &<\left(\frac{2\xi+1}{\xi+1}\right)^{1-\lambda}\left(\frac{2}{2\xi+1}\right)^{\lambda}\left[\frac{(2\xi+1)\gamma}{2\sin(2\xi+1)\frac{\gamma}{2}}\right]^{\lambda}\\ &\leqslant\left(\frac{2\xi+1}{\xi+1}\right)^{1-\lambda}\left(\frac{2}{2\xi+1}\right)^{\lambda}\left[\frac{3\gamma}{2\sin\frac{3\gamma}{2}}\right]^{\lambda}<1.26,\end{aligned}$$

for $0\leqslant\xi\leqslant 1$ and $n\geqslant 18$.

Suppose now that $\theta\geqslant\mu\cdot\frac{\gamma}{2}$, $\mu\geqslant 3$. Then,

$$\begin{aligned}U(\theta,\xi,\gamma)&\leqslant\left[\frac{\sin\theta}{\sin\left(\theta-\frac{\gamma}{2}\right)}\right]^{1-\lambda}\left(\frac{\gamma}{\sin\theta}\right)^{\lambda}\leqslant\left[\frac{\sin\mu\frac{\gamma}{2}}{\sin(\mu-1)\frac{\gamma}{2}}\right]^{1-\lambda}\left[\frac{\gamma}{\sin\mu\frac{\gamma}{2}}\right]^{\lambda}\\ &<\left(\frac{\mu}{\mu-1}\right)^{1-\lambda}\left(\frac{2}{\mu}\right)^{\lambda}\left[\frac{\mu\gamma}{2\sin\mu\frac{\gamma}{2}}\right]^{\lambda}=p(\mu).\end{aligned}$$

Hence,　$p(3) < 1.2472$,　$p(5) < 0.9678$,　$p(9) < 0.7915$

and the assertions of the Lemma follow.

Similarly, when $\frac{3\gamma}{2} \leqslant \theta \leqslant \frac{5\gamma}{2}$ we write $\theta = \frac{3\gamma}{2} + \xi\gamma$, $0 \leqslant \xi \leqslant 1$, and we see that

$$V(\theta,\xi,\gamma) \leqslant \left[\frac{\theta}{\theta - \frac{\gamma}{2} - \xi\gamma}\right]^{1-\lambda} \left(\frac{\gamma}{\sin\theta}\right)^{\lambda} \leqslant \left(\frac{5}{2}\right)^{1-\lambda} \left(\frac{2}{3}\right)^{\lambda} \left[\frac{3\gamma}{2\sin\frac{3\gamma}{2}}\right]^{\lambda} < 1.8477.$$

On the other hand, for $\theta > \frac{5\gamma}{2}$,

$$V(\theta,\xi,\gamma) \leqslant \left[\frac{\sin\theta}{\sin\left(\theta - \frac{3\gamma}{2}\right)}\right]^{1-\lambda} \left(\frac{\gamma}{\sin\theta}\right)^{\lambda} < \left[\frac{\sin\frac{5\gamma}{2}}{\sin\gamma}\right]^{1-\lambda} \left[\frac{\gamma}{\sin\frac{5\gamma}{2}}\right]^{\lambda}$$

$$\leqslant \frac{5}{2}\left(\frac{4}{25}\right)^{\lambda} \left[\frac{5\gamma}{2\sin\frac{5\gamma}{2}}\right]^{\lambda} < 1.6495.$$

This completes the proof of the lemma.

Now, in view of the Lemma above we have

$$\left(\cos\frac{\theta}{2}\right)^{2-2\lambda} - 0.7196\, U(\theta,\xi,\gamma) > \begin{cases} 0.059, & \frac{\gamma}{2} \leqslant \theta \leqslant \frac{5\gamma}{2}, \\ 0.19, & \frac{5\gamma}{2} \leqslant \theta \leqslant \frac{9\gamma}{2}, \\ 0.017, & \frac{9\gamma}{2} \leqslant \theta \leqslant \frac{\pi}{2}, \end{cases}$$

and

$$\left(\cos\frac{\theta}{2}\right)^{2-2\lambda} - 0.2879\, V(\theta,\xi,\gamma) > 0.054.$$

These in combination with (2.32) and (2.33) complete the proof of (2.23).　□

References

[1] Askey R. Positive Jacobi polynomial sums. Tôhoku Math. Journ. 1972, 24: 109—119.

[2] Askey R. Orthogonal polynomials and special functions. Regional Conf.

Lect. Appl. Math., 21, SIAM, Philadelphia, 1975.

[3] Askey R. Problems which interest and/or annoy me. J. Comp. Appl. Math., 1993, 48: 3—15.

[4] Askey R, Fitch. J. Integral representations for Jacobi polynomials and some applicatios. J. Math. Anal. Appl., 1969, 16: 411—437.

[5] Askey R. Gasper G. Positive Jacobi polynomial sums Ⅱ. Amer. J. Math., 1976, 98: 709—737.

[6] Brown G, Koumandos S, Wang K. Positivity of more Jacobi polynomial sums. Math. Proc. Camb. Phil. Soc., 1996, 119(4): 681—694.

[7] Burnside W S, Panton A W. Theory of Equations, Dublin University Press, 1886.

[8] Char B W, Geddes K O, Gonnet G H, Leong B L, Monagan M B, Watt S M. Maple V first leaves, a tutorial introduction to Maple V and library reference manual. Springer-Verlag, 1992.

[9] Erdélyi A, Magnus W, Oberhettinger F, Tricomi F G. Higher transcendental functions. Volume 2, McGraw-Hill, New York, 1953.

[10] Fejér L. Über die Laplacesche Reihe. Math. Annalen, 1909, 67: 76—109. Gesammelte Arbeiten I, 503—537, Birkhäuser Verlag, Basel and Stuttgart, 1970.

[11] Feldheim E. On the positivity of certain sums of ultraspherical polynomials. J. Analyse Math., 1963, 11: 275—284. (edited with additional notes by G. Szegö), also in G. Szegö collected papers Vol. 3, Birkhäuser Boston, 1982, 821—830.

[12] Gasper G. Non-negative sums of cosine, ultraspherical and Jacobi polynomials. J. Math. Anal. Appl., 1969, 26: 60—68.

[13] Gasper G. Positive sums of the classical orthogonal polynomials. SIAM J. Math. Anal., 1977, 8: 423—447.

[14] Lorch L, Muldoon M E, Szego P. Some monotonicity properties of Bessel functions. SIAM J. Math. Anal., 1973, 4: 385—392.

[15] Vilenkin, N Ja. Some relations for Gegenbauer functions. Uspekhi Mat. Nauk., 1958, 13(3): 167—172.

[16] Watson G N. A Treatise on the Theory of Bessel Functions. 2nd ed., Cambridge University Press, 1944.

Journal of the London Mathematical Society，1999，59(2)：939－954.

关于某些基本 Legendre 多项式和的正性

On the Positivity of Some Basic Legendre Polynomial Sums①

Abstract The positivity of the partial sums of the series of Legendre polynomials is a classical result of Fejér. The paper presents corresponding results for both even and odd Legendre polynomials with best possible leading constants. These results have applications to demonstrating positivity of Cotes numbers at Jacobi abscissas.

§ 1. Introduction

A classical result of Fejér [7, 8] states that

$$\sum_{k=0}^{n} P_k(\cos\theta) \geqslant 0, \quad 0 \leqslant \theta \leqslant \pi, \tag{1.1}$$

Where P_k is the Legendre polynomial of degree k. This result has been used in [8] in the proof of $(C, 2)$ summability of Laplace series. Several extensions of (1.1) in the case of Gegenbauer polynomials have been derived by Fejér [9], Feldheim [10] and Szegö [13]. Further results of this kind concerning positive sums of Jacobi polynomials have been proved by Askey [1], Askey and Fitch [2], Askey and Gasper

① This research was supported by a grant from the Australian Research Council.

本文与 Brown G 和 Koumandos S 合作.

Received：1994-06-15；Revised：1997-09-30.

[3] and Gasper [11].

In this paper our aim is to establish some new inequalities concerning basic sums of Legendre polynomials.

We consider the Legendre polynomial sums defined by

$$S_n(\theta)=\frac{1}{2}+\sum_{k=1}^{n}P_{2k}(\cos\theta),$$

$$T_n(\theta)=\frac{\sqrt{15}}{45}+\sum_{k=1}^{n}P_{2k-1}(\cos\theta),$$

$$U_n(\theta)=\frac{1}{2}+\sum_{k=1}^{n}P_{4k}(\cos\theta),$$

for $0\leqslant\theta\leqslant\pi$ and $n\in\mathbf{N}$.

Our main results are the following.

Theorem 1

$$S_n(\theta)\geqslant 0 \quad \text{for} \quad 0\leqslant\theta\leqslant\pi \quad \text{and} \quad n\in\mathbf{N},$$

$$T_n(\theta)\geqslant 0 \quad \text{for} \quad 0\leqslant\theta\leqslant\frac{\pi}{2} \quad \text{and} \quad n\in\mathbf{N}.$$

Theorem 2

$$U_n(\theta)>0 \quad \text{for} \quad 0\leqslant\theta\leqslant\pi \quad \text{and} \quad n\in\mathbf{N}.$$

Observe that the leading constants in S_n and T_n in Theorem 1 are best possible, for $S_1=0$ when $\cos\theta=0$ and $T_2=0$ when $\cos\theta=\frac{1}{\sqrt{15}}$. More to the point we are able to use the sharp results on S_n and U_n to establish positivity of Cotes numbers at new Jacobi abscissas. See [4, 12].

§ 2. Proof of the main results

We define

$$R_n(\theta)=\frac{1}{\pi}\int_0^\theta\frac{\sin\left(n+\frac{1}{2}\right)t}{\sin t(2\cos t-2\cos\theta)^{\frac{1}{2}}}\mathrm{d}t,\quad 0<\theta<\frac{\pi}{2}$$

and

$$Q_n(\theta)=\frac{1}{\pi}\int_0^\theta\frac{\cos\left(n+\frac{1}{2}\right)t}{\cos t(2\cos t-2\cos\theta)^{\frac{1}{2}}}\mathrm{d}t,\quad 0<\theta<\frac{\pi}{2}.$$

By application of the Dirichlet-Mehler formula

$$P_k(\cos\theta)=\frac{2}{\pi}\int_0^\theta\frac{\cos\left(k+\frac{1}{2}\right)t}{(2\cos t-2\cos\theta)^{\frac{1}{2}}}\mathrm{d}t,\quad 0<\theta<\pi$$

(as can be found in [14]) and using a direct summation, we obtain

$$S_n(\theta)=-\frac{1}{2}+\frac{1}{4\cos\frac{\theta}{2}}+R_{2n+1}(\theta),\quad 0<\theta<\frac{\pi}{2},\tag{2.1}$$

$$T_n(\theta)=\frac{\sqrt{15}}{45}-\frac{1}{4\cos\frac{\theta}{2}}+R_{2n}(\theta),\quad 0<\theta<\frac{\pi}{2}.\tag{2.2}$$

We see that in order to prove Theorem 1 we need (only!) to find appropriate estimates of the integral $R_n(\theta)$. It will become clear that we must estimate also the integral $Q_n(\theta)$ in order to prove Theorem 2. The proofs will be broken down into a sequence of lemmas which will demonstrate the relative difficulty of the estimates for S_n, T_n and U_n.

It is convenient to fix notation and write $\gamma=\gamma_n=\frac{\pi}{n+\frac{1}{2}}$ and $\alpha=\gamma_{49}=\frac{2\pi}{99}$. Our first lemma gives the main part of the necessary estimate for S_n.

Lemma 1 For $n\geqslant 49$ and $7\gamma<\theta\leqslant\frac{\pi}{2}$,

$$R_n(\theta)>\frac{0.2273}{\sin\frac{\theta}{2}}-\frac{0.1304}{\sin\theta}.\tag{2.3}$$

Proof Write $h(t)=\sin^2 t(\cos t-\cos\theta)$ and observe that $h(t)$ as a function of t in $(0,\theta)$, for $0<\theta\leqslant\frac{\pi}{2}$, attains its absolute maximum at $t_0=\arccos\left(\frac{1}{3}\cos\theta+\frac{1}{3}(3+\cos^2\theta)^{\frac{1}{2}}\right)$. In fact we are not concerned with the precise value of t_0 but rather with the information that $h(t)$ is increasing in $(0,t_0)$ and decreasing in (t_0,θ). By considering the integral of a regularly oscillating sine curve multiplied by the reciprocal of

$h(t)^{\frac{1}{2}}$ and choosing the integer $k \geqslant 4$ such that

$$(2k-1)\gamma < \theta \leqslant (2k+1)\gamma, \tag{2.4}$$

we infer that

$$R_n(\theta) > \sum_{\mu=0}^{3} (-1)^{\mu} I_{\mu}(\theta) - J_0(\theta), \tag{2.5}$$

where

$$I_{\mu}(\theta) = \frac{1}{\pi}\int_0^{\gamma} \frac{\sin\left(\frac{\pi}{\gamma}t\right)\mathrm{d}t}{\sin(\mu\gamma+t)[2\cos(\mu\gamma+t)-2\cos\theta]^{\frac{1}{2}}},$$

$$J_0(\theta) = \frac{1}{\pi}\int_0^{\theta-(2k-1)\gamma} \frac{\sin\left(\frac{\pi}{\gamma}t\right)\mathrm{d}t}{\sin[(2k-1)\gamma+t][2\cos[(2k-1)\gamma+t]-2\cos\theta]^{\frac{1}{2}}}.$$

The point is that the sine curve has positive arches at both ends of the interval where we replace the integrand by zero.

It is easy to see that

$$\frac{1}{2\pi\sin\frac{\theta}{2}}\int_0^{\pi}\frac{\sin t}{\mu\pi+t}\mathrm{d}t < I_{\mu}(\theta) < \frac{1}{2\pi\sin\frac{\theta}{2}}\frac{(\mu+1)\alpha}{\sin(\mu+1)\alpha}p_{\mu}(\gamma,\theta)\int_0^{\pi}\frac{\sin t}{\mu\pi+t}\mathrm{d}t, \tag{2.6}$$

where

$$p_{\mu}(\gamma,\theta) = \left(1-\frac{\sin^2\left[\frac{(\mu+1)\gamma}{2}\right]}{\sin^2\frac{\theta}{2}}\right)^{-\frac{1}{2}}.$$

From the fact that

$$\int_0^{\pi}\frac{\sin t}{\mu\pi+t}\mathrm{d}t = (-1)^{\mu}\{\mathrm{Si}[(\mu+1)\pi]-\mathrm{Si}[\mu\pi]\},$$

we obtain the estimates

$$I_0(\theta) > \frac{0.294\,7}{\sin\frac{\theta}{2}}, \qquad I_2(\theta) > \frac{0.040\,8}{\sin\frac{\theta}{2}}, \tag{2.7}$$

and

$$I_1(\theta) < \frac{0.069\,2}{\sin\frac{\theta}{2}}p_1(\gamma,\theta), \qquad I_3(\theta) < \frac{0.029\,4}{\sin\frac{\theta}{2}}p_3(\gamma,\theta). \tag{2.8}$$

In fact

$$p_1(\gamma,\ \theta)\leqslant\left[1-\frac{\sin^2\alpha}{\sin^2(3.5\alpha)}\right]^{-\frac{1}{2}}=1.0442\cdots,$$

$$p_3(\gamma,\ \theta)\leqslant\left[1-\frac{\sin^2 2\alpha}{\sin^2(3.5\alpha)}\right]^{-\frac{1}{2}}=1.2218\cdots,$$

so that, for $7\gamma<\theta\leqslant\frac{\pi}{2}$,

$$I_1(\theta)<\frac{0.0723}{\sin\frac{\theta}{2}},\qquad I_3(\theta)<\frac{0.0359}{\sin\frac{\theta}{2}}. \tag{2.9}$$

Now we turn to estimates of $J_0(\theta)$ and, for future reference, it is as well to note that we may relax the condition $k\geqslant 4$ to $k\geqslant 2$ for this limited purpose.

In fact we have

$$J_0(\theta)<\frac{1}{\sqrt{2}\pi}\frac{1}{\sin[(2k-1)\gamma]}\cdot\left(\frac{1}{\sin\frac{\theta+(2k-1)\gamma}{2}}\right)^{\frac{1}{2}}\varphi_n(\theta),$$

where

$$\varphi_n(\theta)=\int_0^{\xi}\frac{\sin\left(\frac{\pi}{\gamma}t\right)}{\left[2\sin\frac{\theta-(2k-1)\gamma-t}{2}\right]^{\frac{1}{2}}}\mathrm{d}t,\quad \xi=\min\{\gamma,\theta-(2k-1)\gamma\}.$$

If $0<\xi\leqslant\frac{15}{16}\gamma$, that is, $(2k-1)\gamma<\theta\leqslant\left(2k-\frac{1}{16}\right)\gamma$, then

$$\varphi_n(\theta)<\left[\frac{\frac{15}{32}\alpha}{\sin\left(\frac{15}{32}\alpha\right)}\right]^{\frac{1}{2}}\left(\frac{\gamma}{\pi}\right)^{\frac{1}{2}}\int_0^{\frac{\pi}{\gamma}\xi}\frac{\sin\left(\frac{\pi}{\gamma}\xi-t\right)}{\sqrt{t}}\mathrm{d}t.$$

We define the function

$$I(x)=\int_0^x\frac{\sin(x-t)}{\sqrt{t}}\mathrm{d}t,\quad 0\leqslant x\leqslant\pi.$$

We can easily verify that

$$I(x)=\sqrt{2\pi}\left\{\sin xC\left(\overline{\frac{2x}{\pi}}\right)-\cos xS\left(\overline{\frac{2x}{\pi}}\right)\right\},$$

where

$$C(z)=\int_0^z\cos\left(\frac{\pi}{2}t^2\right)\mathrm{d}t$$

and

$$S(z)=\int_0^z \sin\left(\frac{\pi}{2}t^2\right)\mathrm{d}t.$$

It can be easily checked that the function $I(x)$ attains its maximum in $[0, \pi]$ at $x=2.297\,439\cdots$ and $\max\limits_{0\leqslant x\leqslant\pi} I(x)=2.378\,93\cdots$. The verification of this fact is particularly simple using numerical methods.

Hence we get

$$J_0(\theta)<\frac{0.302\,1}{\sin[(2k-1)\gamma]}\cdot\left(\frac{\gamma}{\sin\dfrac{\theta+(2k-1)\gamma}{2}}\right)^{\frac{1}{2}}. \tag{2.10}$$

If $\frac{15}{16}\gamma<\xi\leqslant\gamma$, that is, $\left(2k-\frac{1}{16}\right)\gamma<\theta\leqslant(2k+1)\gamma$, then

$$\varphi_n(\theta)<\left(\frac{\alpha}{\sin\alpha}\right)^{\frac{1}{2}}\left(\frac{\gamma}{\pi}\right)^{\frac{1}{2}}\cdot\int_0^{\frac{15}{16}\pi}\frac{\sin\left(\frac{15}{16}\pi-t\right)}{\sqrt{t}}\mathrm{d}t=2.025\,5\cdot\left(\frac{\gamma}{\pi}\right)^{\frac{1}{2}}$$

and hence

$$J_0(\theta)<\frac{0.257\,2}{\sin[(2k-1)\gamma]}\cdot\left[\frac{\gamma}{\sin\dfrac{\theta+(2k-1)\gamma}{2}}\right]^{\frac{1}{2}}. \tag{2.11}$$

Combining the two cases above we get for $7\gamma<\theta\leqslant\frac{\pi}{2}$,

$$J_0(\theta)<\frac{0.130\,4}{\sin\theta}. \tag{2.12}$$

The combination of (2.5)(2.7)(2.9) and (2.12) gives (2.3) and completes the proof of the lemma. □

We summarise the progress made so far in the next lemma which indicates that more work will be needed for the estimate of $T_n(\theta)$.

Lemma 2 Assume that $n\geqslant25$. Then

$$S_n(\theta)>0, \quad 7\gamma_{2n+1}<\theta\leqslant\frac{\pi}{2},$$

$$T_n(\theta)>0, \quad 7\gamma_{2n}<\theta\leqslant\frac{5\pi}{12}.$$

Proof Applying Lemma 1 and taking into account (2.1), we obtain, for $7\gamma_{2n+1}<\theta\leqslant\frac{\pi}{2}$,

$$S_n(\theta)\sin\theta > -0.5\sin\theta + 0.5\sin\frac{\theta}{2} + 0.4546\cos\frac{\theta}{2} - 0.1304 \geqslant 0.04.$$

From Lemma 1 and (2.2), we obtain, for $7\gamma_{2n} < \theta \leqslant \frac{\pi}{2}$,

$$T_n(\theta)\sin\theta > \frac{\sqrt{15}}{45}\sin\theta - 0.5\sin\frac{\theta}{2} + 0.4546\cos\frac{\theta}{2} - 0.1304.$$

The function on the right-hand side of the previous inequality decreases in the interval $\left(0, \frac{1}{2}\pi\right)$ but becomes negative. Because it is positive for $\theta = \frac{5\pi}{12}$ the lemma is proved. □

Let us proceed to dispose of small values of θ, proving the case where $\frac{5\pi}{12} < \theta \leqslant \frac{\pi}{2}$ for $T_n(\theta)$ in Lemma 6 below.

Lemma 3 Assume that $n \geqslant 49$. If $0 < \theta \leqslant 7\gamma$, then

$$R_n(\theta) > \begin{cases} 0.7867, & 0 < \theta \leqslant 3\gamma, \\ 0.6303, & 3\gamma < \theta \leqslant 7\gamma. \end{cases}$$

Proof We maintain the notations of the proof of Lemma 1.

Case 1 $0 < \theta \leqslant \gamma$. We have in this case

$$R_n(\theta) > \frac{1}{2\pi\sin\frac{\theta}{2}}\int_0^{\frac{\pi}{\gamma}\theta}\frac{\sin t}{t}\mathrm{d}t > \frac{\pi\theta}{2\pi\sin\frac{\theta}{2}}\ \frac{1}{\gamma}\left\{1 - \frac{1}{18}\left(\frac{\pi}{\gamma}\theta\right)^2\right\} > 4.$$

Case 2 $\gamma < \theta \leqslant 2\gamma$. We have

$$R_n(\theta) = I_0(\theta) - J_0(\theta).$$

Now

$$J_0(\theta) <$$

$$\frac{1}{\sqrt{2\pi}}\ \frac{1}{\pi}\left(\frac{\frac{\gamma}{2}}{\sin\frac{\gamma}{2}}\right)^{\frac{1}{2}}\left(\frac{\gamma}{\sin\frac{\theta+\gamma}{2}}\right)^{\frac{1}{2}}\frac{1}{\sin\gamma}\cdot\int_0^{\frac{\pi}{\gamma}(\theta-\gamma)}\frac{\sin\left[\frac{\pi}{\gamma}(\theta-\gamma)-t\right]}{\sqrt{t}}\mathrm{d}t.$$

If $\gamma \leqslant \theta \leqslant 1.5\gamma$, then

$$I\left(\frac{\pi}{\gamma}(\theta-\gamma)\right) \leqslant I\left(\frac{\pi}{2}\right) = 1.9549\cdots$$

and hence by (2.7)

$$R_n(\theta) > \frac{0.2947}{\sin\frac{\theta}{2}} - \frac{0.25}{\sin\gamma} > 2.$$

If $1.5\gamma < \theta \leqslant 1.84\gamma$ then by the fact that $\max\limits_{0\leqslant x\leqslant \pi} I(x) = 2.37893\cdots$, we obtain

$$R_n(\theta) > \frac{0.2947}{\sin\frac{\theta}{2}} - \frac{2.3789}{\sqrt{2\pi}\pi}\left[\frac{\gamma}{\sin(1.25\gamma)}\ \frac{\frac{\gamma}{2}}{\sin\frac{\gamma}{2}}\right]^{\frac{1}{2}} \frac{1}{\sin\gamma} > 0.7867.$$

If $1.84\gamma < \theta \leqslant 2\gamma$ then $I\left(\frac{\pi}{\gamma}(\theta-\gamma)\right) \leqslant I(0.84\pi) = 2.27875\cdots$. Hence

$$R_n(\theta) > \frac{0.2947}{\sin\frac{\theta}{2}} - \frac{2.2788}{\sqrt{2\pi}\pi}\left[\frac{\gamma}{\sin(1.42\gamma)}\ \frac{\frac{\gamma}{2}}{\sin\frac{\gamma}{2}}\right]^{\frac{1}{2}} \frac{1}{\sin\gamma} > 0.7877.$$

Case 3 $2\gamma < \theta \leqslant 3\gamma$. Now we have

$$R_n(\theta) > I_0(\theta) - I_1(\theta)$$

and

$$I_1(\theta) < \frac{1}{\pi}\left(\frac{\gamma}{\sin 2\gamma}\right)\left[\frac{1}{\sin(1.5\gamma)}\ \frac{1}{\sin\frac{\gamma}{2}}\right]^{\frac{1}{2}} \int_0^1 \frac{\sin(\pi t)}{(1+t)\sqrt{1-t}}\mathrm{d}t.$$

It is easy to see that

$$\int_0^1 \frac{\sin(\pi t)}{(1+t)\sqrt{1-t}}\mathrm{d}t = 0.65907\cdots.$$

Hence we get

$$I_1(\theta) < \frac{0.1216}{\gamma};$$

therefore

$$R_n(\theta) > \frac{0.2947}{\sin\frac{\theta}{2}} - \frac{0.1216}{\gamma} > 1.$$

Case 4 $3\gamma < \theta \leqslant 4\gamma$. Now

$$R_n(\theta) > I_0(\theta) - I_1(\theta) - J_0(\theta).$$

Noticing in this case that

$$\frac{\sin\gamma}{\sin\frac{\theta}{2}} \leqslant \frac{\sin\gamma}{\sin(1.5\gamma)} \leqslant 0.6672$$

and applying (2.8) we get

$$I_1(\theta)<\frac{0.0929}{\sin\dfrac{\theta}{2}}.$$

By (2.10) and (2.11) for $k=2$ we have

$$J_0(\theta)<0.3021\left[\frac{\gamma}{\sin(3\gamma)}\right]^{\frac{3}{2}}\cdot\frac{1}{\gamma}\leqslant\frac{0.0587}{\gamma}.$$

Hence we get

$$R_n(\theta)>\frac{0.0422}{\gamma}\geqslant 0.6649.$$

Case 5 $4\gamma<\theta\leqslant 5\gamma$. Now

$$R_n(\theta)>I_0(\theta)-I_1(\theta)-I_3(\theta)$$

and

$$\begin{aligned}I_3(\theta)&=\frac{1}{\pi}\int_0^{\gamma}\frac{\sin\left(\dfrac{\pi}{\gamma}t\right)\mathrm{d}t}{\sin(3\gamma+t)[2\cos(3\gamma+t)-2\cos\theta]^{\frac{1}{2}}}\\&<\frac{1}{\sqrt{2\pi}\pi}\cdot\frac{1}{\sin(3\gamma)}\cdot\left[\frac{\gamma}{\sin(3.5\gamma)}\cdot\frac{\dfrac{\gamma}{2}}{\sin\dfrac{\gamma}{2}}\right]^{\frac{1}{2}}I(\pi)<\frac{0.0409}{\gamma}.\end{aligned}$$

Then by (2.8)

$$I_1(\theta)<\frac{0.0801}{\sin\dfrac{\theta}{2}}.$$

Hence we have

$$R_n(\theta)>\frac{0.2146}{\sin\dfrac{\theta}{2}}-\frac{0.0409}{\gamma}>\frac{0.0449}{\gamma}\geqslant 0.7075.$$

Case 6 $5\gamma<\theta\leqslant 7\gamma$. It follows from (2.8) that

$$I_1(\theta)<\frac{0.0756}{\sin\dfrac{\theta}{2}}.$$

Applying (2.10) and (2.11) for $k=3$ we get

$$J_0(\theta)<\begin{cases}0.0277\gamma^{-1}, & 5\gamma<\theta\leqslant 5\dfrac{15}{16}\gamma,\\[2ex] 0.0226\gamma^{-1}, & 5\dfrac{15}{16}\gamma<\theta\leqslant 7\gamma.\end{cases}$$

Thereforc wc have

$$R_n(\theta)>\frac{0.2191}{\sin\frac{\theta}{2}}-J_0(\theta)>0.04\gamma^{-1}\geqslant 0.6303.$$

The proof of the lemma is complete. □

Lemma 4 If $1\leqslant n\leqslant 24$ then

$$S_n(\theta)\geqslant 0,\quad T_n(\theta)\geqslant 0,\quad 0\leqslant\theta\leqslant\frac{\pi}{2}.$$

Proof We set $x=\cos\theta$. Then, $S_n(\theta)$ and $T_n(\theta)$ are polynomials of degree $2n$ and $2n-1$ in x, which we denote by $s_n(x)$ and $t_n(x)$ respectively. It is trivial that $s_1(x)\geqslant 0$, $s_2(x)>0$, $0\leqslant x\leqslant 1$, and $t_1(x)>0$, $t_2(x)\geqslant 0$, $0\leqslant x\leqslant 1$.

For $3\leqslant n\leqslant 24$ we observe that

$$s_n(0)>0 \quad\text{and}\quad t_n(0)>0.$$

Then the positivity of $s_n(x)$ and $t_n(x)$ in $[0, 1]$ follows from the absence of zeros of these polynomials in $[0, 1]$. The application of Sturm's theorem [5, §89] reduces the verification of this fact to a straightforward computation.

In order to facilitate our calculation we have used Maple V (see [6]). Details of the programs used may be obtained from the second author. □

The proof of the positivity of $S_n(\theta)$ is now complete and, in order to achieve the proof of Theorem 1, it remains only to check $T_n(\theta)$ for $n\geqslant 25$ in the interval $\left(\frac{5\pi}{12}, \frac{\pi}{2}\right]$. In fact we go back to the method of Lemma 1 and remove fewer arches.

Lemma 5 For $n\geqslant 49$ and $\frac{5\pi}{12}<\theta\leqslant\frac{\pi}{2}$.

$$R_n(\theta)>\frac{0.2406}{\sin\frac{\theta}{2}}-\frac{0.0703}{\sin\theta}. \tag{2.13}$$

Proof As in the proof of Lemma 1 we take the integer k such that $(2k-1)\gamma<\theta\leqslant(2k+1)\gamma$, noting on this occasion that $k\geqslant 10$. we are thus able to make the estimate

$$R_n(\theta) > \sum_{\mu=0}^{7}(-1)^{\mu}I_{\mu}(\theta) - \sum_{\mu=0}^{2}(-1)^{\mu}J_{\mu}(\theta), \tag{2.14}$$

where previous notation is retained and we have written also for $\mu=1,2$,

$$J_{\mu}(\theta) = \frac{1}{\pi}\int_0^{\gamma}\frac{\sin\left(\frac{\pi}{\gamma}t\right)\mathrm{d}t}{\sin[(2k-1-\mu)\gamma+t][2\cos[(2k-1-\mu)\gamma+t]-2\cos\theta]^{\frac{1}{2}}}.$$

Note first that

$$p_1(\gamma,\ \theta)<\left[1-\frac{\sin^2\alpha}{\sin^2\left(\frac{5}{24}\pi\right)}\right]^{-\frac{1}{2}}=1.005\ 4\cdots,$$

$$p_3(\gamma,\ \theta)<\left[1-\frac{\sin^2(2\alpha)}{\sin^2\left(\frac{5}{24}\pi\right)}\right]^{-\frac{1}{2}}=1.022\ 3\cdots$$

and hence by (2.8)

$$I_1(\theta)<\frac{0.069\ 6}{\sin\frac{\theta}{2}},\qquad I_3(\theta)<\frac{0.030\ 1}{\sin\frac{\theta}{2}}. \tag{2.15}$$

To estimate $I_4(\theta)-I_5(\theta)$ and $I_6(\theta)-I_7(\theta)$ we write

$$q_{\mu}(\gamma,\ \theta)=\frac{\sin(\mu\gamma)}{\sin[(\mu+1)\gamma]}\left\{\frac{\cos(\mu\gamma)-\cos\theta}{\cos[(\mu+1)\gamma]-\cos\theta}\right\}^{\frac{1}{2}}.$$

It is not hard to prove that

$$I_4(\theta)-I_5(\theta)>[1-q_5(\gamma,\ \theta)]I_4(\theta),$$

$$I_6(\theta)-I_7(\theta)>[1-q_7(\gamma,\ \theta)]I_6(\theta).$$

It is then easy to see that

$$I_4(\theta)>\frac{0.022\ 6}{\sin\frac{\theta}{2}},\qquad I_6(\theta)>\frac{0.015\ 6}{\sin\frac{\theta}{2}},$$

and

$$q_5(\gamma,\ \theta)\leqslant\frac{\sin(5\alpha)}{\sin(6\alpha)}\left\{\frac{\cos(5\alpha)-\cos\left(\frac{5}{12}\pi\right)}{\cos(6\alpha)-\cos\left(\frac{5}{12}\pi\right)}\right\}^{\frac{1}{2}}=0.853\ 0\cdots,$$

$$q_7(\gamma,\ \theta)\leqslant\frac{\sin(7\alpha)}{\sin(8\alpha)}\left\{\frac{\cos(7\alpha)-\cos\left(\frac{5}{12}\pi\right)}{\cos(8\alpha)-\cos\left(\frac{5}{12}\pi\right)}\right\}^{\frac{1}{2}}=0.904\ 6\cdots.$$

Hence, we obtain

$$I_4(\theta)-I_5(\theta)>\frac{0.0033}{\sin\frac{\theta}{2}},\qquad I_6(\theta)-I_7(\theta)>\frac{0.0015}{\sin\frac{\theta}{2}}.\tag{2.16}$$

Going back to (2.10) and (2.11) we deduce that, for $\frac{5\pi}{12}<\theta\leqslant\frac{\pi}{2}$,

$$J_0(\theta)<\begin{cases}\dfrac{0.3075}{\sin\theta}\left[\dfrac{\gamma}{\sin\frac{1}{2}(\theta+(2k-1)\gamma)}\right]^{\frac{1}{2}}, & (2k-1)\gamma<\theta\leqslant\dfrac{2k-1}{16}\gamma\\[2ex] \dfrac{0.2684}{\sin\theta}\left[\dfrac{\gamma}{\sin\frac{1}{2}(\theta+(2k-1)\gamma)}\right]^{\frac{1}{2}}, & \left(2k-\dfrac{1}{16}\right)\gamma<\theta\leqslant(2k+1)\gamma.\end{cases}\tag{2.17}$$

Moreover we have

$$J_1(\theta)-J_2(\theta)>[1-q_{2k-2}(\gamma,\theta)^{-1}]J_1(\theta).$$

In the case $(2k-1)\gamma<\theta\leqslant\left(2k-\frac{1}{16}\right)\gamma$ we have

$$\begin{aligned}q_{2k-2}(\gamma,\theta)^{-1}\leqslant&\left[\cos\alpha+\sin\alpha\cot\left(\frac{5}{12}\pi-\frac{31}{16}\alpha\right)\right]\cdot\\&\left\{\left[\cos\frac{\alpha}{2}+\sin\frac{\alpha}{2}\cot\left(\frac{5}{12}\pi-\frac{31}{32}\alpha\right)\right]\cdot\frac{\sin\left(\frac{15}{32}\alpha\right)}{\sin\left(\frac{31}{32}\alpha\right)}\right\}^{\frac{1}{2}}\\=&0.7158\cdots.\end{aligned}$$

Accordingly

$$J_1(\theta)-J_2(\theta)>0.2842J_1(\theta).$$

We have for $(2k-1)\gamma<\theta\leqslant\left(2k-\frac{1}{16}\right)\gamma$,

$$\begin{aligned}J_1(\theta)>&\frac{1}{\pi}\frac{1}{\sin\theta}\left(\frac{\gamma}{2\pi\sin\frac{\theta+(2k-1)\gamma}{2}}\right)^{\frac{1}{2}}\int_0^{\pi}\frac{\sin t}{\left(\frac{15}{16}\pi+t\right)^{\frac{1}{2}}}\mathrm{d}t\\=&\frac{0.1206}{\sin\theta}\cdot\left(\frac{\gamma}{\sin\frac{\theta+(2k-1)\gamma}{2}}\right)^{\frac{1}{2}}.\end{aligned}$$

Hence we get

$$J_1(\theta)-J_2(\theta)>\frac{0.0343}{\sin\theta}\left(\frac{\gamma}{\sin\frac{\theta+(2k-1)\gamma}{2}}\right)^{\frac{1}{2}}.\tag{2.18}$$

Combining (2.17) and (2.18) we get

$$J_0(\theta)-J_1(\theta)+J_2(\theta)<\frac{0.273\,2}{\sin\theta}\cdot\left[\frac{\alpha}{\sin\left(\frac{5}{12}\pi-\frac{15}{32}\alpha\right)}\right]^{\frac{1}{2}}=\frac{0.070\,3}{\sin\theta}.$$

In the case $\left(2k-\frac{1}{16}\right)\gamma<\theta\leqslant(2k+1)\gamma$ it follows from (2.17) that

$$J_0(\theta)-J_1(\theta)+J_2(\theta)<J_0(\theta)<\frac{0.069\,5}{\sin\theta}.$$

Combining the two cases, we get

$$J_0(\theta)-J_1(\theta)+J_2(\theta)<\frac{0.070\,3}{\sin\theta}. \tag{2.19}$$

Now substituting (2.7)(2.15)(2.16) and (2.19) into (2.14) we get (2.13) and this completes the proof of the lemma.

Lemma 6 If $n\geqslant 25$ and $\frac{5\pi}{12}<\theta\leqslant\frac{\pi}{2}$ then $T_n(\theta)>0$.

Proof Let

$$F(\theta)=\frac{\sqrt{15}}{45}\sin\theta-0.5\sin\frac{\theta}{2}+0.481\,2\cos\frac{\theta}{2}-0.070\,3.$$

Then F is decreasing for $0\leqslant\theta\leqslant\frac{\pi}{2}$ and $F\left(\frac{\pi}{2}\right)>0$. The combination of (2.2) and Lemma 5 now gives the result. □

At this stage the proof of Theorem 1 is complete and we turn to the proof of Theorem 2. As well as estimating $R_n(\theta)$ (our existing estimates are more than adequate) we need to estimate the analogous cosine integral, as the next lemma shows.

Lemma 7

$$2U_n(\theta)=R_{4n+1}(\theta)+Q_{4n+1}(\theta)+Q_0(\theta)+\frac{1}{4\cos\frac{\theta}{2}}-1. \tag{2.20}$$

Proof Using the Dirichlet-Mehler formula we have

$$U_n(\theta)=\frac{1}{2}+\frac{1}{\pi}\int_0^\theta\frac{1}{(2\cos t-2\cos\theta)^{\frac{1}{2}}}\cdot\frac{\sin\left(4n+\frac{5}{2}\right)t-\sin\frac{5}{2}t}{\sin 2t}\mathrm{d}t.$$

Since

$$\frac{\sin\left(4n+\frac{5}{2}\right)t-\sin\frac{5}{2}t}{\sin 2t}=$$

$$\frac{\sin\left(4n+\frac{3}{2}\right)t}{2\sin t}+\frac{\cos\left(4n+\frac{3}{2}\right)t}{2\cos t}-2\cos\frac{t}{2}+\frac{1}{4\cos\frac{t}{2}}+\frac{\cos\frac{t}{2}}{2\cos t},$$

$$P_0(\cos\theta)=\frac{2}{\pi}\int_0^\theta\frac{\cos\frac{t}{2}\mathrm{d}t}{(2\cos t-2\cos\theta)^{\frac{1}{2}}}=1$$

and

$$\frac{1}{4\pi}\int_0^\theta\frac{1}{\cos\frac{t}{2}}\frac{\mathrm{d}t}{(2\cos t-2\cos\theta)^{\frac{1}{2}}}=\frac{1}{8}\frac{1}{\cos\frac{\theta}{2}}=\frac{1}{2}R_0(\theta),$$

we obtain (2. 20). Note also that

$$Q_0(\theta)=\frac{1}{2\sqrt{\cos\theta}}.\qquad\square$$

Assume that $n\geqslant 12$ and write $\delta=\delta_n=\frac{\pi}{4n+\frac{3}{2}}$. we note that $\delta\leqslant\alpha$.

Define $\varphi_0=0$, $\varphi_k=\left(k-\frac{3}{4}\right)\delta$, $k=1,2,\cdots,2n+1$, $\varphi_{2n+2}=\left(2n+\frac{3}{4}\right)\delta=\frac{\pi}{2}$. Write

$$h_\varphi(t)=\frac{1}{\sin t(2\sin t-2\sin\varphi)^{\frac{1}{2}}}\quad\text{for}\quad 0<\varphi<t\leqslant\frac{\pi}{2}.$$

Write

$$f(\varphi)=Q_{4n+1}\left(\frac{\pi}{2}-\varphi\right)=\frac{1}{\pi}\int_\varphi^{\frac{\pi}{2}}h_\varphi(t)\sin\left(\frac{\pi}{\delta}t-\frac{\pi}{4}\right)\mathrm{d}t,\quad 0<\varphi\leqslant\frac{\pi}{2},$$

$$f_0(\varphi)=Q_0\left(\frac{\pi}{2}-\varphi\right)=\frac{1}{\pi}\int_\varphi^{\frac{\pi}{2}}h_\varphi(t)\sin\left(\frac{t}{2}+\frac{\pi}{4}\right)\mathrm{d}t,\quad 0<\varphi\leqslant\frac{\pi}{2}.$$

We note that

$$f_0(\varphi)=\frac{1}{2}\frac{1}{\sin^{\frac{1}{2}}\varphi},\qquad 0<\varphi\leqslant\frac{\pi}{2}.$$

Define for $0<\varphi\leqslant\varphi_k$, $k=1,2,\cdots,2n+1$.

$$a_k(\varphi)=(-1)^{k-1}\frac{1}{\pi}\int_{\varphi_k}^{\varphi_{k+1}}h_\varphi(t)\sin\left(\frac{\pi}{\delta}t-\frac{\pi}{4}\right)\mathrm{d}t.$$

We observe that

$$a_k(\varphi) = \frac{1}{\pi}\int_0^{\delta} \sin\left(\frac{\pi}{\delta}t\right)h_{\varphi}(\varphi_k + t)\mathrm{d}t, \quad k = 1,2,\cdots,2n,$$

$$a_{2n+1}(\varphi) = \frac{1}{\pi}\int_0^{\frac{\delta}{2}} \sin\left(\frac{\pi}{\delta}t\right)h_{\varphi}(\varphi_{2n+1} + t)\mathrm{d}t.$$

By the periodicity of the function $\sin\left(\left(\frac{\pi}{\delta}\right)t\right)$ we get for $\varphi_k < \varphi \leqslant \varphi_{k+1}$, $k=0,1,2,\cdots,2n$,

$$f(\varphi) = \frac{1}{\pi}\int_{\varphi}^{\varphi_{k+1}} h_{\varphi}(t)\cdot\sin\left(\frac{\pi}{\delta}t - \frac{\pi}{4}\right)\mathrm{d}t + \sum_{j=k+1}^{2n}(-1)^{j-1}a_j(\varphi) + a_{2n+1}(\varphi)$$

Where as a convention we let $\sum\limits_{j=2n+1}^{2n} = 0$.

Since the function $h_{\varphi}(t)\left(\varphi \leqslant t \leqslant \frac{\pi}{2}\right)$ is decreasing we conclude that

(1) If $\varphi_{2k} < \varphi \leqslant \varphi_{2k+1}$, $k=0,1,2,\cdots,n$, then

$$f(\varphi) > \frac{1}{\pi}\int_{\varphi}^{\varphi_{2k+1}} h_{\varphi}(t)\cdot\sin\left(\frac{\pi}{\delta}t - \frac{\pi}{4}\right)\mathrm{d}t. \tag{2.21}$$

(2) If $\varphi_{2k-1} < \varphi \leqslant \varphi_{2k}$, $k=1,2,\cdots,n$, then

$$f(\varphi) > -a_{2k}(\varphi) \geqslant -a_{2k}(\varphi_{2k}). \tag{2.22}$$

(3) If $\varphi_{2n+1} < \varphi \leqslant \varphi_{2n+2} = \frac{\pi}{2}$ then $f(\varphi) \geqslant 0$.

We define for $k=0,1,2,\cdots,n$,

$$b_k(\varphi) = -\frac{1}{\pi}\int_{\varphi}^{\varphi_{2k+1}} h_{\varphi}(t)\sin\left(\frac{\pi}{\delta}t - \frac{\pi}{4}\right)\mathrm{d}t, \quad \varphi_{2k} \leqslant \varphi \leqslant \varphi_{2k+1}.$$

Then we see that

$$b_k(\varphi) = \frac{1}{\pi}\int_{\varphi-\varphi_{2k}}^{\delta} h_{\varphi}(\varphi_{2k} + t)\sin\left(\frac{\pi}{\delta}t\right)\mathrm{d}t, \quad \varphi_{2k} \leqslant \varphi \leqslant \varphi_{2k+1}.$$

Since

$$h_{\varphi}(\varphi_{2k}+t) < \frac{1}{\sqrt{2}}\frac{1}{\sin\varphi}\cdot\frac{1}{\sqrt{\cos\left(\varphi+\frac{t}{2}\right)}}\frac{1}{\sqrt{t-(\varphi-\varphi_{2k})}}\left(\frac{\frac{\delta}{2}}{\sin\frac{\delta}{2}}\right)^{\frac{1}{2}},$$

$$\varphi-\varphi_{2k} < t \leqslant \delta,$$

we get

$$b_k(\varphi)<\frac{1}{\sqrt{2\pi}}\frac{1}{\pi}\cdot\left(\frac{\frac{\delta}{2}}{\sin\frac{\delta}{2}}\right)^{\frac{1}{2}}\cdot\frac{1}{\sqrt{\sin\varphi}}\left[\frac{\delta}{\sin\varphi\cos\left(\varphi+\frac{\delta}{2}\right)}\right]^{\frac{1}{2}}\cdot$$

$$\int_0^{\left(1-\frac{\varphi-\varphi_{2k}}{\delta}\right)\pi}\sin\left[\left(1-\frac{\varphi-\varphi_{2k}}{\delta}\right)\pi-t\right]\frac{\mathrm{d}t}{\sqrt{t}}.$$

We know that $\max\limits_{0\leqslant x\leqslant\pi}\int_0^x\frac{\sin(x-t)}{\sqrt{t}}\mathrm{d}t=2.378\,93\cdots$. Therefore,

$$b_k(\varphi)<\frac{0.302\,1}{\sqrt{\sin\varphi}}\left\{\frac{2\delta}{\sin\left(2\varphi+\frac{\delta}{2}\right)-\sin\frac{\delta}{2}}\right\}^{\frac{1}{2}},\quad \varphi_{2k}\leqslant\varphi\leqslant\varphi_{2k+1}. \tag{2.23}$$

From this it follows that

$$b_1(\varphi)<\frac{0.271\,2}{\sqrt{\sin\varphi}}. \tag{2.24}$$

Also, if $2\leqslant k\leqslant n-2$ then the condition $\varphi_{2k}\leqslant\varphi\leqslant\varphi_{2k+1}$ implies that $\sin\left(2\varphi+\frac{\delta}{2}\right)\geqslant\sin(7\delta)$. Hence

$$b_k(\varphi)\leqslant\frac{0.302\,1}{\sqrt{\sin\varphi}}\left\{\frac{2\delta}{\sin(7\delta)-\sin\frac{\delta}{2}}\right\}^{\frac{1}{2}}\leqslant\frac{0.170\,6}{\sqrt{\sin\varphi}}. \tag{2.25}$$

Now we can easily prove the following.

Lemma 8 Let $V_n(\theta)=Q_{4n+1}(\theta)+Q_0(\theta)$. If $n\geqslant12$, then

$$V_n(\theta)>\begin{cases}\dfrac{0.228\,8}{\sqrt{\cos\theta}}, & \dfrac{5}{2}\delta<\theta\leqslant\dfrac{9}{2}\delta,\\[2mm] \dfrac{0.329\,4}{\sqrt{\cos\theta}}, & \dfrac{9}{2}\delta<\theta\leqslant\dfrac{\pi}{2}-\dfrac{9}{4}\delta,\\[2mm] 0.606\,5, & \dfrac{\pi}{2}-\dfrac{9}{4}\delta<\theta\leqslant\dfrac{\pi}{2}.\end{cases}$$

Proof By (2.21) we see that if $0<\varphi\leqslant\varphi_1$, then

$$f(\varphi)>\frac{1}{\pi}\int_{\varphi}^{\frac{\delta}{4}}h_\varphi(t)\sin\left(\frac{\pi}{\delta}t-\frac{\pi}{4}\right)\mathrm{d}t.$$

However

$$f_0(\varphi)=\frac{1}{\pi}\int_{\varphi}^{\frac{\delta}{4}}h_\varphi(t)\sin\left(\frac{t}{2}+\frac{\pi}{4}\right)\mathrm{d}t+\frac{1}{\pi}\int_{\frac{\delta}{4}}^{\frac{\pi}{2}}h_\varphi(t)\sin\left(\frac{t}{2}+\frac{\pi}{4}\right)\mathrm{d}t,$$

so we get

$$V_n\left(\frac{\pi}{2}-\varphi\right)>\frac{1}{\pi}\int_{\frac{\delta}{4}}^{\frac{\pi}{2}}h_\varphi(t)\sin\left(\frac{t}{2}+\frac{\pi}{4}\right)\mathrm{d}t$$

$$>\frac{\cos\frac{\delta}{4}}{\left(\sin\frac{\delta}{4}\right)^{\frac{1}{2}}}\frac{1}{\sqrt{2}\pi}\sin\left(\frac{\pi}{4}+\frac{\delta}{8}\right)>1,\quad 0<\varphi\leqslant\frac{\delta}{4},$$

If $\varphi_1<\varphi\leqslant\varphi_2$, it follows from (2.22) and (2.24) that

$$f(\varphi)>-a_2(\varphi_2)=-b_1(\varphi_2)>-\frac{0.2712}{\sqrt{\sin\varphi_2}}.$$

However

$$f_0(\varphi)=\frac{0.5}{\sqrt{\sin\varphi}}.$$

Hence

$$V_n\left(\frac{\pi}{2}-\varphi\right)>\frac{0.2288}{\sqrt{\sin\varphi_2}}\geqslant 0.8127,\qquad \frac{\delta}{4}<\varphi\leqslant\frac{5}{4}\delta.$$

When $\varphi_2<\varphi\leqslant\varphi_3$ we make use of (2.21) and (2.24) to get

$$f(\varphi)>-\frac{0.2712}{\sqrt{\sin\varphi}}.$$

Then we get

$$V_n\left(\frac{\pi}{2}-\varphi\right)>\frac{0.2288}{\sqrt{\sin\varphi}}\geqslant 0.6065\quad\text{for}\quad \frac{5}{4}\delta<\varphi\leqslant\frac{9}{4}\delta.$$

Combining the above results we get

$$V_n(\theta)>0.6065\quad\text{for}\quad \frac{\pi}{2}-\frac{9}{4}\delta\leqslant\theta\leqslant\frac{\pi}{2}.$$

On the other hand for $\varphi_{2k-1}<\varphi\leqslant\varphi_{2k}(k=2,3,\cdots,n-2)$ it follows from (2.22) that

$$f(\varphi)>-a_{2k}(\varphi_{2k})=-b_k(\varphi_{2k}). \tag{2.26}$$

Hence for $k=2,3,\cdots,n-2$ from (2.21)(2.25) and (2.26) we have

$$f(\varphi)>-\frac{0.1706}{\sqrt{\sin\varphi}},\qquad \varphi_{2k-1}<\varphi\leqslant\varphi_{2k+1}.$$

Therefore

$$V_n(\theta)>\frac{0.3294}{\sqrt{\cos\theta}},\qquad \frac{9}{2}\delta\leqslant\theta<\frac{\pi}{2}-\frac{9}{4}\delta.$$

If $\varphi_{2n-3}<\varphi\leqslant_{2n-1}$, then $\sin\left(2\varphi+\frac{\delta}{2}\right)-\sin\frac{\delta}{2}\geqslant\sin(3\delta)-\sin\frac{\delta}{2}$.

Finally, $b_{n-1}(\varphi)<\frac{0.3021}{\sqrt{\sin\varphi}}\cdot\left\{\frac{2\delta}{\sin 3\delta-\sin\frac{\delta}{2}}\right\}^{\frac{1}{2}}\leqslant\frac{0.2712}{\sqrt{\sin\varphi}}$.

Hence we conclude that $V_n(\theta)>\frac{0.2288}{\sqrt{\cos\theta}}, \quad \frac{5}{2}\delta<\theta\leqslant\frac{9}{2}\delta$.

The proof is complete. □

Lemma 9 If $n\geqslant 12\left(0<\theta\leqslant\frac{\pi}{2}\right)$, then $U_n(\theta)>0.01755$.

Proof If $\frac{5}{2}\delta<\theta\leqslant 7\delta$ then by Lemma 3, Lemma 7 and Lemma 8 we have

$$R_{4n+1}(\theta)>0.6303, \quad Q_{4n+1}(\theta)+Q_0(\theta)>0.2288.$$

Hence

$$2U_n(\theta)>0.6303+0.2288+0.25-1=0.1091.$$

When $7\delta<\theta\leqslant\frac{\pi}{2}-\frac{9}{4}\delta$, we use Lemma 1 and Lemma 8 to get

$$R_{4n+1}(\theta)>\frac{0.2273}{\sin\frac{\theta}{2}}-\frac{0.1304}{\sin\theta}, \quad Q_{4n+1}(\theta)+Q_0(\theta)>\frac{0.3294}{\sqrt{\cos\theta}}.$$

Write

$$G(\theta)=\frac{0.2273}{\sin\frac{\theta}{2}}-\frac{0.1304}{\sin\theta}+\frac{0.3294}{\sqrt{\cos\theta}}+\frac{0.25}{\cos\frac{\theta}{2}}-1.$$

From Lemma 7 we get $\quad 2U_n(\theta)>G(\theta), \quad 7\delta<\theta\leqslant\frac{\pi}{2}-\frac{9}{4}\delta$.

It can be easily checked that

$$\min_{0<\theta<\frac{\pi}{2}} G(\theta)=G(0.94664\cdots)=0.0497\cdots.$$

Hence we get $\quad 2U_n(\theta)>0.0497, \quad 7\delta<\theta\leqslant\frac{\pi}{2}-\frac{9}{4}\delta$.

When $\frac{\pi}{2}-\frac{9}{4}\delta<\theta\leqslant\frac{\pi}{2}$ by Lemma 7, Lemma 1 and Lemma 8 we have

$$2U_n(\theta)>\frac{0.2273}{\sin\frac{\theta}{2}}-\frac{0.1304}{\sin\theta}+\frac{0.25}{\cos\frac{\theta}{2}}-0.3935>0.13$$

for

$$\frac{\pi}{2}-\frac{9}{4}\delta<\theta<\frac{\pi}{2}.$$

Finally, let $0<\theta\leqslant\frac{5}{2}\delta$. We simply have

$$Q_{4n+1}(\theta)+Q_0(\theta)>\left(1-\frac{1}{\cos\left(\frac{5}{4}\delta\right)}\right)Q_0(\theta)\geqslant-\frac{0.001\,6}{\sqrt{\cos\theta}}\geqslant-0.001\,6.$$

From Lemma 3 and Lemma 7 we get

$$2U_n(\theta)>0.786\,7-0.001\,6+0.25-1=0.035\,1.$$

The proof of the lemma is complete. □

Lemma 10 $U_n(\theta)>0$ for $n=1,2,\cdots,11$.

Proof We set $x=\cos\theta$, then $U_n(\theta)$ is a polynomial of degree $4n$ in x which we denote by $u_n(x)$. We observe that $u_n(0)>0$, $n=1,2,\cdots,11$ and the positivity of the polynomials $u_n(x)$ in $[0,1]$, $n=1,2,\cdots,11$, follows from the absence of zeros of these polynomials in $[0,1]$, as can be directly verified. In fact, as in Lemma 4 we apply the method of Sturm sequences to obtain the desired result. Details of the program used may also be obtained from the second author. □

The combination of Lemma 9 with Lemma 10 yields Theorem 2.

Remark 1 Let us consider the more general inequalities

$$\frac{1}{2}+\sum_{k=1}^{n}P_{2vk}(\cos\theta)>0,\quad v\in\mathbf{N}^*.\tag{2.27}$$

The cases $v=1,2$ are settled by Theorem 1 and Theorem 2 respectively.

It seems that (2.27) is true when $v=3,4,5$ for all n and $0<\theta<\pi$. For these cases, by applying the method of Sturm sequences as in Lemma 4 and Lemma 10 we can establish (2.27) for $n=1,2,\cdots,10$.

The proof of (2.27) when $v=3,4,5$ for $n\geqslant 11$, in the same way as in Theorem 1 and Theorem 2, appears to be very complicated, so another method will have to be developed to prove (2.27) when $v=3,4,\cdots$ for all n. It would also be very interesting to find some applications of these inequalities.

References

[1] Askey R. Positive Jacobi polynomial sums. Tôhoku Math. J., 1972, 24: 109—119.

[2] Askey R, Fitch J. Integral representations for Jacobi polynomials and some

applications. J. Math. Anal. Appl., 1969, 26: 411—437.

[3] Askey R, Gasper G. Positive Jacobi polynomial sums Ⅱ. Amer. J. Math., 1976, 98: 709—737.

[4] Brown G, Koumandos S, Wang K. Positivity of Cotes numbers at more Jacobi abscissas. Monatsh. Math., 1996, 122: 9—19.

[5] Burnside W S, Panton A W. Theory of Equations. Dublin University Press, 1886.

[6] Char B W, Geddes K O, Gonnet G H, Leong B L, Monagan M B, Watt S M. Maple V first leaves: a tutorial introduction to Maple V and library reference manual. Springer, 1992.

[7] Fejér L. Sur le développement d'une fonction arbitraire suivant les fonctions de Laplace. C. R. Acad. Sci. Paris, 1908: 224—227; Gesammelte Arbeiten I. Birkhäuser, Basel, 1970: 319—322.

[8] Fejér L. Über die Laplacesche Reihe. Math. Ann., 1909, 67: 76—109; Gesammelte Arbeiten I. Birkhäuser, Basel, 1970: 503—537.

[9] Fejér L. Ultrasphärikus polynomok összegéről. Mat. Fiz. Lapok, 1931, 38: 161—164; Gesammelte Arbeiten Ⅱ. Birkhäuser, Basel, 1970: 421—423.

[10] Feldheim E. On the positivity of certain sums of ultraspherical polynomials. J. Anal. Math, 1963, 11: 275—284; reprinted in G. Szegö, Collected papers-Vol. 3. Birkhäuser, Boston, 1982: 821—830.

[11] Gasper G. Positive sums of the classical orthogonal polynomials. SIAM J. Math. Anal. 1977, 8: 423—447.

[12] Koumandos S. Some positive Cotes numbers for the Jacobi weight function. Computational methods and function theory. 1994. Series in Approximations and Decompositions 5, World Scientific Publishing, 1995: 197—205.

[13] Szegö G. Ultrasphaerikus polynomok összegéről. Mat. Fiz. Lapok, 1938, 45: 36 — 38; reprinted in G. Szegö, Collected papers-Vol. 2. Birkhäuser, Boston, 1982: 700—703.

[14] Szegö G, Orthogonal polynomials. American Mathematical Society Colloquium Publications 23, 4th edn, American Mathematical Society, Providence, RI, 1975.

Ramanujan Journal，2007，14(3)：471－507.

Vietoris 不等式之推广（Ⅰ）

Extensions of Vietoris's inequalities (Ⅰ)①

Abstract Let $\beta_0=0.308\ 443\cdots$ denote the Littlewood-Salem-Izumi number, i. e. , the unique solution of the equation

$$\int_0^{\frac{3}{2}\pi}\frac{\cos t}{t^\beta}\mathrm{d}t=0.$$

In this paper it is proved that if $a_0\geqslant a_1\geqslant\cdots\geqslant a_n>0$ and $a_{2k}\leqslant\left(1-\frac{\beta_0}{k}\right)a_{2k-1}$, $k\geqslant 1$ then for all $\theta\in(0,\pi)$,

$$\sum_{k=0}^{n}a_k\cos k\theta>0,$$

and furthermore, the number β_0 is best possible in the sense that for any $\beta\in(0,\beta_0)$,

$$\lim_{n\to+\infty}\min_{\theta\in(0,\pi)}\sum_{k=0}^{n}c_k(\beta)\cos k\theta=-\infty,$$

where the coefficients $c_k(\beta)$ are defined as

$$c_0(\beta)=c_1(\beta)=1,\quad c_{2k}(\beta)=c_{2k+1}(\beta)=\left(1-\frac{\beta}{k}\right)c_{2k-1}(\beta),\quad k\geqslant 1.$$

---

① This research was supported by a grant from the Australian Research Council. The Dai F was also supported in part by the NSERC Canada under grant G121211001. The Wang K Y was also supported in part by the NSF of China under grant 10471010.

本文与 Brown G 和 Dai Feng 合作.

Received：2005-04-14；Accepted：2005-06-07.

Results for the sine sums are obtained as well.

These results generalize and sharpen the well known trigonometric inequalities of Vietoris.

Keywords Vietoris's inequalities; Positive trigonometric sums.

§ 1. Introduction and main results

In 1958 L. Vietoris [11] proved that if $a_0 \geqslant a_1 \geqslant \cdots \geqslant a_n > 0$ and $a_{2k} \leqslant \left(1-\frac{1}{2k}\right)a_{2k-1}$, $k \geqslant 1$, then for all $n \geqslant 1$ and $\theta \in (0, \pi)$,

$$\sum_{k=1}^{n} a_k \sin k\theta > 0 \tag{1.1}$$

and

$$\sum_{k=0}^{n} a_k \cos k\theta > 0, \tag{1.2}$$

which extend both the Fejér-Jackson-Gronwall inequality [9],

$$\sum_{k=1}^{n} \frac{\sin k\theta}{k} > 0, \quad 0 < \theta < \pi,$$

and W. H. Young's inequality [12],

$$1 + \sum_{k=1}^{n} \frac{\cos k\theta}{k} > 0, \quad 0 < \theta < \pi.$$

As shown by Vietoris himself, his result is actually equivalent to the following

Theorem A (Vietoris [11]) If

$$a_0 = a_1 = 1, \quad a_{2k} = a_{2k+1} = \left(1 - \frac{1}{2k}\right)a_{2k-1}, \quad k \in \mathbf{N},$$

then for any $n \in \mathbf{N}$ and $\theta \in (0, \pi)$,

$$\sum_{k=1}^{n} a_k \sin k\theta > 0, \quad \sum_{k=0}^{n} a_k \cos k\theta > 0.$$

These inequalities of Vietoris have turned out to be very useful. In fact, it was shown by Askey and Steinig [4] in 1974 that they can be applied to yield various new results, including improved estimates for the localization of zeros of a class of trigonometric polynomials and new positive sums of ultraspherical polynomials. Furthermore, in his report

[3] Askey showed that one of the problems suggested by these inequalities leads to the derivation of certain hypergeometric summation formulas.

Due to the importance of Vietoris's inequalities, any extension of these inequalities would be of interest. One recent work in this direction was done by Brown and Yin [6] and their main result can be stated as follows.

Theorem B (Brown and Yin [6]) If $\alpha\in(-1, 2]$ and

$$a_0(\alpha)=a_1(\alpha)=1,\quad a_{2k}(\alpha)=a_{2k+1}(\alpha)=\left(1-\frac{1}{2k+\alpha}\right)a_{2k-1},\quad k\in\mathbf{N},$$

then for all $\theta\in(0, \pi)$ and $n\geqslant 1$,

$$S_n^\alpha(\theta)=\sum_{k=0}^{n}a_k(\alpha)\cos k\theta>0.$$

In the case $\alpha=1$, Theorem B was previously proved by Brown and Hewitt [5].

The best possible extension of the Brown-Yin inequality (Theorem B) has been obtained recently by the current authors in [8], where we proved the following

Theorem C ([8]) Let $S_n^\alpha(\theta)$ be as defined in Theorem B and let α_0 $:=2.3308\cdots$ be the unique solution of the equation

$$\min_{x\in(0,\pi)}S_6^\alpha(x)=0.$$

Then following statements are true:

(1) For any $-1\leqslant\alpha\leqslant\alpha_0$,

$$S_n^\alpha(\theta)\geqslant 0,\quad \theta\in(0,\pi),\quad n\in\mathbf{N},$$

with equality holding for some $x\in(0, \pi)$ only when $\alpha=\alpha_0$ and $n=6$.

(2) For any $\alpha>\alpha_0$, there exists a positive integer $N(\alpha)$ depending only on α such that for all $\theta\in(0, \pi)$,

$$S_n^\alpha(\theta)>0,\quad \text{whenever } n\geqslant N(\alpha),$$

and moreover, the smallest positive integer $N(\alpha)$ here is greater than 6 whenever $\alpha>\alpha_0$, and tends $t_0+\infty$ as $\alpha\to+\infty$.

In the present paper we will extend Vietoris's inequality (1.2) in another best possible way. To be precise, for $\beta \in (0, 1)$, we define, $c_j(\beta)$, $j \geqslant 0$, inductively, by

$$c_0(\beta) = c_1(\beta) = 1, \quad c_{2k}(\beta) = c_{2k+1}(\beta) = \left(1 - \frac{\beta}{k}\right) c_{2k-1}(\beta), \quad k \geqslant 1, \tag{1.3}$$

and for $n \geqslant 1$, define

$$T_n^\beta(\theta) := \sum_{k=0}^{n} c_k(\beta) \cos k\theta. \tag{1.4}$$

Our main result in this paper is stated in the following

Theorem 1.1 For $0 < \theta < \pi$ and $n \geqslant 1$, the inequality,

$$T_n^\beta(\theta) = \sum_{k=0}^{n} c_k(\beta) \cos k\theta > 0, \tag{1.5}$$

is true whenever $\beta_0 \leqslant \beta < 1$, where $\beta_0 \in (0.308\,443, 0.308\,444)$ is the unique solution of the equation

$$\int_0^{\frac{3\pi}{2}} \frac{\cos t}{t^\beta} \mathrm{d}t = 0. \tag{1.6}$$

Furthermore, β_0 is best possible in the sense that for any $\beta \in (0, \beta_0)$ we have

$$\lim_{n \to +\infty} \min_{\theta \in (0, \pi)} T_n^\beta(\theta) = -\infty.$$

As a consequence, we have, by Abel's transform,

Corollary 1.2 Let $\beta_0 \in (0.308\,443, 0.308\,444)$ be defined as in Theorem 1.1 and let $\{a_k\}_{k=0}^n$ be a sequence of real numbers satisfying $a_0 \geqslant a_1 \geqslant \cdots \geqslant a_n > 0$ and $a_{2k} \leqslant \left(1 - \frac{\beta_0}{k}\right) a_{2k-1}$, $k \geqslant 1$. Then for $\theta \in (0, \pi)$ we have

$$\sum_{k=0}^{n} a_k \cos k\theta > 0.$$

It is of some historical interest that the critical constant β_0 is the Littlewood-Salem-Izumi number which is critical for the uniform boundedness of the partial sums, $1 + \sum_{k=1}^{n} k^{-\beta} \cos k\theta$. It was shown by Brown, Wang and Wilson in [7] that this same constant is also critical for positivity of these partial sums. Although we have $c_{2m+2}(\beta) = O(m^{-\beta})$ it would have been unreasonable to conjecture the truth of Theorem 1.1 on this basis.

Since Vietoris's inequality (1.2) implies or is related to many other inequalities (see for example [3] and [4]), we believe that some of these other inequalities can also be extended due to our result here. But we shall not concern ourselves with this problem in the present paper because our arguments have already been very long.

We organize this paper as follows. Section 2 contains a series of useful lemmas (Lemmas 2.1～2.5), without proofs except for Lemma 2.1. Assuming these lemmas, we proceed with the proof of Theorem 1.1 in Sect. 3. In Sect. 4, we prove some new estimates related to gamma functions, which play an important role in the proofs of Lemmas 2.2～2.5 and which seem to be of independent interest. More precisely, in Lemma 4.1 we prove a new estimate involving the asymptotic expansion of the function $\frac{\Gamma(x+1-\beta)}{\Gamma(x+1)}$ as $x \to +\infty$ and the advantage of this new estimate lies in the fact that we have a much better control of the remainder. Furthermore, in Lemma 4.3 of Sect. 4, we obtain an important asymptotic expression for the function $\mu(x)$ appearing in the well known formula (see [2, p. 24])

$$\Gamma(x) = \sqrt{2\pi} x^{x-\frac{1}{2}} \mathrm{e}^{-x+\mu(x)}, \quad x > 0$$

and as a result, we correct an error in the celebrated book [2] by E. Artin. Based on the estimates in Sect. 4, we deduce two more useful expressions for the sums $T_n^\beta(\theta)$ in Sect. 5. Lemmas 2.2～2.5 are then proved in Sect. 6. Finally, in Sect. 7 we conclude the paper with some remarks concerning the corresponding sine sums. Proofs of these remarks are also given there.

A different proof of Theorem 1.1 was obtained independently by S. Koumandos in [10] in the same issue of this journal.

§ 2. Statements of the lemmas

We first define the following two functions:

$$\gamma_m^\beta(\theta) = \Gamma(1-\beta)(2\sin\theta)^{1-\beta}\sum_{k=0}^{m} c_{2k}(\beta)\cos\left(2k+\frac{1}{2}\right)\theta, \tag{2.1}$$

$$\mu_m^\beta(\theta) = \frac{\Gamma(1-\beta)(\sin\theta)^{1-\beta}}{2^\beta\cos\frac{\theta}{2}} c_{2m+2}(\beta)\cos(2m+2)\theta. \tag{2.2}$$

We will keep the notations γ_m^β and μ_m^β throughout the rest of the paper.

For the proof of Theorem 1.1, we need several lemmas.

Lemma 2.1 For $\theta\in(0, \pi)$, we have

$$T_{2m+1}^\beta(\theta) = \frac{2^\beta\cos\frac{\theta}{2}}{\Gamma(1-\beta)(\sin\theta)^{1-\beta}}\gamma_m^\beta(\theta),$$

$$T_{2m+2}^\beta(\theta) = \frac{2^\beta\cos\frac{\theta}{2}}{\Gamma(1-\beta)(\sin\theta)^{1-\beta}}(\gamma_m^\beta(\theta)+\mu_m^\beta(\theta)).$$

Proof This is an immediate consequence of (1.3) and (1.4).

Lemma 2.2 For $m\geqslant 100$ and $\beta\in(0.308\,443, 0.308\,444)$,

$$c_{2m+2}(\beta) \leqslant \frac{0.999\,7}{\Gamma(1-\beta)\left(m+\frac{5}{4}\right)^\beta}. \tag{2.3}$$

Lemma 2.3 For $\beta\in(0.308\,443, 0.308\,444)$, $m\geqslant 100$ and $\theta\in\left(\frac{\pi}{4(m+1)}, \frac{\pi}{2}\right]$, we have

$$\gamma_m^\beta(\theta) > 0.221\,8\theta - 0.059\,7\theta^2 - \frac{0.230\,2\theta^2}{\left(\left(m+\frac{3}{4}\right)\theta\right)^{\beta+1}} +$$

$$\left(\frac{\theta}{\sin\theta}\right)^\beta\left(\int_0^{\left(2m+\frac{3}{2}\right)\theta}\frac{\cos t}{t^\beta}\mathrm{d}t + 0.059\,1\theta\cdot\int_{\left(2m+\frac{3}{2}\right)\theta}^{+\infty}\frac{\cos t}{t^{1+\beta}}\mathrm{d}t\right).$$

Lemma 2.4 For $\beta\in(0, 1)$, we have

$$\gamma_m^\beta\left(\frac{3\pi}{4m+3}\right) = \int_0^{\frac{3}{2}\pi}\frac{\cos t}{t^\beta}\mathrm{d}t + O(m^{-1}),$$

as $m\to+\infty$.

Lemma 2.5 Let $\varphi\in\left(0, \frac{\pi}{2}\right]$, $m\geqslant 100$ and $\beta\in(0.308\,443, 0.308\,444)$, we have

$$\gamma_m^\beta(\pi-\varphi) \geqslant 1.311\,6\cos(0.484\,5+0.191\,6\varphi) - \left(\frac{\varphi}{\sin\varphi}\right)^\beta\int_{\left(2m+\frac{3}{2}\right)\varphi}^{+\infty}\frac{\sin t}{t^\beta}\mathrm{d}t +$$

$$0.0591\varphi\cdot\left(\frac{\varphi}{\sin\varphi}\right)^{\beta}\int_{(2m+\frac{3}{2})\varphi}^{+\infty}\frac{\sin t}{t^{1+\beta}}\mathrm{d}t+$$

$$0.0391\varphi^{2}\cdot\left(\frac{\varphi}{\sin\varphi}\right)^{\beta}\int_{(2m+\frac{3}{2})\varphi}^{+\infty}\frac{\sin t}{t^{2+\beta}}\mathrm{d}t+$$

$$\frac{\varphi^{2}}{\left(\left(2m+\frac{3}{2}\right)\varphi\right)^{1+\beta}}\left[0.3748+\frac{0.2484+0.4136\varphi}{\left(2m+\frac{3}{2}\right)\varphi}\right].$$

The proofs of Lemmas 2. 2～2. 5 are quite long, therefore we postpone them to Sect. 6. For the moment, we take these for granted and proceed with our proof of Theorem 1. 1.

§ 3. Proof of Theorem 1. 1

First, we prove the second assertion of the theorem, which is quite easy. We use Lemmas 2. 1 and 2. 4 to obtain that for $\beta\in(0, 1)$,

$$T_{2m+1}^{\beta}\left(\frac{3\pi}{4m+3}\right)=\frac{2^{\beta}}{\Gamma(1-\beta)}\left(\frac{4}{3}\right)^{1-\beta}\left(m^{1-\beta}\int_{0}^{\frac{3}{2}\pi}\frac{\cos t}{t^{\beta}}\mathrm{d}t+O(m^{-\beta})\right).\tag{3.1}$$

Since $c_{2m+2}(\beta)=O(m^{-\beta})$ as $m\to+\infty$, on account of (2. 2), the same identity is also true for T_{2m+2}^{β}. Therefore, noting the fact that for $0<\beta<\beta_0$,

$$\int_{0}^{\frac{3}{2}\pi}\frac{\cos t}{t^{\beta}}\mathrm{d}t<0,$$

we deduce that for $0<\beta<\beta_0$,

$$\lim_{n\to+\infty}\min_{\theta\in(0,\pi)}T_{n}^{\beta}(\theta)=-\infty,$$

which is the same as the desired.

Next, we point out that it will suffice to prove (1. 5) in the case when $\beta=\beta_0$. For we may write, by Abel's transform,

$$T_{n}^{\beta}(\theta)=\sum_{k=0}^{n-1}\left(\frac{c_k(\beta)}{c_k(\beta_0)}-\frac{c_{k+1}(\beta)}{c_{k+1}(\beta_0)}\right)T_{k}^{\beta_0}(\theta)+\frac{c_n(\beta)}{c_n(\beta_0)}T_{n}^{\beta_0}(\theta),$$

and furthermore, for $\beta\in(\beta_0, 1)$, it's easy to verify that

$$\frac{c_k(\beta)}{c_k(\beta_0)}=\frac{c_{k+1}(\beta)}{c_{k+1}(\beta_0)},$$

$$\frac{c_k(\beta)}{c_k(\beta_0)}-\frac{c_{k+1}(\beta)}{c_{k+1}(\beta_0)}=\frac{c_k(\beta)}{c_k(\beta_0)}\left(1-\frac{k+1-2\beta}{k+1-2\beta_0}\right)>0,$$

according as k is even or odd.

Finally, let us prove the positivity of $T_n^{\beta_0}(\theta)$. Mechanical computation using Maple 7 shows① that for $0\leqslant n\leqslant 200$ and $\theta\in(0,\pi)$,

$$T_n^{0.308}(\theta)>0.$$

By Abel's transform and the fact that $\beta_0\in(0.308\,443, 0.308\,444)$, it follows that for $0\leqslant n\leqslant 200$ and $\theta\in(0,\pi)$ we have $T_n^{\beta_0}(\theta)>0$. So, it remains to prove the positivity of $T_{2m+1}^{\beta_0}(\theta)$ and $T_{2m+2}^{\beta_0}(\theta)$ for $m\geqslant 100$ and $\theta\in(0,\pi)$.

For convenience, throughout the rest of the proof, we will always assume $\beta=\beta_0$, $\beta_1=0.308\,443$ and $\beta_2=0.308\,444$ so that $\beta\in(\beta_1,\beta_2)$, and instead of γ_m^β and μ_m^β we simply write γ_m and μ_{m}.

Before starting our proof it's worthwhile to point out that all our numerical calculations below will have the accuracy up to 10^{-4} and that we will always use the approximate value $\beta_1=0.308\,443$ of β to compute the integrals $\int_a^b\frac{\cos t}{t^{i+\beta}}\mathrm{d}t$, $\int_a^b\frac{\sin t}{t^{i+\beta}}\mathrm{d}t$, $(i=0,1,2,1\leqslant a<b\leqslant+\infty)$, which is sensible due to the following facts:

$$\left|\int_a^b\frac{\cos t}{t^{i+\beta}}\mathrm{d}t-\int_a^b\frac{\cos t}{t^{i+\beta_1}}\mathrm{d}t\right|\leqslant 4.770\,8\times 10^{-6},\qquad (3.2)$$

$$\left|\int_a^b\frac{\sin t}{t^{i+\beta}}\mathrm{d}t-\int_a^b\frac{\sin t}{t^{i+\beta_1}}\mathrm{d}t\right|\leqslant 4.770\,8\times 10^{-6}.\qquad (3.3)$$

For the proof of (3.2), without loss of generality we may assume $b<+\infty$. We then write

$$\int_a^b\frac{\cos t}{t^{i+\beta}}\mathrm{d}t-\int_a^b\frac{\cos t}{t^{i+\beta_1}}\mathrm{d}t=\int_{\beta_1+i}^{\beta+i}\int_a^b\frac{\ln t}{t^\alpha}\cos t\,\mathrm{d}t\,\mathrm{d}\alpha.$$

Since the function $g(t):=\frac{\ln t}{t^\alpha}$ is strictly increasing on $[1, \mathrm{e}^{\frac{1}{\alpha}}]$ and strictly decreasing on $[\mathrm{e}^{\frac{1}{\alpha}}, +\infty)$, applying the second mean value the-

① For details of the computation the reader can contact the third named author.

orem for integrals twice, we obtain

$$\left|\int_a^b \frac{\ln t}{t^\alpha}\cos t\mathrm{d}t\right| \leqslant \frac{4}{\mathrm{e}\alpha},$$

and hence

$$\left|\int_a^b \frac{\cos t}{t^{i+\beta}}\mathrm{d}t - \int_a^b \frac{\cos t}{t^{i+\beta_1}}\mathrm{d}t\right| \leqslant \int_{\beta_1+i}^{\beta+i} \frac{4}{\mathrm{e}\alpha}\mathrm{d}\alpha \leqslant \frac{4}{\mathrm{e}\beta_1}(\beta_2-\beta_1) = 4.770\,8\times 10^{-6},$$

which is (3.2). Obviously, the proceeding argument with "cos" replaced by "sin" remains valid. This gives (3.3).

The proof of the positivity of T_n^β is very long, therefore we shall break it into two parts.

3.1 Case I: $0<\theta<\frac{\pi}{2}$

In this case, by Lemma 2.1, it will suffice to prove that for $m\geqslant 100$ and $\theta\in\left(0, \frac{\pi}{2}\right)$,

$$\gamma_m(\theta)>0 \tag{3.4}$$

and

$$\gamma_m(\theta)+\mu_m(\theta)>0. \tag{3.5}$$

For simplicity, we let

$$p_m(\theta)=\int_0^{\left(2m+\frac{3}{2}\right)\theta}\frac{\cos t}{t^\beta}\mathrm{d}t,$$

$$q_m(\theta)=\int_{\left(2m+\frac{3}{2}\right)\theta}^{+\infty}\frac{\cos t}{t^{1+\beta}}\mathrm{d}t. \tag{3.6}$$

By (2.1) and (2.2), it's obvious that both (3.4) and (3.5) are true when $\theta\in\left(0, \frac{\pi}{4(m+1)}\right]$. So, it remains to prove (3.4) and (3.5) in the case when $\theta\in\left[\frac{\pi}{4(m+1)}, \frac{\pi}{2}\right]$. To this end, we consider the following three cases:

Case 1 $\frac{\pi}{4(m+1)}\leqslant\theta\leqslant\frac{3\pi}{4m+3}\leqslant 0.023\,4$.

In this case, it's obvious that

$$q_m(\theta)\geqslant\int_{\frac{\pi}{2}}^{+\infty}\frac{\cos t}{t^{1+\beta}}\mathrm{d}t=-0.364\,7$$

and

$$\left(\left(m+\frac{3}{4}\right)\theta\right)^{1+\beta} \geqslant \left(\frac{\pi}{4}\cdot\frac{100.75}{101}\right)^{1+\beta} = 0.7266.$$

So, from Lemma 2.3, it follows that

$$\gamma_m(\theta) > 0.2218\theta - 0.0597\theta\times 0.0234 - \frac{0.2302\times 0.0234\theta}{0.7266} + \left(\frac{\theta}{\sin\theta}\right)^{\beta} p_m(\theta) - \left(\frac{0.0234}{\sin(0.0234)}\right)^{\beta_2}\times 0.3647\times 0.0591\,\theta$$

$$= 0.1914\theta + \left(\frac{\theta}{\sin\theta}\right)^{\beta} p_m(\theta). \tag{3.7}$$

This proves (3.4) in this case since, by the definition of $\beta=\beta_0$, $p_m(\theta)$ is always nonnegative.

For the proof of (3.5), different arguments are needed for each of the intervals

$$\frac{\pi}{4(m+1)}\leqslant\theta\leqslant\frac{2.4\pi}{4m+3},\ \frac{2.4\pi}{4m+3}\leqslant\theta\leqslant\frac{2.91\pi}{4m+4},\ \frac{2.91\pi}{4(m+1)}\leqslant\theta\leqslant\frac{3\pi}{4m+3}.$$

For $\frac{\pi}{4(m+1)}\leqslant\theta\leqslant\frac{2.4\pi}{4m+3}$, we have

$$p_m(\theta) \geqslant \min_{\frac{403\pi}{808}\leqslant z\leqslant 1.2\pi}\int_0^z \frac{\cos t}{t^{\beta}}\mathrm{d}t = \int_0^{1.2\pi}\frac{\cos t}{t^{\beta}}\mathrm{d}t = 0.2670.$$

On the other hand, by (2.2) and (2.3),

$$|\mu_m(\theta)| \leqslant \left(\frac{1}{2}\right)^{\beta_1}\frac{\theta}{\cos\frac{\theta}{2}}\frac{0.9997}{\left(\left(m+\frac{5}{4}\right)\theta\right)^{\beta}}$$

$$\leqslant \left(\frac{1}{2}\right)^{\beta_1}\frac{0.0234}{\cos\frac{0.0234}{2}}\frac{0.9997}{(0.25\pi)^{\beta_2}} = 0.0204.$$

Combining these with (3.7) gives

$$\gamma_m(\theta)+\mu_m(\theta) \geqslant 0.1914\theta + 0.2670 - 0.0204 > 0.$$

For $\frac{2.4\pi}{4m+3}\leqslant\theta\leqslant\frac{2.91\pi}{4m+4}$, we set $z=\left(2m+\frac{3}{2}\right)\theta$ so that $1.2\pi\leqslant z\leqslant 1.455\pi$. Since, by (2.3),

$$\frac{\Gamma(1-\beta)(\sin\theta)^{1-\beta}}{2^{\beta}\cos\frac{\theta}{2}}c_{2m+2}(\beta)$$

$$\leqslant \frac{\theta}{2^{\beta}\cos\frac{\theta}{2}} \frac{0.9997}{\left(\left(m+\frac{5}{4}\right)\theta\right)^{\beta}} \leqslant \frac{0.0234}{2^{\beta_1}\cos\frac{0.0234}{2}} \frac{0.9997}{(0.6\pi)^{\beta_1}} = 0.0155,$$

we obtain from (2.2),

$$\mu_m(\theta) > 0.0155\cos\left(2m+\frac{3}{2}\right)\theta - 0.0155\times 2\left|\sin\left(\left(2m+\frac{7}{4}\right)\theta\right)\sin\frac{\theta}{4}\right|$$

$$\geqslant 0.0155\cos z - 0.0078\theta,$$

which, together with (3.6) and (3.7), implies

$$\gamma_m(\theta)+\mu_m(\theta) \geqslant 0.1836\theta + \int_0^z \frac{\cos t}{t^{\beta}}\mathrm{d}t + 0.0155\cos z$$

$$\geqslant 0.1836\theta + \min_{1.2\pi\leqslant u\leqslant 1.455\pi}\left\{\int_0^u \frac{\cos t}{t^{\beta}}\mathrm{d}t + 0.0155\cos u\right\}.$$

Mechanical calculation by Maple 7 shows that the second term in the right side of this last inequality has a positive lower bound 0.0040.

Finally, for $\frac{2.91\pi}{4(m+1)} \leqslant \theta \leqslant \frac{3\pi}{4m+3}$, using the fact that $p_m(\theta)\geqslant 0$ and combining (3.7) with (2.2) and (2.3), we deduce

$$\gamma_m(\theta)+\mu_m(\theta) \geqslant 0.1914\theta + \frac{\theta}{2^{\beta_1}\cos\frac{\theta}{2}} \frac{0.9997}{\left(\left(m+\frac{5}{4}\right)\theta\right)^{\beta_1}}\cos(1.455\pi)$$

$$\geqslant 0.1914\theta + \frac{\theta}{2^{\beta_1}\cos\frac{0.0234}{2}} \frac{0.9997}{\left(\frac{1.455\pi}{2}\right)^{\beta_1}}\cos(1.455\pi)$$

$$= 0.1032\theta > 0.$$

We remind the reader that it's here that the exact value of β has to be used so that $p_m(\theta)\geqslant 0$ and that any numerical calculation of the integral $\int_0^z \frac{\cos t}{t^{\beta}}\mathrm{d}t$ using approximate value of β will cause trouble.

Case 2 $\frac{3\pi}{4m+3} \leqslant \theta \leqslant 0.3$.

In this case, the following estimates are obvious:

$$q_m(\theta) \geqslant \int_{\frac{5}{2}\pi}^{+\infty} \frac{\cos t}{t^{1+\beta}}\mathrm{d}t = -0.0647,$$

$$\left(\left(m+\frac{3}{4}\right)\theta\right)^{\beta+1} \geqslant (0.75\pi)^{1+\beta_1} = 3.0691.$$

Hence, applying Lemma 2.3 gives

$$\gamma_m(\theta) > 0.2218\theta - 0.0597 \times 0.3\theta - \frac{0.2302 \times 0.3\theta}{3.0691} -$$
$$\left(\frac{0.3}{\sin(0.3)}\right)^{\beta_2} \times 0.0591 \times 0.0647\theta + p_m(\theta)$$
$$= 0.1776\theta + p_m(\theta), \tag{3.8}$$

which is positive and therefore gives (3.4).

For the proof of (3.5), we rewrite

$$\mu_m(\theta) = \frac{\Gamma(1-\beta)(\sin\theta)^{1-\beta}}{2^\beta \cos\frac{\theta}{2}} c_{2m+2}(\beta)\cos\left(2m+\frac{3}{2}\right)\theta -$$
$$\frac{\Gamma(1-\beta)(\sin\theta)^{1-\beta}}{2^{\beta-1}\cos\frac{\theta}{2}} c_{2m+2}(\beta)\sin\left(\left(2m+\frac{7}{4}\right)\theta\right)\sin\frac{\theta}{4}.$$

The absolute value of the second term in the right hand side of this last equality is less than

$$\frac{\theta^2}{2^{\beta+1}\cos\frac{\theta}{2}}\frac{0.9997}{\left(\left(m+\frac{5}{4}\right)\theta\right)^\beta} \leqslant \frac{0.3\theta}{2^{\beta_1+1}\cos\frac{0.3}{2}}\frac{0.9997}{(0.75\pi)^{\beta_1}} = 0.09402\theta.$$

So, by (3.8),

$$\gamma_m(\theta) + \mu_m(\theta)$$
$$\geqslant 0.0835\theta + p_m(\theta) + \frac{\Gamma(1-\beta)(\sin\theta)^{1-\beta}}{2^\beta \cos\frac{\theta}{2}} c_{2m+2}(\beta)\cos\left(2m+\frac{3}{2}\right)\theta.$$

It will be shown that

$$p_m(\theta) + \frac{\Gamma(1-\beta)(\sin\theta)^{1-\beta}}{2^\beta \cos\frac{\theta}{2}} c_{2m+2}(\beta)\cos\left(2m+\frac{3}{2}\right)\theta \geqslant 0, \tag{3.9}$$

from which (3.5) will follow.

To prove (3.9), without loss of generality, we may assume $\cos\left(2m+\frac{3}{2}\right)\theta<0$ so that $\left(2m+\frac{3}{2}\right)\theta\geqslant\frac{5}{2}\pi$ and

$$\frac{\Gamma(1-\beta)(\sin\theta)^{1-\beta}}{2^\beta\cos\frac{\theta}{2}} c_{2m+2}(\beta)\cos\left(2m+\frac{3}{2}\right)\theta$$
$$\geqslant \frac{\theta}{2^\beta\cos\frac{\theta}{2}}\frac{0.9997}{\left(\left(m+\frac{5}{4}\right)\theta\right)^\beta}\cos\left(2m+\frac{3}{2}\right)\theta$$

$$\geqslant \frac{0.3}{\cos\frac{0.3}{2}} \frac{0.9997}{\left(\left(m+\frac{3}{2}\right)\theta\right)^{\beta_1}} \cos\left(2m+\frac{3}{2}\right)\theta$$

$$\geqslant 0.3034 \frac{\cos\left(2m+\frac{3}{2}\right)\theta}{\left(\left(2m+\frac{3}{2}\right)\theta\right)^{\beta_1}}, \tag{3.10}$$

otherwise (3.9) is obvious since both terms in the sum are nonnegative. We define

$$F(z) = \frac{0.3034\cos z}{z^{\beta_1}} + \int_{\frac{3}{2}\pi}^{z} \frac{\cos t}{t^{\beta}}\mathrm{d}t.$$

It then follows from (3.10) that

$$p_m(\theta) + \frac{\Gamma(1-\beta)(\sin\theta)^{1-\beta}}{2^{\beta}\cos\frac{\theta}{2}} c_{2m+2}(\beta)\cos\left(2m+\frac{3}{2}\right)\theta \geqslant \min_{z\geqslant\frac{5}{2}\pi} F(z).$$

If $2.5\pi\leqslant z\leqslant 30.5\pi$ then mechanical calculation using Maple 7 shows $F(z)\geqslant 0.1133$; if $z\geqslant 30.5\pi$, then

$$F(z) \geqslant -\frac{0.3034}{(30.5\pi)^{\beta}} + \int_{1.5\pi}^{29.5\pi} \frac{\cos t}{t^{\beta}}\mathrm{d}t = -0.0743 + 0.3636 = 0.2893.$$

Hence in either case we have $F(z)>0$, and consequently, (3.9) follows.

Case 3 $0.3<\theta<\frac{\pi}{2}$.

In this case, we have $\left(2m+\frac{3}{2}\right)\theta\geqslant 19\pi$. It then follows that

$$p_m(\theta) \geqslant \int_{1.5\pi}^{17.5\pi} \frac{\cos t}{t^{\beta}}\mathrm{d}t = 0.3204$$

and

$$q_m(\theta) \geqslant \int_{18.5\pi}^{+\infty} \frac{\cos t}{t^{\beta+1}}\mathrm{d}t = -0.0049.$$

Therefore, by Lemma 2.3,

$$\gamma_m(\theta) > 0.2218\theta - 0.0597\times\frac{\pi}{2}\times\theta - \frac{0.2302\times\frac{\pi}{2}\times\theta}{(100.75\times 0.3)^{1+\beta_1}} +$$

$$p_m(\theta) - 0.0591\times 0.0049\times\left(\frac{\pi}{2}\right)^{\beta}\theta$$

$$= 0.1235\theta + p_m(\theta)$$

$$\geqslant 0.1235\theta + 0.3204, \qquad (3.11)$$

and, by (2.2),

$$|\mu_m(\theta)| \leqslant \frac{1}{2^{\beta_1}\cos\frac{\pi}{4}} \frac{0.9997}{(101.25)^{\beta_1}} = 0.2748.$$

Both (3.4) and (3.5) then follow.

3.2 Case Ⅱ: $\frac{\pi}{2} \leqslant \theta < \pi$.

It follows from Lemma 2.1 that for $\pi - \frac{\pi}{4(m+1)} \leqslant \theta < \pi$ both $T_{2m+1}^{\beta}(\theta)$ and $T_{2m+2}^{\beta}(\theta)$ are positive. Now we assume $\theta \in \left[\frac{\pi}{2}, \pi - \frac{\pi}{4(m+1)}\right]$ and set $\varphi = \pi - \theta$ so that $\varphi \in \left[\frac{\pi}{4(m+1)}, \frac{\pi}{2}\right]$. By Lemma 2.1, it will then be sufficient to show that for $\varphi \in \left[\frac{\pi}{4(m+1)}, \frac{\pi}{2}\right]$ and $m \geqslant 100$, both of the following inequalities are true:

$$\gamma_m(\pi - \varphi) > 0, \qquad (3.12)$$

$$\gamma_m(\pi - \varphi) + \mu_m(\pi - \varphi) > 0. \qquad (3.13)$$

We first observe that

$$\mu_m(\pi - \varphi) \geqslant$$

$$\frac{\varphi}{\sin\frac{\varphi}{2}} \frac{0.9997}{\left(\left(2m + \frac{5}{2}\right)\varphi\right)^{\beta}} \cos\left(2m + \frac{3}{2}\right)\varphi - \frac{\varphi^2}{2\sin\frac{\varphi}{2}} \frac{0.9997}{\left(\left(2m + \frac{5}{2}\right)\varphi\right)^{\beta}}$$

$$(3.14)$$

in the case when $\cos\left(2m + \frac{3}{2}\right)\varphi < 0$. Indeed, this is a simple consequence of Lemma 2.2 and the following formula of trigonometry:

$$\cos(2m+2)\varphi - \cos\left(2m + \frac{3}{2}\right)\varphi = -2\sin\left(\left(2m + \frac{7}{4}\right)\varphi\right)\sin\frac{\varphi}{4}.$$

For the proofs of (3.12) and (3.13), we consider the following three cases:

Case 1 $\frac{\pi}{4(m+1)} \leqslant \varphi \leqslant \frac{1.5\pi}{2m + \frac{3}{2}} \leqslant 0.0234.$

In this case, it is easy to verify the following estimates:

$$0.0591\varphi\left(\frac{\varphi}{\sin\varphi}\right)^{\beta}\int_{(2m+\frac{3}{2})\varphi}^{+\infty}\frac{\sin t}{t^{1+\beta}}\mathrm{d}t$$

$$\geqslant 0.0591\times 0.0234\times\left(\frac{0.0234}{\sin(0.0234)}\right)^{\beta}\int_{\pi}^{+\infty}\frac{\sin t}{t^{1+\beta}}\mathrm{d}t$$

$$=-0.0003,$$

$$0.0391\varphi^{2}\left(\frac{\varphi}{\sin\varphi}\right)^{\beta}\int_{(2m+\frac{3}{2})\varphi}^{+\infty}\frac{\sin t}{t^{2+\beta}}\mathrm{d}t$$

$$\geqslant 0.0391\times 0.0234^{2}\times\left(\frac{0.0234}{\sin(0.0234)}\right)^{\beta}\int_{\pi}^{+\infty}\frac{\sin t}{t^{2+\beta}}\mathrm{d}t$$

$$=-0.1034\times 10^{-5},$$

$$\frac{\varphi^{2}}{\left(\left(2m+\frac{3}{2}\right)\varphi\right)^{1+\beta}}\left[0.3748+\frac{0.2484+0.4136\varphi}{\left(2m+\frac{3}{2}\right)\varphi}\right]$$

$$\leqslant\frac{(0.0234)^{2}}{\left(\frac{403}{808}\pi\right)^{1+\beta}}(0.3748+0.2484+0.4136\times 0.0234)$$

$$=0.0002.$$

Therefore, from Lemma 2.5 it follows that

$$\gamma_{m}(\pi-\varphi)\geqslant 1.3116\cos(0.4845+0.1916\times 0.0234)-$$

$$0.0003-0.0002-\left(\frac{\varphi}{\sin\varphi}\right)^{\beta}\int_{(2m+\frac{3}{2})\varphi}^{+\infty}\frac{\sin t}{t^{\beta}}\mathrm{d}t$$

$$=1.1574-\left(\frac{\varphi}{\sin\varphi}\right)^{\beta}\int_{(2m+\frac{3}{2})\varphi}^{+\infty}\frac{\sin t}{t^{\beta}}\mathrm{d}t. \tag{3.15}$$

Observing that for $\frac{\pi}{4(m+1)}\leqslant\varphi\leqslant\frac{1.5\pi}{2m+\frac{3}{2}}$,

$$\left(\frac{\varphi}{\sin\varphi}\right)^{\beta}\int_{(2m+\frac{3}{2})\varphi}^{+\infty}\frac{\sin t}{t^{\beta}}\mathrm{d}t$$

$$\leqslant\left(\frac{0.0234}{\sin 0.0234}\right)^{\beta_{2}}\times\max_{\frac{403\pi}{808}\leqslant u\leqslant 1.5\pi}\int_{u}^{+\infty}\frac{\sin t}{t^{\beta}}\mathrm{d}t=0.1159,$$

we obtain

$$\gamma_{m}(\pi-\varphi)\geqslant 1.1574-0.1159=1.0415, \tag{3.16}$$

which gives (3.12) in this case.

For the proof of (3.13), set $z=\left(2m+\frac{3}{2}\right)\varphi$. Without loss of generality, we may assume $z\in\left[\frac{\pi}{2},\frac{3}{2}\pi\right]$ so that $\cos z<0$ and, by (3.14),

$$\begin{aligned}\mu_m(\pi-\varphi)&\geqslant\frac{0.0234}{\sin\frac{0.0234}{2}}\times\frac{0.9997}{z^{\beta_1}}\cos z-\frac{0.0234^2}{2\sin\frac{0.0234}{2}}\times\frac{0.9997}{(0.5\pi)^{\beta}}\\&=1.9995\times\frac{\cos z}{z^{\beta_1}}-0.0204,\end{aligned}\tag{3.17}$$

otherwise $\mu_m(\pi-\varphi)\geqslant-0.0204$ and by (3.16), there's nothing to prove.

Observing

$$\begin{aligned}&\left(\left(\frac{\varphi}{\sin\varphi}\right)^{\beta}-1\right)\int_{\left(2m+\frac{3}{2}\right)\varphi}^{+\infty}\frac{\sin t}{t^{\beta}}\mathrm{d}t\\&\leqslant\left(\left(\frac{0.0234}{\sin(0.0234)}\right)^{\beta_2}-1\right)\times\max_{\frac{403\pi}{808}\leqslant u\leqslant\frac{3}{2}\pi}\int_u^{+\infty}\frac{\sin t}{t^{\beta}}\mathrm{d}t<10^{-5},\end{aligned}$$

by (3.15) and (3.17), we obtain

$$\begin{aligned}\gamma_m(\pi-\varphi)+\mu_m(\pi-\varphi)&\geqslant1.1370-\int_z^{+\infty}\frac{\sin t}{t^{\beta}}\mathrm{d}t+\frac{1.9995\cos z}{z^{\beta_1}}\\&\geqslant1.1370-\int_z^{21\pi}\frac{\sin t}{t^{\beta}}\mathrm{d}t+\frac{1.9995\cos z}{z^{\beta_1}},\end{aligned}$$

which, by mechanical calculation, has a positive lower bound 0.1262 when $z\in\left[\frac{\pi}{2},\frac{3}{2}\pi\right]$. This proves (3.13).

Case 2 $\frac{3\pi}{4m+3}\leqslant\varphi\leqslant\frac{1}{2}$.

Mechanical calculation shows that $\int_{1.5\pi}^{3\pi}\frac{\sin t}{t^{1+\beta}}\mathrm{d}t>0$ and $\int_{1.5\pi}^{3\pi}\frac{\sin t}{t^{2+\beta}}\mathrm{d}t<0$. Hence, in this case, we have the following estimates:

$$\begin{aligned}&0.0591\varphi\cdot\left(\frac{\varphi}{\sin\varphi}\right)^{\beta}\int_{\left(2m+\frac{3}{2}\right)\varphi}^{+\infty}\frac{\sin t}{t^{1+\beta}}\mathrm{d}t\\&\geqslant0.0591\times0.5\times\left(\frac{0.5}{\sin0.5}\right)^{\beta}\int_{3\pi}^{+\infty}\frac{\sin t}{t^{1+\beta}}\mathrm{d}t\\&=-0.0015,\end{aligned}$$

$$0.0391\varphi^2 \cdot \left(\frac{\varphi}{\sin\varphi}\right)^\beta \int_{(2m+\frac{3}{2})\varphi}^{+\infty} \frac{\sin t}{t^{2+\beta}}dt$$

$$\geqslant 0.0391 \times 0.5^2 \times \left(\frac{0.5}{\sin 0.5}\right)^\beta \int_{1.5\pi}^{+\infty} \frac{\sin t}{t^{2+\beta}}dt$$

$$= -9.3 \times 10^{-5},$$

$$\frac{\varphi^2}{\left(\left(2m+\frac{3}{2}\right)\varphi\right)^{1+\beta}}\left[0.3748 + \frac{0.2484 + 0.4136\varphi}{\left(2m+\frac{3}{2}\right)\varphi}\right]$$

$$\leqslant \frac{0.5^2}{(1.5\pi)^{1+\beta}}\left(0.3748 + \frac{0.2484 + 0.4136 \times 0.5}{1.5\pi}\right)$$

$$= 0.0155,$$

$$1.3116\cos(0.4845 + 0.1916 \times \varphi)$$

$$\geqslant 1.3116\cos(0.4845 + 0.1916 \times 0.5) = 1.0969.$$

We then use Lemma 2.5 to obtain

$$\gamma_m(\pi-\varphi) \geqslant 1.0969 - 0.0015 - 0.0156 - \left(\frac{\varphi}{\sin\varphi}\right)^\beta \int_{(2m+\frac{3}{2})\varphi}^{+\infty} \frac{\sin t}{t^\beta}dt$$

$$= 1.0798 - \left(\frac{\varphi}{\sin\varphi}\right)^\beta \int_{(2m+\frac{3}{2})\varphi}^{+\infty} \frac{\sin t}{t^\beta}dt, \tag{3.18}$$

which, together with the fact that

$$\left(\frac{\varphi}{\sin\varphi}\right)^\beta \int_{(2m+\frac{3}{2})\varphi}^{+\infty} \frac{\sin t}{t^\beta}dt \leqslant \left(\frac{0.5}{\sin 0.5}\right)^{\beta_2} \int_{2\pi}^{+\infty} \frac{\sin t}{t^\beta}dt = 0.5696,$$

yields

$$\gamma_m(\pi-\varphi) \geqslant 1.0798 - 0.5696 = 0.5102. \tag{3.19}$$

This gives (3.12).

For the proof of (3.13), without loss of generality, we may assume $\cos\left(2m+\frac{3}{2}\right)\varphi<0$ so that, by (3.14),

$$\mu_m(\pi-\varphi)$$

$$\geqslant \frac{0.9997 \times 0.5}{\sin 0.25} \frac{\cos\left(\left(2m+\frac{3}{2}\right)\varphi\right)}{\left(\left(2m+\frac{3}{2}\right)\varphi\right)^\beta} - \frac{0.5^2 \times 0.9997}{2\sin(0.25) \cdot \left(\left(2m+\frac{3}{2}\right)\varphi\right)^\beta}$$

$$= \frac{2.0204\cos\left(\left(2m+\frac{3}{2}\right)\varphi\right)}{\left(\left(2m+\frac{3}{2}\right)\varphi\right)^\beta} - \frac{0.5051}{\left(\left(2m+\frac{3}{2}\right)\varphi\right)^\beta}, \tag{3.20}$$

otherwise, we have

$$\mu_m(\pi-\varphi)\geqslant-\frac{0.5051}{(1.5\pi)^\beta}=-0.3131,$$

and hence, (3.13) is a consequence of (3.19). Now set $z=\left(2m+\frac{3}{2}\right)\varphi$ so that $z\geqslant 2.5\pi$. It then follows from (3.18) and (3.20) that

$$\begin{aligned}&\gamma_m(\pi-\varphi)+\mu_m(\pi-\varphi)\\&\geqslant 1.0798-\left[\left(\frac{0.5}{\sin(0.5)}\right)^\beta-1\right]\int_{2\pi}^{+\infty}\frac{\sin t}{t^\beta}\mathrm{d}t-\int_z^{+\infty}\frac{\sin t}{t^\beta}\mathrm{d}t+\\&\quad\frac{2.0204\cos z}{z^\beta}-\frac{0.5051}{z^\beta}\\&=1.0725-\int_z^{+\infty}\frac{\sin t}{t^\beta}\mathrm{d}t+\frac{2.0204\cos z}{z^\beta}-\frac{0.5051}{z^\beta}=:F(z).\end{aligned}$$

If $2.5\pi\leqslant z\leqslant 31\pi$ then

$$F(z)\geqslant 1.0731-\int_z^{31\pi}\frac{\sin t}{t^\beta}\mathrm{d}t-\frac{2.0204}{z^\beta}-\frac{0.5051}{z^\beta}\geqslant 0.0614,$$

as mechanical calculation shows; if $z>31\pi$ then

$$F(z)\geqslant 1.0725-\int_{32\pi}^{+\infty}\frac{\sin t}{t^\beta}\mathrm{d}t-\frac{2.0204}{(31\pi)^\beta}-\frac{0.5051}{(31\pi)^\beta}=0.2161.$$

Hence, in both cases we have $F(z)>0$. This proves (3.13).

Case 3 $\frac{1}{2}\leqslant\varphi\leqslant\frac{\pi}{2}$.

In this case $\left(2m+\frac{3}{2}\right)\varphi\geqslant 100=31.831\pi$, hence the following estimates hold:

$$\begin{aligned}&1.3116\cos(0.4845+0.1916\times\varphi)\\&\geqslant 1.3116\cos(0.4845+0.1916\times 0.5\pi)\geqslant 0.9274,\end{aligned}$$

$$\left(\frac{\varphi}{\sin\varphi}\right)^\beta\int_{(2m+\frac{3}{2})\varphi}^{+\infty}\frac{\sin t}{t^\beta}\mathrm{d}t\leqslant(0.5\pi)^\beta\int_{32\pi}^{+\infty}\frac{\sin t}{t^\beta}\mathrm{d}t=0.2773,$$

$$\begin{aligned}&0.0591\varphi\cdot\left(\frac{\varphi}{\sin\varphi}\right)^\beta\int_{(2m+\frac{3}{2})\varphi}^{+\infty}\frac{\sin t}{t^{1+\beta}}\mathrm{d}t\\&\geqslant 0.0591\times\left(\frac{\pi}{2}\right)^{1+\beta}\int_{31\pi}^{+\infty}\frac{\sin t}{t^{1+\beta}}\mathrm{d}t=-0.0003,\end{aligned}$$

$$0.0391\varphi^2\cdot\left(\frac{\varphi}{\sin\varphi}\right)^\beta\int_{(2m+\frac{3}{2})\varphi}^{+\infty}\frac{\sin t}{t^{2+\beta}}\mathrm{d}t$$

$$\geqslant 0.0391\times\left(\frac{\pi}{2}\right)^{2+\beta}\int_{31\pi}^{+\infty}\frac{\sin t}{t^{2+\beta}}\mathrm{d}t=-2.9\times 10^{-6},$$

$$\frac{\varphi^2}{\left(\left(2m+\frac{3}{2}\right)\varphi\right)^{1+\beta}}\left[0.3748+\frac{0.2484+0.4136\varphi}{\left(2m+\frac{3}{2}\right)\varphi}\right]$$

$$\leqslant\frac{(0.5\pi)^2}{100^{1+\beta}}\left(0.3748+\frac{0.2484+0.4136\times 0.5\pi}{100}\right)=0.0023.$$

It then follows from Lemma 2.5 that

$$\gamma_m(\pi-\varphi)\geqslant 0.9274-0.2773-0.0003-0.0025=0.6473. \tag{3.21}$$

On the other hand, by (2.2) and (2.3), we have

$$|\mu_m(\pi-\varphi)|\leqslant\frac{\varphi}{\sin\frac{\varphi}{2}}\frac{0.9997}{\left(\left(2m+\frac{5}{2}\right)\varphi\right)^{\beta}}\leqslant\frac{0.9997\times\frac{\pi}{2}}{100^{\beta}\times\sin\frac{\pi}{4}}=0.5366.$$

Putting these together, we obtain both (3.12) and (3.13).

§ 4. Some estimates related to gamma functions

Before stating our main results in this section, we need to introduce some necessary notations:

$$A(\beta):=\frac{1}{2}\beta^2-\frac{1}{4}\beta,\quad B(\beta):=-\frac{1}{96}\beta-\frac{1}{8}\beta^2+\frac{1}{6}\beta^3, \tag{4.1}$$

$$e(\beta):=\frac{1}{64}\beta+\frac{95}{128}\beta^2-\frac{23}{384}\beta^3-\frac{1}{192}\beta^4+\frac{5}{96}\beta^5+\frac{1}{48}\beta^6. \tag{4.2}$$

We will keep these notations throughout the rest of the paper.

Our main results in this section can now be stated as follows.

Lemma 4.1 For $x\geqslant 100$ and $\beta\in(0, 1)$, we have

$$\frac{\Gamma(x+1-\beta)}{\Gamma(x+1)}=$$

$$\frac{1}{\left(x+\frac{1}{4}\right)^{\beta}}+\frac{A(\beta)}{\left(x+\frac{1}{4}\right)^{1+\beta}}+\frac{\frac{1}{2}A^2(\beta)+B(\beta)}{\left(x+\frac{1}{4}\right)^{2+\beta}}+\frac{R(x)}{\left(x+\frac{1}{4}\right)^{3+\beta}}, \tag{4.3}$$

where $R(x)\equiv R(x, \beta)$ satisfies

$$|R(x)-e(\beta)| \leqslant 0.0301. \tag{4.4}$$

Lemma 4.2 Let $u(x)=\dfrac{\Gamma(x+1-\beta)}{\Gamma(x+1)}$. Then for $x \geqslant 10$ and $\beta \in (0, 1)$, the following inequalities are true:

$$-\beta u(x)\left(\frac{1}{x+1-\beta}+\frac{1}{(x+1-\beta)^2}\right)<u'(x)<0; \tag{4.5}$$

$$0<u''(x)<u(x)\frac{\beta}{(x+1-\beta)^2}\left(1+\beta+\frac{2+2\beta}{x+1-\beta}+\frac{\beta}{(x+1-\beta)^2}\right); \tag{4.6}$$

$$u'''(x)<0. \tag{4.7}$$

For Gamma functions we have the following identity (see [2, p. 24, (3.9)])

$$\Gamma(x)=\sqrt{2\pi}x^{x-\frac{1}{2}}\mathrm{e}^{-x+\mu(x)}, \quad x>0, \tag{4.8}$$

where

$$\mu(x)=\sum_{n=0}^{+\infty}g(x+n), \quad g(x)=\sum_{k=1}^{+\infty}\frac{1}{(2k+1)(2x+1)^{2k}}. \tag{4.9}$$

It is known that (see [2, p. 22])

$$\mu(x)<\frac{1}{12x}. \tag{4.10}$$

An incorrect statement that there exists a $\theta \in (0, 1)$ independent of x such that

$$\mu(x)=\frac{\theta}{12x}$$

was also given there. The proof there does not yield this result and it will be shown in our Lemma 4.3 below that the parameter θ does depend on x.

For the proof of Lemma 4.1, we need to use the identity (4.8) and some properties of the function $\mu(x)$. The known properties (4.9) and (4.10) of $\mu(x)$ are not sufficient for our purpose. We shall prove the following

Lemma 4.3 For $x \geqslant 10$,

$$\mu(x)=\frac{1}{12x}+\frac{\theta(x)}{x^3},$$

where

$$-0.0065<\theta(x)<0. \tag{4.11}$$

Proof　We start with the first identity in (4.9). We then obtain

$$\mu(x)=\sum_{n=0}^{+\infty}\int_{n-\frac{1}{2}}^{n+\frac{1}{2}}(g(x+n)-g(x+y))\mathrm{d}y+\int_{-\frac{1}{2}}^{+\infty}g(x+y)\mathrm{d}y.$$

Since (see [2, p. 21])

$$g''(x)=\frac{1}{2x^2(x+1)^2},$$

it follows that

$$\begin{aligned}&-\sum_{n=0}^{+\infty}\int_{n-\frac{1}{2}}^{n+\frac{1}{2}}(g(x+n)-g(x+y))\mathrm{d}y\\&=\int_0^{\frac{1}{2}}\int_0^y\int_{-t}^t\Big(\sum_{n=0}^{+\infty}g''(x+n+\tau)\Big)\mathrm{d}\tau\,\mathrm{d}t\,\mathrm{d}y\\&<\int_0^{\frac{1}{2}}\int_0^y\int_{-t}^t\left(\sum_{n=0}^{+\infty}\frac{1}{2(x+n+\tau)^4}\right)\mathrm{d}\tau\,\mathrm{d}t\,\mathrm{d}y\\&<\int_0^{\frac{1}{2}}\int_0^y\int_{-t}^t\left[\frac{1}{2\left(x-\frac{1}{2}\right)^4}+\int_0^{+\infty}\frac{\mathrm{d}s}{2\left(x+s-\frac{1}{2}\right)^4}\right]\mathrm{d}\tau\,\mathrm{d}t\,\mathrm{d}y\\&=\frac{1}{24}\left[\frac{1}{2\left(x-\frac{1}{2}\right)^4}+\frac{1}{6\left(x-\frac{1}{2}\right)^3}\right]\\&=\frac{1}{144\left(x-\frac{1}{2}\right)^3}+\frac{1}{48\left(x-\frac{1}{2}\right)^4}.\end{aligned}$$

On the other hand, invoking the second identity in (4.9), we obtain

$$\int_{-\frac{1}{2}}^{+\infty}g(x+y)\mathrm{d}y=\sum_{k=1}^{+\infty}\frac{1}{2(2k-1)(2k+1)(2x)^{2k-1}}>\frac{1}{12x}+\frac{1}{240x^3}.$$

Putting these together, we get

$$\mu(x)>\frac{1}{12x}+\frac{1}{240x^3}-\frac{1}{144\left(x-\frac{1}{2}\right)^3}-\frac{1}{48\left(x-\frac{1}{2}\right)^4}.$$

But for $x\geqslant 10$,

$$\begin{aligned}&\frac{1}{240x^3}-\frac{1}{144\left(x-\frac{1}{2}\right)^3}-\frac{1}{48\left(x-\frac{1}{2}\right)^4}\\&=\frac{-1}{x^3}\left(\frac{-1}{240}+\frac{1}{144}\left(1+\frac{1}{2x-1}\right)^3+\frac{1}{48}\left(1+\frac{1}{2x-1}\right)^3\frac{2}{2x-1}\right)\end{aligned}$$

$$\geqslant \frac{-1}{x^3}\left(\frac{-1}{240}+\frac{1}{144}\left(1+\frac{1}{19}\right)^3+\frac{1}{48}\left(1+\frac{1}{19}\right)^3\frac{2}{19}\right)>-\frac{0.0065}{x^3}.$$

It then follows that

$$\mu(x)>\frac{1}{12x}-\frac{0.0065}{x^3},$$

which, together with (4.10), implies the desired assertion. □

Proof of lemma 4.1 For simplicity, we define

$$\varphi(x)=\left(x+\frac{1}{4}\right)^{\beta}\frac{\Gamma(x+1-\beta)}{\Gamma(x+1)}. \tag{4.12}$$

We then use Lemma 4.3 to obtain

$$\varphi(x)=\left(\frac{x+1-\beta}{x+1}\right)^{x+\frac{1}{2}}\left(\frac{x+\frac{1}{4}}{x+1-\beta}\right)^{\beta}e^{\beta+\frac{\beta}{12(x+1-\beta)(x+1)}+\psi(x)},$$

where

$$\psi(x)=\psi(x,\beta)=\frac{\theta(x+1-\beta)}{(x+1-\beta)^3}-\frac{\theta(x+1)}{(x+1)^3},$$

and by (2.5)

$$\frac{-0.0065}{(x+1-\beta)^3}<\psi(x)<\frac{0.0065}{(x+1)^3}. \tag{4.13}$$

It follows that

$$\begin{aligned}\ln\varphi(x)&=\left(x+\frac{1}{2}\right)\ln\left(1-\frac{\beta}{x+1}\right)+\beta\ln\left(1-\frac{\frac{3}{4}-\beta}{x+1-\beta}\right)+\\&\quad\beta+\frac{\beta}{12(x+1-\beta)(x+1)}+\psi(x)\\&=-\sum_{k=1}^{+\infty}\frac{\beta^k}{k(x+1)^{k-1}}\left(1-\frac{\frac{1}{2}}{x+1}\right)-\sum_{k=1}^{+\infty}\frac{\beta}{k}\left(\frac{\frac{3}{4}-\beta}{x+1-\beta}\right)^k+\\&\quad\beta+\frac{\beta}{12(x+1-\beta)(x+1)}+\psi(x)\\&=\sum_{k=1}^{+\infty}\left[\left(\frac{\beta^k}{2k}-\frac{\beta^{k+1}}{k+1}\right)\frac{1}{(x+1)^k}-\frac{\beta}{k}\left(\frac{\frac{3}{4}-\beta}{x+1-\beta}\right)^k\right]+\\&\quad\frac{\beta}{12(x+1-\beta)(x+1)}+\psi(x)\end{aligned}$$

$$= \left(\frac{\beta}{12(x+1-\beta)(x+1)} + \sum_{k=1}^{2}\cdots\right) + \left(\psi(x) + \sum_{k=3}^{+\infty}\cdots\right)$$

$$=: \left(\frac{\beta}{12(x+1-\beta)(x+1)} + \sum_{k=1}^{2}\cdots\right) + r_1(x), \tag{4.14}$$

where

$$r_1(x) := \psi(x) + \sum_{k=3}^{+\infty}\left[\left(\frac{\beta^k}{2k} - \frac{\beta^{k+1}}{k+1}\right)\frac{1}{(x+1)^k} - \frac{\beta}{k}\left[\frac{\frac{3}{4}-\beta}{x+1-\beta}\right]^k\right]. \tag{4.15}$$

By elementary calculation, we have

$$\sum_{k=1}^{2}\left[\left(\frac{\beta^k}{2k} - \frac{\beta^{k+1}}{k+1}\right)\frac{1}{(x+1)^k} - \frac{\beta}{k}\left[\frac{\frac{3}{4}-\beta}{x+1-\beta}\right]^k\right] + \frac{\beta}{12(x+1-\beta)(x+1)}$$

$$= \frac{A(\beta)}{x+\frac{1}{4}} + \frac{B(\beta)}{\left(x+\frac{1}{4}\right)^2} + r_2(x), \tag{4.16}$$

where

$$r_2(x) = \frac{h_1(\beta)}{\left(x+\frac{1}{4}\right)^2(x+1)} + \frac{h_2(\beta)}{\left(x+\frac{1}{4}\right)^2(x+1-\beta)} + \frac{h_3(\beta)}{\left(x+\frac{1}{4}\right)^2(x+1)^2} + \frac{h_4(\beta)}{\left(x+\frac{1}{4}\right)^2(x+1-\beta)^2} \tag{4.17}$$

and

$$h_1(\beta) = -\frac{3}{64} + \frac{9}{32}\beta + \frac{3}{32}\beta^2 + \frac{1}{2}\beta^3, \quad h_2(\beta) = \frac{1}{12}\left(\frac{3}{4}-\beta\right)^2,$$

$$h_3(\beta) = \frac{9}{16}\beta^2\left(\frac{1}{4} - \frac{1}{3}\beta\right), \qquad h_4(\beta) = -\frac{1}{2}\beta\left(\frac{3}{4}-\beta\right)^4.$$

Now substituting (4.16) into (4.14) yields

$$\ln\varphi(x) = \frac{A(\beta)}{x+\frac{1}{4}} + \frac{B(\beta)}{\left(x+\frac{1}{4}\right)^2} + r_1(x) + r_2(x).$$

It then follows that

$$\varphi(x) = \exp\left[\frac{A(\beta)}{x+\frac{1}{4}} + \frac{B(\beta)}{\left(x+\frac{1}{4}\right)^2} + r_1(x) + r_2(x)\right]$$

$$= 1 + \frac{A(\beta)}{x+\frac{1}{4}} + \frac{B(\beta)}{\left(x+\frac{1}{4}\right)^2} + r_1(x) + r_2(x) +$$

$$\frac{1}{2}\left[\frac{A(\beta)}{x+\frac{1}{4}} + \frac{B(\beta)}{\left(x+\frac{1}{4}\right)^2} + r_1(x) + r_2(x)\right]^2 +$$

$$\sum_{k=3}^{+\infty}\frac{1}{k!}\left[\frac{A(\beta)}{x+\frac{1}{4}} + \frac{B(\beta)}{\left(x+\frac{1}{4}\right)^2} + r_1(x) + r_2(x)\right]^k$$

$$= 1 + \frac{A(\beta)}{x+\frac{1}{4}} + \frac{\frac{1}{2}A^2(\beta)+B(\beta)}{\left(x+\frac{1}{4}\right)^2} + \frac{R(x)}{\left(x+\frac{1}{4}\right)^3},$$

where

$$R(x) = A(\beta)B(\beta) + r(x)\left(x+\frac{1}{4}\right)^3 + \frac{B^2(\beta)}{2x+\frac{1}{2}} + A(\beta)r(x)\left(x+\frac{1}{4}\right)^2 +$$

$$B(\beta)r(x)\left(x+\frac{1}{4}\right) + \frac{1}{2}r(x)^2\left(x+\frac{1}{4}\right)^3 +$$

$$\left(x+\frac{1}{4}\right)^3\sum_{k=3}^{+\infty}\frac{1}{k!}\left[\frac{A(\beta)}{x+\frac{1}{4}} + \frac{B(\beta)}{\left(x+\frac{1}{4}\right)^2} + r(x)\right]^k. \qquad (4.18)$$

$$r(x) = r_1(x) + r_2(x).$$

This gives (4.3) on account of (4.12).

Now it remains to prove (4.4). By (4.13) (4.15) and (4.17), we have

$$\left|\left(x+\frac{1}{4}\right)^3 r(x) - D(\beta)\right| \leqslant 0.0191,$$

where

$$D(\beta) = \frac{1}{64}\beta + \frac{71}{96}\beta^2 - \frac{1}{12}\beta^3 + \frac{1}{12}\beta^4.$$

Then from (4.18), (4.4) will follow. The verification of these estimates contains many tedious numerical calculations, therefore we omit the details. □

Proof of Lemma 4.2 We start with the identity (see [2, p. 17]),

$$\frac{\mathrm{d}}{\mathrm{d}x}\ln\Gamma(x)=-\gamma-\frac{1}{x}+\sum_{k=1}^{+\infty}\left(\frac{1}{k}-\frac{1}{k+x}\right),$$

where $\gamma=0.57\cdots$ denotes Euler's constant. We then obtain

$$\begin{aligned}u'(x)&=u(x)\frac{\mathrm{d}}{\mathrm{d}x}(\ln\Gamma(x+1-\beta)-\ln\Gamma(x+1))\\&=-u(x)\sum_{k=0}^{+\infty}\frac{\beta}{(x+1+k-\beta)(x+1+k)}\end{aligned}\tag{4.19}$$

and

$$\begin{aligned}u''(x)=&\,u(x)\Big(\sum_{k=0}^{+\infty}\frac{\beta}{(x+1+k-\beta)(x+1+k)}\Big)^2+\\&u(x)\sum_{k=0}^{+\infty}\Big(\frac{1}{(x+1+k-\beta)^2}-\frac{1}{(x+1+k)^2}\Big)\\=&\,u(x)\Big(\sum_{k=0}^{+\infty}\frac{\beta}{(x+1+k-\beta)(x+1+k)}\Big)^2+\\&u(x)\sum_{k=0}^{+\infty}\Big(\frac{\beta}{(x+1+k-\beta)(x+1+k)}\Big(\frac{1}{x+1+k-\beta}+\frac{1}{x+1+k}\Big)\Big).\end{aligned}\tag{4.20}$$

Now (4.5) is a consequence of (4.19) and the following estimates:

$$\begin{aligned}&\sum_{k=0}^{+\infty}\frac{1}{(x+1+k-\beta)(x+1+k)}\\&\leqslant\frac{1}{(x+1-\beta)(x+1)}+\int_0^{+\infty}\frac{\mathrm{d}\tau}{(x+1+\tau-\beta)(x+1+\tau)}\\&<\frac{1}{(x+1-\beta)(x+1)}+\frac{1}{x+1-\beta}<\frac{1}{x+1-\beta}+\frac{1}{(x+1-\beta)^2};\end{aligned}$$

while (4.7) follows directly from (4.20) and (4.5).

For the proof of (4.6), we observe that

$$\begin{aligned}&\sum_{k=0}^{+\infty}\Big(\frac{\beta}{(x+1+k-\beta)(x+1+k)}\Big(\frac{1}{x+1+k-\beta}+\frac{1}{x+1+k}\Big)\Big)\\&<\sum_{k=0}^{+\infty}\frac{2\beta}{(x+1+k-\beta)^3}<\frac{2\beta}{(x+1-\beta)^3}+\frac{\beta}{(x+1-\beta)^2}.\end{aligned}\tag{4.21}$$

and

$$\begin{aligned}&\Big(\sum_{k=0}^{+\infty}\frac{\beta}{(x+1+k-\beta)(x+1+k)}\Big)^2\\&=\Big(\frac{u'(x)}{u(x)}\Big)^2\quad(\text{by }(4.19))\end{aligned}$$

$$\leqslant \left(\frac{\beta}{x+1-\beta}+\frac{\beta}{(x+1-\beta)^2}\right)^2. \quad (\text{by } (4.5)).$$

These combined with (4.21) and (4.20) give (4.6). □

§ 5. Two fundamental lemmas

For the statements of our main results in this section, we need to introduce the following notations:

$$f(\theta):=2\Gamma(1-\beta)\cos\left(\frac{\pi}{2}\beta+\left(\frac{1}{4}-\frac{\beta}{2}\right)\theta\right)\sin\left(\left(\frac{1}{4}-\frac{\beta}{2}\right)\theta\right),$$

$$h(\theta):=\Gamma(1-\beta)\sin\left(\frac{\pi}{2}\beta\right)\left(\left(\frac{\theta}{\sin\theta}\right)^{\beta}-1\right),$$

$$p_m(\theta):=\int_0^{\left(2m+\frac{3}{2}\right)\theta}\frac{\cos t}{t^{\beta}}\mathrm{d}t, \quad q_m(\theta):=\int_{\left(2m+\frac{3}{2}\right)\theta}^{+\infty}\frac{\cos t}{t^{1+\beta}}\mathrm{d}t.$$

Henceforth we will always keep these notations.

Recall that

$$\gamma_m^{\beta}(\theta)=\Gamma(1-\beta)(2\sin\theta)^{1-\beta}\sum_{k=0}^{m}c_{2k}(\beta)\cos\left(2k+\frac{1}{2}\right)\theta.$$

The following lemma will play a fundamental role in our later proof of Lemmas 2.3 and 2.4 in Sect. 2.

Lemma 5.1 For $\beta\in(0, 1)$ and $\theta\in(0, \pi)$, we have

$$\gamma_m^{\beta}(\theta)=f(\theta)+\left(\frac{\theta}{\sin\theta}\right)^{\beta}p_m(\theta)-h(\theta)-2\theta A(\beta)\left(\frac{\theta}{\sin\theta}\right)^{\beta}q_m(\theta)-R_m(\theta), \tag{5.1}$$

where for $m\geqslant 100$ the remainder term $R_m(\theta)$ satisfies the following conditions: for $\theta\in\left(0, \frac{\pi}{2}\right]$ and $\beta\in(0, 1)$,

$$|R_m(\theta)|\leqslant C_{\beta}\frac{1}{(m\theta)^{\beta}}\left(\theta+\frac{1}{m^2}\right) \tag{5.2}$$

while for $\theta\in(0, \pi)$ and $\beta\in(0.308\,443, 0.308\,444)$,

$$|R_m(\theta)|$$

$$\leqslant\frac{1.615\,0\theta}{\left(\left(m+\frac{3}{4}\right)\sin\theta\right)^{\beta}}\left[\frac{0.008\,0}{m+\frac{3}{4}}+\frac{0.017\,2}{\left(m+\frac{3}{4}\right)^2\sin\theta}+\frac{0.051\,9\theta}{(m+1)\sin\theta}\right]. \tag{5.3}$$

Proof For simplicity, we define

$$\gamma_{\infty}^{\beta}(\theta)=\Gamma(1-\beta)(2\sin\theta)^{1-\beta}\sum_{k=0}^{+\infty}c_{2k}(\beta)\cos\left(2k+\frac{1}{2}\right)\theta.$$

By (1.3), the coefficients $c_{2k}(\beta)$ can be expressed in terms of gamma functions:

$$c_{2k}(\beta)=\frac{\Gamma(k+1-\beta)}{\Gamma(1-\beta)\Gamma(k+1)},\quad k\geqslant 0.$$

It then follows that for a complex number z with $|z|\leqslant 1$ and $z\neq 1$,

$$\sum_{k=0}^{+\infty}c_{2k}(\beta)z^{k}=(1-z)^{-1+\beta}.$$

This last equality implies

$$\gamma_{\infty}^{\beta}(\theta)=\Gamma(1-\beta)\sin\left(\frac{\pi}{2}\beta+\left(\frac{1}{2}-\beta\right)\theta\right).$$

Now we rewrite

$$\gamma_{m}^{\beta}(\theta)=\Gamma(1-\beta)\sin\left(\frac{\pi}{2}\beta+\left(\frac{1}{2}-\beta\right)\theta\right)-R_{m,1}(\theta), \tag{5.4}$$

where

$$R_{m,1}(\theta)=\Gamma(1-\beta)(2\sin\theta)^{1-\beta}\sum_{k=m+1}^{+\infty}c_{2k}(\beta)\cos\left(2k+\frac{1}{2}\right)\theta.$$

For simplicity, set

$$u(x)=\frac{\Gamma(x+1-\beta)}{\Gamma(x+1)}. \tag{5.5}$$

Then since

$$\cos\left(2k+\frac{1}{2}\right)\theta=\frac{\theta}{\sin\theta}\int_{k-\frac{1}{4}}^{k+\frac{3}{4}}\cos(2\theta t)\,\mathrm{d}t,$$

it follows that

$$\begin{aligned}R_{m,1}(\theta)&=(2\sin\theta)^{1-\beta}\frac{\theta}{\sin\theta}\Bigg[\sum_{k=m+1}^{+\infty}\int_{k-\frac{1}{4}}^{k+\frac{3}{4}}\left(u(k)-u\left(t-\frac{1}{4}\right)\right)\cos(2\theta t)\,\mathrm{d}t+\\&\qquad\int_{m+\frac{3}{4}}^{+\infty}u\left(t-\frac{1}{4}\right)\cos(2\theta t)\,\mathrm{d}t\Bigg]\\&=:(2\sin\theta)^{1-\beta}\frac{\theta}{\sin\theta}(\sigma_m(\theta)+I_m(\theta)),\end{aligned} \tag{5.6}$$

where

$$\sigma_m(\theta):=\sum_{k=m+1}^{+\infty}\int_{k-\frac{1}{4}}^{k+\frac{3}{4}}\left(u(k)-u\left(t-\frac{1}{4}\right)\right)\cos(2\theta t)\,\mathrm{d}t,$$

$$I_m(\theta) := \int_{m+\frac{3}{4}}^{+\infty} u\left(t-\frac{1}{4}\right)\cos(2\theta t)\,\mathrm{d}t. \tag{5.7}$$

Observing (see [1, p. 50, Exercise 19])

$$\sin\left(\frac{\pi}{2}\beta\right)\Gamma(1-\beta) = \int_0^{+\infty} \frac{\cos t}{t^\beta}\mathrm{d}t,$$

we obtain from Lemma 4.1

$$I_m(\theta) = \frac{\sin\left(\frac{\pi}{2}\beta\right)\Gamma(1-\beta)}{(2\theta)^{1-\beta}} - \frac{1}{(2\theta)1-\beta}\int_0^{(2m+\frac{3}{2})\theta} \frac{\cos t}{t^\beta}\mathrm{d}t +$$

$$A(\beta)(2\theta)^\beta\int_{(2m+\frac{3}{2})\theta}^{+\infty} \frac{\cos t}{t^{1+\beta}}\mathrm{d}t + J_m(\theta), \tag{5.8}$$

where

$$J_m(\theta) = \left(\frac{1}{2}A^2(\beta) + B(\beta)\right)(2\theta)^{1+\beta}\int_{(2m+\frac{3}{2})\theta}^{+\infty} \frac{\cos t}{t^{2+\beta}}\mathrm{d}t +$$

$$\int_{m+\frac{3}{4}}^{+\infty} \frac{\cos(2\theta t)}{t^{3+\beta}}R\left(t-\frac{1}{4}\right)\mathrm{d}t \tag{5.9}$$

and $R(t)$ is the same function as in Lemma 4.1.

Now combining (5.4)(5.6) and (5.8), we obtain (5.1) with

$$R_m(\theta) := (2\sin\theta)^{1-\beta}\frac{\theta}{\sin\theta}(J_m(\theta) + \sigma_m(\theta)). \tag{5.10}$$

So it remains to prove $R_m(\theta)$ satisfies (5.2) for $\beta\in(0, 1)$ and (5.3) for $\beta\in(0.308\,443, 0.308\,444)$.

It follows from (4.5) that for $\beta\in(0, 1)$ and $t\geqslant 10$,

$$|u'(t)| \leqslant C_\beta t^{-1-\beta}.$$

This, together with (5.7), implies

$$|\sigma_m(\theta)| \leqslant C_\beta m^{-\beta}.$$

On the other hand, invoking Lemma 4.1 and the second mean value theorem for the integrals in (5.9) yields

$$|J_m(\theta)| \leqslant C_\beta \theta^{1+\beta}\left(\frac{1}{m\theta}\right)^{2+\beta}.$$

Now (5.10) combined with these last two inequalities gives that for $\beta\in(0, 1)$ and $\theta\in\left(0, \frac{\pi}{2}\right]$,

$$|R_m(\theta)| \leqslant C_\beta\left(\frac{1}{m\theta}\right)^\beta\left(\theta + \frac{1}{m^2}\right),$$

which is (5.2).

For the proof of (5.3), we observe that for $\beta\in(0.308\ 443,\ 0.308\ 444)$, the following are true:

$$A(\beta)=\frac{\beta^2}{2}-\frac{\beta}{4}=-0.029\ 5,\quad B(\beta)=-\frac{\beta}{96}-\frac{\beta^2}{8}+\frac{\beta^3}{6}=-0.010\ 2,$$

$$e(\beta)=\frac{1}{64}\beta+\frac{95}{128}\beta^2-\frac{23}{384}\beta^3-\frac{1}{192}\beta^4+\frac{5}{96}\beta^5+\frac{1}{48}\beta^6=0.073\ 8,$$

$$|R(t)|\leqslant e(\beta)+0.030\ 1=0.103\ 9,\quad t\geqslant 100\quad (\text{by}(4.4)),\tag{5.11}$$

$$u(x)=\frac{1}{\left(x+\frac{1}{4}\right)^{\beta}}-\frac{0.029\ 5}{\left(x+\frac{1}{4}\right)^{1+\beta}}-\frac{0.009\ 8}{\left(x+\frac{1}{4}\right)^{2+\beta}}+\frac{R(x)}{\left(x+\frac{1}{4}\right)^{3+\beta}},$$

$$x\geqslant 100\quad (\text{by }(4.3)).\tag{5.12}$$

It will be shown that for $\beta\in(0.308\ 443,\ 0.308\ 444)$, $m\geqslant 100$ and $\theta\in(0,\ \pi)$,

$$|J_m(\theta)|\leqslant\frac{0.008\ 0}{\left(m+\frac{3}{4}\right)^{1+\beta}}\tag{5.13}$$

and

$$|\sigma_m(\theta)|\leqslant\frac{0.017\ 2}{\left(m+\frac{3}{4}\right)^{2+\beta}\sin\theta}+\frac{0.051\ 9\theta}{(m+1)^{1+\beta}\sin\theta}.\tag{5.14}$$

Then by (5.10), it will follow that

$$|R_m(\theta)|\leqslant\frac{2^{1-\beta}\theta}{\left(\left(m+\frac{3}{4}\right)\sin\theta\right)^{\beta}}\left[\frac{0.008\ 0}{m+\frac{3}{4}}+\frac{0.017\ 2}{\left(m+\frac{3}{4}\right)^2\sin\theta}+\frac{0.051\ 9\theta}{(m+1)\sin\theta}\right],$$

which is (5.3) since $2^{1-\beta}=1.615\ 0$. So the proof of (5.3) reduces to the proofs of (5.13) and (5.14).

We first prove (5.13), which is relatively easier. Recall that

$$\begin{aligned}&J_m(\theta)\\&=\left(\frac{1}{2}A^2(\beta)+B(\beta)\right)(2\theta)^{1+\beta}\int_{\left(2m+\frac{3}{2}\right)\theta}^{+\infty}\frac{\cos t}{t^{2+\beta}}\mathrm{d}t+\int_{m+\frac{3}{4}}^{+\infty}\frac{\cos(2\theta t)}{t^{3+\beta}}R\left(t-\frac{1}{4}\right)\mathrm{d}t\\&=-0.024\ 2\theta^{1+\beta}\int_{\left(2m+\frac{3}{2}\right)\theta}^{+\infty}\frac{\cos t}{t^{2+\beta}}\mathrm{d}t+\int_{m+\frac{3}{4}}^{+\infty}\frac{\cos(2\theta t)}{t^{3+\beta}}R\left(t-\frac{1}{4}\right)\mathrm{d}t.\end{aligned}\tag{5.15}$$

By (5.11), the absolute value of the second integral in (5.15) is dominated by

$$\int_{m+\frac{3}{4}}^{+\infty}\frac{0.1039}{t^{3+\beta}}\mathrm{d}t=\frac{0.1039}{(2+\beta)\left(m+\frac{3}{4}\right)^{2+\beta}}=\frac{0.045}{\left(m+\frac{3}{4}\right)^{2+\beta}}\leqslant\frac{0.0005}{\left(m+\frac{3}{4}\right)^{1+\beta}}.$$

On the other hand, obviously, the absolute value of the first term in (5.15) is estimated by

$$0.0242\theta^{1+\beta}\int_{(2m+\frac{3}{2})\theta}^{+\infty}\frac{1}{t^{2+\beta}}\mathrm{d}t=\frac{0.0242}{(1+\beta)2^{1+\beta}\left(m+\frac{3}{4}\right)^{1+\beta}}=\frac{0.0075}{\left(m+\frac{3}{4}\right)^{1+\beta}}.$$

Putting these together, we obtain (5.13).

To prove (5.14), we start with (5.7). We rewrite the term $\int_{k-\frac{1}{4}}^{k+\frac{3}{4}}\left(u(k)-u\left(t-\frac{1}{4}\right)\right)\cos(2\theta t)\mathrm{d}t$ as follows:

$$\begin{aligned}
&\int_{-\frac{1}{2}}^{\frac{1}{2}}(u(k)-u(k+t))\cos\left(\left(2k+\frac{1}{2}\right)\theta+2\theta t\right)\mathrm{d}t\\
=&\int_{0}^{\frac{1}{2}}\int_{0}^{t}(u'(k-s)-u'(k+s))\mathrm{d}s\cos\left(\left(2k+\frac{1}{2}\right)\theta-2\theta t\right)\mathrm{d}t+\\
&\int_{0}^{\frac{1}{2}}\int_{0}^{t}u'(k+s)\mathrm{d}s\left(\cos\left(\left(2k+\frac{1}{2}\right)\theta-2\theta t\right)-\right.\\
&\left.\cos\left(\left(2k+\frac{1}{2}\right)\theta+2\theta t\right)\right)\mathrm{d}t\\
=&-\int_{0}^{\frac{1}{2}}\int_{0}^{t}\int_{-s}^{s}u''(k+h)\cos\left(\left(2k+\frac{1}{2}\right)\theta-2\theta t\right)\mathrm{d}h\,\mathrm{d}s\,\mathrm{d}t+\\
&2\int_{0}^{\frac{1}{2}}\int_{0}^{t}u'(k+s)\sin\left(\left(2k+\frac{1}{2}\right)\theta\right)\sin(2\theta t)\mathrm{d}s\,\mathrm{d}t.
\end{aligned}$$

It then follows that

$$\begin{aligned}
\sigma_m(\theta)=&-\int_{0}^{\frac{1}{2}}\int_{0}^{t}\int_{-s}^{s}\left(\sum_{k=m+1}^{+\infty}u''(k+h)\cos\left(\left(2k+\frac{1}{2}\right)\theta-2\theta t\right)\right)\mathrm{d}h\,\mathrm{d}s\,\mathrm{d}t+\\
&2\int_{0}^{\frac{1}{2}}\int_{0}^{t}\left(\sum_{k=m+1}^{+\infty}u'(k+s)\sin\left(\left(2k+\frac{1}{2}\right)\theta\right)\right)\sin(2\theta t)\mathrm{d}s\,\mathrm{d}t. \qquad (5.16)
\end{aligned}$$

Recall that

$$u(x)=\frac{\Gamma(x+\beta-1)}{\Gamma(x+1)},$$

and that, by Lemma 2.3, $u''(x)$ is strictly decreasing on $[10, +\infty)$.

Applying Abel's transform gives that for $0<h<\frac{1}{2}$,

$$\left|\sum_{k=m+1}^{+\infty} u''(k+h)\cos\left(\left(2k+\frac{1}{2}\right)\theta-2\theta t\right)\right| \leqslant \frac{u''(m+1+h)}{\sin\theta}.$$

This last term does not exceed

$$\frac{u(m+1+h)}{\sin\theta}\ \frac{\beta}{(m+1)^2}\left(1+\beta+\frac{2+2\beta}{m+1}+\frac{\beta}{(m+1)^2}\right)$$

(by (4.6) applied to $x=m+h+1$)

$$\leqslant \frac{u\left(m+\frac{1}{2}\right)}{\sin\theta}\ \frac{0.411\,6}{(m+1)^2} \text{ (since } u \text{ is strictly decreasing on } [10, +\infty))$$

$$\leqslant \frac{0.411\,6}{(m+1)^2\sin\theta}\left[\frac{1}{\left(m+\frac{3}{4}\right)^{\beta}}+\frac{0.103\,9}{\left(m+\frac{3}{4}\right)^{3+\beta}}\right] \quad \text{(by (5.12))}$$

$$\leqslant \frac{0.411\,601}{(m+1)^{2+\beta}\sin\theta}.$$

It then follows that

$$\int_0^{\frac{1}{2}}\int_0^t\int_{-s}^s\left|\sum_{k=m+1}^{+\infty} u''(k+h)\cos\left(\left(2k+\frac{1}{2}\right)\theta-2\theta t\right)\right| \mathrm{d}h\,\mathrm{d}s\,\mathrm{d}t$$

$$\leqslant \frac{0.411\,601}{(m+1)^{2+\beta}\sin\theta}\int_0^{\frac{1}{2}}\int_0^t\int_{-s}^s \mathrm{d}h\,\mathrm{d}s\,\mathrm{d}t$$

$$= \frac{0.017\,2}{(m+1)^{2+\beta}\sin\theta}. \tag{5.17}$$

On the other hand, since, by Lemma 4.2, u' is negative and increasing,

$$\left|\sum_{k=m+1}^{+\infty} u'(k+s)\sin\left(\left(2k+\frac{1}{2}\right)\theta\right)\right| \leqslant \frac{|u'(m+1+s)|}{\sin\theta},$$

which, by (4.5) and (5.12), is estimated by

$$\frac{\beta}{\sin\theta}\left(\frac{1}{m+1}+\frac{1}{(m+1)^2}\right)u(m+1) \leqslant \frac{0.311\,5}{\sin\theta(m+1)^{1+\beta}}.$$

Observing

$$2\int_0^{\frac{1}{2}}\int_0^t \sin(2\theta t)\,\mathrm{d}s\,\mathrm{d}t = \frac{1}{2\theta}\left(\frac{\sin\theta}{\theta}-\cos\theta\right)$$

$$= \frac{1}{2\theta}\sum_{k=1}^{+\infty}\frac{(-1)^{k-1}\theta^{2k}}{(2k-1)!(2k+1)} < \frac{\theta}{6},$$

we obtain, for $\theta \in (0, \pi)$,

$$\left|2\int_0^{\frac{1}{2}}\int_0^t\left(\sum_{k=m+1}^{+\infty} u'(k+s)\sin\left(\left(2k+\frac{1}{2}\right)\theta\right)\right)\sin(2\theta t)\,\mathrm{d}s\,\mathrm{d}t\right|$$

$$< \frac{0.0519\theta}{(m+1)^{1+\beta}\sin\theta}. \tag{5.18}$$

Now the combination of (5.18) and (5.17) with (5.16) gives (5.14). This completes the proof. □

Parallel to Lemma 5.1, we have the following

Lemma 5.2 For $\varphi \in \left(0, \frac{\pi}{2}\right)$, we have

$$\gamma_m(\pi-\varphi) = \Gamma(1-\beta)\cos\left(\frac{\pi}{2}\beta + \left(\frac{1}{2}-\beta\right)\varphi\right) - \left(\frac{\varphi}{\sin\varphi}\right)^{\beta}\left(\int_{(2m+\frac{3}{2})\varphi}^{+\infty}\frac{\sin t}{t^{\beta}}dt + 2A(\beta)\varphi\int_{(2m+\frac{3}{2})\varphi}^{+\infty}\frac{\sin t}{t^{1+\beta}}dt + 4\left(\frac{1}{2}A(\beta)^2 + B(\beta)\right)\varphi^2\int_{(2m+\frac{3}{2})\varphi}^{+\infty}\frac{\sin t}{t^{2+\beta}}dt\right) - 2\widetilde{R}_m(\varphi), \tag{5.19}$$

where, for $m \geqslant 100$ and $\beta \in (0.308443, 0.308444)$, $\widetilde{R}_m(\varphi)$ satisfies

$$|\widetilde{R}_m(\varphi)| \leqslant \frac{\varphi}{\sin\varphi}\frac{2}{\left(\left(2m+\frac{3}{2}\right)\sin\varphi\right)^{\beta}}\left(\frac{0.0172}{\left(m+\frac{3}{4}\right)^2} + \frac{0.0519\varphi}{m+1}\right) + \frac{0.0835\varphi^{1-\beta}}{\left(m+\frac{3}{4}\right)^{2+\beta}}. \tag{5.20}$$

Proof The proof is completely analogous to that of Lemma 5.1. We sketch it as follows. By (5.4), we obtain

$$\gamma_m^{\beta}(\pi-\varphi) = \Gamma(1-\beta)\cos\left(\frac{\pi}{2}\beta + \left(\frac{1}{2}-\beta\right)\varphi t\right) - R_{m,1}(\pi-\varphi),$$

where

$$R_{m,1}(\pi-\varphi) := (2\sin\varphi)^{1-\beta}\sum_{k=m+1}^{+\infty}u(k)\sin\left(2k+\frac{1}{2}\right)\varphi,$$

and u is defined by (5.5).

Parallel to (5.6), the remainder term $R_{m,1}(\pi-\varphi)$ can be rewritten as follows:

$$R_{m,1}(\pi-\varphi) = (2\sin\varphi)^{1-\beta}\frac{\varphi}{\sin\varphi}(\tilde{\sigma}_m(\varphi) + \widetilde{I}_m(\varphi)),$$

where

$$\tilde{\sigma}_m(\varphi) = \sum_{k=m+1}^{+\infty} \int_{k-\frac{1}{4}}^{k+\frac{3}{4}} \left(u(k) - u\left(t - \frac{1}{4}\right)\right) \sin(2\varphi t)\mathrm{d}t,$$

$$\tilde{I}_m(\varphi) = \int_{m+\frac{3}{4}}^{+\infty} u\left(t - \frac{1}{4}\right) \sin(2\varphi t)\mathrm{d}t.$$

For the first term $\tilde{\sigma}_m$, like (5.14), we have

$$|\tilde{\sigma}_m(\varphi)| < \frac{0.017\,2}{\left(m + \frac{3}{4}\right)^{2+\beta} \sin\varphi} + \frac{0.051\,9\varphi}{(m+1)^{1+\beta} \sin\varphi}.$$

It then follows that

$$(2\sin\varphi)^{1-\beta} \frac{\varphi}{\sin\varphi} |\tilde{\sigma}_m(\varphi)|$$

$$< \frac{\varphi}{\sin\varphi} \frac{2}{\left(\left(2m + \frac{3}{2}\right)\sin\varphi\right)^{\beta}} \left(\frac{0.017\,2}{\left(m + \frac{3}{4}\right)^{2}} + \frac{0.051\,9\varphi}{m+1}\right).$$

For the second term $\tilde{I}_m$, parallel to (5.8), we have

$$\tilde{I}_m(\varphi) = \int_{m+\frac{3}{4}}^{+\infty} \frac{\sin(2\varphi)}{t^{\beta}} \mathrm{d}t + A(\beta) \int_{m+\frac{3}{4}}^{+\infty} \frac{\sin(2\varphi t)}{t^{1+\beta}} \mathrm{d}t +$$

$$\left(\frac{1}{2}A(\beta)^2 + B(\beta)\right) \int_{m+\frac{3}{4}}^{+\infty} \frac{\sin(2\varphi t)}{t^{2+\beta}} \mathrm{d}t +$$

$$\int_{m+\frac{3}{4}}^{+\infty} R\left(t - \frac{1}{4}\right) \frac{\sin(2\varphi t)}{t^{3+\beta}} \mathrm{d}t.$$

By (5.11), it's easy to verify that

$$(2\sin\varphi)^{1-\beta} \frac{\varphi}{\sin\varphi} \left| \int_{m+\frac{3}{4}}^{+\infty} R\left(t - \frac{1}{4}\right) \frac{\sin(2\varphi t)}{t^{3+\beta}} \mathrm{d}t \right|$$

$$\leqslant 2\varphi^{1-\beta} \left(\frac{\pi}{4}\right)^{\beta} \int_{m+\frac{3}{4}}^{+\infty} \frac{0.103\,9}{t^{3+\beta}} \mathrm{d}t = \frac{0.083\,5\varphi^{1-\beta}}{\left(m + \frac{3}{4}\right)^{2+\beta}}.$$

Now putting these together, we obtain (5.19) with

$$\tilde{R}_m(\varphi) = (2\sin\varphi)^{1-\beta} \frac{\varphi}{\sin\varphi} \left(\tilde{\sigma}_m(\varphi) + \int_{m+\frac{3}{4}}^{+\infty} R\left(t - \frac{1}{4}\right) \frac{\sin(2\varphi t)}{t^{3+\beta}} \mathrm{d}t\right)$$

satisfying (5.20). □

§ 6. Proofs of Lemmas 2.2~2.5

Now we are ready to prove the lemmas left in Sect. 2.

Proof of lemma 2.2 By (5.12), we have, for $\beta \in (0.308\,443,$

0. 308 444),

$$
\begin{aligned}
&\Gamma(1-\beta)c_{2k}(\beta)\\
&= u(m+1)\\
&= \frac{1}{\left(m+\frac{5}{4}\right)^{\beta}} - \frac{0.0295}{\left(m+\frac{5}{4}\right)^{1+\beta}} - \frac{0.0098}{\left(m+\frac{5}{4}\right)^{2+\beta}} + \frac{R(m+1)}{\left(m+\frac{5}{4}\right)^{3+\beta}},
\end{aligned}
$$

with R satisfying (5. 11). It then follows that

$$
\Gamma(1-\beta)c_{2k}(\beta) \leqslant \frac{1}{\left(m+\frac{5}{4}\right)^{\beta}}\left[1-\frac{0.0295}{m+\frac{5}{4}}+\frac{0.1039}{\left(m+\frac{5}{4}\right)^{3}}\right] \leqslant \frac{0.9997}{\left(m+\frac{5}{4}\right)^{\beta}},
$$

provided $m\geqslant 100$. This is (2. 3). □

Proof of Lemma 2. 4 For simplicity, put $\theta_m=\frac{3\pi}{4m+3}$. We use Lemma 5. 1 with $\theta=\theta_m$. Then working directly from the definition, it's easy to verify the following estimates:

$$
f(\theta_m) = 2\Gamma(1-\beta)\cos\left(\frac{\pi}{2}\beta+\left(\frac{1}{4}-\frac{\beta}{2}\right)\theta_m\right)\sin\left(\left(\frac{1}{4}-\frac{\beta}{2}\right)\theta_m\right)=O(m^{-1}),
$$

$$
h(\theta_m) = \Gamma(1-\beta)\sin\left(\frac{\pi}{2}\beta\right)\left(\left(\frac{\theta_m}{\sin\theta_m}\right)^{\beta}-1\right)=O(m^{-2}),
$$

$$
q_m(\theta_m) = \int_{(2m+\frac{3}{2})\theta_m}^{+\infty}\frac{\cos t}{t^{1+\beta}}\mathrm{d}t = \int_{\frac{3}{2}\pi}^{+\infty}\frac{\cos t}{t^{1+\beta}}\mathrm{d}t,
$$

$$
|R_m(\theta_m)| = O(m^{-1}) \quad (\text{by}(5.2)),
$$

$$
\left(\frac{\theta_m}{\sin\theta_m}\right)^{\beta}p_m(\theta_m) = \left(\frac{\theta_m}{\sin\theta_m}\right)^{\beta}\int_0^{(2m+\frac{3}{2})\theta_m}\frac{\cos t}{t^{\beta}}\mathrm{d}t = \int_0^{\frac{3}{2}\pi}\frac{\cos t}{t^{\beta}}\mathrm{d}t+O(m^{-2}).
$$

Lemma 2. 4 then follows. □

Proof of Lemma 2. 3 We use Lemma 5. 1. By the definition, we have

$$
f(\theta) = 2.6232\cos(0.4845+0.0958\theta)\sin(0.0958\theta),
$$

$$
h(\theta) = 0.6109\left(\left(\frac{\theta}{\sin\theta}\right)^{\beta}-1\right).
$$

Noticing that for $0<x<1$,

$$
\cos x > 1-\frac{1}{2}x^2, \quad \sin x > x-\frac{1}{6}x^3,
$$

we obtain, for $0<\theta<\pi$,

$$f(\theta) > 0.221\,76\theta - 0.011\,66\theta^2 - 0.001\,49\theta^3.$$

On the other hand, for $\theta \in \left(0, \frac{\pi}{2}\right)$,

$$\left(\frac{\theta}{\sin\theta}\right)^{\beta} - 1 < \left(1 - \frac{1}{6}\theta^2\right)^{-\beta} - 1 < \frac{1}{6}\beta\theta^2 + \frac{\beta(1+\beta)}{72}\frac{\theta^4}{1 - \frac{1}{6}\theta^2}$$

$$< 0.051\,4\theta^2 + 0.009\,5\theta^4,$$

and hence,

$$h(\theta) < 0.031\,4\theta^2 + 0.005\,8\theta^4.$$

So, for $\theta \in \left(0, \frac{\pi}{2}\right)$,

$$f(\theta) - h(\theta) > 0.221\,8\theta - 0.059\,7\theta^2.$$

It follows from (5.3) that for $\theta \in \left[\frac{\pi}{4(m+1)}, \frac{\pi}{2}\right]$,

$$|R_m(\theta)| \leqslant \frac{1.615\,0\theta}{\left(\left(m+\frac{3}{4}\right)\sin\theta\right)^{\beta}}\left(\frac{0.008\,0}{m+\frac{3}{4}} + \frac{0.017\,2}{\left(m+\frac{3}{4}\right)^2\sin\theta} + \frac{0.051\,9\theta}{(m+1)\sin\theta}\right)$$

$$\leqslant \frac{\theta^2}{\left(\left(m+\frac{3}{4}\right)\theta\right)^{1+\beta}} \times$$

$$\left(\left(\frac{\pi}{2}\right)^{\beta} \times 1.615 \times 0.008 + 1.615 \times 0.051\,9 \times \left(\frac{\pi}{2}\right)^{\beta+1} +\right.$$

$$\left.1.615 \times 0.017\,2 \times \left(\frac{\pi}{2}\right)^{1+\beta} \times \frac{4}{\pi} \times \frac{101}{100.75}\right)$$

$$= \frac{0.230\,2\theta^2}{\left(\left(m+\frac{3}{4}\right)\theta\right)^{1+\beta}}.$$

Now invoking Lemma 5.1 we find that

$$\gamma_m^{\beta}(\theta) = f(\theta) - h(\theta) + \left(\frac{\theta}{\sin\theta}\right)^{\beta}(p_m(\theta) - 2\theta A(\beta)q_m(\theta)) - R_m(\theta)$$

$$> 0.221\,8\theta - 0.059\,7\theta^2 - \frac{0.230\,2\theta^2}{\left(\left(m+\frac{3}{4}\right)\theta\right)^{1+\beta}} +$$

$$\left(\frac{\theta}{\sin\theta}\right)^{\beta}(p_m(\theta) + 0.059\,1\theta \cdot q_m(\theta)),$$

as desired. □

Proof of Lemma 2.5 We use Lemma 5.2. For $\beta \in (0.308\,443,$

0.308 444), we have

$$2A(\beta)=-0.0591,\quad 2A(\beta)^2+4B(\beta)=-0.0391,$$

$$\Gamma(1-\beta)\cos\left(\frac{\pi}{2}\beta+\left(\frac{1}{2}-\beta\right)\varphi\right)\geqslant\Gamma(1-\beta_1)\cos\left(\frac{\pi}{2}\beta_2+\left(\frac{1}{2}-\beta_2\right)\varphi\right)$$

$$=1.3116\cos(0.4845+0.1916\varphi),$$

where $\beta_1=0.308\,443$, $\beta_2=0.308\,444$. On the other hand, it follows from (5.20) that

$$|\widetilde{R}_m(\varphi)|\leqslant\frac{\varphi^2}{\left(\left(2m+\frac{3}{2}\right)\varphi\right)^{1+\beta}}\left(0.3748+\frac{0.2484+0.4136\varphi}{\left(2m+\frac{3}{2}\right)\varphi}\right).$$

Now Lemma 2.5 is a direct consequence of Lemma 5.2. □

§7. Concluding remarks on the sine sums

We end this paper with some remarks concerning the sine sums:

$$\sigma_n^\beta(\theta)=\sum_{k=1}^{n}c_k(\beta)\sin k\theta,\quad n\in\mathbf{N},\ \theta\in(0,\pi),$$

with the coefficients $c_k(\beta)$ defined by (1.3). It follows from Vietoris's inequality (1.1) that for $\beta\in\left[\frac{1}{2},1\right)$,

$$\sigma_n^\beta(\theta)>0,\quad n\in\mathbf{N},\ \theta\in(0,\pi).$$

However, for $\beta\in\left(0,\frac{1}{2}\right)$, things are different. In fact, we have

Theorem 7.1 Let $\beta_0=0.308\,443\cdots$ denote the Littlewood-Salem-Izumi number and let $\sigma_n^\beta(\theta)$ denote the sine sums defined above. Then the following statements are true:

(1) For $\beta\in[\beta_0,1)$ and odd n we have

$$\sigma_n^\beta(\theta)>0,\quad \theta\in(0,\pi).$$

(2) For $\beta\in\left(0,\frac{1}{2}\right)$ and even n,

$$\min_{\theta\in\left(\pi-\frac{\pi}{n},\pi\right)}\sigma_n^\beta(\theta)<0 \tag{7.1}$$

whereas for $\beta\in[\beta_0,1)$ and even n,

$$\sigma_n^\beta(\theta)>0,\quad \theta\in\left(0,\pi-\frac{\pi}{n}\right). \tag{7.2}$$

Proof (1) From the definition, it's easy to obtain the following identity:

$$\sigma_{2m+1}^{\beta}(\theta)=\cot\left(\frac{\theta}{2}\right)T_{2m+1}^{\beta}(\pi-\theta), \tag{7.3}$$

where T_n^{β} is defined by (1.4). The conclusion of (1) then follows directly from Theorem 1.1.

(2) By (1.3), we have, for $\varphi\in\left(0, \frac{\pi}{2}\right)$,

$$\sigma_{2m}^{\beta}(\pi-\varphi)=\sum_{k=1}^{m}(k-\beta)c_{2k-1}(\beta)\left(\frac{\sin(2k-1)\varphi}{k-\beta}-\frac{\sin(2k\varphi)}{k}\right).$$

Since for $\beta\in\left(0, \frac{1}{2}\right)$ and each $k\geqslant 1$,

$$\lim_{\varphi\to 0}\frac{1}{\varphi}\left(\frac{\sin(2k-1)\varphi}{k-\beta}-\frac{\sin(2k\varphi)}{k}\right)=\frac{2k-1}{k-\beta}-2<0,$$

the first assertion of (2) follows.

Now it remains to prove the second assertion of (2). By Abel's transform, it's enough to prove it in the case $\beta=\beta_0$. Numerical calculation using Maple 7 shows that for $1\leqslant m\leqslant 100$,

$$T_{2m}^{0.308}(\theta)>0, \quad \theta\in\left(0, \pi-\frac{\pi}{2m}\right).$$

By Abel's transform, the same inequalities with 0.308 replaced by β_0 must be also true. So we need only to prove (7.2) with $n=2m+2$ and $m\geqslant 100$. Henceforth, we always assume $\beta=\beta_0$.

We start with the identity (7.3). We then obtain from Lemma 2.1,

$$\sigma_{2m+2}^{\beta}(\pi-\theta)=\frac{2^{\beta}\sin\frac{\theta}{2}}{\Gamma(1-\beta)(\sin\theta)^{1-\beta}}(\gamma_m^{\beta}(\theta)-\tau_m^{\beta}(\theta)),$$

where

$$\tau_m^{\beta}(\theta):=\frac{\Gamma(1-\beta)(\sin\theta)^{1-\beta}}{2^{\beta}\sin\frac{\theta}{2}}c_{2m+2}(\beta)\sin(2m+2)\theta. \tag{7.4}$$

So it's sufficient to prove that for $\theta\in\left[\frac{\pi}{2m+2}, \pi-\frac{\pi}{2m+2}\right]$,

$$\gamma_m^{\beta}(\theta)-\tau_m^{\beta}(\theta)>0. \tag{7.5}$$

If $\sin(2m+2)\theta<0$ then (7.5) is obvious since both γ_m^{β} and $-\tau_m^{\beta}$ are positive. So in what follows we may assume $\sin(2m+2)\theta\geqslant 0$. Under this assumption, by Lemma 2.2, we have

$$\tau_m(\theta)\leqslant\frac{1.9994\cos\frac{\theta}{2}}{\left(\left(2m+\frac{5}{2}\right)\sin\theta\right)^{\beta}}\sin(2m+2)\theta. \tag{7.6}$$

On the other hand, it follows from (3.7)(3.8) and (3.11) that

$$\gamma_m^{\beta}(\theta)\geqslant\begin{cases}p_m(\theta), & \text{if } \frac{\pi}{2m+2}\leqslant\theta\leqslant\frac{\pi}{2},\\ 0.1235\theta+0.3204, & \text{if } 0.3\leqslant\theta\leqslant\frac{\pi}{2},\end{cases} \tag{7.7}$$

where

$$p_m(\theta)=\int_0^{\left(2m+\frac{3}{2}\right)\theta}\frac{\cos t}{t^{\beta}}\mathrm{d}t.$$

We first prove (7.5) for $\theta\in\left[\frac{\pi}{2m+2},1\right]$. If $\frac{2\pi}{2m+2}\leqslant\theta\leqslant\frac{3\pi}{2m+2}\leqslant 0.05$ then, by (7.6),

$$\begin{aligned}\tau_m^{\beta}(\theta)&\leqslant\frac{1.9994}{\left(\left(2m+\frac{3}{2}\right)\theta\right)^{\beta}}\left(\frac{0.05}{\sin(0.05)}\right)^{\beta}\sin(2m+2)\theta\\&=\frac{1.9997}{\left(\left(2m+\frac{3}{2}\right)\theta\right)^{\beta}}\left(\sin\left(2m+\frac{3}{2}\right)\theta+2\cos\left(\left(2m+\frac{7}{4}\right)\theta\right)\sin\frac{\theta}{4}\right)\\&\leqslant\frac{1.9997}{\left(\left(2m+\frac{3}{2}\right)\theta\right)^{\beta}}\sin\left(2m+\frac{3}{2}\right)\theta+\frac{0.05}{\left(\left(2m+\frac{3}{2}\right)\theta\right)^{\beta}},\end{aligned}$$

and hence, by (7.7),

$$\begin{aligned}\gamma_m^{\beta}(\theta)-\tau_m^{\beta}(\theta)&\geqslant\min_{\frac{201.5}{202}\times 2\pi\leqslant z\leqslant 3\pi}\left(\int_0^z\frac{\cos t}{t^{\beta}}\mathrm{d}t-\frac{1.9997\sin z}{z^{\beta}}-\frac{0.05}{z^{\beta}}\right)\\&=:\min_{\frac{201.5}{202}\times 2\pi\leqslant z\leqslant 3\pi}F_1(z);\end{aligned}$$

similarly, if $\frac{4\pi}{2m+2}\leqslant\theta\leqslant\frac{19\pi}{2m+2}\leqslant 0.3$, then

$$\tau_m^{\beta}(\theta)\leqslant\frac{2.0087}{\left(\left(2m+\frac{3}{2}\right)\theta\right)^{\beta}}\sin\left(2m+\frac{3}{2}\right)\theta+\frac{0.3010}{\left(\left(2m+\frac{3}{2}\right)\theta\right)^{\beta}},$$

and $\gamma_m^{\beta}(\theta)-\tau_m^{\beta}(\theta)\geqslant\min_{\frac{201.5}{202}\times 4\pi\leqslant z\leqslant 19\pi}\left(\int_0^z\frac{\cos t}{t^{\beta}}\mathrm{d}t-\frac{2.0087\sin z}{z^{\beta}}-\frac{0.3010}{z^{\beta}}\right)$

$$=: \min_{\frac{201.5}{202}\times 4\pi\leqslant z\leqslant 19\pi} F_2(z);$$

if $\frac{20\pi}{2m+2}\leqslant\theta\leqslant\frac{31\pi}{2m+2}\leqslant\frac{1}{2}$, then

$$\tau_m^\beta(\theta)\leqslant\frac{2.0255}{\left(\left(2m+\frac{3}{2}\right)\theta\right)^\beta}\sin\left(2m+\frac{3}{2}\right)\theta+\frac{0.5051}{\left(\left(2m+\frac{3}{2}\right)\theta\right)^\beta},$$

and

$$\gamma_m^\beta(\theta)-\tau_m^\beta(\theta)\geqslant\min_{\frac{201.5}{202}\times 20\pi\leqslant z\leqslant 31\pi}\left(\int_0^z\frac{\cos t}{t^\beta}\mathrm{d}t-\frac{2.0255\sin z}{z^\beta}-\frac{0.5051}{z^\beta}\right)$$

$$=: \min_{\frac{201.5}{202}\times 20\pi\leqslant z\leqslant 31\pi} F_3(z);$$

if $\frac{32\pi}{2m+2}\leqslant\theta\leqslant 1$, then

$$\tau_m^\beta(\theta)\leqslant\frac{2.1087}{\left(\left(2m+\frac{3}{2}\right)\theta\right)^\beta}\sin\left(2m+\frac{3}{2}\right)\theta+\frac{1.0434}{\left(\left(2m+\frac{3}{2}\right)\theta\right)^\beta},$$

and

$$\gamma_m^\beta(\theta)-\tau_m^\beta(\theta)\geqslant\min_{z\geqslant\frac{201.5}{202}\times 32\pi}\left(\int_0^z\frac{\cos t}{t^\beta}\mathrm{d}t-\frac{2.1087\sin z}{z^\beta}-\frac{1.0434}{z^\beta}\right)$$

$$=: \min_{z\geqslant\frac{201.5}{202}\times 32\pi} F_4(z);$$

Mechanical computation using Maple 7 shows that

$$\min_{\frac{201.5}{202}\times 2\pi\leqslant z\leqslant 3\pi} F_1(z),\quad \min_{\frac{201.5}{202}\times 4\pi\leqslant z\leqslant 19\pi} F_2(z),\quad \min_{\frac{201.5}{202}\times 20\pi\leqslant z\leqslant 31\pi} F_3(z),\quad \min_{z\geqslant\frac{201.5}{202}\times 32\pi} F_4(z)$$

are all positive. This proves (7.5) in the case when $\theta\in\left[\frac{\pi}{2m+2}, 1\right]$ and $\sin(2m+2)\theta\geqslant 0$.

Next, for $\theta\in\left[1, \frac{\pi}{2}\right]$, we use (7.6) to obtain

$$\tau_m^\beta(\theta)\leqslant\frac{1.9994\cos(0.5)}{(202.5\sin 1)^\beta}=0.3597,$$

and we use the second estimate in (7.7) to obtain

$$\gamma_m^\beta(\theta)\geqslant 0.1235+0.3204=0.4439.$$

Inequality (7.5) then follows.

Finally, we prove (7.5) for $\theta\in\left[\frac{\pi}{2}, \pi-\frac{\pi}{2m+2}\right]$. Set $\varphi=\pi-\theta$ so that $\varphi\in\left[\frac{\pi}{2m+2}, \frac{\pi}{2}\right]$. If $\varphi\in\left[\frac{\pi}{2m+2}, \frac{1}{2}\right]$ then, by (7.6),

$$\gamma_m^\beta(\pi-\varphi)\leqslant\frac{1.9994\sin\frac{\varphi}{2}}{((2m+2.5)\varphi)^\beta}\left(\frac{\varphi}{\sin\varphi}\right)^\beta\leqslant\frac{1.9994\sin 0.25}{\pi^\beta}\left(\frac{0.5}{\sin 0.5}\right)^\beta$$
$$=0.3520,$$

and hence, by (3.16) and (3.19),

$$\gamma_m^\beta(\pi-\varphi)-\tau_m^\beta(\pi-\varphi)\geqslant 0.5102-0.3520>0;$$

if $\varphi\in\left[\frac{1}{2}, \frac{\pi}{2}\right]$ then, by (7.6),

$$\tau_m^\beta(\pi-\varphi)\leqslant\frac{1.9994\times\sin\frac{\pi}{4}}{(202.5\times\sin 0.5)^\beta}=0.3447,$$

and by (3.21), $\quad\gamma_m^\beta(\pi-\varphi)-\tau_m^\beta(\pi-\varphi)\geqslant 0.6473-0.3447>0.$

Therefore for all $\varphi\in\left[\frac{\pi}{2m+2}, \frac{\pi}{2}\right]$ we have

$$\gamma_m^\beta(\pi-\varphi)-\tau_m^\beta(\pi-\varphi)>0.$$

This completes the proof of (7.2).

As an immediate consequence of Theorem 7.1, we have

Corollary 7.2 Let $\beta_0=0.308443\cdots$ be the unique solution of (1.6) and let $\{a_k\}_{k=0}^n$ be a sequence of real numbers satisfying $a_1\geqslant a_2\geqslant\cdots\geqslant a_n>0$ and $a_{2k}\leqslant\left(1-\frac{\beta_0}{k}\right)a_{2k-1}$, $k\geqslant 1$. Then

$$\sum_{k=1}^n a_k\sin k\theta>0$$

if either n is odd and $\theta\in(0, \pi)$ or n is even and $\theta\in\left(0, \pi-\frac{\pi}{n}\right)$.

Acknowledgments

The authors would like to express their sincere gratitude to Professor G. Gasper for his proposal to publish this paper and the Koumandos paper [10] in the same issue of the Ramanujan Journal.

References

[1] Andrews G E, Askey R, Roy R. Special Functions. Encyclopedia of Mathematics and its Applications, Vol. 71. Cambridge University Press, Cambridge, 1999.

[2] Artin E. The Gamma Function. Holt, Rinehart and Winston, New York, 1964.

[3] Askey R. Vietoris's inequalities and hypergeometric series. In: Recent Progress in Inequalities, 1996, Math. Appl. 430, 63—76. Kluwer Academic, Dordrecht, 1998.

[4] Askey R, Steinig J. Some positive trigonometric sums. Trans. Am. Math. Soc., 1974, 187: 295—307.

[5] Brown G, Hewitt E. A class of positive trigonometric sums. Math. Ann., 1984, 268: 91—122.

[6] Brown G. Yin Q. Positivity of a class of cosine sums. Acta Sci. Math., (Szeged) 2001, 67: 221—247.

[7] Brown G, Wang K Y, Wilson D C, Positivity of some basic cosine sums. Math. Proc. Camb. Philos. Soc., 1993, 114(3): 383—391.

[8] Brown G, Dai F, Wang K Y, Extensions of Vietoris's inequalities Ⅱ. Adv. Math., 2005, 34(2): 253—255.

[9] Fejér L, Einige Sätze. die sich auf das Vorzeichen einer ganzen rationalen Funktion beziehen; nebst Anwendungen dieser Sätze auf die Abschnitte und Abschnittesmittelwerte von ebenen und räumlichen harmonischen Entwicklungen und von beschränkten Potenzreihen. Monatsh, Math. Phys., 1928, 35: 305—344.

[10] Koumandos S. An extension of Vietoris's inequalities. Ramanujan J., 2007, 14(1): 1—38.

[11] Vietoris L. Uber das Vorzeichen gewisser trigonometricher Summen. Sitzungsber, Öst. Acad. Wiss. Math. -Natur. Kl. S. -B. Ⅱ, 1958, 167: 125—135.

[12] Young W H. On certain series of Fourier. Proc. Lond. Math. Soc., 1913, 11: 357—366.

数学进展，2005，34(2)：253—255.

Vietoris 不等式之推广(Ⅱ)①

Extensions of Vietoris's Inequalities (Ⅱ)

It was shown by Vietoris[1] in 1958 that for $n\geqslant 1$ and $x\in(0,\pi)$,

$$\sum_{k=0}^{n} c_k \cos kx > 0, \tag{1}$$

where $c_0=c_1=1$ and

$$c_{2k}=c_{2k+1}=\frac{2k-1}{2k}c_{2k-1}, \qquad k\in\mathbf{N}.$$

Here the coefficients can be expressed in terms of gamma functions:

$$c_{2k}=c_{2k+1}=\frac{\Gamma(k+0.5)}{\Gamma(0.5)\Gamma(k+1)}, \qquad k\in\mathbf{N}. \tag{2}$$

One way to extend Vietoris's inequality (1) is via taking the coefficients in (1) to be

$$d_{2k}(\beta)=d_{2k+1}(\beta)=\frac{\Gamma(k+1-\beta)}{\Gamma(1-\beta)\Gamma(k+1)}, \qquad k\in\mathbf{N} \tag{3}$$

with $\beta<1$. In this direction, the current authors[4] recently have proved the following result:

For $0<x<\pi$ and $n\geqslant 1$,

$$S_n^{\beta}(x)=\sum_{k=0}^{n} d_k(\beta)\cos kx > 0,$$

---

① Foundation item: This work was supported by a grant from Australian Research Council. Wang K Y was also supported by NSFC.

本文与 Brown G 和戴峰合作.

Received: 2005-01-10.

whenever $\beta \geqslant \beta_0$, where $\beta_0 \in (0.308\,443, 0.308\,444)$ is the unique solution of the equation

$$\int_0^{\frac{3\pi}{2}} \frac{\cos t}{t^\beta} \mathrm{d}t = 0.$$

Moreover, the number β_0 is best possible in the sense that for any $\beta < \beta_0$,

$$\lim_{n \to +\infty} \min_{x \in (0,\pi)} S_n^\beta(x) = -\infty.$$

In the present work, we will extend Vietoris's inequality (1) in another way. More precisely, for $\alpha > -1$ and $k \geqslant 0$, we let

$$c_{2k}(\alpha) = c_{2k+1}(\alpha) = \frac{\Gamma\left(\frac{2+\alpha}{2}\right)\Gamma\left(k+\frac{1+\alpha}{2}\right)}{\Gamma\left(\frac{1+\alpha}{2}\right)\Gamma\left(k+\frac{2+\alpha}{2}\right)} \tag{4}$$

and define for $n \geqslant 1$,

$$T_n^\alpha(x) = \sum_{k=0}^{n} c_k(\alpha) \cos kx. \tag{5}$$

Our main purpose is to find suitable conditions on α under which

$$T_n^\alpha(x) > 0$$

holds true for $n \in \mathbf{N}$ and $x \in (0, \pi)$.

It can be shown that the equation

$$\psi(\alpha) := \min_{x \in (0,\pi)} T_6^\alpha(x) = 0 \tag{6}$$

has a unique solution α_0 in $(-1, +\infty)$, and moreover, by mechanical computation, $\alpha_0 \in (2.330\,8, 2.330\,9)$.

One of our main results reads as follows.

Theorem 1 For $-1 < \alpha \leqslant \alpha_0 = 2.330\,8\cdots$,

$$T_n^\alpha(x) := \sum_{k=0}^{n} c_k(\alpha) \cos kx > 0, \quad x \in (0,\pi), \quad n \in \mathbf{N},$$

where the equality holds for some $x \in (0, \pi)$ if and only if $n = 6$ and $\alpha = \alpha_0$. Moreover, for $\alpha > \alpha_0$,

$$\min_{x \in (0,\pi)} T_6^\alpha(x) < 0.$$

It is clear that Theorem 1 for $\alpha = 0$ corresponds to Vietoris's inequality (1). We remark that Theorem 1 for $\alpha = 1$ is due to Brown and Hewitt[2] while for $\alpha = 2$ is due to a recent paper[3] by Brown and Yin.

As an immediate consequence of Theorem 1, we obtain, by Abel's transform,

Corollary Let $\alpha_0=2.3308\cdots$ be as above and let $\{a_k\}_{k=0}^{+\infty}$ be a sequence of real numbers satisfying $a_0\geqslant a_1\geqslant\cdots\geqslant a_n>0$ and $(2k+\alpha_0)a_{2k}\leqslant(2k-1+\alpha_0)a_{2k-1}$, $k\geqslant1$. Then we have

$$\sum_{k=0}^{n} a_k\cos kx>0,\quad x\in(0,\pi),\quad n\in\mathbf{N},$$

where the equality holds for some $x\in(0,\pi)$ if and only if $n=6$ and

$$a_0=a_1=1,\qquad a_2=a_3=\frac{1+\alpha_0}{2+\alpha_0},$$

$$a_4=a_5=\frac{(1+\alpha_0)(3+\alpha_0)}{(2+\alpha_0)(4+\alpha_0)},$$

$$a_6=\frac{(1+\alpha_0)(3+\alpha_0)(5+\alpha_0)}{(2+\alpha_0)(4+\alpha_0)(6+\alpha_0)}.$$

Our second result can be sated as follows.

Theorem 2 For any $\alpha>-1$, there's an $M=M(\alpha)$ depending only on α such that for $n\geqslant M$ and $x\in(0,\pi)$,

$$T_n^{\alpha}(x):=\sum_{k=0}^{n} c_k(\alpha)\cos kx>0. \tag{7}$$

Furthermore, given an integer $n\geqslant2$ there is an $\gamma=\gamma(n)$ depending only on n such that for $\alpha>\gamma(n)$,

$$\min_{x\in(0,\pi)} T_n^{\alpha}(x)<0.$$

Remark If we denote by $N(\alpha)$ the smallest positive integer for which (7) holds for all $x\in(0,\pi)$ and $n\geqslant N(\alpha)$. Then Theorem 2 above means that $N(\alpha)$ is finite for each fixed $\alpha>-1$, however,

$$\min_{\alpha\to+\infty} N(\alpha)=+\infty.$$

It would be interesting if one could find a better upper estimate of $N(\alpha)$ for each specific α. However, as was demonstrated in our proof of Theorem 1 where a better upper estimate for $N(2.3309)$ is given, technically this seems to be very difficult.

We end this announcement with a result concerning the sine sums. It follows by [Vietoris, 1] and a simple use of Abel's transform that for

$-1<\alpha\leqslant 0$, $n\geqslant 1$ and $x\in(0,\pi)$,

$$\sigma_n^{\alpha}(x):=\sum_{k=1}^{n} c_k(\alpha)\sin kx > 0.$$

However, things are different for $\alpha>0$. In fact, we have

Theorem 3 For the sine sums $\sigma_n^{\alpha}(x)$ defined above, the following are true:

(i) The conclusions of Theorems 1 and 2 remain valid with $T_n^{\alpha}(x)$ replaced by $\sigma_{2m+1}^{\alpha}(x)$.

(ii) For $\alpha>-1$ there exists an $M=M(\alpha)$ such that for $m\geqslant M(\alpha)$ and $x\in\left(0,\ \pi-\frac{\pi}{2m+2}\right)$,

$$\sigma_{2m+2}^{\alpha}(x)>0. \tag{8}$$

(iii) For $\alpha>0$ and any $m\geqslant 2$,

$$\min_{x\in\left(\pi-\frac{\pi}{2m+2},\pi\right)} \sigma_{2m+2}^{\alpha}(x)<0.$$

References

[1] Vietoris L. Uber das Vorzeichen gewisser trigonometricher Summen. Sitzungsber, Öst. Acad. Wiss. Math.-Natur. Kl. S.-B. II, 1958, 167: 125—135.

[2] Brown G, Hewitt E. A class of positive trigonometric sums. Math. Ann., 1984, 268: 91—122.

[3] Brown G, Yin Qinghe. Positivity of a class of cosine sums. Acta Sci. Math. (Szeged), 2001, 67: 221—247.

[4] Brown G, Dai Feng, Wang Kunyang. Extensions of Vietoris's inequalities I. Advances in Mathematics(China), 2005, 34(1): 125—127.

[5] Erdélyi and etal. Higher Transcendental Functions, Volume 1. McGraw-Hill Book Company, INC. 1953.

Math. Proc. Camb. Phil. Soc., 2007, 142: 219—232.

关于正的余弦和①

On Positive cosine Sums

Abstract For $\alpha\in(0, 1)$ and $x\in[0, \pi]$, we define $S_0(x, \alpha)=1$ and

$$S_n(x,\alpha):=1+\sum_{k=1}^{n}a_k(\alpha)\cos kx, \quad n\in\mathbf{N}^*,$$

where

$$a_k(\alpha)=\frac{\Gamma(k+1-\alpha)}{\Gamma(2-\alpha)k!}, \quad k\in\mathbf{N}^*.$$

Also, we set, for $\alpha\in(0, 1)$,

$$N(\alpha)=\inf\{N\geqslant 0 \mid \min_{x\in[0,\pi]} S_n(x, \alpha)>0 \text{ whenever } n\geqslant N\},$$

where it is agreed that $\inf\varnothing=\infty$. It is shown in this paper that

$$N(\alpha)=\begin{cases}0, & \text{if } \alpha^*<\alpha<1,\\ 8, & \text{if } \alpha=\alpha^*,\\ \infty, & \text{if } 0<\alpha\leqslant\alpha_0,\end{cases}$$

and $8\leqslant N(\alpha)<+\infty$ if $\alpha_0<\alpha<\alpha^*$, where $\alpha^*:=0.335\ 42\cdots$ is the unique solution $\alpha\in(0, 1)$ of the equation

$$\min_{x\in[0,\pi]} S_7(x, \alpha)=0,$$

① Brown G was Supported by the Australian Research Council. Dai Feng was also partially supported by the NSERC Canada under grant G121211001.

Wang Kunyang was also partially supported by NSF of China under the grant 10071007.

Received: 2005-11-21.

while $\alpha_0=0.308\,443\cdots$ is the unique solution $\alpha\in(0,1)$ of the equation

$$\int_0^{\frac{3}{2}\pi}\frac{\cos t}{t^\alpha}\mathrm{d}t=0.$$

In the case $\alpha^*<\alpha<1$, this extends a result of Brown, Wang and Wilson (1993).

§ 1. Introduction

First, we introduce some necessary notations. Given $0<\alpha<1$, we set

$$a_k(\alpha)=\frac{\Gamma(k+1-\alpha)}{\Gamma(2-\alpha)k!},\qquad k\in\mathbf{N}^* \tag{1.1}$$

and define

$$S_n(x,\alpha):=1+\sum_{k=1}^{n}a_k(\alpha)\cos kx,\quad x\in[0,\pi],\quad n\in\mathbf{N}^*. \tag{1.2}$$

Throughout this paper, we denote by α_0 the unique solution $\alpha\in(0,1)$ of the equation

$$\int_0^{\frac{3}{2}\pi}\frac{\cos t}{t^\alpha}\mathrm{d}t=0,$$

and by α^* the unique solution $\alpha\in(0,1)$ of the equation

$$\min_{x\in[0,\pi]}S_7(x,\alpha)=0. \tag{1.3}$$

Numerical evaluation shows that $\alpha_0=0.308\,443\cdots$ and $\alpha^*=0.335\,42\cdots$.

For convenience, we set, for $\alpha\in(0,1)$,

$$N(\alpha):=\inf\{N\geqslant 0\mid \min_{x\in[0,\pi]}S_n(x,\alpha)>0\text{ whenever }n\geqslant N\}, \tag{1.4}$$

where it is agreed that $\inf\varnothing=\infty$.

Our main result in this paper, which concerns the positivity of the partial sums $S_n(x,\alpha)$, can be stated as follows.

Theorem 1.1

(1) If $0.335\,42\cdots=\alpha^*\leqslant\alpha<1$ then

$$S_n(x,\alpha)\geqslant 0,\qquad x\in[0,\pi],\qquad n\in\mathbf{N}^*,$$

with equality being true for some $x\in[0,\pi]$ if and only if $\alpha=\alpha^*$ and $n=7$.

(2) If $0.308\,443\cdots=\alpha_0<\alpha<\alpha^*=0.335\,42\cdots$ then

$$8\leqslant N(\alpha)\leqslant N_0\cdot(\alpha-\alpha_0)^{-\frac{1}{\alpha_0(1-\alpha_0)}}=N_0\cdot(\alpha-\alpha_0)^{-4.688\,1\cdots},$$

Where N_0 is a positive absolute constant

(3) If $\alpha=\alpha_0=0.308\ 443\cdots$ then

$$\limsup_{n\to+\infty}(\min\{S_n(x,\ \alpha_0)\mid x\in[0,\ \pi]\})=-\frac{\alpha_0}{1-\alpha_0}=-0.446\ 014\cdots \tag{1.5}$$

(4) If $0<\alpha<\alpha_0$ then

$$\liminf_{n\to+\infty}(\min\{S_n(x,\ \alpha)\mid x\in[0,\ \pi]\})=-\infty.$$

Notice that, by the definition (1.1),

$$\left(1-\frac{\alpha}{k+1}\right)a_k(\alpha)=a_{k+1}(\alpha),\qquad k\in\mathbf{N}. \tag{1.6}$$

Thus, by Theorem 1.1 and a summation by parts, we deduce the following:

Corollary 1.2

(1) If $\{c_k\}_{k=0}^{+\infty}$ is a sequence of positive numbers satisfying $c_0\geqslant c_1$ and

$$\left(1-\frac{\alpha^*}{k+1}\right)c_k\geqslant c_{k+1},\qquad k\in\mathbf{N},$$

where $\alpha^*=0.335\ 42\cdots$ is the unique solution $\alpha\in(0,\ 1)$ of the equation (1.3), then

$$\sum_{k=0}^{n}c_k\cos kx\geqslant 0,\quad x\in[0,\pi],\quad n\in\mathbf{N},$$

with equality being true for some $x\in[0,\ \pi]$ if and only if $n=7$ and $c_k=c_0a_k(\alpha^*)$ for $0\leqslant k\leqslant 7$ and some positive number c_0.

(2) If $0.308\ 443\cdots=\alpha_0<\alpha<\alpha^*=0.335\ 42\cdots$, and $\{c_k\}_{k=0}^{+\infty}$ is a sequence of positive numbers satisfying $c_0\geqslant c_1$,

$$\left(1-\frac{\alpha}{k+1}\right)c_k\geqslant c_{k+1},\qquad k\in\mathbf{N},$$

then we have

$$\sum_{k=0}^{n}c_k\cos kx\geqslant\min_{0\leqslant j<N(\alpha)}\left(\sum_{k=0}^{j}c_k\cos kx\right),\quad x\in[0,\pi],\quad n\in\mathbf{N},$$

where $N(\alpha)$ is defined by (1.4).

Remarks 1) It is of some historical interest that the critical constant α_0 is the Littlewood-Salem-Izumi number which is critical for the uniform boundedness of the partial sums, $1+\sum_{k=1}^{n}k^{-\alpha}\cos kx$. It was shown by Brown, Wang and Wilson in [5] that this same constant is also critical for positivity of these partial sums, that is, the inequality

$$1+\sum_{k=1}^{n} k^{-\alpha}\cos kx > 0,\quad n\in\mathbf{N},\quad x\in[0,\pi] \tag{1.7}$$

is true if and only if $\alpha\geqslant\alpha_0$. Although we have $a_k(\alpha)=O(k^{-\alpha})$ it would have been unreasonable to conjecture the truth of Theorem 1.1 on this basis. Note that for $0<\alpha<1$ and $k\geqslant1$,

$$\frac{(k+1)^{-\alpha}}{k^{-\alpha}}=\left(1-\frac{1}{k+1}\right)^{\alpha}\leqslant1-\frac{\alpha}{k+1}=\frac{a_{k+1}(\alpha)}{a_k(\alpha)}. \tag{1.8}$$

Thus, a summation by parts shows that our result, Theorem 1.1 (1), implies (1.7) when $\alpha\geqslant\alpha^*:=0.335\ 42\cdots$ (here α^* is bigger than the critical constant α_0). We point out that it is Equation (1.8) that motivates the considerations in the present paper.

2) Obviously, from Theorem 1.1, we conclude that

$$N(\alpha)\begin{cases}=0, & \text{if } \alpha^*<\alpha<1,\\ =8, & \text{if } \alpha=\alpha^*,\\ <+\infty, & \text{if } \alpha_0<\alpha<\alpha^*,\\ =+\infty, & \text{if } 0<\alpha\leqslant\alpha_0,\end{cases}$$

and moreover, $\lim\limits_{\alpha\to\alpha_0+}N(\alpha)=N(\alpha_0)=+\infty$. This means the Corollary 1.2 (2) is useful only insofar as $\alpha>\alpha_0$ is bounded away from α_0. In that case we may employ Lemma 2.2 to find a better bound for $N(\alpha)$. For example, our proof in Section 4 (see Lemma 4.1) shows that $N(\alpha_1)=N(\alpha^*)=8$, where $\alpha_1=0.335\ 42$ is slightly smaller than the critical constant α^*.

3) As a product of our proof of Theorem 1.1, we have the following asymptotic estimate of $S_n(x,\alpha)$ (see Lemma 2.2), which seems to be of independent interest: for $\alpha\in(0,1)$, we have

$$S_n(x,\alpha)=-\frac{\alpha}{1-\alpha}+\frac{\sin\left(\frac{\pi\alpha}{2}+\frac{(1-\alpha)x}{2}\right)}{(1-\alpha)\left(2\sin\frac{x}{2}\right)^{1-\alpha}}-\frac{x^{\alpha}}{2\Gamma(2-\alpha)\sin\frac{x}{2}}\int_{(n+\frac{1}{2})x}^{+\infty}\frac{\cos t}{t^{\alpha}}\mathrm{d}t+O(n^{-\alpha}), \tag{1.9}$$

uniformly for $x\in(0, \pi]$, as $n\to+\infty$. The point in expression (1.9) is that for $n\geqslant 101$, the remaining term $O(n^{-\alpha})$ can be controlled by a very small, computable constant.

4) After this paper was completed, we learnt that the following result was obtained recently by S. Koumandos and St. Ruscheweyh [6, equation (3)]: the inequality

$$T_n^{\mu}(x):=\sum_{k=0}^{n}\frac{\Gamma(2\mu+k)}{\Gamma(2\mu)k!}\cos kx>0, \quad x\in[0,\pi], \quad n\in\mathbf{N} \quad (1.10)$$

holds if and only if $0<\mu\leqslant\mu_0:=\frac{1-\alpha_0}{2}=0.345\ 778\cdots$ (1.10) is the key to all results in the paper [6]. As was shown in [6], (1.10) alone does not imply (1.7), but can be used to prove the following weaker inequality:

$$\frac{1}{1-\alpha}+\sum_{k=1}^{n}\frac{\cos kx}{k^{\alpha}}>0, \quad x\in[0,\pi], \quad n\in\mathbf{N}, \quad 0<\alpha\leqslant\alpha_0.$$

On the other hand, it was also pointed out in [6] that (1.7) implies (1.10) in the sharper version

$$T_n^{\mu}(x)\geqslant 1-2\mu, \quad x\in[0, \pi], \quad n\in\mathbf{N}, \quad (1.11)$$

for $0<\mu\leqslant\mu_1$, where $\mu_1=0.307\ 51\cdots$ is the largest number μ for which the sequence $\left\{\frac{\Gamma(2\mu+k)k^{\alpha_0}}{k!}\right\}_{k=1}^{+\infty}$ is decreasing. Note that

$$T_n^{\mu}(x)=2\mu S_n(x, 1-2\mu)+1-2\mu. \quad (1.12)$$

Thus, on the one hand, our result, Theorem 1.1(1), implies that the sharper inequality (1.11) holds if and only if $1-2\mu\geqslant\alpha^*$, i.e., $0<\mu\leqslant\mu_2:=\frac{1-\alpha^*}{2}=0.332\ 28\cdots$. On the other hand, however, the inequality (1.10) of Koumandos and Ruscheweyl implies that the inequality

$$S_n(x, \alpha)\geqslant-\frac{\alpha}{1-\alpha}, \quad x\in[0, \pi], \quad n\in\mathbf{N}$$

holds whenever $\alpha\geqslant\alpha_0$.

We organize this paper as follows. Our main goal in Section 2 is to prove Lemma 2.2, which gives an asymptotic estimate of $S_n(x, \alpha)$ and will play a fundamental role in our proof of Theorem 1.1. For technical reasons, we prove Statements (2)~(4) of Theorem 1.1 first in Section 3. After that, following almost the same idea, we prove Theorem 1.1 (1) with more technical details in the final section, Section 4.

Finally, we refer the reader to [2, 3] for more background information.

§ 2. Fundamental lemmas

We start with the following useful lemma, which was proved recently in [4, Lemma 4.1].

Lemma 2.1 For $u \geqslant 100$ and $\alpha \in (0, 1)$, we have

$$\frac{\Gamma(u+1-\alpha)}{\Gamma(u+1)}=\frac{1}{\left(u+\frac{1}{4}\right)^{\alpha}}+\frac{A(\alpha)}{\left(u+\frac{1}{4}\right)^{1+\alpha}}+\frac{\frac{1}{2}A^2(\alpha)+B(\alpha)}{\left(u+\frac{1}{4}\right)^{2+\alpha}}+\frac{R(u, \alpha)}{\left(u+\frac{1}{4}\right)^{3+\alpha}},$$

where

$$A(\alpha):=\frac{1}{2}\alpha^2-\frac{1}{4}\alpha, \quad B(\alpha):=-\frac{1}{96}\alpha-\frac{1}{8}\alpha^2+\frac{1}{6}\alpha^3 \tag{2.1}$$

and

$$|R(u, \alpha)-e(\alpha)| \leqslant 0.0301, \tag{2.2}$$

with

$$e(\alpha):=\frac{1}{64}\alpha+\frac{95}{128}\alpha^2-\frac{23}{384}\alpha^3-\frac{1}{192}\alpha^4+\frac{5}{96}\alpha^5+\frac{1}{48}\alpha^6. \tag{2.3}$$

The point of Lemma 2.1 is that for a given $\alpha \in (0, 1)$, both $A(\alpha)$ and $B(\alpha)$ are computable and the remainder term $R(u, \alpha)$ can be well approximated by $\frac{e(\alpha)}{\left(u+\frac{1}{4}\right)^{3+\alpha}}$ which is also computable.

For the rest of this paper, we set, for $\alpha \in (0, 1)$ and $n \geqslant 100$,

$$\rho_n(\alpha)=|A(\alpha)|+\frac{\left|\frac{1}{2}A^2(\alpha)+B(\alpha)\right|}{n+\frac{5}{4}}+\frac{|e(\alpha)|+0.0301}{\left(n+\frac{5}{4}\right)^2}. \tag{2.4}$$

The following lemma, which gives an asymptotic estimate of the partial sums $S_n(x, \alpha)$, will play a fundamental role in the proof of

Theorem 1.1.

Lemma 2.2 For $0<\alpha<1$, $0<x\leqslant\pi$ and $n\geqslant 100$,

$$S_n(x,\alpha)=-\frac{\alpha}{1-\alpha}+\frac{\sin\left(\frac{\pi\alpha}{2}+\frac{(1-\alpha)x}{2}\right)}{(1-\alpha)\left(2\sin\frac{x}{2}\right)^{1-\alpha}}-$$

$$\frac{x^{\alpha}}{2\sin\frac{x}{2}}\frac{1}{\Gamma(2-\alpha)}\int_{(n+\frac{1}{2})x}^{+\infty}\frac{\cos t}{t^{\alpha}}\mathrm{d}t+r_n(x,\alpha), \qquad (2.5)$$

where $r_n(x,\alpha)$ satisfies

$$|r_n(x,\alpha)|\leqslant\frac{x}{\sin\frac{x}{2}}\frac{1}{4\Gamma(2-\alpha)\left(n-\frac{1}{2}\right)^{\alpha}}+\frac{\rho_n(\alpha)}{\alpha\Gamma(2-\alpha)\left(n+\frac{1}{4}\right)^{\alpha}}. \qquad (2.6)$$

For the proof of Lemma 2.2, we need the following.

Lemma 2.3 For $0<x\leqslant\pi$ and $0<\alpha<1$, we have

$$S(x,\alpha):=1+\sum_{k=1}^{+\infty}a_k(\alpha)\cos kx=\frac{\sin\left(\frac{\pi\alpha}{2}+(1-\alpha)\frac{x}{2}\right)}{(1-\alpha)\left(2\sin\frac{x}{2}\right)^{1-\alpha}}-\frac{\alpha}{1-\alpha}.$$

Proof We note that, by the definition (1.1), for a complex number z,

$$(1-z)^{\alpha-1}=1+(1-\alpha)\sum_{k=1}^{+\infty}a_k(\alpha)z^k,$$

where the equality holds if and only if the last series converges. Thus, setting $z=\mathrm{e}^{\mathrm{i}x}$ with $0<x\leqslant\pi$, we deduce

$$1+\sum_{k=1}^{+\infty}a_k(\alpha)\cos kx=\operatorname{Re}\left\{\frac{(1-\mathrm{e}^{\mathrm{i}x})^{\alpha-1}}{1-\alpha}\right\}-\frac{\alpha}{1-\alpha}$$

$$=\frac{\left(2\sin\frac{x}{2}\right)^{\alpha-1}}{1-\alpha}\operatorname{Re}\{\mathrm{e}^{\mathrm{i}(\alpha-1)\frac{x-\pi}{2}}\}-\frac{\alpha}{1-\alpha}=\frac{\sin\left(\frac{\pi\alpha}{2}+(1-\alpha)\frac{x}{2}\right)}{(1-\alpha)\left(2\sin\frac{x}{2}\right)^{1-\alpha}}-\frac{\alpha}{1-\alpha},$$

which completes the proof.

Now we are in a position to prove Lemma 2.2.

Proof of Lemma 2.2 By Lemma 2.1 and the definition (1.1), we obtain

$$a_k(\alpha)=\frac{1}{\Gamma(2-\alpha)\left(k+\frac{1}{4}\right)^{\alpha}}+\frac{c_k(\alpha)}{\Gamma(2-\alpha)\left(k+\frac{1}{4}\right)^{1+\alpha}},$$

where

$$c_k(\alpha)=A(\alpha)+\frac{\frac{1}{2}A^2(\alpha)+B(\alpha)}{k+\frac{1}{4}}+\frac{R(k,\ \alpha)}{\left(k+\frac{1}{4}\right)^{2}}. \tag{2.7}$$

It then follows that

$$\sum_{k=n+1}^{+\infty}a_k(\alpha)\cos kx=\frac{1}{\Gamma(2-\alpha)}\sum_{k=n+1}^{+\infty}\frac{\cos kx}{\left(k+\frac{1}{4}\right)^{\alpha}}+h_n(x,\alpha), \tag{2.8}$$

where

$$h_n(x,\alpha)=\frac{1}{\Gamma(2-\alpha)}\sum_{k=n+1}^{+\infty}\frac{c_k(\alpha)\cos kx}{\left(k+\frac{1}{4}\right)^{1+\alpha}}. \tag{2.9}$$

Note that

$$\begin{aligned}\frac{1}{\Gamma(2-\alpha)}\sum_{k=n+1}^{+\infty}\frac{\cos kx}{\left(k+\frac{1}{4}\right)^{\alpha}}&=\frac{1}{\Gamma(2-\alpha)2\sin\frac{x}{2}}\sum_{k=n+1}^{+\infty}\int_{(k-\frac{1}{2})x}^{(k+\frac{1}{2})x}\frac{\cos t}{\left(k+\frac{1}{4}\right)^{\alpha}}\mathrm{d}t\\&=g_n(x,\alpha)+\frac{x^{\alpha}}{\Gamma(2-\alpha)2\sin\frac{x}{2}}\int_{(n+\frac{1}{2})x}^{+\infty}\frac{\cos t}{t^{\alpha}}\mathrm{d}t,\end{aligned}$$

where

$$g_n(x,\alpha)=\frac{x^{\alpha}}{\Gamma(2-\alpha)2\sin\frac{x}{2}}\sum_{k=n+1}^{+\infty}\int_{(k-\frac{1}{2})x}^{(k+\frac{1}{2})x}\left[\frac{1}{\left(\left(k+\frac{1}{4}\right)x\right)^{\alpha}}-\frac{1}{t^{\alpha}}\right]\cos t\mathrm{d}t. \tag{2.10}$$

Therefore, by (2.8), we obtain

$$\sum_{k=n+1}^{+\infty}a_k(\alpha)\cos kx=\frac{x^{\alpha}}{\Gamma(2-\alpha)2\sin\frac{x}{2}}\int_{(n+\frac{1}{2})x}^{+\infty}\frac{\cos t}{t^{\alpha}}\mathrm{d}t+g_n(x,\alpha)+h_n(x,\alpha).$$

This last equality together with Lemma 2.3 implies

$$S_n(x,\alpha)=S(x,\alpha)-\sum_{k=n+1}^{+\infty}a_k(\alpha)\cos kx$$

$$= \frac{\sin\left(\alpha\frac{\pi}{2}+(1-\alpha)\frac{x}{2}\right)}{(1-\alpha)\left(2\sin\frac{x}{2}\right)^{1-\alpha}} - \frac{x^\alpha}{\Gamma(2-\alpha)2\sin\frac{x}{2}}\int_{\left(n+\frac{1}{2}\right)x}^{+\infty}\frac{\cos t}{t^\alpha}\mathrm{d}t -$$

$$\frac{\alpha}{1-\alpha}+r_n(x,\alpha),$$

where

$$r_n(x,\alpha)=-g_n(x,\alpha)-h_n(x,\alpha). \tag{2.11}$$

Therefore, it remains to show that $r_n(x,\alpha)$ satisfies (2.6). This can be achieved by a straightforward calculation. In fact, by (2.10), we have

$$|g_n(x,\alpha)|$$

$$\leqslant \frac{x^\alpha}{\Gamma(2-\alpha)2\sin\frac{x}{2}}\sum_{k=n+1}^{+\infty}\left(\int_{\left(k-\frac{1}{2}\right)x}^{kx}+\int_{kx}^{\left(k+\frac{1}{2}\right)x}\right)\left|\frac{1}{\left(\left(k+\frac{1}{4}\right)x\right)^\alpha}-\frac{1}{t^\alpha}\right|\mathrm{d}t$$

$$\leqslant \frac{x^\alpha}{\Gamma(2-\alpha)2\sin\frac{x}{2}}\sum_{k=n+1}^{+\infty}\left(\frac{\alpha}{\left(\left(k-\frac{1}{2}\right)x\right)^{1+\alpha}}\cdot\frac{3}{4}x\cdot\frac{1}{2}x+\right.$$

$$\left.\frac{\alpha}{(kx)^{1+\alpha}}\cdot\frac{1}{4}x\cdot\frac{1}{2}x\right)$$

$$\leqslant \frac{x}{\Gamma(2-\alpha)4\sin\frac{x}{2}}\sum_{k=n+1}^{+\infty}\frac{\alpha}{\left(k-\frac{1}{2}\right)^{1+\alpha}}$$

$$\leqslant \frac{x}{\Gamma(2-\alpha)4\sin\frac{x}{2}\left(n-\frac{1}{2}\right)^\alpha}. \tag{2.12}$$

But on the other hand, by (2.9)(2.7) and (2.4), we have

$$|h_n(x,\alpha)|\leqslant\frac{1}{\Gamma(2-\alpha)}\sum_{k=n+1}^{+\infty}\frac{|c_k(\alpha)|}{\left(k+\frac{1}{4}\right)^{1+\alpha}}\leqslant\frac{\rho_n(\alpha)}{\Gamma(2-\alpha)\alpha\left(n+\frac{1}{4}\right)^\alpha}. \tag{2.13}$$

Now combining (2.13) and (2.12) with (2.11), we obtain (2.6) and therefore complete the proof. □

§ 3. Proof of Theorem 1.1 (2)~(4)

For convenience, we set, for $\alpha\in(0,1)$ and $x\in(0,\pi]$,

$M(x, \alpha)=$

$$\frac{1}{(1-\alpha)\left(2\sin\frac{x}{2}\right)^{1-\alpha}}\sin\left(\frac{\pi\alpha}{2}+\frac{(1-\alpha)x}{2}\right)-\frac{x^{\alpha}}{2(1-\alpha)\sin\frac{x}{2}}\sin\frac{\alpha\pi}{2}. \tag{3.1}$$

Since (see [1, p. 50, exercise 19])

$$\sin\frac{\pi\alpha}{2}=\frac{1}{\Gamma(1-\alpha)}\int_0^{+\infty}\frac{\cos t}{t^{\alpha}}\mathrm{d}t,$$

it follows by Lemma 2.2 that for $n\geqslant 100$, $\alpha\in(0, 1)$ and $x\in(0, \pi]$,

$$S_n(x,\alpha)=-\frac{\alpha}{1-\alpha}+M(x,\alpha)+\frac{x^{\alpha}}{2\Gamma(2-\alpha)\sin\frac{x}{2}}\int_0^{\left(n+\frac{1}{2}\right)x}\frac{\cos t}{t^{\alpha}}\mathrm{d}t+r_n(x,\alpha), \tag{3.2}$$

where $r_n(x, \alpha)$ satisfies (2.6).

Our proof of Theorem 1.1 (2)～(4) will be based on (3.1) and the following lemma.

Lemma 3.1　For a given $\alpha\in(0, 0.4)$, the function $M(\cdot, \alpha)$ is positive on $(0, \pi]$ and moreover

$$M(x, \alpha)=\frac{x^{\alpha}}{2}\cos\frac{\pi\alpha}{2}+O(x^{\alpha+1}),\ \text{as } x\to 0^{+}. \tag{3.3}$$

Proof　For simplicity, we set

$$L(x, \alpha):=\frac{2(1-\alpha)\sin\frac{x}{2}}{x^{\alpha}}M(x, \alpha). \tag{3.4}$$

Then by (3.1), we have

$$\begin{aligned}L(x, \alpha)&=\left(\frac{2\sin\frac{x}{2}}{x}\right)^{\alpha}\sin\left[\alpha\frac{\pi}{2}+(1-\alpha)\frac{x}{2}\right]-\sin\left(\alpha\frac{\pi}{2}\right)\\&=2\left(\frac{2\sin\frac{x}{2}}{x}\right)^{\alpha}\cos\left(\left(\frac{\pi}{2}-\frac{x}{4}\right)\alpha+\frac{x}{4}\right)\sin\frac{(1-\alpha)x}{4}-\\&\quad\sin\frac{\alpha\pi}{2}\left[1-\left(\frac{2\sin\frac{x}{2}}{x}\right)^{\alpha}\right].\end{aligned} \tag{3.5}$$

Notice that this last equality implies that for any fixed $x\in(0, \pi]$, $L(x, \cdot)$ is a decreasing function of $\alpha\in(0, 1)$. Thus, by numerical evaluation, we obtain, for $0<\alpha\leqslant 0.4$,

$$\frac{L(x,\ \alpha)}{x}\geqslant\frac{L(x,\ 0.4)}{x}>0.07,\qquad x\in(0,\ \pi],$$

which implies the positivity of $M(\cdot,\ \alpha)$ on $(0,\ \pi]$.

To show (3.3), we note that, by (3.5),

$$L(x,\ \alpha)=\left(1-\frac{\alpha}{24}x^2+O(x^4)\right)\left(\sin\frac{\pi\alpha}{2}+\frac{1-\alpha}{2}x\cos\frac{\pi\alpha}{2}+O(x^2)\right)-\sin\frac{\pi\alpha}{2}$$

$$=\frac{(1-\alpha)x}{2}\cos\frac{\pi\alpha}{2}+O(x^2),\qquad \text{as } x\to 0^+.$$

It then follows by (3.4) that

$$M(x,\ \alpha)=\frac{x^\alpha}{2(1-\alpha)\sin\frac{x}{2}}L(x,\ \alpha)=\frac{x^\alpha}{2}\cos\frac{\pi\alpha}{2}+O(x^{\alpha+1}),\qquad \text{as } x\to 0^+,$$

which proves (3.3). This completes the proof of the lemma. □

Now we are in a position to prove Theorem 1.1 (ii)~(iv).

Proof of Theorem 1.1 (2). First, we show that for $\alpha_0<\alpha<\alpha^*$,

$$N(\alpha)\geqslant 8. \tag{3.6}$$

Let $x_0\in[0,\ \pi]$ be such that $S_7(x_0,\ \alpha^*)=0$. Then, for the proof of (3.6), by the definition (1.4), it is sufficient to prove

$$S_7(x_0,\ \alpha)\leqslant 0. \tag{3.7}$$

Suppose (3.7) is not true. Then we would deduce

$$\min_{0\leqslant j\leqslant 6} S_j(x_0,\ \alpha)<0$$

since, by a summation by parts, we have

$$0=S_7(x_0,\alpha^*)=\sum_{j=0}^{6}\left(\frac{a_j(\alpha^*)}{a_j(\alpha)}-\frac{a_{j+1}(\alpha^*)}{a_{j+1}(\alpha)}\right)S_j(x_0,\alpha)+\frac{a_7(\alpha^*)}{a_7(\alpha)}S_7(x_0,\alpha),$$

and since, by (1.6),

$$\frac{a_j(\beta_1)}{a_j(\beta_2)}\geqslant\frac{a_{j+1}(\beta_1)}{a_{j+1}(\beta_2)},\qquad j\in\mathbf{N},\qquad \text{whenever } \beta_2<\beta_1, \tag{3.8}$$

where it is agreed that $a_0(\beta)=1$ for $0<\beta<1$. Suppose $j_{x_0}\in\{0,\ 1,\ \cdots,\ 6\}$ is the largest integer for which $S_{j_{x_0}}(x_0,\ \alpha)<0$. Then a summation by parts shows that

$$\sum_{j=j_{x_0}+1}^{7} a_j(\alpha^*)\cos jx$$

$$=-\frac{a_{j_{x_0}+1}(\alpha^*)}{a_{j_{x_0}+1}(\alpha)}S_{j_{x_0}}(x_0,\alpha)+\sum_{j=j_{x_0}+1}^{6}\left(\frac{a_j(\alpha^*)}{a_j(\alpha)}-\frac{a_{j+1}(\alpha^*)}{a_{j+1}(\alpha)}\right)S_j(x_0,\alpha)+$$

$$\frac{a_7(\alpha^*)}{a_7(\alpha)} S_7(x_0, \alpha) > 0.$$

This means that

$$S_7(x_0, \alpha^*) > \min_{0 \leqslant j \leqslant 6} S_j(x_0, \alpha^*). \tag{3.9}$$

However, on the other hand, numerical evaluation shows that for $\alpha_1 = 0.33542 (< \alpha^*)$,

$$\min\{S_j(x, \alpha_1) \mid 0 \leqslant j \leqslant 6, x \in [0, \pi]\} > 0,$$

which, by a summation by parts, implies that $S_j(x_0, \alpha^*) > 0$ for $0 \leqslant j \leqslant 6$. Thus, by (3.9), we conclude

$$S_7(x_0, \alpha^*) > 0,$$

which contradicts the definition of α^*. This proves (3.7) and hence (3.6).

Next, we show that there exists an absolute constant $N_0 > 0$ such that for $\alpha_0 < \alpha < \alpha^*$ and all $x \in [0, \pi]$,

$$S_n(x, \alpha) > 0, \qquad \text{whenever } n \geqslant N_0 \cdot (\alpha - \alpha_0)^{-\frac{1}{\alpha_0(1-\alpha_0)}}, \tag{3.10}$$

which, by the definition (1.4), will imply

$$N(\alpha) \leqslant N_0 \cdot (\alpha - \alpha_0)^{-\frac{1}{\alpha_0(1-\alpha_0)}}.$$

To show (3.10), note, first, that by the definition (1.2), (3.10) is obvious for $x \in \left[0, \frac{\pi}{2n}\right]$. For the rest of the proof, unless otherwise stated, we will assume $\frac{\pi}{2n} \leqslant x \leqslant \pi$, $\alpha_0 < \alpha < \alpha^*$ and $n \geqslant 100$.

Since $\left(n + \frac{1}{2}\right) x > \frac{\pi}{2}$, it follows that

$$\int_0^{(n+\frac{1}{2})x} \frac{\cos t}{t^\alpha} \mathrm{d}t \geqslant \int_0^{\frac{3\pi}{2}} \frac{\cos t}{t^\alpha} \mathrm{d}t.$$

Thus, by (3.1) and the positivity of $M(\cdot, \alpha)$, we obtain

$$\begin{aligned} S_n(x, \alpha) &\geqslant \frac{x^\alpha}{2\sin\frac{x}{2}} \frac{1}{\Gamma(2-\alpha)} \int_0^{\frac{3\pi}{2}} \frac{\cos t}{t^\alpha} \mathrm{d}t - \frac{\alpha}{1-\alpha} + r_n(x, \alpha) \\ &\geqslant C_1 \left(\frac{1}{x}\right)^{1-\alpha} \int^{\frac{3\pi}{2}} \frac{\cos t}{t^\alpha} \mathrm{d}t - C_2, \end{aligned} \tag{3.11}$$

where and below, C_1, C_2, C_3 denote positive absolute constants.

Note, however, that by the definition of α_0,

$$\int_0^{\frac{3\pi}{2}} \frac{\cos t}{t^\alpha}\mathrm{d}t = \int_0^{\frac{3\pi}{2}} \frac{\cos t}{t^\alpha}\mathrm{d}t - \int_0^{\frac{3\pi}{2}} \frac{\cos t}{t^{\alpha_0}}\mathrm{d}t$$
$$\geqslant (\alpha-\alpha_0) \min_{0.308\leqslant\beta\leqslant 0.4}\left(\int_0^{\frac{3\pi}{2}} \frac{\cos t}{t^\beta}\ln\frac{1}{t}\mathrm{d}t\right)$$
$$\geqslant 0.28(\alpha-\alpha_0).$$

Thus, substituting this last inequality into (3.11), we obtain

$$S_n(x, \alpha)\geqslant 0.28C_1\cdot(\alpha-\alpha_0)\left(\frac{1}{x}\right)^{1-\alpha}-C_2,$$

which is positive provided that

$$x<\varepsilon\cdot(\alpha-\alpha_0)^{\frac{1}{1-\alpha_0}},$$

where

$$\varepsilon=\frac{1}{2}\mathrm{e}^{-4\mathrm{e}^{-1}}\left(\frac{0.28C_1}{C_2}\right)^2.$$

Thus, it remain to consider the case $\varepsilon\cdot(\alpha-\alpha_0)^{\frac{1}{1-\alpha_0}}:=x_0<x\leqslant\pi$. To this end, on the one hand, note that, $x_0<x\leqslant\pi$,

$$\frac{x^\alpha}{2\sin\frac{x}{2}}\left|\int_{(n+\frac{1}{2})x}^{+\infty}\frac{\cos t}{t^\alpha}\mathrm{d}t\right|\leqslant\frac{\pi}{2}x^{\alpha-1}\frac{2}{\left(\left(n+\frac{1}{2}\right)x\right)^\alpha}\leqslant\frac{\pi}{n^\alpha x_0}.$$

On the other hand, however, numerical evaluation shows that for $\alpha\in[0.3, 0.4]$ and all $x\in(0, \pi]$,

$$\frac{\sin\left(\frac{\pi\alpha}{2}+\frac{(1-\alpha)x}{2}\right)}{(1-\alpha)\left(2\sin\frac{x}{2}\right)^{1-\alpha}}-\frac{\alpha}{1-\alpha}\geqslant 0.28.$$

Therefore, using Lemma 2.2, we conclude that for $x_0<x\leqslant\pi$,

$$S_n(x, \alpha)\geqslant 0.28-\frac{C_3}{n^\alpha x_0}=0.28-\frac{C_3}{n^\alpha\varepsilon\cdot(\alpha-\alpha_0)^{\frac{1}{1-\alpha_0}}}=:\gamma,$$

the number γ being positive provided that

$$n\geqslant\left(\frac{C_3}{0.28\varepsilon}+1\right)^{\frac{1}{\alpha_0}}\left(\frac{1}{\alpha-\alpha_0}\right)^{\frac{1}{\alpha_0(1-\alpha_0)}}.$$

In summary, we have shown that

$$\min_{x\in[0,\pi]} S_n(x, \alpha)>0$$

whenever

$$n \geqslant \max\left\{100, \left(\frac{C_3}{0.28\varepsilon}+1\right)^{\frac{1}{\alpha_0}}\left(\frac{1}{\alpha-\alpha_0}\right)^{\frac{1}{\alpha_0(1-\alpha_0)}}\right\}.$$

This proves (3.10) and therefore, completes the proof of Theorem 1.1 (2).

Proof of Theorem 1.1 (3). Set

$$\theta_n=\frac{3\pi}{2n+1}.$$

First, we show that

$$\lim_{n\to+\infty} S_n(\theta_n, \alpha_0)=-\frac{\alpha_0}{1-\alpha_0}, \tag{3.12}$$

from which the inequality

$$\limsup_{n\to+\infty}(\min\{S_n(x, \alpha_0) \mid x\in[0, \pi]\}) \leqslant -\frac{\alpha_0}{1-\alpha_0} \tag{3.13}$$

will follow immediately. In fact, by (3.1) and the definition of α_0, we have

$$\begin{aligned} &S_n(\theta_n,\alpha_0) \\ &= M(\theta_n,\alpha_0)+\frac{\theta_n^{\alpha_0}}{2\sin\frac{\theta_n}{2}}\frac{1}{\Gamma(2-\alpha_0)}\int_0^{\frac{3\pi}{2}}\frac{\cos t}{t^{\alpha_0}}\mathrm{d}t-\frac{\alpha_0}{1-\alpha_0}+r_n(\theta_n,\alpha_0) \\ &= M(\theta_n, \alpha_0)-\frac{\alpha_0}{1-\alpha_0}+r_n(\theta_n, \alpha_0). \end{aligned}$$

(3.13) then follows since, by (2.6) and (3.3),

$$\lim_{n\to+\infty} M(\theta_n, \alpha_0)=\lim_{n\to+\infty} r_n(\theta_n, \alpha)=0.$$

Next, we show

$$\liminf_{n\to+\infty}(\min\{S_n(x, \alpha_0) \mid x\in[0, \pi]\}) \geqslant -\frac{\alpha_0}{1-\alpha_0}, \tag{3.14}$$

which together with (3.13) will imply the desired equality (1.5).

Since for any $y \geqslant \frac{\pi}{2}$,

$$\int_0^y \frac{\cos t}{t^{\alpha_0}}\mathrm{d}t \geqslant \int_0^{\frac{3\pi}{2}} \frac{\cos t}{t^{\alpha_0}}\mathrm{d}t = 0,$$

it follows that for any positive integer n and $x\in[0, \pi]$,

$$\int_0^{(n+\frac{1}{2})x} \frac{\cos t}{t^{\alpha_0}}\mathrm{d}t \geqslant 0.$$

Thus, by (3.1) and the positivity of $M(\cdot, \alpha)$, we conclude

$$S_n(x, \alpha_0) \geqslant -\frac{\alpha_0}{1-\alpha_0} + r_n(x, \alpha_0), \quad \text{for all } x \in [0, \pi]. \tag{3.15}$$

Since, by (2.6),

$$\lim_{n \to +\infty} r_n(x, \alpha_0) = 0, \quad \text{uniformly for } x \in (0, \pi],$$

(3.14) follows from (3.15). This completes the proof of (3).

Proof of Theorem 1.1 (4). Again, we set

$$\theta_n = \frac{3\pi}{2n+1}.$$

It then follows by (3.1), (2.6) and the positivity of $M(\cdot, \alpha)$ that

$$S_n(\theta_n, \alpha) = \frac{\theta_n^\alpha}{2\Gamma(2-\alpha)\sin\frac{\theta_n}{2}} \int_0^{\frac{3\pi}{2}} \frac{\cos t}{t^\alpha} \mathrm{d}t - \frac{\alpha}{1-\alpha} + o(1), \quad \text{as } n \to +\infty. \tag{3.16}$$

Since for $0 < \alpha < \alpha_0$,

$$\int_0^{\frac{3\pi}{2}} \frac{\cos t}{t^\alpha} \mathrm{d}t < 0,$$

it follows that

$$\lim_{n \to +\infty} S_n(\theta_n, \alpha) = -\infty,$$

which implies

$$\lim_{n \to +\infty} \inf(\min\{S_n(x, \alpha) \mid x \in [0, \pi]\}) = -\infty.$$

This completes the proof of (4). □

§ 4. Proof of Theorem 1.1 (1)

For simplicity, we set $\alpha_1 = 0.335\ 42$ throughout this section, and define, for $\alpha \in (0, 1)$,

$$m_k(\alpha) = \min_{x \in [0, \pi]} S_k(x, \alpha), \ k \in \mathbf{N}.$$

The following lemma is the key to the proof of Theorem 1.1 (1).

Lemma 4.1 The number $m_n(\alpha_1)$ is positive if n is a positive integer not equal to 7.

Proof

First, by numerical evaluation, we have

$$m_n(\alpha_1)\begin{cases}>0, & \text{if } 1\leqslant n\leqslant 6,\\ =-1.285\times 10^{-5}, & \text{if } n=7,\\ \geqslant 0.000\,78 & \text{if } 8\leqslant n\leqslant 100.\end{cases}$$

Thus, it remains to show Lemma 4.1 for $n\geqslant 101$. To this end, we need two observations:

Observation 1 For $n\geqslant 101$ and $x\in(0, \pi]$,

$$\frac{\alpha_1}{1-\alpha_1}+|r_n(x, \alpha_1)|\leqslant\frac{0.059\,1x}{\sin\frac{x}{2}}+0.532\,42, \tag{4.1}$$

where $r_n(x, \alpha)$ is the same as in Lemma 2.2.

To prove this observation, we use (2.6) to obtain

$$\begin{aligned}&\frac{\alpha_1}{1-\alpha_1}+|r_n(x, \alpha_1)|\\ \leqslant&\frac{x}{\Gamma(2-\alpha_1)4\sin\frac{x}{2}(100)^{\alpha_1}}+\frac{\rho_n(\alpha_1)}{\alpha_1\Gamma(2-\alpha_1)(101)^{\alpha_1}}+\frac{\alpha_1}{1-\alpha_1}\\ =&\frac{0.059\,1x}{\sin\frac{x}{2}}+0.702\,62\rho_n(\alpha_1)+0.504\,71,\end{aligned} \tag{4.2}$$

where $\rho_n(\alpha)$ is defined by (2.4). However, by (2.1)~(2.4), we deduce

$$\begin{aligned}\rho_n(\alpha_1)&=|A(\alpha_1)|+\frac{|A^2(\alpha_1)+B(\alpha_1)|}{n+\frac{5}{4}}+\frac{|e(\alpha_1)|+0.030\,1}{\left(n+\frac{5}{4}\right)^2}\\ &=0.027\,60+\frac{0.010\,89}{n+\frac{5}{4}}+\frac{0.116\,77}{\left(n+\frac{5}{4}\right)^2}\\ &\leqslant 0.027\,60+\frac{0.010\,89}{102}+\frac{0.116\,77}{102^2}\\ &\leqslant 0.027\,71,\end{aligned}$$

which together with (4.2) implies (4.1).

The next observation is a simple exercise in calculus.

Observation 2 If $\alpha\in(0, 1)$, k is a nonnegative integer and $y\geqslant 2k\pi+\frac{\pi}{2}(k\in\mathbf{Z})$, then

$$\frac{1}{\Gamma(1-\alpha)}\int_y^{+\infty}\frac{\cos t}{t^\alpha}\mathrm{d}t\leqslant J_k(\alpha):=\frac{1}{\Gamma(1-\alpha)}\int_{2k\pi+\frac{3\pi}{2}}^{+\infty}\frac{\cos t}{t^\alpha}\mathrm{d}t.$$

With the above two observations, we can prove Lemma 4.1 for $n \geqslant 101$ easily. In fact, using Lemma 2.2 and Observation 1, we obtain, for $n \geqslant 101$,

$$S_n(x,\ \alpha_1) \geqslant \frac{\sin\left[\alpha_1 \frac{\pi}{2} + (1-\alpha_1)\frac{x}{2}\right]}{(1-\alpha_1)\left(2\sin\frac{x}{2}\right)^{1-\alpha_1}} - \frac{x^{\alpha_1}}{2(1-\alpha_1)\sin\frac{x}{2}} \frac{1}{\Gamma(1-\alpha_1)} \int_{(n+\frac{1}{2})x}^{+\infty} \frac{\cos t}{t^{\alpha_1}} \mathrm{d}t - \frac{0.0591x}{\sin\frac{x}{2}} - 0.53242. \tag{4.3}$$

On the other hand, using Observation 2 with $k=0, 1$ and 15 respectively, we deduce, for $n \geqslant 101$ and $\frac{\pi}{2n} \leqslant x \leqslant \pi$,

$$\frac{1}{\Gamma(1-\alpha)} \int_{(n+\frac{1}{2})x}^{+\infty} \frac{\cos t}{t^{\alpha_1}} \mathrm{d}t \leqslant D(x), \tag{4.4}$$

where the function D is defined by

$$D(x) := \begin{cases} J_0(\alpha_1) = 0.43081\cdots, & \text{if } x \in [0,\ 0.11]; \\ J_1(\alpha_1) = 0.32838\cdots, & \text{if } x \in (0.11,\ 1]; \\ J_{15}(\alpha_1) = 0.15769\cdots, & \text{if } x \in (1,\ \pi]. \end{cases} \tag{4.5}$$

Thus, combining (4.3) ~ (4.5), we conclude, for $n \geqslant 101$ and $x \in \left[\frac{\pi}{2n},\ \pi\right]$,

$$S_n(x,\ \alpha_1) \geqslant \frac{\sin\left[\alpha_1 \frac{\pi}{2} + (1-\alpha_1)\frac{x}{2}\right]}{(1-\alpha_1)\left(2\sin\frac{x}{2}\right)^{1-\alpha_1}} - \frac{D(x)x^{\alpha_1}}{2(1-\alpha_1)\sin\frac{x}{2}} - \frac{0.0591x}{\sin\frac{x}{2}} - 0.53242,$$

which, by numerical evaluation, is bigger than 0.04 for all $x \in (0,\ \pi]$.

Finally, note that it is obvious that, by the definition (1.2), $S_n(\cdot,\ \alpha_1)$ is positive on $\left[0,\ \frac{\pi}{2n}\right]$. This completes the proof. □

Now we are in a position to prove Theorem 1.1 (1).

Proof of Theorem 1.1 (1). The proof runs along the same lines as that of Theorem 1.1 (2). In fact, the positivity of $m_j(\alpha^*)$ for $0 \leqslant j \leqslant 6$

follows from Lemma 4.1 by a summation by parts, while the equality $m_7(\alpha^*)=0$ follows by the definition of α^*.

Thus, it remains to show that for all $x\in[0, \pi]$,

$$S_k(x, \alpha^*)>0, \quad k=8, 9, \cdots. \tag{4.6}$$

Note that (4.6) follows again from Lemma 4.1 by a summation by parts if $S_7(x, \alpha)>0$. Thus, without loss of generality, we may assume $S_7(x, \alpha)<0$. Then, using a summation by parts once again, we obtain, for $k\geqslant 8$,

$$\sum_{j=8}^{k} a_j(\alpha^*)\cos kx = -\frac{a_8(\alpha^*)}{a_8(\alpha)}S_7(x,\alpha)+$$

$$\sum_{j=8}^{k-1}\left(\frac{a_j(\alpha^*)}{a_j(\alpha)}-\frac{a_{j+1}(\alpha^*)}{a_{j+1}(\alpha)}\right)S_j(x,\alpha)+\frac{a_k(\alpha^*)}{a_k(\alpha)}S_k(x,\alpha).$$

Since it is assumed that $S_7(x, \alpha)<0$, by (3.8) and Lemma 4.1 it follows that

$$\sum_{j=8}^{k} a_j(\alpha^*)\cos kx > 0,$$

which means

$$S_k(x, \alpha^*)>S_7(x, \alpha^*)\geqslant 0.$$

This proves Theorem 1.1 (1). □

References

[1] Andrews G E, Askey R, Roy R. Special functions. Encyclopedia Math. Appl., 71. Cambridge Univ. Press, Cambridge, 1999.

[2] Askey R. Vietoris's inequalities and hypergeometric series. Recent Progr. Ineq., 1996: 63—76. Math. Appl., 1998, 430.

[3] Askey R, Steinig J. Some positive trigonometric sums. Trans. Amer. Math. Soc., 1974, 187: 295—307.

[4] Brown G, Dai F, Wang K Y. Extensions of Vietoris's inequalities (Ⅰ). Ramanujan J., 2007, 14(3): 471—507.

[5] Brown G, Wang K Y, Wilson D C. Positivity of some basic cosine sums. Math. Proc. Camb. Phil. Soc., 1993, 114(3): 383—391.

[6] Koumandos S, Ruscheweyh ST. Positive gegenbauer polynomial sums and applications to starlike functions. Constr. Approx., 2006, 23(2): 197—210.

四、大学数学系分析类课程内容的改进

Ⅳ. Improvement of the University Analysis Courses

数学通报，1997，(10)：33－36；(11)：45－47.

极限、实数、指数函数

——对于改编高中理科用数学教材的建议

Limits，Real Numbers，Exponential Functions：a Suggestion on Modifying the School Text Book

§0. 前　言

当今流行的大学微积分教材中普遍存在两个逻辑漏洞.

1. 先在实数系上建立微积分理论，然后再定义实数或永远不讨论这个问题. 其实初中代数课本第二册([1]，151 页）早已明确定义：**“……无限不循环小数，又叫作无理数”**. 只是没有极限概念不可能理解这个极好的定义. 只有 20 世纪 50 年代引入的苏联菲赫金哥尔茨的《微积分学教程》[2]一开篇在“绪论”中就讲实数. 但此书主要是用 Dedekind 的分划来定义实数，不能直接与初中的讲法接轨. 而且 D 氏方法的局限性是无法推广应用于处理没有顺序关系的距离空间的完备化的问题(关于距离空间的完备化，可参阅任何一本关于泛函分析的教科书).

2. 承袭中学对于指数函数的描述(不是定义). 高中代数课本上册([3]) 第一章讲幂函数、指数函数和对数函数，就是做这种描述. 此书第 95 页写道“在科学技术中常常使用以无理数 e＝2.718 28…为底的对数，以 e 为底的对数叫作自然对数.”大学的课本，就是在这些描述的基础上，来讨论指数函数的各种性质的. 结果陷入更深的逻辑漏洞. 例如，以默认对数函数的连续性来证明指数函数连续. 只有在讲完 Taylor 展开，人们才明白指数函数是什么.

当今中学数学教材存在的问题是，内容的量和质都不够(这给中学腾出了大量时间搞应试训练)，不适应科技发展需要，不利于学生素质的提高. 可喜的是，1996 年 5 月出版的《全日制普通高级中学数学教学大纲(供试验用)》规定：限定选修课安排在高三. 其中供理科选用部分，每周 4 课时，共 104 课时，内容是

(1) 概率与统计(18 课时)，(2)极限(14 课时)，(3)导数与微分(20 课时)，(4)积分(14 课时)，(5)复数(18 课时).

而且，这些内容也将在高考的命题范围之内.

本文的目的是，提出一个关于极限、实数、指数函数的简明的讲稿，添补微积分教材中的两个漏洞，尽快理顺微积分课程的逻辑系统，与初中数学接轨. 文章分三节. §1 讲有理数列的极限. §2 讲实数的定义和实数系的完备性，完全与初中二年级时给出的实数的定义一致. §3 讲区间上的连续函数的一些特性并借以给出指数函数和对数函数的确切定义.

特建议将本文§1，§2 编入上述大纲限定选修课供理科选用部分的第 2 部分内容当中；而将本文§3 编入第 3 部分内容当中，可将该部分的标题(导数与微分)改为“初等函数及其导数”，这样更确切. 大学理工科高考的深度应该掌握在 heiti 区分学生是否具有学习高等数学的能力的程度. 大学微积分课，原则上不必重复高中的内容.

§1. 数列的极限

我们现在只在有理数范围内讨论. 用 $\mathbf{N}$ 代表自然数集，用 $\mathbf{Q}$ 代表有理数集.

定义 1(数列)　设 f 是从 $\mathbf{N}$ 到 $\mathbf{Q}$ 内的映射，则称 f 为(有理)数列，记作 $f=\{f(n)\}_{n=1}^{+\infty}$. 当然，也可用其他记号，如 a_n，b_n 等来代表 $f(n)$. 注意，f 的值域 $f(\mathbf{N})=\{f(n)\mid n\in\mathbf{N}\}$ 与数列是两回事.

定义 2(数列的极限)　设 $f=\{f(n)\}_{n=1}^{+\infty}$ 是(有理)数列，$l\in\mathbf{Q}$. 如果对于任意的 $k\in\mathbf{N}$，总找得到 $n_k\in\mathbf{N}$，使得当 $n>n_k$ 时 $|f(n)-l|<\frac{1}{k}$，那么，就说数列 f 收敛到极限 l，记作 $\lim\limits_{n\to+\infty}f(n)=l$.

如果对于任意的 $k\in\mathbf{N}$，总找得到 $n_k\in\mathbf{N}$，使得当 $n>n_k$ 时，$f(n)>k$，那么，就说数列 f 发散到极限 $+\infty$(读作正无穷)，记作

$$\lim_{n\to+\infty}f(n)=+\infty.$$

如果对于任意的 $k\in\mathbf{N}$，总找得到 $n_k\in\mathbf{N}$，使得当 $n>n_k$ 时，$f(n)<-k$，那么，就说数列 f 发散到极限 $-\infty$（读作负无穷），记作

$$\lim_{n\to+\infty}f(n)=-\infty.$$

不收敛的情形都叫作发散.

定义 3（基本列—Cauchy 列） 设 $f=\{f(n)\}_{n=1}^{+\infty}$ 是数列. 如果对于任意的 $k\in\mathbf{N}$，总找得到 $n_k\in\mathbf{N}$，使得当 m，$n>n_k$ 时，$|f(m)-f(n)|<\frac{1}{k}$，那么，就说数列 f 是基本列（或 Cauchy 列）.

定理 1 收敛数列必是基本列.

证 设数列 f 收敛到 l. 那么，不管 $k\in\mathbf{N}$ 多大，总找得到 $n_k\in\mathbf{N}$，使得当 $n>n_k$ 时，$|f(n)-l|<\frac{1}{2k}$. 于是，当 m，$n>n_k$ 时，

$$|f(m)-f(n)|<|f(m)-l|+|l-f(n)|<\frac{1}{k}. \qquad \square$$

定义 4 设 f 和 g 都是数列. 如果 $\lim\limits_{n\to+\infty}(f(n)-g(n))=0$，那么就说 f 和 g 等价.

显然，数列的等价关系具有反身性、对称性和传递性.

定义 5（子列） 设 f 和 g 都是数列，g 只取自然数值且严格增，即对于一切 $n\in\mathbf{N}$，有 $\mathbf{N}\ni g(n)<g(n+1)$. 那么，称数列 $\{f(g(n))\}_{n=1}^{+\infty}$ 为 f 的子列，记作 $f\circ g$.

定理 2 若 f 是基本列，则它的子列与它等价.

显然.

定理 3 若 f 和 g 分别收敛到 a 和 b，且 c，$d\in\mathbf{Q}$，那么

$$\lim_{n\to+\infty}(cf(n)+dg(n))=ca+db, \qquad \lim_{n\to+\infty}(f(n)g(n))=ab.$$

如果还知 $b\neq0$，那么

$$\lim_{n\to+\infty}\frac{f(n)}{g(n)}=\frac{a}{b}.$$

注 因为 $b\neq0$，当 n 充分大时必有 $g(n)b>0$ 成立，所以 $\frac{f(n)}{g(n)}$ 对于大的 n 有定义. 那么，谈到它的极限，总把前有限项使 $g(n)=0$ 者略去不要.

定理的证明平庸，略. $\square$

§2. 实数

用**Z**代表整数集. 使用习惯了的十进制计数法(虽然二进制更简单).

定义 6(标准数列) 设数列$\{a_n\}_{n=1}^{+\infty}$满足: $a_n\in\{0, 1, 2, \cdots, 9\}$, $n\in\mathbf{N}$ 且对于任何 $k\in\mathbf{N}$ 都找得到 $n>k$ 使 $a_n<9$. 定义小数 $f(n)=0.a_1a_2\cdots a_n$, $n\in\mathbf{N}$. 对于任何 $m\in\mathbf{Z}$, 称数列$\{m+f(n)\}_{n=1}^{+\infty}$为标准(数)列.

定理 4 标准列是基本列.

证 设$g=\{m+f(n)\}_{n=1}^{+\infty}$是标准列, 其中 $f(n)=0.a_1a_2\cdots a_n$, $n\in\mathbf{N}$. 对于任给的 $k\in\mathbf{N}$, 当 μ, $\nu\in\mathbf{N}$ 且 μ, $\nu>k$ 时显然有

$$|g(\mu)-g(\nu)|=|f(\mu)-f(\nu)|<10^{-k}<\frac{1}{k}. \qquad \square$$

定理 5 任给一个基本列, 存在唯一一个标准列与之等价.

证 设 $f=\{f(n)\}_{n=1}^{+\infty}$是基本列.

对于 $k\in\mathbf{N}$, 存在 $n_k\in\mathbf{N}$ 使得当 μ, $\nu\geqslant n_k$ 时, $|f(\mu)-f(\nu)|<10^{-k-1}$. 归纳地保证所选择的 n_k 满足 $n_k<n_{k+1}$, $k\in\mathbf{N}$. 把 $f(n_k)$写成不以 9 为循环节的小数. 由于$|f(n_k)-f(n_{k+1})|<10^{-k-1}$, 那么 $f(n_k)$和 $f(n_{k+1})$的小数表示中位于小数点后第 k 位以前的数字都是一样的. 也就是说, $f(n_k)$的表达式具有如下形状:

$$f(n_k)=m+0.a_1a_2\cdots a_kb_{k+1}^kb_{k+2}^k\cdots.$$

当小数 $0.a_1a_2\cdots$不以 9 为循环节时, 定义$g(k)=m+0.a_1a_2\cdots a_k$, $k\in\mathbf{N}$.

若小数 $0.a_1a_2\cdots$以 9 为循环节, 则有两种可能性. 一是, 所有的 a_k 全部为 9, 这时我们定义$g(k)=m+1$. 另一种可能的情况是, 存在 $N\in\mathbf{N}$, 使得 $a_N<9$. 而当 $k>N$ 时恒有 $a_k=9$. 在这种情况下, 定义:

$$当\ 1\leqslant k<N\ 时, b_k=a_k, b_N=a_N+1, 当\ k>N\ 时, b_k=0.$$

并定义$g(k)=m+0.b_1b_2\cdots b_k$.

那么, 在任何情况下, $g=\{g(k)\}_{k=1}^{+\infty}$都是标准列, 并且

$$|f(n_k)-g(k)|<10^{-k}.$$

从而, g 与 f 的子列$\{f(n_k)\}_{k=1}^{+\infty}$等价. 根据定理 2, g 与 f 等价. 另外,

容易看出，不相同的两个标准列是不可能等价的.

定义 7（实数） 设 $f=\{m+0.a_1a_2\cdots a_n\}_{n=1}^{+\infty}$ 是标准列，其中 $m\in\mathbf{Z}$，$a_n\in\{0,1,\cdots,9\}$ 且诸 a_n 不以 9 为循环节. 那么把十进小数

$$\overline{f}=m+0.a_1a_2\cdots$$

叫作与 f 对应的**实数**. 全体实数所成的集合(如我们已知的)记作 $\mathbf{R}$.

我们看到，标准列与实数之间建立了一一对应.

定理 6 任给标准列 $\{f(n)\}_{n=1}^{+\infty}$，$\{g(n)\}_{n=1}^{+\infty}$. 那么

(1) $\{f(n)+g(n)\}_{n=1}^{+\infty}$，$\{f(n)g(n)\}_{n=1}^{+\infty}$ 和 $\{-f(n)\}_{n=1}^{+\infty}$ 都是基本列.

(2) 如果存在 $N\in\mathbf{N}$ 使得当 $n>N$ 时，$f(n)\neq 0$，那么 $\left\{\dfrac{1}{f(n+N)}\right\}_{n=1}^{+\infty}$ 也是基本列.

证 结论(1)是明显的. 我们来证结论(2).

设 $f(n)=m+0.a_1a_2\cdots a_n$，其中，$m\in\mathbf{Z}$，$0.a_1a_2\cdots$ 是不以 9 为循环节的小数. 并设 $f(N)\neq 0$. 若 $m\geqslant 0$，则当 $n>N$ 时，

$$f(n)\geqslant f(N)>0.$$

若 $m<0$，则 $f(n)\leqslant -1+0.a_1a_2\cdots a_n$. 由于诸 a_k 不全为 9，可设 $a_l\neq 9$. 那么当 $n>l$ 时，

$$f(n)\leqslant f(l)+10^{-l}\leqslant -10^{-l}.$$

从而，

$$|f(n)|\geqslant -10^{-l}.$$

总之，在任何情况下，都存在正数 $\delta\in\mathbf{Q}$ 以及 $l\in\mathbf{N}$，使得当 $n>l$ 时，

$$|f(n)|\geqslant\delta.$$

那么，当 μ，$\nu>l+N$ 时，

$$\left|\frac{1}{f(\mu)}-\frac{1}{f(\nu)}\right|\leqslant\delta^{-2}\,|f(\mu)-f(\nu)|.$$

由此可见，结论(2)成立. □

定义 8（实数的四则运算） 设 f,g 都是标准列. 那把与 $\{f(n)+g(n)\}_{n=1}^{+\infty}$ 等价的标准列记作 $f+g$；把与 $\{f(n)g(n)\}_{n=1}^{+\infty}$ 等价的标准列记作 fg. 规定 $\overline{f}$ 与 $\overline{g}$ 的和与乘积如下：

$$\overline{f}+\overline{g}=\overline{f+g}；\quad \overline{f}\,\overline{g}=\overline{fg}.$$

并把与 $\{-f(n)\}_{n=1}^{+\infty}$ 等价的标准列记作 $-f$，把 -1 与 $\overline{f}$ 的乘积规定为

$-\overline{f}=\overline{-f}$. 把 $\overline{f}+(-\overline{g})$ 叫作 $\overline{f}$ 与 $\overline{g}$ 的差，记作 $\overline{f}-\overline{g}$. 如果存在 $N\in\mathbf{N}$ 使得当 $n>N$ 时，$f(n)\neq 0$，那么把与 $\left\{\frac{1}{f(n+N)}\right\}_{n=1}^{+\infty}$ 等价的标准列记作 $\frac{1}{f}$. 规定

$$\frac{1}{\overline{f}}=\overline{\left(\frac{1}{f}\right)}.$$

定义 9(实数的大小)　给定实数 $r=m+0.a_1a_2\cdots$，其中 $m\in\mathbf{Z}$. 如果 $m\geqslant 0$，而 m 以及诸 a_n 不全是零，那么称 r 为正数，记作 $r>0$；如果 m 以及诸 a_n 全是零，那么称 r 为零，记作 $r=0$；如果 $m<0$，那么称 r 为负数，记作 $r<0$. 若实数 a，b 满足 $a-b>0$，则说 a 大于 b，记作 $a>b$，或说 b 小于 a，记作 $b<a$. 把“不大于”记作“$\leqslant$”，“不小于”记作“$\geqslant$”.

注　定义 7 完全与我们熟知的关于有理数和无理数的定义一致，形式上可能不同的只是我们总把实数写成不超过它的最大整数与一个非负小数的和. 定义 8 规定的运算也与 $\mathbf{Q}$ 中熟知的运算毫无抵触，熟知的算律也仍旧成立. 而且，对于无限小数完全可以像有限小数(即以 0 为循环节的无限)小数那样进行四则运算. 例如，我们可以动笔算出

$$\sqrt{2}\pi=(1.414\,213\cdots)(3.141\,592\,\cdots)=4.442\,88\cdots,$$

$$4-\pi=0.858\,407\cdots,\quad -\pi=-4+0.858\,407\cdots.$$

现在我们可以说，§1 中在有理数集 $\mathbf{Q}$ 中进行的关于数列的讨论，在实数集 $\mathbf{R}$ 中全部适用. 以后我们的讨论将完全在实数集 $\mathbf{R}$ 中进行.

定理 7　$\mathbf{R}$ 是完备的，也就是说，在 $\mathbf{R}$ 中，基本列一定收敛.

证　设 $f=\{f(n)\}_{n=1}^{+\infty}$ 是 $\mathbf{R}$ 中的基本列. 把实数 $f(n)$ 写成不以 9 为循环节的十进小数

$$f(n)=m_n+0.a_1^na_2^n\cdots,\ m_n\in\mathbf{Z},\ a_k^n\in\{0,1,2,\cdots,9\},\ k\in\mathbf{N}.$$

令有理数

$$g(n)=m_n+0.a_1^na_2^n\cdots a_n^n,\qquad n\in\mathbf{N}.$$

按定义 8 和定义 9，

$$0\leqslant f(n)-g(n)<10^{-n}.$$

于是，有理数列 $g=\{g(n)\}_{n=1}^{+\infty}$ 与数列 f 等价. 把与 g 等价的标准列记作 $h=\{h(n)\}_{n=1}^{+\infty}$，把与 h 对应的实数记为 $\overline{h}$. 根据定义 8 和定义 9，

$$0\leqslant\overline{h}-h(n)<10^{-n}.$$

那么，根据定义 2(对于实数列的情形)，

$$\lim_{n\to+\infty} h(n)=\bar{h}.$$

从而，与 h 等价的实数列 f 收敛到 $\bar{h}\in\mathbf{R}$. □

定义 10(实数的级数表示) 设实数 r 表示成不以 9 为循环节的十进小数

$$r=m+0.a_1a_2\cdots.$$

那么记号

$$r=m+\sum_{n=1}^{+\infty}a_k10^{-k}$$

叫作 r 的十进级数表示.

现在可以说，我们对于八年级学到的“无限不循环小数”终于有了透彻的理解.

在这节的最后，我们来说一说实数在实直线(或实数轴)上的稠密性.

定理 8 (有理数和无理数都在 **R** 中稠密) 设 $r\in\mathbf{R}$，$\delta>0$. 那么，一定找得到有理数 a 和无理数 b，使得

$$a\in(r-\delta,\ r+\delta),\ b\in(r-\delta,\ r+\delta).$$

证 取 $n\in\mathbf{N}$ 充分大，使 $\frac{\pi}{n}<\delta$. 若 $r\in\mathbf{Q}$，则取 $a=r+\frac{1}{n}$，$b=r+\frac{\pi}{n}$ 便合乎定理的要求.

设 r 是无理数. 依定义 10 表之为十进级数

$$r=m+\sum_{k=1}^{+\infty}r_k10^{-k}.$$

取 $a=10^{-n}+m+\sum_{k=1}^{n}r_k10^{-k}$, $b=\pi10^{-n}+\sum_{k=1}^{n}r_k10^{-k}$. 那么，$a$ 是有理数，b 是无理数，并且

$$r<a<b<a+\delta.$$ □

§ 3. 连续函数、指数函数和对数函数

我们一般地用 I 代表 **R** 中的区间，可取下列九种形式：

$(a,\ b)$，$[a,\ b)$，$(a,\ b]$，$[a,\ b]$，$(-\infty,\ b)$，$(-\infty,\ b]$，

$(a, +\infty)$，$[a, +\infty)$，$(-\infty, +\infty)=\mathbf{R}$，
其中$-\infty<a<b<+\infty$.

定义 11　设 f 是 I 上的实函数，即从 I 到 $\mathbf{R}$ 的映射．设 $x_0\in I$. 如果对于任给的 $\varepsilon>0$ 总找得到 $\delta>0$，使得当 $|x-x_0|<\delta$ 且 $x\in I$ 时，

$$|f(x)-f(x_0)|<\varepsilon,$$

那么就说 f 在 x_0 处连续．当 x_0 是 I 的右(左)端点时，也把在 x_0 连续叫左(右)连续．如果 f 在 I 的每点处都连续，就说 f 在 I 上连续，记作 $f\in C(I)$（可理解 C 为 Continuous 的首字母）.

定义 12　说 I 上的函数 f 严格增(减)，指的是当 x，$y\in I$，$x<y$ 时，必有 $f(x)<f(y)$($f(x)>f(y)$).

显然，若 f 在 I 上严格增(减)，则 f 具有定义在 $f(I)$(f 的值域)上的反函数，记作 f^{-1}.

定理 9(连续函数取遍中间值)　设 $f\in C(I)$，x，$y\in I$. 若 $f(x)=a<c<f(y)=b$，则存在 ξ 介于 x，y 之间，使得 $f(\xi)=c$.

证　先考虑 $c=0$ 的情形.

不失一般性，可认为 $x<y$，$a<b$. 记 $I_0=[x, y]$. 把 I_0 二等分成 $\left[x, \frac{x+y}{2}\right]$和$\left[\frac{x+y}{2}, y\right]$. 若 $f\left(\frac{x+y}{2}\right)=0$，取$\frac{x+y}{2}=\xi$ 就使得$f(\xi)=0$. 设 $f\left(\frac{x+y}{2}\right)\neq 0$. 那么，当 $f\left(\frac{x+y}{2}\right)<0$ 时，取 $x_1=\frac{x+y}{2}$，$y_1=y$，$I_1=[x_1, y_1]$，使得 $f(x_1)<0<f(y_1)$. 若 $f\left(\frac{x+y}{2}\right)>0$，则取 $x_1=x$，$y_1=\frac{x+y}{2}$，$I_1=[x_1, y_1]$. 仍使 $f(x_1)<0<f(y_1)$成立．然后在 I_1 上实施上述讨论．即把 I_1 二等分，若 f 在 I_1 的中点取零值，则讨论可终止．否则，分成的两个小区间中必有一个，记作 $I_2=[x_2, y_2]$，使得 $f(x_2)<0<f(y_2)$成立．继续重复这个手续．若有限次后得一区间之中点，为 f 之零点，则讨论中止．否则，得一列区间 $I_n=[x_n, y_n]$，$n\in\mathbf{N}$. 其中，I_n是 I_{n-1}的一半．我们看到

$$x_n\leqslant x_{n+1}\leqslant y_{n+1}\leqslant y_n, \qquad y_n-x_n=2^{-n}(b-a).$$

那么，按定义 4，数列$\{x_n\}_{n=1}^{+\infty}$和$\{y_n\}_{n=1}^{+\infty}$等价．对于任何 $k\in\mathbf{N}$，当 $\mu>\nu>\frac{k}{b-a}$时，

$$0\leqslant x_{\mu}-x_{\nu}<y_{\nu}-x_{\nu}=2^{-\nu}(b-a)<\frac{1}{k}.$$

故按定义 3，$\{x_n\}_{n=1}^{+\infty}$是基本列．那么，根据定理 7，它收敛到一个实数 ξ．显然，与之等价的$\{y_n\}_{n=1}^{+\infty}$亦收敛到 ξ．然而，$f(x_n)<0<f(y_n)$并且 $f\in C(I)$．所以

$$f(\xi)=\lim_{n\to+\infty}f(x_n)\leqslant 0\leqslant \lim_{n\to+\infty}f(y_n)=f(\xi).$$

由此得到 $f(\xi)=0$．我们对于 $c=0$ 的情形证明了定理．

现设 $c\neq 0$．令 $g(x)=f(x)-c$，并对 g 使用已证的事实，就得到定理的结论．

定理 9 的推论 区间上的连续函数的值域是区间．

这是定理 9 的改写．

定理 10 设 $f\in C(I)$且 f 严格增(或严格减)．那么反函数

$$f^{-1}\in C(f(I)).$$

证 设 $f\in C(I)$严格增．记 $J=f(I)$．据上述推论，J 是区间．f^{-1}作为从区间 I 到区间 J 的严格增映射 f 的逆映射，当然是严格增的．

设 $y_0\in J$，$f^{-1}(y_0)=x_0\in I$．我们来证明 f^{-1}在 y_0 连续．

对于任给的 $\varepsilon>0$，如果 $x_0+\varepsilon\in I$，我们记 $f(x_0+\varepsilon)=y_1\in J$；如果 $x_0-\varepsilon\in I$，我们记 $f(x_0-\varepsilon)=y_2\in J$．分三种情况来定义正数 δ：当 $x_0+\varepsilon\in I$ 且 $x_0-\varepsilon\in I$ 时，定义 $\delta=\min(y_1-y_0,\ y_0-y_2)$；当 $x_0+\varepsilon\in I$ 且 $x_0-\varepsilon\notin I$ 时，定义 $\delta=y_1-y_0$；当 $x_0+\varepsilon\notin I$ 且 $x_0-\varepsilon\in I$ 时，定义 $\delta=y_0-y_2$．当 $x_0+\varepsilon\notin I$ 且 $x_0-\varepsilon\notin I$ 时定义 $\delta=1$．

由于 f 严格增，所以 $\delta>0$．并且，当 $x\notin(x_0-\varepsilon,\ x_0+\varepsilon)$但 $x\in I$ 时，$f(x)\notin(y_0-\delta,\ y_0+\delta)$．我们断定，当 $y\in J$，$|y-y_0|<\delta$ 时，

$$|f^{-1}(y)-f^{-1}(y_0)|<\varepsilon. \quad \square$$

现在我们来定义指数函数．

引理 11 设 $x\in\mathbf{R}$，$n\in\mathbf{N}$．定义

$$S_n(x)=\sum_{k=0}^{n}\frac{1}{k!}x^k.$$

若 $q=2([|x|]+1)$，m，$n\in\mathbf{N}$，$n>q$，则

$$|S_{n+m}(x)-S_n(x)|\leqslant q^q 2^{-n}.$$

(我们用$[x]$表示不大于 x 的最大整数．)

证 显然，当 $k\geqslant q$ 时，$k!>q^{k-q}$并且 $2|x|<q$．所以，当 $n\geqslant$

q 时，

$$|S_{n+m}(x)-S_n(x)|\leqslant\sum_{k=n+1}^{n+m}\frac{1}{k!}|x|^k\leqslant q^q\sum_{k=n+1}^{n+m}2^{-k}\leqslant q^q2^{-n}.\quad\Box$$

根据这个引理，我们得知，对于任何实数 x，$\{S_n(x)\}_{n=1}^{+\infty}$ 都是基本列.

定义 13(指数函数)　把数列 $\{S_n(x)\}_{n=1}^{+\infty}$ 的极限记作 e^x，叫作 e 的 x 次幂；作为 x 的函数，e^x 叫作以 e 为底的指数函数.

定理 12　对于任何 x，$y\in\mathbf{R}$，

$$\mathrm{e}^x\mathrm{e}^y=\mathrm{e}^{x+y}.$$

证　设 S_n 如引理 10 所定义. 令

$$f_n(x,\ y)=S_{2n}(x+y)-S_n(x)S_n(y).$$

我们有

$$\begin{aligned}S_{2n}(x+y)&=\sum_{\mu=0}^{2n}\frac{1}{\mu!}(x+y)^\mu=\sum_{\nu=0}^{2n}\sum_{\mu=\nu}^{2n}\frac{1}{\nu!}x^\nu\frac{1}{(\mu-\nu)!}y^{\mu-\nu}\\&=\sum_{\nu=0}^{2n}\sum_{\mu=0}^{2n-\nu}\frac{1}{\nu!}x^\nu\frac{1}{\mu!}y^\mu\\&=\Big(\sum_{\nu=0}^{n}\sum_{\mu=0}^{n}+\sum_{\nu=0}^{n-1}\sum_{\mu=n+1}^{2n-\nu}+\sum_{\nu=n+1}^{2n}\sum_{\mu=0}^{2n-\nu}\Big)\frac{1}{\nu!}x^\nu\frac{1}{\mu!}y^\mu.\end{aligned}$$

于是我们得到

$$f_n(x,y)=\Big(\sum_{\nu=0}^{n-1}\sum_{\mu=n+1}^{2n-\nu}+\sum_{\nu=n+1}^{2n}\sum_{\mu=0}^{2n-\nu}\Big)\frac{1}{\nu!}x^\nu\frac{1}{\mu!}y^\mu.$$

从而，

$$\begin{aligned}|f_n(x,\ y)|\leqslant&S_{n-1}(|x|)(S_{2n}(|y|)-S_n(|y|))+\\&S_n(|y|)(S_{2n}(|x|)-S_n(|x|)).\end{aligned}$$

由于 $\{S_n(u)\}_{n=1}^{+\infty}$ 收敛，令 $n\to+\infty$ 就得到

$$\lim_{n\to+\infty}f_n(x,\ y)=0=\lim_{n\to+\infty}S_{2n}(x+y)-\lim_{n\to+\infty}S_n(x)\lim_{n\to+\infty}S_n(y).$$

由此得到所需的结果.　$\Box$

定理 13　令 $f(x)=\mathrm{e}^x$，$x\in\mathbf{R}$. 那么，

(1) f 严格增；　(2) f 在 $\mathbf{R}$ 上连续，值域是 $(0,\ +\infty)$.

证　根据 S_n 的定义(见引理 11)知道，如果 $x\geqslant0$，$n\in\mathbf{N}$，那么，

$$\begin{aligned}1+x+\frac{1}{2}x^2\leqslant S_{n+2}(x)&=1+x+x^2\sum_{k=0}^{n}\frac{1}{(k+2)!}x^k\\&\leqslant1+x+x^2S_n(x).\end{aligned}$$

令 $n\to+\infty$，得

$$1+x+\frac{1}{2}x^2\leqslant e^x\leqslant 1+x+x^2e^2.$$

由此得到

$$e^x\geqslant e^0=1,\ (1-x^2)e^x\leqslant 1+x.$$

用定理 12，得 $e^{-x}=\frac{1}{e^x}$. 这样，我们首先证明了 $f(x)>0$.

对于 $y>0$，$x\in\mathbf{R}$，根据上面的结果，有

$$e^{x+y}-e^x=e^x(e^y-1)>e^x(1+y)>0.$$

那么我们证明了结论(1).

最后，我们已推得，当 $0<x<1$ 时，

$$1+x<e^x\leqslant\frac{1}{1-x},\quad \frac{1}{1-x}>e^{-x}>1+x.$$

由此可知，对于任何 $\varepsilon>0$，只要 $|x|<\frac{\varepsilon}{1+\varepsilon}$ 就有

$$1-\varepsilon<1-|x|\leqslant e^x\leqslant\frac{1}{1-|x|}<1+\varepsilon.$$

那么，据定义 11，f 在 $x=0$ 处连续. 再使用等式

$$e^{x+y}-e^x=e^x(e^y-e^0)$$

就证得 f 在 $\mathbf{R}$ 上连续. 由于 $f(k)>1+k$，$f(-k)<\frac{1}{1+k}$ 对于一切 $k\in\mathbf{N}$ 成立，我们得到结论(2). □

定义 14(对数函数) 函数 e^x 的反函数叫作以 e 为底的对数(或自然对数)函数，记作 $\ln x$.

设 $a>0$，$n\in\mathbf{N}$. 我们早已知道，a 的 n 次幂指的是 n 个 a 相乘的积，记作 a^n. 由于 $a=e^{\ln a}$，而根据定理 12，n 个 $a=e^{\ln a}$ 相乘的结果是 $e^{(n\ln a)}$，所以必有

$$a^n=e^{(n\ln a)},\ \ln(a^n)=n\ln a.$$

为书写简便，以后总把肩膀上的乘积外面的括号略掉. 定义 $f(a)=a^n$，$a>0$. 那么从等式

$$f(a+b)-f(a)=b\sum_{k=0}^{n-1}(a+b)^{n-1-k}b^k,\ (a,\ b>0)$$

可知，f 在 $(0,+\infty)$ 上连续、严格增、值域为 $(0,+\infty)$. 于是，根据

定理 9，f 的反函数 f^{-1} 是$(0, +\infty)$上的严格增连续函数，叫作 n 次根函数(我们只谈正数的正根)．记 $f^{-1}(u)=u^{\frac{1}{n}}$．由于 $u^{\frac{1}{n}}=\mathrm{e}^{\ln(u^{\frac{1}{n}})}$，那么根据定理 12，得到

$$u=\mathrm{e}^{n\ln(u^{\frac{1}{n}})},\ \ln u=n\ln(u^{\frac{1}{n}}),\ u^{\frac{1}{n}}=\mathrm{e}^{\frac{1}{n}\ln u},\quad (u>0).$$

根据这些结果，我们进一步导出，对于 $a>0$ 和正有理数 $r=\frac{m}{n}$，$m, n\in\mathbf{N}$，

$$a^r=(a^{\frac{1}{n}})^m=\mathrm{e}^{m\ln(a^{\frac{1}{n}})}=\mathrm{e}^{r\ln a}.$$

根据这个事实，我们做出下述定义.

定义 15(以 a 为底的指数函数)　设 $a>0$，$x\in\mathbf{R}$. 把 a 的 x 次幂规定为

$$a^x=\mathrm{e}^{(x\ln a)}.$$

作为 x 的函数，它叫作以 a 为底的指数函数.

我们看到，如果 $a>1$，那么 $\ln a>0$. 于是 a^x 是 x 的严格增连续函数，值域是$(0, +\infty)$. 如果 $0<a<1$，那么 $\ln a<0$. 则 a^x 是 x 的严格减连续函数，值域还是$(0, +\infty)$. 于是我们做出下述定义.

定义 16(以 a 为底的对数函数)　设 $a>0$，$a\neq1$. 把以 a 为底的指数函数的反函数叫作以 a 为底的对数函数，记做 $\log_a x$.

注　根据定义，我们得知到对于任意的 x，$y\in\mathbf{R}$，$a>0$，

$$(a^y)^x=\mathrm{e}^{x\log_a(a^y)}=\mathrm{e}^{xy\log_a a}=a^{xy}=(a^y)^x.$$

参考文献

[1] 九年义务教育三年制初级中学教科书，代数，第二册．北京：人民教育出版社，1993.

[2] Г. М. 菲赫金哥尔茨．《微积分学教程》，叶彦谦，等译．北京：人民教育出版社，1956.

[3] 高级中学课本，代数，上册．北京：人民教育出版社，1995.

数学教育学报，1999，8(3)：95—98.

关于 Riemann 积分理论的本质缺陷及以 Lebesgue 积分理论取代之的看法[①]

On the Essential Shortage of the Integral Theory of Riemann and the Opinion to Replace it by the Theory of Lebesgue

摘要 简要论述了 Riemann 积分理论的本质缺陷是不承认 σ-可加性，而 Lebesgue 积分理论的出发点是 σ-可加性；在大学数学系本科基础课中以 Lebesgue 积分取代 Riemann 积分是历史的必然.

关键词 Riemann 积分；Lebesgue 积分；缺陷.

为较完整地说明问题，先介绍“可数”的概念和 Riemann 积分的定义.

§1. 可数集

众所周知，有限个元素组成的集合，它的元素是数(shǔ)得清的，当然应该叫作“可数集”. 同时，我们有办法一个一个地“清点”全体自然数所成的集合 **N**. 只要从一开始，一个比一个多一地数(shǔ)下去，虽然永无休止，但却一个也不会漏掉. 因此，我们也把 **N** 叫作可数集. 把这种常识抽象成数学概念，得下述定义.

定义 1 可与 **N** 的一个子集作成一一对应的集合叫作**可数集**，空集也叫可数集.

显然，任何无限的可数集都可作成与 **N** 的一一对应. 因此，我们认为，它与 **N** 有同样多个元素.

---

① 教育部资助的教学改革研究项目.

定义 2 规定无限可数集所含的元素的数目为$\aleph_0$，读如“阿列夫零”.

若集合A有$\aleph_0$个元，也就是说在A与$\mathbf{N}$之间可建立一一对应，那么，把A中与自然数n对应的元记作a_n(或$a(n)$或a^n等)，就可把A写成$\{a_n \mid n\in\mathbf{N}\}$的形式.

以后，谈到可数个东西的时候，要么指的是有限个(包括零个)，要么指的是$\aleph_0$个.

例 1 有理数的全体$\mathbf{Q}$是可数集. 对于任意的$n\in\mathbf{N}$，我们把n维Euclid空间$\mathbf{R}^n$中的每个坐标都是有理数的点叫作**有理点**，把有理点的全体记作$\mathbf{Q}^n$，它也是可数集.

例 2 实数集$\mathbf{R}$不是可数集.

证 设不然. 那么，可把$\mathbf{R}$写成$\{r_n \mid n\in\mathbf{N}\}$.

令$I_1=[0, 1]$. 把I_1三等分成三个闭区间，其中至少有一个不含r_1，记这样的一个闭区间为I_2，它包含于I_1，且长度为$\frac{1}{3}$. 把I_2三等分成三个闭区间，其中至少有一个不含r_2，记这样的一个闭区间为I_3，它包含于I_2，且长度为$\frac{1}{3^2}$. 无限地继续这一步骤，得到$\aleph_0$个闭区间，它们的全体记作$\{I_n \mid n\in\mathbf{N}\}$. 那么，对于每个$n\in\mathbf{N}$，$r_n\notin I_n$，$I_{n+1}\subset I_n$，并且$I_n$的长度为$\frac{1}{3^{n-1}}$. 由区间套定理知，存在实数$r\in\bigcap\limits_{n=1}^{+\infty}I_n$. 那么，$r$不可能属于$\{r_n \mid n\in\mathbf{N}\}$. 荒谬. □

§ 2. Riemann 积分的定义

定义 3 设$n\in\mathbf{N}$，$m\in\mathbf{N}$，$k=(k_1, k_2, \cdots, k_n)\in\mathbf{Z}^n$. 称集合

$$Q_m(k)=\{x\in\mathbf{R}^n \mid 2^{-m}k_j\leqslant x_j\leqslant 2^{-m}(k_j+1),\ j=1, 2, \cdots, n\}$$

为$\mathbf{R}^n$中的第m级第k方块. 规定它的体积为$|Q_m(k)|=2^{-mn}$(边长的n次幂).

定义 4 设f是定义在$\mathbf{R}^n$的非空有界集D上的实值函数. 令

$$\tilde{f}(x)=\begin{cases} f(x), & x\in D, \\ 0, & x\in\mathbf{R}^n\setminus D. \end{cases}$$

设$m\in\mathbf{N}$. 令

$$\overline{S}_m(f) = \sum_{k \in \mathbf{Z}^n} \sup\{\widetilde{f}(x) \mid x \in Q_m(k)\} |Q_m(k)|,$$

$$\underline{S}_m(f) = \sum_{k \in \mathbf{Z}^n} \inf\{\widetilde{f}(x) \mid x \in Q_m(k)\} |Q_m(k)|.$$

分别称 $\overline{S}_m(f)$ 与 $\underline{S}_m(f)$ 为 f(在 D 上)的第 m(Darboux)上和与第 m(Darboux)下和.

注 1

(1) 定义 4 中的和 $\sum\limits_{k \in \mathbf{Z}^n}$ 实为有限和，因为只有有限个 $k \in \mathbf{Z}^n$ 使所加的数不是零.

(2) 如果 f 无上界，那么 $\overline{S}_m(f)=+\infty$，如果 f 无下界，那么

$$\underline{S}_m(f)=-\infty.$$

(3) $\forall m \in \mathbf{N}$，$\underline{S}_m(f) \leqslant \underline{S}_{m+1}(f) \leqslant \overline{S}_{m+1}(f) \leqslant \overline{S}_m(f)$，$\underline{S}_m(f) < +\infty$，$\overline{S}_m(f) > -\infty$.

定义 5 设 f 定义在 $\mathbf{R}^n$ 的有界集 D 上，分别称

$$\overline{I}(f) = \lim_{m \to +\infty} \overline{S}_m(f) \quad 和 \quad \underline{I}(f) = \lim_{m \to +\infty} \underline{S}_m(f)$$

为 f(在 D 上)的上积分与下积分，如果 $\overline{I}(f)=\underline{I}(f)=I$，那么称 I 为 f(在 D 上)的积分，记作 $I=\int_D f(x)\mathrm{d}x=\int_D f$. 当 $I \in \mathbf{R}$ 时，称 f 在 D 上可积. 把在有界集 D 上可积的函数的全体记作 $\mathfrak{R}(D)$.

注 2 在定义 5 中，没有限定 f 有界，所以 $I=+\infty$ 或 $-\infty$ 的情形可能发生.

从定义 5 知，f 在 D 上可积的必要条件是 f 在 D 有界. 用二进网格 $\{Q_m(k) \mid k \in \mathbf{Z}^n, m \in \mathbf{N}\}$ 处理涉及分划的问题，是近代实分析理论中的一个基本技巧. 我们借助这个网格来定义 Riemann 积分，与原始的定义是等价的，然而却使得叙述变得简洁.

定义 6 设 D 是 $\mathbf{R}^n$ 的有界集，χ_D 表示 D 的特征函数，即

$$\chi_D(x) = \begin{cases} 1, & x \in D, \\ 0, & x \notin D. \end{cases}$$

如果 $\chi_D \in \mathfrak{R}(D)$，那么称 D 为 Jordan 可测集，简称为可测集. 称 $\int_D \chi_D$ 为 D 的测度，记之为 $|D|$.

对于无界集，我们如下规定它的可测性及测度. 设 $N \in \mathbf{N}$，$A_N =$

$\{x\in\mathbf{R}^n \mid |x_j|\leqslant N,\ j=1,2,\cdots,n\}$，若 $\forall N\in\mathbf{N}$，$E\cap A_N$ 可测，则说 E 可测，且定义 $|E|=\lim\limits_{N\to+\infty}|E\cap A_N|$.

注 3 名称"Jordan 可测集"源于法国数学家 C. Jordan 在 1893 年发表的《分析教程》第二版中的"容度"一词.

例 3 $\mathbf{R}^n$ 中的单点集是可测集，且测度为零.

设 $D\subset\mathbf{R}^n$，$x\in D$. 若存在正数 r，使得 $\{y\in\mathbf{R}^n \mid |x-y|<r\}\subset D$，就把 x 叫作 D 的内点，把 D 的一切内点所成的集记作 $\mathring{D}$，称之为 D 的内部. 使 $D=\mathring{D}$ 的集 D 叫作开集，(规定 $\varnothing=\mathring{\varnothing}$).

例 4 设 $\mathbf{Q}^n$ 为 n 维有理点集，即 $\mathbf{Q}^n=\{x\in\mathbf{R}^n \mid x_j\in\mathbf{Q},\ j=1,2,\cdots,n\}$. 那么，对于 $\mathbf{R}^n$ 的任意的有界集 D，只要 $\mathring{D}=\varnothing$，集合 $\mathbf{Q}^n\cap D$ 就不是 Jordan 可测的.

事实上 $\mathring{D}\neq\varnothing$ 蕴含着，存在充分大的 $j\in\mathbf{N}$ 和适当的 $k\in\mathbf{Z}^n$，使小方块 $Q:=Q_j(k)\subset D$. 显然，$|Q|>0$. 那么，记函数 $\chi_{\mathbf{Q}^n\cap D}=f$，必有

$$\forall m\in\mathbf{N},\quad \overline{S}_m(f)\geqslant|Q|>0,\quad \underline{S}_m(f)=0.$$

故 f 不可积，即 $\mathbf{Q}^n\cap D$ 不可测.

§ 3. Riemann 积分理论的本质缺陷是不承认 σ-可加性

设 E_j，$(j\in\mathbf{N})$ 是 $\mathbf{R}^n$ 中的两两不相交的可数个可测集. 对于任意的 $m\in\mathbf{N}$，令 $F_m=\bigcup\limits_{j=1}^{m}E_j$. 由定义可知，$F_m$ 是可测集，且 $|F_m|=\sum\limits_{j=1}^{m}|E_j|$. 更一般地，有限个可积函数的和仍是可积函数，且和函数的积分等于各被加函数的积分的和. 这种性质，叫有限可加性.

如果代替"有限个"以"$\aleph_0$"个，仍成立相应的结论，那么相应的性质就叫作 σ-可加性.

从例 3 和例 4，我们已看到，Riemann 积分理论(所涉及的 Jordan 测度)不具有 σ-可加性. 下面我们做些进一步的分析.

例 5 设 $\mathbf{Q}^n\cap[0,1]^n=\{r_j \mid j\in\mathbf{N}\}$，$\delta>0$. 定义

$$B_j(\delta)=\{x\in\mathbf{R}^n \mid |x-r_j|<2^{-j-1}\delta\},\qquad G(\delta)=\bigcup_{j=1}^{+\infty}B_j(\delta).$$

那么，$G(\delta)$是$\mathbf{R}^n$中的有界开集，它一定可以表示成彼此互不重叠(但可能相交)的$\aleph_0$个(如定义3所说的)方块的并. 我们说，当$\delta<1$时，$G(\delta)$不可测.

证 记$f=\chi_{G(\delta)}$. 对于每个$m\in\mathbf{N}$，定义

$$I_m=\{k\in\mathbf{Z}^n\mid Q_m(k)\subset G(\delta)\},\qquad J_m=\{k\in\mathbf{Z}^n\mid Q_m(k)\cap G(\delta)\neq\varnothing\}.$$

显然，I_m是有限集. 若$k\in I_m$，即$Q_m(k)\subset G(\delta)$，则由有限覆盖定理知，存在$N(k)\in\mathbf{N}$使$Q_m(k)\subset\bigcup\limits_{j=1}^{N(k)}B_j(\delta)$. 令$M=\max\{N(k)\mid k\in I_m\}$. 那么$\sum\limits_{k\in I_m}|Q_m(k)|\leqslant\sum\limits_{j=1}^{M}|B_j(\delta)|\leqslant\sum\limits_{j=1}^{M}(2^{-j}\delta)^n$. 由此推出$\underline{S}_m(f)=\sum\limits_{k\in I_m}|Q_m(k)|\leqslant\delta^n$. 另一方面，由$\bigcup\limits_{k\in J_m}Q_m(k)\supset(0,1)^n$知$\overline{S}_m(f)\geqslant\sum\limits_{k\in J_m}|Q_m(k)|\geqslant 1$. 那么，根据定义5，当$\delta<1$时有界开集$G(\delta)$不是Jordan可测集. □

下面我们把例5所说的话翻译成积分的语言.

例6 设G是例5所说的相应于$\delta=0.4$的有界开集. 把它表示成$\aleph_0$个互不重叠的方块的并：$G=\bigcup\limits_{j=1}^{\infty}Q_j$. 令

$$E_1=Q_1,\ E_{j+1}=Q_{j+1}\setminus\Big(\bigcup_{i=1}^{j}Q_i\Big),\quad j\in\mathbf{N}.$$

那么，诸E_j可测且两两不交，$|E_j|=|Q_j|$. 而且$G=\bigcup\limits_{j=1}^{+\infty}E_j$. 定义$f_j=\chi_{E_j}$，$j\in\mathbf{N}$. 那么，

$$\forall j\in\mathbf{N},\quad f_j\in\Re(G),\quad \int_G f_j(x)\mathrm{d}x=|E_j|.$$

但是，$\chi_G=\sum\limits_{j=1}^{+\infty}f_j\notin\Re(G)$.

我们想一想，集合的测度规定的是它的大小尺寸. 有限个彼此互不相交的有测度的集合合起来是有大小的，它的大小就是这些集的测度的总和. 这合乎我们的常识，反映了客观规律. 当我们对客观世界的认识从“有限”发展到“$\aleph_0$”之后，是不是应该承认$\aleph_0$个彼此互不相交的有测度的集合合起来是有大小的，它的大小就是这些集的测度的总和呢? 也就是说，我们的认识是不是也应该从“有限可加性”这个初等水平发展到

σ-可加性的水平呢？

在这个问题上 Riemann 的理论是停留在初等水平的，它不承认测度的 σ-可加性，这是它的本质缺陷.

§4. 创造条件，逐步以 Lebesgue 积分取代 Riemann 积分

20 世纪初由法国数学家 H. L. Lebesgue 建立起来的积分理论，克服了 Riemann 积分理论的缺陷，恰恰以承认 σ-可加性为其出发点. 正如 M. 克莱因[1](第四册，123 页)所说：

> “Cantor 曾证明，直线上的任一开集 U 必是一族可数个两两不相交的开区间的并集. Borel 利用 Cantor 的结果，不再用有穷个区间包围 U 去逼近 U 的方法，而是提出把一个有界开集的各个构成区间的长度的总和，作为这个开集的测度.”

同样的思想也适用于 $\mathbf{R}^n$. 容易证明，$\mathbf{R}^n$ 中的非空开集 G 必是 $\aleph_0$ 个形如 $Q_m(k)$ 的两两不相重叠(即无公共内点)的二进方块的并. 把这些方块的体积之和作为 G 的测度，是非常自然的事. 这正是 Lebesgue 理论的出发点. Lebesgue 就是如此在他的老师 Borel 及其他前辈工作的基础上建立起他的理论的.

因此，上面例 4～例 6 中发生的问题，在 Lebesgue 积分理论中是不存在的，例 4 和例 5 的毛病被 Lebesgue 测度的定义克服掉了. 而例 6 的问题被从定义(σ 可加性)导出的单调收敛定理(即以意大利人 B. Levi 的名字命名的定理)轻而易举地解决了.

我们愿意再举个例，说明 Lebesgue 积分理论基于 σ 可加性所表现的巨大优越性. 这是[2]第 57 页的习题 25 的一部分.

例 7　设 $f(x)=\begin{cases} x^{-\frac{1}{2}}, & 0<x<1, \\ 0, & x\notin(0,1). \end{cases}$ 把 $\mathbf{Q}$ 排成数列 $\{r_n\}_{n=1}^{+\infty}$. 定义

$$g(x)=\sum_{n=1}^{+\infty} 2^{-n} f(x-r_n).$$

那么，g 在 $\mathbf{R}$ 的任何有内点的区间上都是无界的. 显然，Riemann 积分

(即使是瑕积分)无法处理g. 可是按 Lebesgue 积分的控制收敛定理(它也是σ可加性的结果),

$$\int_{\mathbf{R}} g(x)\mathrm{d}x=\sum_{k=1}^{+\infty}2^{-n}\int_{\mathbf{R}} f(x-r_n)\mathrm{d}x=2.$$

Lebesgue 积分理论的重要性和优越性现已举世公认. 以至于法国著名数学教育家 J. Dieudonné 说([3], 159 页):

> "……这里明显地没有微积分教程中一个古老的题目. 即'黎曼积分'. 人们大概会感觉到:如果不是它的有权威的名字,它老早就该没落下去了,因为对于任何一位从事研究工作的数学家来说(带着对黎曼天才的应有尊敬),十分清楚,现今这一'理论'的重要性在测度与积分的一般理论中,最多不过是一普通的有趣的练习(参看 13.9 问题 7). 只有那种学究传统的顽固保守主义才会把它冻结成课程的正规部分,长时间以后必将失去它的历史重要性."

我理解这段话有两个意思. 第一个意思是,按照科学发展的客观规律,Lebesgue 积分取代 Riemann 积分是理所当然的;第二个意思是说 Lebesgue 积分目前尚未取代 Riemann 积分原因有二:其一是由于或许 Riemann 比 Lebesgue 在人们心中更伟大,其二是**学究传统的顽固保守主义的束缚**. 我认为传统的束缚(把那些伤人的定语去掉)是最糟糕的. 我们的改革,恰恰就是要打破那些落后于时代的传统的束缚. 事实上这也是最难的.

我们现行的课程体系中,Riemann 积分在一、二年级讲,讲得很细,算得很多;而 Lebesgue 积分在三年级讲,72 学时,讲得糙,算得少,甚至根本不算(因为在 Riemann 积分中算多了). 使教师和学生在心理上潜在地形成 Riemann 积分比 Lebesgue 积分更重要或更基本的印象. 这不能不说是一个违背科学发展规律的错误. 其结果,不仅是学生,就是教师,对于 Lebesgue 积分的理解都是马马虎虎的. 如今现实的局面是,许多数学系的教师只懂 Riemann 积分而不懂(或不真正懂) Lebesgue 积分. 虽然时有"新三高"的提法(把"实变函数论"等三门课的地位提高的提法),但实际上,只要 Riemann 积分不让位,Lebesgue 积分就不可能被摆在数学系基础课的地位.

Riemann 积分不让位的一个现实的原因是,以往的 Lebesgue 积分被讲得太难. 从“有限可加”到 σ 可加，人们的认识要从“有限”提高到 $\aleph_0$. 困难是不可避免的. 教师的职责应是把难的东西经过自己反复的思考消化，尽可能化解得易为学生接受，而不该是躲避困难，故步自封.

用 Lebesgue 积分取代 Riemann 积分，对于数学系一、二年级的学生是否可行?

我认为，重要的问题是怎样教. 如果能循序渐进，因材施教，不要求每个学生都得 90 分，那就没什么行不通. 倒是我们的教师必须先提高自己.

关于 Lebesgue 积分比 Riemann 积分难多少，S. Saks 在 60 多年前阐述 Lebesgue 积分的定义时说过(转译自[4]，第 2，3 页)：

> “此外，Lebesgue 的方法不仅更为一般，而且从某种观点来说，它比 Riemann-Darboux 的方法更为简单. 因为它不同时引入上积分和下积分这样两个作为极值的积分.”

我国有些学者，例如匡继昌教授[5]等，早就提出过用 Lebesgue 积分取代 Riemann 积分的问题，有的已打算开始实施试验. 我愿加入这个向传统的束缚挑战的行列，为推进数学教育的发展而不遗余力.

参考文献

[1] 克莱因·M. 北京大学数学系数学史翻译组，译. 古今数学思想. 上海：上海科学技术出版社，1981.

[2] Folland G. B. Real Analysis. John Wiley and Sons，New York，1984.

[3] 迪厄多内 J. 现代分析基础，第一卷. 北京：科学出版社，1982.

[4] Saks S. Theory of the Integral (English translation by L. C. Young). Hafner Publish Company，New York，1937.

[5] 匡继昌. 寻求数学分析改革突破口的思考与实践，数学教育学报，1997，6(2)：81—84.

Abstract It is shown briefly that the essential shortage of the integral theory of Riemann is to deny σ-additivity which is the starting point of the theory of Lebesgue; the replacement of Riemann's theory by Lebesgue's in the basic courses of the mathematical department of universities must be historically necessary.

Keywords Riemann Integral; Lebesgue Integral; Shortage.

高等数学研究，2001，4(3)：13－14.

谈指数函数的定义①

——在大学数学分析课中妥善定义指数函数

On the Definition of Exponential Functions—Defining Exponential Functions Well in the University Mathematical Analysis Cources

对于如何定义指数函数，我们提出下述看法，与同行们商讨.

§1. 承袭前人的结果，不必重复对事物的认识过程

在数学发展的历史上，对于对数函数的研究比对指数函数的研究来得还早，这似乎是不合逻辑的反常现象，但却是事实.

16 世纪，当人们对于指数概念的了解还不很完全的时候，由于天文学和航海事业的需要，英国数学家 J. Napier(1550—1617)使用三角公式，花了约 20 年时间，制作了精密的对数表. 于 1614 年出版了《Mirifici Logarithmorum Canonic Description》(《奇妙的对数定律说明书》)(参阅[1])一书. 他的工作不涉及指数函数的概念，当时关于非正整数指数的概念还是模糊的. 另一位英国数学家和天文学家 H. Briggs(1561—1630)是 Napier 的追随者和合作者，继承 Napier 未竟的事业. 他们于 1624 年合作出版了《Arithmetica Logarithmica》(《对数算术》)一书. 为了纪念 Briggs，以 10 为底的对数(常用对数)常被称为 Briggs 对数. 在他们那个年代无理数 e 尚未被发现.

第一个明确地阐明对数是幂运算的逆运算的是大数学家 L. Euler(1707—1783)，生于瑞士. 自然对数的底也是他发现的，并以他的名字

① 本文与张培恒合作.

收稿日期：2001-01-11.

的首字母的小写形式 e 来表示．英格兰数学家 B. Taylor(1685—1731)于 1715 年提出函数的级数表示公式，在这个公式下，e^x 的值将由 x 代入下面的级数中而得到

$$e^x = \sum_{n=0}^{+\infty} \frac{x^n}{n!}. \tag{1.1}$$

如今我们怎样向学生讲授对数概念呢？恐怕没有人会从 Napier 和 Briggs 的著作《Arithmetica Logarithmica》讲起．我们都不会重复历史上前人走过的路，而是直截了当地先讲指数，再作为其逆运算引入对数.

§ 2. 现在流行的教科书中，对指数函数定义的讲法之缺点

上一段的历史事实说明，我们在承袭前人对于指数、对数的研究结果时，没有重复前人走过的老路，而是走了一个“捷径”．但是在怎样向学生讲授指数函数这个问题上，现行的教科书中的讲法却是保守的，我们来看指数函数

$$f(x)=a^x(a>0,\ a\neq 1),\ x\in\mathbf{R} \tag{2.1}$$

通常是怎样定义的．先考虑 x 取自然数的情形，那么 $f(x)$ 是 x 个 a 连乘之积，这时 $f(x)$ 叫作 a 的 x 次幂．其次考虑 x 取自然数的倒数 $x=\frac{1}{n}(n\in\mathbf{N})$ 的情形，这时把 $f(x)$ 规定为一个其 $n\left(=\frac{1}{x}\right)$ 次幂为 a 的正数，并把它叫作是 a 的 n 次方根．进一步 $f(x)$ 的定义可推广到 x 为有理数的情形．注意，这里暗藏着一个正数的 n 次方根存在的问题．试问，在定义之初，一个正数一定有 n 次方根这件事是不是容易说得清楚？定义了 $n(n\in\mathbf{N})$ 次方根之后，接着，考虑 x 是正分数 $\frac{n}{m}(m,\ n\in\mathbf{N})$ 的情形．$a^{\frac{n}{m}}$ 被定义为 $(a^{\frac{1}{m}})^n$．其后对于 x 是有理数的情形，给出 $f(x)$ 的定义．最后，如何过渡到 x 是无理数的情形呢？是不是要通过 x 作为有理数列的极限来过渡呢？这样的过渡是很干净利索的吗？这样定义的指数函数其连续性是不是已蕴含于定义本身了呢？处理诸如此类逻辑上的细节是很省事的吗？

这样一个定义指数函数的思想过程，所涉及的指数函数的基本性质(或特征)乃是

$$a^x a^y = a^{x+y}, \quad a>0,\ x,\ y \in \mathbf{R}. \tag{2.2}$$

关于这个算律，中学生已经记得清清楚楚了，就像他们已经清清楚楚记得勾股定理一样.

那么，这样一个定义指数函数的方法，使学生从中学的认识水平提高到怎样的程度呢?

我们认为这样讲授指数函数至少有三个缺点.

(1) 细节麻烦，逻辑不清，叙述啰嗦. 如上所述，正数的方根的存在性，只能承认，说不清楚. 正数(当然可以是无理数)的“无理数次幂”也不是三言两语说得明白的.

(2) 与第一个缺点相关联，在讨论指数函数的连续性时发生逻辑错误.

(3) 按上述方法恐怕很难，甚至不可能把指数函数的定义域扩充到复平面上. 也就是说，对于虚数 $\mathrm{i}(\mathrm{i}^2=-1)$，$f(\mathrm{i})=a^{\mathrm{i}}$ 的定义恐怕只能重新规定.

上述第(2)个缺点，已在有些教科书中表现出来. 在这些书中，如上“定义”了指数函数之后，随之“证明”指数函数是严格单调的，然后定义它的反函数 $\log_a x$. 当证明 $f(x)=a^x\,(a>0)$ 在 $x=0$ 处的连续性时，论述如下：

$\forall X>0$，要使 $|f(x)-f(0)|<X$，只需

$$1-X=f(0)-X<f(x)<f(0)+X=1+X,$$

此式等价于 x 介于 $f^{-1}(1-X)=\log_a(1-X)$ 和 $f^{-1}(1+X)=\log_a(1+X)$ 之间. 于是取 $W=\min\{|\log_a(1+X)|,\ |\log_a(1-X)|\}$ 就保证当 $|x|<W$ 时，$|f(x)-f(0)|<X$.

请问：这是证明吗? 它的根据是什么? 能说这不是逻辑错误吗?

顺便说一说，在多数通用的教材中都在显要的地位上叙述“两个重要的极限”，其中之一就是数列 $\left\{\left(1+\frac{1}{n}\right)^n\right\}$ 当 $n\to+\infty$ 时的极限，并把这个极限为数 e 的定义. 这种做法具有很长的历史，是有些背景的. 而证明这个数列收敛，大多是考虑它的单调有界性. 如果引申一下，进一步考虑数列 $\left\{\left(1+\frac{x}{n}\right)^n\right\}$，其中 x 是复数，譬如说 $x=\mathrm{i}$，$(\mathrm{i}^2=-1)$，那么前述考察单调有界的方法就不可用了. 因此，我们认为这个所谓的

“重要的极限”不宜作为定义指数函数的前导方式，而只宜作为一个习题.

§3. 对引入指数函数定义的建议

把公式(1.1)作为指数函数 e^z 的定义，即定义

$$f(z)=\sum_{n=0}^{+\infty}\frac{1}{n!}z^n=\lim_{m\to+\infty}\sum_{n=0}^{m}\frac{1}{n!}z^n, z\in\mathbf{C}.$$

那么 $f(1)=e$，并记 $f(z)=e^z$. 在复变函数论的课程中(例如[2])就是这样做的. 然后非常简便地(利用数列极限的理论)得到 $f(x+y)=f(x)f(y)$(参阅[3])以及 f 的连续性. 当 $x\in\mathbf{R}$ 时，f 在 $\mathbf{R}$ 上的严格增加性质是明显的，它的反函数，记为 $f^{-1}=\ln x$ 进而对于任意的 $a>0$，定义以 a 为底的指数函数 $a^x=f(x\log_e a)=e^{x\ln a}$ 及其反函数.

这个定义十分简明扼要，一扫§2中所述的诸项缺点，能把学生的认识水平真正从中学时代的知其然而上升到大学时代的知其所以然.

参考文献

[1] 梁宗巨. 世界数学史简编. 沈阳：辽宁人民出版社，1980：157.

[2] [苏]И. И. 普里瓦诺夫. 复变函数引论，北京大学数学力学系分析教研组，译. 上册. 上海：商务印书馆，1953：78.

[3] 王昆扬. 简明数学分析. 北京：高等教育出版社，2001：37.

高等数学研究，2011，14(4)：7—8.

谈谈 Lebesgue 数钱[①]

On Counting Coins by Lebesgue

摘要 用 Lebesgue 关于数钱的比喻，解释了 Lebesgue 积分的本质优点.

关键词 积分；Lebesgue；Riemann；有限加性；σ 加性.

谈到 Lebesgue 积分理论时，人们自然地把它与 Riemann 积分理论加以比较. 在中文的教科书上，多见到引述 Lebesgue 数钱的比喻，用以表述两种积分方法的区别.

我曾力图寻找 Lebesgue 自己述说的数钱的比喻的原文，可惜没找到，中文的传说倒是俯拾皆是. 下面引 3 段从网上录下的文字.

(1) 一次随手拿了一本《实变函数论》，北京大学周民强写的. 在绪论里介绍了 Rieman 积分的局限性，同时也引出了 Lebesgue 积分. 其中说到 Lebesgue 给自己的积分和 Rieman 的区别时，他举了一个例子：假如你要还钱给人家，要先数一下自己钱包里有多少钱. 一种数法就是，把钱全拿出来，按照面额大小分类，再把各部分总数求和. 这种方法就是 Lebesgue 积分的思想. 另外一种就是，按照一定次序逐项相加，比如按照取出钱的次序来算总额. 这就是 Rieman 积分的思想.

① 分析类课程建设国家级教学团队经费支持.

(2) 勒贝格自己是这样解释他所创立的积分：假如桌子上有一大堆不同面值的硬币，你会怎么来数呢？

黎曼是这样做的：把这一大堆硬币任意分成很多小堆，每一堆的硬币数量足够少(这样才能“数清”)，然后把每一小堆的硬币总值数好，最后加起来.

勒贝格这样做：把一大堆硬币按照不同面值分开，每一堆的硬币面值都是一样的. 用每一堆硬币共同的面值乘这一堆的硬币数. 最后求和，即得.

(3) Riemann积分与 Lebesgue 积分就是两种不同的数钱方式.

既然 Lebesgue 的比喻如此深入人心，就有必要谈谈如何正确地、深入地理解这个比喻，并进一步正确地掌握 Lebesgue 积分.

一个实例

我们设计一个数钱的实例，看看用 Lebesgue 的办法该怎样数，用 Riemann 的办法又该怎样数.

区间(0，1)中的分数(俗称为“有理数”)总共有$\aleph_0$ 个. 也就是说，这些分数的全体所成的集合，权且记之为 A，与正整数集 $\mathbf{N}^*$ 之间可建立一一对应(即可逆映射). 那么，A 的元素可以排成一列(排列的方式当然不是唯一的). 假设 A 的元素已经排成序列$\{a_k\}_{k=1}^{+\infty}$.

定义　$I_k=(a_k-4^{-k},\ a_k+4^{-k})\cap(0,\ 1)$. 那么，$I_k$ 是开区间，长度不超过 $2\cdot4^{-k}$. 定义

$$G=\bigcup_{k=1}^{+\infty}I_k.$$

那么，G 是(0，1)的开子集，并且 $A\subset G$. 于是，G 必是$\aleph_0$ 个两两不交的开区间 J_k 的并集. 即

$$G=\bigcup_{k=1}^{+\infty}J_k,\qquad \text{当 } m\neq n \text{ 时，} J_m\cap J_n=\varnothing.$$

① 任取正整数 m，定义集合 $G_m=\bigcup\limits_{k=1}^{m}J_k$. 用 χ_E 代表集合 E 的特征函数，即

$$\text{当 } x\in E \text{ 时，} \chi_E(x)=1\text{，当 } x\notin E \text{ 时，} \chi_E(x)=0.$$

现在计算函数 χ_{G_m} 在[0，1]上的积分. 按照 Lebesgue 的比喻，函数 χ_{G_m} 被看作是口袋里的"钱"，它的积分就是这些钱的总值. 钱 χ_{G_m} 由单一的"面值"为 1 的"硬币"组成.

先用 Riemann 的办法来数.

实际上，Riemann 必须把整个区间[0，1]想象成一条含金的金属棒. 他把这个棒切割成有限段，譬如说 n 段，分别记作 Δ_k，$k=1,2,\cdots,n$. 小段 Δ_k 的含金量，就是它与 G_m 的交集 $\Delta_k\cap G_m$ 的"长度"：这里 Riemann 只能在两种极端的情形下来计算 $\Delta_k\cap G_m$ 的长度，第一种是当 $\Delta_k\subset G_m$ 时，这个长度就是 Δ_k 自己的长度，记作 $|\Delta_k|$，这是这个小区间中所含面值为 1 的"硬币"的数量，由于面值为 1，价值就是 $|\Delta_k|$. 第二种情形就是 $\Delta_k\cap G_m=\varnothing$(空集)的情形. 此时小段 Δ_k 中根本不含钱，价值以 0 计.

然而还有第三种情形，即 $\Delta_k\setminus G_m\neq\varnothing$ 且 $\Delta_k\cap G_m\neq\varnothing$ 的情形. 这时 Riemann 无法判断这一段的价值，但知道它的价值一定介于 0 和 $|\Delta_k|$ 之间. 所以可以随意取一个这样的值来近似. 然后把全部结果加起来，作为总钱数的"近似值". 这样数的误差不超过第三种情形的区间长度之和.

关键是，Riemann 把这个棒切割得尽可能细. 把 Δ_k，$k=1,2,\cdots,n$ 的长度的最大值记作 λ(积分法中的"分法的模"). 当 $m=1$ 时，记 $\alpha=1$；当 $m>1$ 时，记诸区间 J_k，$k=1,2,\cdots,m$ 之间的距离的最小值为 α. 即

$$\alpha=\min\{|x-y|\mid x\in J_s,\ y\in J_t,\ 1\leqslant s\neq t\leqslant m\}.$$

注意，当分法的模 $\lambda<\alpha$ 时，使得第三种情形出现的区间，也就是使 $\Delta_k\setminus G_m\neq\varnothing$ 且 $\Delta_k\cap G_m\neq\varnothing$ 的小段 Δ_k 顶多只有 $2m$ 段. 这些小段的长度总和必定不超过 $2m\lambda$. 可见，λ 越小，数钱的误差越小. 从理论上来说，"当 $\lambda\to0$ 时，就得到精确结果"(此实为定义). 显然，这个结果就是 $\sum\limits_{k=1}^{m}|J_k|$.

用 Lebesgue 的办法数就简单多了. 他一下子把面值为 1 的钱 G_m 都拿出来，口袋里已经空无一文了，马上得到结果 $|G_m|=\sum\limits_{k=1}^{m}|J_k|$.

② 现在转而计算函数 χ_G 的积分，也就是说，"数钱 χ_G".

Lebesgue 的数法与①中一样，一下子从口袋里掏出来就得到结果

$$|G|=\sum_{k=1}^{+\infty}|J_k|.$$

Riemann 的数法可就麻烦大了. 不管把[0，1]分成多短的**有限**小段 Δ_k，$k=1,2,\cdots,n$，这时每小段 Δ_k 都含 G 的点，这是因为，任何一个小区间 Δ_k，一定含有分数，所以第二种情形不发生；小段 Δ_k 使第三种情形发生的唯一的条件是 $\Delta_k \setminus G\neq\varnothing$. 这样的小段的全体一定覆盖 $(0,1)\setminus G$. 因此，这样的小段 Δ_k 的长度总和一定不小于 $(0,1)\setminus G$ 的测度，即

$$|(0,1)\setminus G|=1-|G|\geqslant 1-\sum_{k=1}^{+\infty}|I_k|\geqslant 1-2\sum_{k=1}^{+\infty}4^{-k}=\frac{1}{3}.$$

所以，用 Riemann 的办法不管怎么数，不能保证误差小于$\frac{1}{3}$. 结论是，Riemann 的数法失效. 换言之，**Riemann 的办法不能量度开集 G 的大小.**

结论

Riemann 积分只承认有限加性，不承认 σ 加性(或叫作 $\aleph_0$ 加性)；Lebesgue 积分以 σ 加性为出发点. 拙文[1]曾论及此事.

Lebesgue 积分理论本质上比 Riemann 积分理论提高了一步. 当然，比喻不能代替推理和证明，只具启发作用. Lebesgue 不说他的积分论比 Riemann 的高明，只戏称“数钱”方式不同而已. 谦恭之态，人之常情也.

参考文献

[1] 王昆扬. 关于 Riemann 积分理论的本质缺陷及以 Lebesgue 积分取代之的看法. 数学教育学报，1999，8(3)：95－98.

Abstract　The essential advantage of Lebesgue integration is explained by the metaphor of Lebesgue on counting coins.

Keywords　integration; Lebesgue; Riemann; finite additivity; σ-additivity.

数学通报(英文版)，2009，(增刊)：92－96.

实数的十进表示

Decimal Representation of Real Numbers

Abstract The problem “to represent real numbers by decimal fractions” is rigorously investigated. Based on the knowledge about rational numbers originally represented as fractions the concept of standard sequence is introduced and is applied to prove that every real number represented by a decimal fraction is just the limit of the standard sequence equivalent to the decimal fraction itself. Then the completeness of the real number set **R** is proved.

Keywords Rational numbers; Decimal fractions; Representation of real numbers; Fundamental sequences; Standard sequences.

§ 1. Introduction

If people ask the question “What is the ‘time’?” I am unable to answer. And I even think it is not necessary to answer such a question. Similarly, if people ask the question “What is a ‘real number’?” I am also unable to answer, and I even think it is not necessary to answer.

But the question “ **How do you represent a real number mathematically**?” must be answered clearly. Because any concept in mathematics has significance and can be used when only it gets a clear mathematical representation.

The history on this topic can be traced back to 19 century. At that time many mathematicians such as W. R. Hamilton, K. Weierstrass, C. Méray, R. Dedekind and G. Cantor had discussed this topic. German mathematician Dedekind, J. W. R, (1831—1916) defined irrational numbers as the cuts of rational numbers. His theory does round more widely.

Dedekind's idea can be traced to the year 1858 when he gave lectures on calculus and realized that the real number system had no logical foundation (see [2]). His theory was published in 1872 [3].

It was pointed out in [2] that "though Dedekind's theory of irrational numbers , with minor modifications, …, is logically satisfactory, Cantor criticized it because cuts do not appear naturally in analysis". The author of [2] also said "the irrational number, logically defined, is an intellectual monster, …".

In [4] there is a note said that in Weyl's work "Der Circulus vitiosus in der heutigen Begründung der Analysis" a vicious circle in Dedekind's argument was pointed out. But I could not find Weyl's original paper.

The main idea with the notion "standard sequence" in the present article appeared first in a paper in Chinese journal SHUXUE TONGBAO, No. 10 and No. 11, 1997, and then had been collected in my Chinese text book [5]. Essentially, my argument is based on Cantor's idea.

Most recent work on this topic may be found in Tao's book "Analysis I" [5]. The following sentence in Tao's book is worth to point out: "But to get the reals from the rationals is to pass from a 'discrete' system to a 'continuous' one, and requires the introduction of a somewhat different notion—that of a limit." Tao's argument on real numbers is also based on Cantor's idea, I think.

The purpose of the present article is to discuss rigorously the decimal representation of real numbers.

All students have well known the decimal representation of integers since they studied in elementary schools. In China, the concept of real numbers has been introduced to students since the second year of study in the junior high school [1]. The students are taught that

"Any periodical decimal fraction is a rational number; and any non-periodical decimal fraction is an irrational number. Both rational and irrational numbers are all called real numbers."

But since then there is no further explanation to this important concept, and the cognition to this concept of students has only been kept in the perceptual stage. This situation is kept until the students entering universities.

But even in the mathematical departments of some universities, most students still can not understand more then they have accepted in middle school. In traditional text books of university level, the real number is explained by the Dedekind "cut" which is apparently quite different then what the students have learned about from the middle school text book. By my opinion, Dedekind "cut" is not good for representing a real number, and is not suitable to teach students.

§ 2. Rational numbers and sequence of rational numbers

We declare that rational numbers are assumed to be understood well. This is our starting point. In this section we are concerned with only rational numbers.

Here, any integer is represented in the decimal system. A rational number is represented as a ratio of two integers m and n ($n \neq 0$). Denote by $\mathbf{N}^*$ the set of all positive integers, and by $\mathbf{Z}$ the set of all integers, by $\mathbf{Q}$ the set of all rational numbers. A rational number can also be called a fraction which is written as $\frac{m}{n}$ where $m \in \mathbf{Z}$ and $n \in \mathbf{N}^*$. This gives just the well known representation of rational numbers. In what follows we will call this representation of rational numbers **original rep-**

resentation.

There is a special form of original representation. When a rational number r has original representation $r=\frac{m}{n}$ with $m\in\mathbf{Z}$ and $n=10^k$ ($k\in\mathbf{N}^*$), then r can be written as a **finite decimal number** $p+0.a_1a_2\cdots a_k$ where $p\in\mathbf{Z}$, $a_j\in\{0, 1, 2, 3, 4, 5, 6, 7, 8, 9\}$, $j=1,2,\cdots,k$. So, any finite decimal number is also an original representation of a rational number.

Definition 1 (Sequence of rational numbers) If f is a mapping from the set of positive integers $\mathbf{N}^*$ to $\mathbf{Q}$, and the elements of the range of f are arranged naturally in the same order as the positive integers, i. e. the number $f(n)$ is regarded ahead of $f(n+1)$ for every $n\in\mathbf{N}^*$, then f is called a sequence of rational numbers and is written as $f=\{f(n)\}_{n=1}^{+\infty}$. Each $f(n)$ is called a term (the n-th term) of f. A sequens can also be written in an expansion form as

$$f(1),\ f(2),\ \cdots.$$

Definition 2 (Limit of a sequence of rational numbers) Let $f=\{f(n)\}_{n=1}^{+\infty}$ be a sequence of rational numbers. Assume $l\in\mathbf{Q}$. If for every $k\in\mathbf{N}^*$ there exists an integer $n_k\in\mathbf{N}^*$ (depending generally on k such that for all $n>n_k$),

$$|f(n)-l|<\frac{1}{k}$$

then the sequence f is said to be convergent to l, and we write

$$\lim_{n\to+\infty}f(n)=l.$$

Definition 3 (Fundamental sequence) Let $f=\{f(n)\}_{n=1}^{+\infty}$ be a sequence of rational numbers. If $\forall k\in\mathbf{N}^*$, $\exists n_k\in\mathbf{N}^*$, such that

$$|f(m)-f(n)|<\frac{1}{k},\qquad \forall m,\ n>n_k,$$

then the sequence f is said to be fundamental. A fundamental sequence is also called a Cauchy sequence.

Theorem 1 Any convergent sequence is fundamental.

We need the concept of subsequence. Let $f=\{f(n)\}_{n=1}^{+\infty}$ be a sequence of rational numbers. Its any infinitely many terms by keeping the original order in f consists a new sequence which is called a subsequence of f.

Definition 4 Let f and g be two sequences. If $\lim_{n\to+\infty}(f(n)-g(n))=0$ then f and g are said to be equivalent. We use the notation $f\sim g$ to denote that f is equivalent to g.

Theorem 2 Any subsequence of a fundamental sequence f is still fundamental and is equivalent to f.

Theorem 3 If a sequence f converges to limit $l\in\mathbf{Q}$, then any sequence equivalent to f converges to the same limit.

Theorem 4 If f is a fundamental sequence then its any subsequence is equivalent to f.

Theorem 5 Assume sequence f and sequence g converge to $a\in\mathbf{Q}$ and $b\in\mathbf{Q}$, respectively. Assume c, $d\in\mathbf{Q}$. Then

$$\lim_{n\to+\infty}(cf(n)+dg(n))=ca+db,\quad \lim_{n\to+\infty}(f(n)g(n))=ab.$$

Moreover, if $b\neq 0$ then

$$\lim_{n\to+\infty}\frac{f(n)}{g(n)}=\frac{a}{b}.$$

Here we note that when $b\neq 0$ then all $g(n)\neq 0$ except a finite number of terms, and when we are concerning with limit we may assume that all $g(n)\neq 0$, then for all $n\in\mathbf{N}^*$, $\frac{f(n)}{g(n)}\in\mathbf{Q}$.

The proofs of these theorems are all easy exercises of using the corresponding definitions.

At last we make the following definition. A sequence $\{f(n)\}_{n=1}^{+\infty}$ is said to be bounded from up if there is a number a such that $f(n)<a$ for all $n\in\mathbf{N}^*$; A sequence $\{f(n)\}_{n=1}^{+\infty}$ is said to be bounded from below if there is a number b such that $f(n)>b$ for all $n\in\mathbf{N}^*$; A sequence is said to be bounded if it is both bounded from up and below.

It is obvious that any fundamental sequence must be bounded.

§ 3. Definitions of decimal fractions and standard sequences

We now rewrite the definition of decimal fraction which appears in the middle school text book with a little unessential modification.

Definition 5(Decimal fraction) Let $k \in \mathbf{N}^*$, $a_k \in \{0, 1, \cdots, 9\}$, and for any $N \in \mathbf{N}^*$, there exists $k > N$ such that $a_k < 9$. Let $p \in \mathbf{Z}$. We call the notation

$$A := p + 0.a_1a_2\cdots \tag{3.1}$$

a decimal fraction (or just a **decimal**, simply). When $p=0$ in (3.1), A is written as $A=0.a_1a_2\cdots$ shortly. Every decimal fraction is regarded as **a representation of a real number.** In this sense each decimal is just called a **real number.** The set of all real numbers is denoted by **R.**

Remark 1 Before further discussion a decimal of form (3.1) is only a symbol. Although $0.a_1a_2\cdots$ was taught to students in school as "infinite decimal fraction", no more explanation was given in middle school. Note that (3.1) is currently a symbol. The symbol "+" inside (between p and $0.a_1a_2\cdots$) does not mean anything right now. Of course, later on when we establish an appropriate arithmetic construction on **R**, the symbol "+" will denote a real operation of addition.

Definition 6(Standard sequence) Let $A := p + 0.a_1a_2\cdots$ be a decimal. Define

$$A_n = p + 0.a_1a_2\cdots a_n = p + \sum_{k=1}^{n} \frac{a_k}{10^k}, \quad n \in \mathbf{N}^*.$$

The sequence of rational numbers

$$\overline{A} := \{A_n\}_{n=1}^{+\infty} \tag{3.2}$$

is called **the sequence equivalent to the decimal** A. Any sequence equivalent to a decimal is called a **standard sequence.**

Remark 2 Definition 6 is important. It gives an explanation for the "symbol" (3.1) by known concepts. Expression (3.2) is a sequence of rational numbers with original representation. We know the meaning of (3.2) exactly. Now we transfer (3.1) to (3.2). In fact we may regard (3.1) as (3.2). They give the same thing.

It is obvious that any standard sequence is fundamental.

Definition 7 (Periodical decimal) Let $m\in\mathbf{N}^*$, a_1, a_2, $\cdots$, a_m be numbers belong to $\{0, 1, 2, \cdots, 9\}$ and at least one of them be not 9. Suppose $p\in\mathbf{Z}$. We call

$$A := p+0.a_1a_2\cdots a_m a_1 a_2\cdots a_m\cdots \tag{3.3}$$

a **periodical decimal fraction** or periodical decimal, simply. The ordered digits $a_1a_2\cdots a_m$ is called a period of A. And we write A simply as

$$A=p+0.\dot{a}_1\dot{a}_2\cdots\dot{a}_m.$$

If b_1, b_2, $\cdots$, $b_\mu\in\{0, 1, 2, \cdots, 9\}$ $(\mu\in\mathbf{N}^*)$, then the decimal

$$B := p+0.b_1b_2\cdots b_\mu a_1a_2\cdots a_m a_1a_2\cdots a_m\cdots \tag{3.4}$$

is also called **periodical decimal fraction** (periodical decimal, simply) with period $a_1a_2\cdots a_m$, and is written as

$$B=p+0.b_1b_2\cdots b_\mu\dot{a}_1\dot{a}_2\cdots\dot{a}_m.$$

Remark 3 For a periodical decimal fraction with period 0 we has no reason to omit its period to write it in a "finite" form currently. Any finite decimal is the original representation of a rational number, but a periodical decimal fraction with period 0 is right now a new defined "symbol". Only when we prove that it represents the same rational number which the corresponding finite decimal represents originally, we can then make them equal. For example, only when we have proved that 0.000 ⋯ is exactly the decimal representation of the number 0 (or the number represented by 0 originally), we then write $0=0.000\cdots$.

Remark 4 By Definition 7, a periodical decimal fraction may be expressed in different notation with a different period. For example, the periodical decimal fraction

$$0.01010101\cdots$$

may be written as $0.\dot{0}\dot{1}$ and may also be written as $0.0\dot{1}\dot{0}$. Also, $0.\dot{2}$ and $0.\dot{2}\dot{2}$ denote same periodical decimal fraction. However, as a convention, the single digit 9 may never be a period.

§ 4. Decimal representation of rational numbers

We now give a clear explanation for the sentence that appeared in the school text book: "periodical decimal fraction is called rational number".

Theorem 5 Any standard sequence equivalent to a periodical decimal fraction converges to a rational number.

Proof Assume standard sequence (3.2) is equivalent to a periodical decimal fraction (3.3) with a period $a_1a_2\cdots a_m$. Let $B_k=A_{mk}$, $k\in\mathbf{N}^*$. We see that $\overline{B}:=\{B_k\}_{k=1}^{+\infty}$ is a subsequence of $\overline{A}$. And

$$B_k=p+\sum_{j=1}^{k}0.a_1a_2\cdots a_m\left(\frac{1}{10^m}\right)^{j-1}=p+0.a_1a_2\cdots a_m\sum_{j=0}^{k-1}\left(\frac{1}{10^m}\right)^j.$$

Using the formula $(1-q)\sum_{j=0}^{k-1}q^j=1-q^k$, we get (for $q=10^{-m}$)

$$B_k=p+0.a_1a_2\cdots a_m\frac{1-10^{-mk}}{1-10^{-m}}.$$

Hence we get

$$\lim_{k\to+\infty}B_k=p+\frac{0.a_1a_2\cdots a_m}{1-10^{-m}}=p+\frac{a_1a_2\cdots a_m}{10^m-1},$$

where $a_1a_2\cdots a_m$ denotes a decimal integer. Since $\overline{B}$ is a subsequence of $\overline{A}$ which is standard and hence fundamental, we conclude that they have the same limit. That is

$$\lim_{n\to+\infty}A_n=p+\frac{a_1a_2\cdots a_m}{10^m-1}.$$

By the same argument we may prove that any standard sequence (2) equivalent to a periodic decimal of form (4) converges to the limit

$$p+0.b_1b_2\cdots b_\mu+10^{-\mu}\frac{a_1a_2\cdots a_m}{10^m-1}\in\mathbf{Q}.$$

Theorem 6 Let $r=\frac{m}{n}\in\mathbf{Q}$, where $m\in\mathbf{Z}$, and $n\in\mathbf{N}^*$. Then there exists a periodical decimal fraction of form (3.3) or (3.4) such that the standard sequence equivalent to the decimal converges to r.

Proof Without loss of generality we may assume $1\leqslant m<n$. Then

we have

$10m=a_1n+r_1$, $a_1\in\{0, 1, 2, 3, 4, 5, 6, 7, 8, 9\}$, $r_1\in\mathbf{Z}$, $0\leqslant r_1<n$,

$10_{r_1}=a_2n+r_2$, $a_2\in\{0, 1, 2, 3, 4, 5, 6, 7, 8, 9\}$, $r_2\in\mathbf{Z}$, $0\leqslant r_2<n$

……

$10_{r_n}=a_{n+1}n+r_{n+1}$, $a_{n+1}\in\{0, 1, 2, 3, 4, 5, 6, 7, 8, 9\}$, $r_{n+1}\in\mathbf{Z}$, $0\leqslant r_{n+1}<n$,

Since r_1, r_2, $\cdots$, r_{n+1} are all integers taking values from the set $\{0, 1, \cdots, n-1\}$ which consists of n elements, there are at least two indices μ and ν, $1\leqslant\mu<\nu\leqslant n+1$, such that $r_\mu=r_\nu$. Then we get

$$m=0.a_1a_2\cdots a_\mu n+10^{-\mu}r_\mu;\qquad r_\mu=0.a_{\mu+1}a_{\mu+2}\cdots a_\nu n+10^{\mu-\nu}r_\nu.$$

Since $r_\mu=r_\nu$ we get from the second equality that

$$r_\mu=\frac{0.a_{\mu+1}a_{\mu+2}\cdots a_\nu}{1-10^{\mu-\nu}}n.$$

Hence we get

$$r=\frac{m}{n}=0.a_1a_2\cdots a_\mu+10^{-\mu}\frac{0.a_{\mu+1}a_{\mu+2}\cdots a_\nu}{1-10^{\mu-\nu}}.$$

Since $r_\mu<n$ we see that the digits $a_{\mu+1}$, $a_{\mu+2}$, $\cdots$, a_ν can not be all 9. Then from the proof of Theorem 5 we see that r is the limit of a standard sequence which is equivalent to the periodical decimal fraction

$$A:=0.a_1a_2\cdots a_\mu\dot{a}_{\mu+1}\dot{a}_{\mu+2}\cdots\dot{a}_\nu.$$

By Theorem 5 and Theorem 6 we conclude that for any rational number r there exists unique periodical decimal fraction A (note that we do not allow the single digit 9 to be a period) such that the standard sequence equivalent to A converges to r. Now we make the following definition to complete the representation by decimal fractions for rational numbers.

Definition 8 For any rational number $r\in\mathbf{Q}$ we define the periodical decimal fraction A to be its decimal representation if the standard sequence equivalent to A converges to r.

Now, each rational number has representations of two different kinds. By the original representation, any rational number r can be written as fraction of form $\frac{m}{n}$. This form is obviously not unique. By the representation of periodical decimal fractions A, r is just the limit of the standard sequence equivalent to A. The representation A is absolutely unique. From now on we write

$$\frac{m}{n}=A,$$

if $\frac{m}{n}$ is an original representation of the rational number $r\in\mathbf{Q}$ and A is the decimal representation of r.

Now every rational number obtains a new form of representation, called decimal representation. It is obvious that the arithmetic construction on $\mathbf{Q}$ which is built via the original representation is now naturally transferred to be built via the decimal representation. For example, given two periodical decimals $A=p+0.a_1a_2\cdots$ and $B=q+0.b_1b_2\cdots$, if the rational numbers represented by A and B have original representations s and t respectively, then we know their sum $r:=s+t$ by the addition between the original rational numbers very well. So, there is unique periodical decimal C which represents the number represented originally by r. Hence the sum of A and B must be the periodical decimal C. This shows that if $\overline{A}=\{A_n\}_{n=1}^{+\infty}$ and $\overline{B}=\{B_n\}_{n=1}^{+\infty}$ are the standard sequences equivalent to A and B respectively, then the fundamental sequence $\{A_n+B_n\}_{n=1}^{+\infty}$ is equivalent to the standard sequence which is equivalent to the periodical decimal C (with $C=r$ in the sense of representing same rational number).

So, our next task is to generalize the arithmetic construction on $\mathbf{Q}$ via the decimal representation obtained by the above way to the whole $\mathbf{R}$. Before the generalization we need to prove a crucial theorem in the next section. Here we remember that a rational number represented by a periodical decimal is just the limit of the standard sequence equivalent to this decimal.

§ 5. A crucial theorem

The following theorem has essential meaning in our arguments.

Theorem 7 Given any fundamental sequence (of rational numbers represented originally) there exists a unique standard sequence equivalent to it.

Proof Assume $f=\{f(n)\}_{n=1}^{+\infty}$ is a fundamental sequence, where each $f(n)\in\mathbf{Q}$ is originally represented. So, f is bounded. There is a positive integer M such that for all $n\in\mathbf{N}^*$, $|f(n)|<M$. Now, let $A^n=p_n+0.a_1^na_2^n\cdots$ be the decimal representation of $f(n)$, i. e.

$$f(n)=\lim_{k\to+\infty}A_k^n$$

where $A_k^n=p_n+0.a_1^na_2^n\cdots a_k^n$. We see that $p_n\leqslant f(n)<p_n+1$. Hence we get

$$|p_n|\leqslant M+1$$

for all $n\in\mathbf{N}^*$. So, the integers p_n can only take finite many (say, $2M+2$) different values. Hence we may find a subsequence f_1 from f such that the decimal representations of all terms $f_1(n)$ ($n\in\mathbf{N}^*$) of f_1 have same integer part, i. e. there is a common integer p such that

$$f_1(n)=p+0.b_1^nb_2^n\cdots$$

for all $n\in\mathbf{N}^*$, in decimal form. Next, since $b_1^n\in\{0, 1, 2, 3, 4, 5, 6, 7, 8, 9\}$ we may choose a subsequence f_2 from f_1 such that

$$f_2(n)=p+0.a_1c_2^nc_3^n\cdots$$

with a common $a_1\in\{0, 1, 2, 3, 4, 5, 6, 7, 8, 9\}$ for all $n\in\mathbf{N}^*$, in decimal form. Continuing this argument we get a sequence of sequences of rational numbers $\{f_k\}$ and a sequence of integer numbers p, a_1, a_2, $\cdots$ such that f_1 is a subsequence of f, f_{k+1} is a subsequence of f_k for all $k\in\mathbf{N}^*$; and every term of f_k has the decimal form

$$f_k(n)=p+0.a_1a_2\cdots a_kc_{k+1}^nc_{k+2}^n\cdots.$$

We know the meaning of the above equality is that the standard sequence equivalent to the decimal fraction $p+0.a_1a_2\cdots a_kc_{k+1}^nc_{k+2}^n\cdots$ converges to the rational number $f_k(n)$. So, if we write $A_k=p+0.a_1a_2\cdots a_k$

which is a rational number in original form, then we get

$$0 \leqslant f_k(n) - A_k \leqslant 10^{-k}, \quad n \in \mathbf{N}^*.$$

Since each $f_k(k)$ is a term of the sequence f and its position in f must be in front of $f_{k+1}(k+1)$ strictly, we see that $\{f_k(k)\}_{k=1}^{+\infty}$ is a subsequence of f. Since f is fundamental so $\{f_k(k)\}_{k=1}^{+\infty}$ is also. Then from the above inequality we see that the sequence $\{A_k\}_{k=1}^{+\infty}$ is equivalent to f.

If for any $n \in \mathbf{N}^*$ there exists $k > n$ such that $a_k < 9$, in other words, $0.a_1a_2\cdots$ with no period 9, then $p + 0.a_1a_2\cdots$ is a decimal fraction and $\{A_k\}_{k=1}^{+\infty}$ is just the standard sequence equivalent to this decimal fraction by Definition 6.

In the case when all a_k are equal to 9, then $\{A_k\}_{k=1}^{+\infty}$ is obviously equivalent to the sequence with constant term $p+1$ which is the standard sequence equivalent to the decimal $(p+1) + 0.00\cdots$ where $p+1$ is a real sum of the integers p and 1. Now assume $a_k < 9$ and all $a_n = 9$ when $n > k$. Then we may easily check that the standard sequence equivalent to the decimal $p + 0.a_1a_2\cdots(a_k+1)\dot{0}$ (here $a_k + 1$ is a real sum of the integer a_k and 1) is equivalent to $\{A_k\}_{k=1}^{+\infty}$ and hence equivalent to f. □

§ 6. Arithmetic construction on the whole R

Now we generalize the arithmetic construction on **Q** represented via decimals to the whole **R**.

Definition 9 (Positive, Negative numbers and Zero) Let A be a decimal fraction as in (3.1). If $p=0$ and $a_k = 0$, $\forall k \in \mathbf{N}^*$, then A is called zero and write $A=0$, if p is a positive integer or $p=0$ but there is $k \in \mathbf{N}^*$ such that $a_k \neq 0$ then A is said to be positive and write $A>0$; if p is a negative integer then A is called to be negative and write $A<0$. The notation $A \neq 0$ means $A>0$ or $A<0$.

We have already proved that any periodical decimal represents a rational number in the sense that it is just the limit of the standard sequence equivalent to itself. Hence Definition 10 consistent with the corresponding known definitions for rational numbers.

Theorem 8 Let both $\overline{A} := \{A_n\}_{n=1}^{+\infty}$, $\overline{B} := \{B_n\}_{n=1}^{+\infty}$ be standard, A and B be the decimal fractions (which are called real numbers by Definition 5) equivalent to $\overline{A}$ and $\overline{B}$, respectively. Then

(1) $\{A_n+B_n\}_{n=1}^{+\infty}$, $\{A_nB_n\}_{n=1}^{+\infty}$ and $\{-A_n\}_{n=1}^{+\infty}$ are all fundamental.

(2) If $A\neq 0$ then there exists $N\in\mathbf{N}^*$ such that for $A_n\neq 0$ for all $n>N$, and $\left\{\frac{1}{A_{n+N}}\right\}_{n=1}^{+\infty}$ is also fundamental.

Proof of this theorem is straightforward and we omit it.

Definition 11(Arithmetic operations for real numbers) Let $\overline{A} := \{A_n\}_{n=1}^{+\infty}$ and $\overline{B} := \{B_n\}_{n=1}^{+\infty}$ be standard sequences equivalent to real numbers A and B, respectively. Write the standard sequence equivalent to $\{\overline{A}_n+\overline{B}_n\}_{n=1}^{+\infty}$ as $\overline{A+B}$; and the standard sequence equivalent to $\{\overline{A}_n\overline{B}_n\}_{n=1}^{+\infty}$ as $\overline{AB}$. Define the sum of real numbers A and B to be the number equivalent to the standard sequence $\overline{A+B}$ and write it as $A+B$; and define the product of A and B to be the number equivalent to the standard sequence $\overline{AB}$, and write it as AB. Write $-1\cdot A$ as $-A$, and define $A-B$ to be $A+(-B)$, called the difference of A and B.

Moreover, if there exists $N\in\mathbf{N}^*$ such that for all $n>N$, $f(n)\neq 0$, then we write the standard sequence equivalent to $\left\{\frac{1}{A_{n+N}}\right\}_{n=1}^{+\infty}$ as $\overline{A^{-1}}$. Define the reciprocal of the number A to be the real number equivalent to the standard sequence $\overline{A^{-1}}$ and write it as A^{-1}.

Remark 5 In the definition of A^{-1}, the value of the integer N is not essential, i. e. any value which makes $A_k\neq 0$ for all $k>N$ can be taken and yields the same result.

It is easy to check that the above operations defined on **R** satisfy the all familiar laws satisfied by the same known operations defined on **Q**. These are the commutative law for addition and multiplication, associative law for addition and multiplication, distributive law of the multiplication with respect to the addition, and so on.

Definition 12(The order in R) If real numbers A, B are such that $A-B>0$, then we say A is bigger than B and write $A>B$, or we say B

is smaller than A and write $B<A$. Denote by "$\leqslant$" "not bigger", by "$\geqslant$" "not smaller". Furthermore, define the absolute value of a real number A to be itself when it is nonnegative, to be $-A$ when it is negative, and write the absolute value of A as $|A|$.

Now we have defined the concepts of zero, positivity and negativity and absolute value for all real numbers (or their decimal representations, precisely), have established an arithmetic construction on whole **R** including operations of addition, substitution, multiplication, division, and including the order relation between real numbers. All these are consistent with known concepts for rational numbers. And by using the subtraction and the absolute value we can define the **distance** between real numbers (decimals) A and B as $d(A, B):=|A-B|$. It is easy to check the following properties of the distance:

① $\forall A, B\in\mathbf{R}$, $d(A, B)=d(B, A)\geqslant 0$, $d(A, B)=0$ if and only if $A=B$.

② $\forall A, B, C\in\mathbf{R}$, $d(A, B)\leqslant d(A, C)+d(B, C)$ (triangle inequality).

And for rational numbers the new defined distance is consistent with the known distance defined via original representation.

Here we note that for two rational numbers we can calculate the distance between them by their original representation and also can by their decimal representation. Both ways lead to the same result. But when we treat with a decimal (real number) and a fraction (originally represented rational number), we must use the decimal representation of them. For example, when we calculate the distance between a decimal $A=p+0.a_1a_2\cdots$ and $A_n=p+0.a_1a_2\cdots a_n$—the originally represented rational number which is the n-th term of the standard sequence equivalent to A we must use the decimal representation of $A_n=p+0.a_1a_2\cdots a_n\dot{0}$. Then we get

$$d(A, A_n)=0.00\cdots 0a_{n+1}a_{n+2}\cdots<10^{-n}.$$

Here the number on the most right side is written back in original form again.

Now we may generalize the definition of the limit (Definition 2) from the case of sequences of rational numbers to the case of real numbers. So the whole discussion on **Q** in section 2 applies completely the real numbers. All definitions (Def. 1～5) can be restated for real numbers without other change. As an example, we may state the definition of convergence of a sequence of real numbers as follows.

Let $\{A_n\}_{n=1}^{+\infty}$ be a sequence of real numbers, A be a real number. If $\forall k \in \mathbf{N}^*$, $\exists N \in \mathbf{N}^*$, such that $|A_n - A| < \frac{1}{k}$, for all $n > N$ ($n \in \mathbf{N}^*$), then the sequence $\{A_n\}_{n=1}^{+\infty}$ is said to converge to A. This fact is written as

$$\lim_{n \to +\infty} A_n = A.$$

Also, Theorem 1～4 are all applied to **R**.

§ 7. Conclusions

The following theorem shows the meaning of the representation for real numbers.

Theorem 9 If $\overline{A} = \{A_n\}_{n=1}^{+\infty}$ is a standard sequence then the real number A equivalent to $\overline{A}$ is the limit of $\overline{A}$, i. e.

$$A = \lim_{n \to +\infty} A_n.$$

This is a direct result of the inequality $d(A, A_n) \leqslant 10^{-1}$.

Theorem 10 **R** is complete. In other words, any fundamental sequence of real numbers must converge to a real number.

Proof Let $f = \{f(n)\}_{n=1}^{+\infty}$ be a fundamental sequence of real numbers represented as decimals. Let $f(n) = p_n + 0.a_1^n a_2^n \cdots$ in decimal form. Define $g(n) = p_n + 0.a_1^n a_2^n \cdots a_n^n$ which is a rational number in the original form. Then we get a sequence of rational numbers $g = \{g(n)\}_{n=1}^{+\infty}$. Since $0 \leqslant f(n) - g(n) < 10^{-n}$, the sequence g is equivalent to f and hence is also fundamental.

By the crucial theorem (Theorem 7) there is a (unique) standard sequence $\overline{A} := \{A_n\}_{n=1}^{+\infty}$ which is equivalent to g and hence to f. So, by Theorem 9, the real number A which is equivalent to the standard sequence $\overline{A}$ is the limit of $\overline{A}$ and hence is the limit of the fundamental sequence f.

References

[1] Chinese text book "Algebra" Vol. Ⅱ for the junior high school. People's Education Press, Beijing, 1993.

[2] Morris Kleine. Mathematical thought from ancient to modern times. Oxford Univ. Press, New York, 1972.

[3] Richard Dedekind. Stetigkeit und irrationale Zahlen, 1872 (English translation: Essays on the Theory of Numbers, Dover (reprint), 1963).

[4] Boyer C B. The History of the Calculus and Its Conceptual Development. Chinese translation by Tang Sheng, Fudan University Press, 2007.

[5] Wang Kunyang. Concise Mathematical Analysis (Chinese). High Education Press, Beijing, 2001.

[6] Terence Tao. Analysis I. Hindustan Book Agency (India), 2006.

实数的十进表示

摘要 严格地研究了用十进数表示实数的问题. 基本有理数本原地表示为分数的知识，引入了标准列的概念，并用以证明每个用十进数表示的实数恰是与此十进数等价的标准列的极限. 然后证明了实数集 **R** 的完备性.

关键词 有理数；十进数；实数的十进表示；基本列；标准列.

五、纪念导师

V.

In Memory of My Supervisor

钱伟长，总主编．王元，主编．20世纪中国知名科学家学术成绩概览，数学卷，第三分册．北京：科学出版社，2012：59—66.

孙永生

Sun Yongsheng

孙永生(1929-01-22—2006-03-22)，河北沧县人．实函数论专家．1952年毕业于北京师范大学．1954年赴苏联学习，1958年毕业于莫斯科大学，获物理-数学副博士学位．其后一直在北京师范大学任教．1965～1982年担任数学系副主任，1983～1993年担任该系学术委员会主任．1990～1994年任国家教委首届高等学校数学力学教学指导委员会委员．曾任科学出版社《现代数学丛书》和北京师范大学《现代数学丛书》编辑委员，国际学术期刊《Approximation Theory and Its Applications》《East Journal on Approximations》编辑委员，以及《数学季刊》《数学研究》的编辑委员．在北京师范大学数学系历任副教授、教授，1981年由国务院学位委员会批准为北京师范大学首批博士生导师．被宁夏大学、河南大学、河北师范大学聘为兼职教授.

著有《函数逼近论》上、下册(下册与房艮孙合著)；撰写学术论文80余篇，大多收入《孙永生文集》；翻译《逼近论的极值问题》《正交级数》(与王昆扬合译)，编写教材《泛函分析》及其他著作等共7种.

§1. 自幼勤奋、好学

孙永生的祖父是中医，家里有田产，在当地是富裕户．伯父孙震涛在20世纪20年代毕业于北京师范大学化学系，毕业后做中学教师．父亲孙震营读过几年小学，在农村生活．孙永生七八岁时在村里念过一年小学，一年私塾．1937年七七事变后随父母离乡．1940～1941年在沧县县城住了一年，毕业于县立牛市街小学，后考入沧县师范初中部. 1941 年以后父母又搬回乡下，他在校住宿，生活很艰苦，伙食极差，经常吃不饱. 孙永生埋头读书. 那时他对数学产生了浓厚兴趣，对课堂上讲的内容不满足，就利用课余时间自学. 三年内不但超前学完了全部初中数学，而且自学了部分高中数学科目，读了金品、叔桂超合编的《几何学》，严济慈的《几何证题法》，做了大量几何难题. 1944 年考入北平市立高中. 该校是原先的河北省立第 17 中学，基础扎实. 孙永生利用该校较好的师资和设备条件，更加埋头念书. 他的学业在班上经常名列前茅，对数学则情有独钟. 高一时热衷于初等平面几何，自学过王邦珍、R. Johnson 等人的初等近世几何的著作，他自学过林鹤一、Dickson 的初等方程式论的书. 他在高中不但养成了自学习惯，而且也形成了自己的一套自学方法. 高中毕业后，他被学校保送进了北平师范学院数学系. 大一的课程除了微积分、立体解析几何、初等方程式论等数学科目，还有大量非专业课程，周学时高达 40. 他对那些枯燥无味的课程一律采取应付态度，集中精力自学了竹内端三的《极限论》和 Dedekind 的《Irrational Number and Continuity》. 孙永生埋头读书，从不参加社团活动. 但是他有正义感，对国民党政权的腐朽深为不满，对其疯狂压制学生的爱国民主运动的种种暴行深恶痛绝. 大学读书环境的日趋恶劣使他深感不安. 到了 1947 年年底，他对所处的环境越来越感到厌恶. 经过和高中时期的同学王鸿书、许忠(时在唐山工学院)反复密商，终于下决心于 1948 年 3 月 23 日结伴离校奔赴华北解放区. 是年 4 月到达正定市，进了华北大学第二部政治班第 12 班学习，接受初步的革命人生观教育. 1948 年 8 月政治班结业后，他和十几位同班学友被分配到设在井陉的华北大学工学院学习业务. 华北大学工学院是华北解放区培养工程技术人员的最高学府. 工学院领导热情欢迎这十几位新学员，特别批准孙永生、李讲、吴沙凡三人学习数学. 为此，成立了一个数学自修组，自学微积分和微分方程. 到 1948 年 12 月中旬，华北人民政府从华北大学工学院抽调十名学员赴平津前线，准备于平津解放后参加接管国民党的工矿企业. 孙永生参加了这项工作. 他随大队于 12 月底到达北平市郊，俟北平解放后参加了接管华北钢铁公司的工作. 接管工作结束后，于 1949 年 5 月又回到井陉. 1949 年 8 月，华北大学工学院迁到北平，孙永生返回北京师范大学数学系，从一年级重读. 中华人民共和国成立后的北京师范大学数学系由傅种孙任系主任. 孙永生仅仅读了三年，在 1952 年 8 月提前一年毕业. 在这三年里政治运动很频繁，孙永生是青年团干部，社会工作重，课外活动很多. 他不满足于课堂上安排的那几门基础课，很想利用课余时间多念点书，但苦于时间太少，仅仅自学了熊庆来的《高等算学分析》和哈代的《Pure Mathematics》等分析类的著作. 到毕业时，他对微分方程、概率、拓扑学、泛函分析、函数论几乎一无所知. 好在他从中学时期就养成了自学习惯，形成了自己的一套自修方法，养成了锲而不舍、知难而进的勤奋好学的精神.

毕业后，孙永生被分配到北京俄文专修学校(以下简称俄专)留苏预备部做了一个学期的文化教员. 1953 年 1 月孙永生被调回北京师范大学数学系，协助张禾瑞举办“师范师专算术代数课讲习班”. 孙永生突击翻译了苏联教材《算术》《代数和初等函数》作为讲习班教材. 借此机会他自学了张禾瑞的《近世代数基础》和施伯纳(樊𡒄译)的《解析几何与代数》. 1953 年 7 月系里推荐孙永生和袁兆鼎参加留苏预备考试. 他们被录取，遂于 1953 年 10 月到俄专留苏预备部. 经过 10 个月的准备，1954 年 9 月踏上了赴莫斯科的旅途.

§2. 在莫斯科大学的研究成果

1954 年 9 月，孙永生被分配到莫斯科大学力学数学系，他的指导老师是当时的青年专家 Stechkin 教授. Stechkin 为他选定了专业方向：实函数逼近论，制订了一个一年半的学位课程计划. 首先要补上常微分方程论和数学物理方程两门大学课程. 学位课程则包括函数论(实变与复变)、泛函分析、三角级数论、函数逼近论以及发散级数求和理论.

1955 年秋季，孙永生参加了 Steklov 研究所为硕士生组织的“拓扑学和泛函分析”讨论班. 他在莫斯科大学选听了著名学者 Menshov 教授主讲的发散级数求和和 Bari 教授主讲的三角级数论，经常参加由 Menshov，Bari 主持的函数论讨论班. 这个讨论班在莫斯科大学力学数学系举行，每周一次，吸引了莫斯科市的许多函数论方向的专家学者. 这些课程和学术活动，不但引导他进入了现代函数论的重要领域，而且莫斯科大学浓厚的学术氛围，俄罗斯分析分派的严谨、创新、精细的学术风格，Bari，Menshov 等一批当代数学大师的学者风范和人格魅力，使孙永生为之倾倒，其印象终生难忘.

1956 年 1 月我国学者曾远荣、徐利治两位教授去莫斯科大学，参加在那里举行的国际泛函分析讨论会. 孙永生在会上结识了徐利治，帮助他和 Stechkin 建立了联系. 是年 6 月我国派了一个庞大的代表团赴莫斯科参加全苏第三届数学大会，会场就设在莫斯科大学力学数学系. 孙永生等也参加了大会. 他在这里结识了陈建功、程民德两位教授，帮助他们和苏联同行建立联系.

1956 年上半年，Stechkin 在讨论班上提出了几个问题，鼓励孙永生去作. Stechkin 的指导相当具体，不但提出问题、介绍问题的背景和有关文献，而且阐述自己对解决问题的一些思路，并预测可能的结果. 这对刚想跨进科研门槛的研究生非常宝贵，可以引导他们较快地进入工作状态. 不过，孙永生在开始阶段工作并不顺利. 到 1956 年底，他证得了一些结果，但主要是模仿、推广，没有体现出创造性. 孙永生自己对此颇为苦恼. 1957 年年初，他读了 V. Dzjadyk 1953 年发表的一个工作，解决了当 $0<r<1$ 时周期的 Weyl 可微类 $\widetilde{W}^r_\infty$ 利用 n 阶三角多项式的最佳逼近常数的精确计算问题. 这个结果为完全解决法国数学家 J. Favard 在 1936 年提出的 Weyl 可微类及其共轭类的最佳逼近常数精确估计问题迈出了重要一步，因而在当时苏联函数逼近论学者中间颇受重视. 但是，到 1957 年，对于 $\widetilde{W}^r_\infty$ 的共轭类，以及 $r>1$ 的情形，问题都有待解决. 由于 $r>0$ 时，若 r 不是整数，函数类及其共轭类的卷积核没有了对称性，因此对它验证 Markov 准则变得十分困难. 孙永生本打算解决 $r>1$ 的情形，但是花费了半年时间，克服了很多困难，仅得到了部分结果($r>6$). 但 Stechkin 对这项结果很满意. 孙永生的副博士学位论文经 Nikolskii，Dzjadyk 两位教授的审查，颇受赞赏，于 1958 年 2 月顺利通过了答辩，得到了物理-数学副博士学位. 他这一时期的主要研究结果分别发表在苏联的重要杂志 Dokl. Akad. Nauk USSR，1958，118(2)：247－250，Izv. Akad. Nauk USSR Math. Series，1959，23(1)：67－92 以及 Izv. Akad. Nauk USSR Math. Series，1961，25：143－152 上.

§3. 在北京师范大学的研究成果及对学生的悉心栽培

1958 年 3 月孙永生从莫斯科回到北京师范大学. 他希望尽快地组织起实变函数讨论班，把研究工作继续下去. 但由于政治运动等，直到 1962 年年初实变函数讨论班才运转起来，当时，总共有三个成员(包括陆善镇和张阳春). 他们选择了三角 Fourier 分析和实变逼近论作为研究方向. 孙永生继续研究 Weyl 类的三角逼近问题，陆善镇则专攻多元 Fourier 分析. 这个讨论班存在了三年半，到 1964 年 10 月因社会主义教育运动(即"四清运动")而停顿. 经过"文化大革命"，直到 1978 年年底讨论班才恢复. 到现在，这个当年仅有三个成员的讨论班，已经发展成拥有 7 位教授的具有较高素质的研究梯队，在三个研究方向上卓有成效地开展着工作. 陆善镇、丁勇、杨大春三位教授主持着 Hardy-Herz 空间和 Calderon-Zygmnud 算子论方向；孙永生、房艮孙、刘永平三位教授主持着逼近论的极值问题和信息复杂性理论方向；王昆扬、刘永平两位教授主持着多元逼近、球面调和分析和特殊函数的交叉问题方向. 这些方向和当前国际上的主流学派息息相通，密切接轨. 经过了 20 年的工作，有了较深厚的学术积累，做出了在国内外有影响的工作.

1958 年以来，孙永生在函数逼近论方向发表学术论文(包括与他人合作的)达 80 多篇，出版专著一部，论文集一部，其他著作(教科书，译著)七种. 下面扼要介绍孙永生以及孙永生与他的学生们合作的主要工作.

3.1 切彼晓夫最佳逼近

1958～1962 年，孙永生完全解决了在 Weyl 意义下的可微周期函数类(即是具有非整数阶有界导数的周期函数)及其共轭类借助于次数不超过 n 的三角多项式子空间的最佳一致逼近的逼近常数的精确估计问题. 微分指标 $r>1$ 时得到了基本函数类及其共轭类上的精确解，而 $\frac{1}{2}\leqslant r<1$ 时得到了共轭类上的精确解. 这一工作是和苏联数学家 V. Dzjadyk 的工作互相独立进行的，二者的方法不同，而且孙永生对 $\frac{1}{2}\leqslant r<1$ 时共轭情形的结果比 Dzjadyk 的工作早 12 年发表. 这项结果得到了国际同行

的高度评价. 苏联函数逼近论著名学者 N. Korneichuk 院士于 1974 年在苏联科学院 Steklov 数学研究所的一次学术报告中称赞这一结果"是对逼近论的极值理论的重要贡献". A. Timan、N. Korneichuk 以及 V. Tikhomirov 等国际著名的逼近论学者在他们关于函数逼近论的著作中多次引述这一结果. 这项结果曾用三种文字(中文、俄文、英文)发表. 中文稿《用三角多项式近迫周期可微函数》, 发表在《数学学报》(1962, 12(2): 181—202); 英文稿发表在《Chinese Mathematica Acta》(1963, 3: 196—217), 俄文稿发表在《Scientia Sinica》(1962, 6(11): 1 455—1 474). 还发表了[1].

3.2 周期函数类上线性算子的逼近

1979～1980 年, 孙永生研究正阶 Cesàro 平均对周期连续函数和周期可积函数分别在一致尺度及 L_1 尺度下的逼近误差[2], 得到了该误差的 Jackson 不等式的最佳常数的强渐近估计, 证明了在两种尺度下的 Jackson 逼近常数具有相同的渐近展开的主项.

在这期间, 孙永生还研究了具有正核的多项式型周期卷积算子的逼近性能, 得到了很广泛的结果. 他引入了一类称为弱正核的卷积算子以代替正的多项式核, 并且对弱正型算子在周期的 Sobolev 类上的整体逼近建立了广泛的结果[3]. 这项成果包含了从 20 世纪 40 年代以来苏联学派的逼近论学者 S. Nikolskii, S. Stechkin 等人在这一方向上给出的一些具体结果, 而且本质地改进了 Motornyi 发表在 1974 年的工作. 至今这仍是关于该问题的最好结果.

3.3 Landau 不等式

S. Stechkin 在 20 世纪 70 年代初发现经典的 Landau 不等式的精确常数问题和逼近论的几个重要极值问题的精确解有深刻联系, 引起了人们对该问题的兴趣. 孙永生在 1984 年把 Landau-Kolmogorov 不等式拓广到由任意常系数的线性微分算子确定的 Sobolev 类上[6], 他的结果包含了 20 世纪 70 年代以来国际上发表的几个著名结果(美国的 S. Karlin, 1976; 加拿大的 Sharma 和 Tzimbalario, 1977; 荷兰的 Ter-morsche, 1985). 孙永生利用该结果完整地解决了微分算子 D 在由一个仅仅具有实特征根的线性微分算子确定的 Sobolev 类上的最优恢复问题. 该问题是美国数学家 C. Micchelli 在 1977 年提出, 他只得到了部分解.

3.4 宽度

逼近论中宽度问题的研究肇端于 A. Kolmogorov 发表于 1936 年的开创性工作. 但是在这以后直到 20 世纪 50 年代末，这一问题的研究基本上处于停滞状态. 20 世纪 60～70 年代经过 V. Tikhomirov，K. Babenko，N. Korneichuk，C. Micchelli，A. Pinkus 等国际上著名逼近论学者的大力工作而成了逼近论发展的一个热点，以其问题的艰深，和现代数学的许多分支的广泛联系而吸引了国际上许多数学家的注意. 孙永生在 1980～1982 年注意到这一方向，奋力拼搏，阅读、整理了大量文献，吸收了该方向的最新成果，接近了国际前沿. 1983 年，他发表了自共轭常系数线性微分算子的 Green 函数关于指标(∞, p)的 Kolmogorov n 宽度的精确估计及极子空间构造的结果([4])，这个结果回答了 C. Micchelli 和 A. Melkman 在 1977 年提出的关于全正核宽度的一个猜想. 在 1983～1985 年孙永生和黄达人合作继续这一方向的研究，在由常系数线性微分算子确定的周期 Sobolev 类上的宽度和单边宽度的精确估计问题的研究中得到了系统的结果. 其中一部分主要结果发表在[5]、《科学通报》等期刊上[7]，并且曾在 1983 年美国 A & M 大学举行的国际第四届逼近论大会上，以及 1985 年在杭州举行的中美逼近论联合讨论会上做过报告.

3.5 平均宽度和最优插值

从 1985 年起，孙永生开始研究定义在全实轴上的光滑函数类的最优恢复问题. 他首先提出了利用基数是$\aleph_0$ 的线性信息族对上述函数类的最优插值的研究. 这一课题把最优恢复问题从利用有限基数的线性信息对以紧集为定义域的，紧致的函数集为恢复对象的情况实质性地推进了一步. 1985～1989 年，他连续发表了《On optimal interpolation for a differentiable function class (I)》[8]、《可微函数类的最优恢复》[9]，以及《一个卷积类上的最优插值问题》[10]等三篇论文，把对一致范数情形的研究逐步展开并引向深入；稍后和李淳合作研究 L_2 范数情形，完全解决了 Sobolev 类 $W_2^r(\mathbf{R})$在 $L_2(\mathbf{R})$内借助于可数个线性信息的最优恢复问题[11][12]. 1989 年 9 月，苏联著名的函数逼近论学者 V. Tikhomirov 和 N. Kormeichuk 来我国访问. Tikhomirov 介绍了他对 $\mathbf{R}$ 上的 Sobolev 类的宽度问题的一些思考，特别介绍了他基于 Shannon-

Kolmogorov 熵的思想提出的关于无限维线性子空间的平均维数的概念. 他以此为基础建立了全实轴上的 Sobolev 类的平均宽度概念. Tikhomirov 了解到孙永生的工作和他的想法关系十分密切，对孙的工作极感兴趣. 通过和俄罗斯学派的交流，孙永生把自己的想法和 Tikhomirov 的想法结合起来，带领博士生展开了深入系统的工作. 1989～1993 年，在他和黄达人、李淳合作已经得到的一些基本结果的基础上，他的博士生陈迪荣、刘永平、房艮孙等得到了一系列进一步的深刻结果，不但完全解决了 $\mathbf{R}$ 上 Sobolev 类 $W_2^r(\mathbf{R})$ 在 $L_p(\mathbf{R})$ 内的平均 Kolmogorov 宽度和最优插值的精确解，而且更进一步地引入了一个新的函数类：Sobolev-Wiener 类 $W_{pq}^r(\mathbf{R})$，在 $p \geqslant q \geqslant 1$ 条件下圆满解决了 $W_{pq}^r(\mathbf{R})$ 在 $L_q(\mathbf{R})$ 内的平均宽度和最优插值问题，从而建立起了在 $\mathbf{R}$ 上和 T 上完全类似的结果. 这一轮工作成果十分丰硕，和俄罗斯 Tikhomirov 学派的工作并驾齐驱，互有交叉，其中对 Sobolev-Wiener 类的研究，以及对全实轴上以 PF 密度为核的卷积类的研究，是孙永生和他的学生们的独特的工作. 此外，陈迪荣在其博士论文中部分地证实了 Pinkus 关于 B 核宽度的第二猜想. 至今，这仍是在该问题研究中的最好结果.

孙永生和他的学生们的这一轮工作(见[13]～[20])得到了国际、国内同行的重视和好评. V. Tikhomirov 教授于 1991 年致函孙永生，信中写道："您和我以及我们的学生们几乎同时提出了旨在实现全实轴上逼近的广泛计划，这给我留下了非常深刻的印象."他在 1995 年俄刊《Mathematical Survey》上发表文章，其中提到"孙永生的学术集体独立于我们，提出了与我们的想法相近的思想. 他们是通过研究全实轴上的最优插值问题实现其想法的."德国 Erlangen 大学 H. Berens 教授在评价陈迪荣的博士论文时写道："在逼近论中宽度问题被认为是难度最大的一个领域……在中国，孙永生教授是这一方向研究中起带头作用的分析学家，他和他的学生们作了出色的工作，特别是在宽度和最优恢复的结合方面."

徐利治教授指出(《数学研究与评论》，1991，11(4))："以孙永生为代表的北京师范大学逼近论研究集体，多年来已经在中国结出丰硕成果，并引起国际同行的注意……孙永生、黄达人与房艮孙所建立的某些周期卷积类上宽度估计的精确结果，大大拓广了苏联学派和 Pinkus 等外国学者的工作，确实具有深刻的理论意义和价值.”

3.6 退休之前，孙永生把研究的注意力转到多元问题，特别关注逼近论与数值分析和连续问题的信息复杂性问题的交叉，他和他的学生汪成咏、汪和平、蒋艳杰、罗俊波等人在多元最优求积公式、线性算子方程的近似解法及其计算复杂性、在带概率测度的 Banach 空间内的宽度和随机逼近等广泛的方向上取得了一批新的结果.

孙永生于 1999 年 4 月退休. 当时他说：“我就要成为科教战线上一个退役的老兵了，但我并不感到孤独，因为有我们的学术集体在.”事实上，他一如既往地关心学术研究和研究生培养，继续参加讨论班活动，及中俄学术交流活动. 特别有两件值得提一提. 一是他在身患肺癌的情况下，仍与他的学生合作翻译了俄罗斯 Kashin 与 Saakian 的专著《正交函数》，遗憾的是他没有来得及看到此书的出版. 另一件事就是他协助王昆扬对于博士生戴峰关于球面上的宽度研究给予了非常重要和具体的指导，这才使戴峰的论文被评为 2004 年全国优秀博士论文.

孙永生在逼近论方向的研究成果和陆善镇在实 Hardy 空间方向的研究成果曾获得 1988 年国家教委科技进步奖一等奖和 1989 年国家第四届自然科学奖四等奖. 他的专著《函数逼近论》(上、下册)为我国函数逼近论专业方向研究生广泛使用，在 1992 年获国家教委颁发的高校出版社优秀学术著作特等奖. 他共培养了 14 名博士，18 名硕士，另外还有 34 名访问学者和进修教师. 这当中已有 9 人晋升为教授，4 人被批准为博士生导师. 他和陆善镇获得了 1993 年北京市高等学校优秀教学成果二等奖.

§4. 崇高的人格

孙永生一生孜孜不倦地致力于数学学术研究，切实地做到了学为人师. 同时孙永生教授，热爱教育事业，热爱学生，淡泊名利，以自己的行为给学生做榜样，也做到了行为世范.

北京师范大学校领导曾经想推荐他申报中国科学院院士，可是他说：“我已疾病缠身，还要这个名誉做什么?”接着又说“院士是一个岗位，不单纯是名誉.”

孙永生曾从 1965 年到 1982 年担任数学系副主任的工作，他的一位同事回忆，改革开放之初，孙永生主持数学系的科研工作，当时有两个机会，一个是教育部给北京师范大学数学系一个去美国作学术访问的教授名额，另一个是给北京师范大学数学系一个国务院数学学科评审组的委员的名额．这两个机会被戏称为“甜活”，却被孙永生很自然地安排给了系里的其他教授．

孙永生在担任系领导工作近 20 年之后，为集中精力做好数学系的学术建设，提出辞去副系主任职务．他在辞职申请中写道：“匹夫虽愚，志不可夺．”他还说过“有所为，有所不为”的话．

孙永生曾多年担任本科生基础课的教学工作．曾经为 1977 级、1978 级本科生和 1982 级本科生讲授泛函分析．并编写了《泛函分析讲义》．还曾为 1984 级本科生讲授数学分析，为 1985 级本科生讲授实变函数论．孙永生讲课总是准备得十分充分，把深刻的内容，讲得生动、容易．他每次上课前要提前半小时来到数学楼前的小树林中，把要讲的内容梳理一遍，上课时是不看讲稿的．有的学生说：“儒雅的先生、精彩的讲授使艰深的数学变得优美而引人入胜，原来抽象得近乎神学的数学变得像诗歌音乐般美妙．”还有的学生说：“一个在通常教科书里非常烦琐的证明，在先生的讲解中，显得那么简单而生动．”

在没有课时津贴的 1984～1985 年，孙永生的教学工作量超额 983 学时，居全系之首．孙永生常为研究生翻译俄文文献，仅 Tikhomirov 的一篇《谐波和样条：做为逼近和恢复的最优工具》，他给学生翻译的手稿就多达 70 页．

要做到像孙永生那样，没有渊博的知识不行，没有对于学生的挚爱及崇高的奉献精神也是绝对做不到的．

1987 年，由于孙永生与其他老一辈学者多年的辛勤努力，北京师范大学数学系基础数学被评为国家重点学科．可惜在 21 世纪初的评选中，这个重点学科丢失了．2006 年 1 月，在孙永生临终前，他对看望他的同事、朋友及院领导说的话是“在我即将离开这个世界的时候……我有一个愿望，北京师范大学要把基础数学重点学科拿回来．”可以告慰孙先生的是，2007 年下半年，他的这个遗愿实现了．

§5. 孙永生的主要论文

[1] 孙永生. 周旋函数的最佳一致近迫和最佳线性近迫. 数学学报，1962，12(3)：301—319.

[2] 孙永生. 关于 Cesàro 算子的逼近常数. 数学学报，1981，24(4)：516—537.

[3] 孙永生. 关于周期函数用线性算子的平均逼近. 数学学报，1982，25(3)：561—577.

[4] 孙永生. 一个广义样条函数类上的极值问题和有关的宽度问题. 中国科学，1983，12A(8)：677—688.

[5] Sun Yongsheng，Huang Daren. On one-sided approximation of class of smooth functions $\Omega_p^{2r+\sigma}$. ATA，1984，1(1)：19—35.

[6] 孙永生. 线性微分算子的 Landau-Kolmogorov 型不等式. 科学通报，1985，30(7)：481—484.

[7] 孙永生，黄达人. 广义 Bernoulli 核的宽度. 科学通报，1985，30(10)：728—731.

[8] Sun Yongsheng. On optimal interpolation for a differentiable function class (I). ATA，1986，2(4)：49—54.

[9] 孙永生. 可微函数类的最优恢复. 科学通报，1986，30(11)：809—812.

[10] 孙永生. 一个卷积类上的最优插值问题. 科学通报，1988，33(13)：964—967.

[11] Sun Yongsheng，Li Chun. Best approximation of differentiable functions by cardinal-splines of higher order. (Russian) Matem Zametki，1990，48(4)：100—109. (Translation in Math. Notes，1990，48(3—4)：1 038—1 044.)

[12] Sun Yongsheng，Li Chun. Optimal recovery for $W_2^r(\mathbf{R})$ in $L_2(\mathbf{R})$. Acta Math. Sinica (New series)，1991，7(4)：309—323.

[13] Chen Dirong，Sun Yongsheng. Hardy-Littlewood-Polya inequality for a linear differential operator and some related optimal problems. ATA，1992，8(1)：50—58.

[14] Sun Yongsheng. Average n-width of a point set in Hilbert space. Chinese Science Bulletin, 1992, 37(14): 1 153—1 157.

[15] Sun Yongsheng, Liu Yongping. Optimal recovery of the Sobolev-Wiener class of smooth functions by double sampling. Constructive Approximation, 1993, 9(4): 391—405.

[16] Sun Yongsheng, Wang Chengyong. μ-Average n-widths on the Wiener space. J. Complexity, 1994, 10(4): 428—436.

[17] Sun Yongsheng, Wang Chengyong. Average error bounds of best approximation of continuous functions on the Wiener space. J. Complexity, 1995, 11(1): 74—104.

[18] Sun Yongsheng, Wang Heping. Representation and approximation of multivariate periodic functions with bounded mixed moduli of smoothness. Proc. of Steklova Math. Ins., 1997, 219(4): 350—371.

[19] Sun Yongsheng, Liu Yongping, Chen Dirong. Extremal problems in approximation theory for some classes of smooth functions defined on $\mathbf{R}^d$. 北京师范大学学报(自然科学版), 1999, 35(增刊): 79—144.

[20] Sun Yongsheng. Optimal subspaces for a periodic convolution class with a B-kernel. East J. on Approx., 2000, 6(1): 53—63.

附录

Appendix

论文和著作目录

Bibliography of Papers and Works

论文目录

[序号] 作者. 论文题目. 杂志名称，年份，卷(期)：起页—止页.

[1] 王昆扬. 二元连续周期函数用其 Marcinkiewicz 型和强性逼近的估计式. 北京师范大学学报(自然科学版)，1981，(1)：7—22.

[2] 王昆扬. 多重 de la Vallee Poussin 方形余项的估计. 数学年刊，1982，3(6)：789—802.

[3] 王昆扬. 关于多重 Fourier 级数的线性求和. 北京师范大学学报(自然科学版)，1983，(2)：1—12.

[4] 王昆扬. 多重共轭 Fourier 级数的强求和(研究通讯). 科学通报，1983，28(23)：1 470.

[5] 罗俊波，王昆扬. 二元连续周期函数用其 Fourier 级数的 Marcinkiewicz 型线性平均逼近. 北京师范大学学报(自然科学版)，1984，(1)：37—48.

[6] 王昆扬. 二元周期函数用其 Marcinkiewicz 型平均逼近与绝对求和. 数学年刊，1984，5A(2)：133—152.

[7] 王昆扬. 二重 Fourier 级数及其共轭级数的 M 型(H，q)求和. 数学学报，1984，27(6)：811—816.（王昆扬. 二重 Fourier 级数的 M-型(H，q)求和(研究通讯). 科学通报，1982，27(14)：894.）

[8] 王昆扬. 多重共轭 Fourier 级数的(C，l)求和. 科学通报，1984，

29(24)：1 473－1 477.（Wang Kunyang.（C，l）summability of multiple conjugate Fourier series. Chinese Science Bulletin，1985，30(2)：160－165.）

[9] 王昆扬. 关于 Walsh-Fourier 级数的几乎处处收敛问题的一点注记. 北京师范大学学报(自然科学版)，1985，(1)：11－16.

[10] Wang Kunyang. Generalized spherical Riesz means of multiple Fourier series. 数学进展，1985，14(1)：77－80.

[11] Wang Kunyang. Approximation for continuous periodic functions of several variables and its conjugate functions by Riesz means on a set of total measure. 数学进展，1985，14(1)：81－84.（Research announcements）；Approximation Theory and Its Application，1985，1(4)：19－56(全文).

[12] 王昆扬. 多重共轭 Fourier 级数的强求. 数学学报，1986，29(2)：156－175.

[13] 王昆扬. 多重 Fourier 级数及其线性平均. 数学季刊，1986，2(2)：59－68.

[14] Wang Kunyang. Generalized spherical Riesz means of multiple Fourier series. Acta Math. Sinica，New Series，1986，2(2)：178－202.

[15] 王昆扬. 一类奇异积分算子的逼近性质. 北京师范大学学报(自然科学版)，1987，(4)：1－8.

[16] 王昆扬. 再论多重共轭 Fourier 级数的强求和. 科学通报，1987，32(15)：1 124－1 127.（Wang Kunyang. Further discussion on strong summability of multiple conjugate Fourier series. Chinese Science Bulletin，1988，33(12)：971－972.）

[17] 王昆扬. 多重 Fourier 级数及其线性平均. 数学季刊，1987，2(2)：59－68.

[18] 陆善镇，王昆扬. H^p 实空间变理论对逼近问题的应用. 见：全国第五届函数逼近论会议论文集，河南大学学报，1988，(增刊)：49－52.

[19] 王昆扬. 关于多重 Fourier 级数的收敛性. 北京师范大学学报(自

然科学版)，1988，(2)：18—27.

[20] Wang Kunyang. On the approximation of L^2-functions by their spherical Fourier sums on set of full measure. Approximation Theory and Its Applications，1989，5(4)：69—77.

[21] Wang Kunyang. Strong uniform approximation by Bochner-Riesz means. Multivariate Approximation Theory (Oberwolfach)，1989，4：337—342.

[22] Muller D，Wang Kunyang. On the rate of convergence of certain summability methods for Fourier integrals of L^2-functions. Artiv für Mathematik，1991，29(2)：261—276.

[23] Wang Kunyang. Equiconvergent operator of Cesàro means on sphere and its applications. 北京师范大学学报(自然科学版)，1993，29(2)：143—154.

[24] Wang Kunyang. Pointwise convergence of Cesàro means on sphere. 北京师范大学学报(自然科学版)，1993，29(2)：158—164.

[25] Brown G，Wang Kunyang，Wilson D C. Positivity of some basic consine sums. Mathematical Proceedings of the Cambridge Philosophical Society，1993，114(3)：383—391.

[26] Wang Kunyang，Brown G. Approximation by Bochner-Riesz means and Hardy summability. 北京师范大学学报(自然科学版)，1994，30(2)：163—169.

[27] Wang Kunyang，Zhang Pu. Strong uniform approximation on sphere. 北京师范大学学报(自然科学版)，1994，30(3)：321—328.

[28] Riemenschneider S，Wang Kunyang. Approximation theorems of Jackson type on the sphere (Research Announcements). 数学进展，1995，24(2)：184—186.

[29] Brown G，Wang Kunyang. Jacobi polynomial estimates and Fourier-Laplace Convergence (Research Announcements). 数学进展，1995，24(6)：563—565.

[30] Brown G，Koumandos S，Wang Kunyang. Positivity of Cotes numbers at more Jacobi abscissas. Monatshefte Fŭr Mathematik，

1996，122(1)：9—19.

[31] Brown G，Koumandos S，Wang Kunyang. Positivity of more Jacobi polynomial sums. Mathematical Proceedings of the Cambridge Philosophical Society，1996，119(4)：681—694.

[32] 王昆扬. 极限　实数　指数函数. 数学通报，1997，(10)：33—36；(11)：45—47.

[33] Wang Kunyang，Ma Bolin. Maximal translation operator on sphere. 北京师范大学学报(自然科学版)，1997，33(1)：43—48.

[34] Brown G，Wang Kunyang. An extension of the Fejer-Jackson inequality. Journal of Australasian Mathematical Society，1997，62A(1)：1—12.

[35] Brown G，Wang Kunyang. Jacobi polynomial estimates and Fourier-Laplace convergence. The Journal of Fourier Analysis and Applications，1997，3(6)：705—714.

[36] 张伦传，王昆扬. 素 AW^*-代数 A 的本原性的证明. 北京师范大学学报(自然科学版)，1998，34(3)：289—291.

[37] Tang Pin，Wang Kunyang. Strong approximation of de la Vallee Poussin type on the sphere. 北京师范大学学报(自然科学版)，1998，34(2)：162—167.

[38] Brown G，Koumandos S，Wang Kunyang. Positivity of basic sums of ultraspherical polynomials. Analysis，1998，18：313—331.

[39] Brown G，Wang Kunyang. On a conjecture of F. Mmoricz. Proceedings of the American Mathematical Society，1998，126(12)：3 527—3 537.

[40] Chen Wengu，Wang Kunyang. Some positive cotes numbers for Jacobi abscissas. Mathematical Proceedings of the Cambridge Philosophical Society，1998，123(2)：193—198.

[41] 洪勇，王昆扬. 球面上的变阶分数次积分. 数学年刊，1999，20A(1)：27—32.

[42] 王昆扬. 关于 Riemann 积分理论的本质缺陷及以 Lebesgue 积分理

论取代之的看法. 数学教育学报，1999，8(3)：95—98.
[43] 王昆扬. 怎样讲实数理论. 高等数学研究，1999，(1)：1—5.
[44] Wang Kunyang. Some recent progress in Fourier-Laplace analysis on the sphere. 北京师范大学学报(自然科学版)，1999，35(增刊)：23—44.
[45] Brown G，Koumandos S，Wang Kunyang. On the positivity of some basic Legendre polynomial sums. Journal of the London Mathematical Society，1999，59(2)：939—954.
[46] Dai Feng，Wang Kunyang. On approximation of functions on sphere. Approximation Theory and Its Applications，1999，15(4)：50—59.
[47] Wang Kunyang，Zhang Xirong. Approximation of locally integrable functions on the ereal line. Укр. Мат. Журн，1999，51(11)：1 549—1 561.
[48] Ван Куньян，Теляковский С А. Дифференциальные Свойства Сумм Одного Класса Тригонометрических Рядов. Вестн. Моск. Ун-Та.，1999，(1)：24—29.
[49] 戴峰，王昆扬. Fourier-Laplace 级数强可和点的刻画. 北京师范大学学报(自然科学版)，2000，36(6)：718—721.
[50] 许贵桥，王昆扬，孙宇峰，连丽霞. Grün wald 插值算子的加权 L_1 收敛速度. 河北科技大学学报，2000，21(3)：14—20.
[51] Dai Feng，Wang Kunyang. On the Fourier-Laplace a. e. Convergence. Approximation Theory and Its Applications (Ukraine). 2000：72—81.
[52] Wang Kunyang，Wang Sheng. Some constructive properties of functions defined on the sphere. Approximation Theory and Its Applications，2000，16(1)：26—35.
[53] 王昆扬. 给中学生讲好微积分基本知识. 数学通报，2001，(6)：39—41.
[54] 王昆扬，张培恒. 谈指数函数的定义——在大学数学课程中妥善定义指数函数. 高等数学研究，2001，4(3)：13—14，28.

[55] 张奠宙，王昆扬. 面向21世纪“高师教学改革计划”数学课题组结题会议纪要. 数学教育学报，2001，10(3)：43—49.

[56] 戴峰，王昆扬. Fourier-Laplace分析中用连续模给出的几乎处处收敛条件. 北京师范大学学报(自然科学版)，2002，38(4)：440—444.

[57] 王昆扬，刘永平. 现代数学家孙永生. 见：王淑芳，邵红英，主编：师范之光. 北京：北京师范大学出版社，2002：595—599.

[58] 王昆扬.《孙永生》，见：程民德，主编：中国现代数学家传(第五卷). 江苏教育出版社，2002：405—418.

[59] Dai Feng，Wang Kunyang，Yu Chunwu. On a conjecture of Ditzian and Runovskii. Journal of Approxiamtion Theory，2002，118：202—204.

[60] 戴峰，王昆扬. Fourier-Laplace级数的缺项平均在Lebesgue点处的收敛性. 数学学报，2003，46(4)：729—732.（Dai Feng，Wang Kunyang. Summability of Fourier-Laplace series with the method of Lacunary arithmetical means at Lebesgue points. Acta Mathematica Sinica，English Series，2001，17(3)：489—496.）

[61] 余纯武，陈莘萌，王昆扬，戴峰. 球面函数逼近算法研究. 中国科学，2003，33E(10)：924—933.（Yu Chunwu，Chen Xinmeng，Wang Kunyang，Dai Feng. Approximation of Hardy space on the unit sphere. Science in China，2003，45E(4)：309—320.）

[62] Wang Kunyang，Yu Dan. On the absolute (C，α) convergence for functions of bounded variation. Journal of Approximation Theory，2003，123(2)：300—304.

[63] 郇中丹，王昆扬，执笔. 全国数学教育实践调研会纪要. 数学通报，2004，(1)：45—46.

[64] Lin Qincheng，Wang Kunyang. Convergence rate of Fourier-Laplace series of L^2-functions. Journal of Approximation Theory，2004，128(1)：103—114.

[65] Dai Feng，Wang Kunyang. A note on the equivalences between the averages and the K-functionals related to the Laplacian. Journal

of Approximation Theory，2004，130：38—45.

[66] Brown G，Dai Feng，Wang Kunyang. Extensions of Vietoris's inequalities Ⅰ. 数学进展，2005，34(1)：125—127.

[67] Brown G，Dai Feng，Wang Kunyang. Extensions of Vietoris's inequalities Ⅱ. 数学进展，2005，34(2)：253—255.

[68] Dai Feng，Wang Kunyang. Strong approximation by Cesàro means with critical index in the Hardy spaces $H^p(S^{d-1})$. Acta Mathematica Sinica，English Series，2005，21(2)：439—448.

[69] 王昆扬. 甘肃省天水地区部分中学教师座谈会记录. 数学通报，2005，44(3)：4.

[70] 王昆扬，郇中丹. 培养高素质数学人才的研究与实践. 见：教育部高等教育司编：发展创新改革(下)，北京：高等教育出版社，2005：11—13.

[71] 王昆扬. 谈中学数学课程中的"实数". 数学通报，2006，45(12)：14—16.

[72] Wang Kunyang. A note on a paper of H. Alzer and S. Koumandos. Journal of Inequalities in Pure and Applied Mathematics，2006，7(2)：Article ID 46.

[73] Wang Kunyang. Professor Sun Yongsheng. Analysis in Theory and Applications，2006，22(2)：101—104.

[74] 王昆扬，张英伯. 您在我们的心中永生. 数学通报，2006，45(4)：2—3.

[75] Brown G，Dai Feng，Wang Kunyang. On positive cosine sums. Mathematical Proceedings of the Cambridge Philosophical Society，2007，142(2)：219—232.

[76] Brown G，Dai Feng，Wang Kunyang. Extensions of Vietoris's inequalities Ⅰ. The Ramanujan Journal，2007，14(3)：471—507.

[77] Huang Rong，Wang Kunyang. A covering lemma on the unit sphere and application to the Fourier-Laplace convergence. Acta Mathematica Sinica，English Series，2007，23(7)：1 327—1 332.

[78] Wang Kunyang, Dai Feng. Recent progress on spherical harmonic approximation made by BNU research group. Analysis in Theory and Applications, 2007, 23(1): 50—63.

[79] Chen Xirong, Dai Feng, Wang Kunyang. Estimations of the remainder of spherical harmonic series. Mathematical Proceedings of the Cambridge Philosophical Society, 2008, 145(1): 243—255.

[80] Dai Feng, Wang Kunyang. Convergence rate of spherical harmonic expansions of smooth functions. Journal of Mathematical Analysis and Applications, 2008, 348(1): 28—33.

[81] Dai Feng, Wang Kunyang. A survey of polynomial approximation on the sphere. International Journal of Wavelets, Multiresolution and Information Processing, 2009, 7(6): 749—771.

[82] Wang Kunyang. Decimal Representation of Real Numbers. 数学通报(英文版), 2009, (增刊): 92—96.

[83] Dai Feng, Huang Hongwei, Wang Kunyang. Approximation by the Bernstein-Durrmeyer operator on a simplex. Constructive Approximation, 2010, 31(3): 289—308.

[84] Huang Hongwei, Wang Kunyang. On the widths of Sobolev's classes. Journal of Complexity, 2011, 27(2): 201—220.

[85] 王昆扬. 《孙永生》. 见：钱伟长，总主编，王元，数学分卷主编：20世纪中国知名科学家学术成就概览，数学卷，第3分册. 北京：科学出版社，2012：59—66.

[86] 王昆扬. 坚持高的学术水平，坚持因材施教. 见：北京市教育委员会高等教育处、人事处、师资管理办公室，北京市高等学校师资培训中心，编：高校名师的教学视野，国家级高等学校教学名师谈教学丛书，第1卷. 北京：首都师范大学出版社，2012：37—50.

[87] 王昆扬. 谈谈Lebesgue数钱. 高等数学研究，2011，14(4)：7—8.

[88] Huang Hongwei, Wang Kunyang. On the widths of Sobolev's classes. International Journal of Wavelets, Multiresolute and Information Processing, 2014, 12(5): Article ID 1461004.

著作

[序号] 著者，译者. 书名. 出版地：出版社，出版年份.

[1] Зорич B. A. 著. 蒋铎，王昆扬，周美珂，邝荣雨. 译. 数学分析(第一卷第一、二分册). 北京：高等教育出版社，1987，1989.

[2] 陆善镇，王昆扬. Bochner-Riesz 平均. 北京：北京师范大学出版社，1988.

[3] 陆善镇，王昆扬. 实分析. 北京：北京师范大学出版社，第 1 版，1997；第 2 版，2006.

[4] Wang Kunyang，Li luoqing. Harmonic analysis and approximation on the unit sphere. Beijing：Science Press，2000.

[5] 王昆扬. 数学分析专题研究(专升本教材). 北京：高等教育出版社，2001.

[6] 王昆扬. 简明数学分析(面向 21 世纪课程教材). 北京：高等教育出版社，2001.

[7] 张奠宙，王昆扬，主编. 高等师范教育面向 21 世纪教学内容和课程体系改革成果丛书，八，数学. 北京：北京师范大学出版社，2001.

[8] Архипов Г. И.，Садовничий В. А.，Чубариков В. Н. 著. 王昆扬，译，数学分析讲义. 北京：高等教育出版社，2006.

[9] Зорич B. A. 著. 蒋铎，王昆扬，周美珂，邝荣雨，译，周美珂，校. 数学分析，第 1～2 卷(第 4 版). 北京：高等教育出版社，2006.

[10] Кашин Б. С.，Саакян А. А. 著. 孙永生，王昆扬，译. 正交级数. 北京：北京师范大学出版社，2007.

[11] 孙永生，王昆扬. 泛函分析讲义(北京高等教育精品教材)，第 2 版. 北京：北京师范大学出版社，2007.

[12] Tao T. 著. 王昆扬，译. 陶哲轩实分析. 北京：人民邮电出版社，2008.

[13] 郇中丹，刘永平，王昆扬. 简明数学分析，第 2 版(普通高等教育“十一五”国家级规划教材). 北京：高等教育出版社，2009.

[14] 王昆扬. 实变函数论讲义(北京高等教育精品教材立项项目). 北京：高等教育出版社，2011.

[15] 王昆扬. 实数的十进表示. 北京：科学出版社，2011.

[16] 王昆扬. 数学分析简明教程. 北京：高等教育出版社，2015.

后记

Postscript by the Chief Editor

在2003年年底，北京师范大学出版社的新一届领导将5位先生（王世强、孙永生、严士健、王梓坤、刘绍学）的文集列入《北京师范大学数学家文库》出版计划，文集在2005年8～10月出版．在2007年4月至2008年9月出版了3位已经逝世的先生（汤璪真、白尚恕、范会国）的文集．2011年11月启动出版在20世纪30年代出生的先生文集．2018年初启动出版在20世纪40年代出生的先生文集．

北京师范大学数学科学学院在20世纪40年代出生的博士生导师共5人，以年龄大小为序：王昆扬、郑学安、陈木法、张英伯、房艮孙，至少有以下特点．

陈木法教授是中国科学院院士和发展中国家科学院院士，其余4人均任教授二级岗位．

前3人是“老五届”，即1966～1970届本科毕业生．后两人是“老三届”，即1966～1968届的高中毕业生（不是“老三届”的初中毕业生），高中毕业后到黑龙江生产建设兵团劳动锻炼．

严士健教授对我说过，数学科学学院20世纪40年代的教授水平，已经超过他们．

数学系“老五届”最初留系工作的十多位毕业生均已先后提升，或调到校部机关，或调到外校．因此，在1966～1970年毕业生中有5位教师在数学系任教，有两位教师是教育部寄存在数学系的教师，1971～1975年无毕业的学生留系，这是一种不合理的师资结构且明显断档．

1978年3月，我国恢复招收研究生．10月，数学系共招收硕士研究生 11 人，其中概率方向 3 人：陈木法、郑小谷和唐守正；分析方向两人：王昆扬和罗俊波；代数与数理逻辑方向 6 人：罗里波、沈复兴、张英伯、王成德、孙晓岚和程汉生．毕业后先后有 6 人留数学系任教，王昆扬老师是其中之一．

值得一提的是，数学系第一届博士生共 6 人，5 人均是 1978 级硕士生，他们是陈木法、沈复兴、唐守正、王昆扬和郑小谷．毕业后除唐守正外，其余 5 人留数学系任教，开启了本系博士毕业生留数学系的先河．之后，数学系教师队伍的学历结构逐渐发生重大变化，由选留硕士生，进而选留博士生和公开招聘教师，补充数学系的教师队伍．从 1999 年 8 月起，数学系新进教师必须具有博士学位．

王昆扬文集的前言略显简单，这里略作补充．王老师于 1981 年后在北京师范大学数学系任教，1985 年 7 月任副教授，1991 年 9 月任教授，1993 年 12 月任博士生导师，2014 年 6 月退休．1998～2007 年任北京市第 9～10 届政治协商会议委员会委员，2000～2004 年任中国数学会教育工作委员会主任，2001～2005 年任教育部教学指导委员会数学分委员会委员，2001～2009 年任《数学进展》编委，2006 年任《数学研究与评论》和《Analysis in Theory and Applications》编委．

王昆扬老师长期从事基础数学的教学和研究．编写过本科生教材和发表教学改革论文(见附录)．主持完成教育部师范司教改重点项目 1 项．两次主持国家理科基地创建名牌课程项目，2005～2009 年主持数学分析精品课程．获 1997 年宝钢优秀教师奖，2000 年北京师范大学首届本科优秀教学奖特等奖和最受学生欢迎的本科教学十佳教师称号，2001 年宝钢优秀教师特等奖，2001 年全国模范教师称号，2002 年全国普通高等学校优秀教材二等奖，2003 年北京市教学名师奖，2005 年北京市优秀教学成果奖一等奖，2008 年第四届高等学校教学名师奖．2009 年负责的分析类课程教学团队被评为国家级优秀教学团队．

王昆扬文集的出版，得到了数学科学学院党委书记唐仲伟教授，前院长李增沪教授和现任院长王恺顺教授的大力支持，同时得到了北京师范大学出版社的大力支持，在此表示衷心的感谢．

华罗庚教授说："一个人最后余下的就是一本选集．"(龚昇论文选集，中国科学技术大学出版社，2008)这些选集的质量反映了我们学院某一学科，或几个学科，或学科群的整体学术水平．而将北京师范大学数学科学学院著名数学家、数学教育家和科学史专家论文进行整理和选编出版，是学院学科建设的一项重要的和基础性的工作，是学院的基本建设之一．它对提高学院的知名度和凝聚力，激励后人，有着重要的示范作用．当然，这项工作还在继续做下去，收集和积累数学科学学院各种资料的工作还在继续进行．

主编　李仲来

2018-11-08